Oxidative Stress and Redox Regulation

Ursula Jakob • Dana Reichmann
Editors

Oxidative Stress and Redox Regulation

Springer

Editors
Ursula Jakob
Department of Molecular, Cellular and Developmental Biology
University of Michigan
Ann Arbor, MI
USA

Dana Reichmann
Department of Biological Chemistry
The Alexander Silberman Institute of Life Science
The Hebrew University of Jerusalem
Givat Ram Campus
Jerusalem
Israel

ISBN 978-94-007-5786-8 ISBN 978-94-007-5787-5 (eBook)
DOI 10.1007/978-94-007-5787-5
Springer Dordrecht Heidelberg New York London

Library of Congress Control Number: 2013933705

Printed on acid-free paper

Springer is part of Springer Science+Business Media (www.springer.com)

I dedicate this book to my beloved partner and companion, Dori Reichmann, and to my remarkable parents, Ludmila and Semion Garber, for their unfailing love, endless support, and continuous enthusiasm to understand my research. I also dedicate this book to my amazing daughter, Noya, who keeps asking the question "Why?" which drives our life as researchers.

Dana Reichmann

I dedicate this book to my husband Jim and my children Alina and Adrian – every day I am amazed by their generosity, wisdom, humor, and love.

Ursula Jakob

Preface

Reactive oxygen species (ROS) have long been branded as undesired cytotoxic compounds to be avoided at all costs. This view has changed with the realization that ROS, which are constantly produced in all aerobic organisms, play essential roles in a plethora of signaling processes that act to control cell differentiation, proliferation, and apoptosis. It is now clear that organisms need to achieve a fine-tuned balance between oxidants and antioxidants (i.e., redox homeostasis) in order to function properly. At the heart of many cellular ROS-controlled processes are redox-sensitive proteins. These proteins use reversible posttranslational modifications, particularly on functionally and/or structurally important thiol groups, to regulate their activity and the activity of the pathway that they are part of. Based on extensive research particularly in the last decade, a picture is now emerging that cells harbor a highly complex network of redox-sensitive processes, which are constantly adjusting the cell's metabolism and physiology to the prevailing redox conditions of the environment. Alterations in a cell's redox homeostasis are involved in initiating stem cell differentiation, are associated with tumor development, and are considered to be a critical component in aging and age-related diseases.

The realization that redox-regulated processes influence almost every area of modern biology and will be inevitably encountered by many non-redox biologists provided a key motivation in the assembly of this book. We aim to provide a concise overview of the current state of the art in redox biology for a wide range of people from the novice to the expert. We start with a description of the chemical basis of thiol oxidation followed by a summary of the major enzymatic and non-enzymatic players involved in ROS detoxification and redox homeostasis in bacteria and eukaryotes. The second part of the book introduces state-of-the-art experimental and computational methodologies that are currently used in redox biology and discusses their advantages and limitations. We decided to focus on three major methodologies that nicely complement each other: (1) redox proteomics, which provides means to discover novel redox-sensitive proteins and pathways in the cell and is a useful approach to monitor changes in redox homeostasis during distinct physiological or pathological states; (2) fluorescence imaging of ROS in multicellular organisms using a variety of highly specific detection methods, which allow temporal and spatial tracking of ROS in live organisms; and (3) redox bioinformatics to analyze, characterize, and eventually predict features of redox-sensitive and redox-regulated proteins in cells. Integration of these methods will eventually allow us to draw a global and system-wide picture about redox homeostasis and redox regulation in the living organism. The final, and the most applied, part of this book details biological examples of oxidative stress and redox regulation in a variety of different pro- and eukaryotic systems. Some chapters illustrate the importance of ROS in cell signaling, epigenetics, and stem cell differentiation, while others provide expert summaries and discussions about the role of redox regulation in infectious diseases, cancer, and aging. The closing chapter summarizes recent drug developments, which are aimed to modulate specific redox pathways in different human disease pathologies.

We are tremendously grateful to each of our contributors for their enthusiasm and willingness to provide the chapters for this book. Without them, this book would have not been possible. It was great fun putting this book together, and we hope that our book will infect the readers with the same enthusiasm about this fascinating research field that we experience every day in working with redox-regulated proteins.

Dana Reichmann
Ursula Jakob

Contents

Chapter 1
The Chemistry of Thiol Oxidation and Detection

Mauro Lo Conte and Kate S. Carroll

Abstract The thiol functional group of the amino acid cysteine can undergo a wide array of oxidative modifications and perform a countless number of physiological functions. In addition to forming covalent cross-links that stabilize protein structure and functioning as a powerful nucleophile in many enzyme active sites, cysteine appears to be the principal actor in redox signaling, functioning as a regulatory reversible molecular switch. It is increasingly appreciated that the thiol group of cysteine in subset of proteins undergoes oxidative modification in response to changes in the intracellular redox environment. To understand these complex but critical biological phenomena, the chemistry of the thiol functionality and related oxidation products must also be taken into consideration. Selective methods to monitor and quantify discrete cysteine modifications will be central to understanding their regulatory and pathophysiologic function. Accordingly, this chapter focuses on the chemical feature of thiol oxidation and on selective methods for detecting oxidants and individual cysteine chemotypes.

Keywords Thiol oxidation • Cysteine • Reactive oxygen species • Reactive nitrogen species • Reactive sulfur species

1.1 Introduction

The thiol functional group of the amino acid cysteine can undergo a wide array of oxidative modifications and perform a countless number of physiological functions. In addition to forming covalent cross-links that stabilize protein structure and

---

This chapter is dedicated to the memory of Professor William S. Allison.

M. Lo Conte • K.S. Carroll (✉)
Department of Chemistry, The Scripps Research Institute, Jupiter, FL 33458, USA
e-mail: kcarroll@scripps.edu

U. Jakob and D. Reichmann (eds.), *Oxidative Stress and Redox Regulation*,
DOI 10.1007/978-94-007-5787-5_1,

functioning as a powerful nucleophile in many enzyme active sites, cysteine appears to be the principal actor in redox signaling, functioning as a regulatory reversible molecular switch. It is increasingly appreciated that the thiol group of cysteine in subset of proteins undergoes oxidative modification in response to changes in the intracellular redox environment. To understand these complex, but pivitol biological phenomena, the chemistry of the thiol functionality and related oxidation products must be taken into consideration. Along these lines, selective methods to monitor and quantify discrete cysteine modifications are key to understanding their regulatory and pathophysiologic function. This chapter focuses on the chemistry of thiol oxidation and detection.

1.2 The Biological Chemistry of Thiols

The cysteine side chain is generally considered the most potent nucleophile of all amino-acid side chains under physiological conditions. This notable level of reactivity is due to the presence of a thiol functional group. Thiol is a sulfur analogue of alcohol, but the smaller difference in electronegativity between the sulfur atom and the hydrogen atom makes the S–H bond less polarized than the O–H bond, leading to a diminished propensity to form hydrogen bonds. In contrast, thiols are much more acidic in comparison to alcohols, and this property can be explained by the weakness of the S–H bond and the greater likelihood that the negative charge will be distributed within sulfur 3d orbitals. Cysteine can be considered a triprotic acid in which the pK_a of thiol group has been determined to be 8.2 (Tajc et al. 2004). In glutathione, the most abundant low molecular weight thiol in the cytosol, cysteine residue has a pK_a of 9.1 (Tang and Chang 1996). In general, therefore, thiols are mild acids, but the protein microenvironment can dramatically influence the pKa value. The presence of a positively charged residue, such as lysine or arginine (Copley et al. 2004), as well as the formation of a hydrogen bond (Wang et al. 2001), may increase thiol acidity by 3–4 orders of magnitude. The reactivity of thiols is correlated with its pKa value (Szajewski and Whitesides 1980). In cysteines, a thiolate side chain becomes a stronger nucleophile and readily reacts with oxidants and electrophilic species, although interactions with specific residues or metals can also stabilize the thiolate form. With its remarkable reactivity, the cysteine thiol group can play a key biological role in catalysis and serve as an important site for many post-translational modifications. Considering the propensity of thiols to undergo oxidative reactions, the need for methods and chemical tools to monitor both reduced and oxidized cysteine residues has become clear. The key challenge for effective detection of bio-functional groups is summarized by the concept of chemoselectivity (Trost 1983) and, more specifically, in bioorthogonality (Bertozzi 2011). Indeed, chemists and biologists have sought to identify highly selective and facile reactions to detect biomolecules in living systems without interfering with native biochemical processes. Moreover, the relatively low abundance of cysteine, in comparison to other amino acids, combined with its remarkable nucleophilicity

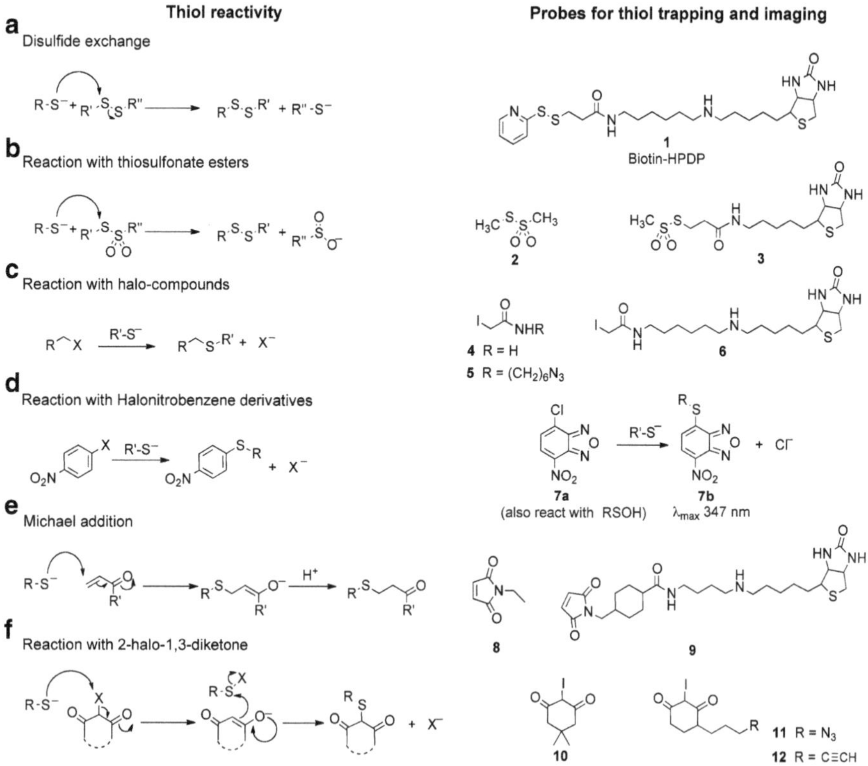

Fig. 1.1 Chemical reactivity and detection of thiols

has made cysteine the most common target for selective protein bio-conjugation (Chalker et al. 2009), creating fertile ground for the development of site-specific strategies for protein modification. In this context, the thiol reacts as a soft nucleophile with alkyl and aryl halides, carbonyl, phosphoryl, and sulfonyl groups as well as with unsaturated compounds (Fig. 1.1a–f). There are a large number of reagents that selective modify thiols, even in the presence of other strong nucleophiles such as lysine or histidine (Fig. 1.1, **1–12**). Such thiol conjugations can be further subdivided into two categories: reversible and irreversible.

To date, the most well studied example of cysteine modification is disulfide formation between two thiol groups and, thus it is no coincidence that one of the first strategies to detect thiols was inspired by the process of thiol-disulfide exchange (Fig. 1.1a). Bearing in mind that disulfide exchange is an equilibrium reaction (see Sect. 1.4.3), a series of specific disulfides were designed in order to favor one direction over the other. The well-known 5,5′-dithiobis-(2-nitrobenzoate) (DTNB) or Ellman's reagent is perhaps most common disulfide-based reagent used to monitor and quantify protein free thiols (Ellman 1959). The notably low pKa of the 5-mercapto-2-nitrobenzoate leaving group (4.4, Little and Brocklehurst 1972) ensures that this reaction occurs rapidly and, at the same time, does not perturb naturally occurring disulfides in the protein of interest.

Thiopyridyl disulfides are another class of common disulfide-exchange based detection reagents (Olsen et al. 2004). Thiopyridyl disulfides such as **1** (Fig. 1.1) react rapidly with thiols to afford an electro-stabilized compound, pyridine 2-thione, which can be followed by colorimetric analysis (i.e., absorption at 420 nm). Based on the same concept, a series of selenyl-sulfides (Gamblin et al. 2004) and thiosulfonates (Kenyon and Bruice 1997) have also been developed to selectively functionalize cysteine side chains. For example, methyl methane thiosulfonate (**2**) is the classic thiol-targeted reagent used to block free proteins thiols in the biotin switch assay (Jaffrey et al. 2001). All of these reagents (e.g., **1–3**) react with the thiol side chain to form a mixed disulfide that can subsequently be reduced to regenerate the free cysteine residue. The reversible nature of the disulfide linkage is acceptable in some circumstances, but in other situations this feature can be problematic. As a result, specific alkylating agents have been designed to irreversibly modify the cysteine side chain (**4–11**). Alkyl halides react with thiols to yield stable thioethers (Fig. 1.1c). Such reagents are usually not exclusive in their reaction with thiols; however, with careful control of pH, α-halo carbonyl compounds (**4–6**) can provide a satisfactory degree of bioorthogonality. Iodoacetamide (**4**) is employed in standard protein digestion protocols to block cysteine residues (Shevchenko et al. 1996). Despite the popularity of iodoacetamide, side-reactions with lysine, methionine, tyrosine, and histidine residues have been observed (Nielsen et al. 2008). In such cases, this issue can be resolved by using chloroacetamide (Weerapana et al. 2008).

Another unusual class of organo-halides is halo-nitrophenyl derivatives (Fig. 1.1d and 7). The strong electro-withdrawing nitro group makes haloarene extremely reactive toward thiols, although other sulfur species may also be modified in the presence of similar compounds (see Sect. 1.4). Thiolates are good Michael donors and the addition reaction to an α,β-unsaturated system is an alternative way to alkylate cysteine residues (Fig. 1.1e). Historically, maleimides have been the most widely used tools in biochemistry for the modification of thiols. The popularity of *N*-ethyl maleimide (**8**) is due to its good cell permeability and the ease with which the ethyl group can be elaborated by other substituents (e.g., **9**). However, what must be considered is that the reaction is only selective toward the thiol group at pH 7 (Crankshaw and Grant 1996). Vinyl sulfones are an alternative to the use of maleimide derivatives. Although they offer the advantage of yielding a single stereoisomer as a product (i.e., in theoretical models, maleimides can yield a mixture of two diastereoisomers), vinyl sulfones may cross react with the ε-NH_2 of lysine (Masri and Friedman 1988). A selective new reagent for thiol modification reported by Carroll's research group is based on the α-halo 1,3-diketone scaffold (Fig. 1.1f). In this reaction, 2-iodo-5, 5-dimethyl-1, 3-cyclohexandione (**10**, distinguished from the aforesaid α-halo carbonyl compounds by virtue of its secondary halogen) and related compounds (e.g., **11, 12**) react with a thiol to give the sulfenyl iodide species; this intermediate is rapidly attacked by the 1,3-diketone carbon nucleophile to afford a stable thioether (Seo and Carroll 2011).

1.3 Two-Electron Oxidants

Reactive oxygen species (ROS) are a family of molecules that are continuously generated in cells as consequence of aerobic life. Traditionally, their production has been strictly associated with oxidative stress, aging, and disease. However, growing evidence indicates that controlled generation of ROS contributes to physiological intracellular signaling events. ROS mediate redox modification of various biomolecules *via* two-electron oxidation or radical-based reactions, and they react in particular with cysteine residues. The term ROS encompasses a wide class of activated oxygen molecules which, can undergo further reaction with nitrogen or sulfur compounds, to produce reactive nitrogen species (RNS) and reactive sulfur species (RSS). Here, we focus on two-electron oxidants without mentioning the radical species, which will be described in subsequent chapters of this book.

1.3.1 Hydrogen Peroxide

A handful of enzymes generate H_2O_2 directly. For example, significant quantities of H_2O_2 are produced in the peroxisome by the action of enzymes, such as D-amino acid oxidase or uric acid oxidase. More generally, however, H_2O_2 is formed in cells *via* the disproportionation of superoxide ($O_2^{\bullet-}$). The major sources of $O_2^{\bullet-}$ production are the electron transport chain (ETC.) complex in the mitochondria (Murphy 2009) and the family of NADPH oxidases (NOXs) (Bedard and Krause 2007). Superoxide disproportionation may occur spontaneously (7.6×10^5 M^{-1} s^{-1}, Kutala et al. 2008) or may be catalyzed by superoxide dismutase (SOD) (Forman and Fridovich 1973; Fukai and Masuko 2011). Finally, several studies have showed the importance of monoamino oxidase (MAO) as an important source of H_2O_2 in neurons (Maker et al. 1981) and in relationship to neurodegenerative diseases (Andersen 2004). Hydrogen peroxide has the ability to move between different cellular compartments. Although H_2O_2 has long been thought to diffuse freely across membranes, its dipole moment of 2.2×10^{-18} C·m (Cohen and Pickett 1981) is only slightly higher than that of water (1.9×10^{-18} C·m), rendering its passive diffusion similarly limited. Recent studies have demonstrated the implication of aquaporin water channels as specific mediators of H_2O_2 passage across membranes (Bienert et al. 2007) and the involvement of these channels in intracellular signaling (Miller et al. 2010).

The unusual reactivity of H_2O_2 is generally attributed to relatively low O–O bond energy (47 kcal/mol) – hence, the ease with which it is homolytically or heterolytically cleaved (Bach et al. 1996). H_2O_2 is a strong oxidant and may behave as an electrophile as well a nucleophile depending on the nature of the reactant. In cells, the principal targets of H_2O_2-mediated oxidation are thiols by means of two electron nucleophilic substitution. A thiol with a low pK_a value reacts rapidly to yield sulfenic acid and water (Eq. 1.1).

$$R - SH + HO - OH \rightarrow R - S - OH + H_2O \qquad (1.1)$$

Although glutathione (GSH) is the most abundant low molecular weight thiol, GSH contributes little to the reduction of H_2O_2 in cells. In fact, the total protein thiol concentration in human cells is at least two times higher than GSH (Hansen et al. 2009). At steady state, cellular H_2O_2 is kept in check by enzymes of the peroxiredoxin (Prx) family (Winterbourn and Hampton 2008). The rate of H_2O_2 reduction catalyzed by Prx is exceptionally fast ($1 \times 10^7\ M^{-1}\ s^{-1}$ at pH 7), but this reactivity cannot be explained solely by the low pK_a of the peroxidative cysteine (~5.7, Rhee and Woo 2011). More generally, the reactivity of peroxides (RO–OH) shows an inverse relationship with the pKa of the leaving group RO (for this reason CH_3CO_3H is much more reactive than H_2O_2). The presence of specific proton-donating moieties within the protein microenvironment can serve to stabilize the poor hydroxide-leaving group and thus, accelerate the reaction rate. Along these lines, a recent series of experimental and computational studies by Nagy et al. suggests that two highly conserved arginine residues contribute to reactivity of Prx toward H_2O_2 *via* hydrogen bonding (Nagy et al. 2011).

Although the recombinant form of proteins may undergo H_2O_2-mediated thiol oxidation in the test tube, in many cases, it has been difficult to reconcile relatively low oxidation rates observed *in vitro* with intracellular regulation of enzymatic activity. For example, the rate constant measured for inactivation of recombinant protein tyrosine phosphatase 1B (PTB1B) is $\sim 10\ M^{-1}\ s^{-1}$ (Denu and Tanner 1998), which would appear to be incompatible with intracellular signaling events. With respect to this long-standing issue, recent work by Gates and co-workers has proposed that more reactive species, such as peroxymonocarbonate and peroxymonophosphate, oxidize PTPs in cells (Zhou et al. 2011). Peroxymonocarbonate, which may be spontaneously generated from bicarbonate anion in the presence of localized high concentrations of H_2O_2 (Eq. 1.3), can increase the rate of PTB1B oxidation by a factor of 20; the elevated reactivity is easily explained by the higher pK_a value of the hydrogen carbonate leaving group.

$$H_2O + CO_2 \rightleftharpoons H_3O^+ + HCO_3^- \tag{1.2}$$

$$HCO_3^- + H_2O_2 \rightarrow H_2O + HO_2COOH^- \tag{1.3}$$

Peroxymonophosphate is even more powerful and reacts with PTB1B more than 7,000 times faster than H_2O_2, although the generation and existence of such a species *in vivo* remains to be evaluated. One very speculative possibility is that peroxymonophosphate could be biologically accessible *via* direct enzymatic phosphorylation of H_2O_2 (LaButti et al. 2007). On the other hand, the proximity of proteins to the source of ROS/RNS can also have an obvious influence on target selectivity and rates of thiol oxidation within the cell (Chen et al. 2008; Paulsen et al. 2012).

Besides thiol oxidation, H_2O_2 can also easy react *via* one electron reduction with transition metals such as iron and copper. This reaction, known as the Fenton reaction, generates hydroxyl radical (Eq. 1.4; where L = ligand), a highly reactive

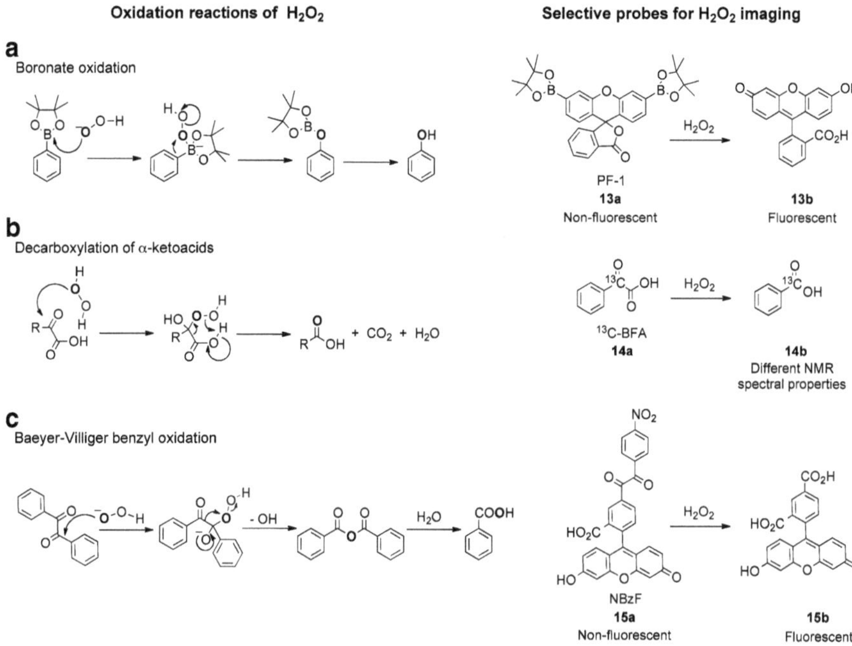

Fig. 1.2 Probes for the detection of hydrogen peroxide

species that can indiscriminately damage different biological target molecules such as DNA, proteins, or lipids *via* radical-mediated reactions and contributes significantly to the development of oxidative stress (Prousek 2007).

$$Fe^{2+} + H_2O_2 \rightarrow LFe^{3+} + HO^{\bullet} + HO^{-} \tag{1.4}$$

Although H_2O_2 has a relatively short half-life in cells ($t_{1/2} = 10^{-5}$ s), high variations in concentration can form locally and temporarily in cells (Giorgio et al. 2007). Molecular imaging is a powerful method for real-time monitoring of H_2O_2 in biological systems. Traditional tools for ROS detection, such 2′-7′-dichlorodihydroflurescein, are not specific for H_2O_2, but recent years has seen the development of many chemoselective probes, in particular by Chang's research group (Lippert et al. 2011). Because of its ambiphilic reactivity, H_2O_2 can be chemically differentiated from other ROS. With a pKa value of about 11, H_2O_2 is also a good nucleophile owing to the α-effect of adjacent nonbonding orbitals on its oxygen atoms (Jencks and Carriulo 1960), particularly in the deprotonated form. H_2O_2-mediated deprotection of aryl boronates to phenols (Kuivila and Armour 1957, Fig. 1.2a) has served as the starting point for the development of selective probes for H_2O_2 imaging. The installation of boronic esters on the 3′ and 6′ positions of the fluorescein core, for example, produces masked fluorescein (e.g., **13a**) in which the two boronates block the molecule in the closed lactone form, eliminating its absorptive and emissive properties. The reaction with H_2O_2 transforms the boronates into phenols (e.g., **13b**) with a strong increase in fluorescence intensity (Chang et al. 2004). Although boronates should also react with alkyl peroxides (like lipid peroxides), the reaction with H_2O_2 should be faster because hydroxide anion is a better leaving group than alkoxides (RO^-). Obviously, the reaction is accelerated at higher pH levels, and local pH changes can alter the probe response. Based on the same approach, several other probes have been developed with enhanced sensitivity (Dickinson et al. 2011) or which are targeted to specific subcellular compartments (Dickinson and Chang 2008). Other bioorthogonal reactions have inspired alternative probes for selective H_2O_2 detection. Chang's group, for example, has developed a hyperpolarized ^{13}C probe based on oxidative decarboxylation of α-ketoacids (Fig. 1.2b) for ^{13}C MRI resonance imaging of H_2O_2 (**14a**, Lippert et al. 2011). Finally, Nagano's group has taken advantage of the less known Baeyer-Villiger benzil oxidations (Fig. 1.2c) to design a highly sensitive probe for H_2O_2 (**15a**, Abo et al. 2011). Together, these novel chemical tools have revealed fundamental new insights into the role of H_2O_2 in cells (Dickinson et al. 2011; Miller et al. 2010).

1.3.2 Hypo(Pseudo)halous Acid

The heme-containing enzymes myeloperoxidase (MPO) and lactoperoxidase (LPO) use H_2O_2 to oxidize halides (Cl^-, Br^- and I^-) and thiocyanate (SCN^-) into their respective hypohalous acids (HOXs). These species are more reactive than H_2O_2 and are used as antimicrobials by the immune system (Albrich et al. 1981). Although chloride is the most abundant negative electrolyte in living systems (1–60 mM in cell, 100 mM in extracellular fluids), the major substrate of MPO is SCN^-, whose affinity for the enzyme is a 1,000-fold greater than is that of Cl^- (Hawkins 2009). Generally SCN^- has a concentration of 20–250 μM (Van Dalen et al. 1997) but can reach a concentration of 1 mM in saliva, where LPO produces almost exclusively HSCN (Ashby 2008). Br^- is nearly 60-fold more specific for MPO but, with a concentration of 20–100 μM, the HOBr generated from neutrophils is estimated to be ~10% of HOCl production (Chapman et al. 2009). Finally, since the iodide concentration in cells is usually low (~1 μM), the production of hypoiodous acid *in vivo* can be considered negligible. Hypohalous acids are weak acids; their pKa values are 7.6 for HOCl, 8.7 for HOBr, 10.4 for HOI, and 5.3 for HOSCN (Davies et al. 2008). Except for HOSCN, which is completely deprotonated, these species exist as a mixture of their acid and anion forms at physiological pH. HOCl and, to a lesser extent HOBr, are extremely strong oxidants and indiscriminately react with a large variety of functional groups in proteins, nucleotides, and membrane lipids. HSCN is several times less reactive and selectively oxidizes thiols (Lloyd et al. 2008).

HOX species may also interconvert, and this reaction reflects the relative oxidizing strengths of these acids. As a result, if HOCl can oxidize Br^-, I^-, and SCN^-, HOBr reacts solely with I^- and SCN^- (Spalteholz et al. 2005), though only the inter-conversion with SCN^- appears to be relevant *in vivo* (Nagy et al. 2006). HOXs are electrophilic oxidants and undergo two-electron nucleophilic substitution. Thiols react with HOCl and HOBr *via* formation of a very unstable sulfenyl halide (Eqs. 1.5 and 1.6), which readily rearranges to sulfenic acid (Nagy and Ashby 2007a). Otherwise, HOSCN generates a sulfonyl-thiocyanate (Eq. 1.8), which can react with another thiol to yield a disulfide and SCN^- (Eq. 1.9, Ashby and Aneetha 2004). Kinetic studies indicate that reaction proceeds *via* thiolate and hypothiocyanous acid reactants, indicating that this mild oxidant is very selective for proteins with low pK_a thiol residues (Nagy et al. 2009).

$$R-S^- + HO-Cl \rightarrow R-S-Cl + HO^- \tag{1.5}$$

$$R-S^- + HO-Br \rightarrow R-S-Br + HO^- \tag{1.6}$$

$$R-S-X + H_2O \rightarrow R-S-OH + X^- \tag{1.7}$$

$$R-S^- + HO-SCN \rightarrow R-S-SCN + HO^- \tag{1.8}$$

$$R_1-S^- + R-S-SCN \rightarrow R-S-S-R_1 + SCN^- \tag{1.9}$$

It is generally accepted that the sulfur-containing residues (cysteine and methionine) react most rapidly with HOCl (with rate constants of $\sim 3\text{–}4 \times 10^{-8}\ M^{-1}\ s^{-1}$), but strong nucleophilic amines (such as lysine and histidine) are also modified at high HOCl concentrations to yield chloramines (Pattison and Davies 2001).

$$R_1R_2NH + HO-Cl \rightarrow R_1R_2N-Cl + HO^- \tag{1.10}$$

$$R-S^- + R_1R_2N-Cl + H^+ \rightarrow R-S-Cl + R_1R_2NH \tag{1.11}$$

Small-molecule amines such as ammonia or glycine may react with HOCl to form membrane-permeable chloramines (Eq. 1.10), which selectively mediate thiol oxidations (Midwinter et al. 2006). Taurine-chloramine (TauCl) is another biologically relevant chlroamine. This cell-impermeable chloramine is generated especially in neutrophils, in which taurine is the most abundant free "amino acid" (10–30 mM) and also appears to play a scavenger role for chlorinated oxidants (Marcinkiewicz et al. 1995).

Although HOCl plays a fundamental role in destroying a wide range of pathogens, neutrophil-mediated HOCl production has also been implicated in several inflammation-associated diseases. A series of detection methods have been developed in order to shed light on its physiological and pathological roles (Fig. 1.3). These selective probes are based on the major oxidant strength of HOCl in comparison with other ROS. For example, *p*-methoxyphenol is selectively

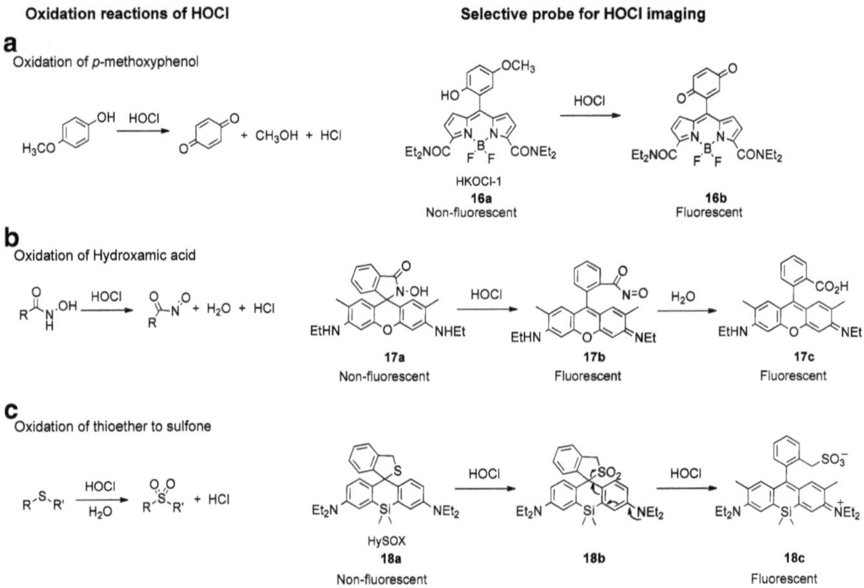

Fig. 1.3 Probes for the detection of hypochlorous acid

oxidized by HOCl (Fig. 1.3a) and shows good stability in the presence of other ROS and RNS. Yang's research group has taken advantage of this reaction to design a BODIPY-based probe (**16a**) that displays a significant increase of fluorescence after HOCl incubation (Sun et al. 2008). By analogy to chloramine formation, Tae's research group has instead developed a fluorescent probe based on HOCl-mediated oxidation of hydroxamic acid (Yang et al. 2009). Following generation of an unstable chlorinated intermediate, hydroxamic acid yields an acyl nitroso compound with elimination of HCl (Fig. 1.3b). Starting from a rhodamine scaffold, Tae's group designed a hydroxyamido derivative enclosed in a stable spirocyclic non-fluorescent form (**17a**), which demonstrates a rapid and selective fluorescent response in the presence of exogenous HOCl in cells as well as in mice. Recently, a thioether-rhodamine derivative was designed as a chemoselective probe for HOCl. In fact, thioethers and thiols are the major target of HOCl-mediated oxidation and, although thiols indiscriminately react with any ROS and RNS, thioethers show selective oxidation (Fig. 1.3c). The thioether group maintains the Si-rhodamine probe in the cyclic non-fluorescent form (**18a**), and the system recovers its highly conjugated structure through the generation of a sulfone (**18b**), followed by elimination of sulfinic acid (which then oxidizes to sulfonic acid) and a robust increase in fluorescence (Koide et al. 2011). Finally, although many efforts have been made, few methods (with the exception of probe **17a**) appear to have the appropriate characteristics for biological application *in vivo*.

1.3.3 Nitric Oxide and Peroxynitrite

NADPH-dependent nitric oxide synthase (NOS) catalyzes the synthesis of nitric oxide (NO) and L-citrulline *via* oxidation of the guanidine group of L-arginine (Alderton et al. 2001). NO, like carbon monoxide (CO) and hydrogen sulfide (H_2S), is a small gaseous signaling molecule implicated in numerous physiological processes (Cary et al. 2006). NO is a free radical species ($^{\bullet}N{=}O$) with a relative long half-life in comparison to other ROS/RNS ($t_{1/2} = 1$ s, Pacher et al. 2007) and can freely diffuse across membranes. Although an *S*-nitrosothiol could be directly generated by reaction with thiyl radicals (Jourd'heuil et al. 2003), most of its activity is mediated by development of intermediary nitrosating agents (Keszler et al. 2010). These nitrosating species are generated *via* a series of oxidation reactions. NO forms nitrous anhydride (N_2O_3) in the presence of oxygen (Eqs. 1.12a and 1.12b), and this unstable species can easy be hydrolyzed (Eq. 1.13) or undergo two-electron nucleophilic substitution by thiols (Goldstein and Czapski 1996) and amines (Caulfield et al. 1998). N_2O_3 is generally accepted as the major nitrosylating agent *in vivo*, although other pathways are proposed to explain *S*-nitrosothiol formation in cells (see Sect. 1.4.2).

$$NO + O_2 \rightarrow NO_2 \tag{1.12a}$$

$$NO + NO_2 \rightarrow O_2N - N = O \tag{1.12b}$$

$$O_2N - N = O + 2OH^- \rightarrow 2NO_2^- + H_2O \tag{1.13}$$

$$O_2N - N = O + R - S^- \rightarrow R - S - NO + NO_2^- \tag{1.14}$$

$$O_2N - N = O + R - NH_2 \rightarrow R - NH - NO + HNO_2 \tag{1.15}$$

$O_2^{\bullet -}$ transforms the relatively unreactive NO into peroxynitrous acid (ONOOH), a strong oxidant (Eq. 1.16). This diffusion-controlled reaction has the largest rate constant known for NO (1×10^{10} M^{-1} s^{-1}) and, *in vivo*, is spatially associated with the sources of superoxide (Radi et al. 2001). ONOOH is a weak acid with a pK_a value of 6.8 (Kissner et al. 1997) and is in equilibrium with its deprotonated form (peroxynitrite, $ONOO^-$) at physiological pH. Both species are strong oxidants and can participate in one- and two-electron oxidation reactions, although HONOOH shows a faster reactivity (Moro et al. 1994) and $ONOO^-$ appears to be a powerful nucleophile as well. Despite its short half-life, ONOOH appears able to diffuse across the membranes (Ferrer-Sueta and Radi 2009) where it can react with a wide group of substrates (Marla et al. 1997). Typically, thiols are oxidized to disulfides by ONOOH (Squadrito and Pryor 1998), but it is still uncertain which intermediate species are involved between the sulfenic acid (R-SOH) and *S*-nitrothiol (R-SNO_2).

Nevertheless, kinetic analysis clearly shows that this reaction involves the pairing of the thiolate/peroxynitrous acid. Other studies have shown that ONOOH can also mediate the *S*-nitrosylation of GSH with elimination of H_2O_2 (van der Vliet et al. 1998). In any case, considering that ONOOH is very unstable and can undergo hemolytic fission to generate the one-electron oxidant, hydroxyl ($^{\bullet}OH$), as well as $^{\bullet}NO_2$ radicals (Szabo et al. 2007), it is difficult to assess which reaction ONOOH participates in directly. ONOOH decomposition is also facilitated by bicarbonate (HCO_3^-). CO_2 and readily undergoes nucleophilic addition by $ONOO^-$ (with a constant rate of $5.8 \times 10^4\ M^{-1}\ s^{-1}$) to yield $^{\bullet}NO_2$ and $CO_3^{\bullet-}$ (Eqs. 1.18 and 1.19). CO_2 also mediates the generation of other oxidant species (Eq. 1.20) and is considered to contribute to the nitration of tyrosine side chains (Squadrito and Pryor 1998).

$$NO + O_2^{\bullet-} \rightarrow ONOOH \tag{1.16}$$

$$ONOOH \rightarrow {}^{\bullet}NO_2 + {}^{\bullet}OH \tag{1.17}$$

$$ONOO^- + CO_2 \rightarrow ONO - OCO_2^- \tag{1.18}$$

$$ONO - OCO_2^- \rightarrow {}^{\bullet}NO_2 + CO_3^{\bullet-} \tag{1.19}$$

$$ONO - OCO_2^- \rightarrow O_2N - OCO_2^- \tag{1.20}$$

The development of powerful tools to detect NO signaling remains an active area of research. The first approach developed to detect NO was based on the *o*-phenylenediamine scaffold which, in the presence of NO and O_2 (Uppu and Pryor 1999), yields the corresponding aryl triazole (Fig. 1.4a). The electronic difference between the electron-rich diamine (**19a**) and the electron-poor triazole (**19b**) leads to a robust fluorescent switch (Kojima et al. 1998). Obviously, this is an indirect approach for NO detection. Specifically, the aryl diamine reacts with N_2O_3 to form *N*-nitrosylamine, which evolves to triazole *via* intramolecular nucleophilic displacement. Recently, Xu's research group has developed an elegant new probe based on the fluorescein scaffold; in this case, the diamine blocks the fluorophore in spirolactame form; reaction with NO leads to the formation of a triazole ring and concomitant fluorescence response (Zheng et al. 2008).

Parallel approaches for direct detection of NO have also been developed. It is well-known that NO has strong affinity towards transition metals, including the ability to bind the heme-group. Taking advantage of this property, several probes have been designed conjugating transition metals to a fluorescent core (Lim and Lippard 2007). This approach was based upon the release of a fluorophore, initially quenched by electron or energy transfer through coordination to a paramagnetic transition-metal center such as Fe(II), Co(II), or Cu(II). The presence of NO causes

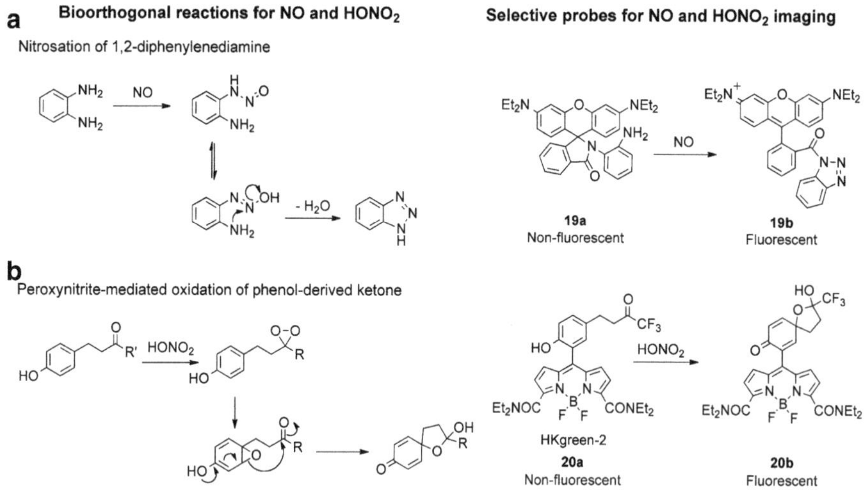

Fig. 1.4 Probes for the detection of nitric oxide and peroxynitrite

the displacement of the metal-complex with concomitant fluorescence activation. Taken together, the two approaches described are perfectly complementary for chemospecific NO detection in living systems.

Largely because of the ongoing difficulty of direct and unambiguous detection, the biological relevance of ONOOH remains controversial. Selective probes for peroxynitrous acid have been inspired based on the similar chemical behavior of $ONOO^-$ and peroxymonosulfate ($HOOSO_3^-$). In fact, $ONOO^-$ can react with ketones to yield dioxiranes (Yang et al. 1999). Furthermore, it is known that internal dioxiranes can intramolecularly oxidize phenol derivatives to quinones and that this reaction is facilitated when the dioxiranes are generated from ketones with electron-withdrawing groups (Yang et al. 2000). Yang's group has demonstrated that $ONOO^-$ can also oxidize anisole-derivative ketones *via* dioxirane *in situ* formation (Fig. 1.4b). Based upon this reaction, several fluorescent probes have been developed in which an anisole-derivative is conjugated with a fluorophore and the peroxynitrite-mediated oxidation selectively allows the release of the fluorescent molecule (e.g., **20a**; Yang et al. 2006; Peng and Yang 2010).

1.3.4 Hydrogen Sulfide

For many years, hydrogen sulfide (H_2S) was solely considered as a toxic gas, but in the last decade its involvement in mammalian cell signaling has become apparent (Li et al. 2011). H_2S is synthesized by two pyridoxal 5′-phosphate-dependent enzymes involved in cysteine metabolism (Chiku et al. 2009): cystathione β-synthase (CBS) and cystathione γ-lyase (CSE). H_2S can also be generated from L-methionine *via* transsulfuration (Bearden et al. 2010) or release from a persulfide (RS–SH) by thiol-disulfide exchange. Like other gaseous signaling molecules, such as NO and CO, H_2S rapidly crosses cell membranes with no specific transporter (Mathai et al. 2009). In fact, although H_2S shows structural similarities with H_2O it does not form strong hydrogen bonds and, with a dipole moment of 0.9 D, H_2S can freely diffuse through the hydrophobic core of biological membranes.

H_2S is a diprotic acid with a pKa_1 of 6.9 and pKa_2 of 11.96; consequently, H_2S is in equilibrium with its deprotonated form (with a H_2S/HS^- of 1:3) at physiological pH (Liu et al. 2011). In many respects, H_2S exhibits a reactivity profile akin to cysteine. It is a powerful nucleophile and, like thiols, can react with electrophiles as well as oxidants. With two-electron oxidants (Eq. 1.21), H_2S forms hydrogen thioperoxide (HSOH – Carballal et al. 2011). HSOH is a very reactive sulfur species and can be oxidized rapidly to $SO_4{}^{2-}$ or undergo reaction with thiols. *In vitro*, the reaction of H_2S with HSOH leads exclusively to the formation of a complex mixture of polysulfides ($HS_X{}^-$ with x = 2–8) (Eq. 1.22), which have been proposed to mediate the *S*-sulfhydration of proteins (Nagy and Winterbourn 2010).

$$HS^- + H_2O_2 \rightarrow HS - OH + H_2O \qquad (1.21)$$

$$HS - OH + n\,HS^- \rightarrow HS - S_n^- + H_2O \qquad (1.22)$$

The biological functions of H_2S are still unclear, but growing evidence suggests a role for H_2S in the regulation of cardiovascular (Elsey et al. 2010) and gastrointestinal (Wang 2010) systems. Moreover, its biologically relevant levels are still debated, and a wide range of concentrations has been reported (Kabil and Banerjee 2010). To investigate such issues, a number of research groups have focused their efforts on the development of small-molecule probes for H_2S detection (Fig. 1.5). H_2S can be considered a strong reducing agent and this property has been widely exploited in organic synthesis to reduce aromatic nitro (Lin and Lang 1980) and azido (N_3) (Pang et al. 2009) groups to aniline. Chang's research group developed a selective probe for H_2S detection by installing an N_3 group in the 6′ position of a rhodamine core (**21a**), which locks the fluorophore in the spirocyclic non-fluorescent form. Reduction to the amine releases the fluorescence signal (**21b**) and such probes display high selectivity for H_2S over other biologically relevant reactive sulfur, oxygen, and nitrogen species (Lippert et al. 2011). Subsequently, Wang's research group used a similar approach to design a sulfonylazide dansyl derivative whereby the difference in electronegativity between the azide and amine group triggers a change in electronic and, thus, fluorescent properties (Liu et al. 2011). Both probes appear quite promising given that azido group is quite inert and has shown considerable compatibility for application in living systems.

As previously mentioned, H_2S exhibits a similar reactivity as compared to cysteine. Theoretically, H_2S can be considered a non-substituted thiol and can give nucleophilic attack twice, whereas other thiols, such as cysteine, can participate in just one nucleophilic attack. Exploring this property, Xian's research group

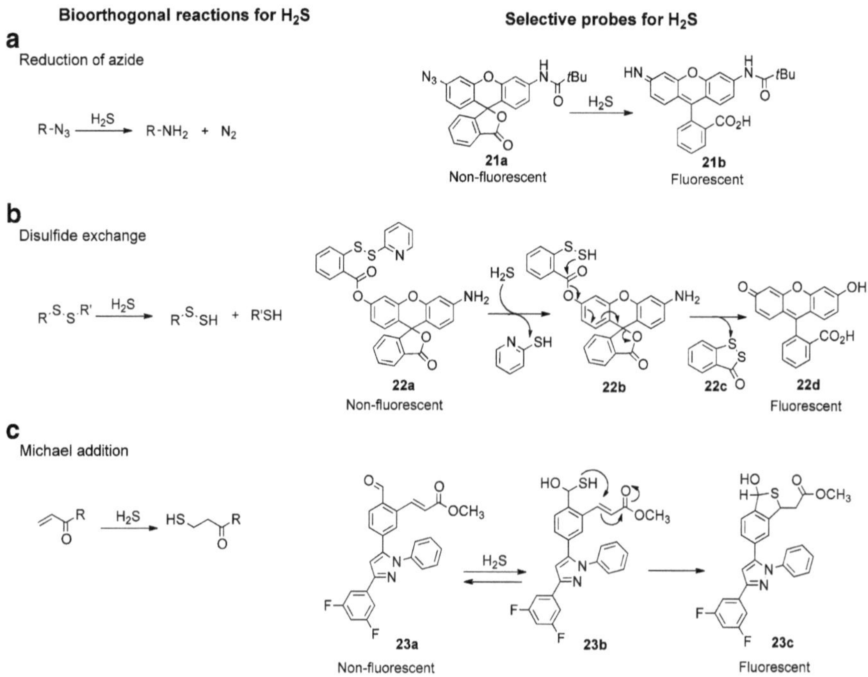

Fig. 1.5 Probes for the detection of hydrogen sulfide

hypothesized that a probe with two electrophilic centers could be selective for H_2S detection and developed a thiosalicylic acid ester of fluorescein in which the thiol group is functionalized as thiopyridyl disulfide (**22a**). Both thiols and H_2S readily yield disulfide exchange with activated disulfide, but only the disulfide intermediate from H_2S (**22b**) can rearrange, *via* nucleophilic attack on the intramolecular phenyl ester, leading to benzodithiolone (**22c**) and fluorescein (**22d**) products (Liu et al. 2011). Along these lines, Chuan's research group has developed a selective probe for H_2S using a Michael acceptor and an aldehyde group as two electrophilic centers (**23a**). H_2S forms a thioacetal with the aldehyde group (**23b**) followed by an intramolecular Michael addition with the internal α,β-unsaturated compound (**23c**). Such H_2S-specific rearrangement leads to a strong increase in fluorescence. Using this probe, Chuan's group has succeeded for the first time in imaging enzymatic H_2S production in living cells (Qian et al. 2011a).

1.4 Oxidative Modification of Protein Cysteine Residues

The cysteine side chain, with its high nucleophilic capacity, appears to be the principal target of ROS/RNS in cells. The sulfur atom of cysteine may assume a wide range of oxidation states (i.e., −2 to +4, Fig. 1.6) and each form exhibits a distinct

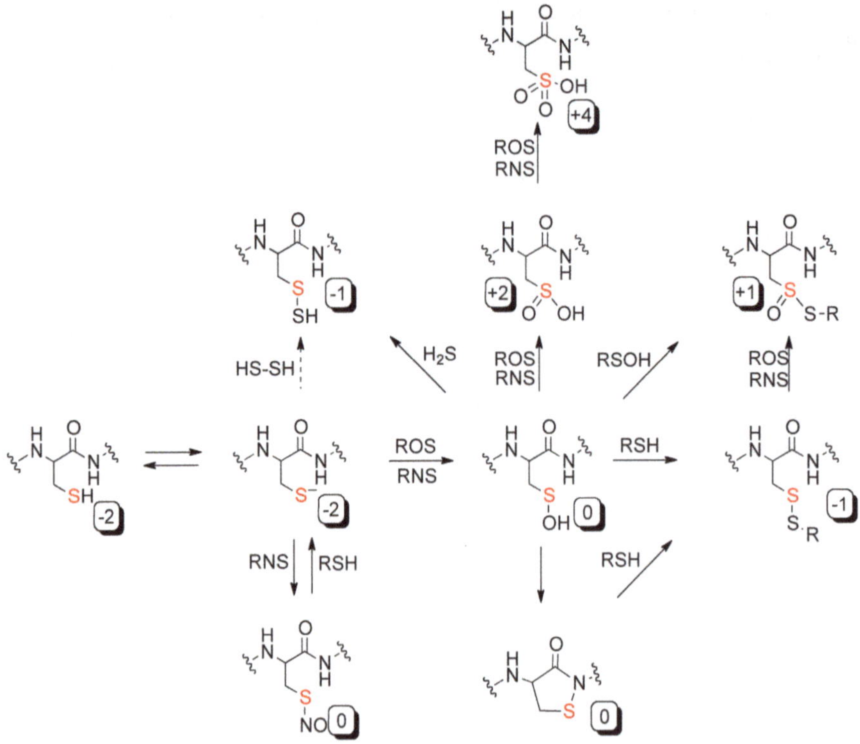

Fig. 1.6 Biologically relevant cysteine oxidation states. Oxidation number correspond to sulfur atom denoted in red

chemical reactivity. In the thiolate form, sulfur undergoes oxidation to generate a sulfenic acid, and this oxoform can be considered as a central species among thiol modifications. Sulfenic acid may be reduced to a disulfide by reaction with intra- and inter-molecular thiols or further oxidized to sulfinic acid at high ROS/RNS concentrations. In some cases, the sulfenic acid can lead to the formation of sulfenamide and thiosulfinate ester groups. The cysteine reactivity landscape becomes more complex given that the thiolate may react with RNS and RSS to form *S*-nitrosothiol and persulfide, respectively. Moreover, depending upon the nature of the protein microenvironment, many of these modifications are reactive and can interconvert with one another. In order to highlight the possible role and significance of each modification, their distinct physical and chemical properties are outlined below.

1.4.1 Sulfenic Acid

Sulfenic acids (RSOH) are directly formed *in vivo* by the oxidation of thiols with two-electron oxidants such as H_2O_2, ONOOH, or alkyl peroxides and many enzymes form SOH intermediates during their catalytic cycles (Lim et al. 2011). In addition, RSOH can be generated from the hydrolysis of sulfonyl-halides (Nagy and Ashby 2007a), sulfonyl-thiocyanates (Ashby and Aneetha 2004), and thiosulfinates (Nagy and Ashby 2007b). Hydrolysis of *S*-nitrosothiols can also yield RSOH (Percival et al. 1999), but the reaction has a high activation barrier (Moran et al. 2011). The thiyl radical ($RS^•$) can also yield sulfenic acid in the presence of hydroxyl radicals ($^•OH$); however, this reaction is not likely to be widespread in a biological setting. Otherwise, sulfenic acids are usually obtained in organic synthesis from the thermolysis of sulfoxides (Sivaramakrishnan et al. 2005) or thiosulfinates (Block and O'Conner 1974).

The S–O bond length in RSOH is distinctly longer than that of sulfoxides, indicating that the tautomeric structure, R–S(O)H, can be excluded (Goto et al. 1997). RSOH is a weak acid but, because of their high reactivity, their pKa has been difficult to measure. In the case of some stable small-molecule sulfenic acids, an acidity of two to three orders of magnitude lower than the corresponding thiols has been found (Heckel and Pfleiderer 1983). This trend could quite realistically offer an estimation of pKa value for other sulfenic acids. For example, the pKa value of a high-hindered triptycenyl sulfenic species was recently reported to be 12.5 (McGrath et al. 2010), about three times higher than a typical thiol. Although they are usually defined as transient species, intramolecular hydrogen bonds (Heckel and Pfleiderer 1983) and steric hindrance (Nakamura 1983) can play major roles in stabilizing protein RSOH. More generally, hydrogen bonds in apolar protein microenvironments, and the absence of vicinal thiols can all stabilize sulfenic acids (Claiborne et al. 1993).

Sulfenic acids exhibit potent electrophilic (Fig. 1.7) and relative weak nucleophilic (Fig. 1.8) reactivity. In small-molecule RSOHs, this dual behavior can lead to self-condensation in which one sulfur atom functions as a nucleophile and the second as an electrophile to yield a thiosulfinate ester (Fig. 1.7a). In this reaction, intermolecular hydrogen bonding mediates the self-condensation of RSOH, thereby reducing the free energy of activation for thiosulfinate formation thermodynamically preferred over the acid (Davis et al. 1986). On the other hand, self-condensation of RSOH only competes with thiol-based reduction of sulfenic acid at high pH (Nagy and Ashby 2007a, b) and is otherwise considered negligible.

As indicated above, thiols can reduce sulfenic acid to form a disulfide (Fig. 1.8a) and this condensation represents an important biological reaction. Moreover, this reaction allows the recycling of sulfenic acid in as much as the disulfide can be reduced to thiol through the action of cellular-reducing agents such as glutathione (GSH), glutaredoxin (Grx) and thioredoxin (Trx). Dithiols (such as DTT) and Trx can also directly reduce RSOH back to the thiol form (Poole et al. 2004). Thiols are also used as trapping agents to demonstrate the formation of aleatory sulfenic acid in the thermolysis of sulfoxides (Kamiya et al. 1973). Several inorganic molecules, such as $NaAsO_2$ (Saurin et al. 2004), hydrazine, and NaN_3 (Allison 1976), have been reported to reduce RSOH to the thiol; however, their selectivity has not been sufficiently explored. Although sodium ascorbate (Fig. 1.9a) is generally considered a specific reducing agent for *S*-nitrosothiols (Turell et al. 2008), it can also reduce

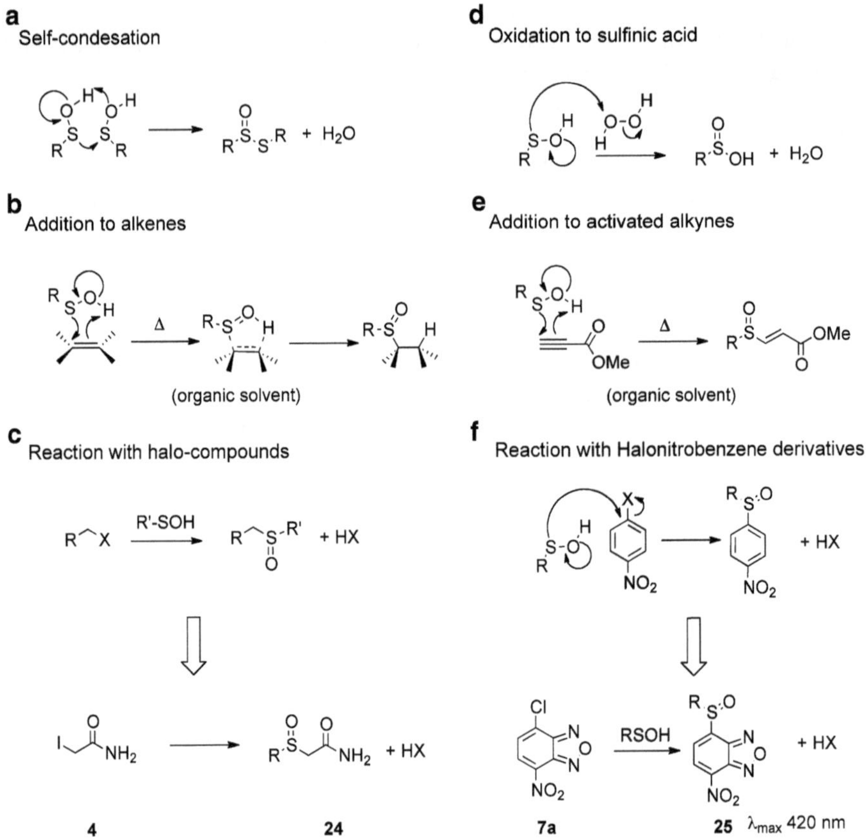

Fig. 1.7 Sulfenic acid reactivity: S atom as a nucleophile

RSOH through an addition/elimination mechanism (You et al. 1975; Fig. 8b); recently, the ability of sodium ascorbate to reduce sulfenic acid *in vivo* has also been proposed (Monteiro et al. 2007).

Sulfenic acids can also be oxidized by two-electron oxidants (Fig. 1.7b) or by O_2 in the presence of trace metal ions. The oxidation of non-hindered thiols leads only to the disulfide. Considering that sulfenic acids can be intermediates in disulfide formation, it is worth noting that oxidation of sulfenic acid is slower, compared to its condensation rate with thiols (Luo et al. 2004). Kinetic studies have showed that the rate constant for oxidation of cysteine sulfenic acid is approximately two to three orders of magnitude slower than cysteine (Hugo et al. 2009). Two pathways may be hypothesized for H_2O_2 oxidation of RSOH: first, a concerted mechanism mediated by formation of hydrogen bond and, second, the direct participation of a sulfenate anion (RSO^-). The pH profile of H_2O_2-mediated oxidation of RSOH indicates that the sulfenate is the reacting species and therefore, RSOH generated from a very low pKa cysteine should be more susceptible to irreversible oxidation to sulfinic acid.

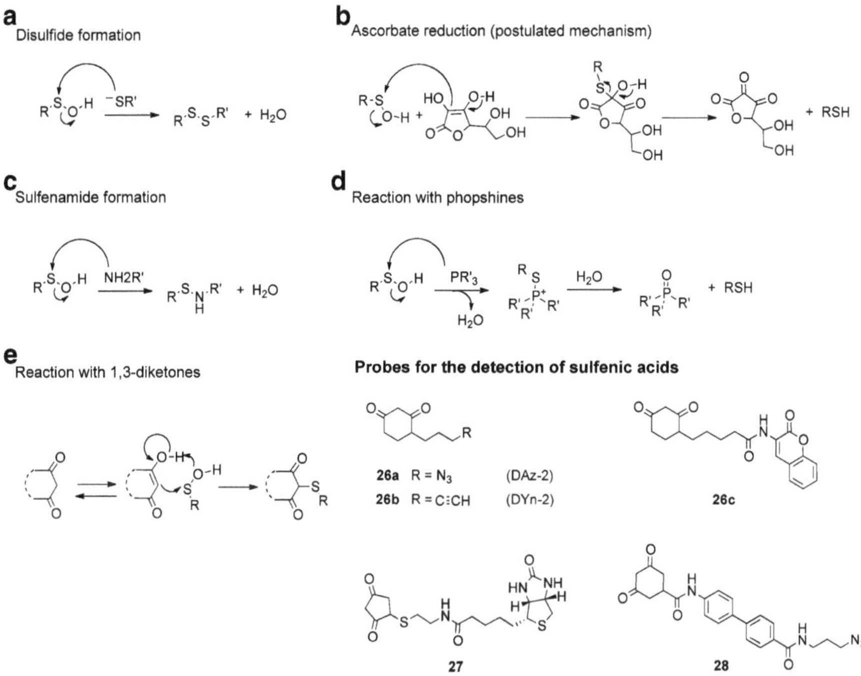

Fig. 1.8 Sulfenic acid reactivity: S atom as an electrophile

Reactive cysteine residues in specific proteins may readily oxidize to sulfenic acid. This reversible thiol modification plays a significant role in protein regulation and cell signaling (Paulsen and Carroll 2010). As the first (and sometimes only) oxidation product, RSOH is a key "marker" for ROS/RNS-sensitive cysteine residues and, as such, there is clear evidence of its role in cell signaling (Klomsiri et al. 2011; Roos and Messens 2011). As a result, interest in the RSOH detection has grown rapidly in the last decade and the reactivity of protein sulfenic acids continues to be explored. Sulfenic acids react as weak nucleophiles with alkyl or aryl halides, alkenes, and alkynes (Fig. 1.7c–e). Although sulfenate ions (RSO^-) should be an ambident nucleophile, alkylation with halides always yields the sulfoxide (Hogg and Robinson 1979). This reaction should be taken in account when blocking protein thiols with iodoacetamide since alkylation of sulfenic acid (Fig. 1.7e, **24**) becomes relevant at reagent concentrations typically employed in such procedures. The aromatic nucleophilic substitution of halonitroarene has been employed to detect the sulfenic acid group in proteins (Fig. 1.7f). For example, the electrophilic reagent 7-chloro-4-nitrobenzo-2-oxa-1,3-diazole (NBD-Cl) reacts with both sulfenic acids and thiols to yield covalent adducts (**25** and Fig. 1.7f) that are distinguishable from each other with respect to mass and their UV–vis absorption profiles (Ellis and Poole 1997). Conversely, alkenes (Kingsbury and Cram 1960) and electron-deficient alkynes (Goto et al. 1997) have long been used in organic synthesis in organic

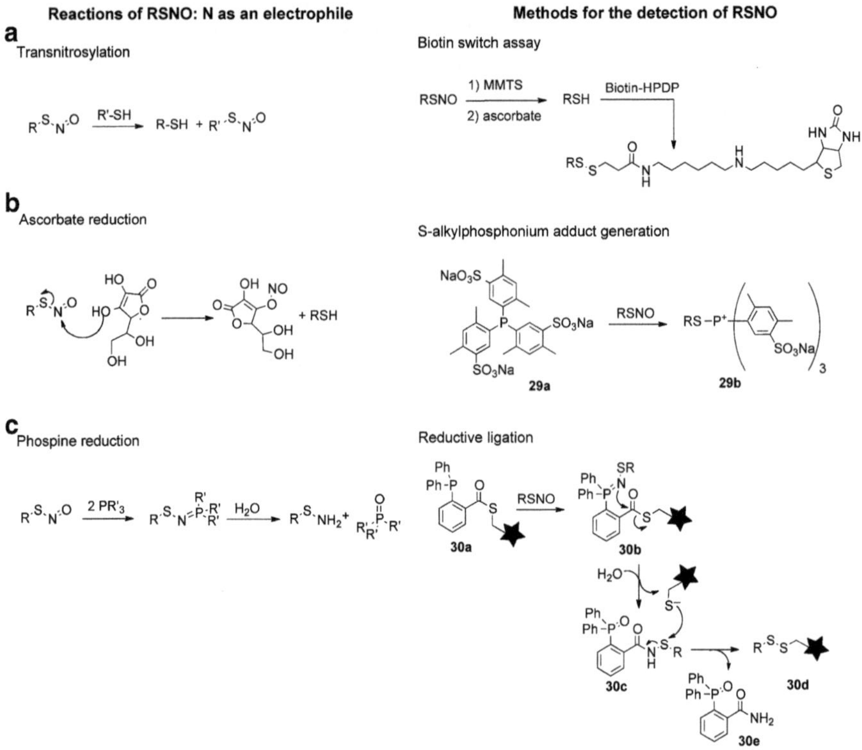

Fig. 1.9 Chemical reactivity and detection of *S*-nitrosothiols

solvent at high temperature for to trap transient sulfenic acid species (Fig. 1.7c–d). Such nucleophilic addition reactions occur through a concerted, cyclic process – the reverse of the elimination of sulfenic acids from sulfoxides – and should therefore proceed through the same transition state *via* regiospecific Markovnikov addition (Fig. 1.7c).

The sulfur atom in sulfenic acid also functions as an electrophile (Fig. 1.8a–e). Primary and secondary amines react with RSOH to yield sulfenamides (Fig. 1.8c, Allison et al. 1973). This reaction is quite slow and, in a biological context, only makes sense when this reaction takes place intramolecularly. In fact, in some proteins sulfenic acid can react with the nitrogen backbone of the neighboring residue to yield a five-member cyclic sulfenamide (see Sect. 1.4.4). Although RSOH could also react with intramolecular lysine residues, this species has been rarely detected (Raftery et al. 2001) and its formation *in vivo* remains unclear. Sulfenic acids also react with phosphorous III compounds to yield trialkylphosphine oxide and the corresponding thiol *via* hydrolysis of the intermediate phosphonium (Fig. 1.8d, Goto et al. 2003). To date, coupling of RSOH with cyclic 1,3-diketones is the only reaction that has demonstrated high selectivity for this cysteine oxoform.

Cyclic 1,3-diketones react with sulfenic acid in their enol-form, likely through hydrogen bond-mediated direct substitution (Fig. 1.8e).

5,5-dimethyl-1,3-cyclohexandione, commonly known as dimedone, is often used trapping agent for detecting protein sulfenic acid modifications (Benitez and Allison 1974). However, this reagent is not ideal for proteomic studies since it lacks an affinity and/or detection handle. In recent years, several advantageous derivatives of dimedone have been developed. The first approach was direct conjugation of 1,3-cyclohexandione with a biotin or a fluorophore (Fig. 1.8, **25**) moiety (Poole et al. 2005). More recently, Carroll's research group has functionalized the 1,3-cyclohexandione scaffold with an azide (**26a**) (Leonard et al. 2009) or alkyne (**26b**) (Paulsen et al. 2012) chemical reporter, known as DAz-2 and DYn-2, respectively. These small probes are membrane-permeable and enable detection and trapping of protein sulfenic acids directly in cells. After cellular tagging, lysates are generated and coupled to detection reagents *via* Staudinger ligation or Huisgen azide-alkyne 1,3-dipolar cycloaddition (also known as click chemistry). Carroll's group has also reported the development of a heavy-isotope coded dimedone analogue, which in combination with **10** (see Sect. 1.2) represents a powerful new approach to quantify the extent of sulfenic acid modification at individual protein cysteine residues (Seo and Carroll 2011). Furdui's research group has reported a new effective probe for sulfenic acids based on a related scaffold, 1,3-cyclopentadione (**27**, Qian et al. 2011b). Finally, Carroll's group has recently reported tri-functional probes consisting of a dimedone-like warhead, chemical reporter, and binding module (**28**) that enhances the detection of sulfenic acid in specific classes of signaling proteins (e.g., PTPs, Leonard et al. 2011). In parallel research, antibodies that recognize the protein-dimedone adduct have also been developed (Seo and Carroll 2009; Maller et al. 2011). Carroll and co-workers have applied such antibodies to visualize sulfenic acid modifications in tumor cells and profile growth factor-dependent changes in thiol oxidation among breast cancer cell lines (Seo and Carroll 2009).

1.4.2 Nitrosothiol

NO and its metabolites may react with the side chain of cysteine to yield *S*-nitrosothiols (Fig. 1.9, RSNO). This modification can lead to significant changes in the structure and function of proteins (Denninger and Marletta 1999). *S*-nitrosylation represents an important signaling mechanism and many proteins have been identified as *S*-nitrosylation targets (Jaffrey et al. 2001). S–NOs may be formed through the direct reaction of thiols with N_2O_3 and less commonly with ONOOH (see Sect. 1.3.3) or by trans-nitrosylation with an NO-donor (Fig. 1.9a). Trans-nitrosylation is also the most important reaction in which RS–NOs are involved in biology and appears to the principal mechanism upon which NO-signaling is based (Tsikas et al. 1999). RS–NOs are exceptionally labile (Kashiba-Iwatsuki et al. 1997) and easy react with other thiols by trans-nitrosylation or, in some cases, by disulfide formation. These two possible pathways are due to the concomitant presence of two electrophilic centers on the *S*-nitrosothiol: one is the sulfur atom and the second is the nitrogen atom of the NO group. Sodium ascorbate (Fig. 1.9b) or Cu I (Dicks and Williams 1996) can also restore the thiol group from RS–NOs. Although the mechanism of ascorbate reduction is not entirely clear, the formation of *O*-nitrosoascorbic acid (Holmes and Williams 2000) appears quite accepted. The ascorbate-mediated reaction is used in the biotin switch assay. This technique is the most popular method for detecting *S*-nitrosothiols in proteins from cell extracts (Jaffrey et al. 2001) and involves three steps: (1) free thiols are blocked with a "thiol-specific reagent" (see Sect. 1.2) such as MMTS; (2) nitrosothiols are reduced by sodium ascorbate to thiols; and (3) nascent thiols are then conjugated to biotin with N-[6-(biotinamido)hexyl]-30-(20-pyridyldithio)-propionamide (biotin-HPDP, **1**) and analyzed by Western blot (Fig. 1.9).

Along with issues inherent to indirect or subtractive methods of detection, another challenge for this technique is the selective reduction of RSNOs with sodium ascorbate. It is well-known that sodium ascorbate can give false-positive signals for *S*-nitrosothiols when detected by the biotin switch assay (Huang and Chen 2006). In fact, ascorbate can also reduce sulfenic acids (see Sect. 1.4.1) and disulfides, albeit less efficiently. Furthermore, the ascorbate-mediated reduction of RS–NOs is quite slow and some nitrosothiols cannot be reduced during the incubation time. The reaction rate of reduction can be increased using catalytic amounts of Cu^+ (Kirsch et al. 2009). Copper promotes the direct de-nitrosylation of RS–NOs while ascorbate maintains the minimum concentration of Cu(I) necessary to catalyze the reaction, reducing Cu^{2+} to Cu^+ (Eqs. 1.23 and 1.24). However, it is critical to note that Cu^+ may also lead to oxidation of protein thiols.

$$\mathrm{RS-NO + Cu^+ \rightarrow R-S^- + {}^{\bullet}NO + Cu^{2+}} \tag{1.23}$$

$$\mathrm{Cu^{2+} + Asc^- \rightarrow Cu^+ + Asc^{\bullet -}} \tag{1.24}$$

In light these limitations the biotin switch assay has been superseded by direct and selective bioorthogonal reactions for RS–NOs in recent years. It is known that triaryl phosphines react with RS–NOs to yield an aza-ylide intermediate, which is hydrolyzed to phosphine oxide in water (Fig. 1.9c, Haake 1972). Xian's lab has employed such phosphine-mediated reactions to selectively modify *S*-nitrosothiols. In these studies, an aza-ylide nucleophile is proposed, and in the presence of an intramolecular electrophilic group (such as an ester), the *S*-nitrosothiols can be converted into a sulfenamide by intramolecular acyl transfer (Wang and Xian 2008). Although the resulting sulfenamide is more stable than the *S*-nitrosothiol, it is not suitable for RS–NO detection in proteins. In particular, the sulfenamide is reduced to the thiol in the presence of excess phosphine reagent. Consequently, Xian's research group has developed triarylphosphines (**30a**) with a thioester as a trap for the aza-ylide intermediate (**30b**). This kind of reagent reacts with RS–NOs to generate a sulfenamide (**30c**), which immediately undergoes nucleophilic attack of the thiol leaving-group (formed after the intramolecular acyl transfer) to yield a disulfide (**30d**), which is still prone to reduction, but is more stable than an S–N bond.

This approach appears promising and has been used to detect protein RS–NOs in cell extracts (Zhang et al. 2010). However, the relative instability of the disulfide link remains as a significant challenge for robust application of this approach *in vitro* and especially in cells.

Several other phosphine reagents have been developed to detect RS–NOs (see the review: Wang and Xian 2011). King's group reported the first water-soluble phosphine capable of direct labeling of *S*-nitrosothiols (Bechtold et al. 2010). Tris(4,6-dimethyl-3-sulfonatophenyl)-phosphine trisodium salt hydrate (TXPTS) appears to form a stable sterically hindered *S*-alkylphosphonium adduct, which can be detected by ^{31}P-NMR. However, the selectivity of all these reagents must be careful investigated. In fact, disulfides (see Sect. 1.4.3) and sulfenic acids (see Sect. 1.4.1) are also reduced by phosphines. On the other hand, triarylphosphines do not appear to be fast disulfide-reducing agents (Saxon and Bertozzi 2000) and the presence of electron-withdrawing substituents should further reduce their reactivity (Wang and Xian 2008). In contrast, Davis's group has shown that a disulfide formed between a short thio-peptide and methyl 4-thio-2-nitrobenzoate reacts very rapidly ($\sim$15 min) with several phosphines, including triphenylphosphine (Chalker et al. 2011). This reaction can be explained by the low pKa of methyl 4-thio-2-nitrobenzoate, which works as an excellent leaving group during the phosphine attack. Along these lines, some disulfides could exhibit cross-reactivity with the phosphine reagents developed for RS–NOs detection (Bechtold et al. 2010). Finally, cross-reactivity with sulfenic acids remains poorly understood. Although the selectivity of triarylphospine probes for RS–NOs detection over sulfenic acids was recently suggested (Li et al. 2012), however, the use of PTB1B as a model appears contradictory. In fact, when PTB1B is exposed to H_2O_2 the catalytic cysteine residue forms a stable, cyclic sulfenamide (Salmeen et al. 2003), which is predicted to be less reactive toward phosphines.

1.4.3 Disulfides

The formation of a disulfide bond between two cysteine residues or between a cysteine residue and a low-molecular-weight thiol may also have a significant impact on the structure and function of the macromolecule (Fan et al. 2009; Wouters et al. 2010). Protein disulfide generation is generally mediated by sulfenic acid formation (Eq. 1.25, see Sect. 1.4.1) or by thiol-disulfide exchange (Eq. 1.26). Thiols can also react with sulfenyl halides (Eq. 1.27), sulfenyl thiocyanates (Eq. 1.28), and thiosulfinate esters (Eq. 1.29) to form disulfide. Moreover, the reaction of *S*-nitrosothiols (Eq. 1.30) with thiols can sometimes proceed with release of nitroxyl (HNO) and disulfide formation (Hogg 2002).

$$R - S - OH + R_1 - S^- \rightarrow R - S - S - R_1 + OH^- \tag{1.25}$$

$$R - S - SR_1 + R_2 - S^- \rightarrow R - S - S - R_2 + R_1 - S^- \tag{1.26}$$

$$R-S-X+R_1-S^- \rightarrow R-S-S-R_1+X^- \tag{1.27}$$

$$R-S-SCN+R_1-S^- \rightarrow R-S-S-R_1+SCN^- \tag{1.28}$$

$$R-S-S\,(O)\,R_1+R_2-S^- \rightarrow R-S-S-R_2+R_1-S-O^- \tag{1.29}$$

$$R-S-NO+R_1-S^- \rightarrow R-S-S-R_1+NO^- \tag{1.30}$$

Thiol-disulfide exchange is one of the most important sulfur-based reactions in biology and can regulate the structure and activity of proteins that contain regulatory cysteines (Ilbert et al. 2007). Although this reaction is spontaneous, it can also be catalyzed by transition metals (Arisawa and Yamaguchi 2008). The driving force of the reaction is the relatively low activation energy required for thiols to break disulfide bonds (Fernandes and Ramos 2004). The thiol–disulfide exchange reaction is a bimolecular nucleophilic substitution reaction (SN2) in which the attacking nucleophile is the thiolate. The reaction proceeds with a linear transition state and this geometry has a significant impact on the reaction rate. Consequently, in proteins, cysteines that are situated in a position in which a linear transition state is difficult to achieve will be much less reactive.

The rate of protein thiol–disulfide exchange reactions is influenced by several additional factors. These include the pKa of the thiolate and the leaving-group thiol, nearby charged amino acid residues, and steric factors. When a thiolate anion attacks an asymmetrical disulfide ($R_1S–SR_2$), the best leaving group will be the thiol with lowest pKa (Jensen et al. 2009). Moreover, since thiol-disulfide exchange involves negatively charged species, the reaction rate can be influenced by electrostatic factors, such as negative charges adjacent to the reaction center (Bulaj et al. 1998). The formation of an intramolecular disulfide is typically more efficient than the formation of intermolecular disulfides. Consequently, the incubation of disulfide with dithiothreitol (DTT) always leads to the oxidation of DTT. In biology, thiol-disulfide exchange is the basis of the mechanism of action of many enzymes. Another thiol/disulfide exchange process that deserves special consideration is *S*-glutathionylation of cysteine in proteins. Although *S*-glutathionylation was thought to protect cysteine from irreversible oxidation (Thomas et al. 1995), it was later shown that this modification affects the catalytic activity for several enzymes, suggesting a regulatory role (Demasi et al. 2003).

Disulfides also react with strong oxidants to yield thiosulfinate and thiosulfinate esters. However, this reaction is significantly slower than oxidation of a thiol or sulfenic acid and thus, would take place only at extremely elevated, cytotoxic concentrations of ROS/RNS. Disulfides are reduced to thiols by a number of inorganic compounds, including sodium borohydride ($NaBH_4$) and sodium cyanoborohydride ($NaBH_3CN$). In biochemistry, dithiols and phosphite esters are the most common reducing agents. It has been demonstrated that sodium ascorbate can also reduce a disulfide (Holmes and Williams 2000), but this reaction appears to be catalyzed by Copper(II). Chemical approaches to the detection of disulfides are largely indirect. The initial step consists of blocking free thiols with an irreversible alkylating agent (see Sect. 1.2), followed by a disulfide reduction (e.g., with tris(2-carboxyethyl)phosphine – TCEP). Finally, the nascent thiols are trapped with the usual reagents (see Sect. 1.2). Two important considerations are: (1) owing to a pKa of ~8 for most thiols, the blocking step must be carried out at neutral pH or higher for this reaction to be efficient; (2) commonly used reducing agents also reduce other reversible cysteine oxoforms, such as sulfenic acids, nitrosothiols, sulfenamides, and thiosulfinates. Direct detection methods exist only for *S*-glutathionylation (RS–SG), which can be monitored through reduction of this bond by glutaredoxin (Lind et al. 2002), radiolabeled with ^{35}S-GSH (Fratelli et al. 2002) or biotinylated glutathione ethyl ester (BioGEE, Sullivan et al. 2000). Alternately, an immunochemical approach using specific antibodies against the protein-glutathione adduct may be employed (Dalle-Donne et al. 2003).

1.4.4 Sulfenamide

A sulfenamide is a compound containing a trivalent nitrogen atom bonded to divalent sulfur and is formally derived from the condensation of sulfenic acid with an amine (Allison et al. 1973). Although a sulfenic acid and sulfenyl halide may react with any nucleophilic amine to generate a sulfenamide, in general, this reaction is several orders of magnitude slower then the equivalent reaction with a thiol. Nonetheless, interest in such modifications has grown since the discovery of cyclic sulfenamide formation in the active site of protein tyrosine phosphatase 1B (PTP1B) (Salmeen et al. 2003). The sulfenic acid intermediate produced by oxidation of the PTP1B catalytic cysteine is rapidly converted into the sulfonamide form. The crystalline structure of oxidized PTP1B reveals that the sulfenamide is characterized by an isothiazolidinone ring formed by the binding of sulfur of the cysteine to the backbone nitrogen of the adjacent serine residue (van Montfort et al. 2003). This species appears to be generated by nucleophilic attack of the main-chain amide nitrogen on the electrophilic sulfur in sulfenic acid with elimination of water. Beyond PTP1B, few other proteins have been found to form cyclic sulfonamide modifications (Lee et al. 2007) and, as a result, it is difficult to say whether such modification has a wider role in biology.

Sulfenamides appear to have similar a reactivity profile akin to sulfenic acid and undergo attack by several nucleophilic species. For example they react with thiols to yield disulfides (Nti-Addae et al. 2011) and, consequently sulfenamide formation may mediate the *S*-glutathionylation of proteins such as PTP1B (den Hertog et al. 2005). Sulfenamides can also undergo trans-amination reactions (Craine and Raban 1989); considering that the only biologically relevant sulfenamide known to date is cyclic, however, this reaction is unlikely occur in cells because the equilibrium of the reaction is always shifted toward the cyclic form. Sulfenamides can be reduced to thiols by phosphite esters, which also react with disulfides, sulfenic acids, and *S*-nitrosothiols. The cyclic sulfonamide within PTP1B is often cited for its ability

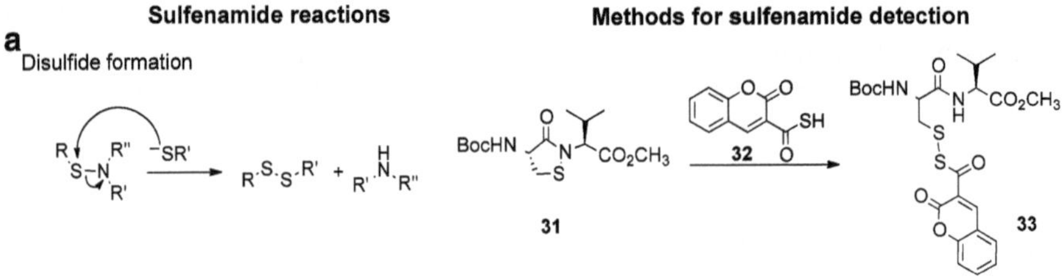

Fig. 1.10 Chemical reactivity and detection of sulfenamides

to protect the reactive thiol against irreversible oxidation to sulfinic and sulfonic acid. This statement can be rationalized by the lesser nucleophilic character of the sulfur atom in the cyclic sulfonamide as compared to the sulfenic acid. In aqueous basic solutions of H_2O_2, the sulfenamide is readily oxidized to sulfinamide and sulfonamide (Kharasch et al. 1946), but at physiological pH such oxidation is negligible. In fact, although sulfinamide formation has been reported in short peptides exposed to ROS (Shetty and Neubert 2009), no evidence of sulfinamide or sulfonamide formation has been observed in PTP1B. This is true even in the presence of high H_2O_2 concentrations, as PTP1B exclusively undergoes sulfinic and sulfonic acid modification, ascribed to the direct oxidation of the cysteine sulfenic acid intermediate (Held et al. 2010).

The development of specific probes for sulfenamides is made more difficult by the similarity with sulfenic acid reactivity (Fig. 1.10b). Shiau et al. have used a series of relatively weak nucleophilic thiols to identify compounds that react with sulfonamides, but not disulfides (Fig. 1.10). They identified a coumarin thioacid (**32**), which reacts with a model sulfenamide (**31**) to generate the fluorescent disulfide (**33**) and exhibits no cross-reactivity with disulfides (Shiau et al. 2006). Unfortunately, this study provided no data regarding reactivity of **32** with sulfenic acid; nonetheless, it is almost certain that the coumarin-derivative should also react with sulfenic acids and related thiosulfinates.

1.4.5 Thiosulfinate Ester

Thiosulfinate esters can be formed by the condensation of two sulfenic acids (see Sect. 4.1) or by oxidation of disulfides (Eq. 1.31). In a biological context, the auto-condensation of sulfenic acids appears to have poor relevance in proteins owing to steric hindrance, while it should be quite likely for small molecules such as cysteine or glutathione. The direct oxidation of disulfide to the thiosulfinate would require very high, localized concentrations of ROS/RNS and only biologically relevant for glutathione in its oxidized state (Giles et al. 2002). Interestingly, the thiosulfinate group forms as a reaction intermediate during sulfiredoxin (Srx)-mediated reduction of cysteine sulfinic acid in Prxs; however, this modification is generated through a completely different mechanism (see Sect. 1.4.5).

The thiosulfinate ester is a highly reactive species and can readily undergo the nucleophilic attack by thiols to yield a disulfide and a sulfenic acid (Nagy et al. 2007). Glutathione disulfide *S*-monoxide has consequently been proposed as a mediator in *S*-glutathionylation (Huang et al. 2007). Thiosulfinates can also undergo hydrolysis (Eq. 1.32), a reaction that is in equilibrium with self-condensation of sulfenic acids (Nagy and Ashby 2007a, b). Formally, thiosulfinates can be oxidized to sulfonate esters (Eq. 1.33), but this should take place only in the presence of high concentrations of very strong oxidants such as those found in the phagosome. Finally, disproportionation of aromatic thiosulfinate esters has been reported to yield a disulfide and a thiosulfonate ester *via* a complex radical mechanism (Eq. 1.34, Poudrel and Cole 2001).

$$R - S - SR + H_2O_2 \rightleftarrows R - S - S(O)R + H_2O \tag{1.31}$$

$$R - S - S(O)R + H_2O \rightleftarrows 2\,R - S - OH \tag{1.32}$$

$$R - S - S(O)R + H_2O_2 \rightarrow R - S - S(O)_2\,R + H_2O \tag{1.33}$$

$$2\,R - S - S(O)R \rightarrow RS - S - R + R - S - S(O)_2\,R \tag{1.34}$$

Except for the Prx-thiosulfinate intermediate, no other thiosulfinates are known in proteins. This may be due, at least in part, to the difficulty of finding a selective strategy for detecting this thiol modification. The thiosulfinate ester, like a thiosulfonate ester (Schank et al. 2007) may react with dimedone (and its derivatives) to yield a dimedone-thioether and a sulfenic acid, which in turn reacts with a second molecule of dimedone. Rabinkov's group has developed a simple spectrophotometric assay to determine the presence of allicin (a natural thiosulfinate ester present in garlic) and of allinase activity by the reaction between 4-mercaptopyridine (4-MP) and allicin (Miron et al. 2002). The selectivity of this technique is based on the high reactivity of the thiosulfinate in comparison to the disulfide, but its application with proteins in complex biological samples appears since 4-MP should also react with the more abundant sulfenic acid.

1.4.6 *Sulfinic Acid*

Sulfenic acid can be further oxidized to sulfinic acid (RSO_2H) in the presence of excess ROS/RNS (see Sect. 1.4.1). RSO_2H is a relatively stable species as compared to RSOH, and with a pK_a value of ~2, is always deprotonated at physiological pH (Burkhard et al. 1959). Sulfinic acids can be also generated by disproportionation

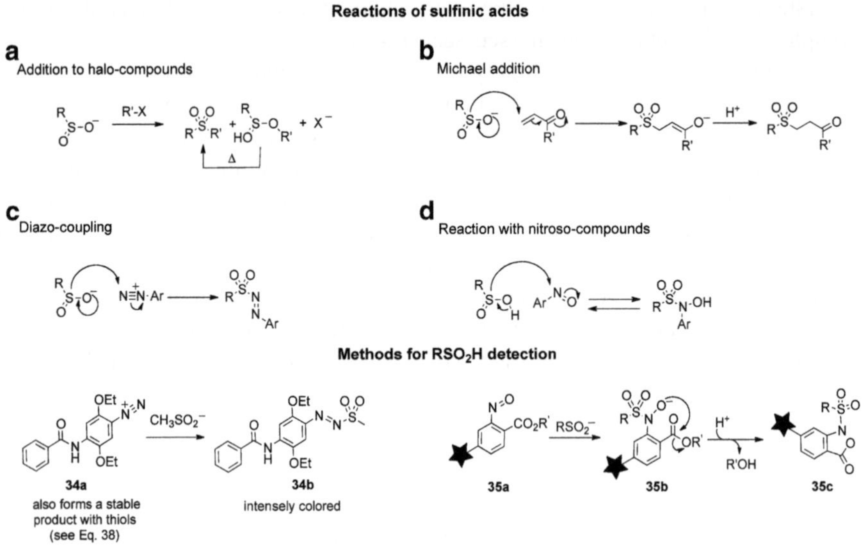

Fig. 1.11 Chemical reactivity and detection of sulfinic acids

of sulfenic acid (Eq. 1.35, Abraham et al. 1983) or by nucleophilic displacement of thiosulfonate esters (Eqs. 1.36 and 1.37, Harpp et al. 1979).

$$2\,R - S - OH \rightarrow R - SH + RS_2OH \tag{1.35}$$

$$R - S - S(O)_2\,R_1 + HO^- \rightarrow RSO_2^- + R_1 - S - OH \tag{1.36}$$

$$R - S - S(O)_2\,R_1 + R_2 - S^- \rightarrow RSO_2^- + R_1 - S - S - R_1 \tag{1.37}$$

In the deprotonated form, the sulfinate group (RSO_2^-) is an ambident anion and the negative charge is delocalized between the sulfur and the two oxygen atoms. RSO_2^- behaves mainly as a soft nucleophile (Reddie and Carroll 2008) and can react with many electrophilic species such as halides (Fig. 1.11a) (Khamis et al. 2010) and α,β-unsaturated compounds (Fig. 1.11b) (Ogata et al. 1970). Sulfinic acids can therefore be alkylated by iodoacetamide as well as ethylmaleimide, although these reactions are slower in comparison to thiols. In all these reactions, *S*-attack is generally favored and leads to the thermodynamically more stable sulfone. In the presence of strong electrophiles, the sulfonyl ester can be kinetically generated (Baidya et al. 2010), but this unstable species slowly rearranges to a sulfone (Fig. 1.11a). Sulfinic acids can oxidize further to sulfonic acid, but neither of these oxoforms can be reduced directly by thiols, hence the term “irreversible”. In recent years, a biological role for RSO_2H has emerged, for example, in both Parkinson’s disease protein DJ-1 (Blackinton et al. 2009) and in matrilysin (MMP-7) activation (Fu et al. 2001), although the best-known cysteine sulfinyl modification is occurs within the Prx family (Wood et al. 2003). Finally, the discovery of sulfiredoxin, an ATP-dependent protein that specifically reduces sulfinic acid in Prx, has opened the door to an additional layer of redox regulation (Jönsson et al. 2008) and increased interest in detecting this specific modification.

Antibodies against individual proteins with sulfinyl/sulfonyl modifications have been developed (see Sect. 1.4.7). At the moment, however, no selective chemical probes are available to detect sulfinic acid formation in proteins. On the basis of early physical-organic studies (Ritchie et al. 1961), the use of diazonium salts (Fig. 1.11c, **34**) as a trapping agent for protein sulfinic acids was recently proposed (Jacob and Ba 2011). Although this approach has proven to be effective for colorimetric detection of methane sulfinic acid (Babbs and Gale 1987), several complications are present in this system. Diazonium salts are highly unstable species that readily decompose under neutral aqueous conditions. Moreover, considering the potential cross-reactivity with tyrosine (Hooker et al. 2004), as well as with cysteine (Patt and Patt 2002), future application of this technique with proteins may be limited. Whereas avoiding basic conditions can minimize cross-reactivity with tyrosine, cysteine appears to react with diazonium salts under a wide range of pHs (Eq. 1.38) to yield a stable sulfenyldiazene species (Lo Conte and Carroll, unpublished data).

$$R - SH + Ar - N_2^+ \rightarrow R - S - N = N - Ar + H^+ \tag{1.38}$$

As an alternative, the Carroll research group has been working to developing a new selective ligation strategy using C-nitroso compounds, which condense with sulfinic acids to yield N-sulfonyl hydroxylamines (Fig. 1.11d). N-sulfonyl hydroxylamines are unstable at neutral or basic pH (Darchen and Moinet 1976). As a result, in order to convert the product into a stable adduct, we can take advantage of the nucleophilic behavior of hydroxylamine. In the presence of an electrophilic center (e.g., a carboxylic ester) on the C-nitroso compound (**35a**), N-sulfonyl hydroxylamine (**35b**) undergoes intramolecular cyclization to form a stable N-sulfonyl benzisoxazolone adduct (**35c**). Although thiols can also target the C-nitroso group, selectivity for sulfinic acid is ensured by the fact that reaction with the thiol leads to an unstable N-sulfenyl hydroxylamine linkage, which is readily cleaved in the presence of additional thiols; this chemical approach is currently under being adapted for detection of protein sulfinic acids *in vitro* (Lo Conte and Carroll 2012).

1.4.7 Sulfonic Acid

Sulfonic acid represents the highest oxidation state for the cysteine sulfur atom and no biological pathway is known to reduce this cysteine oxoform. Sulfinic acid can be further oxidized to sulfonic acid by strong oxidizing species such as peroxynitrous acid and hydrogen peroxide (Eq. 1.39, see Sect. 1.4.6). Sulfonic acid can also be generated by the disproportionation of sulfinic acid (Eq. 1.40). As alluded to above, when an essential enzyme active site cysteine is oxidized to sulfonic acid the catalyst is irreversibly inhibited. However, it has also been proposed that sulfonic acid modification of Prx could enhance its proposed chaperone activity (Lim et al. 2008). This modification may also target proteins for degradation. For example, the oxidation of N-terminal cysteine residues to sulfonic acid can mediate arginylation and degradation of GTPase-activating proteins (Tasaki and Kwon 2007).

$$R - SO_2H + H_2O_2 \rightarrow R - SO_3H + H_2O \tag{1.39}$$

$$4\,R - SO_2H + H_2O \rightarrow RSO_3H + R - S(O)_2 - S - R \tag{1.40}$$

Sulfonic acid is one of the strongest organic acids. Cysteic acid, with a pKa of −3, is always present in the deprotonated form as sulfonate at physiological pH (Chang et al. 2010). Sulfonic acid is a very poor nucleophile and its derivatives are usually obtained by activation as the sulfonyl chloride. No facile chemical methods are available to trap and tag the sulfonic acid modification in proteins. Although base-mediated β-elimination of cysteine sulfonic acid to dehydroalanine and sulfite could provide a starting point for assay design (Dai et al. 2005), the concomitant β-elimination of phosphoserine and phosphothreonine (McLachlin and Chait 2003) represents a serious limitation of such as approach. To date, antibodies are the most common technique for detecting the sulfonic acid form of a specific protein (Woo et al. 2003). Unfortunately, this approach is not easily applied to all proteins, and the resulting antibodies typically exhibit the same affinity for sulfinic and sulfonic acid cysteine oxoforms. Recently, an innovative mass spectroscopy assay was developed to selective enrich and identify peptides containing cysteine sulfonic acid. This approach is based on ionic affinity capture using polyarginine-coated nanodiamonds that exhibit good specificity even in the presence of phosphopeptides (Chang et al. 2010). The technique was applied to selectively enriched sulfopeptides obtained from tryptic digests of over-oxidized BSA.

1.4.8 Sulfhydration

The process by which H_2S mediates *S*-sulfhydration of proteins is still unclear. For example, the direct reaction of cysteine side chains with H_2S (Mustafa et al. 2011) appears quite unlikely without any intermediary oxidant species. Persulfides can be generated by the direct reaction of sulfenic acids (Kabil and Banerjee 2010), sulfenyl halides, sulfenyl-thiocyanates, or thiosulfenyl esters with H_2S (Eqs. 1.41, 1.42, 1.43 and 1.44). Although less reactive, disulfides can also generate persulfides *via* disulfide exchange with H_2S (Eq. 1.45). Finally hydrogen persulfide (HSSH) has recently been proposed as the physiological sulfhydration agent in cells

(Nagy and Winterbourn 2010). HSSH is generated by reaction of H_2S with two-electron oxidants and can easy react with low pKa thiols (Eq. 1.46).

$$R - SOH + HS^- \rightarrow R - S - S^- + H_2O \quad (1.41)$$

$$R - S - X + HS^- \rightarrow R - S - S^- + HX \quad (1.42)$$

$$R - S - SCN + HS^- \rightarrow R - S - S^- + HSCN \quad (1.43)$$

$$R - S - S(O)R_1 + 2\,HS^- \rightarrow 2\,R - S - S^- + H_2O \quad (1.44)$$

$$R - S - S - R_1 + HS^- \rightarrow R - S - S^- + R_1S^- \quad (1.45)$$

$$R - S^- + HS - SH \rightarrow R - S - S^- + HS^- \quad (1.46)$$

Persulfides show intermediate behavior between thiol and disulfide reactivity. In fact, the terminal sulfur is ambiphilic and can behave as an electrophile or a nucleophile. For example, persulfides can undergo oxidation reactions to yield a wide range of products or may also undergo disulfide exchange (Eqs. 1.47 and 1.48), following the same rules described in Sect. 1.4.3. If the internal sulfur atom of a persulfide group has a pKa lower than 6.9 (corresponding to the pKa of hydrogen sulfide), the reaction may proceed *via* trans-sulfhydration; alternately, the result is disulfide formation and the release of hydrogen sulfide. The terminal sulfur, however, can also directly attack a disulfide or react with a sulfenic acid to generate a trisulfide (Eqs. 1.49 and 1.50). Trisulfides are widely distributed in the biological world, although they are not a common post-translational modification, and the number of proteins in which a trisulfide has been unambiguously identified is small (Nielsen et al. 2011).

$$R - S - SH + R_1S^- \rightarrow R - S - S - R_1 + HS^- \quad (1.47)$$

$$R - S - SH + R_1S^- \rightarrow R_1 - S - S^- + RS^- \quad (1.48)$$

$$R - S - S^- + R_1 - S - S - R_1 \rightarrow R - S - S - S - R_1 + R_1S^- \quad (1.49)$$

$$R - S - S^- + R_1 - S - OH \rightarrow R - S - S - S - R_1 + OH^- \quad (1.50)$$

In the presence of transition metals, persulfides can readily decompose to generate ROS, and such reactions could mediate the anti-microbial and anti-cancer properties of some natural products (Chatterji et al. 2005). The terminal sulfur is a strong nucleophile and can react with a huge range of electrophiles. The richness of its chemistry makes the persulfide group a versatile reagent for the incorporation of sulfur along many metabolic pathways (Mueller 2006).

In the last decade, interest in H_2S signaling has grown rapidly, but relatively few proteins have been found to be sulfhydrates (Krishnan et al. 2011). This is due to the difficultly of monitoring such modifications uniquely. The Snyder research group has reported an assay for proteins persulfides, which is a modification of the classic biotin switch assay (Mustafa et al. 2009). In this method, MMTS is employed to block free thiols and any persulfides are reacted with biotin-HPDP (possibly forming a disulfide or trisulfide linkage) and analyzed by Western blot. Applying this approach to complex cell lysates, several proteins were reported as targets of H_2S signaling. A key feature of this approach requires that the MMTS reagent only modify thiols and not persulfides. Unfortunately, no rationale or data was provided by Mustafa et al. to validate this claim. In fact, our own research group has recently determined that MMTS does react with persulfides to yield a trisulfide linkage as expected from the nucleophilic nature of sulfane sulfur (Pan and Carroll, unpublished data). Therefore, considering that the chemical mechanism and selectivity remain unknown, results obtained using this approach should be interpreted with caution.

1.5 Conclusions

In the last decade, awareness of the number of proteins containing redox-sensitive cysteine has grown significantly. Cysteine can assume a wide range of oxidative states in response to dynamic changes in intracellular redox potentials. To understand the role of such protein modifications, methods have been developed to distinguish between different cysteine oxidation states. The role of the biologist is not simply to choose the most appropriate technique, but also to maintain a critical view of possible artifacts. Opting for an *in vitro* approach can prove unsatisfactory because many cysteine oxoforms are reactive and unstable outside of the cellular milieu. The use of indirect methods, by which all free thiols are trapped, must be preceded by the awareness that these thiol-trapping agents can react inefficiently or be incompletely selective, resulting in under- or over-estimation. The same considerations must be taken into account when using reducing agents to identify specific modifications. Indeed, the selectivity and efficiency of such reagents requires careful evaluation. Furthermore, in many cases, interconversion of cysteine redox states dramatically increases the challenges of studying an individual modification. Finally, given that the local microenvironment can exert a remarkable influence on protein thiolate reactivity, the exploration of probes and chemical methods exclusively in low-molecular-weight model systems may prove unsatisfactory. Although more complex, a recommended course for future research is the identification of selective reactions that can be employed to monitor cysteine oxidation *in situ*, directly in cells. Obviously, this presents a formidable challenge, but given the vast number of biological processes that thiol oxidation plays a significant regulatory role, it should be well worth the effort.

References

Abo M, Urano Y, Hanaoka K, Terai T, Komatsu T, Nagano T (2011) Development of a highly sensitive fluorescent probe for hydrogen peroxide. J Am Chem Soc 133:10629–10637

Abraham RT, Benson LM, Jardine I (1983) Synthesis and pH-dependent stability of purine-6-sulfenic acid, a putative reactive metabolite of 6-thiopurine. J Med Chem 26:1523–1526

Albrich JM, McCarthy CA, Hurst JK (1981) Biological reactivity of hypochlorous acid: implications for microbicidal mechanisms of leukocyte myeloperoxidase. Proc Natl Acad Sci USA 78:210–214

Alderton WK, Cooper CE, Knowles RG (2001) Nitric oxide synthases: structure, function and inhibition. Biochem J 357:593–615

Allison WS (1976) Formation and reactions of sulfenic acids in proteins. Acc Chem Res 9:293–299

Allison WS, Benitez LV, Johnson CL (1973) The formation of a protein sulfenamide during the inactivation of the acyl phosphatase activity of oxidized glyceraldehyde-3-phosphate dehydrogenase by benzylamine. Biochem Biophys Res Commun 52:1403–1409

Andersen JK (2004) Oxidative stress in neurodegeneration: cause or consequence? Nat Rev Neurosci 5:s18–s25

Arisawa M, Yamaguchi M (2008) Transition-metal-catalyzed synthesis of organosulfur compounds. Pure Appl Chem 80:993–1003

Ashby MT (2008) Inorganic chemistry of defensive peroxidases in the human oral cavity. J Dent Res 87:900–914

Ashby MT, Aneetha H (2004) Reactive sulfur species: aqueous chemistry of sulfenyl thiocyanates. J Am Chem Soc 126:10216–10217

Babbs CF, Gale MJ (1987) Colorimetric assay for methanesulfinic acid in biological samples. Anal Biochem 163:67–73

Bach RB, Ayala PY, Schlegel HB (1996) A reassessment of the bond dissociation energies of peroxides. An ab initio study. J Am Chem Soc 118:12758–12765

Baidya M, Kobayashi S, Mayr H (2010) Nucleophilicity and nucleofugality of phenylsulfinate ($PhSO_2^-$): a key to understanding its ambident reactivity. J Am Chem Soc 132:4796–4805

Bearden SE, Beard Jr RS, Pfau JC (2010) Extracellular transsulfuration generates hydrogen sulfide from homocysteine and protects endothelium from redox stress. Am J Physiol Heart Circ Physiol 299:H1568–H1576

Bechtold E, Reisz JA, Klomsiri C, Tsang AW, Wright MW, Poole LB, Furdui CM, King SB (2010) Water-soluble triarylphosphines as biomarkers for protein *S*-nitrosation. ACS Chem Biol 5:405–414

Bedard K, Krause K-H (2007) The NOX family of ROS-generating NADPH oxidases: physiology and pathophysiology. Physiol Rev 87:245–313

Benitez LV, Allison WS (1974) The inactivation of the acyl phosphatase activity catalyzed by the sulfenic acid form of glyceraldehyde 3-phosphate dehydrogenase by dimedone and olefins. J Biol Chem 249:6234–6243

Bertozzi CR (2011) A decade of bioorthogonal chemistry. Acc Chem Res 44:651

Bienert GP, Møller ALB, Kristiansen KA, Schulz A, Møller IM, Schjoerring JK, Jahn TP (2007) Specific aquaporins facilitate the diffusion of hydrogen peroxide across membranes. J Biol Chem 282:1183–1192

Blackinton J, Lakshminarasimhan M, Thomas KJ, Ahmad R, Greggio E, Raza AS, Cookson MR, Wilson MA (2009) Formation of a stabilized cysteine sulfinic acid is critical for the mitochondrial function of the parkinsonism protein DJ-1. J Biol Chem 284:6476–6485

Block E, O'Conner J (1974) The chemistry of alkyl thiosulfinate esters. VI. Preparation and spectral studies. J Am Chem Soc 96:3921–3929

Bulaj G, Kortemme T, Goldenberg DP (1998) Ionization-reactivity relationships for cysteine thiols in polypeptides. Biochemistry 37:8965–8972

Burkhard RK, Sellers DE, DeCou F, Lambert JL (1959) The pKa's of aromatic sulfinic acids. J Org Chem 24:767–769

Carballal S, Trujillo M, Cuevasanta E, Bartesaghi S, Möller MN, Folkes LK, García-Bereguiaín MA, Gutiérrez-Merino C, Wardman P, Denicola A, Radi R, Alvarez B (2011) Reactivity of hydrogen sulfide with peroxynitrite and other oxidants of biological interest. Free Radic Biol Med 50:196–205

Cary SPL, Winger JA, Derbyshire ER, Marletta MA (2006) Nitric oxide signaling: no longer simply on or off. Trends Biochem Sci 31:231–239

Caulfield JL, Wishnok JS, Tannenbaum SR (1998) Nitric oxide-induced deamination of cytosine and guanine in deoxynucleosides and oligonucleotides. J Biol Chem 273:12689–12695

Chalker JM, Bernardes GJL, Lin YA, Davis BG (2009) Chemical modification of proteins at cysteine: opportunities in chemistry and biology. Chem Asian J 4:630–640

Chalker JM, Gunnoo SB, Boutureira O, Gerstberger SC, Fernández-González M, Bernardes GJL, Griffin L, Hailu H, Schofield CJ, Davis BG (2011) Methods for converting cysteine to dehydroalanine on peptides and proteins. Chem Sci 2:1666–1676

Chang MCY, Pralle A, Isacoff EY, Chang CJ (2004) A selective cell-permeable optical probe for hydrogen peroxide in living cells. J Am Chem Soc 126:15392–15393

Chang YC, Huang CN, Lin CH, Chang HC, Wu CC (2010) Mapping protein cysteine sulfonic acid modifications with specific enrichment and mass spectrometry: an integrated approach to explore the cysteine oxidation. Proteomics 10:2961–2971

Chapman ALP, Skaff O, Senthilmohan R, Kettle AJ, Davies MJ (2009) Hypobromous acid and bromamine production by neutrophils and modulation by superoxide. Biochem J 417:773–781

Chatterji T, Keerthi K, Gates KS (2005) Generation of reactive oxygen species by a persulfide (BnSSH). Bioorg Med Chem Lett 15:3921–3924

Chen K, Kirber MT, Xiao H, Yang Y, Keaney JF Jr (2008) Regulation of ROS signal transduction by NADPH oxidase 4 localization. J Cell Biol 181:1129–1139

Chiku T, Padovani D, Zhu W, Singh S, Vitvitsky V, Banerjee R (2009) H2S biogenesis by human Cystathionine gamma-Lyase leads to the novel sulfur metabolites lanthionine and homolanthionine and is responsive to the grade of hyperhomocysteinemia. J Biol Chem 248:11601–11612

Claiborne A, Miller H, Parsonage D, Ross RP (1993) Protein-sulfenic acid stabilization and function in enzyme catalysis and gene regulation. FASEB J 7:1483–1490

Cohen EA, Pickett HM (1981) The dipole moment of hydrogen peroxide. J Mol Spectrosc 87: 582–585

Copley SD, Novak WRP, Babbitt PC (2004) Divergence of function in the thioredoxin fold suprafamily: evidence for evolution of peroxiredoxins from a thioredoxin-like ancestor. Biochemistry 43:13981–13995

Craine L, Raban M (1989) The chemistry of sulfenamides. Chem Rev 89:689–712

Crankshaw MW, Grant GA (1996) Modification at cysteine. Current protocols in protein science. Wiley, New York, 15.1.1

Dai J, Zhang Y, Wang J, Li X, Lu Z, Cai Y, Qian X (2005) Identification of degradation products formed during performic oxidation of peptides and proteins by high-performance liquid chromatography with matrix-assisted laser desorption/ionization and tandem mass spectrometry. Rapid Commun Mass Spectrom 19:1130–1138

Dalle-Donne I, Rossi R, Giustarini D, Colombo R, Milzani A (2003) Actin *S*-glutathionylation: evidence against a thiol-disulphide exchange mechanism. Free Radic Biol Med 35:1185–1193

Darchen A, Moinet CJ (1976) Arenesulphinic acids. Nitroso protecting reagents applicable to some nitrosoarenes. J Chem Soc Chem Commun 20:820a–820a. doi:10.1039/C3976000820A

Davies MJ, Hawkins CL, Pattison DI, Rees MD (2008) Mammalian heme peroxidases: from molecular mechanisms to health implications. Antioxid Redox Signal 10:1199–1234

Davis FA, Jenkins LA, Billmers RL (1986) Chemistry of sulfenic acids. 7. Reason for the high reactivity of sulfenic acids. Stabilization by intramolecular hydrogen bonding and electronegativity effects. J Org Chem 51:1033–1040

Demasi M, Silva GM, Netto LES (2003) 20S Proteasome from Saccharomyces cerevisiae is responsive to redox modifications and is *S*-glutathionylated. J Biol Chem 278:679–685

den Hertog J, Groen A, van der Wijk T (2005) Redox regulation of protein-tyrosine phosphatases. Arch Biochem Biophys 434:11–15
Denninger JW, Marletta MA (1999) Guanylate cyclase and the NO/cGMP signaling pathway. Biochim Biophys Acta 1411:334–350
Denu JM, Tanner KG (1998) Specific and reversible inactivation of protein tyrosine phosphatases by hydrogen peroxide: evidence for a sulfenic acid intermediate and implications for redox regulation. Biochemistry 37:5633–5642
Dickinson BC, Chang CJ (2008) A targetable fluorescent probe for imaging hydrogen peroxide in the mitochondria of living cells. J Am Chem Soc 130:9638–9639
Dickinson BC, Peltier J, Stone D, Schaffer DV, Chang CJ (2011) Nox2 Redox signaling maintains essential cell populations in the brain. Nat Chem Biol 7:106–112
Dicks AP, Williams DL (1996) Generation of nitric oxide from *S*-nitrosothiols using protein-bound Cu2+ sources. Chem Biol 8:655–659
Ellis HR, Poole LB (1997) Novel application of 7-chloro-4-nitrobenzo-2-oxa-1,3-diazole to identify cysteine sulfenic acid in the aphc component of alkyl hydroperoxide reductase. Biochemistry 36:15013–15018
Ellman GL (1959) Tissue sulfhydryl groups. Arch Biochem Biophys 82:70–77
Elsey DJ, Fowkes RC, Baxter GF (2010) Regulation of cardiovascular cell function by hydrogen sulfide (H2S) cell biochemistry and function. Cell Biochem Funct 28:95–106
Fan SW, George RA, Haworth NL, Feng LL, Liu JY, Wouters MA (2009) Conformational changes in redox pairs of protein structures. Protein Sci 18:1745–1765
Fernandes PA, Ramos MJ (2004) Theoretical insights into the mechanism for thiol = disulfide exchange. Chemistry 10:257–266
Ferrer-Sueta G, Radi R (2009) Chemical biology of peroxynitrite: kinetics, diffusion, and radicals. ACS Chem Biol 4:161–177
Forman HJ, Fridovich I (1973) Superoxide dismutase: a comparison of rate constants. Arch Biochem Biophys 158:396–400
Fratelli M, Demol H, Puype M, Casagrande S, Eberini I, Salmona M, Bonetto V, Mengozzi M, Duffieux F, Miclet E, Bachi A, Vandekerckhove J, Gianazza E, Ghezzi P (2002) Identification by redox proteomics of glutathionylated proteins in oxidatively stressed human T lymphocytes. Proc Natl Acad Sci USA 99:3505–3510
Fu X, Kassim SY, Parks WC, Heinecke JW (2001) Hypochlorous acid oxygenates the cysteine switch domain of pro-matrilysin (MMP-7). A mechanism for matrix metalloproteinase activation and atherosclerotic plaque rupture by myeloperoxidase. J Biol Chem 276:41279–41287
Fukai T, Masuko U-F (2011) Superoxide dismutases: role in redox signaling, vascular function, and diseases. Antioxid Redox Signal 15:1583–1603
Gamblin DP, Garnier P, van Kasteren S, Oldham NJ, Fairbanks AJ, Davis BG (2004) Glyco-SeS: selenenylsulfide-mediated protein glycoconjugation-a new strategy in post- translational modification. Angew Chem Int Ed 43:828–833
Giles GI, Tasker KM, Collins C, Giles NM, O'rourke E, Jacob C (2002) Reactive sulphur species: an in vitro investigation of the oxidation properties of disulphide *S*-oxides. Biochem J 364(Pt 2):579–585
Giorgio M, Trinei M, Migliaccio E, Pelicci PG (2007) Hydrogen peroxide: a metabolic by-product or a common mediator of ageing signals? Nat Rev Mol Cell Biol 8:722–728
Goldstein S, Czapski G (1996) Mechanism of the nitrosation of thiols and amines by oxygenated NO solutions: the nature of the nitrosating intermediates. J Am Chem Soc 118:3419–3425
Goto K, Holler M, Okazak R (1997) Synthesis, structure, and reactions of a sulfenic acid bearing a novel bowl-type substituent: the first synthesis of a stable sulfenic acid by direct oxidation of a thiol. J Am Chem Soc 119:1460–1461
Goto K, Shimada K, Nagahama M, Okazaki R, Kawashima T (2003) Reaction of stable sulfenic and selenenic acids containing a bowl-type steric protection group with a phosphine. Elucidation of the mechanism of reduction of sulfenic and selenenic acids. Chem Lett 32:1080–1081
Haake M (1972) Zur desoxygenierung von tritylthionitrit. Tetrahedron Lett 13:33–39

Hansen RE, Roth D, Winther JR (2009) Quantifying the global cellular thiol-disulfide status. Proc Natl Acad Sci USA 106:422–427
Harpp DN, Ash DK, Smith RA (1979) Organic sulfur chemistry. 33. Chemistry of sulfenic sulfonic thioanhydrides. Solvent-dependent sulfur extrusion. J Org Chem 44:4135–4140
Hawkins CL (2009) The role of hypothiocyanous acid (HOSCN) in biological systems. Free Radic Res 43:1147–1158
Heckel A, Pfleiderer W (1983) Lumazinesulfenates – a new class of stable sulfenic acids. Tetrahedron Lett 24:5047–5050
Held JM, Danielson SR, Behring JB, Atsriku C, Britton DJ, Puckett RL, Schilling B, Campisi J, Benz CC, Gibson BW (2010) Targeted quantitation of site-specific cysteine oxidation in endogenous proteins using a differential alkylation and multiple reaction monitoring mass spectrometry approach. Mol Cell Proteomics 9:1400–1410
Hogg N (2002) The biochemistry and physiology of *S*-nitrosothiols. Annu Rev Pharmacol Toxicol 42:585–600
Hogg DR, Robinson A (1979) The ambident nucleophilicity of sulphenate anions. J Chem Soc Perkin Trans 1:1125–1128
Holmes AJ, Williams DLH (2000) Reaction of ascorbic acid with snitrosothiols: clear evidence for two distinct reaction pathways. J Chem Soc Perkin Trans 2:1639–1644
Hooker JM, Kovacs EW, Francis MB (2004) Interior surface modification of bacteriophage MS2. J Am Chem Soc 126:3718–3719
Huang B, Chen C (2006) An ascorbate-dependent artifact that interferes with the interpretation of the biotin switch assay. Free Radic Biol Med 41:562–567
Huang KP, Huang FL, Shetty PK, Yergey AL (2007) Modification of protein by disulfide *S*-monoxide and disulfide *S*-dioxide: distinctive effects on PKC. Biochemistry 46:1961–1971
Hugo M, Turell L, Manta B, Botti H, Monteiro G, Netto LES, Alvarez B (2009) Thiol and Sulfenic Acid Oxidation of AhpE, the One-Cysteine Peroxiredoxin from *Mycobacterium tuberculosis*: Kinetics, Acidity Constants, and Conformational Dynamics. Biochemistry 48:9416–9426
Ilbert M, Horst J, Ahrens S, Winter J, Graf PC, Lilie H, Jakob U (2007) The redox-switch domain of Hsp33 functions as dual stress sensor. Nat Struct Mol Biol 14:556–563
Jacob C, Ba LA (2011) Open season for hunting and trapping post-translational cysteine modifications in proteins and enzymes. Chembiochem 12:841–844
Jaffrey SR, Erdjument-Bromage H, Ferris CD, Tempst P, Snyder SH (2001) Protein *S*-nitrosylation: a physiological signal for neuronal nitric oxide. Nat Cell Biol 3:193–197
Jencks WP, Carriulo JC (1960) Reactivity of nucleophilic reagents towards esters. J Am Chem Soc 82:1778–1786
Jensen KS, Hansen RE, Winther JR (2009) Kinetic and thermodynamic aspects of cellular thiol-disulfide redox regulation. Antioxid Redox Signal 11:1047–1058
Jönsson TJ, Johnson LC, Lowther WT (2008) Structure of the sulphiredoxin-peroxiredoxin complex reveals an essential repair embrace. Nature 451:98–101
Jourd'heuil D, Jourd'heuil FL, Feelisch M (2003) Oxidation and nitrosation of thiols at low micromolar exposure to nitric oxide. Evidence for a free radical mechanism. J Biol Chem 278:15720–15726
Kabil O, Banerjee R (2010) Redox biochemistry of hydrogen sulfide. J Biol Chem 285: 21903–21907
Kamiya T, Teraji T, Saito Y, Hasimoto M, Nakaguchi O, Oka T (1973) Studies on β-lactam antibiotics. I. A novel conversion of penicillins into cephalosporins. Tetrahedron Lett 14: 3001–3004
Kashiba-Iwatsuki M, Kitoh K, Kasahara E, Yu H, Nisikawa M, Matsuo M, Inoue M (1997) Ascorbic acid and reducing agents regulate the fates and functions of *S*-nitrosothiols. J Biochem 122:1208–1214
Kenyon GL, Bruice TW (1997) Novel sulfhydryl reagents. Methods Enzymol 47:407–430
Keszler A, Zhang Y, Hogg N (2010) Reaction between nitric oxide, glutathione, and oxygen in the presence and absence of protein: how are *S*-nitrosothiols formed? Free Radic Biol Med 48:55–64

Khamis G, Stoevaa S, Aleksieva D (2010) Reactivity of sodium arenesulfinates in the substitution reaction to γ-functionalized allyl bromides. J Phys Org Chem 23:461–467

Kharasch N, Potempa SJ, Wehrmeister HL (1946) The sulfenic acids and their derivatives. Chem Rev 39:269

Kingsbury CA, Cram DJ (1960) Studies in stereochemistry. XXXII. Mechanism of elimination of sulfoxides. J Am Chem Soc 82:1810–1819

Kirsch M, Buscher AM, Aker S, Schulz R, de Groot H (2009) New insights into the *S*-nitrosothiol-ascorbate reaction. The formation of nitroxyl. Org Biomol Chem 7:1954–1962

Kissner R, Nauser T, Bugnon P, Lye PG, Koppenol WH (1997) Formation and properties of peroxynitrite as studied by laser flash photolysis, high-pressure stopped-flow technique, and pulse radiolysis. Chem Res Toxicol 10:1285–1292

Klomsiri C, Karplus PA, Poole LB (2011) Cysteine-based redox switches in enzymes. Antioxid Redox Signal 14:1065–1077

Koide Y, Urano Y, Hanaoka K, Terai T, Nagano T (2011) Development of an Si-rhodamine-based far-red to near-infrared fluorescence probe selective for hypochlorous acid and its applications for biological imaging. J Am Chem Soc 133:5680–5682

Kojima H, Nakatsubo N, Kikuchi K, Kawahara S, Kirino Y, Nagoshi H, Hirata Y, Nagano T (1998) Detection and imaging of nitric oxide with novel fluorescent indicators: diaminofluoresceins. Anal Chem 70:2446–2453

Krishnan N, Fu C, Pappin DJ, Tonks NK (2011) H2S-induced sulfhydration of the phosphatase PTP1B and its role in the endoplasmic reticulum stress response. Sci Signal 4:ra86

Kuivila HG, Armour AG (1957) Electrophilic displacement reactions. IX. Effects of substituents on rates of reactions between hydrogen peroxide and benzeneboronic acid. J Am Chem Soc 79:5659–5662

Kutala VK, Villamena FA, Ilangovan G, Maspoch D, Roques N, Veciana J, Rovira C, Kuppusamy P (2008) Reactivity of superoxide anion radical with a perchlorotriphenylmethyl (trityl) radical. J Phys Chem B 112:158–167

LaButti J, Chowdhury G, Reilly TJ, Gates KS (2007) Redox regulation of protein tyrosine phosphatase 1B (PTP1B) by Peroxymonophosphate (=O3POOH). J Am Chem Soc 129:5320–5321

Lee JW, Soonsanga S, Helmann JD (2007) A complex thiolate switch regulates the Bacillus subtilis organic peroxide sensor OhrR. Proc Natl Acad Sci USA 104:8743–8748

Leonard SE, Reddie KG, Carroll KS (2009) Mining the thiol proteome for sulfenic acid modifications reveals new targets for oxidation in cell. ACS Chem Biol 4:783–799

Leonard SE, Garcia FJ, Goodsell DS, Carroll KS (2011) Redox-based probes for protein tyrosine phosphatases. Angew Chem Int Ed 50:4423–4427

Li L, Rose P, Moore PK (2011) Hydrogen sulfide and cell signaling. Annu Rev Pharmacol Toxicol 51:169–187

Li S, Wang H, Xian M, Whorton R (2012) Identification of protein nitrosothiols using phosphine-mediated selective reduction. Nitric Oxide 26:20–26

Lim ML, Lippard SJ (2007) Metal-based turn-on fluorescent probes for sensing nitric oxide. Acc Chem Res 40:41–51

Lim JC, Choi HI, Park YS, Nam HW, Woo HA, Kwon KS, Kim YS, Rhee SG, Kim K, Chae HZ (2008) Irreversible oxidation of the active-site cysteine of peroxiredoxin to cysteine sulfonic acid for enhanced molecular chaperone activity. J Biol Chem 283:28873–28880

Lim JC, You Z, Kim G, Levine RL (2011) Methionine sulfoxide reductase a is a stereospecific methionine oxidase. Proc Natl Acad Sci 108:10472–10477

Lin Y-I, Lang SA Jr (1980) Selective reduction of nitro-heterocycles with sodium sulfide in aqueous p-dioxane. J Heterocycl Chem 17:1273–1275

Lind C, Gerdes R, Hamnell Y, Schuppe-Koistinen I, von Lowenhielm HB, Holmgren A, Cotgreave IA (2002) Identification of *S*-glutathionylated cellular proteins during oxidative stress and constitutive metabolism by affinity purification and proteomic analysis. Arch Biochem Biophys 406:229–240

Lippert AR, New EJ, Chang CJ (2011) Reaction-based fluorescent probes for selective imaging of hydrogen sulfide in living cells. J Am Chem Soc 133:10078–10080
Little G, Brocklehurst K (1972) Kinetics of the reversible reaction of papain with 5,5′-dithiobis-(2-nitrobenzoate) dianion: evidence for nucleophilic reactivity in the un-ionized thiol group of cysteine-25 and for general acid catalysis by histidine-159 of the reaction of the 5-mercapto-2-nitrobenzoate dianion with the papain-5-mercapto-2-nitrobenzoate mixed disulphide. Biochem J 128:475–477
Liu C, Pan J, Li S, Zhao Y, Wu LY, Berkman CE, Whorton AR, Xian M (2011) Capture and visualization of hydrogen sulfide by a fluorescent probe. Angew Chem Int Ed Engl 50: 10327–10329
Lloyd MM, Van Reyk DM, Davies MJ, Hawkins CL (2008) Hypothiocyanous acid is a more potent inducer of apoptosis and protein thiol depletion in murine macrophage cells than hypochlorous acid or hypobromous acid. Biochem J 414:271–280
Lo Conte M, Carroll KS (2012) Chemoselective ligation of sulfinic acids with aryl-nitroso compounds (2012) Angew Chem Inter Ed 51:6502–6505
Luo D, Smith SW, Anderson BD (2004) Kinetics and mechanism of the reaction of cysteine and hydrogen peroxide in aqueous solution. J Pharm Sci 94:304–316
Maker HS, Weiss C, Silides DJ, Cohen G (1981) Coupling of dopamine oxidation (monoamine oxidase activity) to glutathione oxidation *via* the generation of hydrogen peroxide in rat brain homogenates. J Neurochem 36:589–593
Maller C, Schröder E, Eaton P (2011) Glyceraldehyde 3-phosphate dehydrogenase is unlikely to mediate hydrogen peroxide signaling: studies with a novel anti-dimedone sulfenic acid antibody. Antioxid Redox Signal 14:49–60
Marcinkiewicz J, Grabowska A, Bereta J, Stelmaszynska T (1995) Taurine chioramine, a product of activated neutrophils, inhibits in vitro the generation of nitric oxide and other macrophage inflammatory mediators. J Leukco Biol 58:667–674
Marla SS, Lee J, Groves JT (1997) Peroxynitrite rapidly permeates phospholipid membranes. Proc Natl Acad Sci USA 94:14243–14248
Masri MS, Friedman M (1988) Protein reactions with methyl and ethyl vinyl sulfones. J Protein Chem 7:49–54
Mathai JC, Missner A, Kügler P, Saparov SM, Zeidel ML, Lee JK, Pohl P (2009) No facilitator required for membrane transport of hydrogen sulfide. Proc Natl Acad Sci USA 106:16633–16638
McGrath AJ, Garrett GE, Valgimigli L, Pratt DA (2010) The redox chemistry of sulfenic acids. J Am Chem Soc 132:16759–16761
McLachlin DT, Chait BT (2003) Improved beta-elimination-based affinity purification strategy for enrichment of phosphopeptides. Anal Chem 75:6826–6836
Midwinter RG, Cheah F-C, Moskovitz J, Vissers MC, Winterbourn CC (2006) IκB is a sensitive target for oxidation by cell-permeable chloramines: inhibition of NF-κB activity by glycine chloramine through methionine oxidation. Biochem J 396:71–78
Miller EW, Dickinson BC, Chang CJ (2010) Aquaporin-3 mediates hydrogen peroxide uptake to regulate downstream intracellular signaling. Proc Natl Acad Sci USA 107:15681–15686
Miron T, Shin I, Feigenblat G, Weiner L, Mirelman D, Wilchek M, Rabinkov A (2002) A spectrophotometric assay for allicin, alliin, and alliinase (alliin lyase) with a chromogenic thiol: reaction of 4-mercaptopyridine with thiosulfinates. Anal Biochem 307:76–83
Monteiro G, Horta BB, Pimenta DC, Augusto O, Netto LES (2007) Reduction of 1-Cys peroxiredoxins by ascorbate changes the thiol-specific antioxidant paradigm, revealing another function of vitamin C. Proc Natl Acad Sci 104:4886–4891
Moran EE, Timerghazin QK, Kwong E, English AM (2011) Kinetics and mechanism of S-nitrosothiol acid-catalyzed hydrolysis: sulfur activation promotes facile NO+ release. J Phys Chem B 115:3112–3126
Moro MA, Darley-Usmar VM, Goodwin DA, Read NG, Zamora-Pino R, Feelisch M, Radomski MW, Moncada S (1994) Paradoxical fate and biological action of peroxynitrite on human platelets. Proc Natl Acad Sci USA 91(14):6702–6706

Mueller EG (2006) Trafficking in persulfides: delivering sulfur in biosynthetic pathways. Nat Chem Biol 2:185–194
Murphy MP (2009) How mitochondria produce reactive oxygen species. Biochem J 417:1–13
Mustafa AK, Gadalla MM, Sen N, Kim S, Mu W, Gazi SK, Barrow RK, Yang G, Wang R, Snyder SH (2009) H2S signals through protein *S*-sulfhydration. Sci Signal 2:ra72
Mustafa AK, Sikka G, Gazi SK, Steppan J, Jung SM, Bhunia AK, Barodka VM, Gazi FK, Barrow RK, Wang R, Amzel LM, Berkowitz DE, Snyder SH (2011) Hydrogen sulfide as endothelium-derived hyperpolarizing factor sulfhydrates potassium channels. Circ Res 109:1259–1268
Nagy P, Ashby MT (2007a) Reactive sulfur species: kinetics and mechanisms of the oxidation of cysteine by hypohalous acid to give cysteine sulfenic acid. J Am Chem Soc 129:14082–14091
Nagy P, Ashby MT (2007b) Reactive sulfur species: kinetics and mechanism of the hydrolysis of cysteine thiosulfinate ester. Chem Res Toxicol 20:1364–1372
Nagy P, Winterbourn CC (2010) Rapid reaction of hydrogen sulfide with the neutrophil oxidant hypochlorous acid to generate polysulfides. Chem Res Toxicol 23:1541–1543
Nagy P, Beal JL, Ashby MT (2006) Thiocyanate is an efficient endogenous scavenger of the phagocytic killing agent hypobromous acid. Chem Res Toxicol 19:587–593
Nagy P, Lemma K, Ashby MT (2007) Reactive sulfur species: kinetics and mechanisms of the reaction of cysteine thiosulfinate ester with cysteine to give cysteine sulfenic acid. J Org Chem 72:8838–8846
Nagy P, Jameson GN, Winterbourn CC (2009) Kinetics and mechanisms of the reaction of hypothiocyanous acid with 5-thio-2-nitrobenzoic acid and reduced glutathione. Chem Res Toxicol 22:1833–1840
Nagy P, Karton A, Betz A, Peskin AV, Pace P, O'Reilly RJ, Hampton MB, Radom L, Winterbourn CC (2011) Model for the exceptional reactivity of peroxiredoxins 2 and 3 with hydrogen peroxide. A kinetic and computational study. J Biol Chem 286(20):18048–18055
Nakamura N (1983) A stable sulfenic acid, 9-triptycenesulfenic acid: its isolation and characterization. J Am Chem Soc 105:7172–7173
Nielsen ML, Vermeulen M, Bonaldi T, Cox J, Moroder L, Mann M (2008) Iodoacetamide-induced artifact mimics ubiquitination in mass spectrometry. Nat Methods 5:459–460
Nielsen RW, Tachibana C, Hansen NE, Winther JR (2011) Trisulfides in proteins. Antioxid Redox Signal 15:67–75
Nti-Addae KW, Laurence JS, Skinner AL, Stella VJ (2011) Reversion of sulfenamide prodrugs in the presence of free thiol-containing proteins. J Pharm Sci 100:3023–3027
Ogata Y, Sawaki Y, Isono M (1970) Kinetics of the addition of substituted benzenesulfinic acids to p-benzoquinone. Tetrahedron 26:731–736
Olsen JV, Andersen JR, Nielsen PA, Nielsen ML, Figeys D, Mann M, Wisniewski JR (2004) HysTag–a novel proteomic quantification tool applied to differential display analysis of membrane proteins from distinct areas of mouse brain. Mol Cell Prot 3:82–92
Pacher P, Beckman JS, Liaudet L (2007) Nitric oxide and peroxynitrite in health and disease. Physiol Rev 87:315–424
Pang L-J, Wang D, Zhou J, Zhang L-H, Ye X-S (2009) Synthesis of neamine-derived pseudodisaccharides by stereo- and regio-selective functional group transformations. Org Biomol Chem 7:4252–4266
Patt J, Patt M (2002) Reaction of [18F] 4- fluorobenzenediazonium cations with cysteine or the cysteinyl group: preparation of 18F-labeled *S*-aryl-cysteine and a radiolabeled peptide. J Label Compd Radiopharm 45:1229–1238
Pattison DI, Davies MJ (2001) Absolute rate constants for the reaction of hypochlorous acid with protein side chains and peptide bonds. Chem Res Toxicol 14:1453–1464
Paulsen CE, Carroll KS (2010) Orchestrating redox signaling networks through regulatory cysteine switches. ACS Chem Biol 5:47–62
Paulsen CE, Truong TH, Garcia FJ, Homann A, Gupta V, Leonard SE, Carroll KS (2012) Peroxide-dependent sulfenylation of the EGFR catalytic site enhances kinase activity. Nat Chem Biol 8:57–64

Peng T, Yang D (2010) HKGreen-3: a rhodol-based fluorescent probe for peroxynitrite. Org Lett 12:4932–4935
Percival MD, Ouellet M, Campagnolo C, Claveau D, Li C (1999) Inhibition of cathepsin K by nitric oxide donors: evidence for the formation of mixed disulfides and a sulfenic acid. Biochemistry 38:13574–13583
Poole LB, Karplus PA, Claiborne A (2004) Protein sulfenic acids in redox signaling. Annu Rev Pharmacol Toxicol 44:325–347
Poole B, Zeng B-B, Knaggs SA, Yakubu M, King SB (2005) Synthesis of chemical probes to Map sulfenic acid modifications on proteins. Bioconjug Chem 16:1624–1628
Poudrel JM, Cole ER (2001) Interchange reactions of aromatic thiosulfinates. Phosphorus Sulfur Silicon Relat Elem 175:79–86
Prousek J (2007) Fenton chemistry in biology and medicine. Pure Appl Chem 79:2325–2338
Qian Y, Karpus J, Kabil O, Zhang S-Y, Zhu H-L, Banerjee R, Zhao J, He C (2011a) Selective fluorescent probes for live-cell monitoring of sulphide. Nat Commun 2. doi:10.1038/ncomms1506
Qian J, Klomsiri C, Wright MW, King SB, Tsang AW, Poole LB, Furdui CM (2011b) Simple synthesis of 1,3-cyclopentanedione derived probes for labeling sulfenic acid proteins. Chem Commun 47:9203–9205
Radi R, Peluffo G, Alvarez MN, Naviliat M, Cayota A (2001) Unraveling peroxynitrite formation in biological systems. Free Radic Biol Med 30:463–488
Raftery MJ, Yang Z, Valenzuela SM, Geczy CL (2001) Novel intra- and inter-molecular sulfinamide bonds in S100A8 produced by hypochlorite oxidation. J Biol Chem 276:33393–33401
Reddie KG, Carroll KS (2008) Expanding the functional diversity of proteins through cysteine oxidation. Curr Opin Chem Biol 12:746–754
Rhee SG, Woo HA (2011) Multiple functions of peroxiredoxins: peroxidases, sensors and regulators of the intracellular messenger H_2O_2, and protein chaperones. Antioxid Redox Signal 15:781–794
Ritchie CD, Saltiel JD, Lewis ES (1961) The reaction of diazonium salts with nucleophiles. VIII. The formation of diazosulfones and the application of linear free energy equations to diazonium salt reactions. J Am Chem Soc 83:4601–4605
Roos G, Messens J (2011) Protein sulfenic acid formation: from cellular damage to redox regulation. Free Radic Biol Med 51:314–326
Salmeen A, Andersen JN, Myers MP, Meng TC, Hink JA, Tonks NK, Barford D (2003) Redox regulation of protein tyrosine phosphatase 1B involves a sulphenyl-amide intermediate. Nature 423:769–773
Saurin AT, Neubert H, Brennan JP, Eaton P (2004) Widespread sulfenic acid formation in tissues in response to hydrogen peroxide. Proc Natl Acad Sci 101:17982–17987
Saxon E, Bertozzi CR (2000) Cell surface engineering by a modified Staudinger reaction. Science 287:2007–2010
Schank K, Bügler S, Folz H, Schott N (2007) Thioreductones and derivatives. Helvetica Chimica Acta 90:1606–1649
Seo YH, Carroll KS (2009) Profiling protein thiol oxidation in tumor cells using sulfenic acid-specific antibodies. Proc Natl Acad Sci USA 106:16163–16168
Seo YH, Carroll KS (2011) Quantification of protein sulfenic acid modifications using isotope-coded dimedone and iododimedone. Angew Chem Int Ed 50:1342–1345
Shetty V, Neubert TA (2009) Characterization of novel oxidation products of cysteine in an active site motif peptide of PTP1B. J Am Soc Mass Spectrom 20:1540–1548
Shevchenko A, Wilm M, Vorm O, Mann M (1996) Mass spectrometric sequencing of proteins from silver-stained polyacrylamide gels. Anal Chem 68:850–858
Shiau TP, Erlanson DA, Gordon EM (2006) Selective reduction of peptide isothiazolidin-3-ones. Org Lett 8:5697–5699
Sivaramakrishnan S, Keerthi K, Gates KS (2005) A chemical model for redox regulation of protein tyrosine phosphatase 1B (PTP1B) activity. J Am Chem Soc 127:10830–10831
Spalteholz H, Wenske K, Arnhold J (2005) Interaction of hypohalous acids and heme peroxidases with unsaturated phosphatidylcholines. Biofactors 24:67–76

Squadrito GL, Pryor WA (1998) Oxidative chemistry of nitric oxide: the roles of superoxide, peroxynitrite, and carbon dioxide. Free Radic Biol Med 25:392–403
Sullivan DM, Wehr NB, Fergusson MM, Levine RL, Finkel T (2000) Identification of oxidant-sensitive proteins: TNF-alpha induces protein glutathiolation. Biochemistry 39:11121–11128
Sun ZN, Liu FQ, Chen Y, Tam PK, Yang D (2008) A highly specific BODIPY-based fluorescent probe for the detection of hypochlorous acid. Org Lett 10:2171–2174
Szabo C, Ischiropoulos H, Radi R (2007) Peroxynitrite: biochemistry, pathophysiology and development of therapeutics. Nat Rev Drug Discov 6:662–680
Szajewski RP, Whitesides GM (1980) Rate constants and equilibrium constants for thiol-disulfide interchange reactions involving oxidized gluthathione. J Am Chem Soc 102:2011–2026
Tajc SG, Tolbert BS, Basavappa R, Miller BL (2004) Direct determination of thiol pKa by isothermal titration microcalorimetry. J Am Chem Soc 126:10508–10509
Tang SS, Chang GGJ (1996) Kinetic characterization of the endogenous glutathione transferase activity of octopus lens S-crystallin. Biochemistry 119:1182–1188
Tasaki T, Kwon YT (2007) The mammalian N-end rule pathway: new insights into its components and physiological roles. Trends Biochem Sci 32:520–528
Thomas JA, Poland B, Honzatko R (1995) Protein sulfhydryls and their role in the antioxidant function of protein *S*-thiolation. Arch Biochem Biophys 319:1–9
Trost BM (1983) Selectivity: a key to synthetic efficiency. Science 219:245–250
Tsikas D, Sandmann J, Rossa S, Gutzki FM, Frölich JC (1999) Investigations of *S*-transnitrosylation reactions between low- and high-molecular-weight *S*-nitroso compounds and their thiols by high-performance liquid chromatography and gas chromatography–mass spectrometry. Anal Biochem 270:231–241
Turell L, Botti H, Carballal S, Ferrer-Sueta G, Souza JM, Durán R, Freeman BA, Radi R, Alvarez B (2008) Reactivity of sulfenic acid in human serum albumin. Biochemistry 47:358–367
Uppu RM, Pryor WA (1999) Nitrosation of 1,2-phenylenediamine by peroxynitrite/ CO2: evidence for a free radical mechanism. J Am Chem Soc 121:9738–9739
Van Dalen CJ, Whitehouse MW, Winterbourn CC, Kettle AJ (1997) Thiocyanate and chloride as competing substrates for myeloperoxidase. Biochem J 327:487–492
Van der Vliet A, Hoen PA, Wong PS, Bast A, Cross CE (1998) Formation of *S*-nitrosothiols *via* direct nucleophilic nitrosation of thiols by peroxynitrite with elimination of hydrogen peroxide. J Biol Chem 273:30255–30262
Van Montfort RL, Congreve M, Tisi D, Carr R, Jhoti H (2003) Oxidation state of the active-site cysteine in protein tyrosine phosphatase 1B. Nature 423:773–777
Wang R (2010) Hydrogen sulfide: the third gasotransmitter in biology and medicine. Antioxid Redox Signal 12:1061–1064
Wang H, Xian M (2008) Fast reductive ligation of *S*-nitrosothiols. Angew Chem Int Ed Engl 47:6598–6601
Wang H, Xian M (2011) Chemical methods to detect *S*-nitrosation. Curr Opin Chem Biol 15:32–37
Wang P-F, McLeish MJ, Kneen MM, Lee G, Kenyon GL (2001) An unusually low pKa for Cys-282 in the active site of human muscle creatine kinase. Biochemistry 40:11698–11705
Weerapana E, Simon GM, Cravatt BF (2008) Disparate proteome reactivity profiles of carbon electrophiles. Nat Chem Biol 4:405–407
Winterbourn CC, Hampton MB (2008) Thiol chemistry and specificity in redox signaling. Free Radic Biol Med 45:549–561
Woo HA, Kang SW, Kim HK, Yang KS, Chae HZ, Rhee SG (2003) Reversible oxidation of the active site cysteine of peroxiredoxins to cysteine sulfinic acid. Immunoblot detection with antibodies specific for the hyperoxidized cysteine-containing sequence. J Biol Chem 278:47361–47364
Wood ZA, Poole LB, Karplus PA (2003) Peroxiredoxin evolution and the regulation of hydrogen peroxide signaling. Science 300:650–653
Wouters MA, Fan SW, Haworth NL (2010) Disulfides as redox switches: from molecular mechanisms to functional significance. Antioxid Redox Signal 12:53–91

Yang D, Tang YC, Chen J, Wang XC, Bartberger MD, Houk KN, Olson L (1999) Ketone-catalyzed decomposition of peroxynitrite via dioxirane intermediates. J Am Chem Soc 121:11976–11983
Yang D, Wong M-K, Yan Z (2000) Regioselective intramolecular oxidation of phenols and anisoles by dioxiranes generated in situ. J Org Chem 65:4179–4184
Yang D, Wang H-L, Sun Z-N, Chung N-W, Shen J-G (2006) A highly selective fluorescent probe for the detection and imaging of peroxynitrite in living cells. J Am Chem Soc 128:6004–6005
Yang ZYK, Cho HJ, Lee J, Shin I, Tae J (2009) A rhodamine-hydroxamic acid-based fluorescent probe for hypochlorous acid and its applications to biological imagings. Org Lett 11:859–861
You KS, Benitez LV, McConachie WA, Allison WS (1975) The conversion of glyceraldehyde-3-phosphate dehydrogenase to an acylphosphatase by trinitroglycerin and inactivation of this activity by azide and ascorbate. Biochem Biophys Acta 384:317–330
Zhang J, Li S, Zhang D, Wang H, Whorton AR, Xian M (2010) Reductive ligation mediated one-step disulfide formation of *S*-nitrosothiols. Org Lett 12:4208–4211
Zheng H, Shang G-H, Yang S-Y, Gao X, Xu J-G (2008) Fluorogenic and chromogenic rhodamine spirolactam based probe for nitric oxide by spiro ring opening reaction. Org Lett 10:2357–2360
Zhou H, Singh H, Parsons ZD, Lewis SM, Bhattacharya S, Seiner SD, LaButti JN, Reilly TJ, Tanner JJ, Gates KS (2011) The biological buffer bicarbonate/CO_2 potentiates H_2O_2-mediated inactivation of protein tyrosine phosphatases. J Am Chem Soc 133:15803–15805

Chapter 2
Radical Scavenging by Thiols and the Fate of Thiyl Radicals

Christine C. Winterbourn

Abstract Reduced glutathione reacts rapidly with the majority of biological free radicals and is regarded as a good radical scavenging antioxidant. However, the reaction generates thiyl radicals, which are strong, potentially damaging oxidants that need to be removed for the process to be effective. Sequential reactions of the radical with the thiolate anion and oxygen drive the scavenging reaction, producing the disulfide and superoxide radicals. Protein thiols are also good radical scavengers via a similar mechanism, and this is an efficient route for generating intramolecular disulfides, mixed disulfides with glutathione and nitrosothiols. These interactions are relevant not only to antioxidant defence but as part of a network of radical reactions that can regulate the oxidation state of glutathione and the extent of protein S-thiolation as well as acting as a source of superoxide and hydrogen peroxide. This chapter discusses the radical chemistry of glutathione and other thiols and how it could contribute to redox activity in the cell.

Keywords Free radical scavenging • Glutathione • Superoxide • Protein disulfide • Thiyl radical

2.1 Introduction

Thiols are prime biological targets for oxidation and their ability to undergo reversible oxidation and reduction enables them to contribute to many cell functions. Thiol groups are required for the activity of numerous enzymes and they have a major role in the antioxidant defenses of the cell. There are also a large

---

C.C. Winterbourn (✉)
Department of Pathology, University of Otago Christchurch, PO Box 4345, Christchurch, New Zealand
e-mail: christine.winterbourn@otago.ac.nz

U. Jakob and D. Reichmann (eds.), *Oxidative Stress and Redox Regulation*,
DOI 10.1007/978-94-007-5787-5_2,

number of regulatory proteins whose activity can be modified by oxidation of the sulfhydryl group. Thus, sulfhydryl oxidation can affect cell function not only through inactivating thiol enzymes, but also by altering the binding characteristics of molecules involved in signaling pathways (Winterbourn and Hampton 2008; Forman et al. 2004).

Thiol groups are generally reactive with all the oxidants that cells encounter, although reaction rates may differ by many orders of magnitude. The reaction may proceed either by a two electron mechanism to generate the sulfenic acid (or sufenyl halide) as the initial product, as is the case with peroxides, hypochlorous acid and peroxynitrite, or by a one electron free radical pathway to produce the thiyl radical. The ultimate products of the two pathways are not necessarily the same. As a consequence of this plus the difference in reactivity of individual reactive oxygen species (ROS), all ROS will not undergo the same reactions in the cell. Therefore, antioxidant strategies, as well as effects on cell metabolism or regulatory pathways will differ depending on the oxidant involved.

The focus of this chapter is on radical pathways, with non-radical pathways covered more extensively elsewhere in this book. However, it is useful to compare the two pathways. With a few exceptions (notably peroxiredoxins, thiol-containing glutathione peroxidases and the bacterial transcription factor oxyR) the direct reaction of peroxides with thiol groups is slow (Winterbourn and Metodiewa 1999; Winterbourn and Hampton 2008). Antioxidant defense against peroxides is not due to a direct reaction with reduced glutathione (GSH), but is provided by well characterized systems that involve either glutathione peroxidases plus GSH, or peroxiredoxins acting in combination with thioredoxin/thioredoxin reductase (Flohe et al. 2011; Rhee and Woo 2011). These enzymatic systems involve two electron steps that cycle between thiol and disulfide and are ultimately dependent on NADPH for reducing equivalents. Direct reactions between thiols and more reactive oxidants such as hypochlorous acid, chloramines or peroxynitrite are more favorable. These can result in oxidation beyond the sulfenic acid or disulfide, to the sulfinic ($-SO_2H$) or sulfonic acid ($-SO_3H$). Particularly in the case of HOCl, condensation with amino groups in GSH or proteins can give rise to sulfinamides or sulfonamides (Harwood et al. 2006; Raftery et al. 2001). Therefore a range of higher oxidation products is possible.

Thiols react with a wide range of radical species. These are generated by a variety of cellular mechanisms (Fig. 2.1) and include hydroxyl, phenoxyl, alkoxyl, arylamino, peroxyl, semiquinone and carbon centred radicals as exemplified in Table 2.1. Some of the parent compounds that give rise to these radicals occur physiologically, others are drugs or environmental chemicals. Some, such as flavonoids, are themselves radical scavengers and of interest for their potential health benefits as antioxidants. This article examines radical scavenging by GSH and by thiol proteins within the context of antioxidant defense and also as a potential mechanism for regulating redox-sensitive cell functions.

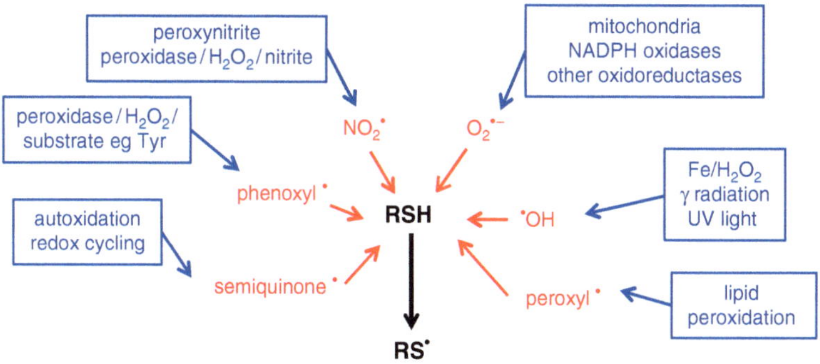

Fig. 2.1 Examples of physiological sources of thiyl radicals. As described in the text and Table 2.1, a majority of physiologically relevant free radicals can react with GSH and protein thiols (*shown as RSH in Figure*) to generate thiyl radicals. Major sources of such radicals are shown. Autoxidation and redox cycling mechanisms apply to many drugs and environmental chemicals including polyphenols, quinones and other aromatic compounds. Other oxidoreductases that generate superoxide radicals include xanthine oxidase, lipoxygenase and nitric oxide synthase. Superoxide reacts only slowly with thiols (*as designated by dashed line*) and these mechanisms are more relevant as a source of hydrogen peroxide for metal-catalyzed or peroxidase- mediated radical generation

Table 2.1 Examples of radical generating systems that cause oxidation of GSH to its thiyl radical

Class of compound	System	References
Tyrosine	Peroxidase	Nakamura et al. (1986), Pichorner et al. (1995)
Phenols	Peroxidase, radiolysis	Ross and Moldeus (1985), D'Arcy Doherty et al. (1986), Subrahmanyam and O'Brien (1985)
Sugars	Radiolysis	Baker et al. (1982)
DNA bases	Radiolysis	Willson (1983)
Nitrogen dioxide and peroxynitrite	Direct reaction	Quijano et al. (1997), Bonini and Augusto (2001)
Aromatic amines, phenothiazines	Peroxidase, autoxidation	Ross and Moldeus (1985), Subrahmanyam and O'Brien (1985), Subrahmanyam et al. (1987)
Ethanol (hydroxyethyl)	Thermal decomposition	Stoyanovsky et al. (1998)
Tocopherol	Radiolysis	Niki et al. (1982)
Hydroxpyrimidines	Autoxidation	Munday and Winterbourn (1989)

2.2 Radical Scavenging by Reduced Glutathione

Radical scavenging by GSH generates the glutathionyl radical (reaction 2.1). In most cases the reaction is fast; for example with tyrosine the rate constant, k, is $2 \times 10^6\ M^{-1}\ s^{-1}$ (Folkes et al. 2011) and with millimolar concentrations typically present inside cells, GSH would be expected to be an effective physiological scavenger. However, the reduction potential of GSH is sufficiently high that scavenging of many radicals is not highly favored on thermodynamic grounds (Wardman and von Sonntag 1995). The reaction is reversible, and in many cases the reverse reaction is at least as fast and the equilibrium lies far to the left (e.g. for acetaminophen $K_1 = 3 \times 10^{-4}$ (Ramakrishna Rao et al. 1990)). For a scavenging reaction to provide antioxidant protection, the products of the reaction must be less reactive or damaging than the initial species. However, as discussed in Sect. 2.7, thiyl radicals are oxidizing species that can undergo a number of potentially damaging reactions with biological molecules. On these grounds, GSH would not appear to have the qualities of a good radical scavenging antioxidant. Yet in experimental systems where radicals are generated from substrates such as acetaminophen and tyrosine, they are efficiently scavenged by GSH (for example Ramakrishna Rao et al. 1990; Ross 1988; Pichorner et al. 1995).

$$R^{\bullet} + GSH \rightleftarrows RH + GS^{\bullet} \tag{2.1}$$

$$GS^{\bullet} + GS^{-} \rightleftarrows GSSG^{\bullet -} \tag{2.2}$$

$$GSSG^{\bullet -} + O_2 \rightarrow GSSG + O_2^{\bullet -} \tag{2.3}$$

There are features of thiyl radical chemistry that enable GSH and other thiols to act as effective scavengers and antioxidants (Wardman and von Sonntag 1995; Winterbourn 1993; Wardman 1995). Most importantly, reaction (2.1) can be kinetically driven in the forward direction by removal of $GS^{\bullet}$ through a rapid reaction with the thiolate anion (GS^{-}) (reaction 2.2). Reaction (2.2) is also an equilibrium. It converts an oxidizing radical ($GS^{\bullet}$) into the disulfide radical anion ($GSSG^{\bullet -}$), which is probably the strongest reductant produced in biological systems (Buettner 1993). The position of equilibrium (2.2) is influenced by the thiolate ion concentration, which is dependent on the pH, the pK_a of the thiol (8.8 for GSH) and the GSH concentration. At pH 7.4 and 5 mM GSH, the ratio of $GS^{\bullet}$ to $GSSG^{\bullet -}$ is 2:1 (Wardman 1995). However, the equilibrium is established rapidly and reactions of $GSSG^{\bullet -}$ may dominate even if its concentration is relatively low. $GSSG^{\bullet -}$ reacts with oxygen irreversibly (reaction 2.3) at a near diffusion-controlled rate ($k = 2 \times 10^8\ M^{-1}\ s^{-1}$). This displaces equilibrium (2.2) to the right and provides the driving force for removing $GS^{\bullet}$. Thus reactions (2.1), (2.2), and (2.3) account for the good scavenging ability of GSH.

Physiologically, this is a major decay route for thiyl radicals. An alternative pathway is for the radical to react with oxygen to form a peroxyl radical (reaction 2.4). This reaction is fast, but it is also reversible and in most situations it contributes less to $GS^{\bullet}$ removal than reaction (2.3). However, it becomes more significant at lower pH or GSH concentration where less thiolate is present. Secondary reactions of the peroxyl radical are the most likely source of the higher oxidation states of glutathione (such as the sulfinic and sulfonic acid) that are minor products in some radical systems (Wefers et al. 1985; Winterbourn and Metodiewa 1994). $GSSG^{\bullet-}$ may undergo alternative reactions to reaction (2.3) (e.g. reduce quinones and heme proteins) when oxygen is limiting. Dimerization of $GS^{\bullet}$ radicals to give the disulfide (GSSG) is of limited significance at the low steady state radical concentrations that are likely to be present physiologically and even under hypoxic conditions it should only be a minor route for radical decay. Indeed, radical transfer to a carbon centre on the peptide, as observed under anerobic conditions (Zhao et al. 1994, 1997) is likely to be more favourable than dimerization.

$$GS^{\bullet} + O_2 \rightleftarrows GSOO^{\bullet} \tag{2.4}$$

From the above argument, it is evident that oxygen plays an important role in enabling GSH to act as a good scavenger without the build up of potentially damaging thiyl radicals. As a corollary, in addition to producing GSSG as a final product, the reaction sequence consumes oxygen and generates superoxide, which will dismutate to hydrogen peroxide (reaction 2.5). Therefore, scavenging by GSH could be regarded as a generator of oxidative stress or a potential source of oxidants involved in redox regulation.

$$2O_2{}^{\bullet-} + 2H^+ \rightarrow H_2O_2 + O_2 \tag{2.5}$$

From the perspective of antioxidant protection, however, this mechanism enables GSH to act as an intermediary for removing radicals and channeling them to superoxide (Fig. 2.2). Superoxide acts as a radical sink and, with superoxide dismutase (SOD) present, the sequence provides a mechanism for a single enzyme to control the effects of radical generation (Munday and Winterbourn 1989; Winterbourn 1993). GSH acting in concert with SOD can therefore provide effective radical scavenging antioxidant activity, with the sum of reactions (2.1), (2.2), (2.3) and (2.5) for radical removal giving the net reaction (2.6). For full antioxidant protection, enzymatic removal of hydrogen peroxide is also required.

$$2R^{\bullet} + 4GSH + O_2 \rightarrow 2RH + 4GSSG + H_2O_2 \tag{2.6}$$

There is a plethora of evidence that superoxide and hydrogen peroxide are produced during radical scavenging by GSH (for example O'Brien 1988; Munday 1994; Pichorner et al. 1995; Stoyanovsky et al. 1995; Ross et al. 1985; Galati et al. 1999). These examples include systems where phenoxyl radicals are generated from tyrosine or dietary flavonoids by a peroxidase plus hydrogen peroxide (reaction 2.7), in which case the hydrogen peroxide generated in the scavenging reaction is re-used by the peroxidase. In this mechanism, referred to as thiol pumping (Ross et al. 1985), much more GSH is oxidized than the initial peroxide added.

$$2RH + H_2O_2 \xrightarrow{\text{peroxidase}} 2R^{\bullet} + 2H_2O \tag{2.7}$$

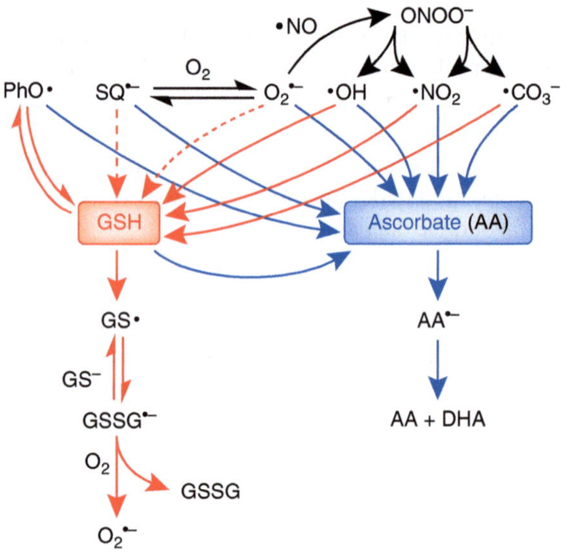

Fig. 2.2 Major aqueous phase radical scavenging pathways in the cell. Pathways show potential interactions with the two main aqueous scavengers, GSH and ascorbate. Note the reversibility and interconversion of many of the reactions. Reactions equivalent to those shown for GSH occur with thioredoxin and glutaredoxin and are also likely with other thiol proteins. A key difference between the glutathione (*red*) and ascorbate (*blue*) pathways is that the former is a radical chain, driven by the reaction of the disulfide radical anion ($GSSG^{\bullet-}$) with oxygen, and producing superoxide, whereas ascorbate generates ascorbyl radicals ($AA^{\bullet-}$). SOD removes superoxide radicals but the hydrogen peroxide generated could be pro-oxidant. *Dashed lines* signify less favourable reactions. *GS*$^{\bullet}$ glutathionyl radical, *PhO*$^{\bullet}$ phenoxyl radical, *SQ*$^{\bullet}$ semiquinone radical, *DHA* dehydroascorbate; reproduced from Winterbourn (2008) with permission

A similar mechanism operates when hydroquinones and hydroxypyrimidines such as dialuric acid autoxidize in the presence of GSH (Winterbourn and Munday 1989, 1990). These compounds undergo superoxide-dependent autoxidation via a semiquinone intermediate. GSH alone, by reducing the semiquinone and generating superoxide, enhances autoxidation and the resultant hydrogen peroxide production, and but with SOD also present, the whole process is inhibited. These are good examples of where both GSH and SOD are required for effective antioxidant protection.

2.3 Reaction of Superoxide with Thiols

The superoxide generated as a result of radical scavenging by GSH could potentially react with more GSH and set up a chain reaction. If this reaction were fast, then large amounts of GSH could be oxidized for each initial radical generated. Data from a number of sources indicate that superoxide does react with GSH. Although a wide range of values for the rate constant is reported in the literature, there is now a consensus that the reaction is relatively slow and at physiological thiol concentrations there is only a short chain (Winterbourn and Metodiewa 1994, 1999). Rate constants of 200 M^{-1} s^{-1} for GSH (Jones et al. 2002) and 68 M^{-1} s^{-1} for N-acetylcysteine (Benrahmoune et al. 2000) have been measured. GSH reacts about 500 times more slowly than ascorbate with superoxide so even if SOD were limiting, this reaction should play a minimal role in superoxide removal or GSH oxidation in the cell. The same argument holds for N-acetylcysteine when it is added to experimental systems as an "antioxidant".

Experimental and theoretical studies with cysteine, GSH and dithiothreitol indicate that the reaction with superoxide is likely to proceed via a complex stabilized by a three electron sulfur – oxygen bond, which breaks down in reaction (2.8) to give a sulfinyl radical (Zhang et al. 1991; Winterbourn and Metodiewa 1994; Cardey and Enescu 2009). The thiyl radical is formed, but via a secondary reaction with the thiol (reaction 2.9) rather than direct electron transfer, and superoxide is regenerated via reactions (2.2) and (2.3). The short chain is probably due to superoxide reacting with the intermediate radicals. Direct electron transfer to give the thiyl radical should be more favorable for the hydroperoxyl radical ($HO_2^{\bullet}$) and may be more relevant at lower pH.

$$O_2^{\bullet-} + GSH \rightarrow [GS \bullet\bullet\bullet O_2H]^{\bullet-} \rightarrow RH + GSO^{\bullet-} \tag{2.8}$$

$$GSO^{\bullet-} + GSH \rightarrow GSOH + GS^{\bullet} \tag{2.9}$$

2.4 Radical Scavenging by Protein Thiols

Protein thiols should undergo similar radical scavenging reactions as GSH. The reactivity of a particular cysteine residue will depend on its pK_a (the thiolate anion is generally more reactive (Wardman and von Sonntag 1995)) and another thiol being accessible so the equivalent of reaction (2.3) can proceed. Cysteine thiyl radicals can be formed directly on proteins exposed to radical generating systems. In addition, cysteine can act as a sink for radicals generated at other sites such as tyrosine, tryptophan or carbon side chains (Schoneich 2008). For example, spin trapping studies with myoglobin have shown intramolecular transfer from an initial tyrosyl radical to cysteine (Witting and Mauk 2001). Radical transfer from tyrosine to cysteine is also seen in ribonucleotide reductase where it is an integral part of its enzymatic mechanism (Holmgren and Sengupta 2010). Generation of a protein thiyl radical has also been observed during the enzymatic activity of mitochondrial NADH dehydrogenase and endothelial nitric oxide synthase (Chen et al. 2005, 2011). With nitric oxide synthase this occurred in the absence of tetrahydrobiopterin cofactor when the enzyme switched from nitric oxide to superoxide production.

One consequence of intramolecular radical transfer is that nitration of tyrosine residues can be inhibited by the presence of neighbouring cysteine residues (Zhang et al. 2005). Tyrosine nitration proceeds by a mechanism that involves combination of a tyrosyl and nitrogen dioxide radical. When this reaction was studied with tyrosyl peptides containing cysteine residues, nitration was suppressed and formation of thiyl radicals due to electron transfer from the tyrosine radical was observed.

Some protein thiols could be much more effective than GSH as radical scavengers. As reaction (2.2) is critical in the scavenging pathway, the presence of a vicinal thiol group should facilitate formation of the disulfide bond and accelerate the reaction. The vicinal thiols in dihydrolipoic acid and thioredoxin have both been shown to scavenge phenoxyl radicals and generate superoxide in the process (Goldman et al. 1995). In contrast to GSH, it was not possible to trap the thiyl radical with these compounds. This implies a rapid reaction of the radical with the vicinal thiol group. We studied this mechanism using the stable tetramethylphenylenediamine (Wurster's blue) radical, following its disappearance by stopped flow spectrophotometry. In this system, dithiothreitol was 5,000 times more efficient than GSH at scavenging the radical (Winterbourn 2003). This observation raises the possibility that proteins containing vicinal thiol groups could be efficient radical scavengers and generators of superoxide. Radical mechanisms could then be a significant route to disulfide formation in these proteins.

If radicals are formed on single cysteine residues on proteins, they would be expected to react with GSH (reactions 2.2 and 2.3) to generate a mixed disulfide, and produce superoxide in the process. This route to glutathionylation of a specific cysteine residue has been observed for nitric oxide synthase (Chen et al. 2011). Glutathionylation by a different radical mechanism can also be facilitated by glutaredoxins. In addition to catalyzing disulfide interchange reactions with glutathione, these thiol proteins react rapidly with glutathionyl radicals, and accelerate oxygen-dependent formation of GSSG (Starke et al. 2003; Gallogly et al. 2008). The proposed mechanism involves an initial reaction between the glutathionyl radical and the active site thiolate of glutaredoxin (the equivalent of reaction 2.2), which is favored because of its low pK_a and because the enzyme has a high affinity for the glutathionyl moiety (Gallogly et al. 2009). An enzyme-$SSG^{\bullet-}$ radical anion would be formed and react rapidly with oxygen (equivalent of reaction 2.3) to give superoxide and glutaredoxin-SSG. This is the catalytic intermediate formed by glutaredoxin in thiol exchange reactions, and is turned over by GSH to regenerate the enzyme thiolate and GSSG. This reaction has the potential to be a significant radical scavenging pathway and source of superoxide in the cell. It may also be a significant route for S-glutathionylation of cell proteins. Glutaredoxins are efficient catalysts of this reaction and in studies with actin, glyceraldehyde-3-phosphate dehydrogenase and protein tyrosine phosphatase 1B, thiyl radical-mediated formation of glutathione adducts was found to be much more efficient than any of the other mechanisms tested (Starke et al. 2003).

Protein glutathionylation has been widely observed in cells under oxidative stress and is proposed as one of the major mechanisms for regulating redox sensitive signalling pathways (Fratelli et al. 2003; Gallogly et al. 2009; Dalle-Donne et al.

2003). The details of how this occurs are not fully understood, although two electron oxidation pathways (e.g. with hydrogen peroxide) or thiol exchange is usually invoked. However, with the addition of GSH to protein thiyl radicals and glutaredoxin-mediated addition of GSH radicals both being facile mechanisms, it is possible that radical mechanisms may play a more prominent role in redox signalling than is generally considered.

2.5 Reactive Nitrogen Species and Nitrosothiol Formation

Nitrosylation of cellular GSH and protein thiols has been observed widely and linked to the biological action of nitric oxide and regulation of cell signaling pathways (Foster et al. 2003). However, in spite the emphasis given to this process as a regulatory mechanism, there is surprisingly little understood about how nitrosothiols are formed from NO. Once one nitrosothiol has formed, it can readily exchange with others, as exemplified by the protein S-thiolation by GSNO. However, NO and GSH do not react directly. Either one or the other must be oxidized. The most commonly cited reaction is between GSH and N_2O_3, which can be formed from the reaction of NO with NO_2. However, calculations by Lancaster (2006, 2008) show that the NO concentrations required for this reaction to be significant (above 20 μM) are unrealistically high for it to be a physiological pathway. In contrast, the reaction between NO and the thiyl radical is fast (Madej et al. 2008), and it can be argued that virtually all *de novo* nitrosothiol formations should occur via this reaction. However, there are caveats here too, as competing reactions will decrease its efficiency. It has been argued that this reaction will compete poorly with hydrogen transfer to the thiyl radical from a carbon centre on the peptide (Hofstetter et al. 2007). In addition, other species that are likely to be present in the systems where thiyl radicals are generated, including the thiolate plus oxygen, are likely to react more efficiently than NO. Therefore, in such situations nitrosation may account for only a small fraction of the thiol modification. Lancaster (2008) has argued further that in the redox signaling field, nitrosothiol formation may be an indicator, but not necessarily the functionally important modification.

Peroxynitrite ($ONOO^-$) is produced in a very fast reaction from superoxide and NO (reaction 2.10). It oxidizes GSH and other thiols both by two electron and radical mechanisms (Ferrer-Sueta and Radi 2009). The direct (two electron) reaction occurs in competition with breakdown of the peroxynitrite, which in physiological media involves reaction with carbon dioxide. This forms an intermediate that decomposes to give carbonate and nitrogen dioxide radicals (reaction 2.11). Both of these react readily with thiols. At high millimolar GSH concentrations, some of the oxidation is likely to be direct, but under most conditions radical-mediated oxidation of GSH and protein thiols should predominate (Quijano et al. 1997). The outcome is typical of any thiyl radical pathway, with disulfides as the major product (reactions 2.12 and 2.13 followed by reactions 2.2 and 2.3). Even when peroxynitrite is generated from NO and superoxide, this is not an efficient mechanism for generating nitrosothiols. However, the pathway does consume oxygen and at maximum efficiency more superoxide is generated than is used to form the initial peroxynitrite (net reaction 2.14). Therefore, in addition to being a route for the formation of GSSG or protein disulfides, this sequence is a physiological mechanism for generating superoxide at the expense of GSH and NO.

$$O_2^{\bullet -} + NO^{\bullet} \rightarrow ONOO^- \quad (2.10)$$

$$ONOO^- + CO_2 \rightarrow \text{intermediate} \rightarrow CO_3^{\bullet -} + NO_2^{\bullet} \quad (2.11)$$

$$CO_3^{\bullet -} + GSH \rightarrow HCO_3^- + GS^{\bullet} \quad (2.12)$$

$$NO_2^{\bullet} + GSH \rightarrow NO_2^- + GS^{\bullet} \quad (2.13)$$

$$NO^{\bullet} + 4GSH + 2O_2 \rightarrow NO_2^- + 2GSSG + H_2O + 2H^+ + O_2^{\bullet -} \quad (2.14)$$

2.6 Interactions Between Glutathione, Ascorbate and Other Radical Scavengers

Although GSH is an effective scavenger in many experimental systems, the question arises as to how efficient this process in relation to other scavenging pathways. Ascorbate is an obvious alternative. It is a better one electron reductant than GSH (Buettner 1993) and scavenges a wide range of radicals including thiyl radicals. Therefore, it could act either by scavenging other radicals directly by intercepting the glutathionyl radical, as shown in Fig. 2.2. The ascorbate radical is relatively stable, does not react with oxygen, and decays primarily by dismutation. Thus scavenging by ascorbate could bypass superoxide production from GSH. Wardman (1995) has considered radical reactions involving ascorbate and GSH from a thermodynamic perspective and calculated that with ascorbate and GSH concentrations in the physiological range, thiyl radicals would preferentially react with ascorbate, with a minor fraction giving rise to superoxide. Sturgeon and co-workers (1998) used a peroxidase system to generate tyrosyl radicals and showed that with 8 mM GSH, both oxygen uptake and thiyl radical formation were inhibited by ascorbate in the 25–100 μM range. It can be concluded from both approaches that ascorbate should dominate over GSH as a radical scavenger under typical intracellular conditions. If this mechanism prevails, there is still synergy between the two antioxidants, as the dehydroascorbate generated can be recycled intracellularly by GSH in a two-electron reduction (Cuddihy et al. 2008). The overall outcome is therefore oxidation of GSH to GSSG.

However, conditions will vary between cell types and within cell compartments, and over time of oxidant exposure as components are consumed, and there are a number of experimental studies in which thiyl radicals have been trapped in cells subjected to free radical stress (for example Kwak et al. 1995; Stoyanovsky et al. 1996). Therefore, it is reasonable to assume that there will be conditions where scavenging by GSH is important, and that relative concentrations of GSH and ascorbate will influence the extent to which GSH undergoes radical-mediated oxidation with resultant superoxide generation. Furthermore, unless cultured cells are supplemented, they almost invariably contain no ascorbate. Most studies of oxidative stress responses and redox signalling have been carried out with cultured cells, in which scavenging by thiols should be more dominant and redox reactions could be different from normal tissues. How this affects signaling mechanisms needs to be addressed using ascorbate-replete cells.

Synergism is likely to be important for other antioxidant activities. Ascorbate, and to a lesser extent GSH, is able to scavenge vitamin E radicals, thereby transferring radicals generated by peroxidation in the lipid phase to the aqueous phase for disposal (Niki et al. 1982). Synergism may also be relevant to the action of dietary or pharmacological antioxidants such as polyphenolics and carotenoids. Even though these compounds are good radical scavengers, dietary intake is unable to increase tissue concentrations sufficiently to impact on total radical scavenging capacity (Halliwell 2007). One way in which they could act as radical scavenging antioxidants (and it should be noted that this mode of action is not unequivocally established), would be to facilitate radical transfer to more abundant species such as ascorbate or GSH.

2.7 Oxidative Reactions of Thiyl Radicals

$GS^{\bullet}$ is an oxidizing species that can react with hydrogen-donating molecules including NADH, polyunsaturated fatty acids, retinol, and ferricytochrome c and other proteins (Schöneich et al. 1992; Forni and Willson 1986a, b; Nauser et al. 2004; Chatgilialoglu and Ferreri 2005; Borisenko et al. 2004). Thus it can initiate lipid peroxidation, generate trans fatty acids through cis-trans isomerization and cause irreversible protein modification. Cysteine thiyl radicals generated on proteins can also undergo intramolecular hydrogen transfer to form carbon centred radicals that give rise to dehydroalanine (Mozziconacci et al. 2011) and other side chain modifications (as reviewed Schoneich 2008). Most of the experimental studies of these reactions have been studied under anaerobic conditions where the reaction of the radical with the thiolate and oxygen is not possible. However, they illustrate the point that without efficient removal of the thiyl radical by the latter mechanism, radical scavenging by thiols is more likely to be damaging rather than protective.

2.8 Detection of Thiyl Radicals

Detection of any short lived radical species in a biological system is difficult and thiyl radicals are no exception. However, they have been detected directly by electron paramagnetic resonance (EPR) in cell free systems and mechanistic studies also provide strong evidence for thiyl radical generation. Recent technological advances have made it possible to detect thiyl radicals in cellular systems (Stoyanovsky et al. 2011). For the most part this involves the use of spin traps, such as DMPO (5,5′-dimethyl-1-pyrroline *N*-oxide), which form radical adducts that can be detected not only by EPR but also by chromatographic, mass spectrometric and immunological methods. Although the initial spin-trapped species is a metastable nitroxide radical, this is readily converted to the stable hydroxylamine or nitrone, which in the case of GSH can be quantified in extracts of cells exposed to a radical stress using HPLC (Stoyanovsky et al. 1996; Cuddihy et al. 2008). A major advance introduced by the Mason group was the introduction of immuno-spin trapping (Mason 2004). In this technique, DMPO is used in an experimental system to trap radicals, then an antibody raised against DMPO is applied to detect radical adducts. Western blotting has been used to detect radical formation on proteins such as myoglobin, and immunoprecipitation plus mass spectrometry to identify the protein modified and pinpoint the radical site that reacted with the DMPO (Bonini et al. 2007; Bhattacharjee et al. 2007). Formation of cysteine radicals on nitric oxide synthase and NADH dehydrogenase have been detected using this approach (Chen et al. 2005, 2011) and the method has wider potential for following thiol oxidation in cells.

2.9 Thiyl Radicals and Redox Regulation

The radical scavenging properties of GSH and other thiols are generally considered in relation to the biological antioxidant defense network. As described in this chapter, thiols perform this function efficiently, provided the oxygen/thiolate mechanism can operate to remove the thiyl radical. However, radical scavenging could have a physiological impact beyond antioxidant protection. The field of redox biology has in recent years evolved from having a sole focus on oxidative damage to encompass oxidative stress and redox regulation, and the current view is of a continuum with no distinct boundaries. The same concept can be applied to thiyl radical reactions, with their capacity both to be regulated by cellular conditions and to be regulators of the proteins with which they interact. They are a source of superoxide and hydrogen peroxide, generated as a result of radical scavenging and able to be regulated by factors such as thiol concentration, oxygen and ascorbate. Thiyl radical reactions are also routes to protein disulfides, glutathionylated proteins and nitrosothiols, which are major players in redox regulation. Currently, the main emphasis is on two electron oxidants as transmitters of redox signals. However, based on the above argument, and the knowledge that major enzymatic generators of reactive oxidants such as NADPH oxidases, nitric oxide synthases, peroxidases and mitochondrial oxidoreductases operate by one electron mechanisms, radical mechanisms warrant more attention.

Acknowledgments The author's work has been supported by the Health Research Council of New Zealand.

References

Baker MZ, Badiello R, Tamba M, Quintiliani M, Gorin G (1982) Pulse radiolytic study of hydrogen transfer from glutathione to organic radicals. Int J Radiat Biol 41:595–602

Benrahmoune M, Therond P, Abedinzadeh Z (2000) The reaction of superoxide radical with N-acetylcysteine. Free Radic Biol Med 29:775–782

Bhattacharjee S, Deterding LJ, Jiang J, Bonini MG, Tomer KB, Ramirez DC, Mason RP (2007) Electron transfer between a tyrosyl radical and a cysteine residue in hemoproteins: spin trapping analysis. J Am Chem Soc 129:13493–13501

Bonini MG, Augusto O (2001) Carbon dioxide stimulates the production of thiyl, sulfinyl, and disulfide radical anion from thiol oxidation by peroxynitrite. J Biol Chem 276:9749–9754

Bonini MG, Siraki AG, Atanassov BS, Mason RP (2007) Immunolocalization of hypochlorite-induced, catalase-bound free radical formation in mouse hepatocytes. Free Radic Biol Med 42:530–540

Borisenko GG, Martin I, Zhao Q, Amoscato AA, Tyurina YY, Kagan VE (2004) Glutathione propagates oxidative stress triggered by myeloperoxidase in HL-60 cells. Evidence for glutathionyl radical-induced peroxidation of phospholipids and cytotoxicity. J Biol Chem 279:23453–23462

Buettner GR (1993) The pecking order of free radicals and antioxidants: lipid peroxidation, a-tocopherol, and ascorbate. Arch Biochem Biophys 300:535–543

Cardey B, Enescu M (2009) Cysteine oxidation by the superoxide radical: a theoretical study. Chemphyschem 10:1642–1648

Chatgilialoglu C, Ferreri C (2005) Trans lipids: the free radical path. Acc Chem Res 38:441–448

Chen YR, Chen CL, Zhang L, Green-Church KB, Zweier JL (2005) Superoxide generation from mitochondrial NADH dehydrogenase induces self-inactivation with specific protein radical formation. J Biol Chem 280:37339–37348

Chen CA, Lin CH, Druhan LJ, Wang TY, Chen YR, Zweier JL (2011) Superoxide induces endothelial nitric-oxide synthase protein thiyl radical formation, a novel mechanism regulating eNOS function and coupling. J Biol Chem 286:29098–29107

Cuddihy SL, Parker A, Harwood DT, Vissers MC, Winterbourn CC (2008) Ascorbate interacts with reduced glutathione to scavenge phenoxyl radicals in HL60 cells. Free Radic Biol Med 44:1637–1644

D'Arcy Doherty M, Wilson I, Wardman P, Basra J, Patterson LH, Cohen GM (1986) Peroxidase activation of 1-naphthol to naphthoxy or naphthoxy- derived radicals and their reaction with glutathione. Chem Biol Interact 58:199–215

Dalle-Donne I, Giustarini D, Rossi R, Colombo R, Milzani A (2003) Reversible S-glutathionylation of Cys 374 regulates actin filament formation by inducing structural changes in the actin molecule. Free Radic Biol Med 34:23–32

Ferrer-Sueta G, Radi R (2009) Chemical biology of peroxynitrite: kinetics, diffusion, and radicals. ACS Chem Biol 4:161–177

Flohe L, Toppo S, Cozza G, Ursini F (2011) A comparison of thiol peroxidase mechanisms. Antioxid Redox Signal 15:763–780

Folkes LK, Trujillo M, Bartesaghi S, Radi R, Wardman P (2011) Kinetics of reduction of tyrosine phenoxyl radicals by glutathione. Arch Biochem Biophys 506:242–249

Forman HJ, Fukuto JM, Torres M (2004) Redox signaling: thiol chemistry defines which reactive oxygen and nitrogen species can act as second messengers. Am J Physiol Cell Physiol 287:C246–C256

Forni LG, Willson RL (1986a) Thiyl and phenoxyl free radicals and NADH. Biochem J 240: 897–903

Forni LG, Willson RL (1986b) Thiyl free radicals and the oxidation of ferrocytochrome c. Biochem J 240:905–907

Foster MW, McMahon TJ, Stamler JS (2003) S-nitrosylation in health and disease. Trends Mol Med 9:160–168

Fratelli M, Demol H, Puype M, Casagrande S, Villa P, Eberini I, Vandekerckhove J, Gianazza E, Ghezzi P (2003) Identification of proteins undergoing glutathionylation in oxidatively stressed hepatocytes and hepatoma cells. Proteomics 3:1154–1161

Galati G, Chan T, Wu B, O'Brien PJ (1999) Glutathione-dependent generation of reactive oxygen species by the peroxidase-catalyzed redox cycling of flavonoids. Chem Res Toxicol 12: 521–525

Gallogly MM, Starke DW, Leonberg AK, Ospina SM, Mieyal JJ (2008) Kinetic and mechanistic characterization and versatile catalytic properties of mammalian glutaredoxin 2: implications for intracellular roles. Biochemistry 47:11144–11157

Gallogly MM, Starke DW, Mieyal JJ (2009) Mechanistic and kinetic details of catalysis of thiol-disulfide exchange by glutaredoxins and potential mechanisms of regulation. Antioxid Redox Signal 11:1059–1081

Goldman R, Stoyanovsky DA, Day BW, Kagan VE (1995) Reduction of phenoxyl radicals by thioredoxin results in selective oxidation of its SH-groups to disulfides. An antioxidant function of thioredoxin. Biochemistry 34:4765–4772

Halliwell B (2007) Dietary polyphenols: good, bad, or indifferent for your health? Cardiovasc Res 73:341–347

Harwood DT, Kettle AJ, Winterbourn CC (2006) Production of glutathione sulfonamide and dehydroglutathione from GSH by myeloperoxidase-derived oxidants and detection using a novel LC-MS/MS method. Biochem J 399:161–168

Hofstetter D, Nauser T, Koppenol WH (2007) The glutathione thiyl radical does not react with nitrogen monoxide. Biochem Biophys Res Commun 360:146–148

Holmgren A, Sengupta R (2010) The use of thiols by ribonucleotide reductase. Free Radic Biol Med 49:1617–1628

Jones CM, Lawrence A, Wardman P, Burkitt MJ (2002) Electron paramagnetic resonance spin trapping investigation into the kinetics of glutathione oxidation by the superoxide radical: re-evaluation of the rate constant. Free Radic Biol Med 32:982–990

Kwak H-S, Yim H-S, Chock PB, Yim MB (1995) Endogenous intracellular glutathionyl radicals are generated in neuroblastoma cells under hydrogen peroxide oxidative stress. Proc Natl Acad Sci USA 92:4582–4586

Lancaster JR Jr (2006) Nitroxidative, nitrosative, and nitrative stress: kinetic predictions of reactive nitrogen species chemistry under biological conditions. Chem Res Toxicol 19:1160–1174

Lancaster JR Jr (2008) Protein cysteine thiol nitrosation: maker or marker of reactive nitrogen species-induced nonerythroid cellular signaling? Nitric Oxide 19:68–72

Madej E, Folkes LK, Wardman P, Czapski G, Goldstein S (2008) Thiyl radicals react with nitric oxide to form S-nitrosothiols with rate constants near the diffusion-controlled limit. Free Radic Biol Med 44:2013–2018

Mason RP (2004) Using anti-5,5-dimethyl-1-pyrroline N-oxide (anti-DMPO) to detect protein radicals in time and space with immuno-spin trapping. Free Radic Biol Med 36:1214–1223

Mozziconacci O, Kerwin BA, Schoneich C (2011) Reversible hydrogen transfer reactions of cysteine thiyl radicals in peptides: the conversion of cysteine into dehydroalanine and alanine, and of alanine into dehydroalanine. J Phys Chem B 115:12287–12305

Munday R (1994) Bioactivation of thiols by one-electron oxidation. Adv Pharmacol 27:237–270

Munday R, Winterbourn CC (1989) Reduced glutathione in combination with superoxide dismutase as an important biological antioxidant defence mechanism. Biochem Pharmacol 38:4349–4352

Nakamura M, Yamazaki I, Ohtaki S, Nakamura S (1986) Characterization of one- and two-electron oxidations of glutathione coupled with lactoperoxidase and thyroid peroxidase reactions. J Biol Chem 261:13923–13927

Nauser T, Pelling J, Schoneich C (2004) Thiyl radical reaction with amino acid side chains: rate constants for hydrogen transfer and relevance for posttranslational protein modification. Chem Res Toxicol 17:1323–1328

Niki E, Tsuchiya J, Tanimura R, Kamiya Y (1982) Regeneration of vitamin E from a-chromanoxyl radical by glutathione and vitamin C. Chem Lett 6:789–792

O'Brien PJ (1988) Radical formation during the peroxidase catalysed metabolism of carcinogens and xenobiotics: the reactivity of these radicals with GSH, DNA, and unsaturated lipid. Free Radic Biol Med 4:169–184
Pichorner H, Metodiewa D, Winterbourn CC (1995) Generation of superoxide and tyrosine peroxide as a result of tyrosyl radical scavenging by glutathione. Arch Biochem Biophys 323:429–437
Quijano C, Alvarez B, Gatti RM, Augusto O, Radi R (1997) Pathways of peroxynitrite oxidation of thiol groups. Biochem J 322:167–173
Raftery MJ, Yang Z, Valenzuela SM, Geczy CL (2001) Novel intra- and inter-molecular sulfinamide bonds in s100a8 produced by hypochlorite oxidation. J Biol Chem 276:33393–33401
Ramakrishna Rao DN, Fischer V, Mason RP (1990) Glutathione and ascorbate reduction of the acetaminophen radical formed by peroxidase. J Biol Chem 265:844–847
Rhee SG, Woo HA (2011) Multiple functions of peroxiredoxins: peroxidases, sensors and regulators of the intracellular messenger HO, and protein chaperones. Antioxid Redox Signal 15:781–794
Ross D (1988) Glutathione, free radicals and chemotherapeutic agents. Mechanisms of free-radical induced toxicity and glutathione- dependent protection. Pharmacol Ther 37:231–249
Ross D, Moldeus P (1985) Generation of reactive species and fate of thiols during peroxidase-catalysed metabolic activation of aromatic amines and phenols. Environ Health Perspect 64:253–257
Ross D, Norbeck K, Moldeus P (1985) The generation and subsequent fate of glutathionyl radicals in biological systems. J Biol Chem 260:15028–15032
Schöneich C (2008) Mechanisms of protein damage induced by cysteine thiyl radical formation. Chem Res Toxicol 21:1175–1179
Schöneich C, Dillinger U, von Bruchhausen F, Asmus KD (1992) Oxidation of polyunsaturated fatty acids and lipids through thiyl and sulfonyl radicals: reaction kinetics, and influence of oxygen and structure of thiyl radicals. Arch Biochem Biophys 292:456–467
Starke DW, Chock PB, Mieyal JJ (2003) Glutathione-thiyl radical scavenging and transferase properties of human glutaredoxin (thioltransferase). Potential role in redox signal transduction. J Biol Chem 278:14607–14613
Stoyanovsky DA, Goldman R, Claycamp HG, Kagan VE (1995) Phenoxyl radical-induced thiol-dependent generation of reactive oxygen species: implications for benzene toxicity. Arch Biochem Biophys 317:315–323
Stoyanovsky DA, Goldman R, Jonnalagadda SS, Day BW, Claycamp HG, Kagan VE (1996) Detection and characterization of the electron paramagnetic resonance-silent glutathionyl-5, 5-dimethyl-1-pyrroline N-oxide adduct derived from redox cycling of phenoxyl radicals in model systems and HL-60 cells. Arch Biochem Biophys 330:3–11
Stoyanovsky DA, Wu D, Cederbaum AI (1998) Interaction of 1-hydroxyethyl radical with glutathione, ascorbic acid and alpha-tocopherol. Free Radic Biol Med 24:132–138
Stoyanovsky DA, Maeda A, Atkins JL, Kagan VE (2011) Assessments of thiyl radicals in biosystems: difficulties and new applications. Anal Chem 83:6432–6438
Sturgeon BE, Sipe HJ JR, Barr DP, Corbett JT, Martinez JG, Mason RP (1998) The fate of the oxidizing tyrosyl radical in the presence of glutathione and ascorbate. Implications for the radical sink hypothesis. J Biol Chem 273:30116–30121
Subrahmanyam VV, O'Brien PJ (1985) Peroxidase catalysed oxygen activation by arylamine carcinogens and phenol. Chem Biol Interact 56:185–199
Subrahmanyam VV, McGirr LG, O'Brien PJ (1987) Glutathione oxidation during peroxidase catalysed drug metabolism. Chem Biol Interact 61:45–59
Wardman P (1995) Reactions of thiyl radicals. In: Packer L, Cadenas E (eds) Biothiols in health and disease. Marcel Dekker Inc., New York
Wardman P, von Sonntag C (1995) Kinetic factors that control the fate of thiyl radicals in cells. Methods Enzymol 251:31–45
Wefers H, Riechmann E, Sies H (1985) Excited species generation in horseradish peroxidase-mediated oxidation of glutathione. J Free Radic Biol Med 1:311–318

Willson RL (1983) Free radical repair mechanisms and the interactions of glutathione and vitamins C and E. In: Nygaard OF, Simic MG (eds) Radioprotectors and anticarcinogens. Academic, New York

Winterbourn CC (1993) Superoxide as an intracellular radical sink. Free Radic Biol Med 14:85–90

Winterbourn CC (2003) Radical scavenging by thiols: biological significance and implications for redox signaling and antioxidant defense. In: Gitler C, Danon A (eds) Cellular implications of redox signaling. Imperial College Press, London

Winterbourn CC (2008) Reconciling the chemistry and biology of reactive oxygen species. Nat Chem Biol 4:278–286

Winterbourn CC, Hampton MB (2008) Thiol chemistry and specificity in redox signaling. Free Radic Biol Med 45:549–561

Winterbourn CC, Metodiewa D (1994) The reaction of superoxide with reduced glutathione. Arch Biochem Biophys 314:284–290

Winterbourn CC, Metodiewa D (1999) Reactivity of biologically important thiol compounds with superoxide and hydrogen peroxide. Free Radic Biol Med 27:322–328

Winterbourn CC, Munday R (1989) Glutathione-mediated redox cycling of alloxan. Mechanisms of superoxide dismutase inhibition and of metal-catalyzed OH. formation. Biochem Pharmacol 38:271–277

Winterbourn CC, Munday R (1990) Concerted action of reduced glutathione and superoxide dismutase in preventing redox cycling of dihydroxypyrimidines, and their role in antioxidant defence. Free Radic Res Commun 8:287–293

Witting PK, Mauk AG (2001) Reaction of human myoglobin and H2O2. Electron transfer between tyrosine 103 phenoxyl radical and cysteine 110 yields a protein-thiyl radical. J Biol Chem 276:16540–16547

Zhang N, Schuchmann H-P, von Sonntag C (1991) The reaction of superoxide radical anion with dithiothreitol: a chain process. J Phys Chem 95:4718–4722

Zhang H, Xu Y, Joseph J, Kalyanaraman B (2005) Intramolecular electron transfer between tyrosyl radical and cysteine residue inhibits tyrosine nitration and induces thiyl radical formation in model peptides treated with myeloperoxidase, H2O2, and NO2-: EPR SPIN trapping studies. J Biol Chem 280:40684–40698

Zhao R, Lind J, Merényi G, Eriksen TE (1994) Kinetics of one-electron oxidation of thiols and hydrogen abstraction by thiyl radicals from a-amino C-H bonds. J Am Chem Soc 116: 12010–12015

Zhao R, Lind G, Merenyi G, Eriksen TE (1997) Significance of the intramolecular transformation of glutathione thiyl radicals to a-aminoalkyl radicals. J Chem Soc Perkin Trans 2(116):569–574

Chapter 3
Redox Homeostasis

Joris Messens, Nicolas Rouhier, and Jean-François Collet

Abstract Multiple factors, including small sulfur-containing molecules and oxidoreductases, are involved in the control of intracellular redox homeostasis. In this chapter, we first review properties and functions of the small sulfur-containing molecules glutathione, mycothiol, bacillithiol and trypanothione. These low molecular weight thiols, which cycle between a reduced and oxidized form, are present at high intracellular concentrations and function as redox buffers to protect cells against oxidative stress conditions. In the second part of this chapter, we focus on the two oxidoreductases, thioredoxin and glutaredoxin. These enzymes are key players in pathways aimed to reduce disulfide bonds in intracellular proteins and to maintain cellular redox homeostasis. We review the general properties of these enzymes and highlight their significant diversity. Finally, we discuss the recent discovery that monothiol glutaredoxins coordinate an iron sulfur cluster, which suggests a novel link between redox and iron homeostasis.

J. Messens (✉)
Department of Structural Biology, Vlaams Instituut voor Biotechnologie (VIB), Pleinlaan 2, B-1050 Brussels, Belgium

Structural Biology Brussels, Vrije Universiteit Brussel (VUB), Pleinlaan 2, B-1050 Brussels, Belgium

Brussels Center for Redox Biology, Brussels, Belgium
e-mail: joris.messens@vib-vub.be

N. Rouhier
Faculté des Sciences, Unité Mixte de Recherches 1136, Interactions Arbres Microorganismes, Lorraine University-INRA, IFR 110 EFABA, Vandoeuvre Cedex, France

J.-F. Collet (✉)
de Duve Institute, Université catholique de Louvain, 75-39 Avenue Hippocrate, B-1200 Brussels, Belgium

Brussels Center for Redox Biology, Brussels, Belgium

WELBIO, B-1200 Brussels, Belgium
e-mail: jfcollet@uclouvain.be

U. Jakob and D. Reichmann (eds.), *Oxidative Stress and Redox Regulation*, DOI 10.1007/978-94-007-5787-5_3,

3.1 Introduction

The redox environment of cellular compartments is tightly controlled by the complementary action of oxidoreductases and high concentrations of sulfur-containing molecules, such as glutathione, which play the role of redox buffers. We review in this chapter the role and function of the low molecular weight thiols glutathione, mycothiol, bacillithiol and trypanothione in maintaining the intracellular redox state under both physiological and oxidative stress conditions. We also focus on thioredoxins and glutaredoxins, the most important classes of oxidoreductases involved in controlling the cellular redox homeostasis in both prokaryotes and eukaryotes.

3.2 Low Molecular Weight Thiols

Most living organisms contain high concentrations of low molecular weight (LMW) thiols that serve as redox buffers to protect the cells against a variety of reactive chemical species, such as reactive oxygen species (ROS), reactive nitrogen species (RNS), reactive electrophilic species (RES), metalloids, and some antibiotics (Dalle-Donne et al. 2008; Roos and Messens 2011). Glutathione is the most ubiquitous of these LMW thiols but certain microorganisms contain high concentrations of other related compounds, such as bacillithiol, mycothiol and trypanothione.

3.2.1 Glutathione

Glutathione (L-γ-glutamyl-L-cysteinylglycine, GSH) is a water-soluble, low molecular weight tripeptide (Fig. 3.1), which is present at millimolar concentrations in nearly all eukaryotic cells and in many bacteria, most of which are Gram-negative. Glutathione is synthesized by the consecutive action of two ATP-dependent enzymes. First, γ-glutamylcysteine synthetase catalyzes the rate-limiting formation of L-γ-glutamylcysteine from glutamic acid and cysteine. Then, glutathione synthetase catalyzes the ligation of L-γ-glutamylcysteine with glycine.

Glutathione cycles between two species, a reduced GSH-form and a disulfide bonded GSSG form. The standard redox potential of the GSH-GSSG couple is −240 mV at pH 7.0 (Meister and Anderson 1983). The reduced form is the most abundant form *in vivo*, and the overall cellular GSH/GSSG ratio ranges from 30:1 to 100:1 (Hwang et al. 1992), which corresponds to a redox potential of approximately −221 to −236 mV (Hwang et al. 1992). The GSH/GSSG ratio is maintained by glutathione reductase, a flavoenzyme that uses the reducing power of NADPH to reduce GSSG back to GSH (Fahey et al. 1978). Noteworthy, the ratio of GSH to GSSG is significantly lower in the secretory pathway, ranging from approximately

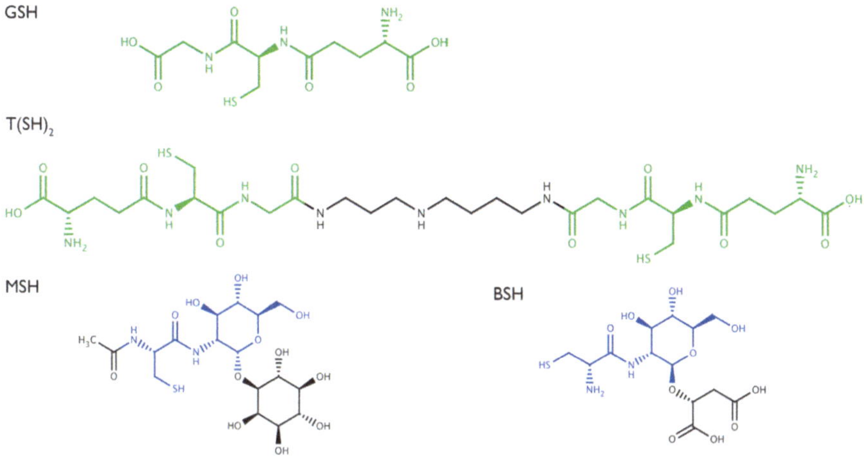

Fig. 3.1 The structure of low molecular weight thiol molecules in their reduced state. Glutathione (GSH): L-γ-glutamyl-L-cysteinylglycine; Dihydrotrypanothione [T(SH)$_2$]: N1,N8-bis(glutathionyl)spermidine; Mycothiol (MSH): N-acetylcysteine amide-linked to 1D-*myo*-inositol 2-acetamido-2-deoxy-α-D-glucopyranoside [GlcN-α(1–1)-Ins] and Bacillithiol (BSH) are shown. Similar chemical motifs are shown in the *same color*. Different chemical motifs are in *black*. Structures were made with MarvinSketch 5.7.1 (ChemAxon)

1:1 to 3:1 (Hwang et al. 1992). This reflects a more oxidizing redox environment in the endoplasmic reticulum, the compartment where disulfide bond formation occurs in eukaryotes [for a review, see Depuydt et al. 2011].

Because of its high concentration and its low redox potential, glutathione, which is essential in eukaryotes but dispensable in bacteria such as *Escherichia coli*, is usually considered as a redox buffer that protects cells against oxidative damages. The thiol moiety of GSH enables this molecule to act as a scavenger reacting with various electrophilic compounds including ROS and RNS. Rate constants at pH 7.4 vary for each oxidant and range between 3×10^7 M^{-1} s^{-1} for NO_2 (Ford et al. 2002), 6.6×10^2 M^{-1} s^{-1} for peroxynitrite (Radi et al. 1991; Koppenol et al. 1992; Zhang et al. 1997; Quijano et al. 1997), 115 M^{-1} s^{-1} for taurine chloramines (Peskin and Winterbourn 2001), 3×10^7 M^{-1} s^{-1} for hypochlorous acid and 1×10^{10} M^{-1} s^{-1} for hydroxyl radicals (Winterbourn and Hampton 2008).

Glutathione is also indirectly involved in peroxide scavenging by participating in the recycling of ascorbate, an electron donor to ascorbate peroxidases (Rouhier et al. 2008), and by supplying some thiol peroxidases, like glutathione peroxidases and peroxiredoxins with reducing equivalents with or without the help of glutaredoxins (Grxs, see below). Glutathione peroxidases and peroxiredoxins are antioxidant enzymes that use a thiol-based chemistry to reduce H_2O_2 and lipid peroxides. However, among the so-called glutathione peroxidases, only the selenocysteine-containing enzymes are truly dependent on glutathione, whereas cysteine-containing enzymes are in fact dependent on thioredoxin (Trx) for their recycling (Navrot et al. 2006). In the peroxiredoxin family, several members have been shown to be recycled by GSH alone or by the GSH/Grx couple, the first reported example being a poplar type II peroxiredoxin (Rouhier et al. 2001). Finally, some members of the glutathione S-transferase family, as well as some Grxs also exhibit a glutathione-dependent peroxidase activity (see the Chap. 8 by Jacquot et al., in this book).

In addition to its role in scavenging oxidizing molecules, glutathione can also covalently modify cysteine residues in a process called S-glutathionylation, although the cellular levels of glutathionylated proteins in non-stressed cells are low (<0.1% of the total protein cysteines) (Hansen et al. 2009). S-glutathionylation is reversible and is assumed either to protect cysteine residues from irreversible oxidation or to regulate proteins, whose cysteines are essential for activity or folding. In general, S-glutathionylation is reversed by the action of Grxs, which constitute a family of glutathione-dependent oxidoreductases that almost exclusively reduce cysteine glutathione adducts formed on proteins in the course of their catalytic cycle or under oxidative stress conditions (see Sect. 3.3.2 for more details). S-glutathionylation is mediated by mechanisms that can be schematically classified as ROS-dependent or ROS-independent (Ghezzi and Di Simplicio 2007). Although the final product is the same, the rates of the reactions are very different. The ROS-independent process occurs via thiol/disulfide exchange between GSSG and the proteins. However, under physiological conditions, the low concentration of GSSG in the cytosol is unlikely to favor the formation of mixed disulfides with protein thiols in this compartment. Moreover, the thermodynamic barrier limits the oxidation of proteins by GSSG, which occurs via a nucleophilic attack of a protein thiolate ($R\text{-}S^-$) on GSSG: the mixed disulfide formed must thus have a redox potential higher than that of the $[GSH]^2/[GSSG]$ couple, which is theoretically possible, but extremely unlikely (Forman et al. 2010). In contrast, in the endoplasmic reticulum, where the GSH/GSSG ratio is lower, the formation of protein-glutathione mixed disulfides is more likely to occur via the direct reaction of a thiolate with GSSG (Townsend 2007; Forman et al. 2010; Chakravarthi et al. 2006). In ROS-dependent processes, sensitive cysteine residues are first oxidized to a sulfenic acid upon ROS exposure. Sulfenic acids are extremely unstable modifications that can be irreversibly oxidized to sulfinic and sulfonic acids (Roos and Messens 2011). Glutathione reacts with sulfenic acid to form a glutathione-protein mixed disulfide, which effectively protects those cysteines against irreversible oxidation. For example, the α-glutamyl transpeptidase is protected against oxidative damage by S-glutathionylation (Dominici et al. 1999). S-glutathionylation is not only a protective mechanism but can also regulate protein function (Dalle-Donne et al. 2008). For instance, the enzyme α-ketoglutarate dehydrogenase is reversibly inactivated by S-glutathionylation in response to alterations in GSH levels in mitochondria (Nulton-Persson et al. 2003). For more examples, we refer to Mieyal et al. (2008). Altogether, the reversible S-glutathionylation of specific proteins has implications for the regulation of cellular homeostasis in health and disease. For instance, changes in the S-glutathionylation state of specific proteins play important roles in diabetes, cardiovascular, lung and neurodegenerative diseases (Dalle-Donne et al. 2008; Mieyal et al. 2008).

The data summarized above indicate that glutathione is a central player in redox homeostasis. However, depending on the organism, the essential function of glutathione might not always be due to its role in thiol-redox control but rather due to its role in iron homeostasis, as recently revealed by work done in the yeast *Saccharomyces cerevisiae* (Kumar et al. 2011).

3.2.2 Trypanothione

Since its discovery in 1985 (Fairlamb et al. 1985), trypanothione has enjoyed a lot of attention (Fig. 3.1). Trypanothione has not been found in prokaryotes but is present in kinetoplastida, which are primitive eukaryotes that parasitize animals and plants. In these organisms, most of the glutathione content is found in the form of a unique thiol, N1, N8-bis(glutathionyl)spermidine, also known as trypanothione. It is a conjugate of two glutathione molecules with one molecule of spermidine, and is synthesized by the ATP-dependent enzyme trypanothione synthetase (Oza et al. 2002).

Oxidized trypanothione (TS_2) is reduced to dihydrotrypanothione [$T(SH)_2$] by the FAD disulfide oxidoreductase trypanothione reductase [reviewed in Flohe et al. 1999]. Whereas the redox potential of the TS_2-$T(SH)_2$ couple (−242 mV) is similar to that of glutathione (Fairlamb and Cerami 1992), trypanothione is more reactive than glutathione in thiol-disulfide exchange reactions under physiological conditions, due to the lower pK_a value of its thiol group (7.4 compared to 8.7 in glutathione) (Krauth-Siegel et al. 2005).

$T(SH)_2$ functions as the donor of reducing equivalents to several enzymes of the parasite, including thioredoxin (Schmidt and Krauth-Siegel 2003), tryparedoxin (Nogoceke et al. 1997; Ludemann et al. 1998), monothiol glutaredoxin-1 (Filser et al. 2008) and ribonucleotide reductase (Dormeyer et al. 2001). $T(SH)_2$ can also spontaneously reduce protein sulfenic acids in the model protein glyceraldehyde-3-phosphate dehydrogenase (Filser et al. 2008).

Tryparedoxin is a distant relative of the thioredoxin superfamily (see below), which has a WCPPCR active site motif and is substantially larger (16 kDa) than most thioredoxins (Krauth-Siegel and Leroux 2012). As such, tryparedoxin is reduced by $T(SH)_2$ (Nogoceke et al. 1997; Gommel et al. 1997) and can reduce glutathione-protein mixed disulfides (Melchers et al. 2007). Tryparedoxin can also transfer reducing equivalents from $T(SH)_2$ to a variety of protein targets, making the tryparedoxin/$T(SH)_2$ couple the determining factor for the intracellular redox state of the parasite (Reckenfelderbaumer and Krauth-Siegel 2002).

In addition to providing reducing equivalents to oxidoreductases, $T(SH)_2$ can also efficiently scavenge hydrogen peroxide, peroxynitrite and radiation-induced radicals (Thomson et al. 2003; Awad et al. 1992; Carnieri et al. 1993). $T(SH)_2$ maintains the redox homeostasis by passing electrons to peroxidases via intermediate shuttle molecules, which can either be tryparedoxin, ascorbate, or even glutathione [reviewed in (Castro and Tomas 2008)].

3.2.3 Mycothiol

Glutathione was thought to be an essential molecule in all living cells until it was found that many Gram-positive bacteria do not produce this compound (Fahey et al. 1978). In the early 1990s, it was shown that glutathione-lacking bacteria *Mycobacterium bovis* and *Streptomyces clavuligerus* produce an alternative low molecular weight thiol that was given the common name "mycothiol" (Spies and Steenkamp 1994; Newton et al. 1995). Mycothiol is the major thiol in most Actinomycetes and is produced at millimolar levels in *Mycobacteria* and *Streptomycetes*. It is comprised of N-acetylcysteine amide-linked to a 1D-*myo*-inosityl 2-acetamido-2-deoxy-α-D-glucopyranoside [GlcN-α(1–1)-Ins] (Fig. 3.1). Mycothiol biosynthesis occurs by linkage of GlcNAc to inositol (Ins), deacetylation to GlcN-Ins, ligation of the latter to L-cysteine, and the transacetylation of the cysteinyl residue by acetyl Coenzyme A (CoASAc) (Newton et al. 2008).

Like glutathione, mycothiol is also present in a reduced (MSH) and a disulfide-bonded (MSSM) state. However, the redox potential of the MSH/MSSM couple is not known, probably due to the limited availability of mycothiol for research purposes. The chemical synthesis of mycothiol with the correct stereochemistry is a challenging task, making bacterial isolation of mycothiol still far more efficient (Ordóñez et al. 2009).

A key property of mycothiol is its high resistance to oxidation by molecular oxygen in the presence of redox metals. For instance, the copper-catalyzed autoxidation of mycothiol is about 30-fold slower than that of cysteine and 7-fold slower than that of glutathione (81), due to the acetyl and GlcN-Ins moieties blocking the amino and carboxyl groups of the cysteine, respectively. In Actinobacteria, mycothiol-disulfide-selective reductase (MTR), a NADPH-dependent flavoenzyme, reduces MSSM back to MSH in order to maintain the intracellular redox homeostasis required for the proper functioning of a variety of biological processes (Rawat and Av-Gay 2007). Interestingly, we showed in *Corynebacterium glutamicum* that mycothiol cannot be replaced by glutathione, which indicates that mycothiol plays a specific role in that organism (Ordóñez et al. 2009). Accordingly, mycoredoxin-1 (Mrx1), the glutaredoxin analog of Actinomycetes, does not function with glutathione but has a strict specificity for mycothiol in a reaction coupled to MTR and NADPH (Ordóñez et al. 2009). These results suggest that glutaredoxins and mycoredoxins have specific binding sites for glutathione and mycothiol, respectively.

Mycothiol serves as a storage form of cysteine in *Mycobacterium smegmatis* (Bzymek et al. 2007) and plays a role in the detoxification of thiol-reactive substances, including formaldehyde, various electrophiles and antibiotics (Newton et al. 2000; Rawat and Av-Gay 2007). Mycothiol S-conjugates derived from electrophiles and antibiotics are then cleaved by mycothiol S-conjugate amidase to release GlcN-Ins, which is used to resynthesize mycothiol, and a mercapturic acid, which is excreted from the cell. Some enzymes depend also on mycothiol for proper functioning, such as formaldehyde dehydrogenase MscR (Misset-Smits et al. 1997), later identified as nitrosomycothiol reductase with a role in the protection against oxidative stress (Vogt et al. 2003), and maleylpyruvate isomerase (Feng et al. 2006; Wang et al. 2007). Moreover, we have recently found that the activity of *C. glutamicum* arsenate reductases 1 and 2 depends on reducing equivalents transferred from mycothiol by mycoredoxin (Ordóñez et al. 2009). Very recently, the identification of a mycothiol S-transferase has been reported in *Mycobacterium smegmatis* and *M. tuberculosis* (Newton et al. 2011). This enzyme is postulated to help protect the cells against antibiotics, as evidenced by the appearance of mercapturic acids in the fermentation broth of producing cultures (Rawat and Av-Gay 2007).

3.2.4 Bacillithiol

Bacillithiol (BshC) was discovered in *Bacillus subtilis* as a molecule disulfide-linked with OhrR. OhrR is a peroxide sensitive transcription factor, which contains a single cysteine residue required for redox sensing (Lee et al. 2007; Fuangthong and Helmann 2002). A thiol compound with the same mass was independently discovered in extracts of *Bacillus anthracis* after labeling with monobromobimane (Nicely et al. 2007).

Bacillithiol is widely found among low-GC Gram-positive bacteria (Firmicutes) and is also sporadically present in more distantly related bacteria, including *Deinococcus radiodurans* (Helmann 2011). The structure of bacillithiol was determined after purification from this latter bacterium (Nicely et al. 2007). Bacillithiol is structurally similar to mycothiol. The inositol group is replaced by L-malate, which makes bacillithiol an α-anomeric glycoside of L-cysteinyl-D-glucosamine with L-malic acid (Fig. 3.1). Bacillithiol biosynthesis requires three enzymes that sequentially couple GlcNAc to malic acid (BshA), deacetylate the GlcNAc-Mal intermediate to generate GlcN-Mal (BshB), and finally couple this latter product to a cysteine to generate bacillithiol. Like mycothiol, bacillithiol is also not available in sufficient amounts to allow biochemical studies; therefore the redox potential of bacillithiol is not known and information on the rate constants of the reaction with ROS and RNS is not available.

By analogy with glutathione and mycothiol, bacillithiol is likely to be a central player in thiol-disulfide homeostasis. It can therefore be anticipated that bacillithiol-containing organisms encode a bacillithiol reductase to mediate the recycling of the oxidized form of bacillithiol as well as oxidoreductases for the reduction of protein-bacillithiol mixed disulfides. The name 'bacilliredoxin' (Brx) has been proposed for such proteins (Helmann 2011). A bacillithiol S-transferase responsible for the conjugation of bacillithiol with various substrates has been recently described in *B. subtilis* (Newton et al. 2011). Moreover, mutants lacking bacillithiol are highly sensitive to fosfomycin (Gaballa et al. 2010), probably due to the inability of FosB, a presumed bacillithiol-S-transferase, to detoxify this antibiotic.

As explained above, protein S-thiolation is thought to protect active site cysteine residues of essential enzymes against irreversible oxidation to sulfinic or sulfonic acids. In *B. subtilis* OhrR, it has been shown that S-bacillithiolation of a single cysteine functions as a redox-switch (Lee et al. 2007; Newton et al. 2009). It leads to the inactivation of the OhrR repressor and derepression of *ohrA* transcription. OhrA is a thiol-dependent peroxiredoxin that converts organic peroxides into organic alcohols and protects the cell against NaOCl. In a recent proteomic study of NaOCl-treated cells using shotgun-LC-MS/MS analysis, several other proteins, including two methionine synthase paralogs (MetE and YxjG), the inorganic pyrophosphatase PpaC, the 3-D-phosphoglycerate dehydrogenase SerA and the thiol-disulfide oxidoreductase YphP were found to be S-bacillithionylated (Chi et al. 2011). The latter enzyme YphP might function as a putative bacilliredoxin that reduces the mixed disulfide between bacillithiol and essential cysteines in proteins.

3.3 A Diversity of Oxidoreductases

Although cells have millimolar concentrations of glutathione and alternatives that serve as redox buffers to prevent the formation of unwanted disulfides, they need more efficient reducing systems to catalyze the reduction of disulfide bonds and maintain the cellular redox homeostasis. The major cellular reducing pathways involve thioredoxins and glutaredoxins.

3.3.1 Thioredoxins

Thioredoxins (Trxs) are ubiquitous antioxidant enzymes that were originally discovered in 1964 in the bacterium *Escherichia coli* as an electron donor for ribonucleotide reductase, an enzyme that provides deoxyribonucleotides required for DNA synthesis and repair (Laurent et al. 1964; Moore et al. 1964). Since their discovery, Trxs have been found to be at the heart of numerous fundamental processes in living organisms, ranging from Archaea to mammals. Trxs are present in various cellular compartments, including the cytosol (Arner and Holmgren 2000), nucleus (Hirota et al. 1997, 1999), mitochondria and plastids (Schürmann and Buchanan 2008), and are found to be attached to the cell membrane (Martin and Dean 1991) or secreted to the extracellular environment (Arner 1999; Xu et al. 2008).

Trxs are best known for their catalytic role in reducing disulfide bonds that form in proteins, either in the course of their catalytic cycle or upon exposure to oxidative stress conditions. Therefore, Trxs are part of the antioxidant defense against peroxides and other ROS. However, Trxs also play other roles in the cell. For instance, Trx proteins have been shown to regulate programmed cell death via denitrosylation (Benhar et al. 2008), to act as growth factor (Powis et al. 2000), to modulate the inflammatory response (Nakamura et al. 2005), to play important roles in the lifecycle of viruses and phages (Holmgren 1989) or to activate antimicrobial peptides (Schroeder et al. 2011).

3.3.1.1 General Properties

Thioredoxins are characterized by a highly conserved structural fold consisting of two motifs (Fig. 3.2a), as described first by Holmgren et al. (1975). The structure of Trx consists of five β-strands surrounded by four α-helices (Holmgren et al. 1975) (Fig. 3.2b). The β-sheets and α-helices can be divided in a N-terminal $\beta_1\alpha_1\beta_2\alpha_2\beta_3$ and a C-terminal $\beta_4\beta_5\alpha_4$ motif connected by the α_3-helix. The β-strands of the N-terminal motif run parallel, and the β-strands of the C-terminal motif run anti-parallel. The α_2 and α_4 helices are located on one side of the central β-sheet while the α_3-helix is located on the opposite side (Fig. 3.2b). The α_3-helix is oriented perpendicularly to helices α_2 and α_4. The catalytic site of Trxs consists of a canonical WCGPC motif that is located on the surface of the protein in a short segment at the N-terminus of the α_2-helix. Many proteins, including glutaredoxins (see below) (Xia et al. 2001) (Fig. 3.2c), protein disulfide isomerases (McCarthy et al. 2000; Tian et al. 2006; Heras et al. 2004), glutathione S-transferases (Reinemer et al. 1991), some thiol-peroxidases (Ren et al. 1997), and the recently characterized protein DsbG (Depuydt et al. 2009; Heras et al. 2004) present a minimal version of the Trx fold. This version, known as the "Trx-fold" (Martin 1995), lacks the β_1-strand and the α_1-helix of Trx.

The WCGPC motif is a landmark feature of Trx proteins. The cysteine residues of this motif are found predominantly reduced *in vivo*, allowing Trxs to break disulfide bonds in oxidized substrate proteins. The reaction can be seen as a transfer of the disulfide bond from the substrate protein to Trx, or as a transfer of electrons from the reduced cysteines of Trx to the substrate protein. The molecular details of the reaction catalyzed by Trx have been the subject of extensive research [see Collet and Messens 2010 for a review]. The reaction starts with a nucleophilic attack of the N-terminal thiol of the WCGPC motif on the disulfide of the target protein, leading to the formation of a mixed disulfide complex between Trx and the target protein. The N-terminal cysteine of the WCGPC motif has a pK_a value of about ~7 (Dyson et al. 1991; Dillet et al. 1998), which is lower than the pK_a value of free cysteines in solution (~9). Therefore, a large fraction of this cysteine is present as a thiolate under physiological conditions, enabling this residue to act as a nucleophile. The low pK_a value of the first cysteine residue results from the stabilization of the negative charge of the thiolate anion through the formation of hydrogen bonds between the sulfur atom and neighboring residues. Once a mixed-disulfide has formed between Trx and its substrate, the C-terminal thiol is activated as a thiolate and attacks the mixed-disulfide (Roos et al. 2009). This results in the release of the reduced substrate protein and oxidized Trx. The recycling of Trx to the reduced form is catalyzed by thioredoxin reductase (TrxR) at the expense of NADPH (Lennon et al. 2000), enabling Trx to start another reaction cycle.

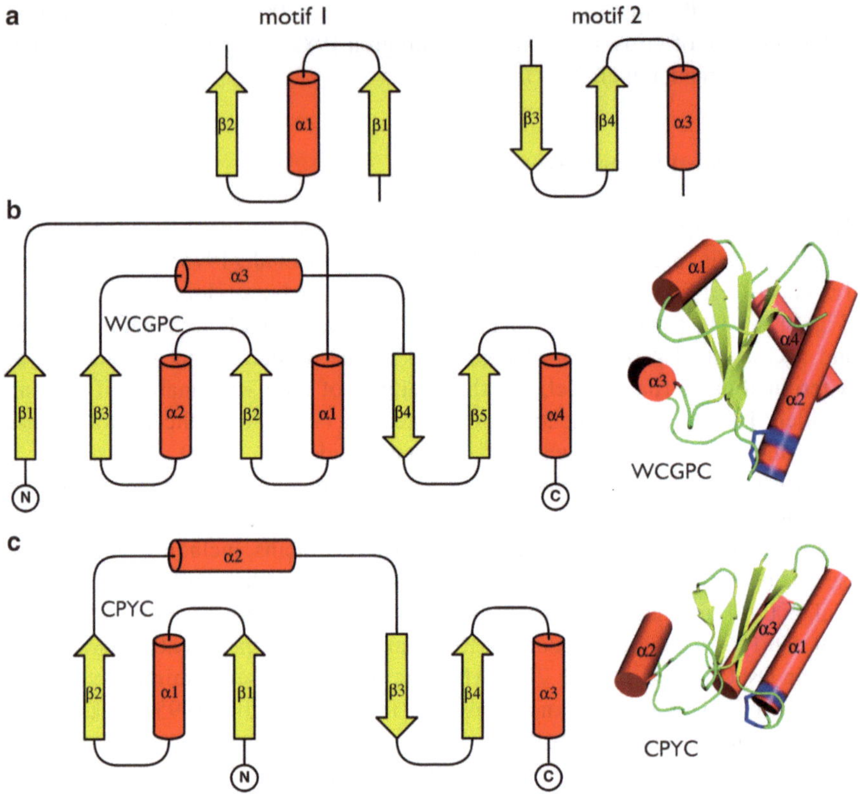

Fig. 3.2 The structures of the Trx-fold proteins, thioredoxin (Trx) and glutaredoxin (Grx). (**a**) The topology of the two motifs that constitute the Trx-fold. (**b**) The secondary and three-dimensional structures (PDB code: 2O7K) (Roos et al. 2007) of *Staphylococcus aureus* Trx are shown. The structure consists of 4 α-helices and a central 5-stranded β-sheet. The conserved WCGPC motif is located at the N-terminus of the α_2-helix. The α-helices are in *red*, the β-strands in *yellow* and the disulfide bond is shown in *blue*. (**c**) The secondary and three-dimensional structures of *E. coli* glutaredoxin are shown (PDB code: 1EGO) (Xia et al. 1992). Glutaredoxins have a Trx-fold consisting of two motifs connected by the α_2-helix. The conserved active site CXXC motif is always located at the same position, at the N-terminal site of an α-helix. The figure was generated using TopDraw from CCP4 suite and MacPyMol (Delano Scientific LLC 2006). An adapted version of the figure was reprinted with the permission of (Collet and Messens 2010)

3.3.1.2 Diversity and Roles of Thioredoxins

Escherichia coli Trx1 (Ec_Trx1), the first identified Trx, is usually considered as the prototype of the thioredoxin superfamily of which members have been identified in most genomes. Here, we briefly review the most prominent Trx proteins to highlight the diversity within this ubiquitous family.

Bacterial Thioredoxins

Ec_Trx1 is the most reducing protein present in the cytoplasm of *E. coli*. Its redox potential value has been estimated to be −270 mV (Krause and Holmgren 1991) but a more recent study suggests that it might be even lower ($E^{o'} = -284$ mV) (Cheng et al. 2007). Ec_Trx1 catalyzes the reduction of several cytoplasmic enzymes that form a disulfide bond in their catalytic cycle, including ribonucleotide reductase, methionine sulfoxide reductases (MsrA, MsrB and fRMsr) and 3-phosphoadenosine 5′-phosphosulfate (PAPS) reductase (Holmgren and Bjornstedt 1995). Ec_Trx1 also provides reducing equivalents to DsbD, an inner membrane protein, which transfers electrons across the membrane and delivers them to periplasmic oxidoreductases (Rietsch et al. 1997). Moreover, Ec_Trx1 is required for the growth of several bacteriophages including T7, M13 and f1 (Russel 1991). For this latter function, which does not require the catalytic cysteine residues, Ec_Trx1 binds to viral DNA polymerase, to mediate the interaction between DNA polymerase, DNA and other replication proteins (Ghosh et al. 2008; Hamdan et al. 2005).

The *E. coli* cytoplasm contains a second thioredoxin, Ec_Trx2 (Miranda-Vizuete et al. 1997). Ec_Trx2, which shares only 28% sequence identity with Ec_Trx1, has a redox potential of −221 mV (El Hajjaji et al. 2009), and is therefore a significantly less reducing enzyme than Ec_Trx1. Like Ec_Trx1, Ec_Trx2 is able to reduce ribonucleotide reductase, DsbD and PAPS reductase. However, Ec_Trx2 has two striking characteristics that distinguish it from Ec_Trx1 and suggest that this protein may have a specific function in *E. coli*. First, the expression of Ec_Trx2 is controlled by OxyR, a transcription factor that controls the response to oxidative stress (Ritz et al. 2000). Second, Ec_Trx2 contains an additional N-terminal domain of 32 amino acids that harbors two CXXC motifs. We found that these additional cysteine residues bind Zn^{2+} with an extremely high affinity (10^{18} M^{-1}) (Collet et al. 2003). The zinc-binding CXXC motifs of Ec_Trx2 are conserved in several other bacterial Trxs, making Ec_Trx2 the prototype of a new zinc-binding Trx family. Noteworthy, we recently showed that the zinc center of Ec_Trx2 fine-tunes its redox and thermodynamic properties (El Hajjaji et al. 2009).

Yeast and Mammalian Thioredoxins

In the budding yeast *Saccharomyces cerevisiae,* there are two cytosolic (Sc_Trx1, Sc_Trx2) and one mitochondrial (Sc_Trx3) thioredoxin (Herrero et al. 2008). A double mutant lacking both *sc_trx1* and *sc_trx2* is viable but exhibits a more severe phenotype than the single mutants, indicating that Sc_Trx1 and Sc_Trx2 are only partially redundant. For instance, mutants lacking both Trxs have a longer S-phase in their cell cycle due to an inefficient reduction of ribonucleotide reductase, are auxotrophic for sulfur amino acids (Muller 1991) due to their inability to reduce PAPS reductase, and are unable to use methionine sulfoxide as a source of organic sulfur due to their inability to reduce methionine sulfoxide reductases.

The function of the mitochondrial Trx, Sc_Trx3, is less defined than that of its cytosolic counterparts. This protein seems to play a role in the defense mechanisms against ROS generated in mitochondria, as suggested by the fact that a mutant lacking Sc_Trx3 is hypersensitive to hydroperoxide (Pedrajas et al. 1999).

Mammalian cells possess two Trx isoforms, ma_Trx1 (~12 kDa) and ma_Trx2 (~18 kDa), which are present in the cytosol and mitochondria, respectively. The ma_Trx1 has also been detected in the nucleus of certain normal and tumor cells, even though it does not have a nuclear localization sequence (Maruyama et al. 1999; Hirota et al. 1997). In addition to the catalytic cysteines present in the CGPC motif, ma_Trx1 contains three extra cysteine residues (Qin et al. 1994). In the human protein, two of these cysteines (C62 and C69) flank the α_3-helix that links both motifs of the Trx-fold (Fig. 3.1a). The third additional cysteine (C73) is located in a turn close to the CGPC active site motif. Several reports have shown that these extra cysteines are involved in regulating the function of Trx via post-translational modifications, such as glutathionylation and S-nitrosylation (Casagrande et al. 2002; Kuster et al. 2006). For instance, C73 has been shown to be *S*-nitrosylated after treating the human protein with S-nitrosoglutathione (Mitchell et al. 2005). This S-nitrosothiol can be transferred from C73 to caspase 3 *in vitro* (Mitchell and Marletta 2005) and *in vivo* (Mitchell et al. 2007). Moreover, under oxidizing conditions, a disulfide can be formed between C69 and C72 (Watson et al. 2003). The formation of this disulfide is predicted to have a profound effect on the structure of Trx and to decrease the rate by which the active site is regenerated by TrxR.

Plant Thioredoxins

The Trx family is particularly important in plants. Plant Trxs have been primarily found to regulate enzymes involved in carbon metabolism but, as suggested by the identification of putative target proteins by proteomic studies, are likely also involved in many other cellular processes, such as photorespiration, lipid metabolism, membrane transport, hormone metabolism, and ATP synthesis (Balmer et al. 2003, 2004; Schürmann and Buchanan 2008). Moreover, plant Trxs also play an important role in sustaining early seedling growth of germinating cereal seeds (Wong et al. 2004).

Plants possess the largest group of Trxs found in all organisms. For instance, recent genomic analyses identified more than 40 Trx genes in higher plants, taking into account atypical variants with modified active sites and multidomain proteins with at least one Trx domain, such as nucleoredoxins (Chibani et al. 2009). The fact that only 28 out of the 46 Trxs found in *Populus trichocarpa* contain the typical WCGPC active site motif illustrates the diversity of the plant Trx family (Chibani et al. 2009).

Plant Trxs with CGPC active site are distributed into nine classes: Trx f, h, m, o, x, y, z, tetratricopeptide domain-containing thioredoxins (TDX) and nucleoredoxins. They are found in several subcellular compartments such as the cytoplasm, the mitochondria and the chloroplasts. Whereas TrxR reduces mitochondrial and cytoplasmic Trxs, Trxs present in chloroplasts are recycled by a ferredoxin-thioredoxin reductase (FTR) [reviewed in Gelhaye et al. 2005]. FTR is almost exclusively found in photosynthetic organisms and uses the reducing power provided by light to reduce Trxs. Although plant Trxs adopt the Trx-fold, some of them present specific characteristics that differentiate them from classical Trxs. For instance, the presence of an additional cysteine residue at position four in poplar Trx *h4* prevents the recycling by TrxR but renders the enzyme glutathione- and Grx-dependent (Koh et al. 2008). In addition, some other Trxs harboring an unusual CXXS active site motif are also recycled by glutathione (Gelhaye et al. 2003).

3.3.2 Glutaredoxins

3.3.2.1 General Properties

Glutaredoxins were first discovered in *Escherichia coli,* when Holmgren identified *E. coli* Grx1 as an alternative electron donor for ribonucleotide reductase in cells lacking Trx1 (Holmgren 1976). Grxs are glutathione-dependent disulfide oxidoreductases that catalyze a variety of thiol-disulfide exchange reactions, including the reduction of protein disulfides, as well as protein glutathionylation and deglutathionylation (Lillig et al. 2008; Mieyal et al. 2008; Gallogly et al. 2009). Grxs are usually small, heat-stable proteins that adopt a typical Trx-fold despite a low sequence homology with Trxs (Eklund et al. 1984; Martin 1995) (Fig. 3.2c). In Grxs, the solvent exposed CXXC active site motif is often replaced by a CXXS motif where the C-terminal active site cysteine is substituted by a Ser [for an overview, see Gallogly et al. 2009; Couturier et al. 2009].

When dithiol Grxs function as general protein disulfide oxidoreductases, they become oxidized upon reducing the protein disulfide, in a way similar to Trx. They are then converted back to the reduced state by glutathione (GSH), which forms a mixed-disulfide with the first cysteine of the catalytic cycle, which is then resolved by a second GSH molecule, releasing reduced Grx and GSSG (Fig. 3.3). As explained above, the intracellular GSH/GSSG ratio is maintained by glutathione reductase at the expense of NADPH. Albeit some Grxs, such as *E. coli* Grx1, catalyze the reduction of protein disulfide bonds, Grxs seem to function mostly as monothiol oxidoreductases for protein deglutathionylation reactions (Fernandes and Holmgren 2004). Under physiological conditions, the thiolate of the active site nucleophilic cysteine attacks the disulfide bond of a S-glutathionylated protein, releasing the protein thiol in the reduced form while becoming itself glutathionylated (Fig. 3.3). Then, GSH attacks the glutathionylated sulfur as explained above, releasing reduced Grx and GSSG. Alternatively, some Grxs were found to be either uniquely regenerated by TrxR or by both, a glutathione- or TrxR-dependent system (Johansson et al. 2004; Zaffagnini et al. 2008).

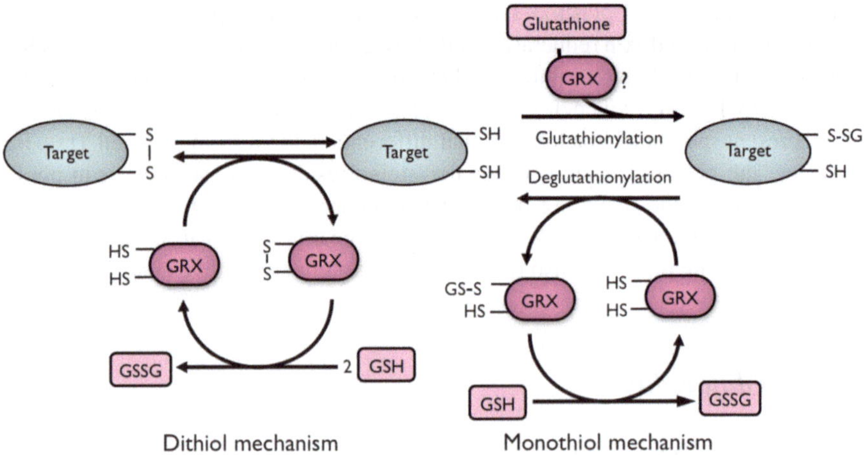

Fig. 3.3 Catalytic mechanisms employed by glutaredoxins. Grxs can reduce both disulfide bonds and protein-glutathione adducts by using either a dithiol or a monothiol mechanism. In the dithiol mechanism, the two cysteines of the active site are required and generally form an intramolecular disulfide bond, which is then reduced by GSH. This is similar to the reduction mechanism used by Trxs, except the latter are reduced by thioredoxin reductases instead of GSH. In the monothiol mechanism, only the catalytic cysteine (*first cysteine of the active site*) is required to reduce the protein-glutathione adducts. The Grx is glutathionylated in the course of the reaction, and is subsequently regenerated by a GSH molecule. Almost no information is available about the glutathionylation process *in vivo,* but it has been proposed that Grxs could also catalyze the glutathionylation of proteins

3.3.2.2 Diversity and Roles of Glutaredoxins

Grxs have been identified in most living organisms, including viruses, bacteria, plants and mammals. For instance, there are four Grxs in *E. coli* (Ec_Grx1 to 4), seven Grxs in *S. cerevisiae* (Sc_Grx1 to 7) and four in human (Hs_Grx1, Grx2, Grx3 and Grx5). Initially, Grxs were classified into a dithiol (with a CPY/FC catalytic motif) and a monothiol (with a CGFS catalytic motif) subgroup based on the active site sequence. However, this classification has now been further refined into six classes, considering the existence of either plant- or bacterial-specific isoforms or of multimodular isoforms containing additional domains (Couturier et al. 2009). The dithiol subgroup is now referred to as class I Grxs while the monothiol Grxs are included into class II. However, several monothiol Grxs with CPYS or CSYS active site motifs also belong to class I (Couturier et al. 2009).

Bacterial Glutaredoxins: The Example of *E. coli* Grxs

Four Grxs (Ec_Grx1, Ec_Grx2, Ec_Grx3 and Ec_Grx4) have been identified in *E. coli*. As stated previously, Ec_Grx1, which is about ten times less abundant than

Ec_Trx1, has been shown to catalyze the reduction of disulfide bonds in various cytoplasmic enzymes. For instance, Ec_Grx1 is an alternate electron donor for PAPS reductase and methionine sulfoxide reductases (Fernandes and Holmgren 2004). Moreover, Ec_Grx1 specifically reduces the disulfide bond that inactivates the transcription factor OxyR (see Sect. 3.2) under oxidative stress conditions (Aslund et al. 1999). As the gene that codes for Ec_Grx1 is induced by OxyR, the OxyR response to oxidative stress is therefore autoregulated [reviewed in Antelmann and Helmann 2011].

Much less is known about the other three *E. coli* glutaredoxins, particularly Ec_Grx3, which exhibits 5% of the activity of Ec_Grx1 with ribonucleotide reductase, and lacks activity with PAPS reductase. The *in vivo* substrates of this protein are unknown. With a molecular mass of about 24 kDa, Ec_Grx2 is unique among *E. coli* Grxs, which typically range in molecular weights from 9 to 14 kDa. Although Ec_Grx2 cannot reduce PAPS reductase or ribonucleotide reductase, this protein, which is up-regulated in stationary phase, is very active in catalyzing deglutathionylation. This is highlighted by the high catalytic activity of the enzyme towards mixed-disulfides between glutathione and artificial or physiological substrates (Vlamis-Gardikas 2008). Ec_Grx4 is the only *E. coli* Grx that belongs to the class of monothiol Grxs. Like other monothiol Grxs, Ec_Grx4 has been shown to bind an iron sulfur cluster (see Sect. 4.3) (Iwema et al. 2009). Its function remains obscure.

Eukaryotic Glutaredoxins: The Example of Plant Grxs

As for the Trx family, comparative genomic analyses revealed the presence of expanded Grx families in higher plants (from 27 to 35 genes), compared to non-photosynthetic organisms, which contain only a limited number of these genes (Couturier et al. 2009). Besides their roles associated with their capacity to bind and transfer Fe-S clusters, which is described in the next section, the established functions for plant Grxs are likely related to their capacity to reduce glutathionylated substrates. Several members from three out of the four Grx classes existing in land plants have been characterized so far. Grxs from class I have been divided into five subclasses (GrxC1, C2, C3, C4 and C5/S12), whereas class II is sub-divided into four subclasses (GrxS14, S15, S16, S17). Because the number of class III Grxs is variable between species (from 13 to 24 isoforms), a precise classification has not yet been achieved (Couturier et al. 2009). In general, if we exclude the targeting sequence, class I and III Grxs have a molecular weight comprised between 10 and 13 kDa. In contrast, the size of class II Grxs ranges from *ca* 12 kDa for GrxS14 and S15 to 53 kDa for GrxS17. The latter is composed of an N-terminal Trx-like domain fused to three Grx domains.

Owing to the large number of Grxs, deciphering the functions of plant Grxs is a daunting task. The present knowledge indicates that Grxs are important for the response of plants to oxidative stress. As already indicated, class I Grxs participate in the regeneration of specific subgroups of antioxidant enzymes, including type

II peroxiredoxins and methionine sulfoxide reductases of the B1 type (Rouhier et al. 2001; Tarrago et al. 2009). These two types of enzymes use a single redox active cysteine that is successively sulfenylated and glutathionylated during the catalytic cycle. Grxs thus contribute to the recycling of the active form by reducing the glutathione adduct. Among class I Grxs, knock-out mutants have only been described for GrxC1 and C2. Whereas the single mutants did not exhibit any phenotype, the double mutant was lethal at an early stage after pollinization. This probably indicates that GrxC1 and GrxC2 have redundant functions, at least for some early developmental stages (Riondet et al. 2011).

Among class II Grxs, the study of knock-out, knock-down or overexpressing lines indicated that seedlings of *A. thaliana* knock-out mutants for GrxS14 (otherwise named AtGrxcp) exhibited a higher sensitivity to oxidative stress treatment, whereas overexpression of the respective ortholog from the arsenic hyperaccumulator *Pteris vittata* increased arsenic resistance (Cheng et al. 2006; Sundaram et al. 2009). Quite similarly, silencing of tomato GrxS16 led to increased sensitivity to oxidative, salt and drought stresses while over-expression had opposite effects (Guo et al. 2010). Arabidopsis mutant plants, in which GrxS17 is disrupted, displayed several phenotypes under high temperature, including defects in proliferation and/or cell cycle control, accumulation of ROS, cellular membrane damages and altered auxin perception. These phenotypes highlight the important role of this protein for temperature-dependent postembryonic growth (Cheng et al. 2011). Last but not least, several studies have implied class III Grxs in developmental or stress response processes, via the regulation of transcription factors of the TGA family (La Camera et al. 2011; Laporte et al. 2011; Ndamukong et al. 2007; Xing and Zachgo 2008).

3.3.2.3 Glutaredoxins and Iron Homeostasis

While Grxs have been identified in the 1970s as an alternative reducing system to Trxs, it has only recently been recognized that several monothiol or dithiol Grxs are able to bind [2Fe-2S] cluster (Fig. 3.4) (Lillig et al. 2005; Rouhier et al. 2007; Bandyopadhyay et al. 2008). The [2Fe-2S] center, which is ligated by the catalytic cysteines of two Grx monomers and two glutathione molecules, can be efficiently transferred to acceptor proteins, as demonstrated for a plant plastidial Grx. Related to this observation, a role for *S. cerevisiae* Grx5, a protein with a CGFS active site, has been suggested as a carrier protein in Fe-S cluster biogenesis in yeast. In a *S. cerevisiae grx5* null mutant, deficient cluster assembly for the mitochondrial Fe-S enzymes aconitase and succinate dehydrogenase has been reported (Rodriguez-Manzaneque et al. 2002; Muhlenhoff et al. 2003). Furthermore, deletion of the zebrafish mitochondrial *grx5*, which also coordinates an iron sulfur cluster, causes hypochromic anemia. As such, the mitochondrial Fe-S cluster assembly machinery is required for heme biosynthesis (Wingert et al. 2005). Moreover, most prokaryotic or eukaryotic Grx orthologs were able to complement the zebrafish *grx5* mutant, which suggests that this function has been conserved throughout evolution (Molina-Navarro et al. 2006).

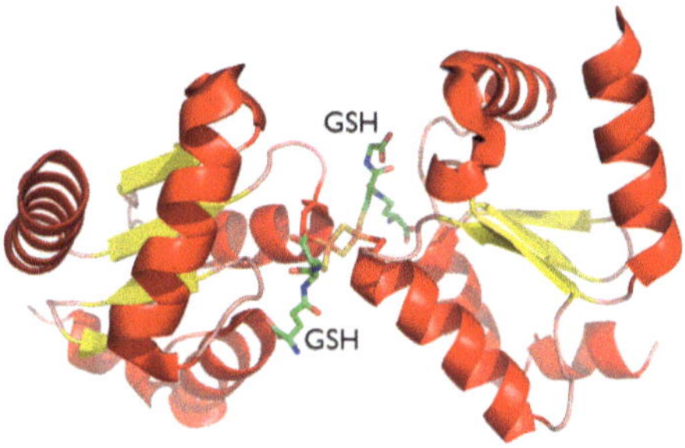

Fig. 3.4 Three-dimensional structure of the dimeric *A. thaliana* GrxC5 bridging a [2Fe-2S] cluster. The α-helices are shown in *red*, and β-strands are shown in *yellow*. The [2Fe-2S] center in *A. thaliana* GrxC5 is bridged by the catalytic cysteine of each monomer (PDB code 3RHC) (Couturier et al. 2011) and by two GSH molecules shown in stick representation. The figure was generated using MacPyMol

Some Fe-S containing Grxs function as redox sensors (Lillig et al. 2005; Rouhier et al. 2007; Couturier et al. 2011). *In vitro,* the [2Fe-2S] clusters on both poplar GrxC1 and human Grx2 are stabilized by GSH and destabilized by GSSG. Therefore, changes in the cellular GSH redox state could serve as a balance to promote or inhibit Grx activity. However, determining the exact ratio between the holo- and apo-protein *in vivo* is technically very challenging and has never been performed, which makes it difficult to differentiate the reductase activity from the capacity to bind Fe-S clusters.

Multidomain monothiol glutaredoxins formed by a Trx-like domain fused to one to three Grx domains have also a function in the nucleus where they are involved in intracellular iron trafficking. In *S. cerevisiae* and *Schizosaccharomyces pombe*, the nucleocytoplasmic class II Grxs (Grx3 and Grx4) regulate iron homeostasis by modifying the function of several transcription factors (Pujol-Carrion et al. 2006; Ojeda et al. 2006; Mercier and Labbe 2009; Jbel et al. 2011). Recently, Hoffman et al. showed that *S. cerevisiae* Grx4 functions as an iron sensor (Hoffmann et al. 2011). The C-terminal domain of this protein binds to the iron-sensing transcription factor Aft1p, whereas its N-terminal Trx domain was found to be essential *in vivo*. Furthermore, Mühlenhoff et al. found that Grx3 and Grx4 have an essential function in intracellular iron trafficking and sensing (Muhlenhoff et al. 2010). Depletion of *grx3* and *grx4* in *S. cerevisiae* specifically impaired all iron-requiring reactions in the cytosol, mitochondria, and nucleus, including the synthesis of Fe-S clusters, heme, and di-iron centers. From all these observations, we can conclude that these Grxs function in the intracellular iron trafficking and sensing through the bridging of a glutathione-containing iron sulfur center.

This class of Grxs forms also a complex with two other proteins, FRA1 and FRA2 (Fe repressor of activation-1 and 2), corresponding to aminopeptidase P- and BolA-like proteins, respectively (Kumanovics et al. 2008). The exact function of the two latter proteins in this complex is not yet clear, since BolA, in particular, was initially described as a protein affecting the morphology and elongation of *E. coli* cells (Aldea et al. 1988). The whole complex is an intermediate that links the mitochondrial Fe-S cluster assembly machinery to the transcription of the iron regulon, which is under the control of the iron-sensing transcription factor Aft1p. The biochemical and spectroscopic characterization of the Grx-BolA heterodimeric complex allowed Li *et al.* to identify three out of the four iron binding sites, which are constituted by the active site cysteine of Grx3 or Grx4, glutathione and a histidine residue from BolA (Li et al. 2009, 2011). The fact that an Fe-S containing Grx homodimer is converted into a Grx-BolA heterodimer upon addition of BolA *in vitro* supports the proposal that BolA could act as an adaptor protein changing the function of Grxs from an Fe-S scaffold/delivery protein to an iron sensor. Based on the strong co-occurrence of *grx* and *bolA* genes and their frequent association in bacterial genomes, it is likely that this interaction is conserved between kingdoms (Couturier et al. 2009; Rouhier et al. 2010). However, the absence of Grx3/4 orthologues with an N-terminal Trx domain in bacteria together with the absence of Aft1p orthologues in non-yeast organisms suggest a diversity of iron sensing mechanisms and transcription factors that might be involved.

3.4 Conclusions

The redox biochemical and biological fields are emerging with the identification of several LMW thiols that function as redox buffers and a variety of oxidoreductases, mostly from the Trx superfamily. However, many cellular mechanisms are still not known. Moreover, further work is required to explore how redox regulation affects complex cellular processes, and to fully characterize the role and function of various Grxs and Trxs involved in redox pathways. How redox and iron homeostasis are interconnected also requires further investigation. To advance our knowledge in redox biochemistry, we need to monitor the redox state and to quantify cellular thiol/disulfide redox buffers *in vivo*. New methods need to be developed to individually quantify both symmetrical and mixed disulfide populations in the correct biological context. Furthermore, for the more recently discovered LMW sulfur-containing compounds, many new targets and mechanisms are waiting to be discovered. Unraveling the role of these sulfur molecules in thiol redox regulation pathways and in the interplay between cytosolic thiol redox status and intra/extracellular stimuli will be one of the challenges for the next decade.

Acknowledgments JM is a group leader of the VIB and JFC is a Chercheur Qualifié of the Belgian FNRS. JM is indebted to JFC for helpful redox discussions over the past 5 years and JFC to JM. This work was supported by the European Research Council (FP7/2007–2013) ERC independent researcher starting grant 282335 – Sulfenic to JFC.

References

Aldea M, Hernandez-Chico C, de la Campa AG, Kushner SR, Vicente M (1988) Identification, cloning, and expression of bolA, an ftsZ-dependent morphogene of *Escherichia coli*. J Bacteriol 170:5169–5176

Antelmann H, Helmann JD (2011) Thiol-based redox switches and gene regulation. Antioxid Redox Signal 14:1049–1063

Arner ES (1999) Superoxide production by dinitrophenyl-derivatized thioredoxin reductase–a model for the mechanism and correlation to immunostimulation by dinitrohalobenzenes. Biofactors 10:219–226

Arner ESJ, Holmgren A (2000) Physiological functions of thioredoxin and thioredoxin reductase. Eur J Biochem 267:6102–6109

Aslund F, Zheng M, Beckwith J, Storz G (1999) Regulation of the OxyR transcription factor by hydrogen peroxide and the cellular thiol-disulfide status. Proc Natl Acad Sci USA 96: 6161–6165

Awad S, Henderson GB, Cerami A, Held KD (1992) Effects of trypanothione on the biological activity of irradiated transforming DNA. Int J Radiat Biol 62:401–407

Balmer Y, Koller A, del Val G, Manieri W, Schurmann P, Buchanan BB (2003) Proteomics gives insight into the regulatory function of chloroplast thioredoxins. Proc Natl Acad Sci USA 100:370–375

Balmer Y, Vensel WH, Tanaka CK, Hurkman WJ, Gelhaye E, Rouhier N, Jacquot JP, Manieri W, Schurmann P, Droux M, Buchanan BB (2004) Thioredoxin links redox to the regulation of fundamental processes of plant mitochondria. Proc Natl Acad Sci USA 101:2642–2647

Bandyopadhyay S, Gama F, Molina-Navarro MM, Gualberto JM, Claxton R, Naik SG, Huynh BH, Herrero E, Jacquot JP, Johnson MK, Rouhier N (2008) Chloroplast monothiol glutaredoxins as scaffold proteins for the assembly and delivery of [2Fe-2S] clusters. EMBO J 27:1122–1133

Benhar M, Forrester MT, Hess DT, Stamler JS (2008) Regulated protein denitrosylation by cytosolic and mitochondrial thioredoxins. Science 320:1050–1054

Bzymek KP, Newton GL, Ta P, Fahey RC (2007) Mycothiol import by *Mycobacterium smegmatis* and function as a resource for metabolic precursors and energy production. J Bacteriol 189:6796–6805

Carnieri EG, Moreno SN, Docampo R (1993) Trypanothione-dependent peroxide metabolism in Trypanosoma cruzi different stages. Mol Biochem Parasitol 61:79–86

Casagrande S, Bonetto V, Fratelli M, Gianazza E, Eberini I, Massignan T, Salmona M, Chang G, Holmgren A, Ghezzi P (2002) Glutathionylation of human thioredoxin: a possible crosstalk between the glutathione and thioredoxin systems. Proc Natl Acad Sci USA 99:9745–9749

Castro H, Tomas AM (2008) Peroxidases of trypanosomatids. Antioxid Redox Signal 10: 1593–1606

Chakravarthi S, Jessop CE, Bulleid NJ (2006) The role of glutathione in disulphide bond formation and endoplasmic-reticulum-generated oxidative stress. EMBO Rep 7:271–275

Cheng NH, Liu JZ, Brock A, Nelson RS, Hirschi KD (2006) AtGRXcp, an Arabidopsis chloroplastic glutaredoxin, is critical for protection against protein oxidative damage. J Biol Chem 281:26280–26288

Cheng Z, Arscott LD, Ballou DP, Williams CH (2007) The relationship of the redox potentials of thioredoxin and thioredoxin reductase from *Drosophila melanogaster* to the enzymatic mechanism: reduced thioredoxin is the reductant of glutathione in *Drosophila*. Biochemistry 46:7875–7885

Cheng NH, Liu JZ, Liu X, Wu Q, Thompson SM, Lin J, Chang J, Whitham SA, Park S, Cohen JD, Hirschi KD (2011) Arabidopsis monothiol glutaredoxin, AtGRXS17, is critical for temperature-dependent postembryonic growth and development via modulating auxin response. J Biol Chem 286:20398–20406

Chi BK, Gronau K, Mader U, Hessling B, Becher D, Antelmann H (2011) S-bacillithiolation protects against hypochlorite stress in Bacillus subtilis as revealed by transcriptomics and redox proteomics. Mol Cell Proteomics 10(11):M111.009506. doi: 10.1074/mcp.M111

Chibani K, Wingsle G, Jacquot JP, Gelhaye E, Rouhier N (2009) Comparative genomic study of the thioredoxin family in photosynthetic organisms with emphasis on Populus trichocarpa. Mol Plant 2:308–322

Collet JF, Messens J (2010) Structure, function, and mechanism of thioredoxin proteins. Antioxid Redox Signal 13:1205–1216

Collet JF, D'Souza JC, Jakob U, Bardwell JC (2003) Thioredoxin 2, an oxidative stress-induced protein, contains a high affinity zinc binding site. J Biol Chem 278:45325–45332

Couturier J, Jacquot JP, Rouhier N (2009) Evolution and diversity of glutaredoxins in photosynthetic organisms. Cell Mol Life Sci 66:2539–2557

Couturier J, Stroher E, Albetel AN, Roret T, Muthuramalingam M, Tarrago L, Seidel T, Tsan P, Jacquot JP, Johnson MK, Dietz KJ, Didierjean C, Rouhier N (2011) Arabidopsis chloroplastic glutaredoxin C5 as a model to explore molecular determinants for iron-sulfur cluster binding into glutaredoxins. J Biol Chem 286:27515–27527

Dalle-Donne I, Milzani A, Gagliano N, Colombo R, Giustarini D, Rossi R (2008) Molecular mechanisms and potential clinical significance of S-glutathionylation. Antioxid Redox Signal 10:445–473

Depuydt M, Leonard SE, Vertommen D, Denoncin K, Morsomme P, Wahni K, Messens J, Carroll K, Collet JF (2009) A periplasmic reducing system protects single cysteine residues from oxidation. Science 326:1109–1111

Depuydt M, Messens J, Collet JF (2011) How proteins form disulfide bonds. Antioxid Redox Signal 15:49–66

Dillet V, Dyson HJ, Bashford D (1998) Calculations of electrostatic interactions and pKas in the active site of *Escherichia coli* thioredoxin. Biochemistry 37:10298–10306

Dominici S, Valentini M, Maellaro E, del Bello B, Paolicchi A, Lorenzini E, Tongiani R, Comporti M, Pompella A (1999) Redox modulation of cell surface protein thiols in U937 lymphoma cells: the role of gamma-glutamyl transpeptidase-dependent H2O2 production and S-thiolation. Free Radic Biol Med 27:623–635

Dormeyer M, Reckenfelderbaumer N, Ludemann H, Krauth-Siegel RL (2001) Trypanothione-dependent synthesis of deoxyribonucleotides by Trypanosoma brucei ribonucleotide reductase. J Biol Chem 276:10602–10606

Dyson HJ, Tennant LL, Holmgren A (1991) Proton-transfer effects in the active-site region of *Escherichia coli* thioredoxin using two-dimensional 1H NMR. Biochemistry 30:4262–4268

Eklund H, Cambillau C, Sjoberg BM, Holmgren A, Jornvall H, Hoog JO, Branden CI (1984) Conformational and functional similarities between glutaredoxin and thioredoxins. EMBO J 3:1443–1449

El Hajjaji H, Dumoulin M, Matagne A, Colau D, Roos G, Messens J, Collet JF (2009) The zinc center influences the redox and thermodynamic properties of *Escherichia coli* thioredoxin 2. J Mol Biol 386:60–71

Fahey RC, Brown WC, Adams WB, Worsham MB (1978) Occurrence of glutathione in bacteria. J Bacteriol 133:1126–1129

Fairlamb AH, Cerami A (1992) Metabolism and functions of trypanothione in the Kinetoplastida. Annu Rev Microbiol 46:695–729

Fairlamb AH, Blackburn P, Ulrich P, Chait BT, Cerami A (1985) Trypanothione: a novel bis(glutathionyl)spermidine cofactor for glutathione reductase in trypanosomatids. Science 227:1485–1487

Feng J, Che Y, Milse J, Yin YJ, Liu L, Ruckert C, Shen XH, Qi SW, Kalinowski J, Liu SJ (2006) The gene *ncgl2918* encodes a novel maleylpyruvate isomerase that needs mycothiol as cofactor and links mycothiol biosynthesis and gentisate assimilation in *Corynebacterium glutamicum*. J Biol Chem 281:10778–10785

Fernandes AP, Holmgren A (2004) Glutaredoxins: glutathione-dependent redox enzymes with functions far beyond a simple thioredoxin backup system. Antioxid Redox Signal 6:63–74

Filser M, Comini MA, Molina-Navarro MM, Dirdjaja N, Herrero E, Krauth-Siegel RL (2008) Cloning, functional analysis, and mitochondrial localization of Trypanosoma brucei monothiol glutaredoxin-1. Biol Chem 389:21–32
Flohe L, Hecht HJ, Steinert P (1999) Glutathione and trypanothione in parasitic hydroperoxide metabolism. Free Radic Biol Med 27:966–984
Ford E, Hughes MN, Wardman P (2002) Kinetics of the reactions of nitrogen dioxide with glutathione, cysteine, and uric acid at physiological pH. Free Radic Biol Med 32:1314–1323
Forman HJ, Maiorino M, Ursini F (2010) Signaling functions of reactive oxygen species. Biochemistry 49:835–842
Fuangthong M, Helmann JD (2002) The OhrR repressor senses organic hydroperoxides by reversible formation of a cysteine-sulfenic acid derivative. Proc Natl Acad Sci USA 99: 6690–6695
Gaballa A, Newton GL, Antelmann H, Parsonage D, Upton H, Rawat M, Claiborne A, Fahey RC, Helmann JD (2010) Biosynthesis and functions of bacillithiol, a major low-molecular-weight thiol in Bacilli. Proc Natl Acad Sci USA 107:6482–6486
Gallogly MM, Starke DW, Mieyal JJ (2009) Mechanistic and kinetic details of catalysis of thiol-disulfide exchange by glutaredoxins and potential mechanisms of regulation. Antioxid Redox Signal 11:1059–1081
Gelhaye E, Rouhier N, Jacquot JP (2003) Evidence for a subgroup of thioredoxin h that requires GSH/Grx for its reduction. FEBS Lett 555:443–448
Gelhaye E, Rouhier N, Navrot N, Jacquot JP (2005) The plant thioredoxin system. Cell Mol Life Sci 62:24–35
Ghezzi P, di Simplicio P (2007) Glutathionylation pathways in drug response. Curr Opin Pharmacol 7:398–403
Ghosh S, Hamdan SM, Cook TE, Richardson CC (2008) Interactions of *Escherichia coli* thioredoxin, the processivity factor, with bacteriophage T7 DNA polymerase and helicase. J Biol Chem 283:32077–32084
Gommel DU, Nogoceke E, Morr M, Kiess M, Kalisz HM, Flohe L (1997) Catalytic characteristics of tryparedoxin. Eur J Biochem 248:913–918
Guo Y, Huang C, Xie Y, Song F, Zhou X (2010) A tomato glutaredoxin gene SlGRX1 regulates plant responses to oxidative, drought and salt stresses. Planta 232:1499–1509
Hamdan SM, Marintcheva B, Cook T, Lee SJ, Tabor S, Richardson CC (2005) A unique loop in T7 DNA polymerase mediates the binding of helicase-primase, DNA binding protein, and processivity factor. Proc Natl Acad Sci USA 102:5096–5101
Hansen RE, Roth D, Winther JR (2009) Quantifying the global cellular thiol-disulfide status. Proc Natl Acad Sci USA 106:422–427
Helmann JD (2011) Bacillithiol, a new player in bacterial redox homeostasis. Antioxid Redox Signal 15:123–133
Heras B, Edeling MA, Schirra HJ, Raina S, Martin JL (2004) Crystal structures of the DsbG disulfide isomerase reveal an unstable disulfide. Proc Natl Acad Sci USA 101:8876–8881
Herrero E, Ros J, Belli G, Cabiscol E (2008) Redox control and oxidative stress in yeast cells. Biochim Biophys Acta 1780:1217–1235
Hirota K, Matsui M, Iwata S, Nishiyama A, Mori K, Yodoi J (1997) AP-1 transcriptional activity is regulated by a direct association between thioredoxin and Ref-1. Proc Natl Acad Sci USA 94:3633–3638
Hirota K, Murata M, Sachi Y, Nakamura H, Takeuchi J, Mori K, Yodoi J (1999) Distinct roles of thioredoxin in the cytoplasm and in the nucleus. A two-step mechanism of redox regulation of transcription factor NF-kappaB. J Biol Chem 274:27891–27897
Hoffmann B, Uzarska MA, Berndt C, Godoy JR, Haunhorst P, Lillig CH, Lill R, Muhlenhoff U (2011) The multidomain thioredoxin-monothiol glutaredoxins represent a distinct functional group. Antioxid Redox Signal 15:19–30
Holmgren A (1976) Hydrogen donor system for *Escherichia coli* ribonucleoside-diphosphate reductase dependent upon glutathione. Proc Natl Acad Sci USA 73:2275–2279
Holmgren A (1989) Thioredoxin and glutaredoxin systems. J Biol Chem 264:13963–13966

Holmgren A, Bjornstedt M (1995) Thioredoxin and thioredoxin reductase. Methods Enzymol 252:199–208

Holmgren A, Soderberg BO, Eklund H, Branden CI (1975) Three-dimensional structure of *Escherichia coli* thioredoxin-S2 to 2.8 Å resolution. Proc Natl Acad Sci USA 72:2305–2309

Hwang C, Sinskey AJ, Lodish HF (1992) Oxidized redox state of glutathione in the endoplasmic reticulum. Science 257:1496–1502

Iwema T, Picciocchi A, Traore DA, Ferrer JL, Chauvat F, Jacquamet L (2009) Structural basis for delivery of the intact [Fe2S2] cluster by monothiol glutaredoxin. Biochemistry 48:6041–6043

Jbel M, Mercier A, Labbe S (2011) Grx4 monothiol glutaredoxin is required for iron limitation-dependent inhibition of Fep1. Eukaryot Cell 10:629–645

Johansson C, Lillig CH, Holmgren A (2004) Human mitochondrial glutaredoxin reduces S-glutathionylated proteins with high affinity accepting electrons from either glutathione or thioredoxin reductase. J Biol Chem 279:7537–7543

Koh CS, Navrot N, Didierjean C, Rouhier N, Hirasawa M, Knaff DB, Wingsle G, Samian R, Jacquot JP, Corbier C, Gelhaye E (2008) An atypical catalytic mechanism involving three cysteines of thioredoxin. J Biol Chem 283:23062–23072

Koppenol WH, Moreno JJ, Pryor WA, Ischiropoulos H, Beckman JS (1992) Peroxynitrite, a cloaked oxidant formed by nitric-oxide and superoxide. Chem Res Toxicol 5:834–842

Krause G, Holmgren A (1991) Substitution of the conserved tryptophan 31 in *Escherichia coli* thioredoxin by site-directed mutagenesis and structure-function analysis. J Biol Chem 266:4056–4066

Krauth-Siegel L, Leroux AE (2012) Low molecular mass antioxidants in parasites. Antioxid Redox Signal 17(4):583–607

Krauth-Siegel RL, Bauer H, Schirmer RH (2005) Dithiol proteins as guardians of the intracellular redox milieu in parasites: old and new drug targets in trypanosomes and malaria-causing plasmodia. Angew Chem Int Ed Engl 44:690–715

Kumanovics A, Chen OS, Li L, Bagley D, Adkins EM, Lin H, Dingra NN, Outten CE, Keller G, Winge D, Ward DM, Kaplan J (2008) Identification of FRA1 and FRA2 as genes involved in regulating the yeast iron regulon in response to decreased mitochondrial iron-sulfur cluster synthesis. J Biol Chem 283:10276–10286

Kumar C, Igbaria A, D'Autreaux B, Planson AG, Junot C, Godat E, Bachhawat AK, Delaunay-Moisan A, Toledano MB (2011) Glutathione revisited: a vital function in iron metabolism and ancillary role in thiol-redox control. EMBO J 30:2044–2056

Kuster GM, Siwik DA, Pimentel DR, Colucci WS (2006) Role of reversible, thioredoxin-sensitive oxidative protein modifications in cardiac myocytes. Antioxid Redox Signal 8:2153–2159

la Camera S, L'Haridon F, Astier J, Zander M, Abou-Mansour E, Page G, Thurow C, Wendehenne D, Gatz C, Metraux JP, Lamotte O (2011) The glutaredoxin ATGRXS13 is required to facilitate Botrytis cinerea infection of Arabidopsis thaliana plants. Plant J 68:507–519

Laporte D, Olate E, Salinas P, Salazar M, Jordana X, Holuigue L (2011) Glutaredoxin GRXS13 plays a key role in protection against photooxidative stress in Arabidopsis. J Exp Bot 63(1):503–515

Laurent TC, Moore EC, Reichard P (1964) Enzymatic synthesis of deoxyribonucleotides. Iv. Isolation and characterization of thioredoxin, the hydrogen donor from *Escherichia coli* B. J Biol Chem 239:3436–3444

Lee J-W, Soonsanga S, Helmann JD (2007) A complex thiolate switch regulates the *Bacillus subtilis* organic peroxide sensor OhrR. Proc Natl Acad Sci USA 104:8743–8748

Lennon BW, Williams CH, Ludwig ML (2000) Twists in catalysis: alternating conformations of *Escherichia coli* thioredoxin reductase. Science 289:1190–1194

Li H, Mapolelo DT, Dingra NN, Naik SG, Lees NS, Hoffman BM, Riggs-Gelasco PJ, Huynh BH, Johnson MK, Outten CE (2009) The yeast iron regulatory proteins Grx3/4 and Fra2 form heterodimeric complexes containing a [2Fe-2S] cluster with cysteinyl and histidyl ligation. Biochemistry 48:9569–9581

Li H, Mapolelo DT, Dingra NN, Keller G, Riggs-Gelasco PJ, Winge DR, Johnson MK, Outten CE (2011) Histidine 103 in Fra2 is an iron-sulfur cluster ligand in the [2Fe-2S] Fra2-Grx3 complex and is required for in vivo iron signaling in yeast. J Biol Chem 286:867–876
Lillig CH, Berndt C, Vergnolle O, Lonn ME, Hudemann C, Bill E, Holmgren A (2005) Characterization of human glutaredoxin 2 as iron-sulfur protein: a possible role as redox sensor. Proc Natl Acad Sci USA 102:8168–8173
Lillig CH, Berndt C, Holmgren A (2008) Glutaredoxin systems. Biochim Biophys Acta 1780:1304–1317
Ludemann H, Dormeyer M, Sticherling C, Stallmann D, Follmann H, Krauth-Siegel RL (1998) Trypanosoma brucei tryparedoxin, a thioredoxin-like protein in African trypanosomes. FEBS Lett 431:381–385
Martin JL (1995) Thioredoxin–a fold for all reasons. Structure 3:245–250
Martin H, Dean M (1991) Identification of a thioredoxin-related protein associated with plasma membranes. Biochem Biophys Res Commun 175:123–128
Maruyama T, Sachi Y, Furuke K, Kitaoka Y, Kanzaki H, Yoshimura Y, Yodoi J (1999) Induction of thioredoxin, a redox-active protein, by ovarian steroid hormones during growth and differentiation of endometrial stromal cells in vitro. Endocrinology 140:365–372
McCarthy AA, Haebel PW, Torronen A, Rybin V, Baker EN, Metcalf P (2000) Crystal structure of the protein disulfide bond isomerase, DsbC, from *Escherichia coli*. Nat Struct Biol 7:196–199
Meister A, Anderson ME (1983) Glutathione. Annu Rev Biochem 52:711–760
Melchers J, Dirdjaja N, Ruppert T, Krauth-Siegel RL (2007) Glutathionylation of trypanosomal thiol redox proteins. J Biol Chem 282:8678–8694
Mercier A, Labbe S (2009) Both Php4 function and subcellular localization are regulated by iron via a multistep mechanism involving the glutaredoxin Grx4 and the exportin Crm1. J Biol Chem 284:20249–20262
Mieyal JJ, Gallogly MM, Qanungo S, Sabens EA, Shelton MD (2008) Molecular mechanisms and clinical implications of reversible protein S-glutathionylation. Antioxid Redox Signal 10: 1941–1988
Miranda-Vizuete A, Damdimopoulos AE, Gustafsson J, Spyrou G (1997) Cloning, expression, and characterization of a novel *Escherichia coli* thioredoxin. J Biol Chem 272:30841–30847
Misset-Smits M, van Ophem PW, Sakuda S, Duine JA (1997) Mycothiol, 1-O-(2′-[N-acetyl-L-cysteinyl]amido-2′-deoxy-alpha-D-glucopyranosyl)-D- myo-inositol, is the factor of NAD/factor-dependent formaldehyde dehydrogenase. FEBS Lett 409:221–222
Mitchell DA, Marletta MA (2005) Thioredoxin catalyzes the S-nitrosation of the caspase-3 active site cysteine. Nat Chem Biol 1:154–158
Mitchell DA, Erwin PA, Michel T, Marletta MA (2005) S-Nitrosation and regulation of inducible nitric oxide synthase. Biochemistry 44:4636–4647
Mitchell DA, Morton SU, Fernhoff NB, Marletta MA (2007) Thioredoxin is required for S-nitrosation of procaspase-3 and the inhibition of apoptosis in Jurkat cells. Proc Natl Acad Sci USA 104:11609–11614
Molina-Navarro MM, Casas C, Piedrafita L, Belli G, Herrero E (2006) Prokaryotic and eukaryotic monothiol glutaredoxins are able to perform the functions of Grx5 in the biogenesis of Fe/S clusters in yeast mitochondria. FEBS Lett 580:2273–2280
Moore EC, Reichard P, Thelander L (1964) Enzymatic synthesis of deoxyribonucleotides. V. Purification and properties of thioredoxin reductase from *Escherichia coli* B. J Biol Chem 239:3445–3452
Muhlenhoff U, Gerber J, Richhardt N, Lill R (2003) Components involved in assembly and dislocation of iron-sulfur clusters on the scaffold protein Isu1p. EMBO J 22:4815–4825
Muhlenhoff U, Molik S, Godoy JR, Uzarska MA, Richter N, Seubert A, Zhang Y, Stubbe J, Pierrel F, Herrero E, Lillig CH, Lill R (2010) Cytosolic monothiol glutaredoxins function in intracellular iron sensing and trafficking via their bound iron-sulfur cluster. Cell Metab 12: 373–385
Muller EG (1991) Thioredoxin deficiency in yeast prolongs S phase and shortens the G1 interval of the cell cycle. J Biol Chem 266:9194–9202

Nakamura T, Nakamura H, Hoshino T, Ueda S, Wada H, Yodoi J (2005) Redox regulation of lung inflammation by thioredoxin. Antioxid Redox Signal 7:60–71

Navrot N, Collin V, Gualberto J, Gelhaye E, Hirasawa M, Rey P, Knaff DB, Issakidis E, Jacquot JP, Rouhier N (2006) Plant glutathione peroxidases are functional peroxiredoxins distributed in several subcellular compartments and regulated during biotic and abiotic stresses. Plant Physiol 142:1364–1379

Ndamukong I, Abdallat AA, Thurow C, Fode B, Zander M, Weigel R, Gatz C (2007) SA-inducible Arabidopsis glutaredoxin interacts with TGA factors and suppresses JA-responsive PDF1.2 transcription. Plant J 50:128–139

Newton GL, Bewley CA, Dwyer TJ, Horn R, Aharonowitz Y, Cohen G, Davies J, Faulkner DJ, Fahey RC (1995) The structure of U17 isolated from *Streptomyces clavuligerus* and its properties as an antioxidant thiol. Eur J Biochem 230:821–825

Newton GL, Av-Gay Y, Fahey RC (2000) A novel mycothiol-dependent detoxification pathway in mycobacteria involving mycothiol S-conjugate amidase. Biochemistry 39:10739–10746

Newton GL, Buchmeier N, Fahey RC (2008) Biosynthesis and functions of mycothiol, the unique protective thiol of Actinobacteria. Microbiol Mol Biol Rev 72:471–494

Newton GL, Rawat M, la Clair JJ, Jothivasan VK, Budiarto T, Hamilton CJ, Claiborne A, Helmann JD, Fahey RC (2009) Bacillithiol is an antioxidant thiol produced in Bacilli. Nat Chem Biol 5:625–627

Newton GL, Leung SS, Wakabayashi JI, Rawat M, Fahey RC (2011) The DinB superfamily includes novel mycothiol, bacillithiol, and glutathione S-transferases. Biochemistry 50: 10751–10760

Nicely NI, Parsonage D, Paige C, Newton GL, Fahey RC, Leonardi R, Jackowski S, Mallett TC, Claiborne A (2007) Structure of the type III pantothenate kinase from Bacillus anthracis at 2.0 A resolution: implications for coenzyme A-dependent redox biology. Biochemistry 46: 3234–3245

Nogoceke E, Gommel DU, Kiess M, Kalisz HM, Flohe L (1997) A unique cascade of oxidoreductases catalyses trypanothione-mediated peroxide metabolism in Crithidia fasciculata. Biol Chem 378:827–836

Nulton-Persson AC, Starke DW, Mieyal JJ, Szweda LI (2003) Reversible inactivation of alpha-ketoglutarate dehydrogenase in response to alterations in the mitochondrial glutathione status. Biochemistry 42:4235–4242

Ojeda L, Keller G, Muhlenhoff U, Rutherford JC, Lill R, Winge DR (2006) Role of glutaredoxin-3 and glutaredoxin-4 in the iron regulation of the Aft1 transcriptional activator in Saccharomyces cerevisiae. J Biol Chem 281:17661–17669

Ordóñez E, van Belle K, Roos G, de Galan S, Letek M, Gil JA, Wyns L, Mateos LM, Messens J (2009) Arsenate reductase, mycothiol, and mycoredoxin concert thiol/disulfide exchange. J Biol Chem 284:15107–15116

Oza SL, Tetaud E, Ariyanayagam MR, Warnon SS, Fairlamb AH (2002) A single enzyme catalyses formation of Trypanothione from glutathione and spermidine in Trypanosoma cruzi. J Biol Chem 277:35853–35861

Pedrajas JR, Kosmidou E, Miranda-Vizuete A, Gustafsson JA, Wright AP, Spyrou G (1999) Identification and functional characterization of a novel mitochondrial thioredoxin system in *Saccharomyces cerevisiae*. J Biol Chem 274:6366–6373

Peskin AV, Winterbourn CC (2001) Kinetics of the reactions of hypochlorous acid and amino acid chloramines with thiols, methionine, and ascorbate. Free Radic Biol Med 30:572–579

Powis G, Mustacich D, Coon A (2000) The role of the redox protein thioredoxin in cell growth and cancer. Free Radic Biol Med 29:312–322

Pujol-Carrion N, Belli G, Herrero E, Nogues A, De la Torre-Ruiz MA (2006) Glutaredoxins Grx3 and Grx4 regulate nuclear localisation of Aft1 and the oxidative stress response in Saccharomyces cerevisiae. J Cell Sci 119:4554–4564

Qin J, Clore GM, Gronenborn AM (1994) The high-resolution three-dimensional solution structures of the oxidized and reduced states of human thioredoxin. Structure 2:503–522

Quijano C, Alvarez B, Gatti RM, Augusto O, Radi R (1997) Pathways of peroxynitrite oxidation of thiol groups. Biochem J 322:167–173
Radi R, Beckman JS, Bush KM, Freeman BA (1991) Peroxynitrite oxidation of sulfhydryls – the cytotoxic potential of superoxide and nitric-oxide. J Biol Chem 266:4244–4250
Rawat M, Av-Gay Y (2007) Mycothiol-dependent proteins in actinomycetes. FEMS Microbiol Rev 31:278–292
Reckenfelderbaumer N, Krauth-Siegel RL (2002) Catalytic properties, thiol pK value, and redox potential of Trypanosoma brucei tryparedoxin. J Biol Chem 277:17548–17555
Reinemer P, Dirr HW, Ladenstein R, Schaffer J, Gallay O, Huber R (1991) The three-dimensional structure of class pi glutathione S-transferase in complex with glutathione sulfonate at 2.3 Å resolution. EMBO J 10:1997–2005
Ren B, Huang W, Åkesson B, Ladenstein R (1997) The crystal structure of seleno-glutathione peroxidase from human plasma at 2.9 Å resolution. J Mol Biol 268:869–885
Rietsch A, Bessette P, Georgiou G, Beckwith J (1997) Reduction of the periplasmic disulfide bond isomerase, DsbC, occurs by passage of electrons from cytoplasmic thioredoxin. J Bacteriol 179:6602–6608
Riondet C, Desouris JP, Montoya JG, Chartier Y, Meyer Y, Reichheld JP (2011) A dicotyledon-specific glutaredoxin GRXC1 family with dimer-dependent redox regulation is functionally redundant with GRXC2. Plant Cell Environ 35(2):360–373
Ritz D, Patel H, Doan B, Zheng M, Aslund F, Storz G, Beckwith J (2000) Thioredoxin 2 is involved in the oxidative stress response in *Escherichia coli*. J Biol Chem 275:2505–2512
Rodriguez-Manzaneque MT, Tamarit J, Belli G, Ros J, Herrero E (2002) Grx5 is a mitochondrial glutaredoxin required for the activity of iron/sulfur enzymes. Mol Biol Cell 13:1109–1121
Roos G, Messens J (2011) Protein sulfenic acid formation: from cellular damage to redox regulation. Free Radic Biol Med 51:314–326
Roos G, Garcia-Pino A, van Belle K, Brosens E, Wahni K, Vandenbussche G, Wyns L, Loris R, Messens J (2007) The conserved active site proline determines the reducing power of *Staphylococcus aureus* thioredoxin. J Mol Biol 368:800–811
Roos G, Foloppe N, van Laer K, Wyns L, Nilsson L, Geerlings P, Messens J (2009) How thioredoxin dissociates its mixed disulfide. PLOS Comput Biol 5:e1000461
Rouhier N, Gelhaye E, Sautiere PE, Brun A, Laurent P, Tagu D, Gerard J, de Fay E, Meyer Y, Jacquot JP (2001) Isolation and characterization of a new peroxiredoxin from poplar sieve tubes that uses either glutaredoxin or thioredoxin as a proton donor. Plant Physiol 127:1299–1309
Rouhier N, Unno H, Bandyopadhyay S, Masip L, Kim SK, Hirasawa M, Gualberto JM, Lattard V, Kusunoki M, Knaff DB, Georgiou G, Hase T, Johnson MK, Jacquot JP (2007) Functional, structural, and spectroscopic characterization of a glutathione-ligated [2Fe-2S] cluster in poplar glutaredoxin C1. Proc Natl Acad Sci USA 104:7379–7384
Rouhier N, Lemaire SD, Jacquot JP (2008) The role of glutathione in photosynthetic organisms: emerging functions for glutaredoxins and glutathionylation. Annu Rev Plant Biol 59:143–166
Rouhier N, Couturier J, Johnson MK, Jacquot JP (2010) Glutaredoxins: roles in iron homeostasis. Trends Biochem Sci 35:43–52
Russel M (1991) Filamentous phage assembly. Mol Microbiol 5:1607–1613
Schmidt H, Krauth-Siegel RL (2003) Functional and physicochemical characterization of the thioredoxin system in Trypanosoma brucei. J Biol Chem 278:46329–46336
Schroeder BO, Wu Z, Nuding S, Groscurth S, Marcinowski M, Beisner J, Buchner J, Schaller M, Stange EF, Wehkamp J (2011) Reduction of disulphide bonds unmasks potent antimicrobial activity of human beta-defensin 1. Nature 469:419–423
Schürmann P, Buchanan BB (2008) The ferredoxin/thioredoxin system of oxygenic photosynthesis. Antioxid Redox Signal 10:1235–1274
Spies HS, Steenkamp DJ (1994) Thiols of intracellular pathogens. Identification of ovothiol A in Leishmania donovani and structural analysis of a novel thiol from Mycobacterium bovis. Eur J Biochem 224:203–213

Sundaram S, Wu S, Ma LQ, Rathinasabapathi B (2009) Expression of a Pteris vittata glutaredoxin PvGRX5 in transgenic Arabidopsis thaliana increases plant arsenic tolerance and decreases arsenic accumulation in the leaves. Plant Cell Environ 32:851–858

Tarrago L, Laugier E, Zaffagnini M, Marchand C, Le Maréchal P, Rouhier N, Lemaire SD, Rey P (2009) Regeneration mechanisms of Arabidopsis thaliana methionine sulfoxide reductases B by glutaredoxins and thioredoxins. J Biol Chem 284:18963–18971

Thomson L, Denicola A, Radi R (2003) The trypanothione-thiol system in Trypanosoma cruzi as a key antioxidant mechanism against peroxynitrite-mediated cytotoxicity. Arch Biochem Biophys 412:55–64

Tian G, Xiang S, Noiva R, Lennarz WJ, Schindelin H (2006) The crystal structure of yeast protein disulfide isomerase suggests cooperativity between its active sites. Cell 124:61–73

Townsend DM (2007) S-glutathionylation: indicator of cell stress and regulator of the unfolded protein response. Mol Interv 7:313–324

Vlamis-Gardikas A (2008) The multiple functions of the thiol-based electron flow pathways of *Escherichia coli*: eternal concepts revisited. Biochim Biophys Acta 1780:1170–1200

Vogt RN, Steenkamp DJ, Zheng R, Blanchard JS (2003) The metabolism of nitrosothiols in the Mycobacteria: identification and characterization of S-nitrosomycothiol reductase. Biochem J 374:657–666

Wang R, Yin YJ, Wang F, Li M, Feng J, Zhang HM, Zhang JP, Liu SJ, Chang WR (2007) Crystal structures and site-directed mutagenesis of a mycothiol-dependent enzyme reveal a novel folding and molecular basis for mycothiol-mediated maleylpyruvate isomerization. J Biol Chem 282:16288–16294

Watson WH, Pohl J, Montfort WR, Stuchlik O, Reed MS, Powis G, Jones DP (2003) Redox potential of human thioredoxin 1 and identification of a second dithiol/disulfide motif. J Biol Chem 278:33408–33415

Wingert RA, Galloway JL, Barut B, Foott H, Fraenkel P, Axe JL, Weber GJ, Dooley K, Davidson AJ, Schmid B, Paw BH, Shaw GC, Kingsley P, Palis J, Schubert H, Chen O, Kaplan J, Zon LI (2005) Deficiency of glutaredoxin 5 reveals Fe-S clusters are required for vertebrate haem synthesis. Nature 436:1035–1039

Winterbourn CC, Hampton MB (2008) Thiol chemistry and specificity in redox signaling. Free Radic Biol Med 45:549–561

Wong JH, Cai N, Balmer Y, Tanaka CK, Vensel WH, Hurkman WJ, Buchanan BB (2004) Thioredoxin targets of developing wheat seeds identified by complementary proteomic approaches. Phytochemistry 65:1629–1640

Xia TH, Bushweller JH, Sodano P, Billeter M, Bjornberg O, Holmgren A, Wuthrich K (1992) NMR structure of oxidized *Escherichia coli* glutaredoxin: comparison with reduced *E. coli* glutaredoxin and functionally related proteins. Protein Sci 1:310–321

Xia B, Vlamis-Gardikas A, Holmgren A, Wright PE, Dyson HJ (2001) Solution structure of *Escherichia coli* glutaredoxin-2 shows similarity to mammalian glutathione-S-transferases. J Mol Biol 310:907–918

Xing S, Zachgo S (2008) ROXY1 and ROXY2, two Arabidopsis glutaredoxin genes, are required for anther development. Plant J 53:790–801

Xu SZ, Sukumar P, Zeng F, Li J, Jairaman A, English A, Naylor J, Ciurtin C, Majeed Y, Milligan CJ, Bahnasi YM, Al-Shawaf E, Porter KE, Jiang LH, Emery P, Sivaprasadarao A, Beech DJ (2008) TRPC channel activation by extracellular thioredoxin. Nature 451:69–72

Zaffagnini M, Michelet L, Massot V, Trost P, Lemaire SD (2008) Biochemical characterization of glutaredoxins from Chlamydomonas reinhardtii reveals the unique properties of a chloroplastic CGFS-type glutaredoxin. J Biol Chem 283:8868–8876

Zhang HW, Squadrito GL, Uppu RM, Lemercier JN, Cueto R, Pryor WA (1997) Inhibition of peroxynitrite-mediated oxidation of glutathione by carbon dioxide. Arch Biochem Biophys 339:183–189

Chapter 4
Sulfenic Acids and Peroxiredoxins in Oxidant Defense and Signaling

Leslie B. Poole, Kimberly J. Nelson, and P. Andrew Karplus

Abstract In recent years, there has been tremendous growth in awareness of the importance of sulfenic acids to biological processes. This is in part related to reactive oxygen species in general and hydrogen peroxide in particular becoming widely recognized as intermediates or second messengers crucial to signaling processes that guide cell growth and development. Thus a picture is emerging of sulfenic acids being important both for defending against oxidative stress and for normal growth. Although the work is challenging due to the fleeting and localized nature of some of the intermediates involved, substantial progress has been made in identifying modified proteins and the roles they play. The landscape of modifications uncovered thus far suggests that we have barely scratched the surface of the many stress-induced and regulatory oxidation events taking place in cells on a regular basis. This chapter recounts the chemistry of sulfenic acids and their effects in modulating function in a few recently-studied proteins. We also discuss cysteine-based peroxidases in more depth, with a focus on the structural and biochemical features of the peroxiredoxin family.

Keywords Redox signaling • Peroxiredoxins • Sulfenic acids • Oxidative stress defenses • Redox regulation

L.B. Poole (✉) • K.J. Nelson
Department of Biochemistry, Wake Forest School of Medicine, Medical Center Boulevard, Winston-Salem, NC 27157, USA
e-mail: lbpoole@wakehealth.edu

P.A. Karplus
Department of Biochemistry and Biophysics, Oregon State University, 2011 Ag Life Sciences Building, Corvallis, OR 97331, USA

U. Jakob and D. Reichmann (eds.), *Oxidative Stress and Redox Regulation*,
DOI 10.1007/978-94-007-5787-5_4,

Abbreviations

ROS	reactive oxygen species
Prx	Peroxiredoxin
Orp1	oxidant resistance protein
NBD-Cl	7-chloro-4-nitrobenzo-2-oxa-1,3-diazole
GAPDH	glyceraldehyde-3-phosphate dehydrogenase
PTP	protein tyrosine phosphatase
SHP	Src homology 2 domain-containing phosphatase
DEP1	density-enhanced phosphatase-1
PTEN	phosphatase and tension homolog
HEK	human embryonic kidney
VEGF	vascular endothelial growth factor
EGF	epidermal growth factor
VEGFR	VEGF receptor
PDGF	platelet-derived growth factor
MEF	mouse embryonic fibroblasts
PKGIα	cGMP-dependent protein kinase
Ohr	organic hydroperoxide resistance protein
C_p	peroxidatic cysteine
C_r	resolving cysteine
Trx	thioredoxin
Grx	glutaredoxin
FF	fully folded
LU	locally unfolded
AhpC	alkyl hydroperoxide reductase C protein
TrxR	thioredoxin reductase
BCP	bacterioferritin comigratory protein
Yap1	yeast activator protein 1
Pap1	activator protein-1 from *Schizosaccharomyces pombe*
Tpx1	thiol peroxidase 1
Nrf-2	NF-E2-related factor
ARE	antioxidant response element
Keap1	Kelch like ECH-associated protein 1
CDK1	cyclin dependent kinase 1

4.1 Introduction and Scope

Reactive oxygen species (ROS) are produced in aerobic organisms by many avenues including as a result of incomplete oxygen reduction by the mitochondrial electron transport chain, as a result of exposure to certain environmental toxins and chemotherapeutics, and even purposefully as part of an immune response or of signaling pathways. It has long been recognized that cellular proteins that remove these damaging species play a critical role in preventing or minimizing mutagenesis and in delaying carcinogenesis and aging (Greenberg and Demple 1988; Imlay 2008; Stadtman 2006). The peroxiredoxin (Prx) family of cysteine-based peroxidases is now thought to be the system responsible for removing the majority of intracellular hydrogen peroxide and other alkyl hydroperoxides (Winterbourn 2008; Adimora et al. 2010). The functions of these enzymes undoubtedly extend well beyond defense, however. It has more recently become apparent that ROS, and hydrogen peroxide in particular, are also produced during normal cell growth by NADPH oxidase enzymes as a part of receptor-mediated signaling pathways that lead to proliferation, differentiation and immune responses (Murphy et al. 2011). Activated cysteine thiol groups (SH) in proteins can be oxidized by peroxide to first form cysteine sulfenic acid (SOH) and, in many cases, then go on to form other reversible species (e.g., disulfides, sulfenamides) or irreversible species (e.g., sulfinic and sulfonic acids). Functionally, SOH has most clearly been established as a catalytic intermediate in cysteine-dependent antioxidant enzymes like Prxs and methionine sulfoxide reductases; however, it has also been shown to serve a role in regulating a range of other transcription factors and signaling proteins (Poole et al. 2004; Paulsen and Carroll 2010; Klomsiri et al. 2011). Here we describe the chemical attributes and summarize the biological roles for sulfenic acids as widespread posttranslational modifications in cellular proteins. We also describe in more detail structural and biochemical characteristics of the peroxiredoxin family as an elegant example of mechanisms by which cysteine oxidation can be used both catalytically and to regulate protein function.

4.2 Cysteine, a Special Amino Acid

A major site of oxidant reactivity that underlies cellular defenses and oxidant-sensitive switches in proteins is the thiol, or sulfhydryl group, of cysteine. This residue has very special physicochemical properties that are responsible for its high prevalence within protein functional sites in spite of its overall low abundance within proteins compared with most other amino acids (Marino and Gladyshev 2010a). These properties include a nominal pK_a value around 8.5 that allows cysteine to function as a nucleophile or an acid/base catalyst, redox properties that bring additional catalytic versatility to enzymes as well as the possibility of forming disulfide bonds that can stabilize proteins or lock in a specific conformation of a protein. Moreover, bioinformatic evidence suggests that the high tendency of this residue to be functional has led to an evolutionary selection against its presence on protein surfaces, leading to the overly simplistic interpretation that, because of its high propensity to be buried in protein structures, this amino acid residue is considerably hydrophobic in nature (Marino and Gladyshev 2010a). Only the very low abundance “21st” amino acid selenocysteine can be functionally interchanged with cysteine in proteins; this feature has been the basis for a bioinformatic approach to identify putative redox-active cysteine residues within the sequences of divergent proteins (Fomenko et al. 2007).

An important chemical product of the oxidation of cysteine thiols is the sulfenic acid, formed when the nucleophilic thiolate anion attacks the terminal oxygen atom of hydroperoxides like hydrogen peroxide, lipid hydroperoxides and peroxynitrite, releasing the alcohol (or water) product. Compared to the rate of reaction with an unactivated thiol or thiolate-containing molecule like glutathione, this chemistry is highly accelerated in the active sites of cysteine-based peroxidases like peroxiredoxins (Parsonage et al. 2005) and NADH peroxidases (Parsonage et al. 1993), and in the oxidant-sensor sites of transcriptional regulators like OxyR (Åslund et al. 1999) and OhrR (Lee et al. 2007). The enhanced reactivity of these proteins toward peroxides reveals the large influence that protein microenvironment can have on the reactivity of particular cysteine residues, supporting second order reaction rates as high as 10^7–10^8 M^{-1} s^{-1}, far higher than the 10–20 M^{-1} s^{-1} expected for small molecule thiolates (Parsonage et al. 2005; Winterbourn 2008; Manta et al. 2009; Hall et al. 2010). The nucleophilic nature of activated cysteine residues also renders them susceptible to alkylation by electrophiles, S-nitrosation, and other types of modifications. A series of bioinformatic analyses have been developed with the goal of identifying features which activate or fine tune the oxidant or electrophile reactivity of given cysteinyl residues; however, this task has been difficult due to the limited number of known redox-regulated proteins and the range of structures and functions exhibited by the proteins reportedly exhibiting sensitivity *in vivo* or *in vitro* (Dennehy et al. 2006; Greco et al. 2006; Salsbury et al. 2008; Sanchez et al. 2008; Codreanu et al. 2009; Marino and Gladyshev 2010b). Early indications based on a functional site profiling and electrostatic analysis of ~50 sites known to be modified to sulfenic acid are that (1) the solvent exposure of modifiable cysteines is not different from the average cysteine, (2) charged residues are underrepresented in the structure near these modifiable sites, and (3) threonine and other polar residues appear to exert a large influence on the cysteine pK_a, thus playing a role in promoting the reactivity of these cysteinyl residues toward oxidants (Salsbury et al. 2008). In the case of the highly peroxide-reactive active site of the widespread and diverse Prx family of proteins, structural and computational analyses have been more informative regarding the active site features which promote peroxide reactivity of these proteins, described in more detail below.

4.3 The Fate of Sulfenic Acids in Proteins

Sulfenic acids are reactive species in solution and in many proteins. In some proteins, however, they are quite stable (e.g., in the cysteine-based peroxidases NADH peroxidase and human PrxVI), and this must be ascribed to features of the protein microenvironment including the presence of stabilizing hydrogen-bonding partners and the lack of nearby thiol groups (Choi et al. 1998; Claiborne

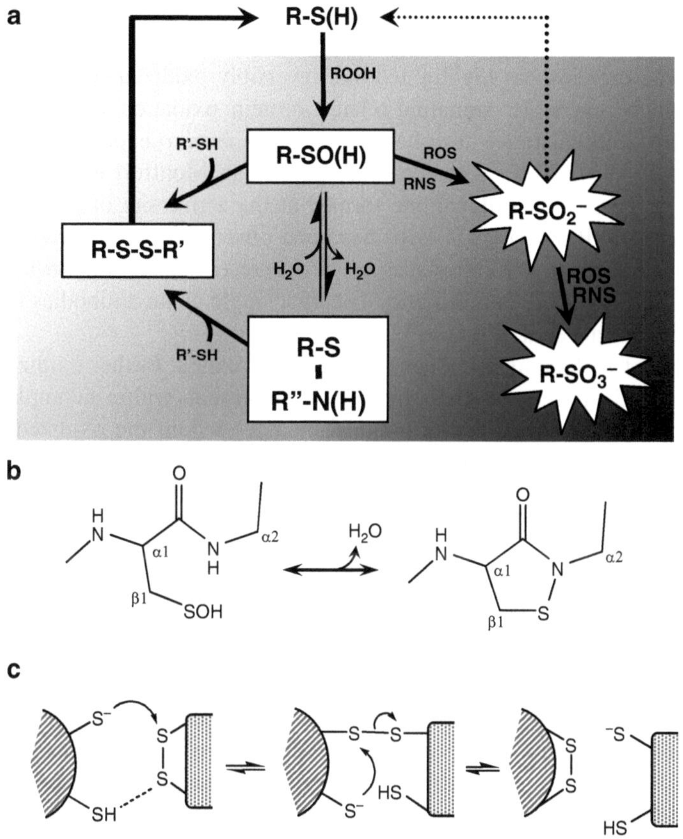

Fig. 4.1 Major redox forms of oxidized cysteinyl residues. (**a**) Sulfenic acids are formed by reactions of cysteine thiolates with oxidants, and particularly hydroperoxides (ROOH). In proteins, sulfenic acids can condense with thiol groups to form intramolecular, intermolecular or mixed disulfides (*left*), undergo a reversible condensation with a proximal amine or amide nitrogen to form a sulfenamide (*middle*), or become further oxidized to sulfinic and perhaps sulfonic acids (*right*). Species in the *middle* and *left* are readily reversible by cellular reductants back to the thiol state. Sulfonic acids are biologically irreversible, as are sulfinic acids, except in the case of the unique repair pathway catalyzed by sulfiredoxin that exists to recover hyperoxidized peroxiredoxins (*dotted line*). (**b**) A cyclic sulfenamide can be formed within protein tyrosine phosphatases like PTP1B through condensation of the cysteine sulfenic acid (*showing* α_1 *and* β_1 *carbons*) with the neighboring main-chain nitrogen of the next residue, serine in PTP1B (side chain attached to the α_2 carbon not shown). (**c**) Thiol-disulfide interchange reactions transfer electrons between dithiol/disulfide centers, effectively transferring the disulfide bond; this mechanism is used by common physiological reductases like thioredoxin and glutaredoxin

et al. 2001). Proximity of a protein sulfenic acid to thiol groups from small molecules, from other proteins, or from other parts of the same protein leads to disulfide bond formation (with loss of one water molecule). This may be the most common fate of cellular protein sulfenic acids, at least when exposed to relatively low levels of ROS where much of the oxidation is reversible (Fig. 4.1). Recent recognition that a sulfenamide (also known as sulfenyl-amide) can form between an amine or amide nitrogen through a condensation reaction with the sulfenic acid in some proteins has highlighted this reversibly oxidized species as another way in which reversible, signaling-related protein oxidation may be sustained (Fig. 4.1a, b). Sulfenamides have been observed in *in vitro* experiments in several proteins (Lee et al. 2007; Salmeen et al. 2003; van Montfort et al. 2003; Yang et al. 2007). The cyclic sulfenamide formed at the active site of protein tyrosine phosphatase 1B (PTP1B) (Fig. 4.1b) has been observed crystallographically and proof of its occurrence and importance *in vivo* has recently been demonstrated using endogenously produced, conformation-specific, single chain antibodies (known as intrabodies) (Haque et al. 2011).

Disulfide or sulfenamide linkages (or those of related, further oxidized forms) have the potential to serve as cross-links between proteins within a complex, and/or to cause or stabilize large conformational changes within the oxidized proteins. These changes may enhance protein stability or make the downstream product(s) of the initial sulfenic acid more accessible to reductants. Disulfide bonds may also readily migrate to other sites containing free thiols through thiol-disulfide interchange reactions, the same type of thiol-disulfide reactivity by which dithiol-containing reductants reductively recycle oxidized proteins (Fig. 4.1c). Thus, any subsequently-generated disulfide bond(s), as downstream products of sulfenic acid formation, may be quite distant from the initial site where the sulfenic acid is formed, including being located on a different protein. This type of disulfide exchange occurs in *Saccharomyces cerevisiae*, for example, during the transfer of oxidizing equivalents from peroxides to Yap1, a stress-activated transcriptional regulator, which occurs via a sulfenic acid initially generated on a reactive cysteine within the Gpx3 glutathione peroxidase homologue Orp1 (i.e., oxidant resistance protein 1) (Delaunay et al. 2002).

Further oxidation of a sulfenic acid as is promoted by high levels of oxidants first generates the sulfinic acid ($R\text{-}SO_2^-$), and then subsequently the sulfonic acid ($R\text{-}SO_3^-$). Nearly all such "hyperoxidized" cysteine residues in proteins are irreversibly damaged in that the thiol group cannot be regenerated in biological systems. The exception to this is in the sulfinic acid-containing Prxs, where a specialized repair protein known as sulfiredoxin can catalyze the ATP-dependent regeneration of the Prx active site thiol (discussed further below) (Biteau et al. 2003; Jönsson et al. 2008b).

In vivo, the fate of sulfenic acids generated in cellular proteins will be highly dependent on the nature and levels of oxidants and reductants within the environment of the protein, as well as the proximity of the sulfenic acid to potential reactive groups. For many proteins that are rapidly oxidized by peroxide, the lifetime of the sulfenic acid that is formed is governed by how fast it forms a disulfide or becomes reduced by a relevant protein (e.g., thioredoxin or glutaredoxin) or small molecule (e.g., glutathione or cysteine) reductant. This allows for the RSOH lifetime to be tuned for functional purposes, with for example, oxidized forms of certain transcriptional regulators purposefully experiencing rather slow turnover so as to optimize the oxidation signal that gives rise to transcriptional activation or derepression (Åslund et al. 1999; Lee et al. 2007; Parsonage et al. 2005; Horta et al. 2010; Reeves et al. 2011).

4.4 Approaches to Assess Cysteine Oxidation Within Proteins

The most straightforward assessment of protein oxidation products can be accomplished when the modification is not susceptible to reductant treatment and has little or no propensity to migrate to other locations during the analysis. Thus, the irreversibly oxidized sulfinic and sulfonic acids, as well as sulfinamide (R-S(O)-NH-R′) and sulfonamide (R-S(O_2)-NH-R′) products that are oxidized beyond the sulfenamide state, can be identified by mass spectrometry and in some cases by peptide mapping techniques (Griffiths et al. 2002; Raftery et al. 2001; Shetty and Neubert 2009; Shetty et al. 2007). Sulfinic and sulfonic acid products at the active sites of Prx and PTP1B proteins have also been characterized crystallographically (Sarma et al. 2005; Schröder et al. 2000; van Montfort et al. 2003), and hyperoxidized forms of mammalian Prxs can be recognized by redox state-sensitive antibodies that do not cross-react with the thiol-containing protein (Woo et al. 2003). Hyperoxidation of Prxs is also often detected by a change in the protein's isoelectric point (due to the more acidic pK_a of the hyperoxidized forms) that causes a shifted mobility in two-dimensional gel electrophoresis (Mitsumoto et al. 2001; Rabilloud et al. 2002; Yang et al. 2002). For human Prx I, the hyperoxidized form is separable from the thiol or sulfenic acid containing forms by reverse-phase HPLC, as was used to quantitatively assess sulfiredoxin activity (Jönsson et al. 2008a). As a point of caution, however, these hyperoxidized protein sites may in some cases be generated from sulfenic acids that become exposed and air oxidized during sample workup and/or analysis (Ellis and Poole 1997a; Fuangthong and Helmann 2002; Shetty et al. 2007). For this reason, chemical trapping of sulfenic acids as early as possible in the analytical protocol used is recommended to minimize such artifacts (Ellis and Poole 1997a; Klomsiri et al. 2010).

In pure proteins amenable to NMR analyses, sulfinic and sulfonic acids can in principle be distinguished from thiols and sulfenic acids using ^{13}C-NMR, although the sulfinic and sulfonic acids are expected to be quite similar to one another in $^{13}C^{\beta}$ chemical shift (Crane et al. 1997). Such analyses were applied to assess the redox states of cysteine within various forms of NADH peroxidase from *Enterococcus faecalis*.

In pure proteins of interest, disulfide-bonded cysteine residues may be accurately assessed as long as great caution is taken to rapidly alkylate free thiols prior to and during denaturation to prevent disulfide "scrambling," which is more of a problem in alkaline pH conditions (Hansen and Winther 2009; Chouchani et al. 2011). Traditional peptide mapping with Edman degradation or MS analyses to sequence peptides, particularly in conjunction with comparisons between oxidized and reduced forms, can often be used to pinpoint the location of the disulfide bond of interest (Baker and Poole 2003; Poole 1996). Methods for detecting sulfenic acids in pure proteins have been more of a challenge to develop, particularly for proteins where this modification is not verifiable by crystallographic or NMR approaches. Because of the transient nature of this species, particularly when proteins are denatured, chemical trapping approaches have been developed that rely on either the nucleophilic or electrophilic properties of this species. The electrophilic compound NBD-Cl (7-chloro-4-nitrobenzo-2-oxa-1,3-diazole) was found to be highly useful in this regard. While this reagent reacts with both thiols and sulfenic acids, the products of these two are readily distinguished by their spectral signatures (with thiol- and sulfenic acid-generated adducts giving absorbance maxima of 420 and 347 nm, respectively), and by the higher mass (by 16 atomic mass units) of the sulfenic acid-derived product (Ellis and Poole 1997a; Poole and Ellis 2002). While this approach has been useful for identifying sulfenic acids in a number of systems (Kettenhofen and Wood 2010; Poole 2008), the nucleophilic compound dimedone (5,5-dimethyl-1,3-cyclohexanedione), which forms a stable thioether bond with cysteine residues containing sulfenic acids, has found wider application. This reagent was used in early experiments with glyceraldehyde-3-phosphate dehydrogenase (GAPDH), taking advantage of the ^{14}C-labeled dimedone that was available at the time; incorporation of the radiolabel was observed with the oxidized, but not reduced, form of the enzyme (Benitez and Allison 1974). As radiolabeled dimedone is no longer commercially available, most subsequent studies using dimedone have relied on the increase in mass (by 140 Da) observed upon covalent attachment of dimedone to proteins (Ellis and Poole 1997a; Kettenhofen and Wood 2010; Willett and Copley 1996). In the past several years, anti-dimedone antibodies have also become available, allowing for yet another way to detect incorporation of this compound into proteins (Seo and Carroll 2009).

For the most part, the methods described above are quite useful when assessing cysteine oxidation within pure proteins, but not for more complex mixtures of proteins as are typically encountered in proteomic experiments. Most strategies for detecting thiol modifications across large proteomes are designed to detect reversible thiol oxidation in general, not distinguishing between disulfides, sulfenic acids and S-nitrosothiols, for example (Leichert and Jakob 2006; Ying et al. 2007; Chouchani et al. 2011; Victor et al. 2012). Recent approaches have, however, been developed to specifically label and detect sulfenic acid formation across proteomes using conjugated derivatives based on dimedone that possess fluorescent or affinity labels (Poole et al. 2005, 2007; Charles et al. 2007; Leonard and Carroll 2011). A major advantage of specifically targeting sulfenic acid is that it is the direct protein product of cysteine modification by peroxides so that one can be confident that the site where the oxidation chemistry is initiated is revealed; this question is more difficult to resolve when subsequent products like disulfide bonds are analyzed. These approaches are beginning to reveal a large and growing list of oxidizable proteins that are important to redox regulation and oxidative stress across a wide range of biological systems.

4.5 Cysteine Sulfenic Acids as Catalytic and Regulatory Components of Enzymes and Transcriptional Regulators

4.5.1 Regulatory Sulfenic Acids in Cellular Proteins

The clearest examples of signaling-related sulfenic acid generation are in the above-mentioned bacterial transcriptional regulators OhrR and OxyR, which sense the presence of peroxides through their reactive cysteine residues, transducing the signal directly to modulate transcriptional activity. Both *Escherichia coli* OxyR and *Xanthomonas campestris* OhrR form intramolecular disulfide bonds upon oxidation; in OxyR, the "locking in" of the significant, oxidation-induced conformational change by disulfide bond formation has been referred to as "fold editing," altering the protein's structural and functional properties (Choi et al. 2001; Panmanee et al. 2006). Interestingly, *Bacillus subtilis* OhrR only has the one cysteine residue that forms sulfenic acid upon exposure to organic peroxides. Without a disulfide-bonding partner within the protein, this sensor site forms either a sulfenamide intermediate or a mixed disulfide with a low molecular weight thiol such as cysteine or the recently-discovered bacillithiol, promoting derepression and suppressing hyperoxidation, which would interfere with reductive recycling of this transcription factor (Lee et al. 2007; Soonsanga et al. 2008; Newton et al. 2009).

Protein sulfenic acids as regulatory features of thiol-dependent enzymes have been well established for many years, particularly through the work of Allison and co-workers, which demonstrated the oxidant sensitivity of the protease papain and the metabolic enzyme GAPDH (Allison 1976). Like papain, another related cysteine protease, cathepsin K, showed evidence of sulfenic acid formation after treatment with S-nitrosoglutathione, suggesting a role for this inhibitory modification in NO-mediated bone resorption (Percival et al. 1999). Cathepsins B and L were also shown to lose 50–75% of their activity after treatment with low levels (10 μM) of H_2O_2 (Headlam et al. 2006). As mentioned above, protein tyrosine phosphatases like PTP1B and PTPα, possess redox-sensitive cysteinyl residues at their active sites (Yang et al. 2007; Tonks 2005); *in vivo* sulfenic acid formation at the active sites of other cysteine-dependent phosphatases like SHP1 and SHP2, DEP1 and PTEN has also been observed after treatment of T-cells with antigen (Michalek et al. 2007), endothelial cells with vascular endothelial growth factor (VEGF) (Oshikawa et al. 2010), and HEK 293 cells with tumor necrosis factor α, respectively (Michalek et al. 2007; Oshikawa et al. 2010; Nelson et al. 2010). Some of these same phosphatases were also shown to be oxidized *in vivo* to sulfenic acid by EGF treatment of A431 cells (Paulsen et al. 2011). Enzymes which use cysteine at the active site to defend against cellular oxidants are also prone to oxidative regulation, and these are discussed in the next section.

Reactive cysteine residues adjacent to or even remote from the active site or binding site have also been shown to modulate the function of a wide variety of signaling-relevant proteins through sulfenic acid formation. Like with phosphorylation, whether oxidation is activating or inhibitory varies with the protein, the site of the modification, and potentially the nature of the downstream oxidation product(s). In the absence of a localized PrxII to remove nearby H_2O_2, the murine VEGF receptor 2 (VEGFR2) is subject to oxidative inhibition of its tyrosine kinase activity through the oxidation of Cys1206 near the C-terminus which forms an intramolecular disulfide with Cys1199 (Kang et al. 2011). In contrast, direct oxidation of epidermal growth factor receptor (EGFR) at Cys797 by H_2O_2 elicited by EGF treatment of A431 cells enhances its tyrosine kinase activity (Paulsen et al. 2011). Another important sulfenic acid modification target of VEGF-induced ROS also appears to be the scaffolding protein IQGAP, which is oxidized preferentially at the leading edge of actively migrating cells (Kaplan et al. 2011). As IQGAP plays a pivotal role in regulating cytoskeletal elements and facilitating localized ROS production, this localized modification potentially serves to promote directional endothelial cell migration. Increasingly, non-receptor kinases are also being identified which undergo oxidative regulation *in vivo* in ways that modulate their activities (reviewed in Cross and Templeton 2006; Klomsiri et al. 2011; Poole and Nelson 2008). A recent study demonstrated that sulfenic acid and disulfide bond generation through platelet derived growth factor (PDGF)-induced ROS occurs in and inhibits serine-threonine protein kinase Akt2 but not Akt1, leading to changes in glucose uptake (Wani et al. 2011b). In a separate study, Akt2 knockout MEFs expressing recombinant Akt2 proteins (wild type and an oxidation insensitive mutant) were also used to demonstrate a role for Akt2 oxidation in promoting cell migration (Wani et al. 2011a). As an example of oxidative activation, disulfide bond formation through initial sulfenic acid formation in cGMP-dependent protein kinase (PKGIα) upon perfusion of rat hearts with H_2O_2 led to intersubunit disulfide bond formation, enzyme relocalization, and an increase in kinase activity of this protein (Burgoyne et al. 2007). Src is another very important nonreceptor, membrane-associated tyrosine kinase which is activated by oxidation at two critical cysteine residues which may form an intrasubunit disulfide bond, although direct evidence for the latter is still lacking; interestingly, *in vitro* oxidation at a different site in this and other Src-family kinases is inhibitory, suggesting that the redox regulation of this class of enzymes may be quite complex (Giannoni et al. 2010). As more methods become available to reveal sites of SOH formation, the list of proteins susceptible to sulfenic acid formation will continue to grow and more specific roles for oxidation to SOH in cellular processes will undoubtedly be established.

4.5.2 *Sulfenic Acids in Catalysis and Regulation of Oxidative Defense Enzymes, Focusing on Prxs*

As examples of sulfenic acid-generating proteins, cysteine-based peroxidases have been much studied, in the 1980s and 1990s with the NADH peroxidase from lactic acid bacteria (Crane et al. 1997; Poole and Claiborne 1989; Yeh et al. 1996)

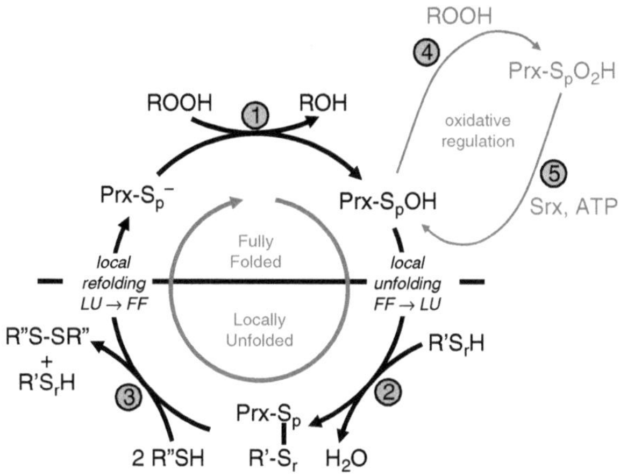

Fig. 4.2 Prx catalytic cycle. The first step (1) of the Prx catalytic cycle is conserved across all Prxs and involves the thiolate form of the C_P (S_P^-) reacting with peroxide to form cysteine sulfenic acid (S_POH). In step 2, the sulfenic acid is resolved through the formation of a disulfide with a resolving thiol (S_rH) from the same subunit, from another Prx subunit, or from an exogenous molecule. In step 3, the Prx is recycled back to the reduced form through a disulfide exchange reaction with Trx or a similar protein or domain. Local unfolding occurs in the C_P-loop (and in 2-Cys Prxs in the region of the protein including C_r) during step 2 so that the disulfide bond can form. Overoxidation of the C_P (step 4) and reduction of the Cys-S_PO_2H by sulfiredoxin (Srx; step 5) depict redox regulation and repair occurring in eukaryotic sensitive 2-Cys Prxs (Adapted from Hall et al. 2009a, b)

and more recently with the widespread Prx proteins (Ellis and Poole 1997a, b; Baker and Poole 2003; Hugo et al. 2009). In these proteins, as well as Ohr (a bacterial organic hydroperoxide resistance enzyme with no structural relationship to Prxs) and non-selenium glutathione peroxidase homologues (Lesniak et al. 2002; Flohé et al. 2011), the peroxidatic cysteine (the C_P) is oxidized to the sulfenic acid by the hydroperoxide substrate, then forms a disulfide bond through condensation with a "resolving" cysteine (the C_r) that is either also present in the peroxidase or provided by another protein or small molecule (Fig. 4.2). The return of the oxidized protein and C_P to the activated thiol state typically involves a thiol-containing reductant like thioredoxin (Trx). Prxs are good models for cysteine chemistry because, with the possible exception of sulfenamides, all of the examples described above of cysteine chemical states (including S-glutathionylated and S-nitrosylated species) are observed in these proteins (Fang et al. 2007; Chae et al. 2012). It may be no coincidence that one group of Prxs (the Prx1/AhpC group) would not be able to form the type of cyclic sulfenamide observed in PTP1B because the residue following the C_P is almost always a proline (99.5% of the members identified from the 2008 release of GenBank) (Nelson et al. 2011).

Any sulfenic acid generated in an active site during catalysis may also impart oxidant sensitivity to that protein through the potential to form the typically irreversible sulfinic and sulfonic acids through further reactions with oxidants. Findings alluded to above showed that some eukaryotic Prxs are indeed sensitive to this type of inactivation, whereas other eukaryotic Prxs and most bacterial Prxs are much less sensitive (Hall et al. 2009a; Wood et al. 2003). The fact that an ATP-dependent repair system (the sulfiredoxins) has evolved to reduce and restore the activity of these easily hyperoxidized Prxs highlights the importance of this redox switch in biology. The nature of when and where Prx hyperoxidation plays an important role is less clear, but it may be critical to the ability of H_2O_2 to serve as a transiently-generated chemical signal or second messenger within cells. One proposal for how this might work is the "floodgate hypothesis" (Hall et al. 2009a; Wood et al. 2003). It suggests that Prxs are generally reduced and abundantly present to rapidly detoxify peroxides when the cell is in a "resting" state, but that during an oxidative signaling event, H_2O_2, superoxide and/or lipid hydroperoxide is locally generated, and the high concentrations at the site of generation lead to inactivation of the nearby Prx molecules before reductants are able to reduce them. This would turn off flux through these enzymes and, in a positive feedback loop, contribute to heightened oxidant concentrations. The built-up oxidants would in turn have the effect of promoting the oxidation of other proteins in the vicinity that would normally not be able to compete with reactive peroxidases. In this way the oxidation signal could influence susceptible sites in a wide range of signaling-relevant proteins. An alternative, but not mutually exclusive, model has also been suggested by the propensity of hyperoxidized Prx proteins in yeast to form large aggregates with chaperone activity; in these organisms, they act like heat shock proteins in helping to ameliorate protein damage due to heat stress (Jang et al. 2004). Such chaperone activity may carry over to higher organisms as well, although this has been a difficult question to address (Moon et al. 2005; Kim et al. 2009; Park et al. 2011). Oxidized Prxs may also behave similarly to the Orp1 protein described above, transmitting the original oxidation signal from the peroxide to other interacting proteins through a series of thiol-disulfide exchange interactions (D'Autreaux and Toledano 2007; Hall et al. 2009a).

The cellular conditions under which Prx hyperoxidation has a significant impact on defense and/or signaling are still being investigated. Direct exposure of cells to low, chronic levels of H_2O_2 can be quite informative in this regard (Phalen et al. 2006; Yang et al. 2002). Recent evidence shows that in mammalian cells, sensitive Prxs are hyperoxidized at even very low levels of H_2O_2 ($<10\ \mu M$) (Baek et al. 2012). These low H_2O_2 levels were not toxic to wild-type cells, but were toxic to cells that were missing sulfiredoxin, verifying that the sulfiredoxin-catalyzed repair process is of critical importance to maintaining the redox homeostasis under these conditions. In a similar vein, studies in yeast showed that normal aging results in an accumulation of the hyperoxidized form of the Prx Tsa1, but that sulfiredoxin expression was upregulated by caloric restriction and blocked that accumulation so as to maintain an effective oxidant defense system (Molin et al. 2011). Strikingly, the addition of an extra copy of the sulfiredoxin gene mimicked the life-extending effects of caloric restriction.

4.6 Active Site Features and Oligomeric Forms Impacting Catalytic Function and Oxidant Hypersensitivity in Prxs

4.6.1 Roles for Conserved Residues in the Specificity and High Catalytic Efficiency of Prxs Toward Hydroperoxide Substrates

All Prxs share a highly conserved active site structure with a nearly universal PXXXTXXC$_P$ sequence and a conserved Arg that is distant in sequence but nearby in the three-dimensional fold. These peroxidases have diverged from a common ancestor of the Trx-fold superfamily that includes Trxs and glutaredoxins (Grxs); comparisons of sequences and structures indicate that the TXXC active site in Prxs directly corresponds to the CXXC active sites in Trxs and Grxs (Fomenko and Gladyshev 2003; Copley et al. 2004). Interestingly, the conserved Arg of the Prxs corresponds in position to a conserved *cis*-Pro of the Trxs and Grxs which, in these proteins, is particularly important for accommodating large protein substrates in the active sites (Copley et al. 2004).

Using a standardized nomenclature for the secondary structural elements (α-helices and β-strands) of Prxs (Hall et al. 2011), the central 5-stranded beta sheet with an initial beta hairpin has two helices on the "back" side (α_1 and α_4) and three on the "front" where the active site is situated (at the N-terminus of helix α_2) (Fig. 4.3). This arrangement of secondary structural elements has been described as a cradle, where the C$_P$-containing central α_2 helix (the "baby") rests upon the β-sheet mattress, flanked by α_3 and α_5 helices that make up the sides of the cradle (Hall et al. 2009c, 2011). This structural arrangement provides stability to the α_2 helix and accommodates the large, though localized, conformational changes which take place during the course of catalysis – conformational changes which vary between Prxs from different subfamilies (see below). These conformational changes are necessary to accomplish the multi-step catalytic cycle of Prxs. At the beginning of the catalytic cycle, the enzyme is in its "fully folded" (FF) state with the deprotonated C$_P$ (thiolate) prepared to encounter the peroxide substrate in the specialized environment of the FF Prx active site (Fig. 4.2). After the oxidation of C$_P$, the sulfenic acid must flip out of the protected pocket of the active site to reach and react with the thiol of the C$_R$, from wherever it comes. This requires an unfolding of the C$_P$-containing loop and helix to generate the "locally unfolded" (LU) state. While details vary somewhat with the different Prxs (Hall et al. 2009c, 2011; Perkins et al. 2012), approximately one turn of the α_2 helix is unraveled, and a disulfide bond is formed between C$_P$ and C$_R$, locking in the LU state. Reductants can then, in principle, attack either sulfur of the disulfide bond to finish the recycling process. However, several studies focused on the Prx1/AhpC group that includes four of the six human Prxs have demonstrated that the physiological reductants attack at the C$_R$ sulfur in these proteins (Budde et al. 2003; Jönsson et al. 2007). The enzyme is reactivated with the refolding of the segment with the regenerated C$_P$ thiol to form the active site in the FF state (Fig. 4.2).

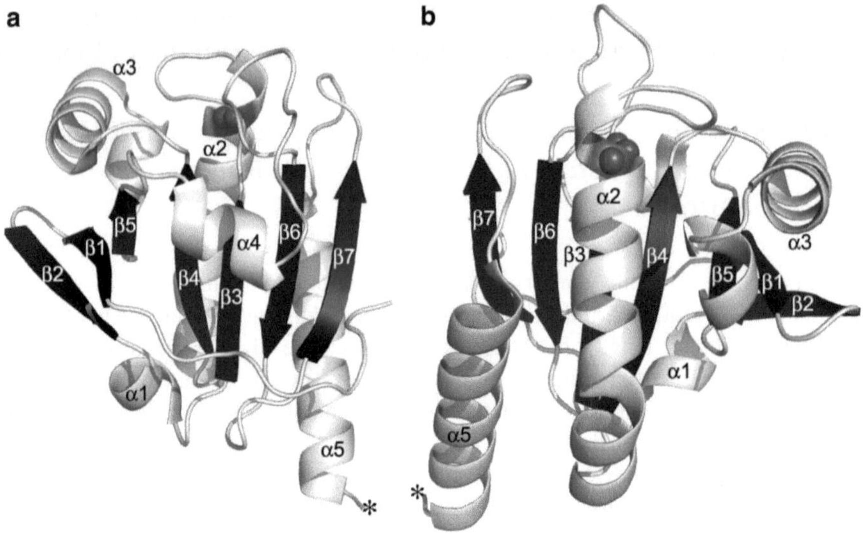

Fig. 4.3 Core secondary structure of Prx proteins. (**a**) Shown is the back view of a representative monomer of a fully folded (FF) Prx (*Salmonella typhimurium* AhpC; PDB entry 1N8J) with the conserved structural elements labeled and β-sheets shown in *black*. The C_p is located in the first turn of helix α_2 and is shown as *spheres*. An *asterisk* represents the position of attachment of the rest of the C-terminal tail that has been hidden in this view (normally absent in other subfamilies); the C-terminus of Prx1 and Prx6 subfamily members extends from here to make extensive interactions with an adjacent monomer. (**b**) The same monomer viewed from the front side of the cradle (related to A by a ~180° rotation around a vertical axis). This and other molecule model figures were prepared using Pymol (DeLano 2002)

Because the thiolate is a much stronger nucleophile than the protonated thiol group, one of the widely recognized features of reactive cysteine residues, including those at the active site of Prxs, is their lowered pK_a relative to that of an unperturbed cysteine ($pK_a \sim 8.5$). For Prxs, pK_a values have been measured by several approaches. A simple one is to take absorbance measurements at 240 nm over a pH range, with the deprotonation event being measureable due to the higher extinction coefficient of the thiolate (Benesch et al. 1955; Kortemme et al. 1996; Roberts et al. 2005). An alternative approach is to measure the pH dependence of catalysis in competition with heme-containing peroxidases; this has the advantage of yielding a function-relevant pK_a that likely specifically reflects the C_p protonation in the FF state (Nelson et al. 2008; Ogusucu et al. 2007). Direct determination of the pH dependence of alkylation rates was not particularly useful, at least for bacterial AhpC, because the C_p in the FF active site is apparently unable to react with iodoacetamide or its derivatives; this means that the pK_a value measured would reflect the C_p in the LU form of the protein (Nelson et al. 2008). Published pK_a values for the C_p residue in the FF state of Prxs have ranged from 5.2 to 6.3 (Ogusucu et al. 2007; Trujillo et al. 2007; Nelson et al. 2008; Hugo et al. 2009; Manta et al. 2009; Horta et al. 2010; Reeves et al. 2011). Using the average pK_a

value of 5.75, about 94% of the C_p residues would be deprotonated at pH 7 as compared with only ~3% if the thiol pK_a were unperturbed. It should be noted, however, that the nucleophilicity of the thiolate decreases as the pK_a is lowered, so it would be a disadvantage to decrease the C_p's pK_a much below the 5.75 value.

During catalysis by Prxs, the two oxygen atoms of the hydroperoxide substrate move apart, with the bond between them ultimately breaking to form the hydroxide/alkoxide and the C_p-sulfenic acid products. The reaction is a classic in-line S_N2 attack by the thiolate nucleophile on the terminal oxygen of ROOH (called O_A here) that weakens the peroxyl -O-O- bond and generates a planar intermediate around O_A, followed by inversion of stereochemistry, bond breakage, and departure of the leaving group. Many of the various ligands found bound to Prxs in crystal structures in fact possess one or two oxygen atoms bound in a way which could mimic the peroxyl oxygens of the substrates or products (Hall et al. 2010). Analysis of these structures, as well as one Michaelis complex with H_2O_2 (PDB identifier 3a2v) and structures with water molecules bound in the active site, together paint a picture of the catalytic process. Interestingly, with all of the structures taken together, the positions of the oxygen atoms from the various structures are all found within the active site distributed along a narrow track (Fig. 4.4a). We interpret these positions as representing a thermodynamic well that stabilizes the positions occupied by substrate oxygens during the bond breaking process. In this way, catalysis can simplistically be viewed as the sliding of the terminal oxygen (O_A) along the S-O_A-O_B track, away from O_B and toward the attacking C_p sulfur, ultimately transferring O_A to the sulfur and releasing the alkoxide (or hydroxide) product (Fig. 4.4b). Close inspection of the hydrogen bonding interactions around the active site indicates that some of the interaction geometries will be stronger with the O_A located partway between O_B and the C_p sulfur, as would be consistent with preferential binding of the enzyme to the transition state (Hall et al. 2010). Such evidence for transition state stabilization fits well with the concept that enzymes accelerate chemical reactions by lowering the activation energy through such preferential interactions with the high-energy transition state, as originally proposed many years ago by Linus Pauling (1946).

Dissection of the hydrogen-bonding and electrostatic interactions around the active site also clarifies specific roles for the active site side chains and backbone interactions, as well as the α_2 helix, not only in activating the C_p sulfur, but also in positioning and activating the incoming peroxide substrate (Fig. 4.4b; Hall et al. 2010). An independent mechanistic analysis of substrate recognition and thiolate activation within the Prx active site offered similar conclusions in this regard, although some aspects of the proposed interactions were different (Ferrer-Sueta et al. 2011). Together, these analyses support the following specific roles for the conserved Prx residues in promoting catalysis: (1) the threonine hydroxyl (on rare occasion substituted by a serine hydroxyl) positions and activates the protonated O_A atom of the incoming peroxide substrate; (2) the arginine positions and interacts with both the active site thiolate and the peroxide substrate through both charge and hydrogen-bonding interactions; (3) two backbone NH groups from the first turn of the α_2 helix help position and activate the peroxide O_A and O_B atoms; and (4) the proline shields the active site and positions the backbone of the next two residues

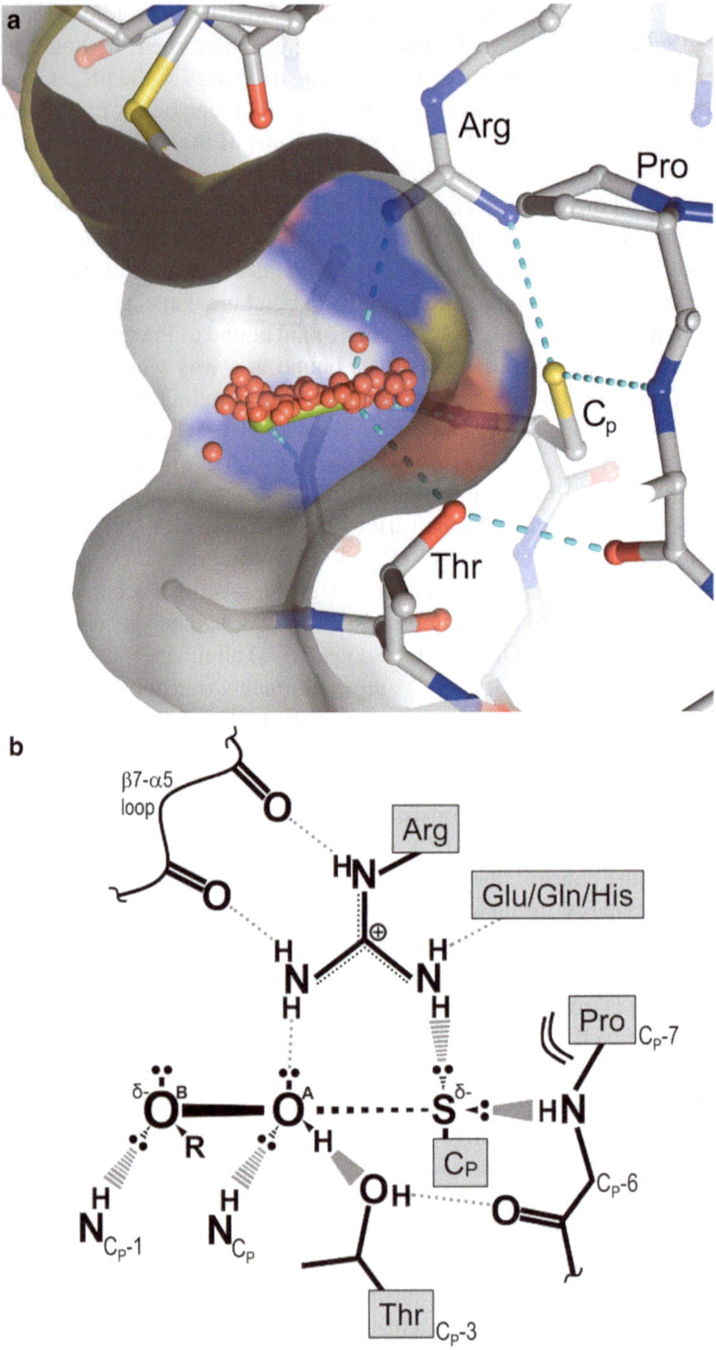

Fig. 4.4 (continued)

to provide additional hydrogen bonding interactions with C_P, and indirectly (via the threonine hydroxyl) with the peroxide O_A (Hall et al. 2010). These analyses have brought considerable insight into the origins of the specificity of Prxs – in other words, how the extremely high catalytic power of the Prx active site toward hydroperoxides is generated without a parallel enhancement of reactivity toward alkylating agents and other oxidants such as chloramines (Peskin et al. 2007; Nelson et al. 2008).

4.6.2 The Structural Basis for Oxidation Sensitivity

As alluded to above, the most abundant eukaryotic Prxs are sensitive to inactivation resulting from hyperoxidation of the C_P sulfenic acid when a second molecule of peroxide outcompetes the resolution reaction and generates the sulfinic acid at C_P. Some bacterial and cyanobacterial Prxs have also exhibited a relatively high degree of oxidant sensitivity (Baker and Poole 2003; Pascual et al. 2010). It has also been seen that a third molecule of peroxide can, in some cases, further oxidize this site to the sulfonic acid form, a redox form which cannot be repaired by sulfiredoxin (Sarma et al. 2005; Seo et al. 2009). The structural basis for this sensitivity has been explored, in particular, within the "typical 2-Cys" (Prx1/AhpC) group of Prxs (see below), where C_R resides near the C-terminus of another monomer within the dimeric (or oligomeric) protein. Comparisons of the structures of representative sensitive versus robust Prxs revealed specific structural features, especially a conserved YF motif within the C-terminal α-helix, correlated with sensitivity, and as described below, provided a rationale for the sensitivity (Wood et al. 2003).

Because the Prx catalytic cycle requires local unfolding of the active site loop and helix to allow C_P-C_R disulfide bond formation, excessive stabilization of the FF form will hinder the resolution reaction and promote hyperoxidation of the longer-lived sulfenic acid within the Prx active site. What was noted by Wood et al. (2003) is that the C-terminal α-helix present in sensitive Prxs packs against the active site C_P-loop to stabilize of the FF form (Fig. 4.5, upper left). In contrast, the architecture and mobility of the protein in the vicinity of the active site in the bacterial AhpC protein

Fig. 4.4 (continued) Interactions in the Prx active site. (**a**) Cutaway view of the active site pocket of a peroxiredoxin (PDB structure 3A2V) with bound H_2O_2 (*green*) overlaid with water molecules (*red spheres*) taken from many other peroxiredoxin structures. Surrounding atoms including the C_p-loop and conserved arginine are shown as *sticks* and *colored* by atom type and select hydrogen bonds are shown as *dashed cyan lines*. (**b**) Schematic drawing of the peroxide-bound active site emphasizing the hydrogen-bonding interactions present. The in-line geometry for the C_p thiolate to attack O_A for the peroxidatic S_N2 reaction (*bold broken line*) and key hydrogen-bonding interactions (*pale broken and dotted lines*) are indicated. The view is as in (**a**). Backbone atoms in the C_p-loop are identified by their residue position relative to C_p. The negative charge on the C_p-thiolate is indicated. Panel **b** is reproduced with permission from the Journal of Molecular Biology (Hall et al. 2010)

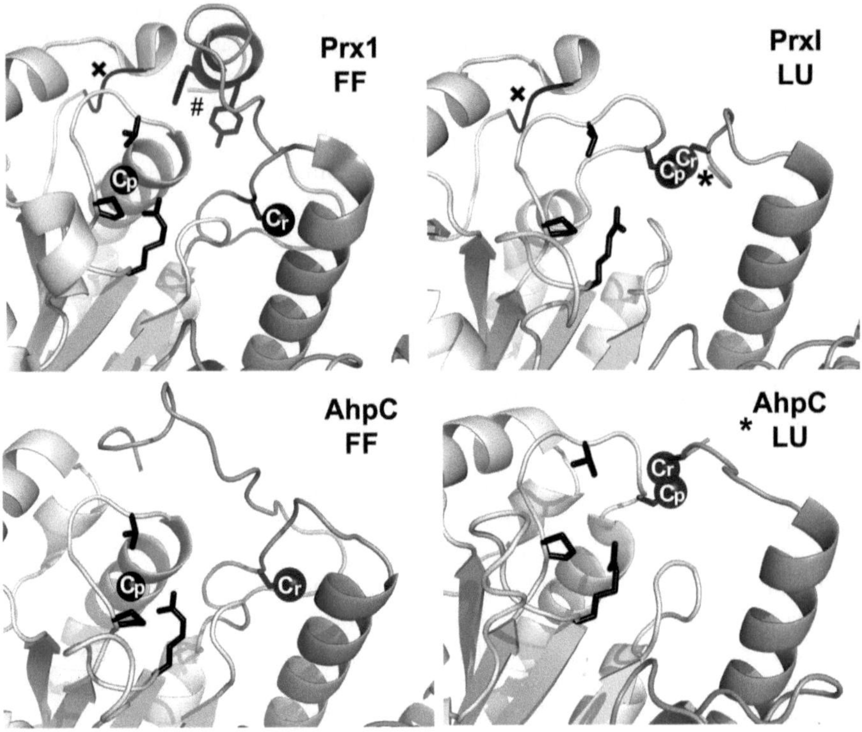

Fig. 4.5 Structural differences between robust and sensitive Prxs identified by Wood et al. (2003). *Ribbon* diagrams of the peroxidatic active site region of sensitive (*top, Rattus norvegicus* PrxI) and robust (*bottom, S. typhimurium* AhpC) structures in the fully folded (FF) and the locally unfolded (LU) conformations. Side chains for C_p and C_r cysteines (with balls emphasizing the sulfur positions), the four universally conserved active site residues (*black*), the YF motif of PrxI (*# in upper left panel*), and the main-chain segment containing the conserved GGLG motif of PrxI (*top panels in black and marked with an "x"*) are shown. The partner subunit contributing the C_r is depicted in a *darker shade of gray*. Structures shown are: *top left*, FF, C52S form of sensitive PrxI (PDB entry 2Z9S); *top right*, LU, disulfide form of sensitive PrxI (1QQ2); *lower left*, FF, C46S form of *St*AhpC (1N8J); and *lower right*, LU, disulfide form of *St*AhpC (1YEP). *Asterisks* marking the ends of the visible C-termini of the LU conformations indicate that many additional C-terminal residues are disordered. Comparison of the upper and lower FF panels (*left*) provides the most dramatic image of how the two sequence features common to sensitive enzymes result in a major difference in the burial of the C_p-containing helix

is noticeably different (Fig. 4.5, lower left), with a shorter C-terminus and greater mobility in all regions near the C_p (as suggested by the higher crystallographic B-factors in these regions). Local unfolding to allow for disulfide bond formation between C_p and C_r is therefore less hindered by the active site microenvironment in AhpC. This explanation of the origins of sensitivity also emphasizes the important role played by C_r in protecting the Prx C_p from overoxidation (Ellis and Poole 1997b; Trujillo et al. 2006).

Validation of this structural explanation for the heightened "redox sensitivity" of certain Prxs has come from mutational and biochemical experiments conducted with several eukaryotic enzymes. Koo et al. demonstrated that a sensitive Prx from *Schizosaccharomyces pombe* became "robust" toward high peroxide levels upon experiencing either single residue substitutions or small deletions at the C-terminus that disrupted the wild type packing interactions (Koo et al. 2002; Wood et al. 2003). Furthermore, simple swapping, through genetic manipulations, of the C-terminal tails of sensitive and robust Prxs from a single organism, *Schistosoma mansoni*, showed that the sensitivity of the enzymes toward oxidative inactivation tracked with the distinct C-terminal sequences (Sayed and Williams 2004). This has been taken as strong evidence that a "tuning" of the sensitivity of Prxs to hyperoxidation has taken place during evolution to match the properties of the specific enzymes to their biological function(s) (Poole et al. 2004).

4.6.3 Quaternary Structures of Prxs

Most Prxs form dimers or in some cases higher order oligomers that are associations of dimers. Only *E. coli* BCP and related proteins in the same subfamily have been shown to exist as stable monomers (Hall et al. 2011; Reeves et al. 2011). Interestingly, Prxs form two types of dimers (Sarma et al. 2005), and these are characteristic of the subfamily to which they belong (see below). A-type dimers, which arguably represent a more ancestral mode of interaction, are formed through a tip-to-tip association at strands β_1 and β_2 (the β-hairpin) and the loops preceding helices α_2, α_3 and α_4 (Fig. 4.6a). B-type dimers are quite distinct, with head-to-tail interactions at the edges of the β-sheets (through the β_7 strands) that generate an extended, 10-stranded β-sheet (Fig. 4.6b). B-type dimers, which form only in proteins in the Prx1 and Prx6 subfamilies (see below) can also form higher order doughnut-shaped structures (typically decamers, but also dodecamers or octamers) through association through the A-interface (Fig. 4.6c). These dimer-dimer interactions were found to be redox-sensitive in proteins like bacterial AhpC and plant and human Prx1-like proteins, and were seen to promote catalysis through stabilization of the FF active site and through a limitation of the dynamic sampling of side chain conformations that strongly affect the C_p pK_a (Barranco-Medina et al. 2008; Parsonage et al. 2005; Wood et al. 2002; Yuan et al. 2010). Larger aggregates of Prx proteins have also been observed and associated with chaperone activity, as mentioned above (Jang et al. 2004); studies of human PrxII using transmission electron microscopy and 3D reconstruction were consistent with the formation (on a holey carbon support film) of dodecamers of the decameric forms that were large, spherical particles with $M_r \sim 2.62$ MDa and an external diameter of ~20 nm (Meissner et al. 2007).

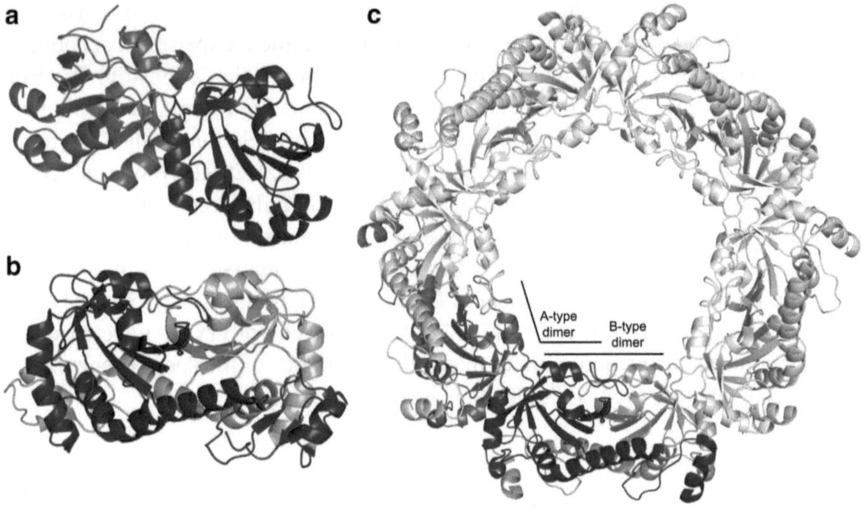

Fig. 4.6 Quaternary structures of Prxs. Many Prxs form dimers either through A-type dimers (**a**) that interact near helix $\alpha3$, or through B-type dimers (**b**) that interact edge to edge to form an extended 10-strand β-sheet. (**c**) Members of the Prx1 and Prx6 subfamily form B-type dimers and many of them form higher order doughnut-like oligomers through the A-interface that can be comprised of 8, 10, or 12 subunits, with the majority being decameric. Structures depicted are for human PrxV (1HD2) (**a**), *Arenicola marina* Prx6 (2V2G) (**b**), and human PrxII (1QMV) (**c**)

4.7 Prx Subfamilies

Prxs are a diverse group of proteins, which can be classified into 6 subfamilies (Prx1, Prx6, Prx5, Tpx, PrxQ, and AhpE) based on sequence similarities, structural homology, and residue conservation in the vicinity of the C_p (Copley et al. 2004; Hall et al. 2011; Nelson et al. 2011). Most species contain multiple Prx proteins from multiple subfamilies; for example, humans have 6 Prxs from 3 subfamilies, *E. coli* has 3 Prxs from 3 subfamilies, *Saccharomyces cerevisiae* has 5 Prxs from 4 subfamilies, and *Arabidopsis thaliana* has 10 Prxs from 4 subfamilies. While the mechanisms for sulfenic acid formation and peroxide reduction, as well as the PXXX(T/S)XXC_p motif, are conserved across all Prxs, other features, including oligomeric state, susceptibility to hyperoxidation, substrate specificity, and the presence and location of the C_r can vary between and within the subfamilies (Hall et al. 2011; Nelson et al. 2011). While the C_r is now known to reside in at least five different locations within the Prx structure (Fig. 4.7a and Lian et al. 2012), in the early 1990s, only a single C_r location was recognized (Chae et al. 1994a). This C_r, near the C-terminus of the Prx, forms a disulfide bond between two chains across the B-type interface (Chae et al. 1994b; Poole 1996) and at that time, the terms "2-Cys" and "1-Cys" were used to distinguish the only two known patterns for resolution of the C_p (Chae et al. 1994a). Later, with the discovery that other Prx proteins formed intrasubunit disulfide bonds using C_r residues in other

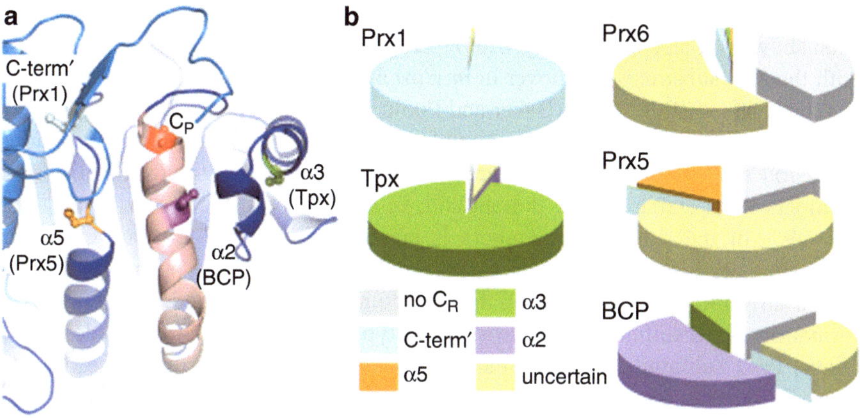

Fig. 4.7 Variability in locations of C_r within and between Prx subfamilies. (**a**) The four prototypical locations for the C_r (labeled by location and the subfamily it is commonly associated with) are mapped onto a composite structure based on *St*AhpC (1YF1). The conserved C_p (*red*) is also shown. The two chains of the B-type dimer are colored in *dark* and *light blue* and helix α2 is colored *pink*. (**b**) Pie charts based on ~3,500 Prx sequences showing the frequency at which the C_r is in a given location for each subfamily (Nelson et al. 2011). Wedges are colored by C_r position consistent with panel A using the same notation: no C_r (*gray*), C-term' (*cyan*), α5 (*orange*), α3 (*green*), α2 (*purple*) and uncertain (*pale yellow*). The exact positions are defined as in Hall et al. (2011). Sequences marked as "uncertain" have additional Cys residues present, but none align exactly with one of the known locations. We suspect most of them are 1-Cys Prxs. This figure is reproduced with permission from the journal Antioxidants and Redox Signaling (Hall et al. 2011)

locations, the term "typical 2-Cys" was used to distinguish the earlier-recognized Prxs containing a C_r in the C-terminus of the protein from "atypical 2-Cys Prxs" with a C_r in any other location. The terms 1-Cys, typical 2-Cys, and atypical 2-Cys are still sometimes used to describe the mechanism by which the protein resolves the sulfenic acid, but they are not particularly useful in distinguishing Prx subfamilies because all subfamilies likely include both 1-Cys and 2-Cys members and because the "atypical 2-Cys" category is not homogeneous (Fig. 4.7b). In all cases, disulfide bond formation requires the localized unfolding of the region surrounding C_P (and often C_r), as described above, promoting the next step of reduction by a thiol-containing disulfide reductase system. The preferred reductant can also vary among and within subfamilies. For many Prxs, Trx serves as the direct reductant, but Grx, AhpF, and small molecule thiols like glutathione have also been shown to reduce particular Prxs.

4.7.1 Prx1 Subfamily

The Prx1 subfamily, also called the Prx1/AhpC subfamily (Nelson et al. 2011), includes highly expressed and widely distributed proteins such as the bacterial alkyl hydroperoxide reductase C protein (AhpC), the tryparedoxin peroxidases, the yeast

TSA1 and 2, and the human PrxI-IV proteins. While members of this subfamily have been shown to react very rapidly with H_2O_2 (catalytic efficiencies of $\sim 10^7\ M^{-1}\ s^{-1}$ with the best substrates), turnover in *in vitro* assays is typically limited by the rate of reduction ($\sim 10^5\ M^{-1}\ s^{-1}$) (Baker and Poole 2003; Horta et al. 2010; Manta et al. 2009; Parsonage et al. 2005). Prx1 subfamily members contain a 40–50 residue extension at the C-terminus compared to the common core structure. All known members form stable B-type dimers and, in the FF form, the resolving cysteine in >99% of the subfamily members is buried within the C-terminal extension approximately 14 Å away from the C_p in the adjacent monomer. Upon oxidation, the region surrounding both the C_p and the C_r become locally unfolded, allowing for the formation of a disulfide bridge across the B-type interface (Fig. 4.5). As described above, Prx1 subfamily members can further associate through the A-interface to form oligomeric, doughnut-like structures of 4–6 dimers. The redox sensitivity of the oligomeric state was first explored in detail for AhpC from *S. typhimurium*. Using analytical ultracentrifugation, reduced AhpC was shown to exist as a stable decamer whereas disulfide bond formation destabilized the decamer, promoting dissociation into dimers at low concentration (Wood et al. 2002). The importance of the decamer interface was further highlighted by studies in which disruption of the decamer interface led to a 100-fold decrease in catalytic efficiency (Parsonage et al. 2005). While eukaryotic Prx1 subfamily members are typically reduced by Trx, many of the bacterial AhpC-like subfamily members have a specific partner flavoprotein reductase, AhpF, encoded within the same operon (Poole 2005). AhpF enzymes have a specially evolved double Trx-fold domain (Roberts et al. 2005; Hall et al. 2009b) that is fused to a C-terminal thioredoxin reductase (TrxR)-like region which, unlike the thioredoxin system, is NADH- rather than NADPH-dependent (Poole and Ellis 1996; Poole et al. 2000).

4.7.2 *Prx6 Subfamily*

The first Prx to be structurally characterized was human PrxVI from which this subfamily takes its name. Prx6 sequences are widely distributed and have sometimes been grouped with Prx1 subfamily members due to their somewhat similar sequences, the formation of B-type dimers, and the presence of a C-terminal extension (Copley et al. 2004). However, in contrast to the Prx1 subfamily, only a subset of Prx6 members forms decamers using the A-type interface. Also, members of this subfamily contain a longer C-terminal extension than observed in the Prx1 subfamily and the large majority appear to be 1-Cys Prxs (i.e., lacking a C_r in the protein) (Nelson et al. 2011). One member of this subfamily, *Aeropyrum pernix* "Tpx", has a cysteine (C213) in the extended C-terminal extension that serves as a C_r (Jeon and Ishikawa 2003), and $\sim$16% of the Prx6 sequences identified in a large-scale bioinformatic analysis of Genbank(nr) contain a cysteine residue in the same location (Nelson et al. 2011). The identity of reductant(s) of Prx6 subfamily members is also largely unknown; some subfamily members are reduced by Trx while others are not. At least some subfamily members may be regenerated by GSTπ and glutathione, or by ascorbate (Hall et al. 2011).

4.7.3 Prx5 Subfamily

The best characterized member of this subfamily is human PrxV, which is less reactive with H_2O_2 (second order rate constant $\sim 10^5$ M^{-1} s^{-1}) than human PrxI or PrxII, but is more reactive with both organic hydroperoxides ($\sim 10^6$ M^{-1} s^{-1}) and peroxynitrite ($\sim 10^7$ M^{-1} s^{-1}) (Trujillo et al. 2007; Manta et al. 2009). Prx5 subfamily members are found in archea, bacteria and higher eukaryotes and are localized to various locations within eukaryotic cells including mitochondria, peroxisomes, cytoplasm, and plant chloroplasts. This family includes a group of bacterial proteins that contain a Grx domain fused to the C-terminus of the Prx5 domain. Although the canonical location of the C_r has been considered to be in helix α_5 as observed in human Prx5, recent analysis shows that a cysteine is found in this location only 17% of the time in this subfamily (Nelson et al. 2011), implying that most Prx5 subfamily members are 1-Cys Prxs.

4.7.4 Tpx Subfamily

Whereas the Tpx subfamily takes its name from the first family member to be biologically characterized, *Escherichia coli* thiol peroxidase, a major confusion is that this same moniker (Tpx) has been used to abbreviate "thioredoxin peroxidase" and is the common name for many Prxs not belonging to this subfamily. Tpx subfamily members are found almost exclusively in bacteria (Nelson et al. 2011) and are comprised of A-type dimers that stay associated in both oxidized and reduced states (Baker and Poole 2003; Hall et al. 2009c). The K_m for cumene hydroperoxide (a bulkier and more hydrophobic substrate than H_2O_2) is more than two orders of magnitude lower than the K_m for H_2O_2 in *E. coli* Tpx (Baker and Poole 2003). For more than 99% of Tpx subfamily members the C_r is located in helix α_3, with the remainder apparently being 1-Cys proteins (Nelson et al. 2011).

4.7.5 PrxQ Subfamily

This subfamily contains the plant PrxQ chloroplast proteins as well as the bacterial BCP proteins and, based on recent bioinformatic analysis, is distributed widely across archea, bacteria, plants, and fungi (Nelson et al. 2011). While the PrxQ subfamily is not as well characterized as other Prx subfamilies, recent structural and kinetic studies have shown that members of this subfamily are rather diverse in terms of oligomeric state, presence and location of the C_r, and kinetic activity, supporting the hypothesis that this subfamily is most similar to the ancestral Prx and that C_r-positions have arisen independently multiple times in the course of evolution (Copley et al. 2004; Hall et al. 2011; Horta et al. 2010; Reeves et al. 2011). PrxQ subfamily members have been shown to exist as either monomers or A-type dimers, and mechanistically can be either 1-Cys or 2-Cys Prxs with the C_r existing in one of two locations: 54% of the subfamily members contain a C_r five residues after the C_p in helix α_2 (Fig. 4.7) while another 13% exhibit a C_r in helix α_3 (the position commonly associated with members of the Tpx subfamily) (Nelson et al. 2011). The catalytic activity of this family also varies widely. *E. coli* BCP exhibits a similar, slow rate of reaction with H_2O_2 and other alkyl hydroperoxides (second order rate constants $\sim 10^4\ M^{-1}\ s^{-1}$) and can be reduced efficiently by a range of reductants including Trx1, Trx2, Grx1, and Grx3 (Reeves et al. 2011). In contrast, a plant PrxQ from *Xylella fastidiosa* has been shown to react much more rapidly with H_2O_2 ($\sim 10^7\ M^{-1}\ s^{-1}$) (Horta et al. 2010). Members of the PrxQ subfamily have also been shown to exhibit both the highest (at −146 mV, *E. coli* BCP) and lowest (at −325 mV, poplar PrxQ) redox potentials reported to date for any Prx (Reeves et al. 2011; Rouhier et al. 2004).

4.7.6 AhpE Subfamily

The AhpE subfamily is comprised of a small number of proteins that have to date only been found in a limited number of gram positive bacterial species closely related to *Mycobacterium tuberculosis* (Nelson et al. 2011). AhpE does not group clearly with any other Prx subfamily, however, it shares ~30% amino acid sequence identity with Prx1 subfamily members and ~25% sequence identity with members of the PrxQ subfamily. *M. tuberculosis* AhpE has been shown to react faster with peroxynitrite ($\sim 10^7\ M^{-1}\ s^{-1}$) than with H_2O_2 ($\sim 10^5\ M^{-1}\ s^{-1}$) and is the only structurally characterized subfamily member to date (Hugo et al. 2009). Although *M. tuberculosis* AhpE is a 1-Cys Prx, other members of this subfamily contain a potential C_r in helix α_2 similar to many PrxQ subfamily members (Nelson et al. 2011).

4.8 Regulation of Prx Activity

4.8.1 Transcriptional Regulation of Prx Expression

Peroxiredoxin expression is frequently up-regulated in response to oxidative stress. In *E. coli*, H_2O_2 activates OxyR through reaction with its sensor site, Cys199, to form a sulfenic acid; subsequent disulfide bond formation with Cys208 leads to activation of the transcription factor, inducing expression of a variety of antioxidant proteins including AhpC and AhpF, KatG catalase, glutathione reductase, and Grx1 (Åslund et al. 1999). Transcriptional activity of OxyR is subsequently modulated by these antioxidants which directly or indirectly decrease H_2O_2 levels in the cell; in addition, Grx1 directly reduces and inactivates OxyR, providing multiple feedback mechanisms to regulate OxyR activity (Åslund et al. 1999). Although OxyR is present in many bacteria, a distinct transcriptional regulator known as PerR regulates Prx expression in *Bacillus subtilis* in response to specific peroxide or general oxidative stress (Mongkolsuk and Helmann 2002). In *Saccharomyces cerevisiae*, the Gpx homolog Orp1 serves as a peroxide sensor by forming a sulfenic acid on Cys36 in the presence of H_2O_2 (Delaunay et al. 2002). Although this protein has its own C_r for intramolecular disulfide bond formation, a kinetic pause at this step allows formation of an intermolecular disulfide bond with yeast activator protein 1 (Yap1), a transcription factor that controls antioxidant gene expression. A second cysteine in Yap1 then attacks the intermolecular disulfide to form an intramolecular disulfide bond within Yap1 and release Orp1, activating transcription and upregulating Prx expression (D'Autreaux and Toledano 2007). In *Schizosaccharomyces pombe*, the Yap1 homologue, activator protein-1 (Pap1), is activated by "low" peroxide levels (below 0.2 mM) in concert with thiol peroxidase 1 (Tpx1), a member of the Prx1 subfamily, and both are inactivated at H_2O_2 levels greater than 1 mM. At this high peroxide level, the Sty1 transcription factor is independently activated through a mechanism involving a bacterial-like His phospho-relay system and potentially promoted by disulfide bond formation either intermolecularly with Tpx1 or intramolecularly through direct oxidation of a cysteine on Sty1 (Day et al. 2012; Karplus and Poole 2012). In *Homo sapiens*, the transcription factor, NF-E2-related factor-2 (Nrf-2) has been shown to regulate the expression of Prx1 and Prx4 as well as the Prx reductants, Trx, TrxR1, and Srx (Bell and Hardingham 2011). Transcriptional regulation of these genes involves a *cis*-acting DNA sequence known as the antioxidant response element (ARE), which recruits the transcription factor Nrf2. Under non-stress conditions, Nrf2 is ubiquitinated and targeted for degradation by Kelch like ECH-associated protein (Keap 1); during oxidative stress, oxidation of one or more cysteine(s) on Keap1 releases Nrf2, allowing for Nrf2 accumulation in the nucleus and subsequent activation of ARE-containing genes (Brigelius-Flohé and Flohé 2011). In parasites, Prxs are very important factors for defending against the immune system, and their expression tends to be regulated by developmental stage and environment (Gretes et al. 2012).

4.8.2 *Regulation of Prx Activity by Phosphorylation and Other Posttranslational Modifications*

Recent studies have identified a group of post-translational modifications (in addition to redox modifications described above) that regulate Prx activity, including phosphorylation, acetylation, and glutathionylation (Parmigiani et al. 2008; Woo et al. 2010; Chae et al. 2012). Most of these studies have focused on human PrxI and PrxII, and it is still too early to completely understand the role some of these modifications play in the cell. In particular, PrxI has been shown to be phosphorylated by cyclin dependent kinase 1 (CDK1) at Thr90 during the mitotic phase in HeLa cells (Chang et al. 2002) and at Tyr194 in response to stimulation by growth factors (e.g. EGF and PDGF), or by antibodies to IgM and CD3 in the case of B-cells and T-cells, respectively (Woo et al. 2010). Both phosphorylation events appear to exhibit specificity for the PrxI isoform over PrxII, and decrease the peroxidase activity of the phosphorylated Prxs, thus allowing for a localized increase in H_2O_2 levels in response to growth factor signaling (Rhee and Woo 2011). Acetylation at the C-terminus of PrxI and PrxII by HDAC6 was reported to increase Prx activity (Parmigiani et al. 2008), perhaps a result of disrupting the packing of the C-terminal helix to promote disulfide bond formation during catalysis and avoid oxidative inactivation. Glutathionylation of the non-catalytic Cys83 has recently been shown to shift the oligomeric state of human PrxI from predominantly decameric to predominantly dimeric structures, eliminating its ability to serve as a chaperone; additional effects of glutathione in recycling or regulating Prx activity likely varies with the specific protein, and adds to the complexity of glutathionylation as a regulatory mechanism for Prxs (Parmigiani et al. 2008; Woo et al. 2010; Chae et al. 2012).

4.9 Outlook

The last two decades have seen a tremendous growth in the recognition of the importance in biology of sulfenic acid derivatives of cysteine. Although much research involving sulfenic acid derivatives does not include the word "sulfenic" in the abstract, Pubmed searches using the word "sulfenic" yield 20, 41, 102 and 199 hits for the 5 year periods of 1990–1994, 1995–1999, 2000–2004, and 2005–2009, respectively. Related to this, hydrogen peroxide has since early in this century transitioned from being simply numbered among the many dangerous oxidative stressors to being in its own right seen as an important and widespread second messenger crucial to fundamental signaling processes that guide normal growth and development. These research areas were initially connected by the recognition of the evolutionary significance of sensitive Prxs (Wood et al. 2003) and their reactivation by sulfiredoxin (Woo et al. 2003; Biteau et al. 2003), along with the discovery of the sulfenamide-inhibited form of protein tyrosine phosphatases (Salmeen et al. 2003; van Montfort et al. 2003), and the identification of widely expressed isoforms of NADPH oxidases that function in growth factor signaling (Oakley et al. 2009). Our understanding of these processes is still rather primitive but the results in hand are enough to predict confidently that discoveries in this area – including the challenging processes of mapping of the temporal and spatial dynamics of the occurrence of sulfenic acids *in vivo* – will play a prominent role in advancing our understanding of basic eukaryotic cell biology and our ability to combat major health problems such as obesity, cancer, and aging.

Acknowledgements The authors thank Andrea Hall for creating Fig. 4.4a. This work was supported by grants from the National Institutes of Health, including RO1 GM050389 to L.B.P. with a subcontract to P.A.K., and R33 CA126659 to L.B.P.

References

Adimora NJ, Jones DP, Kemp ML (2010) A model of redox kinetics implicates the thiol proteome in cellular hydrogen peroxide responses. Antioxid Redox Signal 13:731–743

Allison WS (1976) Formation and reactions of sulfenic acids in proteins. Acc Chem Res 9:293–299

Åslund F, Zheng M, Beckwith J, Storz G (1999) Regulation of the OxyR transcription factor by hydrogen peroxide and the cellular thiol-disulfide status. Proc Natl Acad Sci USA 96: 6161–6165

Baek JY, Han SH, Sung SH, Lee HE, Kim YM, Noh YH, Bae SH, Rhee SG, Chang TS (2012) Sulfiredoxin protein is critical for redox balance and survival of cells exposed to low steady-state levels of H_2O_2. J Biol Chem 287:81–89

Baker LM, Poole LB (2003) Catalytic mechanism of thiol peroxidase from *Escherichia coli*. Sulfenic acid formation and overoxidation of essential CYS61. J Biol Chem 278:9203–9211

Barranco-Medina S, Kakorin S, Lazaro JJ, Dietz KJ (2008) Thermodynamics of the dimer-decamer transition of reduced human and plant 2-cys peroxiredoxin. Biochemistry 47:7196–7204

Bell KF, Hardingham GE (2011) CNS peroxiredoxins and their regulation in health and disease. Antioxid Redox Signal 14:1467–1477

Benesch RE, Lardy HA, Benesch R (1955) The sulfhydryl groups of crystalline proteins. I. Some albumins, enzymes, and hemoglobins. J Biol Chem 216:663–676

Benitez LV, Allison WS (1974) The inactivation of the acyl phosphatase activity catalyzed by the sulfenic acid form of glyceraldehyde 3-phosphate dehydrogenase by dimedone and olefins. J Biol Chem 249:6234–6243

Biteau B, Labarre J, Toledano MB (2003) ATP-dependent reduction of cysteine-sulphinic acid by *S. cerevisiae* sulphiredoxin. Nature 425:980–984

Brigelius-Flohé R, Flohé L (2011) Basic principles and emerging concepts in the redox control of transcription factors. Antioxid Redox Signal 15:2335–2381

Budde H, Flohé L, Hofmann B, Nimtz M (2003) Verification of the interaction of a tryparedoxin peroxidase with tryparedoxin by ESI-MS/MS. Biol Chem 384:1305–1309

Burgoyne JR, Madhani M, Cuello F, Charles RL, Brennan JP, Schroder E, Browning DD, Eaton P (2007) Cysteine redox sensor in PKGIa enables oxidant-induced activation. Science 317: 1393–1397

Chae HZ, Robison K, Poole LB, Church G, Storz G, Rhee SG (1994a) Cloning and sequencing of thiol-specific antioxidant from mammalian brain: alkyl hydroperoxide reductase and thiol-specific antioxidant define a large family of antioxidant enzymes. Proc Natl Acad Sci USA 91:7017–7021

Chae HZ, Uhm TB, Rhee SG (1994b) Dimerization of thiol-specific antioxidant and the essential role of cysteine 47. Proc Natl Acad Sci USA 91:7022–7026

Chae HZ, Oubrahim H, Park JW, Rhee SG, Chock PB (2012) Protein glutathionylation in the regulation of peroxiredoxins: a family of thiol-specific peroxidases function as antioxidants, molecular chaperones and signal modulators. Antioxid Redox Signal 16:506–523

Chang TS, Jeong W, Choi SY, Yu S, Kang SW, Rhee SG (2002) Regulation of peroxiredoxin I activity by Cdc2-mediated phosphorylation. J Biol Chem 277:25370–25376

Charles RL, Schröder E, May G, Free P, Gaffney PR, Wait R, Begum S, Heads RJ, Eaton P (2007) Protein sulfenation as a redox sensor: proteomics studies using a novel biotinylated dimedone analogue. Mol Cell Proteomics 6:1473–1484

Choi HJ, Kang SW, Yang CH, Rhee SG, Ryu SE (1998) Crystal structure of a novel human peroxidase enzyme at 2.0 Å resolution. Nat Struct Biol 5:400–406

Choi H, Kim S, Mukhopadhyay P, Cho S, Woo J, Storz G, Ryu S (2001) Structural basis of the redox switch in the OxyR transcription factor. Cell 105:103–113

Chouchani ET, James AM, Fearnley IM, Lilley KS, Murphy MP (2011) Proteomic approaches to the characterization of protein thiol modification. Curr Opin Chem Biol 15:120–128

Claiborne A, Mallett TC, Yeh JI, Luba J, Parsonage D (2001) Structural, redox, and mechanistic parameters for cysteine-sulfenic acid function in catalysis and regulation. Adv Protein Chem 58:215–276

Codreanu SG, Zhang B, Sobecki SM, Billheimer DD, Liebler DC (2009) Global analysis of protein damage by the lipid electrophile 4-hydroxy-2-nonenal. Mol Cell Proteomics 8:670–680

Copley SD, Novak WR, Babbitt PC (2004) Divergence of function in the thioredoxin fold suprafamily: evidence for evolution of peroxiredoxins from a thioredoxin-like ancestor. Biochemistry 43:13981–13995

Crane EJ 3rd, Vervoort J, Claiborne A (1997) 13C NMR analysis of the cysteine-sulfenic acid redox center of enterococcal NADH peroxidase. Biochemistry 36:8611–8618

Cross JV, Templeton DJ (2006) Regulation of signal transduction through protein cysteine oxidation. Antioxid Redox Signal 8:1819–1827

D'Autreaux B, Toledano MB (2007) ROS as signalling molecules: mechanisms that generate specificity in ROS homeostasis. Nat Rev Mol Cell Biol 8:813–824

Day AM, Brown JD, Taylor SR, Rand JD, Morgan BA, Veal EA (2012) Inactivation of a peroxiredoxin by hydrogen peroxide is critical for thioredoxin-mediated repair of oxidized proteins and cell survival. Mol Cell 45:398–408

Delano WL (2002) The PyMOL molecular graphics system. DeLano Scientific, San Carlos

Delaunay A, Pflieger D, Barrault MB, Vinh J, Toledano MB (2002) A thiol peroxidase is an H2O2 receptor and redox-transducer in gene activation. Cell 111:471–481

Dennehy MK, Richards KA, Wernke GR, Shyr Y, Liebler DC (2006) Cytosolic and nuclear protein targets of thiol-reactive electrophiles. Chem Res Toxicol 19:20–29

Ellis HR, Poole LB (1997a) Novel application of 7-chloro-4-nitrobenzo-2-oxa-1,3-diazole to identify cysteine sulfenic acid in the AhpC component of alkyl hydroperoxide reductase. Biochemistry 36:15013–15018

Ellis HR, Poole LB (1997b) Roles for the two cysteine residues of AhpC in catalysis of peroxide reduction by alkyl hydroperoxide reductase from *Salmonella typhimurium*. Biochemistry 36:13349–13356

Fang J, Nakamura T, Cho DH, Gu Z, Lipton SA (2007) S-nitrosylation of peroxiredoxin 2 promotes oxidative stress-induced neuronal cell death in Parkinson's disease. Proc Natl Acad Sci USA 104:18742–18747

Ferrer-Sueta G, Manta B, Botti H, Radi R, Trujillo M, Denicola A (2011) Factors affecting protein thiol reactivity and specificity in peroxide reduction. Chem Res Toxicol 24:434–450

Flohé L, Toppo S, Cozza G, Ursini F (2011) A comparison of thiol peroxidase mechanisms. Antioxid Redox Signal 15:763–780

Fomenko DE, Gladyshev VN (2003) Identity and functions of CxxC-derived motifs. Biochemistry 42:11214–11225

Fomenko DE, Xing W, Adair BM, Thomas DJ, Gladyshev VN (2007) High-throughput identification of catalytic redox-active cysteine residues. Science 315:387–389

Fuangthong M, Helmann JD (2002) The OhrR repressor senses organic hydroperoxides by reversible formation of a cysteine-sulfenic acid derivative. Proc Natl Acad Sci USA 99: 6690–6695

Giannoni E, Taddei ML, Chiarugi P (2010) Src redox regulation: again in the front line. Free Radic Biol Med 49:516–527

Greco TM, Hodara R, Parastatidis I, Heijnen HF, Dennehy MK, Liebler DC, Ischiropoulos H (2006) Identification of S-nitrosylation motifs by site-specific mapping of the S-nitrosocysteine proteome in human vascular smooth muscle cells. Proc Natl Acad Sci USA 103:7420–7425

Greenberg JT, Demple B (1988) Overproduction of peroxide-scavenging enzymes in *Escherichia coli* suppresses spontaneous mutagenesis and sensitivity to redox-cycling agents in *oxyR*-mutants. EMBO J 7:2611–2617

Gretes MC, Poole LB, Karplus PA (2012) Peroxiredoxins in parasites. Antioxid Redox Signal 17:608–633

Griffiths SW, King J, Cooney CL (2002) The reactivity and oxidation pathway of cysteine 232 in recombinant human alpha 1-antitrypsin. J Biol Chem 277:25486–25492

Hall A, Karplus PA, Poole LB (2009a) Typical 2-Cys peroxiredoxins – structures, mechanisms and functions. FEBS J 276:2469–2477

Hall A, Sankaran B, Poole LB, Karplus PA (2009b) Structural changes common to catalysis in the Tpx peroxiredoxin subfamily. J Mol Biol 393:867–881

Hall A, Parsonage D, Horita D, Karplus PA, Poole LB, Barbar E (2009c) Redox-dependent dynamics of a dual thioredoxin fold protein: evolution of specialized folds. Biochemistry 48:5984–5993

Hall A, Parsonage D, Poole LB, Karplus PA (2010) Structural evidence that peroxiredoxin catalytic power is based on transition-state stabilization. J Mol Biol 402:194–209

Hall A, Nelson K, Poole LB, Karplus PA (2011) Structure-based insights into the catalytic power and conformational dexterity of peroxiredoxins. Antioxid Redox Signal 15:795–815

Hansen RE, Winther JR (2009) An introduction to methods for analyzing thiols and disulfides: reactions, reagents, and practical considerations. Anal Biochem 394:147–158

Haque A, Andersen JN, Salmeen A, Barford D, Tonks NK (2011) Conformation-sensing antibodies stabilize the oxidized form of PTP1B and inhibit its phosphatase activity. Cell 147:185–198

Headlam HA, Gracanin M, Rodgers KJ, Davies MJ (2006) Inhibition of cathepsins and related proteases by amino acid, peptide, and protein hydroperoxides. Free Radic Biol Med 40:1539–1548

Horta BB, de Oliveira MA, Discola KF, Cussiol JR, Netto LE (2010) Structural and biochemical characterization of peroxiredoxin Qbeta from *Xylella fastidiosa*: catalytic mechanism and high reactivity. J Biol Chem 285:16051–16065

Hugo M, Turell L, Manta B, Botti H, Monteiro G, Netto LE, Alvarez B, Radi R, Trujillo M (2009) Thiol and sulfenic acid oxidation of AhpE, the one-cysteine peroxiredoxin from *Mycobacterium tuberculosis*: kinetics, acidity constants, and conformational dynamics. Biochemistry 48:9416–9426

Imlay JA (2008) Cellular defenses against superoxide and hydrogen peroxide. Annu Rev Biochem 77:755–776

Jang HH, Lee KO, Chi YH, Jung BG, Park SK, Park JH, Lee JR, Lee SS, Moon JC, Yun JW, Choi YO, Kim WY, Kang JS, Cheong GW, Yun DJ, Rhee SG, Cho MJ, Lee SY (2004) Two enzymes in one; two yeast peroxiredoxins display oxidative stress-dependent switching from a peroxidase to a molecular chaperone function. Cell 117:625–635

Jeon SJ, Ishikawa K (2003) Characterization of novel hexadecameric thioredoxin peroxidase from *Aeropyrum pernix* K1. J Biol Chem 278:24174–24180

Jönsson TJ, Ellis HR, Poole LB (2007) Cysteine reactivity and thiol-disulfide interchange pathways in AhpF and AhpC of the bacterial alkyl hydroperoxide reductase system. Biochemistry 46:5709–5721

Jönsson TJ, Johnson LC, Lowther WT (2008a) Structure of the sulphiredoxin-peroxiredoxin complex reveals an essential repair embrace. Nature 451:98–101

Jönsson TJ, Murray MS, Johnson LC, Lowther WT (2008b) Reduction of cysteine sulfinic acid in peroxiredoxin by sulfiredoxin proceeds directly through a sulfinic phosphoryl ester intermediate. J Biol Chem 283:23846–23851

Kang DH, Lee DJ, Lee KW, Park YS, Lee JY, Lee SH, Koh YJ, Koh GY, Choi C, Yu DY, Kim J, Kang SW (2011) Peroxiredoxin II is an essential antioxidant enzyme that prevents the oxidative inactivation of VEGF receptor-2 in vascular endothelial cells. Mol Cell 44:545–558

Kaplan N, Urao N, Furuta E, Kim SJ, Razvi M, Nakamura Y, McKinney RD, Poole LB, Fukai T, Ushio-Fukai M (2011) Localized cysteine sulfenic acid formation by vascular endothelial growth factor: role in endothelial cell migration and angiogenesis. Free Radic Res 45: 1124–1135

Karplus PA, Poole LB (2012) Peroxiredoxins as molecular triage agents, sacrificing themselves to enhance cell survival during a peroxide attack. Mol Cell 45:275–278

Kettenhofen NJ, Wood MJ (2010) Formation, reactivity, and detection of protein sulfenic acids. Chem Res Toxicol 23:1633–1646

Kim SY, Jang HH, Lee JR, Sung NR, Lee HB, Lee DH, Park D-J, Kang CH, Chung WS, Lim CO, Yun D-J, Kim WY, Lee KO, Lee SY (2009) Oligomerization and chaperone activity of a plant 2-Cys peroxiredoxin in response to oxidative stress. Plant Sci 177:227–232

Klomsiri C, Nelson KJ, Bechtold E, Soito L, Johnson LC, Lowther WT, Ryu SE, King SB, Furdui CM, Poole LB (2010) Use of dimedone-based chemical probes for sulfenic acid detection: evaluation of conditions affecting probe incorporation into redox-sensitive proteins. Methods Enzymol 473:77–94

Klomsiri C, Karplus PA, Poole LB (2011) Cysteine-based redox switches in enzymes. Antioxid Redox Signal 14:1065–1077

Koo KH, Lee S, Jeong SY, Kim ET, Kim HJ, Song K, Chae H-Z (2002) Regulation of thioredoxin peroxidase activity by C-terminal truncation. Arch Biochem Biophys 397:312–318

Kortemme T, Darby NJ, Creighton TE (1996) Electrostatic interactions in the active site of the N-terminal thioredoxin-like domain of protein disulfide isomerase. Biochemistry 35: 14503–14511

Lee JW, Soonsanga S, Helmann JD (2007) A complex thiolate switch regulates the *Bacillus subtilis* organic peroxide sensor OhrR. Proc Natl Acad Sci USA 104:8743–8748

Leichert LI, Jakob U (2006) Global methods to monitor the thiol-disulfide state of proteins in vivo. Antioxid Redox Signal 8:763–772

Leonard SE, Carroll KS (2011) Chemical 'omics' approaches for understanding protein cysteine oxidation in biology. Curr Opin Chem Biol 15:88–102

Lesniak J, Barton WA, Nikolov DB (2002) Structural and functional characterization of the *Pseudomonas* hydroperoxide resistance protein Ohr. EMBO J 21:6649–6659

Lian FM, Yu J, Ma XX, Yu XJ, Chen Y, Zhou CZ (2012) Structural snapshots of yeast alkyl hydroperoxide reductase Ahp1 peroxiredoxin reveal a novel two-cysteine mechanism of electron transfer to eliminate reactive oxygen species. J Biol Chem 287:17077–17087

Manta B, Hugo M, Ortiz C, Ferrer-Sueta G, Trujillo M, Denicola A (2009) The peroxidase and peroxynitrite reductase activity of human erythrocyte peroxiredoxin 2. Arch Biochem Biophys 484:146–154

Marino SM, Gladyshev VN (2010a) Cysteine function governs its conservation and degeneration and restricts its utilization on protein surfaces. J Mol Biol 404:902–916

Marino SM, Gladyshev VN (2010b) Structural analysis of cysteine S-nitrosylation: a modified acid-based motif and the emerging role of trans-nitrosylation. J Mol Biol 395:844–859

Meissner U, Schröder E, Scheffler D, Martin AG, Harris JR (2007) Formation, TEM study and 3D reconstruction of the human erythrocyte peroxiredoxin-2 dodecahedral higher-order assembly. Micron 38:29–39

Michalek RD, Nelson KJ, Holbrook BC, Yi JS, Stridiron D, Daniel LW, Fetrow JS, King SB, Poole LB, Grayson JM (2007) The requirement of reversible cysteine sulfenic acid formation for T cell activation and function. J Immunol 179:6456–6467

Mitsumoto A, Takanezawa Y, Okawa K, Iwamatsu A, Nakagawa Y (2001) Variants of peroxiredoxins expression in response to hydroperoxide stress. Free Radic Biol Med 30:625–635

Molin M, Yang J, Hanzen S, Toledano MB, Labarre J, Nystrom T (2011) Life span extension and H(2)O(2) resistance elicited by caloric restriction require the peroxiredoxin Tsa1 in *Saccharomyces cerevisiae*. Mol Cell 43:823–833

Mongkolsuk S, Helmann JD (2002) Regulation of inducible peroxide stress responses. Mol Microbiol 45:9–15
Moon JC, Hah YS, Kim WY, Jung BG, Jang HH, Lee JR, Kim SY, Lee YM, Jeon MG, Kim CW, Cho MJ, Lee SY (2005) Oxidative stress-dependent structural and functional switching of a human 2-Cys peroxiredoxin isotype II that enhances HeLa cell resistance to H_2O_2-induced cell death. J Biol Chem 280:28775–28784
Murphy MP, Holmgren A, Larsson NG, Halliwell B, Chang CJ, Kalyanaraman B, Rhee SG, Thornalley PJ, Partridge L, Gems D, Nystrom T, Belousov V, Schumacker PT, Winterbourn CC (2011) Unraveling the biological roles of reactive oxygen species. Cell Metab 13:361–366
Nelson KJ, Parsonage D, Hall A, Karplus PA, Poole LB (2008) Cysteine pK_a values for the bacterial peroxiredoxin AhpC. Biochemistry 47:12860–12868
Nelson KJ, Klomsiri C, Codreanu SG, Soito L, Liebler DC, Rogers LC, Daniel LW, Poole LB (2010) Use of dimedone-based chemical probes for sulfenic acid detection; methods to visualize and identify labeled proteins. Methods Enzymol 473:95–115
Nelson KJ, Knutson ST, Soito L, Klomsiri C, Poole LB, Fetrow JS (2011) Analysis of the peroxiredoxin family: using active-site structure and sequence information for global classification and residue analysis. Proteins 79:947–964
Newton GL, Rawat M, la Clair JJ, Jothivasan VK, Budiarto T, Hamilton CJ, Claiborne A, Helmann JD, Fahey RC (2009) Bacillithiol is an antioxidant thiol produced in Bacilli. Nat Chem Biol 5:625–627
Oakley FD, Abbott D, Li Q, Engelhardt J (2009) Signaling components of redox active endosomes: the redoxosomes. Antioxid Redox Signal 11:1313–1333
Ogusucu R, Rettori D, Munhoz DC, Soares Netto LE, Augusto O (2007) Reactions of yeast thioredoxin peroxidases I and II with hydrogen peroxide and peroxynitrite: rate constants by competitive kinetics. Free Radic Biol Med 42:326–334
Oshikawa J, Urao N, Kim HW, Kaplan N, Razvi M, McKinney R, Poole LB, Fukai T, Ushio-Fukai M (2010) Extracellular SOD-derived H_2O_2 promotes VEGF signaling in caveolae/lipid rafts and post-ischemic angiogenesis in mice. PLOS One 5:e10189
Panmanee W, Vattanaviboon P, Poole LB, Mongkolsuk S (2006) Novel organic hydroperoxide-sensing and responding mechanisms for OhrR, a major bacterial sensor and regulator of organic hydroperoxide stress. J Bacteriol 188:1389–1395
Park JW, Piszczek G, Rhee SG, Chock PB (2011) Glutathionylation of peroxiredoxin I induces decamer to dimers dissociation with concomitant loss of chaperone activity. Biochemistry 50:3204–3210
Parmigiani RB, Xu WS, Venta-Perez G, Erdjument-Bromage H, Yaneva M, Tempst P, Marks PA (2008) HDAC6 is a specific deacetylase of peroxiredoxins and is involved in redox regulation. Proc Natl Acad Sci USA 105:9633–9638
Parsonage D, Miller H, Ross RP, Claiborne A (1993) Purification and analysis of streptococcal NADH peroxidase expressed in *Escherichia coli*. J Biol Chem 268:3161–3167
Parsonage D, Youngblood DS, Sarma GN, Wood ZA, Karplus PA, Poole LB (2005) Analysis of the link between enzymatic activity and oligomeric state in AhpC, a bacterial peroxiredoxin. Biochemistry 44:10583–10592
Pascual MB, Mata-Cabana A, Florencio FJ, Lindahl M, Cejudo FJ (2010) Overoxidation of 2-Cys peroxiredoxin in prokaryotes: cyanobacterial 2-Cys peroxiredoxins sensitive to oxidative stress. J Biol Chem 285:34485–34492
Pauling L (1946) Molecular architecture and biological reactions. Chem Eng News 24:1375
Paulsen CE, Carroll KS (2010) Orchestrating redox signaling networks through regulatory cysteine switches. ACS Chem Biol 5:47–62
Paulsen CE, Truong TH, Garcia FJ, Homann A, Gupta V, Leonard SE, Carroll KS (2011) Peroxide-dependent sulfenylation of the EGFR catalytic site enhances kinase activity. Nat Chem Biol 8:57–64
Percival MD, Ouellet M, Campagnolo C, Claveau D, Li C (1999) Inhibition of cathepsin K by nitric oxide donors: evidence for the formation of mixed disulfides and a sulfenic acid. Biochemistry 38:13574–13583

Perkins A, Gretes MC, Nelson KJ, Poole LB, Karplus PA (2012) Mapping the active site helix-to-strand conversion of CxxxxC Peroxiredoxin Q Enzymes. Biochemistry (in press)

Peskin AV, Low FM, Paton LN, Maghzal GJ, Hampton MB, Winterbourn CC (2007) The high reactivity of peroxiredoxin 2 with H_2O_2 is not reflected in its reaction with other oxidants and thiol reagents. J Biol Chem 282:11885–11892

Phalen TJ, Weirather K, Deming PB, Anathy V, Howe AK, van der Vliet A, Jönsson TJ, Poole LB, Heintz NH (2006) Oxidation state governs structural transitions in peroxiredoxin II that correlate with cell cycle arrest and recovery. J Cell Biol 175:779–789

Poole LB (1996) Flavin-dependent alkyl hydroperoxide reductase from *Salmonella typhimurium*. 2. Cystine disulfides involved in catalysis of peroxide reduction. Biochemistry 35:65–75

Poole LB (2005) Bacterial defenses against oxidants: mechanistic features of cysteine-based peroxidases and their flavoprotein reductases. Arch Biochem Biophys 433:240–254

Poole LB (2008) Measurement of protein sulfenic acid content. Curr Protoc Toxicol Unit 17.2: 1–21

Poole LB, Claiborne A (1989) The non-flavin redox center of the streptococcal NADH peroxidase. II. Evidence for a stabilized cysteine-sulfenic acid. J Biol Chem 264:12330–12338

Poole LB, Ellis HR (1996) Flavin-dependent alkyl hydroperoxide reductase from *Salmonella typhimurium*. 1. Purification and enzymatic activities of overexpressed AhpF and AhpC proteins. Biochemistry 35:56–64

Poole LB, Ellis HR (2002) Identification of cysteine sulfenic acid in AhpC of alkyl hydroperoxide reductase. Methods Enzymol 348:122–136

Poole LB, Nelson KJ (2008) Discovering mechanisms of signaling-mediated cysteine oxidation. Curr Opin Chem Biol 12:18–24

Poole LB, Godzik A, Nayeem A, Schmitt JD (2000) AhpF can be dissected into two functional units: tandem repeats of two thioredoxin-like folds in the N-terminus mediate electron transfer from the thioredoxin reductase-like C-terminus to AhpC. Biochemistry 39:6602–6615

Poole LB, Karplus PA, Claiborne A (2004) Protein sulfenic acids in redox signaling. Annu Rev Pharmacol Toxicol 44:325–347

Poole LB, Zeng BB, Knaggs SA, Yakubu M, King SB (2005) Synthesis of chemical probes to map sulfenic acid modifications on proteins. Bioconjug Chem 16:1624–1628

Poole LB, Klomsiri C, Knaggs SA, Furdui CM, Nelson KJ, Thomas MJ, Fetrow JS, Daniel LW, King SB (2007) Fluorescent and affinity-based tools to detect cysteine sulfenic acid formation in proteins. Bioconjug Chem 18:2004–2017

Rabilloud T, Heller M, Gasnier F, Luche S, Rey C, Aebersold R, Benahmed M, Louisot P, Lunardi J (2002) Proteomics analysis of cellular response to oxidative stress. Evidence for in vivo overoxidation of peroxiredoxins at their active site. J Biol Chem 277:19396–19401

Raftery MJ, Yang Z, Valenzuela SM, Geczy CL (2001) Novel intra- and inter-molecular sulfinamide bonds in S100A8 produced by hypochlorite oxidation. J Biol Chem 276:33393–33401

Reeves SA, Parsonage D, Nelson KJ, Poole LB (2011) Kinetic and thermodynamic features reveal that *Escherichia coli* BCP is an unusually versatile peroxiredoxin. Biochemistry 50:8970–8981

Rhee SG, Woo HA (2011) Multiple functions of peroxiredoxins: peroxidases, sensors and regulators of the intracellular messenger H2O2, and protein chaperones. Antioxid Redox Signal 15(3):781–794

Roberts BR, Wood ZA, Jönsson TJ, Poole LB, Karplus PA (2005) Oxidized and synchrotron cleaved structures of the disulfide redox center in the N-terminal domain of *Salmonella typhimurium* AhpF. Protein Sci 14:2414–2420

Rouhier N, Gelhaye E, Gualberto JM, Jordy MN, de Fay E, Hirasawa M, Duplessis S, Lemaire SD, Frey P, Martin F, Manieri W, Knaff DB, Jacquot JP (2004) Poplar peroxiredoxin Q. A thioredoxin-linked chloroplast antioxidant functional in pathogen defense. Plant Physiol 134:1027–1038

Salmeen A, Andersen JN, Myers MP, Meng TC, Hinks JA, Tonks NK, Barford D (2003) Redox regulation of protein tyrosine phosphatase 1B involves a sulphenyl-amide intermediate. Nature 423:769–773

Salsbury FR Jr, Knutson ST, Poole LB, Fetrow JS (2008) Functional site profiling and electrostatic analysis of cysteines modifiable to cysteine sulfenic acid. Protein Sci 17:299–312

Sanchez R, Riddle M, Woo J, Momand J (2008) Prediction of reversibly oxidized protein cysteine thiols using protein structure properties. Protein Sci 17:473–481

Sarma GN, Nickel C, Rahlfs S, Fischer M, Becker K, Karplus PA (2005) Crystal structure of a novel *Plasmodium falciparum* 1-Cys peroxiredoxin. J Mol Biol 346:1021–1034

Sayed AA, Williams DL (2004) Biochemical characterization of 2-Cys peroxiredoxins from *Schistosoma mansoni*. J Biol Chem 279:26159–26166

Schröder E, Littlechild JA, Lebedev AA, Errington N, Vagin AA, Isupov MN (2000) Crystal structure of decameric 2-Cys peroxiredoxin from human erythrocytes at 1.7 A resolution. Structure 8:605–615

Seo YH, Carroll KS (2009) Profiling protein thiol oxidation in tumor cells using sulfenic acid-specific antibodies. Proc Natl Acad Sci USA 106:16163–16168

Seo JH, Lim JC, Lee DY, Kim KS, Piszczek G, Nam HW, Kim YS, Ahn T, Yun CH, Kim K, Chock PB, Chae HZ (2009) Novel protective mechanism against irreversible hyperoxidation of peroxiredoxin: Nalpha-terminal acetylation of human peroxiredoxin II. J Biol Chem 284:13455–13465

Shetty V, Neubert TA (2009) Characterization of novel oxidation products of cysteine in an active site motif peptide of PTP1B. J Am Soc Mass Spectrom 20:1540–1548

Shetty V, Spellman DS, Neubert TA (2007) Characterization by tandem mass spectrometry of stable cysteine sulfenic acid in a cysteine switch peptide of matrix metalloproteinases. J Am Soc Mass Spectrom 18:1544–1551

Soonsanga S, Lee JW, Helmann JD (2008) Oxidant-dependent switching between reversible and sacrificial oxidation pathways for *Bacillus subtilis* OhrR. Mol Microbiol 68:978–986

Stadtman ER (2006) Protein oxidation and aging. Free Radic Res 40:1250–1258

Tonks NK (2005) Redox redux: revisiting PTPs and the control of cell signaling. Cell 121:667–670

Trujillo M, Mauri P, Benazzi L, Comini M, de Palma A, Flohe L, Radi R, Stehr M, Singh M, Ursini F, Jaeger T (2006) The mycobacterial thioredoxin peroxidase can act as a one-cysteine peroxiredoxin. J Biol Chem 281:20555–20566

Trujillo M, Clippe A, Manta B, Ferrer-Sueta G, Smeets A, Declercq JP, Knoops B, Radi R (2007) Pre-steady state kinetic characterization of human peroxiredoxin 5: taking advantage of Trp84 fluorescence increase upon oxidation. Arch Biochem Biophys 467:95–106

van Montfort RL, Congreve M, Tisi D, Carr R, Jhoti H (2003) Oxidation state of the active-site cysteine in protein tyrosine phosphatase 1B. Nature 423:773–777

Victor KG, Rady JM, Cross JV, Templeton DJ (2012) Proteomic profile of reversible protein oxidation using PROP, purification of reversibly oxidized proteins. PLOS One 7:e32527

Wani R, Bharathi NS, Field J, Tsang AW, Furdui CM (2011a) Oxidation of Akt2 kinase promotes cell migration and regulates G1-S transition in the cell cycle. Cell Cycle 10:3263–3268

Wani R, Qian J, Yin L, Bechtold E, King SB, Poole LB, Paek E, Tsang AW, Furdui CM (2011b) Isoform-specific regulation of Akt by PDGF-induced reactive oxygen species. Proc Natl Acad Sci USA 108:10550–10555

Willett WS, Copley SD (1996) Identification and localization of a stable sulfenic acid in peroxide-treated tetrachlorohydroquinone dehalogenase using electrospray mass spectrometry. Chem Biol 3:851–857

Winterbourn CC (2008) Reconciling the chemistry and biology of reactive oxygen species. Nat Chem Biol 4:278–286

Woo HA, Kang SW, Kim HK, Yang KS, Chae HZ, Rhee SG (2003) Reversible oxidation of the active site cysteine of peroxiredoxins to cysteine sulfinic acid. Immunoblot detection with antibodies specific for the hyperoxidized cysteine-containing sequence. J Biol Chem 278:47361–47364

Woo HA, Yim SH, Shin DH, Kang D, Yu DY, Rhee SG (2010) Inactivation of peroxiredoxin I by phosphorylation allows localized H_2O_2 accumulation for cell signaling. Cell 140:517–528

Wood ZA, Poole LB, Hantgan RR, Karplus PA (2002) Dimers to doughnuts: redox-sensitive oligomerization of 2-cysteine peroxiredoxins. Biochemistry 41:5493–5504

Wood ZA, Poole LB, Karplus PA (2003) Peroxiredoxin evolution and the regulation of hydrogen peroxide signaling. Science 300:650–653

Yang KS, Kang SW, Woo HA, Hwang SC, Chae HZ, Kim K, Rhee SG (2002) Inactivation of human peroxiredoxin I during catalysis as the result of the oxidation of the catalytic site cysteine to cysteine-sulfinic acid. J Biol Chem 277:38029–38036

Yang J, Groen A, Lemeer S, Jans A, Slijper M, Roe SM, den Hertog J, Barford D (2007) Reversible oxidation of the membrane distal domain of receptor PTPalpha is mediated by a cyclic sulfenamide. Biochemistry 46:709–719

Yeh JI, Claiborne A, Hol WG (1996) Structure of the native cysteine-sulfenic acid redox center of enterococcal NADH peroxidase refined at 2.8 Å resolution. Biochemistry 35:9951–9957

Ying J, Clavreul N, Sethuraman M, Adachi T, Cohen RA (2007) Thiol oxidation in signaling and response to stress: detection and quantification of physiological and pathophysiological thiol modifications. Free Radic Biol Med 43:1099–1108

Yuan Y, Knaggs MH, Poole LB, Fetrow JS, Salsbury FR (2010) Conformational and oligomeric effects on the cysteine pK(a) of tryparedoxin peroxidase. J Biomol Struct Dyn 28:51–70

Chapter 5
Fluorescent Imaging of Redox Species in Multicellular Organisms

Yuuta Fujikawa, Bruce Morgan, and Tobias P. Dick

Abstract Redox processes play a crucial role in many aspects of physiology and changes in cellular redox species are increasingly being linked to a wide range of pathological conditions. Redox species can change dynamically at the subcellular compartment-, cell- and tissue-level and different redox species likely convey different biological information. The investigation of redox biology in a living multicellular organism is particularly difficult and is hampered by the lack of tools which offer redox species specificity and the necessary spatial and temporal resolution. In recent years there has been intense development of small organic chemical and genetically encoded fluorescent probes which have vastly improved our ability to investigate cellular redox processes. In this chapter we describe the currently available fluorescent probes, focusing in particular on those which have already been applied to multicellular organisms or those which we believe have the potential for *in vivo* use in the future. We discuss advantages and disadvantages of the different kinds of probes and highlight their major problems and limitations.

Keywords Reactive oxygen species • Redox-sensitive dyes • Redox-sensitive fluorescent proteins

---

The authors Yuuta Fujikawa and Bruce Morgan contributed equally.

Y. Fujikawa • B. Morgan • T.P. Dick (✉)
Division of Redox Regulation, German Cancer Research Center (DKFZ),
DKFZ-ZMBH Alliance, Im Neuenheimer Feld 280, 69120 Heidelberg, Germany
e-mail: t.dick@dkfz.de

U. Jakob and D. Reichmann (eds.), *Oxidative Stress and Redox Regulation*,
DOI 10.1007/978-94-007-5787-5_5,

5.1 Introduction

5.1.1 Why Study Redox Changes In Vivo?

It is becoming increasingly clear that changes in biological redox species are associated with many aspects of human pathology. However, obtaining a detailed understanding of organism redox biology is incredibly difficult. Redox species are regulated at the subcellular compartment-, cell-, and tissue-specific level. Furthermore, many cellular redox species are often independently regulated and change dynamically over time.

Contrary to the classically assumed status of reactive oxygen species (ROS) as unwanted and damaging agents, certain ROS are now known to fulfill crucial roles in many cellular contexts. Hydrogen peroxide (H_2O_2) is specifically produced in response to a range of stimuli, including growth factors and cytokines (Veal et al. 2007). In this context H_2O_2 acts as a second messenger, leading to the post-translational modification of a range of proteins involved in cellular signaling pathways. Thereby, H_2O_2 can influence cellular decisions including proliferation, differentiation, migration and apoptosis (Veal et al. 2007).

Both H_2O_2 and nitric oxide (NO·) are produced in phagocytic immune cells where they contribute to both the killing of invading pathogens and to the regulation of immune signaling pathways, including the pro-inflammatory response (Wink et al. 2011; Grivennikov et al. 2010; Harrison et al. 2011). Inflammation in this context is a crucial part of the immune response; however, prolonged high levels of ROS and reactive nitrogen species (RNS) are associated with chronic inflammation which plays a major role in the initiation and progression of a wide range of pathologies, including atherosclerosis, diabetes and cancer (Drummond et al. 2011; Pashkow 2011; Harrison et al. 2011; Grivennikov et al. 2010). Perturbations in other cellular redox species have also been linked to a range of diseases, for example, sustained alterations of the glutathione system have been associated with malignant, cardiovascular and neurodegenerative disorders (Mieyal et al. 2008; Mieyal and Chock 2012). It goes without saying that redox processes are clearly of huge biomedical interest.

In reality, the actual biomedical significance of redox processes, e.g. protein redox regulation, remains to be demonstrated in the truly *in vivo* context of model animals. It also remains unclear if pharmacological intervention can target cellular redox processes accurately and specifically enough in order to facilitate a beneficial medical outcome. For example, it is becoming clear that even commonly used redox agents such as *N*-acetyl cysteine (NAC) can elicit unexpected outcomes in the context of a multicellular organism (Albrecht et al. 2011).

To date, our understanding of redox processes *in vivo* has been constrained by the lack of appropriate investigative tools and methodology. In particular there are several key difficulties associated with measuring redox processes in the context of multicellular organisms. These include (i) obtaining redox species-specificity, (ii) obtaining subcellular compartment-specificity, (iii) obtaining cell- and tissue-type resolution (iv) obtaining temporal resolution and (v) dealing with the high optical opacity of many model organisms, which impedes standard microscopic investigation.

Redox studies of multicellular organisms have typically relied on disruptive whole organism techniques, but such methods are prone to *post mortem* artifacts and offer little in the way of spatial resolution as evidenced for example by whole organism reduced glutathione (GSH)/oxidized glutathione (GSSG) assays. *In situ* staining (for example of tissue sections) has also been widely used, however, problems include the frequent use of non-specific probes and the observation of late-stage markers of oxidative damage which are indirect and offer little information relating to the initial redox species involved. Cell or tissue culture is more amenable to investigation with redox probes, but at best cell culture only crudely approximates the true *in vivo* environment and must be viewed in light of the typically non-physiological molecular oxygen concentration (de Souza 2007; Atkuri et al. 2007).

In this chapter we focus on the challenge of investigating redox processes taking place inside multicellular organisms. In particular we focus on the organic chemical and genetically-encoded fluorescent probes which are currently available, and discuss their applicability for use in observing biologically relevant redox species in the context of multicellular organisms.

We believe the ultimate goal should be the development of methodologies that allow the non-invasive, real-time observation of defined redox species, with subcellular compartment resolution in living multicellular organisms, even when not optically transparent. At present, in most situations, this goal is far from being realized. The best approximation that is currently available is the imaging of genetically encoded redox probes in transparent model organisms such as *C. elegans, Drosophila* larvae and *Danio rerio* (zebrafish) (Niethammer et al. 2009; Back et al. 2011; Albrecht et al. 2011). For non-transparent organisms including mice, the next best choice is a minimally invasive technique such as intravital imaging, i.e. observation through a window preparation. However, to reach deep tissues, non-transparent organisms must be sacrificed and measurements performed on organ or tissue preparations or tissue sections. While such procedures in principle allow *in situ* redox measurements, there is a high risk that the disruption of organism integrity changes the redox processes of interest. The chemical conservation of the biosensor redox state during dissection is one possible solution to the problem (Albrecht et al. 2011).

5.1.2 *Which Redox Species Are of Particular Interest for In Vivo Measurements?*

We consider there to be two classes of redox species that are particularly worthy of investigation. The first class consists of those oxidants that are primarily produced and involved in signaling and regulation. Most prominently, these are superoxide

($O_2^{\cdot-}$), H_2O_2 and NO·. $O_2^{\cdot-}$ is of interest because in most situations it is the primary ROS produced and a precursor of H_2O_2. Although $O_2^{\cdot-}$ reactivity with thiols is very low, it can react with 4Fe-4S clusters, as seen in the $O_2^{\cdot-}$-responsive bacterial transcription factor SoxR (Winterbourn and Hampton 2008). Thus $O_2^{\cdot-}$ may in principle also play a role in regulating protein function. In terms of redox signaling and redox regulation, H_2O_2 is arguably the most important endogenous oxidant. It is the most stable of the ROS, clearly involved in regulating protein function, and under certain circumstances it is also a precursor of the hydroxyl radical (HO·) and hypochlorous acid (HOCl) (Dickinson and Chang 2011), which along with peroxynitrite ($ONOO^-$) constitute the highly reactive oxygen species (hROS). These species are also of interest as they are very short-lived and indiscriminately react with biological molecules, thus they are likely to be the actual source of ROS-induced cellular damage. NO· is specifically produced by nitric oxide synthases (NOS) and serves as both an autocrine and paracrine signaling molecule, for example in vasodilation and neurotransmission. NO· is known to regulate protein function via post-translational modification. Indeed, S-nitrosylation is reported to affect as many as 3,000 different proteins (Hess and Stamler 2012).

The second class of redox species of fundamental importance comprises redox couples that are central to metabolism and redox homeostasis. Here we focus primarily on the redox couples for which fluorescent probes are currently available, GSH/GSSG and NAD^+/NADH. The glutathione redox couple provides an abundant source of reducing equivalents that protect cells from oxidative stress and are crucial for detoxification of xenobiotics (Grant et al. 1996; Meister 1988). S-glutathionylation is a mechanism for regulating redox-sensitive signaling pathways (Dalle-Donne et al. 2009) and alterations of glutathione homeostasis are known to accompany cell death, senescence, aging and a number of pathophysiological conditions. Likewise, the NAD^+/NADH redox pair has many important roles in the cell. Disrupted NAD^+/NADH homeostasis has been found to be associated with several pathological conditions, including cancer (Zhang et al. 2006).

5.1.3 What Kinds of Redox Probes Exist?

Generally we can distinguish between optical and non-optical redox probes. The most prominent non-optical redox probes are those based on electron paramagnetic resonance (EPR) spectroscopy (Vikram et al. 2010). The main advantage of EPR is the ability to non-invasively image redox species in whole living animals, e.g. mice (Vikram et al. 2010; Shulaev and Oliver 2006). However, the disadvantages of EPR include (i) the low spatial resolution of images (cellular or subcellular compartment imaging is not possible), (ii) the necessity to load organisms with high concentrations of spin-trap probes, which can be toxic and suffer from many of the pharmacokinetic complications described below for chemical fluorescent probes (Shulaev and Oliver 2006) and (iii) the lack of chemical specificity of those EPR probes amenable to *in vivo* application, with the notable exception of NO· specific probes (Yoshimura et al. 1996).

Most optical probes are either based on fluorescence or luminescence. Only relatively few examples of luminescence-based redox probes exist compared to fluorescent probes. In this review we focus primarily on the fluorescent probes, because they have been most widely used for imaging in multicellular organisms and show promise for further improvements.

The vast majority of fluorescent redox probes are either organic chemical or genetically-encoded, and these types are the main focus of this review. However, it should be mentioned that there are now fluorescence-based redox probes, which are neither small organic chemicals nor proteins, namely nanotube- and nanoparticle-based probes designed to react to signaling concentrations of H_2O_2 or NO· (Jin et al. 2010; Casanova et al. 2009; Boghossian et al. 2011).

5.2 Chemical Fluorescent Probes and Their Applications

Chemical fluorescent probes have two key advantages over genetically encoded probes. The first and major advantage is that chemical probes can be applied to almost any sample without the prior need for lengthy genetic manipulation steps. Consequently, chemical probes can be directly applied, for example to human clinical samples, which may ultimately be useful in the context of disease diagnosis.

Chemical fluorescent probes are also especially flexible as their properties can be readily modified via the addition of chemical moieties. This strategy has been used to create novel probe variants with increased specificity towards target molecules and increased membrane permeability. The latter can be achieved by protecting hydroxyl and carboxyl groups with acetyl or acetoxymethyl (AM) groups, thus allowing cells to be more efficiently loaded with fluorescent probes. Chemical modifications can also be used to manipulate spectroscopic properties such as the fluorescence emission wavelength. This can be particularly useful for *in vivo* imaging; near-infrared (NIR) fluorescence, in the range of 650–900 nm enables deeper imaging of animal tissue than visible wavelengths because of low background fluorescence and minimal absorption by hemoglobin and water at these longer wavelengths (Weissleder and Ntziachristos 2003). The exact tissue penetration efficiency of NIR fluorescence is dependent on the particular tissue type examined, but is usually in the order of 1 cm (Lee et al. 2007). In particular, the cyanine fluorophore exhibits strong fluorescence in the NIR range and has been widely used as the basis of fluorescent probes or labeling agents for *in vivo* imaging applications (Klohs et al. 2008).

Chemical fluorescent probes can be broadly categorized into two main classes:

1. Hydro-type fluorescent probes, which are molecules delivered in the non-fluorescent reduced state and which become fluorescent upon oxidation. These are the 'conventional' probes that can react non-specifically and indirectly with a broad range of oxidants.

2. Fluorescent probes selective to particular oxidants. These probes have been designed from the outset to respond specifically to defined oxidant species.

Representative probes and their key properties are summarized in Tables 5.1 and 5.2.

5.2.1 Hydro-Type Fluorescent Probes

In the past, hydro-type chemical fluorescent probes, in particular dichlorodihydrofluorescein, dihydrorhodamine 123 and hydroethidium, have been widely used to report changes in 'cellular redox status' or 'ROS production'. However, fluorescent changes observed with this class of probes can be the result of a wide range of cellular processes and thus great care must be taken when designing and interpreting experiments. Hydro-type fluorescent probes are non-specific, highly susceptible to oxidation by undefined enzymatic processes (Halliwell and Whiteman 2004) and easily excluded from the cell by membrane transporters, including multidrug resistance protein (MDR) and multidrug resistance-associated protein (MRP) transporters (Huai-Yun et al. 1998; Ludescher et al. 1992). Due to the low specificity of these probes and the complexity of the cellular environment, the biological basis of fluorescence changes remains poorly understood in most cases. In fact, it may be difficult to determine if observed fluorescence changes are the result of changes in ROS concentration at all (Wardman 2007). In this section we focus on the three most commonly used hydro-type fluorescent probes, as well as the recently introduced hydrocyanines.

5.2.1.1 2′,7′-Dichlorodihydrofluorescein (DCFH)

DCFH is one of the most commonly used chemical fluorescent redox probes. An acetylated derivative, DCFH-DA, is typically used for intracellular measurements. Intracellular esterases catalyze the deacetylation of DCFH-DA to DCFH, which can then be oxidized to yield the highly fluorescent product 2′,7′-dichlorofluorescein (DCF) (Fig. 5.1a). However, the intracellular process of DCFH oxidation is complex and ill-defined. *In vitro*, DCFH appears to be unreactive toward H_2O_2 and $O_2^{\cdot -}$. DCFH oxidation by H_2O_2 can be facilitated by peroxidase activity (LeBel et al. 1992). DCFH oxidation can also be mediated by ferric iron and cytochrome *c* via the generation of HO· or peroxidase compound I-type oxoferryl species, respectively (LeBel et al. 1992; Lawrence et al. 2003). 5-lipoxygenase (LOX) was also reported as a potent catalyst of DCFH oxidation (Hempel et al. 1999). DCFH oxidation may generally be influenced by the intracellular concentration of heme and heme proteins (Ohashi et al. 2002). Additionally, DCFH is prone to photo-oxidation, which may pose problems when the probe is exposed to high-intensity laser irradiation. In general, DCFH oxidation proceeds via a radical intermediate ($DCF^{\cdot -}$) which is

Table 5.1 Hydro-type fluorescent probes

Name	Target species	Fluorophore	Features	Sensing mechanism	Physiological application	Ref.
DCFH[a]	Nonspecific	Fluorescein	Susceptible to photo-oxidation Non-specific oxidation	Oxidation	✓	1
DHR123[a]	Nonspecific	Rhodamine	Susceptible to photo-oxidation Non-specific oxidation	Oxidation	✓	1
DHE[a]	$O_2^{\cdot-}$ (partially nonspecific)	Ethidium	Highly susceptible to photo-oxidation Non-specific oxidation	Oxidation	✓	2
MitoSOX™ Red[a]	$O_2^{\cdot-}$ (partially nonspecific)	Ethidium	Highly susceptible to photo-oxidation Non-specific oxidation Targeted to mitochondria	Oxidation	✓	2
Hydrocyanines	HO·, $O_2^{\cdot-}$	Cyanine	NIR fluorescence Susceptible to photo-oxidation	Oxidation	✓	3

1. Halliwell and Whiteman (2004), 2. Robinson et al. (2006), 3. Kundu et al. (2009)
[a]Commercially available

Table 5.2 Fluorescent probes selective to oxidants

Name	Target species	Fluorophore(s)	Features	Sensing mechanism	Physiological application	Ref.
BESSo[a]	$O_2^{\cdot -}$	Fluorescein	AM derivative available	Deprotection of BES	✓	1
PG1[a], PC1[a]	H_2O_2	Fluorescein, Resorufin	Slow reaction rate (1 M^{-1} s^{-1})	Deprotection of boronate	✓	2
PF2[a], PF3[a], PE1[a], PY1[a], PO1[a]	H_2O_2	Fluorescein, Rhodol	Different fluorescence wavelength applicable to multicolor imaging	Deprotection of boronate	✓	3
NPF1[a]	H_2O_2	Naphtho-fluorescein	NIR fluorescence	Deprotection of boronate		4
RPF1[a]	H_2O_2	Coumarin and Fluorescein	Ratiometric measurement	Deprotection of boronate		5
PL1[a]	H_2O_2	Naphthalimide	Ratiometric measurement	Deprotection of boronate	✓	6
Mito-PY1[a]	H_2O_2	Rhodol	Targeted to mitochondria	Deprotection of boronate		7
SPGs[a]	H_2O_2	Fluorescein	Controllable localization by SNAP tag expression	Deprotection of boronate		8
PCL-1	H_2O_2	Luciferin	Bioluminescence detection	Deprotection of boronate	✓	9
PF6[a]	H_2O_2	Fluorescein	High intracellular retention	Deprotection of boronate	✓	10
NucPE-1	H_2O_2	Rhodol	Localization to nucleus High background fluorescence	Deprotection of boronate	✓	11
QCy7 based	H_2O_2	Cyanine	NIR fluorescence	Deprotection of boronate	✓	12
NBzF	H_2O_2	Fluorescein	High fluorescence activation upon reaction (∼150 fold)	Baeyer-Villiger reaction with H_2O_2	✓	13
HPF[a]	HO·, $ONOO^-$	Fluorescein	Resistance to photo-oxidation	*O*-dearylation	✓	14
APF[a]	HO·, $ONOO^-$, HOCl	Fluorescein	Resistance to photo-oxidation	*O*-dearylation	✓	14
APC	HO·, $ONOO^-$, HOCl	Calcein	High intracellular retention	*O*-dearylation	✓	15

MitoHR	HO·, $ONOO^-$	Rhodamine	Targeted to mitochondria Resistance to photo-bleaching	*O*-dearylation		16
MitoAR	HO·, $ONOO^-$, HOCl	Rhodamine	Targeted to mitochondria Resistance to photo-bleaching	*O*-dearylation	✓	16
SNAPF	HOCl	Naphtho-fluorescein	NIR fluorescence	*O*-dearylation	✓	17
HySOx	HOCl	Rhodamine	Colorless Resistance to photo-bleaching	Sulfur atom oxidation	✓	18
MMSiR, ws-MMSiR	HOCl	Si-rhodamine	Colorless NIR fluorescence Resistance to photo-bleaching	Sulfur atom oxidation	✓	19
NiSPY1[a]	$ONOO^-$	BODIPY	High fluorescence activation	Nitration		20
HKGreen1	$ONOO^-$	Fluorescein	Low conversion efficiency to fluorescent product after reaction	Reaction with trifluoromethyl ketone		21
HKGreen2	$ONOO^-$	BODIPY	pH insensitivity	Reaction with trifluoromethyl ketone	✓	22
Cy-PSe	$ONOO^-$	Cyanine	NIR fluorescence Reducible by GSH	Selenium atom oxidation	✓	23
FOSCY	HO·, $ONOO^-$, HOCl, $O_2^{\cdot-}$	Cyanine	Static quenching of fluorescence between Cy5 and Cy7	Oxidative decomposition of Cy7 moiety	✓	24

(continued)

Table 5.2 (continued)

Name	Target species	Fluorophore(s)	Features	Sensing mechanism	Physiological application	Ref.
DAF2[a]	NO	Fluorescein	NIR fluorescence Reactive to NO^+	Triazole formation	✓	25
DAF-FM[a]	NO	Fluorescein	DA derivative available Reactive to NO^+	Triazole formation	✓	26
DAR[a]	NO	Rhodamine	Less pH sensitivity than DAF2 Reactive to NO^+	Triazole formation	✓	27
DCl-DA Cal	NO	Calcein	Resistance to photo-bleaching Reactive to NO^+	Triazole formation	✓	28
DAC-P, DAC-S	NO	Cyanine	High intracellular retention Reactive to NO^+	Triazole formation		29
CuFL[a]	NO	Fluorescein	NIR fluorescence Reactive to NO·	Reduction and release of Cu^{2+}	✓	30
Cu_2(FL2A)	NO	Fluorescein	Reactive to NO·	Reduction and release of Cu^{2+}	✓	31
MNIP-Cu	NO	MNIP	High intracellular retention Reactive to NO·	Reduction and release of Cu^{2+}	✓	32
mBCl[a]	GSH	Bimane	Applicable to two photon excitation	Glutathionylation	✓	33

1. Maeda et al. (2007), 2. Miller et al. (2007), 3. Dickinson et al. (2010), 4. Albers et al. (2008), 5. Srikun et al. (2008), 6. Albers et al. (2006), 7. Dickinson and Chang (2008), 8. Srikun et al. (2010), 9. Van de Bittner et al. (2010), 10. Dickinson et al. (2011a), 11. Dickinson et al. (2011b), 12. Karton-Lifshin et al. (2011), 13. Abo et al. (2011), 14. Setsukinai et al. (2003), 15. Izumi et al. (2009), 16. Koide et al. (2007), 17. Shepherd et al. (2007), 18. Kenmoku et al. (2007), 19. Koide et al. (2011), 20. Ueno et al. (2006), 21. Yang et al. (2006), 22. Sun et al. (2009), 23. Yu et al. (2011), 24. Oushiki et al. (2010), 25. Kojima et al. (1998a, c), 26. Kojima et al. (1999), 27. Kojima et al. (2001), 28. Izumi et al. (2009), 29. Sasaki et al. (2005), 30. Lim et al. (2006), 31. McQuade et al. (2010), 32. Ouyang et al. (2008), 33. Rice et al. (1986)

[a]Commercially available

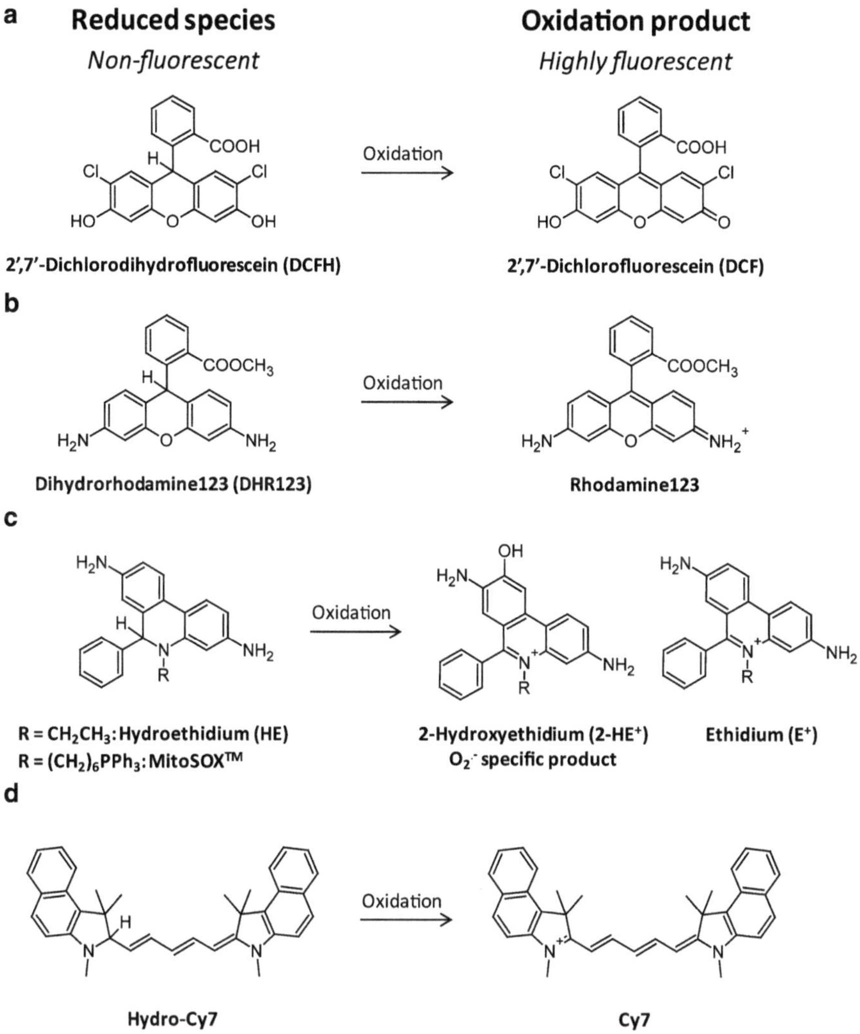

Fig. 5.1 Hydro-type fluorescent probes DCFH (**a**), DHR123 (**b**), DHE and MitoSOX™ Red (**c**), and hydroCy7 (**d**)

further oxidized to DCF with the concomitant generation of $O_2^{\cdot -}$. Thus, DCFH-induced $O_2^{\cdot -}$ formation (and subsequent formation of H_2O_2) likely fuels further DCFH oxidation, potentially leading to an overestimation of ROS production (Wrona et al. 2005). In addition to these considerations, DCFH may also diffuse spontaneously out of cells or be actively removed by membrane transporters. Thus it is often not clear to what extent differences or changes in fluorescence actually reflect intracellular ROS production (Halliwell and Whiteman 2004; Wardman 2007). Despite these limitations, DCFH-DA has been applied widely to the study of redox changes in multicellular organisms, including *C. elegans* and zebrafish (Schulz et al. 2007; Morgan et al. 2010; Kishi et al. 2008; Harding et al. 2003).

5.2.1.2 Dihydrorhodamine123 (DHR123)

DHR123 is based on the rhodamine fluorophore and is frequently used as a ROS probe (Fig. 5.1b). DHR123 readily crosses cellular plasma membranes and thus will likely distribute in the cytosol. DHR123 appears to suffer from many of the problems already described for DCFH and the intracellular reactivity of DHR123 is complex and poorly understood. Like DCFH, DHR123 has been shown to have very low intrinsic reactivity towards H_2O_2 and $O_2^{\cdot -}$ (Hempel et al. 1999). Instead, the reaction of DHR123 with H_2O_2 is dependent on catalysis and was shown to be mediated by peroxidases and oxidases, including xanthine oxidase (Hempel et al. 1999). Consequently, differences or changes in the expression level of peroxidases and oxidases may affect intracellular DHR123 oxidation. Like DCFH, DHR123 is susceptible to photo-oxidation. Following oxidation, the positively charged and hydrophobic product, rhodamine123, accumulates in the mitochondrial matrix due to the negative potential of the mitochondrial inner membrane (Johnson et al. 1981). Therefore, the extent of matrix accumulation may be affected by changes in mitochondrial membrane potential (Johnson et al. 1981). Finally, it should be noted that the level of MRP expression is known to influence the intracellular concentration of rhodamine123 (Ludescher et al. 1992).

5.2.1.3 Hydroethidium (HE) and Mito-HE

Hydroethidium (HE) and Mito-HE (known as MitoSOX™ Red) are commercialized as probes specific for $O_2^{\cdot -}$ (Fig. 5.1c). MitoSOX™ Red is an HE derivative with a triphenylphosphonium (TPP) moiety for mitochondrial targeting. Photo-oxidation of HE is ten times more rapid than that of DCFH and DHR123, leading to high background fluorescence (Buxser et al. 1999). It has been shown that HE can react directly with ferricytochrome c (Benov et al. 1998) and similar to DCFH and DHR123, HE may be oxidized by peroxidases (Wardman 2007). However, recent work suggests that $O_2^{\cdot -}$-mediated HE oxidation can be measured more specifically. It is believed that HE initially interacts with heme proteins to form an ethidium radical ($HE^{\cdot +}$) intermediate, independent of the presence of $O_2^{\cdot -}$ (Robinson et al. 2006). By radical coupling with $O_2^{\cdot -}$ the intermediate can further oxidize to yield 2-hydroxy-ethidium (HO-E^+). In contrast, one electron oxidation by other electron acceptors generates ethidium (E^+), in a reaction possibly mediated by heme proteins (Robinson et al. 2006). Both oxidation products rely on intercalation into DNA to become highly fluorescent. The much faster rate of radical coupling compared to one electron oxidation likely ensures that HO-E^+, rather than E^+, is the dominant product that results from the interaction with $O_2^{\cdot -}$ (Robinson et al. 2006). Thus, for specific detection of $O_2^{\cdot -}$ it is crucial to differentiate between the two different oxidation products. The different excitation wavelengths (385–405 and 480–520 nm for HO-E^+ and E^+, respectively) may allow to discriminate between the two species (Robinson et al. 2006, 2008). However, in nearly all previous studies using HE or MitoSOX™ Red, fluorescence excitation was in the 480–520 nm range, therefore E^+ rather than HO-E^+ was measured, and it remains unclear if $O_2^{\cdot-}$ production was specifically detected. Astrocytes expressing a superoxide dismutase mutant associated with amyotrophic lateral sclerosis were shown to exhibit increased HO-E^+ generation (Robinson et al. 2008).

5.2.1.4 Hydrocyanine-Based Probes

Reduced cyanine-based probes known as hydrocyanines, e.g. Hydro-Cy3 and Hydro-Cy7, fluorescent in the NIR range, were recently employed as ROS probes (Kundu et al. 2009). Hydrocyanines can be obtained by $NaBH_4$-mediated reduction of commercially available cyanine dyes. Similar to other hydro-type probes, reduction disrupts π-conjugation of the cyanine fluorophores, making them non-fluorescent. *In vitro,* direct oxidation by $O_2^{\cdot-}$ or HO· recovered the original cyanine fluorophore (Fig. 5.1d). Hydro-Cy3 was found to be oxidized in live cells and tissue explants in response to angiotensin II and lipopolysaccharide (LPS) stimulation, respectively (Kundu et al. 2009). Hydro-Cy7 oxidation was observed *in vivo* during acute inflammation in LPS-treated mice (Kundu et al. 2009). However, in each case, it remains unclear which ROS are driving probe oxidation. Further probe characterization is required to understand the mechanism of oxidation and to assess specificity and pharmacokinetics.

5.2.2 *Fluorescent Probes Selective to Oxidants*

Recently, chemical fluorescent probes which promise much greater ROS specificity have been developed (Chen et al. 2011). The deliberate design of chemical moieties which react directly and selectively with defined redox species forms the basis for this group of probes. Although these probes should be preferable over hydro-type probes, most of them are not yet commercially available. Several of these probes have so far only been described in one publication and thus remain incompletely characterized, especially in terms of their *in vivo* applicability.

5.2.2.1 $O_2^{\cdot-}$-Selective Fluorescent Probes

The probe 4, 5-dimethoxy-2-nitro-benzenesulfonyl tetrafluorofluorescein (BESSo) becomes fluorescent in the presence of $O_2^{\cdot-}$ (Maeda et al. 2007). Fluorescence is initially quenched because the fluorescein hydroxyl group is protected as a 4, 5-dimethoxy-2-nitro-benzenesulfonyl (BES) ester. Nucleophilic substitution of

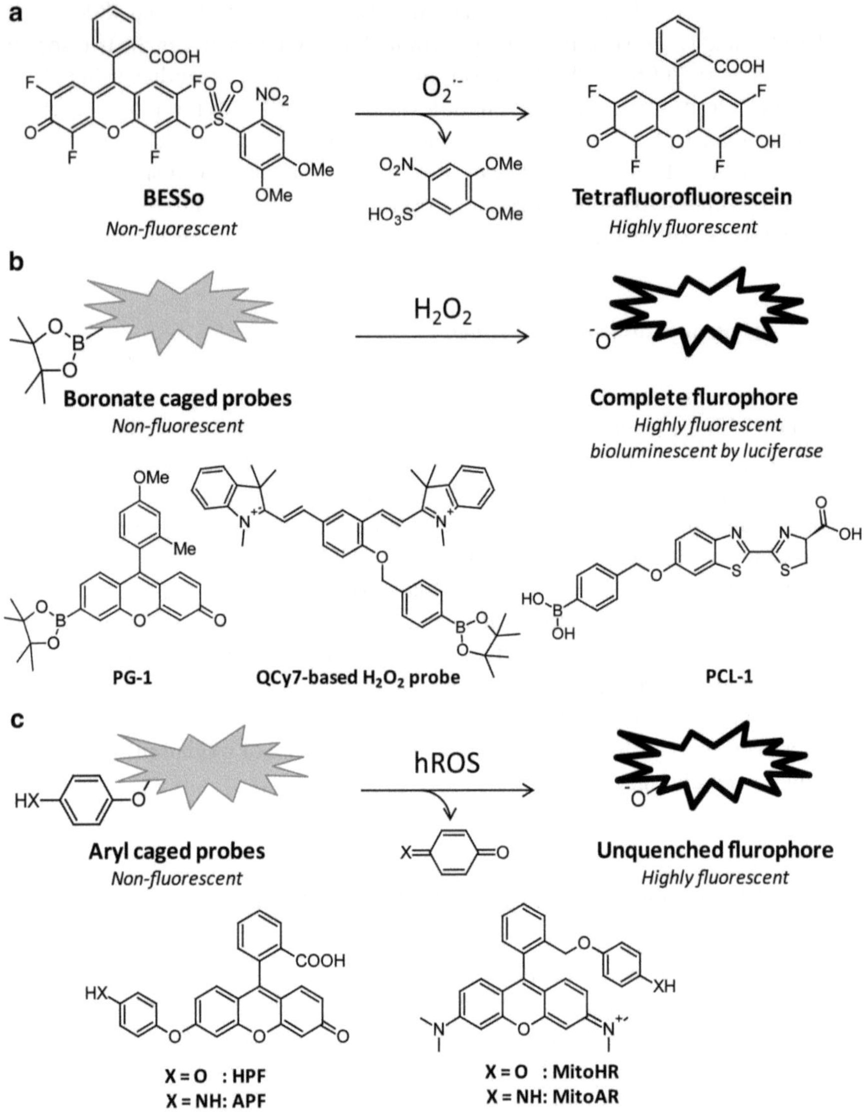

Fig. 5.2 Fluorescent probes selective for superoxide (**a**), hydrogen peroxide (**b**) and hROS (**c**)

the sulfur atom in the BES ester by $O_2^{\cdot -}$ releases the highly fluorescent tetrafluorofluorescein (Fig. 5.2a). *In vitro* experiments suggest that BESSo has little reactivity with a wide range of other redox species, including the dominant cellular nucleophile GSH. An acetoxymethyl derivative of BESSo, which allows for more efficient cellular loading, was used to detect $O_2^{\cdot -}$ generation in activated human neutrophils and following butyric acid treatment of Jurkat T cells (Maeda et al.

2007). The reaction rate of BESSo oxidation by $O_2^{\cdot-}$ has not yet been determined. Thus, the ability of the probe to compete with endogenous $O_2^{\cdot-}$ scavenging systems remains unclear. This aspect may be especially important when studying processes generating low levels of $O_2^{\cdot-}$.

5.2.2.2 H_2O_2-Selective Fluorescent Probes

The selective detection of H_2O_2 has been accomplished by the use of three different chemical strategies. The first strategy utilizes aryl boronate cages as protecting groups. H_2O_2 selectively opens the boronate cage, thus deprotecting the fluorescent dye (Fig. 5.2b). This principle forms the basis for a series of H_2O_2 probes with various spectral and chemical properties, as described below. Probes with a range of different fluorescence emission wavelengths have been created, offering much greater flexibility for making simultaneous measurements with other probes, or potentially for parallel H_2O_2 measurement in more than one subcellular compartment (Albers et al. 2008; Dickinson et al. 2010). Aryl boronate cage-based probes based on fluorescence resonance energy transfer (FRET) (e.g. Ratio Peroxyfluor-1) or internal charge transfer (ICT) (e.g. Peroxy Lucifer-1) have also been generated (Albers et al. 2006; Srikun et al. 2008). These probes enable ratiometric measurement which reduces concerns relating to differential cellular probe accumulation. Further, targeting of these probes to the mitochondrial matrix and nucleus has been demonstrated, thus enabling a certain degree of compartment-specific H_2O_2 detection. (Dickinson and Chang 2008; Srikun et al. 2010; Dickinson et al. 2011a, b). NucPE1, which localizes to the nucleus, has been used for H_2O_2 detection in *C. elegans* (Dickinson et al. 2011b).

A new type of cyanine, QCy7, which fluoresces in the near-infrared region, has also been developed to utilize the boronate-cage protection mechanism for H_2O_2 detection (Karton-Lifshin et al. 2011). π–conjugation in the fluorophore is disturbed by caging with a boronic acid benzyl ether, resulting in fluorescence quenching. The reaction with H_2O_2 drives the release of the caging moiety to generate highly fluorescent, deprotected QCy7. A QCy7-based H_2O_2 probe has already been applied to a mouse model of inflammation, and a fluorescent signal was observed in inflamed tissues (Karton-Lifshin et al. 2011).

The boronate cage protection strategy has also been extended to a bioluminescence-based probe (Van de Bittner et al. 2010). PeroxyCaged Lucifecin-1 (PCL-1) is a firefly luciferin derivative that is unreactive towards luciferase. Its reaction with H_2O_2 leads to the selective removal of the boronate cage, releasing luciferin which then acts normally as a substrate for luciferase, thus producing a bioluminescent readout. PCL-1 was used to show H_2O_2 production in an LNCaP-luc tumor xenograft model following testosterone stimulation (Van de Bittner et al. 2010).

Some boronate-based probes have the disadvantage that they react with H_2O_2 rather slowly, with rate constants of $\sim$1 M^{-1} s^{-1} *in vitro*, as shown for Peroxy Green-1 (PG1) and Peroxy Crimson (PC1). It is not clear what drives the observed fluorescence increase of these probes *in cellulo* given that the highly abundant and efficient H_2O_2 scavenging enzymes (rate constants up to 10^7–10^8 M^{-1} s^{-1}) should almost completely outcompete the probes' reaction with H_2O_2 (Rhee et al. 2010). Further, it should be noted that phenyl boronates were recently shown to react much more efficiently with $ONOO^-$ and HOCl than with H_2O_2 (Sikora et al. 2009), which may be an important consideration when applying boronate-based probes to inflammation models.

The second chemical strategy for the detection of H_2O_2 is the use of a benzil (dibenzoyl) moiety, which selectively reacts with H_2O_2 to form benzoic anhydride via a Baeyer-Villiger reaction and subsequent hydrolysis. The only probe currently based on this mechanism, NBzF, has a much larger fluorescence enhancement ratio than boronate-based probes due to stronger fluorescence quenching prior to the reaction (Abo et al. 2011). NBzF is able to detect physiologically relevant signaling concentrations of H_2O_2 as demonstrated in epidermal growth factor (EGF)-stimulated A431 cells (Abo et al. 2011).

A third strategy for the detection of H_2O_2 involves peroxalate nanoparticles, which carry a peroxylate ester and a fluorescent dye (Lee et al. 2007). One advantage of this probe is light-source independence. The peroxalate ester is thought to react with H_2O_2 specifically, giving rise to a high energy dioxetanedione intermediate which excites the dye to emit in the NIR range. Another advantage of this probe is that the fluorescence emission wavelength can be tuned by the choice of different fluorescent dyes. Fluorescence emission from peroxalate nanoparticles was observed in a mouse inflammation model (Lee et al. 2007).

5.2.2.3 hROS-Selective Fluorescent Probes

2-[6-(4′-hydroxy)phenoxy-3H-xanthen-3-on-9-yl]benzoic acid (HPF) and 2-[6-(4′-amino)phenoxy-3H-xanthen-3-on-9-yl]benzoic acid (APF) have been introduced as probes for highly reactive oxygen species (hROS) (Setsukinai et al. 2003). The intrinsic fluorescence of HPF and APF is quenched by the protection of the phenolic hydroxyl group at the 6′-position of fluorescein by electron-rich 4-hydroxyphenoxyl and 4-aminophenoxyl groups, respectively. Upon reaction with hROS, an O-dearylation reaction releases fluorescein (Fig. 5.2c) (Setsukinai et al. 2003). HPF is known to react selectively with HO· and $ONOO^-$ to become highly fluorescent (Setsukinai et al. 2003). In addition to HO· and $ONOO^-$, APF also reacts with HOCl, thus HOCl production can be infered by comparison between APF and HPF. Both probes are highly resistant to photo-oxidation. Although HPF and APF have clear advantages for detecting hROS, the fluorescent product fluorescein can be transported out of cells. Intracellular retention of APF was improved by the introduction of two imino-diacetic acid groups, yielding APC (Izumi et al. 2009). This simple concept of increasing cellular retention by introducing negative charge to yield a calcein–based structure may serve as a general strategy to increase the sensitivity of fluorescent probes in living cells (Izumi et al. 2009).

MitoHR and MitoAR are based on the rhodamine fluorophore. The positive charge of rhodamine drives localization to the mitochondrial matrix. One advantage of rhodamine-based probes is a general resistance to laser irradiation-induced photo-bleaching, making them especially suitable for fluorescence microscopy (Koide et al. 2007). The intrinsic probe fluorescence is quenched by the electron rich 4-hydroxyl- and 4-aminophenyl ether moieties. Reaction with hROS leads to cleavage of the ether moiety, giving rise to 2-hydroxymethyl tetramethylrhodamine (HMTMR), a highly fluorescent product (Fig. 5.2c) (Koide et al. 2007). MitoHR mainly reacts with HO·, and to a lesser extent with $ONOO^-$ and HOCl, whereas MitoAR reacts with HO· and HOCl, and to a lesser extent with $ONOO^-$. MitoAR was shown to detect myeloperoxidase (MPO)-derived HOCl in HL60 cells (a human promyelocytic leukemia cell line) stimulated with H_2O_2, but not in HeLa cells (which do not express MPO) (Koide et al. 2007).

There are several fluorescent probes considered to detect HOCl specifically, including sulfonaphthoaminophenyl fluorescein (SNAPF), which emits in the NIR region (Shepherd et al. 2007). Similar to the protection mechanism employed in APF, the fluorescence of the unreacted probe is quenched by a 4-aminophenoxyl moiety. In SNAPF, but apparently not APF, this moiety is selectively removed by reaction with HOCl. SNAPF has been used *in vivo* to demonstrate HOCl production in a mouse peritonitis model (Shepherd et al. 2007).

HySox is also designed to specifically detect HOCl and is based on a ring-opening mechanism (Kenmoku et al. 2007). It is initially colorless and non-fluorescent because of a closed ring structure that disrupts π–conjugation in the fluorophore. Upon oxidation of the sulfur atom in the five-membered ring structure, the rhodamine fluorophore is formed. An advantage of this probe is the complete absence of fluorescence in the unreacted molecule. Thus, any increase in fluorescence can be ascribed with a high degree of certainty to HOCl rather than other factors such as unreacted probe accumulation. This property of HySox leads to excellent signal-to-background ratios in contrast to probes that contain the mature fluorophore prior to activation and thus can generate high background signals by simple accumulation (Kenmoku et al. 2007). HySox has been used to detect HOCl production in activated neutrophils (Kenmoku et al. 2007). The same principle has been extended to the NIR fluorophore MMSiR and its water-soluble analogue, wsMMSiR. These are based on the Si-rhodamine NIR fluorophore, and have enabled HOCl detection *in vivo* in a mouse peritonitis model (Koide et al. 2011).

Several fluorescent probes are considered to specifically detect $ONOO^-$. Mechanistically, these probes are based on either nitration (NiSPY-1) (Ueno et al. 2006), formation of a dienone from a trifluoromethyl ketone (HKGreen-1 and HKGreen-2) (Yang et al. 2006; Sun et al. 2009) or selenium oxidation (CyP-Se) (Yu et al. 2011). Of these, HKGreen-2 and CyP-Se, based on BODIPY and cyanine fluorophores, respectively, were applied to demonstrate $ONOO^-$ generation in stimulated macrophages (Sun et al. 2009; Yu et al. 2011). As CyP-Se fluoresces in the NIR region, it may have potential for future applications *in vivo*.

The cyanine-based probe FOSCY-1 consists of two fluorophores, $Cy5SO_3H$ and the Cy7-based IR786S (Oushiki et al. 2010). Fluorescence is quenched by self-stacking between the two hydrophobic fluorophores. Activation of fluorescence is based on the different reactivities of Cy5 and Cy7 to a range of ROS. Upon oxidation of the IR786S moiety by HO·, $ONOO^-$ or HOCl, and to a lesser extent by $O_2^{\cdot-}$, the Cy5 moiety becomes fluorescent due to the loss of intramolecular stacking. FOSCY-1 has been successfully used to detect ROS production in an *in vivo* peritonitis model (Oushiki et al. 2010). Although reactive to a broad range of ROS, the oxidation mechanism of this probe suggests that one advantage may be a very slow rate of photo-oxidation.

5.2.2.4 NO· Selective Fluorescent Probes

NO·-selective fluorescent probes are based either on an *o*-phenylenediamine moiety or a metal complex. The former are indirect NO· probes because they actually react with nitrosonium equivalents (NO^+/N_2O_3) to generate a triazole (Fig. 5.3a). It is assumed that in the presence of molecular oxygen, nitrosonium equivalents are continuously generated from NO· (Ignarro et al. 1993) and can be taken as a proxy for NO·. The diaminofluorescein (DAF) probes are based on this principle (Kojima et al. 1998a, b, 1999). DAF2-DA has been used to detect NO· production in the CA1 region of hippocampus brain slices following stimulation with *N*-methyl-D-aspartic acid (NMDA) (Kojima et al. 1998b). However, DAF2-DA suffers from pH sensitivity and susceptibility to photo-bleaching. DAF-FM was developed to be less sensitive to pH and photo-bleaching than DAF2. Using a di-acetate derivative of DAF-FM, changes in NO· production were observed in the notochord and caudal fin of developing zebrafish (Lepiller et al. 2007). However, it remains to be unambiguously demonstrated that the changes observed with DAF-FM-DA are solely a result of NO· production rather than a differential accumulation of the unreacted probe or fluorescent product. Diaminorhodamine (DAR)-4M, based on the rhodamine fluorophore, is less sensitive to photo-bleaching and pH changes than any of the DAF probes (Kojima et al. 2001). DAR-4M has been used to demonstrate cell layer-specific NO· production in freshly prepared mouse cortical slices upon NMDA stimulation (Imura et al. 2005).

Although the DAF and DAR probes are promising tools for detecting intracellular NO· in cell culture or transparent model organisms, they have the disadvantage of limited cellular retention, thus decreasing their sensitivity for NO· detection. This problem has recently been overcome by the introduction of imino-diacetic acid groups into DAF4, producing dichloro-diamino-calcein (DCl-DA-Cal). DCl-DA-Cal was found to be more sensitive than DAF2 in detecting NOS-generated NO· in bovine aortic endothelial cells upon bradykinin stimulation (Izumi et al. 2009). Cyanine-based diaminocyanine (DAC)-P and DAC-S have also been created to enable NO· detection in the NIR region. DAC-P detected NO· release in isolated rat kidneys perfused with the NO·-generating compound NOC13 (Sasaki et al. 2005).

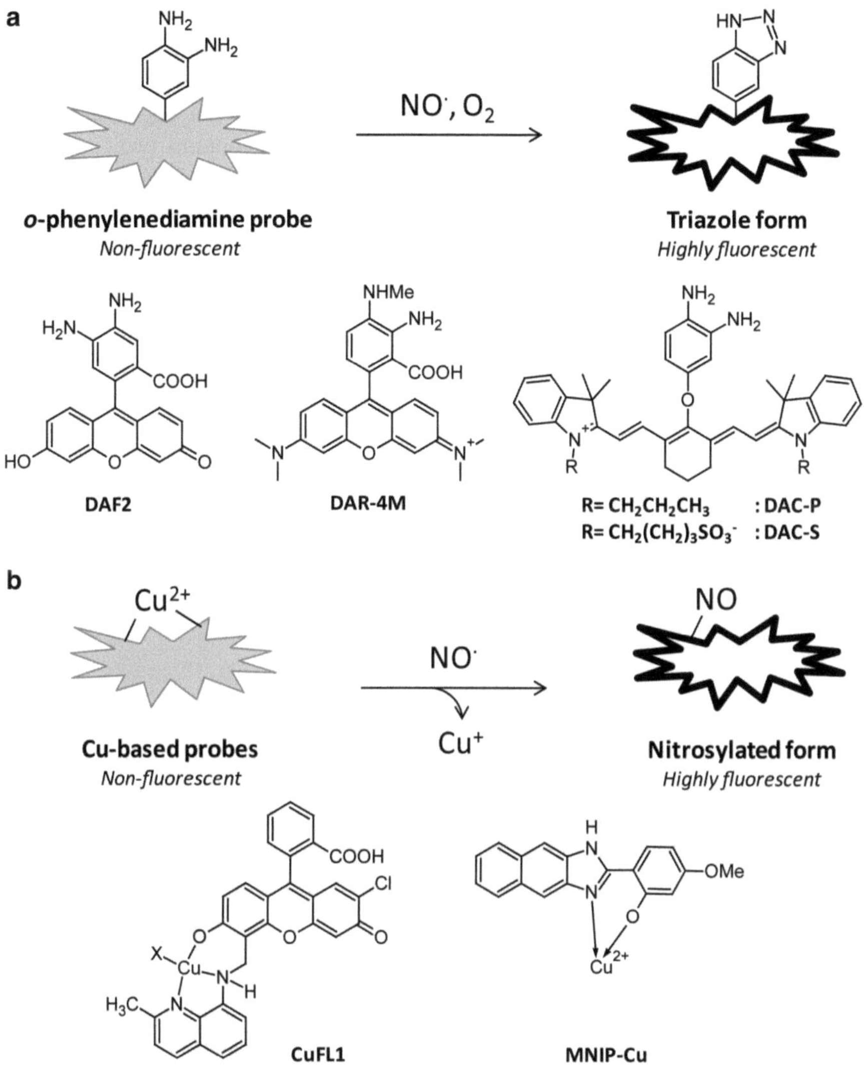

Fig. 5.3 Fluorescent probes selective for nitric oxide based on *o*-phenylenediamine (**a**) or metal complex (**b**)

In contrast to *o*-phenylenediamine-based probes, metal complex-based probes detect the NO· radical directly (Fig. 5.3b). To date, all probes developed in this class consist of a fluorophore ligand coordinating a paramagnetic Cu^{2+} ion, which quenches the fluorescence of the fluorophore. Reduction of Cu^{2+} to Cu^{+} by NO· releases a nitrosylated and highly fluorescent fluorophore ligand. CuFL, a fluorescein-based probe based on this mechanism has been reported to detect NO· generated by constitutive and inducible NOS in SK-N-SH human neuroblastoma cells and Raw264.8 murine macrophages, respectively (Lim et al. 2006). In a direct comparison with DAF2-DA, CuFL generated a brighter signal in response to cNOS activation. The intracellular retention of Cu-based NO· probes has been improved with the development of Cu_2(FL2A), in which two additional negative charges have been introduced (McQuade et al. 2010). Cu_2(FL2E) contains two carboxylic acids protected as ethyl esters which can be cleaved by intracellular esterases. Despite the fact that this probe needs to react with two NO· molecules in order to become fluorescent, its improved cellular retention renders it more sensitive to NO· than CuFL. Application of Cu_2(FL2E) to murine olfactory bulb brain slices successfully detected cell layer-specific NO· generation (McQuade et al. 2010). Another probe utilizing a Cu-based quenching mechanism, MNIP-Cu, is based on the 4-methoxy-2-(1H-naphtho [2,3-d] imidazol-2-yl) phenol (MNIP) fluorophore (Ouyang et al. 2008). Reaction with NO· generates *N*-nitroso MNIP as an NO· reaction-specific fluorescent product. MNIP-Cu does not seem to exhibit acute *in vivo* toxicity. MNIP-Cu injection has been applied to a mouse liver injury model, where a NO·-dependent fluorescent increase was detected in cryosections *in situ* (Ouyang et al. 2008).

5.2.3 Fluorescent Probes Reactive to Thiols

The compartment-specific concentration and redox state of intracellular glutathione is one of the key elements of cellular redox homeostasis. Chemical probes for glutathione are only able to measure relative differences in the concentration of free GSH. Such probes must be membrane-permeable and should be able to discriminate GSH from other thiol-containing molecules, in particular protein thiols and cysteine. Only monochlorobimane (mBCl), i.e. syn-(1-chloroethyl)-1,5-diazabicycla[3.30]acta-3,6,-diene-2,8-dione, is commercially available at this time, although several other fluorescent probes for the detection of intracellular GSH have been reported (Chen et al. 2010). mBCl is essentially non-fluorescent until it forms a strongly fluorescent glutathione-bimane adduct. Adduct formation is catalyzed by intracellular glutathione S-transferases (GSTs) and it is typically assumed that mBCl reacts mainly with cytosolic GSH, due to the abundance of GST activity in this compartment. mBCl is frequently applied to measure relative differences in cytosolic GSH concentration between cell types. However, when making such comparisons it is necessary to ensure that the fluorescent signal has reached a 'plateau' level for all samples. Otherwise, the measurements can be affected by cell-type specific differences in GST activity, which influence the rate of fluorescence accumulation but should not alter the final fluorescence intensity. When mBCl was applied to measure the distribution of GSH concentrations in cancer cells from a primary tumor, substantial variability in mBCl fluorescence was observed (Rice et al. 1986; Cook et al. 1991). Further, cell-type specific mBCl fluorescence was observed by microscopy in both plant tissues and brain slices (Keelan et al. 2001; Meyer and Fricker 2000; Meyer et al. 2001; Sun et al. 2006). As a note of caution, mBCl is routinely used to measure the activity of glutathione conjugate transporters (Ishikawa et al. 1994), whose expression varies between cell types and situations. Thus the process of glutathione-bimane adduct efflux likely influences the intracellular fluorescence to different extents in different cell types. Further, the general specificity of mBCl for glutathione has been questioned (van der Ven et al. 1994).

5.3 Genetically Encoded Redox Probes and Their Applications

Genetically encoded probes have two major advantages over currently available chemical probes. These are (i) precise targeting and (ii) redox species specificity. In principle, genetically encoded probes can be accurately targeted to almost any subcellular location. This is achieved either by the addition of appropriate targeting sequences or via translational fusion to a protein which localizes to the cellular location of interest. Redox species specificity is not an inherent property of all genetically encoded probes but is typically realized by the incorporation of an oxidant-specific protein domain within the probe structure, which serves to couple the probe response to a defined redox species. Current genetically encoded redox probes can be categorized by their mode of operation, into the following four classes:

1. Fluorescent proteins (FPs) that have been engineered to be redox sensitive by placement of a pair of cysteines onto the surface of the beta barrel. There are two archetypes in this class: redox-sensitive yellow fluorescent protein (rxYFP) and reduction-oxidation-sensitive green fluorescent protein (roGFP).
2. Fusion proteins based on the redox-sensitive FPs in class 1. They are engineered redox relays in which a redox enzyme is translationally fused to a redox-sensitive FP in order to catalyze its equilibration with a defined redox pair. This class includes the E_{GSH} probe Grx1-roGFP2 and the H_2O_2 probe roGFP2-Orp1.
3. Probes based on a FP that is not redox sensitive in itself but is responsive to conformational changes in a redox-sensitive protein domain to which it is fused. The best known example being the H_2O_2 probe HyPer.
4. FRET probes which either make use of thiol-containing peptide linkers or redox-sensitive protein domains to modulate the spatial arrangement of two FPs which make up a donor-acceptor FRET pair.

Representatives of the four classes are listed in Table 5.3 and those of particular interest will be discussed briefly in the following subsections. Aspects that have been covered in recent reviews (Meyer and Dick 2010; Morgan et al. 2011; Bjornberg et al. 2006a), in particular molecular mechanisms, thermodynamic considerations, and principles of ratiometric fluorescence measurements, will not be discussed here.

Table 5.3 Genetically encoded redox probes

Type	Name	Ratiometric (pH sensitivity)	FP	Additional sensing domain	Specificity	*In vivo* applications reported	Ref.
I	rxYFP		YFP	none	GSSG/GSH (dependent on endogenous Grx)		1
I	roGFP1	✓ (−)	GFP	none	GSSG/GSH (dependent on endogenous Grx)	✓	2
I	roGFP2	✓ (−)	EGFP	none	GSSG/GSH (dependent on endogenous Grx)	✓	2
II	rxYFP-yGrx1		YFP	Grx1 (*S. cerevisiae*)	GSSG/GSH (E_{GSH})		3
II	Grx1-roGFP2	✓ (−)	EGFP	Grx1 (*H. sapiens*)	GSSG/GSH (E_{GSH})	✓	4
II	roGFP2-Grx1	✓ (−)	EGFP	Grx1 (*H. sapiens*)	GSSG/GSH (E_{GSH})	✓	5
II	roGFP2-Orp1	✓ (−)	EGFP	Orp1 (*S. cerevisiae*)	H_2O_2	✓	6
III	HyPer	✓ (!)	cpYFP	OxyR (*E. coli*)	H_2O_2	✓	7
III	OHSer		cpVenus	OhrR (*X. campestris*)	ROOH (organic hydroperoxides)		8
III	Peredox	✓ (−)	cpGFP	Rex (*Thermus aquaticus*)	NADH/NAD^+		9
III	Frex	✓ (!)	cpYFP	Rex (*B. subtilis*)	NADH		10
IV	HSP-FRET	✓ (?)	CFP/YFP	Hsp-33 (*E. coli*)	? (incl. H_2O_2, NO·, HOCl)		11
IV	CY-RL5 & CY-RL7	✓ (?)	ECFP/EYFP	Peptide linker with 4 Cys	? (incl. GSH/GSSG)		12
IV	Redoxfluor	✓ (?)	Cerulean/Citrine	Peptide linker with 4 Cys (C-terminus of Yap1)	? (incl. H_2O_2, GSH/GSSG, Trx)		13
IV	FRET-MT	✓ (?)	ECFP/EYFP	Metallothionein (MT)	? (incl. NO·)	✓	14

1. Ostergaard et al. (2004), 2. Dooley et al. (2004), Hanson et al. (2004), 3. Bjornberg et al. (2006b), 4. Gutscher et al. (2008), 5. Albrecht et al. (unpublished data), 6. Gutscher et al. (2009), 7. Belousov et al. (2006), 8. Simen Zhao et al. (2010), 9. Hung et al. (2011), 10. Zhao et al. (2011), 11. Waypa et al. (2006), 12. Kolossov et al. (2008, 2011), 13. Yano et al. (2010), 14. Pearce et al. (2000)

In the following discussion of genetically encoded probes we have not included circularly permuted YFP (cpYFP), which has been suggested to be responsive to the presence of $O_2^{\cdot-}$ (Wang et al. 2008). The validity of this suggestion remains controversial (Muller 2009; Schwarzlander et al. 2011; Huang et al. 2011), with the main issue being that cpYFP is inherently highly pH sensitive. Therefore, it is conceivable that a pH response has been misinterpreted as a $O_2^{\cdot-}$ response (Schwarzlander et al. 2011). pH sensitivity must always be taken into account when considering the use of cpYFP or cpYFP-based probes as discussed for HyPer below.

5.3.1 Fluorescent Proteins with an Engineered Dithiol-Disulfide Pair

The founding members of this class are rxYFP (Ostergaard et al. 2004) and the roGFPs 1 and 2 (Hanson et al. 2004; Dooley et al. 2004). Both types of probe are engineered to contain a pair of cysteine residues, C149/C202 in the case of rxYFP and C147/C204 in the case of the two roGFPs. In both instances, the formation of a disulfide bridge between the engineered cysteine residues changes the fluorescent properties of the protein, thus allowing the determination of the redox state of the engineered dithiol-disulfide redox pair. When expressed inside cells, both probes predominantly, if not exclusively, equilibrate with the glutathione redox couple. Importantly, this equilibration is not a spontaneous process, but rather depends on the catalytic activity of endogenous glutaredoxins (Grx) (Ostergaard et al. 2004; Meyer et al. 2007). However, this implies that differences and changes in endogenous Grx expression or activity may affect the state of the probe in a way that does not properly reflect the actual E_{GSH}. This possibility must be kept in mind when comparing different cell types or the same cell type under different conditions.

Applications in multicellular organisms have focused on roGFPs (roGFP1 and roGFP2), rather than rxYFP, for two main reasons: (i) they allow ratiometric measurements and (ii) the measured ratio is resistant to perturbation by changes in pH and halide ion concentration. RoGFP1 is based on wild type GFP and exhibits a redox potential of about −290 mV, whilst roGFP2 is based on the brighter EGFP and has a slightly less negative redox potential of about −280 mV. These low midpoint potentials make both probes suitable for measurements in the cytosol, nucleus, mitochondria, and peroxisomes. However, their utility is limited in the more oxidizing environment of the endoplasmic reticulum (ER). For this reason, roGFP variants with higher midpoint potentials (roGFP1-iX probes) have been developed recently (Lohman and Remington 2008). Attempts to establish the use of these probes in the ER are ongoing (Delic et al. 2010; van Lith et al. 2011).

The first application of roGFPs in a multicellular organism was in the plant *Arabidopsis thaliana*. One early study asked if different cell types in the root display recognizable redox differences; transgenic expression of cytosolic and mitochondrially-targeted roGFP1 under a viral promoter apparently identified increased probe oxidation in the elongation zone of the root tip relative to the other parts of the root (Jiang et al. 2006). Several studies have since applied roGFPs to investigate the impact of stress conditions. Increased oxidation of roGFP2 was observed in response to mechanical wounding of leaves (Meyer et al. 2007). Oxidation of mitochondrial-targeted roGFP2 (mt-roGFP2) was visualized in seedlings exposed to heat, salt and heavy metals (Schwarzlander et al. 2009). Apparent oxidation of cytosolic roGFP1 was observed in response to water stress (Jubany-Mari et al. 2010) and dark-induced senescence in leaves (Rosenwasser et al. 2011). Finally, two *Arabidopsis* studies combined redox imaging with genetic manipulations. Depletion of mitochondrial manganese superoxide dismutase (MnSOD) led to increased oxidation of mitochondrially-targeted roGFP1, but had no significant effect on cytosolic roGFP1 in leaf epidermis cells (Morgan et al. 2008). Further, deletion of all three *Arabidopsis* homologues of the *Plasmodium falciparum* chloroquine-resistance transporter family led to increased cytosolic roGFP1 oxidation, but had no effect on plastid-localized roGFP1. This study showed that the transporters have a role in regulating cytosolic GSH levels, probably by mediating glutathione or γ-glutamylcysteine transport to the cytosol (Maughan et al. 2010).

The first application in mice was the expression of roGFP2 in the liver by adenoviral gene transfer. Hepatic ischemia was found to induce roGFP2 reduction, whilst subsequent reperfusion led to roGFP2 oxidation, the degree of which was dependent on the length of the preceding ischemia and which correlated with subsequent liver damage as assessed by caspase-3 activity and serum levels of damage markers (Haga et al. 2009). So far, two kinds of roGFP-transgenic mice have been reported. First, mice expressing mitochondrially-targeted roGFP2 in dopaminergic neurons under the control of the tyrosine hydroxylase promoter have been used for two photon imaging of midbrain slices. The results indicated that roGFP2 in the substantia nigra is more oxidized than in the ventral tegmental area (Guzman et al. 2010). Second, mice have been created to express roGFP2 under the control of a β-globin mini-promoter, which mediates expression in red blood cells (Xu et al. 2011). The authors concluded from their data that roGFP2 in red blood cells becomes more oxidized with increasing cellular age.

Finally, roGFP1 has been expressed in the mitochondria of body wall muscle cells in *C. elegans* with the aim of observing redox changes linked to the mitochondrial fusion associated proteins EAT-3 and FZO-1. Increased roGFP1 oxidation was observed in strains containing mutant *eat*-3 alleles, but not in *fzo-1* mutants (Johnson and Nehrke 2010).

5.3.2 Fusion Proteins Based on roGFP

The observation that rxYFP and roGFPs require the activity of endogenous Grx to equilibrate with the glutathione redox couple, led to the concept of genetically fusing a Grx directly to the redox-sensitive FP. The advantage is twofold: first, the resulting fusion protein is now a complete glutathione-specific biosensor, independent of endogenous enzymes. Second, the translational fusion increases the effective concentration of Grx by 2–3 orders of magnitude relative to unfused Grx, thus the rate of equilibration with the glutathione redox couple is strongly enhanced. The first example in this class was the fusion of rxYFP to yeast Grx1p, which was used to gain mechanistic insight into the reaction between Grx1p and rxYFP *in vitro* (Bjornberg et al. 2006b). This study found that the fusion increased the rate of GSSG-mediated rxYFP oxidation by a factor of 3,300 while the rate of rxYFP oxidation mediated by other oxidants including cystine remained unchanged. Thus, the Grx1p fusion also dramatically increased the specificity of rxYFP towards glutathione. Based on this principle the ratiometric roGFP2 was fused to human glutaredoxin-1; this construct was used to provide the first real-time measurements of the glutathione redox potential *in cellulo* (Gutscher et al. 2008).

Subsequently the concept of roGFP-based bipartite redox relays was extended to create a probe for H_2O_2 (Gutscher et al. 2009). To this end, the GSSG-sensing Grx domain was replaced with the H_2O_2-sensing oxidant receptor peroxidase 1 (Orp1) protein. Orp1 is well characterized as a sensitive and specific H_2O_2 sensor protein by which yeast cells regulate their transcriptional response to H_2O_2. Orp1 appears to contribute to H_2O_2 tolerance only as a sensor and not as a conventional H_2O_2-scavenging peroxidase (Delaunay et al. 2002). The physiological function of Orp1 is to convert H_2O_2 into a disulfide bond and to relay that disulfide bond to a recipient protein, the transcription factor Yap1, which is activated by oxidation. An adapter protein, Ybp1, is required to bring Orp1 and Yap1 into proximity and position for disulfide exchange. The key idea is to mimic the natural H_2O_2-Orp1-Yap1 relay by creating a H_2O_2-Orp1-roGFP2 relay. A peptide linker between Orp1 and roGFP2 provides proximity and thus abrogates the need for an adaptor protein. Furthermore, roGFP2 replaces Yap1, hence yielding a fluorescent instead of a transcriptional response.

Both Grx1-roGFP2 and roGFP2-Orp1 have been applied as *in vivo* probes in multicellular organisms. First, transgenic expression of cytosolic Grx1-roGFP2 in *A. thaliana* was applied to demonstrate the relative contributions of glutathione reductase and the thioredoxin system to glutathione redox homeostasis (Marty et al. 2009). Second, four different lines of transgenic *Drosophila* have been created to express either Grx1-roGFP2 in the cytosol or in mitochondria or roGFP2-Orp1 in the cytosol or in mitochondria. Of note, the domain order is critical for proper mitochondrial expression in plants and insects. While Grx1-roGFP2 and roGFP2-Grx1 are equivalent probes in terms of specificity, kinetics and dynamic range, in plants and insects only the latter one can be properly targeted to mitochondria (Albrecht et al., unpublished data). Living larvae from each of the four redox probe lines were used in combination with reverse genetics and feeding experiments (Albrecht et al. 2011). Interestingly, E_{GSH} and H_2O_2 levels were found to respond independently of each other, in a subcellular compartment- and tissue-specific manner. The pattern of changes that occurred varied according to the specific chemical treatment or genetic modification studied. In addition a chemical redox conservation method has been developed to prevent artificial roGFP disulfide-bond formation during tissue dissection and paraformaldehyde (PFA) fixation, thereby allowing the imaging of optically-inaccessible tissue from non-transparent adult flies (Albrecht et al. 2011).

5.3.3 Fluorescent Proteins Coupled to Redox-Responsive Proteins Domains

Probes in this class include the H_2O_2 probe HyPer (Belousov et al. 2006), and the very recently introduced NAD^+/NADH probes Peredox (Hung et al. 2011) and Frex (Zhao et al. 2011). All three probes allow ratiometric measurements.

HyPer is based on the bacterial H_2O_2 sensor OxyR, which has been fused to cpYFP. The OxyR domain is selectively oxidized by H_2O_2 to form an intramolecular disulfide bond. The resulting conformational change subsequently changes the fluorescent properties of cpYFP. An improved variant, HyPer2, exhibits a wider dynamic range and a faster response (Markvicheva et al. 2011). Although there is no doubt that HyPer responds to H_2O_2 in a sensitive and specific manner, the significant pH sensitivity, which is inherent to cpYFP, should be given careful attention, especially in the context of *in vivo* studies. The pH sensitivity of the HyPer fluorescence ratio was already reported in the original publication (Supplementary Figure 1 of Belousov et al. 2006), but only recently it has been documented that HyPer measurements can be confounded by pH effects *in cellulo* (Roma et al. 2012). A HyPer mutant with an inactivated H_2O_2-sensing domain (C199S) can be utilized as a ratiometric pH probe (named 'synthetic pH sensor' = SypHer) (Poburko et al. 2011). In conclusion, if HyPer and SypHer are used in parallel experiments it should be possible to discriminate between H_2O_2 and pH mediated ratio changes.

To date, HyPer has been expressed in zebrafish, plants and *C. elegans*. General cytosolic HyPer expression in zebrafish was achieved by injection of HyPer mRNA into oocytes. Interestingly, following tail fin wounding, increased H_2O_2 production was observed at the wound edge, which extended up to 200 μm into the surrounding tissue, creating a gradient of decreasing H_2O_2 away from the wound margin (Niethammer et al. 2009). Very recently a Src family kinase (SFK) was proposed to act as a redox sensor in neutrophils which responds to the wounding-induced H_2O_2 gradient and mediates neutrophil recruitment to the wound site (Yoo et al. 2011). HyPer was also expressed in *A. thaliana*; cytosolic and peroxisomally-targeted HyPer was observed in epidermal leaf cells and stomatal guard cells (Costa et al. 2010) and used to suggest Ca^{2+}-dependent peroxisomal H_2O_2 scavenging.

Finally, HyPer was introduced into *C. elegans*. A plate-reader based method was used to measure HyPer fluorescence. A slight increase in the HyPer fluorescence ratio was observed in a longer lived SOD1 overexpressing strain which was interpreted as an increased steady-state H_2O_2 level. This interpretation was supported by an Amplex red assay performed on worm lysates (Cabreiro et al. 2011). Also HyPer fluorescence was measured in aging *C. elegans* (Back et al. 2011). Differences in HyPer fluorescence ratio were observed dependent upon the region of worm imaged. Further an age-dependent increase in HyPer fluorescence ratio was observed, which was delayed in calorically-restricted worms.

Most recently, two genetically encoded probes for the NAD^+/NADH redox couple have been developed. Both of them are based on the bacterial NADH-binding protein Rex. The Peredox probe (Hung et al. 2011) uses two copies of the *Thermus aquaticus* Rex protein, and interposed between them, the non-ratiometric cpGFP variant T-Sapphire. A tandemly attached mCherry provides the ratiometric reference for signal normalization. Peredox seems to report the NAD^+/NADH ratio. With a pKa of about 5, T-Sapphire should be pH insensitive in the physiological range of most cellular compartments. The Frex probe (Zhao et al. 2011) uses a tandem dimer of the Rex protein from *Bacillus subtilis* in combination with the ratiometric cpYFP. Thus, like other cpYFP-based probes, Frex is pH sensitive. Frex appears to measure changes in NADH concentration, rather than the NAD^+/NADH ratio. As these probes have just been reported in the literature, their *in vivo*/*in situ* application in model animals is a task for the future.

5.3.4 FRET-Based Redox Probes

In comparison to the probes mentioned above, FRET-based redox probes are the least investigated and their overall potential for *in vivo* studies is still to be established. FRET probes are ratiometric by design, yet their dynamic range appears to be quite limited. At least four FRET-based redox probes have been described in the literature. All of them place a redox-sensitive peptide or protein domain between the two FPs that make up the FRET pair.

In the family of RL-FRET probes, an α-helical linker peptide with four cysteine residues separates ECFP from EYFP (Kolossov et al. 2008). Formation of disulfide bonds within the α-helical linker increases FRET efficiency. The mid-point potential of the variant RL7-FRET was determined as −143 mV suggesting that this probe may find applications for measurements in the endoplasmic reticulum (Kolossov et al. 2011). In the 'Redoxfluor' probe Cerulean and Citrine are separated by a tandem repeat of the cysteine-containing 50-amino acid C-terminal peptide of the yeast transcription factor Yap1. Here, disulfide bond formation decreases FRET efficiency (Yano et al. 2010). In HSP-FRET the redox-sensitive heat shock protein Hsp33 links CFP and YFP (Waypa et al. 2006). Oxidation of Zn^{2+}-coordinating thiols in Hsp33 leads to a conformational change (Jakob et al. 1999) which is responsible for a change in FRET efficiency. Finally, FRET-MT places human metallothionein (hMTIIa) between ECFP and EYFP (Pearce et al. 2000). The probe primarily monitors binding and release of metal ions by MT. It was shown to respond to NO· which triggers release of metal ions from MT. FRET-MT seems to the only FRET-based redox probe that has been employed in a transgenic multicellular context. Expression in the mouse lung was achieved by adenoviral somatic gene transfer. The isolated lung was freshly imaged by confocal microscopy and appeared to show hypoxia-induced NO· production (Bernal et al. 2008).

All FRET probes have in common the problem of undefined specificity. Metal ion binding and release by MT is likely influenced by a number of redox- and non-redox processes. The Zn^{2+}-coordinating thiols of Hsp33 are known to react with various oxidants, including H_2O_2, NO· and HOCl. The cysteine-containing peptides of RL-FRET and Redoxfluor seem to equilibrate with GSH/GSSG, but it is unclear which additional redox reactions may occur *in vivo*. It would seem important to define which redox catalysts actually interact with the cysteines in the peptide linkers. Redoxfluor is suggested to communicate with both the glutathione and thioredoxin systems (Oku and Sakai 2012). It remains to be seen how mixed measurements of various redox couples can contribute to the study of physiological processes.

5.4 Conclusions and Outlook

5.4.1 Problems and Considerations

In recent years there has been an explosion of interest in the use of chemical and genetic probes to visualize redox species *in vivo*. There have been many publications using many different redox probes in many different systems. However, in general, very little comparison or cross-checking has been done and many examples of contradictory results exist. In addition, very few examples of established probe application methodology exist. This raises the questions: Which observations and interpretations can be trusted? And what kind of errors can lead to false conclusions? Obviously there are two major sources of human error. First, measurement and/or data processing errors (e.g. the failure to recognize measurement artifacts) and second, interpretation errors, typically based on the belief that the probe is measuring what it is supposed to measure compounded by the lack of appropriate control experiments. Here we give a few examples.

For most fluorescent probes, especially when applied on the level of tissues and organisms, fluctuating signal-to-noise-ratios and auto-fluorescence are serious issues. For example, when analyzing microscopic images based on ratiometric genetic probes, it is critical to calculate the fluorescence emission ratios only for those pixels that show above-background signals in both channels and which are not affected by autofluorescence. Inappropriate image processing procedures can lead to incorrect ratio images that indicate probe oxidation where there is in fact none. Great care should be taken to obtain control images, to identify auto-fluorescence signals and to avoid auto-fluorescent areas altogether.

In many situations and with many of the probes, interpretation of the measurements is difficult. The interpretation of hydro-type probe oxidation is most difficult because it can be influenced by factors unrelated to ROS, and the process of probe oxidation may generate further oxidants. The *in vivo* distribution of small molecule chemical fluorescent probes is usually unpredictable and uncontrolled. Systemic administration may give rise to accumulation in certain locations, leading to areas of increased fluorescence that are unrelated to ROS production. Differential uptake and efflux can lead to the possibility that probe fluorescence is observed at a site different from the site of probe oxidation.

The response of genetically encoded probes should also be interpreted with caution. In the case of the cpYFP-based probes it may be difficult to determine if fluorescence changes are due to changes in pH rather than the redox species of interest. A general concern that applies to probes which utilize two different fluorescent proteins (e.g. FRET probes) is that differences in pH sensitivity, maturation rate and photo-stability can lead to measurement artifacts (Miyawaki 2011).

Interpretation issues can also arise when genetically-encoded probes are adapted for use in other subcellular compartments. For example, roGFPs predominantly equilibrate with the glutathione redox couple, but this requires the presence of Grxs. Nonetheless, roGFP1-iX probes are increasingly being applied with the aim of measuring E_{GSH} in the ER, which is not known to harbor any Grxs, thus it is unclear what is actually measured by these probes. Instead, the ER harbors a protein thiol oxidizing machinery and a host of protein disulfide isomerases (PDIs) involved in thiol oxidation, disulfide reduction and isomerization. At present it not known to what extent roGFP1-iX probes interact with any of these factors. Therefore, the assumption that roGFP1-iX probes report E_{GSH} in the ER is at best premature.

5.4.2 *Future Developments*

Evidently, more detailed probe characterization and especially cross-checking would be of great interest and utility to the redox biology field. It should be possible to compare a panel of different probes (chemical and genetic, e.g. all probes considered to respond to H_2O_2) in the same experimental system and compare the responses under various conditions and treatments. Such experiments should initially be performed in a cell culture setting, but ultimately could be extended to *in vivo* applications.

We are still a long way from obtaining the goal highlighted in Sect. 4.1.1 of being able to image any redox species of interest, in a defined subcellular compartment, with high temporal resolution in a multicellular organism. Below we describe a wish-list of developments which we believe are reasonable to expect in the near future and which are likely to make a substantial contribution to the study of redox biology.

5.4.2.1 Further Development of Chemical Probes

- *Targeting*: very few possibilities exist to target chemical probes to defined subcellular compartments of interest. The recently developed SNAP-tag technology offers the possibility to target chemical probes to specific locations,

with the limitation that this approach relies on prior genetic manipulation. A protein residing in the location of interest is genetically fused to a mutant O6-alkylguanine-DNA alkyltransferase, which reacts specifically and rapidly with benzylguanine (BG) derivatives. Fluorescent probes containing a BG moiety will be retained in the specific location of the SNAP-tag (Srikun et al. 2010). The SNAP-tag principle may be applicable to a range of chemical ROS probes.
- *Reaction rate*: in order to react with oxidants, chemical probes must compete with endogenous oxidant-scavenging systems. At present, ROS-specific chemical probes have very low reaction rates, and fluorescent signals accumulate slowly. Ideally, novel probes would combine ROS specificity with greatly enhanced reaction rates.

5.4.2.2 Further Development of Genetic Probes

- *Multi-color redox imaging*: the development of a redox-sensitive red fluorescent protein (RFP) would greatly enhance the possibilities for making simultaneous measurements of different redox species or of the same redox species in different subcellular compartments.
- *Novel probe specificities*: genetically encoded probes for several key biological redox species are still lacking. Most prominently, probes for the oxidants NO· and $O_2^{\cdot-}$, and for the redox couples NADPH/NADP$^+$, thioredoxin (ox/red) and ascorbate/dehydroascorbate, do not currently exist.

References

Abo M, Urano Y, Hanaoka K, Terai T, Komatsu T, Nagano T (2011) Development of a highly sensitive fluorescence probe for hydrogen peroxide. J Am Chem Soc 133:10629–10637

Albers AE, Okreglak VS, Chang CJ (2006) A FRET-based approach to ratiometric fluorescence detection of hydrogen peroxide. J Am Chem Soc 128:9640–9641

Albers AE, Dickinson BC, Miller EW, Chang CJ (2008) A red-emitting naphthofluorescein-based fluorescent probe for selective detection of hydrogen peroxide in living cells. Bioorg Med Chem Lett 18:5948–5950

Albrecht SC, Barata AG, Grosshans J, Teleman AA, Dick TP (2011) In vivo mapping of hydrogen peroxide and oxidized glutathione reveals chemical and regional specificity of redox homeostasis. Cell Metab 14:819–829

Atkuri KR, Herzenberg LA, Niemi AK, Cowan T (2007) Importance of culturing primary lymphocytes at physiological oxygen levels. Proc Natl Acad Sci USA 104:4547–4552

Back P, de Vos WH, Depuydt GG, Matthijssens F, Vanfleteren JR, Braeckman BP (2011) Exploring real-time in vivo redox biology of developing and aging Caenorhabditis eleganse. Free Radic Biol Med 52(5):850–859

Belousov VV, Fradkov AF, Lukyanov KA, Staroverov DB, Shakhbazov KS, Terskikh AV, Lukyanov S (2006) Genetically encoded fluorescent indicator for intracellular hydrogen peroxide. Nat Methods 3:281–286

Benov L, Sztejnberg L, Fridovich I (1998) Critical evaluation of the use of hydroethidine as a measure of superoxide anion radical. Free Radic Biol Med 25:826–831

Bernal PJ, Leelavanichkul K, Bauer E, Cao R, Wilson A, Wasserloos KJ, Watkins SC, Pitt BR, St Croix CM (2008) Nitric-oxide-mediated zinc release contributes to hypoxic regulation of pulmonary vascular tone. Circ Res 102:1575–1583

Bjornberg O, Ostergaard H, Winther JR (2006a) Measuring intracellular redox conditions using GFP-based sensors. Antioxid Redox Signal 8:354–361

Bjornberg O, Ostergaard H, Winther JR (2006b) Mechanistic insight provided by glutaredoxin within a fusion to redox-sensitive yellow fluorescent protein. Biochemistry 45:2362–2371

Boghossian AA, Zhang J, Barone PW, Reuel NF, Kim JH, Heller DA, Ahn JH, Hilmer AJ, Rwei A, Arkalgud JR, Zhang CT, Strano MS (2011) Near-infrared fluorescent sensors based on single-walled carbon nanotubes for life sciences applications. ChemSusChem 4:848–863

Buxser SE, Sawada G, Raub TJ (1999) Analytical and numerical techniques for evaluation of free radical damage in cultured cells using imaging cytometry and fluorescent indicators. Methods Enzymol 300:256–275

Cabreiro F, Ackerman D, Doonan R, Araiz C, Back P, Papp D, Braeckman BP, Gems D (2011) Increased life span from overexpression of superoxide dismutase in Caenorhabditis elegans is not caused by decreased oxidative damage. Free Radic Biol Med 51:1575–1582

Casanova D, Bouzigues C, Nguyen TL, Ramodiharilafy RO, Bouzhir-Sima L, Gacoin T, Boilot JP, Tharaux PL, Alexandrou A (2009) Single europium-doped nanoparticles measure temporal pattern of reactive oxygen species production inside cells. Nat Nanotechnol 4:581–585

Chen X, Zhou Y, Peng X, Yoon J (2010) Fluorescent and colorimetric probes for detection of thiols. Chem Soc Rev 39:2120–2135

Chen X, Tian X, Shin I, Yoon J (2011) Fluorescent and luminescent probes for detection of reactive oxygen and nitrogen species. Chem Soc Rev 40:4783–4804

Cook JA, Iype SN, Mitchell JB (1991) Differential specificity of monochlorobimane for isozymes of human and rodent glutathione S-transferases. Cancer Res 51:1606–1612

Costa A, Drago I, Behera S, Zottini M, Pizzo P, Schroeder JI, Pozzan T, Lo Schiavo F (2010) H_2O_2 in plant peroxisomes: an *in vivo* analysis uncovers a Ca^{2+}-dependent scavenging system. Plant J 62:760–772

Dalle-Donne I, Rossi R, Colombo G, Giustarini D, Milzani A (2009) Protein S-glutathionylation: a regulatory device from bacteria to humans. Trends Biochem Sci 34:85–96

de Souza N (2007) Too much of a good thing. Nat Methods 4:386

Delaunay A, Pflieger D, Barrault MB, Vinh J, Toledano MB (2002) A thiol peroxidase is an H_2O_2 receptor and redox-transducer in gene activation. Cell 111:471–481

Delic M, Mattanovich D, Gasser B (2010) Monitoring intracellular redox conditions in the endoplasmic reticulum of living yeasts. FEMS Microbiol Lett 306:61–66

Dickinson BC, Chang CJ (2008) A targetable fluorescent probe for imaging hydrogen peroxide in the mitochondria of living cells. J Am Chem Soc 130:9638–9639

Dickinson BC, Chang CJ (2011) Chemistry and biology of reactive oxygen species in signaling or stress responses. Nat Chem Biol 7:504–511

Dickinson BC, Huynh C, Chang CJ (2010) A palette of fluorescent probes with varying emission colors for imaging hydrogen peroxide signaling in living cells. J Am Chem Soc 132:5906–5915

Dickinson BC, Peltier J, Stone D, Schaffer DV, Chang CJ (2011a) Nox2 redox signaling maintains essential cell populations in the brain. Nat Chem Biol 7:106–112

Dickinson BC, Tang Y, Chang Z, Chang CJ (2011b) A nuclear-localized fluorescent hydrogen peroxide probe for monitoring sirtuin-mediated oxidative stress responses in vivo. Chem Biol 18:943–948

Dooley CT, Dore TM, Hanson GT, Jackson WC, Remington SJ, Tsien RY (2004) Imaging dynamic redox changes in mammalian cells with green fluorescent protein indicators. J Biol Chem 279:22284–22293

Drummond GR, Selemidis S, Griendling KK, Sobey CG (2011) Combating oxidative stress in vascular disease: NADPH oxidases as therapeutic targets. Nat Rev Drug Discov 10:453–471

Grant CM, Maciver FH, Dawes IW (1996) Glutathione is an essential metabolite required for resistance to oxidative stress in the yeast Saccharomyces cerevisiae. Curr Genet 29:511–515

Grivennikov SI, Greten FR, Karin M (2010) Immunity, inflammation, and cancer. Cell 140: 883–899

Gutscher M, Pauleau AL, Marty L, Brach T, Wabnitz GH, Samstag Y, Meyer AJ, Dick TP (2008) Real-time imaging of the intracellular glutathione redox potential. Nat Methods 5:553–559

Gutscher M, Sobotta MC, Wabnitz GH, Ballikaya S, Meyer AJ, Samstag Y, Dick TP (2009) Proximity-based protein thiol oxidation by H_2O_2-scavenging peroxidases. J Biol Chem 284:31532–31540

Guzman JN, Sanchez-Padilla J, Wokosin D, Kondapalli J, Ilijic E, Schumacker PT, Surmeier DJ (2010) Oxidant stress evoked by pacemaking in dopaminergic neurons is attenuated by DJ-1. Nature 468:696–700

Haga S, Remington SJ, Morita N, Terui K, Ozaki M (2009) Hepatic ischemia induced immediate oxidative stress after reperfusion and determined the severity of the reperfusion-induced damage. Antioxid Redox Signal 11:2563–2572

Halliwell B, Whiteman M (2004) Measuring reactive species and oxidative damage in vivo and in cell culture: how should you do it and what do the results mean? Br J Pharmacol 142:231–255

Hanson GT, Aggeler R, Oglesbee D, Cannon M, Capaldi RA, Tsien RY, Remington SJ (2004) Investigating mitochondrial redox potential with redox-sensitive green fluorescent protein indicators. J Biol Chem 279:13044–13053

Harding HP, Zhang Y, Zeng H, Novoa I, Lu PD, Calfon M, Sadri N, Yun C, Popko B, Paules R, Stojdl DF, Bell JC, Hettmann T, Leiden JM, Ron D (2003) An integrated stress response regulates amino acid metabolism and resistance to oxidative stress. Mol Cell 11:619–633

Harrison DG, Guzik TJ, Lob HE, Madhur MS, Marvar PJ, Thabet SR, Vinh A, Weyand CM (2011) Inflammation, immunity, and hypertension. Hypertension 57:132–140

Hempel SL, Buettner GR, O'Malley YQ, Wessels DA, Flaherty DM (1999) Dihydrofluorescein diacetate is superior for detecting intracellular oxidants: comparison with 2′,7′-dichlorodihydrofluorescein diacetate, 5(and 6)-carboxy-2′,7′-dichlorodihydrofluorescein diacetate, and dihydrorhodamine 123. Free Radic Biol Med 27:146–159

Hess DT, Stamler JS (2012) Regulation by S-nitrosylation of protein posttranslational modification. J Biol Chem 287(7):4411–4418

Huai-Yun H, Secrest DT, Mark KS, Carney D, Brandquist C, Elmquist WF, Miller DW (1998) Expression of multidrug resistance-associated protein (MRP) in brain microvessel endothelial cells. Biochem Biophys Res Commun 243:816–820

Huang Z, Zhang W, Fang H, Zheng M, Wang X, Xu J, Cheng H, Gong G, Wang W, Dirksen RT, Sheu SS (2011) Response to "A critical evaluation of cpYFP as a probe for superoxide". Free Radic Biol Med 51:1937–1940

Hung YP, Albeck JG, Tantama M, Yellen G (2011) Imaging cytosolic NADH-NAD(+) redox state with a genetically encoded fluorescent biosensor. Cell Metab 14:545–554

Ignarro LJ, Fukuto JM, Griscavage JM, Rogers NE, Byrns RE (1993) Oxidation of nitric oxide in aqueous solution to nitrite but not nitrate: comparison with enzymatically formed nitric oxide from L-arginine. Proc Natl Acad Sci USA 90:8103–8107

Imura T, Kanatani S, Fukuda S, Miyamoto Y, Hisatsune T (2005) Layer-specific production of nitric oxide during cortical circuit formation in postnatal mouse brain. Cereb Cortex 15: 332–340

Ishikawa T, Wright CD, Ishizuka H (1994) GS-X pump is functionally overexpressed in cis-diamminedichloroplatinum (II)-resistant human leukemia HL-60 cells and down-regulated by cell-differentiation. J Biol Chem 269:29085–29093

Izumi S, Urano Y, Hanaoka K, Terai T, Nagano T (2009) A simple and effective strategy to increase the sensitivity of fluorescence probes in living cells. J Am Chem Soc 131:10189–10200

Jakob U, Muse W, Eser M, Bardwell JC (1999) Chaperone activity with a redox switch. Cell 96:341–352

Jiang K, Schwarzer C, Lally E, Zhang S, Ruzin S, Machen T, Remington SJ, Feldman L (2006) Expression and characterization of a redox-sensing green fluorescent protein (reduction-oxidation-sensitive green fluorescent protein) in Arabidopsis. Plant Physiol 141:397–403

Jin H, Heller DA, Kalbacova M, Kim JH, Zhang J, Boghossian AA, Maheshri N, Strano MS (2010) Detection of single-molecule H_2O_2 signalling from epidermal growth factor receptor using fluorescent single-walled carbon nanotubes. Nat Nanotechnol 5:302–309
Johnson D, Nehrke K (2010) Mitochondrial fragmentation leads to intracellular acidification in Caenorhabditis elegans and mammalian cells. Mol Biol Cell 21:2191–2201
Johnson LV, Walsh ML, Bockus BJ, Chen LB (1981) Monitoring of relative mitochondrial membrane potential in living cells by fluorescence microscopy. J Cell Biol 88:526–535
Jubany-Mari T, Alegre-Batlle L, Jiang K, Feldman LJ (2010) Use of a redox-sensing GFP (c-roGFP1) for real-time monitoring of cytosol redox status in Arabidopsis thaliana water-stressed plants. FEBS Lett 584:889–897
Karton-Lifshin N, Segal E, Omer L, Portnoy M, Satchi-Fainaro R, Shabat D (2011) A unique paradigm for a Turn-ON near-infrared cyanine-based probe: noninvasive intravital optical imaging of hydrogen peroxide. J Am Chem Soc 133:10960–10965
Keelan J, Allen NJ, Antcliffe D, Pal S, Duchen MR (2001) Quantitative imaging of glutathione in hippocampal neurons and glia in culture using monochlorobimane. J Neurosci Res 66:873–884
Kenmoku S, Urano Y, Kojima H, Nagano T (2007) Development of a highly specific rhodamine-based fluorescence probe for hypochlorous acid and its application to real-time imaging of phagocytosis. J Am Chem Soc 129:7313–7318
Kishi S, Bayliss PE, Uchiyama J, Koshimizu E, Qi J, Nanjappa P, Imamura S, Islam A, Neuberg D, Amsterdam A, Roberts TM (2008) The identification of zebrafish mutants showing alterations in senescence-associated biomarkers. PLOS Genet 4:e1000152
Klohs J, Wunder A, Licha K (2008) Near-infrared fluorescent probes for imaging vascular pathophysiology. Basic Res Cardiol 103:144–151
Koide Y, Urano Y, Kenmoku S, Kojima H, Nagano T (2007) Design and synthesis of fluorescent probes for selective detection of highly reactive oxygen species in mitochondria of living cells. J Am Chem Soc 129:10324–10325
Koide Y, Urano Y, Hanaoka K, Terai T, Nagano T (2011) Development of an Si-rhodamine-based far-red to near-infrared fluorescence probe selective for hypochlorous acid and its applications for biological imaging. J Am Chem Soc 133:5680–5682
Kojima H, Nakatsubo N, Kikuchi K, Kawahara S, Kirino Y, Nagoshi H, Hirata Y, Nagano T (1998a) Detection and imaging of nitric oxide with novel fluorescent indicators: diaminofluoresceins. Anal Chem 70:2446–2453
Kojima H, Nakatsubo N, Kikuchi K, Urano Y, Higuchi T, Tanaka J, Kudo Y, Nagano T (1998b) Direct evidence of NO production in rat hippocampus and cortex using a new fluorescent indicator: DAF-2 DA. Neuroreport 9:3345–3348
Kojima H, Sakurai K, Kikuchi K, Kawahara S, Kirino Y, Nagoshi H, Hirata Y, Nagano T (1998c) Development of a fluorescent indicator for nitric oxide based on the fluorescein chromophore. Chem Pharm Bull(Tokyo) 46:373–375
Kojima H, Urano Y, Kikuchi K, Higuchi T, Hirata Y, Nagano T (1999) Fluorescent indicators for imaging nitric oxide production. Angew Chem Int Ed Engl 38:3209–3212
Kojima H, Hirotani M, Nakatsubo N, Kikuchi K, Urano Y, Higuchi T, Hirata Y, Nagano T (2001) Bioimaging of nitric oxide with fluorescent indicators based on the rhodamine chromophore. Anal Chem 73:1967–1973
Kolossov VL, Spring BQ, Sokolowski A, Conour JE, Clegg RM, Kenis PJ, Gaskins HR (2008) Engineering redox-sensitive linkers for genetically encoded FRET-based biosensors. Exp Biol Med (Maywood) 233:238–248
Kolossov VL, Spring BQ, Clegg RM, Henry JJ, Sokolowski A, Kenis PJ, Gaskins HR (2011) Development of a high-dynamic range, GFP-based FRET probe sensitive to oxidative microenvironments. Exp Biol Med (Maywood) 236:681–691
Kundu K, Knight SF, Willett N, Lee S, Taylor WR, Murthy N (2009) Hydrocyanines: a class of fluorescent sensors that can image reactive oxygen species in cell culture, tissue, and in vivo. Angew Chem Int Ed Engl 48:299–303

Lawrence A, Jones CM, Wardman P, Burkitt MJ (2003) Evidence for the role of a peroxidase compound I-type intermediate in the oxidation of glutathione, NADH, ascorbate, and dichlorofluorescin by cytochrome c/H_2O_2. Implications for oxidative stress during apoptosis. J Biol Chem 278:29410–29419

Lebel CP, Ischiropoulos H, Bondy SC (1992) Evaluation of the probe 2′,7′-dichlorofluorescin as an indicator of reactive oxygen species formation and oxidative stress. Chem Res Toxicol 5: 227–231

Lee D, Khaja S, Velasquez-Castano JC, Dasari M, Sun C, Petros J, Taylor WR, Murthy N (2007) In vivo imaging of hydrogen peroxide with chemiluminescent nanoparticles. Nat Mater 6:765–769

Lepiller S, Laurens V, Bouchot A, Herbomel P, Solary E, Chluba J (2007) Imaging of nitric oxide in a living vertebrate using a diamino-fluorescein probe. Free Radic Biol Med 43:619–627

Lim MH, Xu D, Lippard SJ (2006) Visualization of nitric oxide in living cells by a copper-based fluorescent probe. Nat Chem Biol 2:375–380

Lohman JR, Remington SJ (2008) Development of a family of redox-sensitive green fluorescent protein indicators for use in relatively oxidizing subcellular environments. Biochemistry 47:8678–8688

Ludescher C, Thaler J, Drach D, Drach J, Spitaler M, Gattringer C, Huber H, Hofmann J (1992) Detection of activity of P-glycoprotein in human tumour samples using rhodamine 123. Br J Haematol 82:161–168

Maeda H, Yamamoto K, Kohno I, Hafsi L, Itoh N, Nakagawa S, Kanagawa N, Suzuki K, Uno T (2007) Design of a practical fluorescent probe for superoxide based on protection-deprotection chemistry of fluoresceins with benzenesulfonyl protecting groups. Chemistry 13:1946–1954

Markvicheva KN, Bilan DS, Mishina NM, Gorokhovatsky AY, Vinokurov LM, Lukyanov S, Belousov VV (2011) A genetically encoded sensor for H_2O_2 with expanded dynamic range. Bioorg Med Chem 19:1079–1084

Marty L, Siala W, Schwarzlander M, Fricker MD, Wirtz M, Sweetlove LJ, Meyer Y, Meyer AJ, Reichheld JP, Hell R (2009) The NADPH-dependent thioredoxin system constitutes a functional backup for cytosolic glutathione reductase in Arabidopsis. Proc Natl Acad Sci USA 106:9109–9114

Maughan SC, Pasternak M, Cairns N, Kiddle G, Brach T, Jarvis R, Haas F, Nieuwland J, Lim B, Muller C, Salcedo-Sora E, Kruse C, Orsel M, Hell R, Miller AJ, Bray P, Foyer CH, Murray JA, Meyer AJ, Cobbett CS (2010) Plant homologs of the Plasmodium falciparum chloroquine-resistance transporter, PfCRT, are required for glutathione homeostasis and stress responses. Proc Natl Acad Sci USA 107:2331–2336

McQuade LE, Ma J, Lowe G, Ghatpande A, Gelperin A, Lippard SJ (2010) Visualization of nitric oxide production in the mouse main olfactory bulb by a cell-trappable copper(II) fluorescent probe. Proc Natl Acad Sci USA 107:8525–8530

Meister A (1988) Glutathione metabolism and its selective modification. J Biol Chem 263: 17205–17208

Meyer AJ, Dick TP (2010) Fluorescent protein-based redox probes. Antioxid Redox Signal 13:621–650

Meyer AJ, Fricker MD (2000) Direct measurement of glutathione in epidermal cells of intact Arabidopsis roots by two-photon laser scanning microscopy. J Microsc Oxf 198:174–181

Meyer AJ, May MJ, Fricker M (2001) Quantitative in vivo measurement of glutathione in Arabidopsis cells. Plant J 27:67–78

Meyer AJ, Brach T, Marty L, Kreye S, Rouhier N, Jacquot JP, Hell R (2007) Redox-sensitive GFP in Arabidopsis thaliana is a quantitative biosensor for the redox potential of the cellular glutathione redox buffer. Plant J 52:973–986

Mieyal JJ, Chock PB (2012) Posttranslational modification of cysteine in redox signaling and oxidative stress: focus on S-glutathionylation. Antioxid Redox Signal 16(6):471–475

Mieyal JJ, Gallogly MM, Qanungo S, Sabens EA, Shelton MD (2008) Molecular mechanisms and clinical implications of reversible protein S-glutathionylation. Antioxid Redox Signal 10: 1941–1988

Miller EW, Tulyathan O, Isacoff EY, Chang CJ (2007) Molecular imaging of hydrogen peroxide produced for cell signaling. Nat Chem Biol 3:263–267
Miyawaki A (2011) Development of probes for cellular functions using fluorescent proteins and fluorescence resonance energy transfer. Annu Rev Biochem 80:357–373
Morgan MJ, Lehmann M, Schwarzlander M, Baxter CJ, Sienkiewicz-Porzucek A, Williams TC, Schauer N, Fernie AR, Fricker MD, Ratcliffe RG, Sweetlove LJ, Finkemeier I (2008) Decrease in manganese superoxide dismutase leads to reduced root growth and affects tricarboxylic acid cycle flux and mitochondrial redox homeostasis. Plant Physiol 147:101–114
Morgan KL, Estevez AO, Mueller CL, Cacho-Valadez B, Miranda-Vizuete A, Szewczyk NJ, Estevez M (2010) The glutaredoxin GLRX-21 functions to prevent selenium-induced oxidative stress in Caenorhabditis elegans. Toxicol Sci 118:530–543
Morgan B, Sobotta MC, Dick TP (2011) Measuring EGSH and H_2O_2 with roGFP2-based redox probes. Free Radic Biol Med 51:1943–1951
Muller FL (2009) A critical evaluation of cpYFP as a probe for superoxide. Free Radic Biol Med 47:1779–1780
Niethammer P, Grabher C, Look AT, Mitchison TJ (2009) A tissue-scale gradient of hydrogen peroxide mediates rapid wound detection in zebrafish. Nature 459:996–999
Ohashi T, Mizutani A, Murakami A, Kojo S, Ishii T, Taketani S (2002) Rapid oxidation of dichlorodihydrofluorescin with heme and hemoproteins: formation of the fluorescein is independent of the generation of reactive oxygen species. FEBS Lett 511:21–27
Oku M, Sakai Y (2012) Assessment of physiological redox state with novel FRET protein probes. Antioxid Redox Signal 16(7):698–704
Ostergaard H, Tachibana C, Winther JR (2004) Monitoring disulfide bond formation in the eukaryotic cytosol. J Cell Biol 166:337–345
Oushiki D, Kojima H, Terai T, Arita M, Hanaoka K, Urano Y, Nagano T (2010) Development and application of a near-infrared fluorescence probe for oxidative stress based on differential reactivity of linked cyanine dyes. J Am Chem Soc 132:2795–2801
Ouyang J, Hong H, Shen C, Zhao Y, Ouyang C, Dong L, Zhu J, Guo Z, Zeng K, Chen J, Zhang C, Zhang J (2008) A novel fluorescent probe for the detection of nitric oxide in vitro and in vivo. Free Radic Biol Med 45:1426–1436
Pashkow FJ (2011) Oxidative stress and inflammation in heart disease: do antioxidants have a role in treatment and/or prevention? Int J Inflamm 2011:514623
Pearce LL, Gandley RE, Han W, Wasserloos K, Stitt M, Kanai AJ, McLaughlin MK, Pitt BR, Levitan ES (2000) Role of metallothionein in nitric oxide signaling as revealed by a green fluorescent fusion protein. Proc Natl Acad Sci USA 97:477–482
Poburko D, Santo-Domingo J, Demaurex N (2011) Dynamic regulation of the mitochondrial proton gradient during cytosolic calcium elevations. J Biol Chem 286:11672–11684
Rhee SG, Chang TS, Jeong W, Kang D (2010) Methods for detection and measurement of hydrogen peroxide inside and outside of cells. Mol Cells 29:539–549
Rice GC, Bump EA, Shrieve DC, Lee W, Kovacs M (1986) Quantitative analysis of cellular glutathione by flow cytometry utilizing monochlorobimane: some applications to radiation and drug resistance in vitro and in vivo. Cancer Res 46:6105–6110
Robinson KM, Janes MS, Pehar M, Monette JS, Ross MF, Hagen TM, Murphy MP, Beckman JS (2006) Selective fluorescent imaging of superoxide in vivo using ethidium-based probes. Proc Natl Acad Sci USA 103:15038–15043
Robinson KM, Janes MS, Beckman JS (2008) The selective detection of mitochondrial superoxide by live cell imaging. Nat Protoc 3:941–947
Roma LP, Duprez J, Takahashi HK, Gilon P, Wiederkehr A, Jonas JC (2012) Dynamic measurements of mitochondrial hydrogen peroxide concentration and glutathione redox state in rat pancreatic beta-cells using ratiometric fluorescent proteins: confounding effects of pH with HyPer but not roGFP1. Biochem J 441(Part 3):971–978
Rosenwasser S, Rot I, Sollner E, Meyer AJ, Smith Y, Leviatan N, Fluhr R, Friedman H (2011) Organelles contribute differentially to reactive oxygen species-related events during extended darkness. Plant Physiol 156:185–201

Sasaki E, Kojima H, Nishimatsu H, Urano Y, Kikuchi K, Hirata Y, Nagano T (2005) Highly sensitive near-infrared fluorescent probes for nitric oxide and their application to isolated organs. J Am Chem Soc 127:3684–3685

Schulz TJ, Zarse K, Voigt A, Urban N, Birringer M, Ristow M (2007) Glucose restriction extends Caenorhabditis elegans life span by inducing mitochondrial respiration and increasing oxidative stress. Cell Metab 6:280–293

Schwarzlander M, Fricker MD, Sweetlove LJ (2009) Monitoring the in vivo redox state of plant mitochondria: effect of respiratory inhibitors, abiotic stress and assessment of recovery from oxidative challenge. Biochim Biophys Acta 1787:468–475

Schwarzlander M, Logan DC, Fricker MD, Sweetlove LJ (2011) The circularly permuted yellow fluorescent protein cpYFP that has been used as a superoxide probe is highly responsive to pH but not superoxide in mitochondria: implications for the existence of superoxide 'flashes'. Biochem J 437:381–387

Setsukinai K, Urano Y, Kakinuma K, Majima HJ, Nagano T (2003) Development of novel fluorescence probes that can reliably detect reactive oxygen species and distinguish specific species. J Biol Chem 278:3170–3175

Shepherd J, Hilderbrand SA, Waterman P, Heinecke JW, Weissleder R, Libby P (2007) A fluorescent probe for the detection of myeloperoxidase activity in atherosclerosis-associated macrophages. Chem Biol 14:1221–1231

Shulaev V, Oliver DJ (2006) Metabolic and proteomic markers for oxidative stress. New tools for reactive oxygen species research. Plant Physiol 141:367–372

Sikora A, Zielonka J, Lopez M, Joseph J, Kalyanaraman B (2009) Direct oxidation of boronates by peroxynitrite: mechanism and implications in fluorescence imaging of peroxynitrite. Free Radic Biol Med 47:1401–1407

Simen Zhao B, Liang Y, Song Y, Zheng C, Hao Z, Chen PR (2010) A highly selective fluorescent probe for visualization of organic hydroperoxides in living cells. J Am Chem Soc 132(48):17065–17067

Srikun D, Miller EW, Domaille DW, Chang CJ (2008) An ICT-based approach to ratiometric fluorescence imaging of hydrogen peroxide produced in living cells. J Am Chem Soc 130:4596–4597

Srikun D, Albers AE, Nam CI, Iavarone AT, Chang CJ (2010) Organelle-targetable fluorescent probes for imaging hydrogen peroxide in living cells via SNAP-Tag protein labeling. J Am Chem Soc 132:4455–4465

Sun X, Shih AY, Johannssen HC, Erb H, Li P, Murphy TH (2006) Two-photon imaging of glutathione levels in intact brain indicates enhanced redox buffering in developing neurons and cells at the cerebrospinal fluid and blood–brain interface. J Biol Chem 281:17420–17431

Sun ZN, Wang HL, Liu FQ, Chen Y, Tam PK, Yang D (2009) BODIPY-based fluorescent probe for peroxynitrite detection and imaging in living cells. Org Lett 11:1887–1890

Ueno T, Urano Y, Kojima H, Nagano T (2006) Mechanism-based molecular design of highly selective fluorescence probes for nitrative stress. J Am Chem Soc 128:10640–10641

van de Bittner GC, Dubikovskaya EA, Bertozzi CR, Chang CJ (2010) In vivo imaging of hydrogen peroxide production in a murine tumor model with a chemoselective bioluminescent reporter. Proc Natl Acad Sci USA 107:21316–21321

van der Ven AJ, Mier P, Peters WH, Dolstra H, van Erp PE, Koopmans PP, van der Meer JW (1994) Monochlorobimane does not selectively label glutathione in peripheral blood mononuclear cells. Anal Biochem 217:41–47

van Lith M, Tiwari S, Pediani J, Milligan G, Bulleid NJ (2011) Real-time monitoring of redox changes in the mammalian endoplasmic reticulum. J Cell Sci 124:2349–2356

Veal EA, Day AM, Morgan BA (2007) Hydrogen peroxide sensing and signaling. Mol Cell 26: 1–14

Vikram DS, Rivera BK, Kuppusamy P (2010) In vivo imaging of free radicals and oxygen. Methods Mol Biol 610:3–27

Wang W, Fang H, Groom L, Cheng A, Zhang W, Liu J, Wang X, Li K, Han P, Zheng M, Yin J, Mattson MP, Kao JP, Lakatta EG, Sheu SS, Ouyang K, Chen J, Dirksen RT, Cheng H (2008) Superoxide flashes in single mitochondria. Cell 134:279–290

Wardman P (2007) Fluorescent and luminescent probes for measurement of oxidative and nitrosative species in cells and tissues: progress, pitfalls, and prospects. Free Radic Biol Med 43:995–1022

Waypa GB, Guzy R, Mungai PT, Mack MM, Marks JD, Roe MW, Schumacker PT (2006) Increases in mitochondrial reactive oxygen species trigger hypoxia-induced calcium responses in pulmonary artery smooth muscle cells. Circ Res 99:970–978

Weissleder R, Ntziachristos V (2003) Shedding light onto live molecular targets. Nat Med 9: 123–128

Wink DA, Hines HB, Cheng RY, Switzer CH, Flores-Santana W, Vitek MP, Ridnour LA, Colton CA (2011) Nitric oxide and redox mechanisms in the immune response. J Leukoc Biol 89: 873–891

Winterbourn CC, Hampton MB (2008) Thiol chemistry and specificity in redox signaling. Free Radic Biol Med 45:549–561

Wrona M, Patel K, Wardman P (2005) Reactivity of 2′,7′-dichlorodihydrofluorescein and dihydrorhodamine 123 and their oxidized forms toward carbonate, nitrogen dioxide, and hydroxyl radicals. Free Radic Biol Med 38:262–270

Xu X, von Lohneysen K, Soldau K, Noack D, Vu A, Friedman JS (2011) A novel approach for in vivo measurement of mouse red cell redox status. Blood 118:3694–3697

Yang D, Wang HL, Sun ZN, Chung NW, Shen JG (2006) A highly selective fluorescent probe for the detection and imaging of peroxynitrite in living cells. J Am Chem Soc 128:6004–6005

Yano T, Oku M, Akeyama N, Itoyama A, Yurimoto H, Kuge S, Fujiki Y, Sakai Y (2010) A novel fluorescent sensor protein for visualization of redox states in the cytoplasm and in peroxisomes. Mol Cell Biol 30:3758–3766

Yoo SK, Starnes TW, Deng Q, Huttenlocher A (2011) Lyn is a redox sensor that mediates leukocyte wound attraction in vivo. Nature 480:109–112

Yoshimura T, Yokoyama H, Fujii S, Takayama F, Oikawa K, Kamada H (1996) In vivo EPR detection and imaging of endogenous nitric oxide in lipopolysaccharide-treated mice. Nat Biotechnol 14:992–994

Yu F, Li P, Li G, Zhao G, Chu T, Han K (2011) A near-IR reversible fluorescent probe modulated by selenium for monitoring peroxynitrite and imaging in living cells. J Am Chem Soc 133: 11030–11033

Zhang Q, Wang SY, Nottke AC, Rocheleau JV, Piston DW, Goodman RH (2006) Redox sensor CtBP mediates hypoxia-induced tumor cell migration. Proc Natl Acad Sci USA 103: 9029–9033

Zhao Y, Jin J, Hu Q, Zhou HM, Yi J, Yu Z, Xu L, Wang X, Yang Y, Loscalzo J (2011) Genetically encoded fluorescent sensors for intracellular NADH detection. Cell Metab 14:555–566

Wang W, Fang H, Groom L, Cheng A, Zhang W, Liu J, Wang X, Li K, Han P, Zheng M, Yin J, Mattson MP, Kao JP, Lakatta EG, Sheu SS, Ouyang K, Chen J, Dirksen RT, Cheng H (2008) Superoxide flashes in single mitochondria. Cell 134:279–290

Wardman P (2007) [illegible] Free Radic Biol Med [illegible]

[illegible]

Chapter 6
Redox Proteomics

Alexandra Müller and Lars I. Leichert

Abstract In the presence of oxidative and nitrosative stress, proteins can undergo oxidative modification. A large variety of such modifications have been identified to date, including carbonylation, oxidation of aromatic amino acids, methionine sulfoxidation, and cysteine oxidation. While unintended oxidative modifications most often lead to the damage of the affected proteins, oxidative modifications also play important roles in cellular redox sensing. Classical redox sensor proteins use reversible oxidative modifications to change their activity in response to a changing redox environment. These redox sensors are the focus of oxidative stress research and have been identified in all three kingdoms of life. They are involved in a wide variety of cellular processes ranging from central energy metabolism over protein quality control to the regulation of the oxidative stress response. Proteomic methods have been used to globally monitor the oxidation state of these redox sensors and to identify novel redox sensitive proteins. These methods can help us in understanding redox regulation and the role of protein oxidation under physiological and pathological conditions. In this chapter, we will provide an overview of the different oxidation products of amino acid side chains in proteins, discuss examples of their physiological relevance and present a selection of global methods to identify them. We put an emphasis on quantitative proteomic methods that are able to identify targets of oxidative modifications down to the amino acid.

Keywords Amino acid oxidation • Cysteine modification • Post-translational modification • Reactive oxygen species • Redox signaling

A. Müller • L.I. Leichert (✉)
Medizinisches Proteom-Center, Redox-Proteomics Group, Ruhr-Universität Bochum, Universitätsstr. 150, ZKF 2.061, 44780 Bochum, Germany
e-mail: lars.leichert@ruhr-uni-bochum.de

U. Jakob and D. Reichmann (eds.), *Oxidative Stress and Redox Regulation*,
DOI 10.1007/978-94-007-5787-5_6,

6.1 Introduction: Reactive Oxygen Species, Oxidative Stress, and Cellular Defense Strategies

6.1.1 Reactive Oxygen Species and Redox Homeostasis

The oxygen-rich atmosphere of present-day earth has emerged some 2.4 billion years ago, when cyanobacteria first started to use water as electron source during photosynthesis (Buick 2008). Thus, molecular oxygen became available as an electron acceptor for metabolic processes such as respiration, which is energetically highly favorable compared to anaerobic fermentation and therefore regarded as a prerequisite for the evolution of multicellular life-forms (Buick 2008; Donoghue and Antcliffe 2010). However, oxygen utilization is a two-edged sword as molecular oxygen is a highly reactive element and its utilization in the respiratory chain and in oxidizing enzymes is accompanied by the generation of reactive oxygen species (ROS) (see Chambers et al. 1985; Turrens 2003 for a comprehensive review). ROS have the potential to damage virtually all cellular macromolecules including DNA, proteins, lipids, and carbohydrates. Three well-studied and common ROS in cellular systems are superoxide ($O_2^{-\cdot}$), hydrogen peroxide (H_2O_2), and the hydroxyl radical ($HO\cdot$). They correspond to the reduction of molecular oxygen by one, two, or three electrons and differ with respect to reactivity, half life, and abundance. Superoxide ($O_2^{-\cdot}$) reacts quite specifically with iron-sulfur clusters and possesses a half life of 0.05 s at 0.1 M concentration (Fridovich 1983). The probably most stable ROS is hydrogen peroxide (H_2O_2) (Pryor 1986). Like superoxide, H_2O_2 is mainly generated as a result of oxygen reduction in the respiratory chain but may also be produced by oxygenases and in the endoplasmic reticulum during disulfide bond formation (Gross et al. 2006; Kuthan and Ullrich 1982; Loschen et al. 1971). $HO\cdot$, which may be generated through biologically relevant Fenton chemistry, i.e. the reduction of hydrogen peroxide catalyzed by redox-cycling metal ions, occurs at very low concentrations and possesses a half life as low as 10^{-9} s (Halliwell and Gutteridge 1989; Sies 1993). Nevertheless, $HO\cdot$ is regarded as the most harmful of these ROS because it reacts instantly with any cellular macromolecules in its vicinity. It is thought that in non-photosynthetic organisms, ROS are predominantly generated in the respiratory chain, whereas in photosynthesizing cells the photosystems are by far the main contributors of partially reduced oxygen (Halliwell and Gutteridge 1989; Ivanov and Khorobrykh 2003).

Because proteins are the most abundant targets of ROS in the cell, organisms have evolved sophisticated defense strategies to protect them against the consequences of oxidative damage (Rinalducci et al. 2008). These cellular defense strategies basically act on three different levels: prevention of ROS formation, interception, and repair or removal of damaged proteins (Fig. 6.1). Prevention of ROS formation includes the chelation of metals that could otherwise drive radical-forming reactions, protection from radiation by specialized pigments, and structural hindrance of radical release by enzymes that are generally prone to radical

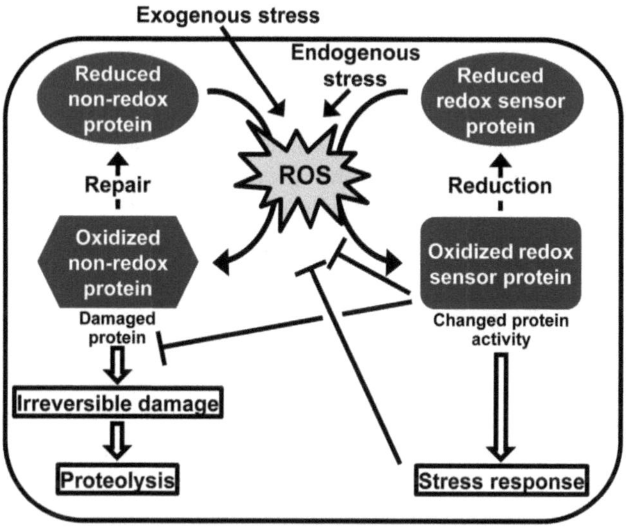

Fig. 6.1 Protein damage and redox regulation in cells exposed to oxidative stress. Endogenous or exogenous oxidative stress leads to the accumulation of reactive oxygen species (ROS) within the cell. These ROS lead to protein oxidation. While typical (non-redox) proteins are damaged by ROS and need to be repaired or degraded, redox sensor proteins alter their activity upon oxidation and often protect cells from oxidative stress. One example would be the redox-regulated chaperone Hsp33, which can prevent protein aggregation under severe oxidative stress conditions. If the redox-sensing protein is a regulator, it can induce different cellular pathways. A prominent example of this mechanism is the induction of the oxidative stress response in *E. coli* following oxidation of the regulator OxyR

formation as for example ribonucleotide reductase (Reichard and Ehrenberg 1983; Sies 1993). When generation inside the cell cannot be prevented or when organisms are exposed to exogenous oxidative stress, cells can remove ROS through enzyme-based decomposition or by the help of small-molecule antioxidants (Halliwell and Gutteridge 1989). Superoxide dismutase converts superoxide to oxygen and hydrogen peroxide. The latter is further dismutated to oxygen and water by catalase or reduced by peroxidase (Chance et al. 1979). These systems are supported by non-enzymatic antioxidants such as glutathione, ascorbic acid, α-tocopherol, flavonoids, and carotenoids. They work in concert with the thioredoxin and the glutaredoxin systems (Holmgren 1989). The thioredoxin/thioredoxin reductase system has been identified in all kingdoms of live. Thioredoxins are small proteins with a highly conserved Cys-Gly-Pro-Cys motif, which is exposed at the surface. They exhibit a highly reducing redox-potential, which allows rapid reduction of oxidized cysteine thiols under conditions of increased oxidative stress. This reduction is accompanied by the formation of a disulfide bond within the Cys-Gly-Pro-Cys motif, which is then re-reduced by thioredoxin reductase in an NADPH-dependent manner. Glutaredoxins are related small redox proteins with a similar active site consisting of two cysteines. In contrast to thioredoxins, however, glutaredoxins are reduced non-enzymatically by the tripeptide glutathione (Holmgren 1989). As long as ROS production does not exceed the capacity of these defense systems, the thiol-disulfide ratio remains fairly stable (Ghezzi et al. 2005). If the exposure to ROS lasts, detoxifying and repairing systems become overwhelmed and irreversibly modified proteins may accumulate under these conditions. These are then subjected to the cellular degradation machinery, which represents the third level of the oxidative stress defense strategies (Grune and Davies 1997; Sies 1993).

6.1.2 Oxidative Stress

However, under certain circumstances, for example under nutrient deficiency, disease, and infection by pathogens, the above mentioned systems may be overwhelmed by ROS, a condition termed oxidative stress (Halliwell and Gutteridge 1989; Sies and Cadenas 1985). This imbalance between ROS generation and proper disposal by cellular defense strategies has been proposed to be one of the underlying mechanisms of aging (Harman 1956). Additionally, a variety of diseases has been linked to oxidative stress for a long time, among them neurological diseases, such as Alzheimer's and Parkinson's, as well as cardiovascular diseases, including heart failure and stroke (Alexandrova and Bochev 2005; Dai et al. 2011; Ogawa et al. 2002). However, in many cases it is still not clear, if oxidative stress is one of the causes or a symptom of these diseases, spawning the development of global and targeted methods to determine the mechanism of cell damage and to study the oxidative damage to biomolecules.

6.1.3 Redox Sensing

Over the past decades it has increasingly been realized that reactive oxygen species not only produce oxidative stress and cause damage of cellular components but also serve as potent signaling molecules involved in important cellular processes, including cell proliferation, differentiation, and apoptosis (D'Autréaux and Toledano 2007; Janssen-Heininger et al. 2008). Therefore the concept of 'redox signaling', which describes entire cascades from sensing of reactive oxygen species down to cellular responses was introduced (Bochner et al. 1984; Proctor and McGinness 1970). This concept is similar to the signaling by nitric oxide (NO·), a radical molecule that was also initially thought to be too unstable and harmful for signaling. While the canonical NO·-sensing is accomplished by a heme center in guanylate cyclase (Craven and DeRubertis 1978; Katsuki et al. 1977), more recently it has been discovered that NO·-signals can be transduced by amino acid modifications, including tyrosine nitration and S-nitrosylation (Ischiropoulos et al. 1992; Stamler et al. 1992; van der Vliet et al. 1994). Similar to oxidative stress, a disturbance of the physiological NO·-levels causes nitrosative stress and uncontrolled modifications, which led to protein damage (Calabrese et al. 2009).

Since the early 1990s a variety of so-called redox sensors has been identified, among them transcription factors like the bacterial OxyR and the eukaryotic NFκB, metabolic enzymes such as glyceraldehyde-3-phosphate dehydrogenase and carbonic anhydrase, as well as molecular chaperones like Hsp33 and peroxiredoxins (Chae et al. 1994; Chai et al. 1991; Jakob et al. 1999; Ravichandran et al. 1994; Staal et al. 1990; Storz et al. 1990; Zheng et al. 1998). What all these proteins have in common is the possession of redox-reactive cysteines, which are specifically oxidized, leading to an alteration of the protein's activity. Once the cause of the oxidation is removed, the cysteines can be reduced by the cellular antioxidant systems and the protein returns to its original redox state and activity (Fig. 6.1). The identification and characterization of these redox sensors is probably just a starting point on our way to understand the full impact of ROS on cellular metabolism, physiology, and signaling. While cysteine, due to the reversibility of thiol modifications in low oxidation states, is a particularly well-suited amino acid for redox sensing, oxidative modifications of other amino acids, such as methionine sulfoxidation have been demonstrated to have functions in redox signaling as well (Vogt 1995; Wong et al. 2008). Hence it is not surprising that in the 'omics' era, a variety of techniques have been developed to globally detect and identify oxidative changes within cellular proteins. The term 'redox proteomics' has emerged to set a boundary to classical proteomic approaches. Redox proteomics "aims to detect and analyze redox-based changes within the proteome both in redox signaling scenarios and in oxidative stress" (Sheehan et al. 2010). This chapter aims to give a comprehensive overview about the most recent proteomic approaches for the large-scale identification of different oxidative protein modifications. In the context of signaling it is important to know which fraction of a redox-sensitive protein is modified and which amino acids are targeted by ROS. Therefore, we will focus on quantitative techniques that allow for the identification of redox-sensitive sites down to the amino acid.

6.2 ROS-Mediated Protein Modifications and Methods for Their Detection

Although virtually all amino acids are vulnerable to oxidation and more than 35 different oxidation products of amino acids have been identified so far, only a few of them have been investigated with respect to their physiological relevance (Madian and Regnier 2010). The observed oxidative modifications can be grouped into two categories: *in vivo* reversible and irreversible modifications. It could be argued that an effective regulative oxidative modification should be reversible *in vivo*, as only reversibility allows for a back-and-forth change and adjustment of the activity of a redox-regulated protein to the changing redox environment. Of the known oxidative amino acid modifications, only the lower oxidation states of cysteine and methionine sulfoxidation fit these criteria and these two amino acids can therefore serve in their corresponding proteins as true ‘redox switches’. Most of the regulative oxidative modifications described so far are cysteine modifications, such as disulfide bond formation or sulfenic acid formation, although recent publications suggest a regulative role for methionine sulfoxidation as well (Abulimiti et al. 2003; Ciorba et al. 1997; Sun et al. 1999). In contrast, amino acid carbonylation and the oxidation of aromatic amino acids is irreversible *in vivo*, and proteins modified in this way typically are rapidly degraded to circumvent harmful aggregation (Nyström 2005). However, recent studies suggest that these irreversible modifications may have a regulatory impact as well (Wong et al. 2008). In the next sections, we will discuss these oxidative protein modifications and proteomic methods, which have been successfully used to study them.

6.2.1 Protein Carbonylation

Protein carbonylation is a post-translational modification resulting from at least four different reactions. The most disruptive event is the oxidation of the protein backbone, leading to the formation of protein fragments with an N-terminal α-ketoacyl amino acid and hence total loss of protein integrity (Stadtman 1990). Additionally, the side-chains of lysine, proline, arginine, and threonine may undergo metal ion-catalyzed oxidation leading to the corresponding ketone or aldehyde derivatives (Fig. 6.2). Carbonylation can also occur by 4-hydroxy-2-nonenal (HNE), which is a product of lipid peroxidation (Dalle-Donne et al. 2003a, b). Finally, lysine, cysteine, and histidine can be glycated or glycoxidated through the conjugation with reducing sugars and their oxidation products, respectively. The latter two reactions are referred to as ‘secondary reactions’ (Dalle-Donne et al. 2006).

Carbonylation is irreversible *in vivo*, and the fate of such modified proteins is degradation by the proteasome in higher organisms or the Lon protease in *E. coli* (Grune et al. 2003; Grune and Davies 1997; Levine 2002). Mildly carbonylated proteins are much more susceptible to degradation than non-modified proteins and it has been suggested that the carbonyl group can act as a recognition tag for the proteasome (Grune et al. 2003; Grune and Davies 1997). Age-related disorders like Parkinson’s disease, Alzheimer’s disease, cancer, and diabetes are associated with an accumulation of carbonylated proteins, which escape the cellular degradation machinery, causing the accumulation of cytotoxic protein aggregates (Dalle-Donne et al. 2003a, b; Levine 2002). Carbonylated proteins therefore often serve as biomarkers for the detection of oxidative stress conditions and several proteomic approaches have been developed to gain insights into the ‘carbonylated proteome’ (Aksenov et al. 2000; Dalle-Donne et al. 2003a, b; Reinheckel et al. 2000). As evidence for a regulatory function of carbonylation emerges (Lee and Helmann 2006; Wong et al. 2008), these methods are constantly improved to enable the site-specific identification of amino acids undergoing carbonylation.

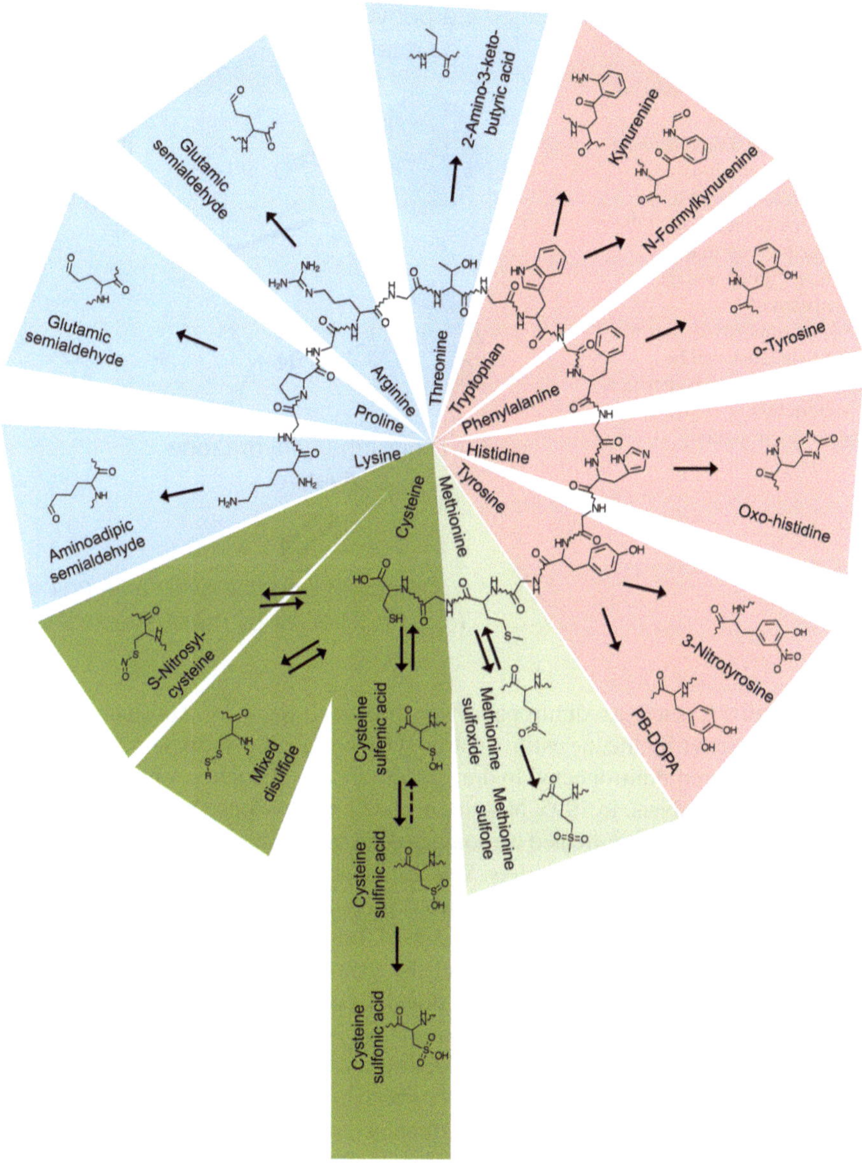

Fig. 6.2 Biologically relevant amino acid side chain modifications caused by reactive oxygen and reactive nitrogen species. *Bidirectional arrows* indicate modifications that are potentially reversible *in vivo*

6.2.1.1 Identification of Carbonylated Proteins

One of the first methods developed to detect carbonylated proteins was the so-called OxyBlot technique. The carbonyl group readily reacts with hydrazine and

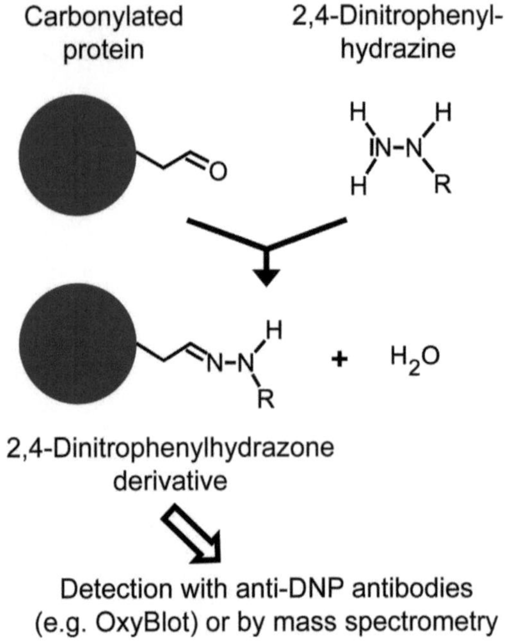

Fig. 6.3 Detection of protein carbonylation by derivatization with 2,4-dinitrophenyl hydrazine ("OxyBlot"). The carbonylated amino acid side chain of a protein is derivatized through a nucleophilic addition of the hydrazin group to the carbonyl group and the formation of a 2,4-dinitrophenylhydrazone. The hydrazone can be detected with antibodies in a western blot or analyzed by mass spectrometry

its derivatives, such as dinitrophenyl hydrazine (Fig. 6.3). This condensation of dinitrophenyl hydrazine with a carbonyl leads to the formation of a stable, intensely colored dinitrophenyl hydrazone product in a reaction well-known from carbohydrate analysis. In 1995, Nakamuro and Goto developed an antibody against dinitrophenyl hydrazine-treated carbonylated BSA, which proved to be specific for the dinitrophenyl hydrazone moiety. Western blots with this antibody allowed them for the first time to identify carbonylated proteins on a 2D gel from rat tissue (Nakamura and Goto 1996). Reinheckel et al. improved this procedure by adding the dinitrophenyl hydrazine in gel, after the separation of the proteins in the first dimension, which made it possible to directly correlate spots from the blot with the stained protein pattern, because there is no isoelectric shift caused when the hydrazone condensation is done after isoelectric focusing (Reinheckel et al. 2000).

Still, the site-specific identification of carbonylated side chains poses a challenge, as the carbonylated form of a given protein is often in low abundance when compared to the unmodified protein. However, Bollineni et al. recently reported a mass spectrometric method, in which dinitrophenyl hydrazine-modified peptides are separated by 2-dimensional liquid chromatography and which makes use of the fact that dinitrophenyl hydrazine can directly act as the matrix for MALDI-based mass spectrometry. In a bottom-up approach they could identify carbonylation-sensitive sites in BSA and β-lactoglobulin (Bollineni et al. 2011).

To further enrich carbonyl-containing proteins it is possible to chemically couple affinity tags to hydrazine. The use of biotin-hydrazine and subsequent affinity-purification with avidin columns to reduce sample complexity made it possible to identify carbonylation sites of rat liver proteins by LC-MS/MS (Mirzaei and Regnier 2005). The same group successfully identified oxidation sites in transferrin by using Girard P reagent, which has the hydrazine linked to a quaternary amine that allowed for enrichment of carbonylated peptides by strong cation exchange chromatography (Mirzaei and Regnier 2006).

Even more sophisticated carbonyl-reactive probes, which combine affinity tags with a so-called element code have been developed. O-ECAT (oxidation-dependent carbonyl-specific element-coded affinity mass tag) specifically labels aldehyde and keto groups of oxidized amino acid side chains by forming an oxime and adds a chelating site for metal ions to the peptide. Differential insertion of metals such as terbium or holmium into the chelating site prior to liquid chromatography allows quantification of the relative oxidation of different samples based on the mass difference of the two metals. With this method oxidation sites in recombinant human serum albumin and *E. coli* RNA polymerase could be identified (Cheal et al. 2009; Lee et al. 2006).

To date these novel mass spectrometry-based methods have been used mainly for model proteins, while experiments on a global level in a physiological context are still missing and the quantification of carbonylations in more complex biological samples seems still challenging.

6.2.2 *Oxidation of Protein-Bound Aromatic Amino Acids*

The aromatic amino acid residues tryptophan, phenylalanine, histidine, and tyrosine are highly prone to oxidation, yielding very different oxidation products (Fig. 6.2) (Guan and Chance 2005). Furthermore, tyrosine is readily nitrated to 3-nitrotyrosine by reactive nitrogen species (Ischiropoulos et al. 1992), which we will discuss here as well. Because of the wide variety of possible modifications and the issue of physiological relevance, site-specific identification of aromatic amino acid modifications in complex samples is still challenging and is therefore, with few exceptions, a relatively unexploited research area to this day.

6.2.2.1 Tryptophan Oxidation

Oxidation of tryptophan may occur by singlet oxygen, derived from UV radiation of molecular oxygen, a so-called type 2 photo-oxidation, or by the reactive nitrogen species peroxynitrite (Ischiropoulos and al-Mehdi 1995; Nakagawa et al. 1977). Although in total 11 different oxidation products of tryptophan have been identified in an *in vitro* system using synthetic peptides, only a few of them have been found *in vivo* (Grosvenor et al. 2010). Among the products found *in vivo* are N-formylkynurenine and kynurenine (Anderson et al. 2002; Taylor et al. 2003b). As antibodies against tryptophan oxidation products are lacking, mass spectrometry is the method of choice for the detection of tryptophan modifications. Oxidation of tryptophan results in mass shifts ranging from +4 Da (kynurenine) to +64 Da (dihydroxy-N-formylkynurenine), which can be used for the identification of modified peptides from MS/MS spectra.

Site-specific identification of oxidized tryptophan was mostly carried out for isolated proteins or protein complexes. In these experiments, the protein of interest is exposed to oxidative stress *in vivo* and then isolated from the complex protein mixture before digestion and LC-MS/MS. This experimental set up was successfully used to identify oxidized tryptophans in histidyl-tRNA synthetase and rat muscle actin (Fedorova et al. 2010; van Dooren et al. 2011). Anderson et al. studied isolated photosystem II from spinach chloroplasts and identified a single tryptophan residue (Trp-352) of CP43 core antenna complex in three distinct oxidation states, namely as kynurenine, oxindolylalanine, and a hydroxylated indole derivative, by tandem mass spectrometry. A role of this tryptophan oxidation as signal for the turnover of photosystem II was proposed (Anderson et al. 2002).

Because of the tell-tale mass shift, the identification of tryptophan oxidation sites in complex samples is possible. Two independent research groups investigated oxidative tryptophan modifications in mitochondrial proteins. Taylor et al. re-evaluated a previously compiled proteomic MS/MS data set of cardiac mitochondria for the occurrence of N-formylkynurenine, making use of the 32 Da mass increase (Taylor et al. 2003a). In total, 51 different N-formylkynurenine-containing peptides from 39 proteins of all mitochondrial respiratory complexes except complex II were discovered upon re-inspection of this data set (Taylor et al. 2003a). The observed tryptophan oxidation of mitochondrial proteins was proposed to be the consequence of exposure to ROS generated during respiration in the mitochondria. Møller and Kristensen studied rice leaf and potato tuber mitochondria and also searched specifically for N-formylkynurenine-containing proteins. 29 peptides containing oxidatively modified tryptophan from 17 different proteins where identified (Møller and Kristensen 2006). Both of these studies identified largely proteins directly involved in cellular respiration, demonstrating that these proteins, often the source of reactive oxygen species themselves, are particularly prone to oxidative damage. However, according to Perdivara et al., some kynurenine and N-formylkynurenine formation can potentially be attributed to artifactual modification during sample processing. A comparison of the extent of kynurenine and N-formylkynurenine occurrence in several antibodies, based on sample handling showed that oxidative tryptophan modifications can be readily identified after separation of the proteins on SDS-PAGE and tryptic in-gel digest but were not detectable after in-solution digest (Perdivara et al. 2010). These results demonstrate the challenges of redox-proteomics, and the need for suitable sample preparation procedures to avoid artifactual oxidative protein modifications.

6.2.2.2 Phenylalanine Oxidation

Oxidation of phenylalanine by $HO^{\cdot}$ or peroxynitrite leads to the formation of ortho-, meta-, and para-tyrosine (Fig. 6.2) (Maskos et al. 1992). The atypical amino acid isomers ortho- and meta-tyrosine are used as biomarkers for the detection of oxidative protein damage associated with diseases such as β-thalassemia and atherosclerosis (Leeuwenburgh et al. 1997; Matayatsuk et al. 2007). Site-specific information about phenylalanine oxidation may be obtained by mass spectrometry, taking advantage of the 16 Da mass increase of tyrosine over phenylalanine. However, a global screening for phenylalanine oxidation has not been reported to date.

6.2.2.3 Histidine Oxidation

Oxidative histidine modification is used as biomarker for oxidative stress (Uchida and Kawakishi 1993). Histidine is particularly prone to metal-catalyzed oxidation leading to 2-oxo-histidine (Fig. 6.2). Histidine oxidation has deleterious effects on proteins: oxidation of histidine in Cu,Zn superoxide dismutase leads to a significant decrease in enzyme activity in aging rats (Maria et al. 1995). But regulatory effects of histidine oxidation have been described as well. In the *Bacillus subtilis* H_2O_2-sensing transcription factor PerR, three histidine and two aspartic acid residues typically coordinate an Fe^{2+}-ion in the regulatory site. Hydroxyl radicals, generated *in situ* through H_2O_2 reduction by the coordinated Fe^{2+}-ion oxidize histidine residues 37 or 91 of PerR to 2-oxo-histidine. The direct consequence is the loss of the metal ion and dissociation of PerR from the DNA followed by the transcription of target genes. This irreversible mechanism allows PerR to sense low levels of hydrogen peroxide *in vivo* (Herbig and Helmann 2001; Lee and Helmann 2006).

Despite its physiological importance, global proteomic screenings for 2-oxo-histidine have not yet been reported. Identification of histidine oxidation sites is complicated by changes in the MS/MS dissociation pattern of peptides containing oxidatively modified histidine due to the lower proton affinity and weaker nucleophilicity of 2-oxo-histidine. This can be particularly challenging in highly oxidized proteins, when several isomers differing in their site of oxidation are present (Bridgewater et al. 2007).

6.2.2.4 Tyrosine Oxidation

Tyrosine residues can be converted to protein-bound 3,4-dihydroxyphenylalanine (PB-DOPA; Fig. 6.2), or if two tyrosines are present in the immediate vicinity, they can be cross-linked to bi-tyrosine (Gross and Sizer 1959; Simpson et al. 1992). Formation of PB-DOPA occurs enzymatically, via metal-catalyzed oxidation, gamma irradiation, or UV light (Ito et al. 1984; Simpson et al. 1993). Similar to phenylalanine and histidine oxidation, oxidative tyrosine modifications can serve as biomarkers for the detection of oxidative stress-associated diseases: PB-DOPA levels were shown to be increased in a number of pathologies including atherosclerosis and lens cataract (Fu et al. 1998; Woods et al. 2003).

A global screening for PB-DOPA was recently published by Lee et al., demonstrating growing interest in this research field (Lee et al. 2010). Lee and co-workers specifically searched for the PB-DOPA modification in LC-MS/MS spectra and identified as many as 67 PB-DOPA-bearing sites in *E. coli* and nine sites in HeLa mitochondria using this approach. These included proteins typically prone to oxidative modification, such as GAPDH and superoxide dismutase.

Sharov and co-workers have developed a derivatization technique transforming PB-DOPA to 6-amino-substituted benzoxazoles using benzylamine derivatives (Sharov et al. 2010). The resulting modified proteins display characteristic fluorescence properties making this method suitable for gel-based identification of PB-DOPA. In addition, the use of substituted benzylamines carrying functional groups to enable enrichment will conceivably facilitate MS-based identification of target proteins.

6.2.2.5 Tyrosine Nitration

In addition to hydroxylation, nitration of tyrosine to 3-nitrotyrosine (3-NT; Fig. 6.2) has been extensively studied. Accumulation of 3-NT is considered a typical marker of peroxynitrite stress (Beal et al. 1997). A variety of techniques for site-specific identification of 3-NT are available, including immunoprecipitation and immunoaffinity chromatography using anti-3-NT antibodies, gel-based techniques with subsequent immunoblotting and mass spectrometry, as well as gel-free mass spectrometry-based techniques (see Abello et al. 2009 for a comprehensive review).

The concentration of 3-NT in human plasma is in the pM to nM range and has been shown to remain fairly constant, even under conditions of disease, physical exercise, and pharmacological treatment. This low abundance in highly complex biological samples complicates its detection, making selection of a suitable methodology most critical for the generation of reliable data (Tsikas 2010). This challenge is further complicated by artifactual tyrosine-nitration, which occurs by acid-catalyzed reactions in the presence of nitrate or nitrite in solution. To avoid this problem, 3-NT can be converted chemically into stable derivatives, preferably during early steps of sample preparation (Tsikas 2010).

A variety of such derivatization techniques, including fluorogenic tagging with ABS (4-(aminomethyl)benzenesulfonic acid), APPD ((3R,4S)-1-(4-(aminomethyl)phenyl-sulfonyl)pyrrolidine-3,4-diol), or benzylamine as well as isotope-coded tagging via iTRAQ after protection of free amino groups and chemical modification was proven to be suitable for identification of 3-NT (Chiappetta et al. 2009; Dremina et al. 2011; Sharov et al. 2008, 2010). APPD derivatization of 3-NT-containing proteins from peroxynitrite-exposed cell lysates was developed recently and makes it now possible to quantify the amount of 3-NT through fluorescence spectrometry or boronate affinity chromatography followed by MS-based proteomics. In addition, localization and quantification of 3-NT has been performed by a clever extension of the iTRAQ chemistry. Specific iTRAQ labeling of 3-NT-bearing sites is facilitated by protection of lysine residues and N-termini of trypsin-digested peptides prior to reduction of the nitro group in 3-NT to an amino group, which can then be iTRAQ modified (Chiappetta et al. 2009).

These methods allowed for site-specific identification of 3-NT even at its typically very low abundance in complex protein samples, such as human uremic plasma and cell lysates (Chiappetta et al. 2009; Dremina et al. 2011).

6.2.3 Methionine Oxidation

Oxidation of protein-bound methionine residues to methionine sulfoxide is one of the few protein oxidation events that are reversible *in vivo* (Fig. 6.2). Methionines that undergo sulfoxidation can thus serve as on/off sensors of oxidative stress in certain proteins. A number of such proteins has been identified, among them calmodulin and glutamine synthetase (Levine et al. 1996; Wolff et al. 1980; Yao et al. 1996). Surface-exposed methionines are especially prone to oxidation, and once oxidized to methionine sulfoxide can be further oxidized to methionine sulfone, a modification that is irreversible *in vivo* (Fig. 6.2) (Levine et al. 2000). The modified amino acid methionine sulfoxide has different physico-chemical characteristics from reduced methionine, which in turn can lead to conformational changes in the affected protein, causing loss of function or altered susceptibility to degradation (Bigelow and Squier 2005). Therefore almost all organisms, from bacteria to human, possess methionine sulfoxide reductases (Msr), which reduce methionine sulfoxide back to methionine in a thioredoxin-dependent manner. Two different classes of methionine sulfoxide reductases, MsrA and MsrB, can be distinguished, based on their stereospecificity towards S- and R-enantiomers of methionine sulfoxide, respectively (see Weissbach et al. 2005 for a comprehensive review).

Methionine sulfoxide reduction activity is required under oxidative stress conditions. Consequently, deletion of *msrA* causes severe phenotypes related to oxidative stress both in prokaryotes and in eukaryotes. Increased sensitivity towards oxidative stress has been observed in *Escherichia coli* and *Saccharomyces cerevisiae msrA* deletion strains (Moskovitz et al. 1995, 1997). Knock-out of *msrA* in mice resulted in a 40% reduced life span, abnormal behavior and significantly increased brain dopamine levels, demonstrating the often observed connection between oxidative stress, aging and neurodegenerative diseases (Oien et al. 2008).

In contrast, *msrA* overexpression in *Drosophila melanogaster* and *S. cerevisiae* extended the life span and increased resistance against oxidative stress, respectively (Moskovitz et al. 1998; Ruan et al. 2002). The observed positive effects of *msrA* overexpression are attributed to an antioxidant effect of methionine oxidation. While reducing oxidized methionines, methionine sulfoxide reductase becomes oxidized, which is then reversed through thioredoxin-based reduction. Hence, methionine oxidation and the subsequent reduction of methionine sulfoxide is thought to be able to act as a sink for reactive oxygen species, protecting amino acids in close vicinity from oxidation (Levine et al. 2000).

In addition to this protective function of methionine oxidation, several studies found a role of methionine sulfoxidation in regulation of enzyme and peptide hormone activity. One of the earliest examples for regulation of protein activity by reversible methionine oxidation was the discovery of the oxidation of Met3 in the voltage-gated potassium channel ShC/B from *Drosophila melanogaster*. Sulfoxidation of this specific methionine residue causes a significant slowdown of channel inactivation, which can be restored by co-expression of *msrA*. The authors proposed that membrane depolarization itself could induce oxidation and that the associated modulation may play a role in learning and memory (Ciorba et al. 1997).

In the case of IκBα (inhibitor of kappa B alpha), which is involved in the immune response, it was shown that oxidation of a methionine residue increased its resistance against proteolytic degradation. This stabilization leads to a decreased activation of the immune response (Agbas and Moskovitz 2009).

However, it is still not clear if the reduction by methionine sulfoxide reductases is a non-specific process on all cellular proteins with exposed methionines or if there are preferred targets of this antioxidant system. To trap potential target proteins, affinity chromatography with immobilized plastidial MSRB1 was carried out in *Arabidopsis thaliana*. Interestingly, 13 out of the 24 identified MRSB1-interacting proteins directly participate in photosynthesis, including the ATP synthase α- and β-subunits, the large RubisCO subunit and GAPDH (Tarrago et al. 2011). These results indicate that methionine sulfoxide reductase activity is of special importance in compartments that are generally exposed to oxidative stress, such as plastids. The authors noted that homologs of several identified MSRB1-interacting proteins were also shown to be methionine sulfoxide reductase substrates in other organisms. These include catalase and GAPDH from *Helicobacter pylori* and *Synechocystis*, respectively (Alamuri and Maier 2006; Sato et al. 2007), indicating that methionine oxidation itself could be targeted to some highly susceptible substrate proteins.

In a recent study, a combination of SILAC (stable isotope labeling with amino acids in cell culture) and COFRADIC (combined fractional diagonal chromatography) was used to identify methionine sulfoxide-containing peptides from complex samples and to quantify their degree of oxidation (Ghesquière et al. 2011). Human Jurkat cells were cultured in medium containing $^{13}C_5$-methionine and treated with hydrogen peroxide prior to cell lysis and tryptic digest. This hydrogen peroxide-treated sample was mixed with a completely oxidized 'reference sample' labeled with $^{12}C_5$ methionine. In contrast to the hydrogen peroxide-treated sample, the 'reference sample' was oxidized *in vitro* after trypsin digest to ensure complete oxidation of all methionines. The subsequently performed COFRADIC method relies on the fact that the retention time of a modified peptide in an LC run is different from the retention time of the unmodified peptide. Thus, the removal of the methionine sulfoxide modifications from a fraction of a first LC run and a subsequent second LC run of this fraction under the same conditions will lead to characteristic peak shifts, which make it possible to identify modified peptides (Fig. 6.4). Subsequent tandem mass spectrometry could identify the oxidation-sensitive methionine of these peptides. Furthermore, by comparison of the peak areas of the light and heavy peptides, which correspond to the fully oxidized reference and the hydrogen peroxide-treated sample, respectively, the degree of sulfoxidation in the cells could be determined. As many as 2,000 oxidation-sensitive methionines from more than 1,600 proteins were identified in this way. As these proteins were not associated with particular pathways or protein classes,

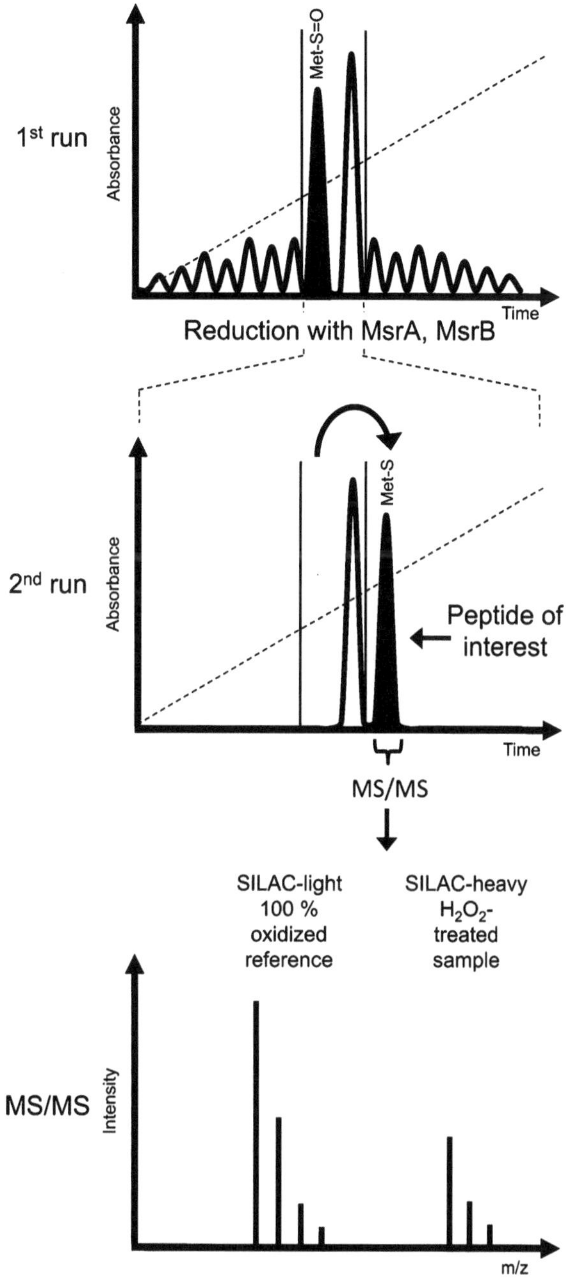

Fig. 6.4 Application of COFRADIC (Combined Fractional Diagonal Chromatography) and SILAC for the identification and quantification of peptides containing methionine sulfoxides. In the first HPLC run, a complex peptide mixture including $^{12}C_5$ methionine and $^{13}C_5$ methionine labeled peptides from a fully oxidized reference and an H_2O_2-treated sample, respectively, is separated. The peptides are collected in fractions and then reduced with methionine sulfoxide reductase A and B, which removes all sulfoxides. A second HPLC run is performed using the same chromatographic conditions. Peptides that were sulfoxidized in the original sample and were reduced by MsrA or MsrB, elute now at a later time from the column and are identified by tandem mass spectrometry. The ratio of the intensities of the $^{12}C_5$ and the $^{13}C_5$ signal in the MS allows for a quantification of the oxidized fraction of the peptide in the H_2O_2-treated sample

it was assumed that hydrogen peroxide-mediated oxidation of methionine is a non-targeted event (Ghesquière et al. 2011). This non-specificity and the susceptibility of methionine-containing peptides to form methionine sulfoxide through air oxidation during sample handling is a particular challenge when studying this modification.

The SILAC/COFRADIC method could also be used by the same group to detect and quantify methionine oxidation in a mouse sepsis model. In mice, the intravenous injection of *Salmonella* leads to a massive oxidative burst event due to the septic shock. Under these conditions 35 oxidatively modified methionines from 27 different proteins were identified in the serum. Most of these proteins, including myoglobin and NADP-dependent malic enzyme are not typical serum proteins and were thought to be released by the onset of necrosis (Ghesquière et al. 2011).

6.2.4 Cysteine Oxidation

Cysteine is one of the least used proteinogenic amino acids, with its fraction ranging between 0.7% in archaea and 2.3% in mammals. One of the reasons seems to be the high reactivity of its terminal thiol group, which brings a high risk of damaging oxidation to proteins that carry this amino acid (Giles et al. 2003). However, this high reactivity is critical for some proteins and can facilitate their structural stabilization by disulfide bond formation, mediate their binding of catalytic metal ions, ensure catalytic activity, and regulate their protein function (Miseta and Csutora 2000).

Depending on the cellular location of a protein, cysteine oxidation can occur under physiological conditions or under pathological stress conditions. In highly oxidizing compartments, such as the bacterial periplasm or the eukaryotic endoplasmic reticulum, protein cysteines are oxidized to form disulfide bonds under physiological conditions. This stabilizes the structure of extra-cellular proteins and helps them to maintain proper function in an environment devoid of active protein-folding systems (Darby and Creighton 1995). In these compartments, oxidoreductases such as protein disulfide isomerases in eukaryotes or the Dsb-system in Gram-negative bacteria introduce disulfide bonds during protein maturation (Depuydt et al. 2011).

In contrast, the cytoplasm is a highly reducing environment and most cysteines are required to be in their free thiol state in this compartment for proper protein function. Here, the thioredoxin and glutaredoxin systems keep cysteines in their reduced state. However, upon oxidative stress, cysteine thiol groups may become oxidatively modified. On the one hand, this protein oxidation often results in protein damage, e.g. through non-native disulfide bond formation, misfolding, or direct blockage of active sites. On the other hand, in so-called redox-regulated proteins, the oxidation can have regulative effects and change the protein's function. Cysteines are well-suited to function in this way as redox-sensing nano-switches, because oxidative thiol modifications in low oxidation states are fully reversible *in vivo* through the thioredoxin and glutaredoxin systems. These reversible modifications include cysteine sulfenic acid (Fig. 6.2), inter- and intramolecular disulfide bonds, and mixed disulfides with free cysteine, glutathione, or other low molecular weight thiols (Dickinson and Forman 2002).

The reactivity of a cysteine residue towards oxidants and other chemicals depends on its solvent exposure and its pK_a. The latter is strongly influenced by the local amino acid environment. The average pK_a of cysteine is 8.5. Because the negatively charged thiolate is the redox-active form, typical cysteines are somewhat protected from oxidative modifications under physiological pH-conditions. However, in specific protein domains or in special local environments, the pK_a may be lowered significantly and the thiol group may exist mainly in its thiolate anion form even at physiological pH. These cysteines are highly prone to oxidation (Lindahl et al. 2011; Salsbury et al. 2008).

6.2.4.1 Formation of Disulfide Bonds

Disulfide bond formation occurs under physiological conditions during oxidative folding in specialized compartments, such as the bacterial periplasm or the eukaryotic endoplasmic reticulum. Disulfide bonds can be formed intramolecularly by two cysteines in the same peptide chain or intermolecularly between two different proteins (Fig. 6.2). Introduction of structural disulfides is achieved by specific oxidoreductases such as the Dsb-system in bacteria and the protein disulfide isomerases in eukaryotes (Depuydt et al. 2011).

Disulfide bond formation can also be observed within the cytoplasm. These disulfide bonds can be formed as part of a catalytic cycle, but under oxidative stress conditions, non-native disulfides may form. Some cytoplasmic proteins are especially prone to the formation of non-native disulfide bonds upon oxidative stress and such disulfides typically need to be removed to restore the protein's activity. This reduction can be carried out by thioredoxin, which leads to the formation of a catalytic intramolecular disulfide bond in thioredoxin. This disulfide bond is subsequently reduced by thioredoxin reductase. Electrons for the reduction of thioredoxin reductase are provided by NAD(P)H (Bindoli et al. 2008).

A well-studied example of a protein prone to form a disulfide bond in its active site is glyceraldehyde-3-phosphate dehydrogenase (GAPDH), which catalyzes the conversion of glyceraldehyde-3-phosphate to 1,3-bisphosphoglycerate during glycolysis. During this reaction, NAD^+ is reduced, yielding NADH. GAPDH possesses two cysteines in its active site, one of which is essential to overcome the high activation energy of the reaction (Harris et al. 1963). Interestingly, these cysteines are not only important for catalysis but are also able to sense oxidative and nitrosative stress through reversible thiol oxidation. The inactivation of GAPDH under these stress conditions is thought to divert glucose to the pentose phosphate pathway, resulting in increased NADPH levels, an important cofactor of the thioredoxin and glutaredoxin systems (Godon et al. 1998; Ralser et al. 2007).

6.2.4.2 Sulfenic Acid Formation

Formation of disulfide bonds within proteins is thought to be preceded by the formation of sulfenic acid when the reaction is initiated by certain oxidants, such as hydrogen peroxide (Fig. 6.2). Generally, cysteine sulfenic acids are considered to be unstable and easily react with other free thiol groups to form disulfides. However, a cysteine sulfenic acid may be stabilized and can have regulatory function, as has been shown in protein tyrosine phosphatase, which catalyzes the dephosphorylation of tyrosine. A reactive cysteine in the active center of this enzyme is reversibly oxidized to sulfenic acid by hydrogen peroxide resulting in inactivation of the enzyme (Meng et al. 2002).

6.2.4.3 Sulfinic and Sulfonic Acid Formation

Persistent oxidative stress may result in the formation of sulfinic acid in proteins (Fig. 6.2). This modification has long been viewed as an irreversible overoxidation product but this assessment dramatically changed with the identification and characterization of sulfiredoxin. This oxidoreductase has been identified in yeast, mammals, and cyanobacteria and catalyzes the regeneration of cysteine sulfinic acid of overoxidized peroxiredoxins (Biteau et al. 2003; Boileau et al. 2011; Chang et al. 2004). Peroxiredoxins are ubiquitous in all kingdoms and belong to the antioxidant enzyme systems, as they catalyze the removal of hydroperoxides and hydrogen peroxide (Hofmann et al. 2002; Wood et al. 2003).

In typical dimeric 2-Cys peroxiredoxins, a sulfenic acid intermediate is formed during the reaction with hydrogen peroxide at an N-terminally located cysteine residue during the reaction with hydroperoxides. The thiol group of a C-terminal cysteine residue from the other subunit reacts with the sulfenic acid to form an intermolecular disulfide bond, which is then reduced by thioredoxin. However, the sulfenic acid intermediate may also be further oxidized to sulfinic acid by excess hydrogen peroxide, which blocks disulfide bond formation and results in the inactivation of the peroxiredoxin (see Wood et al. 2003 for a comprehensive review). Sulfiredoxin can reduce the overoxidized cysteine and thus restore peroxiredoxin activity (Biteau et al. 2003). So far, sulfinic acid reduction by sulfiredoxin has only been demonstrated for peroxiredoxins and a growing body of evidence suggests that sulfiredoxins are highly specific for peroxiredoxin reduction (Woo et al. 2005).

Sulfinic acids can be further oxidized to sulfonic acid (Fig. 6.2), an oxidation event that is thought to be irreversible even in peroxiredoxins and typically leads to the inactivation of the corresponding protein. Sulfonic acid is hence used as a marker of cumulative damage due to excessive oxidative stress. Selective detection of proteins bearing cysteine sulfonic acid can be achieved by nano-diamonds coated with poly-arginine, which specifically bind to the sulfonate groups of over-oxidized proteins (Chang et al. 2010).

6.2.4.4 Formation of Mixed Disulfides by S-Thiolation

Cysteine overoxidation often goes along with protein damage. Glutathione and other low molecular weight thiols can play an important role in preventing this overoxidation both during conditions of oxidative stress and under physiological conditions. In many organisms, the tripeptide glutathione is present at millimolar concentrations in the cytoplasm. Under normal physiological conditions, it has been shown that up to 20% of glutathione exists in a protein-bound state as mixed disulfide (Ghezzi et al. 2005). A protein sulfenic acid that might form when a protein encounters oxidative stressors can react with the free thiol group of glutathione, releasing a water molecule, and forming a glutathione adduct. The mixed disulfide is then protected from overoxidation.

In addition, S-glutathionylation has been shown to modulate the activity of several proteins that are involved in different cellular processes, including signal transduction and metabolism (Rinna et al. 2006; Zaffagnini et al. 2007). Cobalamine-independent methionine synthase (MetE) in *E. coli* is a prominent example of a protein, whose activity is regulated by reversible S-glutathionylation. MetE catalyzes the formation of methionine from homocysteine by the use of methyl-tetrahydrofolate as a methyl group donor. Under oxidative stress conditions, MetE was shown to become S-glutathionylated at cysteine 645 at the entrance of the active site, resulting in a conformational change and thereby the loss of enzyme activity. As a consequence, cells treated with oxidizing agents become methionine auxotroph (Hondorp and Matthews 2004). It has been proposed that the inactivation of MetE by S-glutathionylation protects the enzyme against irreversible oxidation damage with the possibility to restore the activity when the redox-environment is returned to pre-stress conditions.

In eukaryotes, regulation of protein activity by reversible S-glutathionylation seems to be a far more common event than in prokaryotes. Examples of proteins regulated by S-glutathionylation include metabolic enzymes, such as carbonic anhydrase III and α-ketoglutarate dehydrogenase, as well as transcription factors, such as NF-κB and c-Jun. The list of newly identified proteins that become S-glutathionylated is constantly growing (Dalle-Donne et al. 2007). In most organisms, reversibility of S-glutathionylation is achieved by the glutaredoxin system. Not surprisingly, the disturbance of this system is associated with the onset or the progression of neurodegenerative diseases, such as Alzheimer's, Parkinson's and Huntington's disease (Sabens et al. 2012).

Not all organisms use glutathione as a cellular low molecular weight thiol (Zaffagnini et al. 2007). Most Gram-positive bacteria, as for example the soil bacterium *Bacillus subtilis*, do not possess glutathione and although S-thiolation was readily observed, the nature of such modifications remained unclear (Hochgräfe et al. 2005). Interestingly, cysteine biosynthesis is greatly increased upon oxidative stress in *B. subtilis*, suggesting a role of high cysteine levels for oxidative stress protection. Hochgräfe et al. demonstrated, that proteins in *B. subtilis* undergo reversible S-cysteinylation (Hochgräfe et al. 2007). Using a mass spectrometry-based approach, they identified six different proteins that become S-cysteinylated by oxidative stress treatment. One of these proteins is methionine synthase MetE that is regulated in *E. coli* by glutathionylation. This reversible inactivation of MetE in *B. subtilis* is thought to increase cysteine levels by redirecting of synthesis capacity from methionine towards cysteine biosynthesis (Hochgräfe et al. 2007). Hence, though the nature of cysteine modification may differ in particular organisms, it seems that similar metabolic pathways are targeted through the protective inactivation of functionally related proteins. Another common low molecular weight thiol in *B. subtilis* is the recently discovered bacillithiol. Bacillithiol is synthesized from

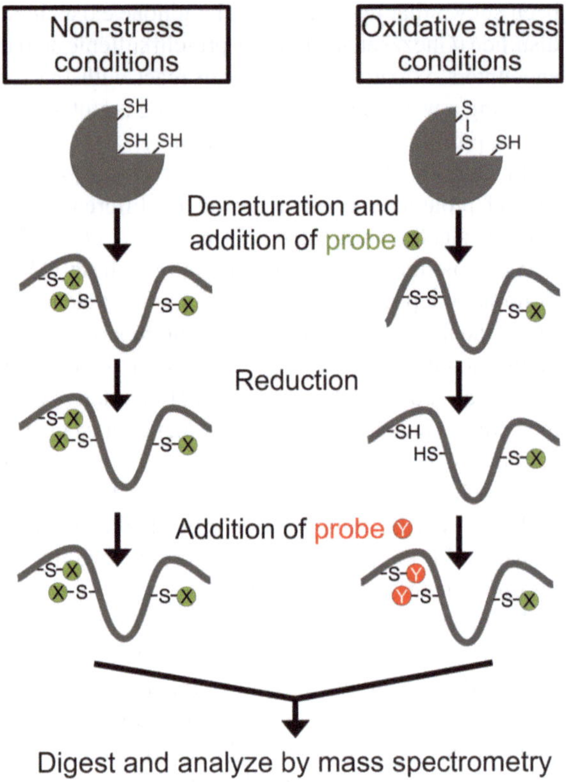

Fig. 6.5 Detection of oxidative cysteine modifications by differential thiol labeling. A hypothetical protein in its reduced (*left*) and oxidized form (*right*) is denatured and a thiol reactive probe "X" is added. This probe blocks all free thiol groups while oxidized cysteines will not be modified. In a subsequent step, reducing agent is added, which reduces the disulfide bond in the oxidized protein. Addition of a second thiol reactive probe "Y" with different properties will then label thiol groups that were initially oxidized. Based on the probes properties, the oxidation state of the protein can be determined, e.g. by radioactivity, fluorescence, or mass difference

L-cysteinyl-D-glucosamine and L-malic acid and functions as a redox buffer similar to glutathione in this organism. It was shown to directly react with protein-bound cysteine residues in regulative events leading to S-bacillithiolation. The anorganic peroxide-sensing transcription factor OhrR in *B. subtilis* is S-bacillithiolated upon hydroperoxide stress, resulting in its release from the operator, which allows transcription of oxidative stress response genes (Lee et al. 2007).

6.2.4.5 Proteomic Identification of Thiol Modifications

The proteomic identification of thiol modifications is typically based on modification of the thiol group with specific probes. The reduced thiolate is highly nucleophilic and is therefore easily modified by nucleophilic substitution or addition reactions. Typical probes used for this purpose are derived from iodoacetamide or maleimide. Because oxidized cysteines typically do not undergo these reactions, this allows for differential thiol trapping: after modification of all free cysteines in a protein sample with one probe, the sample can then be reduced and the now accessible, previously oxidized thiols can be labeled with a different probe in a second labeling step (Fig. 6.5). Thus, reduced and oxidized cysteines can be clearly distinguished and potentially quantified, based on the nature of the probe. When typical thiol reductants such as DTT or TCEP are used in the reduction step, all reversible thiol modifications will be reduced and thus labeled with the second probe.

One of the first approaches to quantify the thiol status by the use of differential thiol trapping was accomplished in *E. coli* using a gel-based approach with unlabeled and ^{14}C-labeled iodoacetamide (Leichert and Jakob 2004). A number of so far unknown redox-regulated proteins were identified by the comparison of control and oxidative stress conditions. The same technique also allowed for the identification of substrates of the disulfide bond forming DsbA-system and the antioxidant thioredoxin system.

The well-known DIGE (differential gel electrophoresis) technique has also been adapted to redox proteomics by the use of derivatized cyanine (Cy) fluorescent dyes, which specifically react with free thiol groups (Chan et al. 2005; Hurd et al. 2007). In a typical workflow, a control sample and an oxidative stress-treated sample are first treated with NEM (N-ethyl maleimide) to block all free thiol groups. DTT is then added, which unspecifically reduces all reversibly oxidized cysteines. Subsequently, two different DIGE dyes (e.g., Cy3 maleimide and Cy5 maleimide) are used to label the formerly oxidized cysteines in the control and the treated sample, respectively. Samples are then mixed and 2-dimensional gel electrophoresis is performed. Due to the use of two distinguishable fluorophores, differences in the thiol redox state of individual proteins can be detected with this so-called redox-DIGE method.

Nevertheless, several control steps are needed, as differences in the relative expression of proteins and their stability, as well as differences in the labeling efficiency of the two dyes might have an influence on spot intensities. Swapping of the two labels and analysis of several gels in parallel can avoid false-positives (Hurd et al. 2007).

Among gel-free methods for the identification of reversible cysteine modifications, the ICAT technique (isotope-coded affinity tag) is well established. This tagging system is composed of a biotin moiety for purification, a linker region that can be isotopically light (light ICAT) or heavy (heavy ICAT), and iodoacetamide. ICAT has initially been developed for quantitative proteomics based on the differential expression of proteins and is typically used with fully reduced samples to gain the highest labeling efficiency (Gygi et al. 1999). But as labeling is based on the reaction of the iodoacetamide moiety with free sulfhydryl groups of cysteines, it also proved useful to detect changes in the oxidation state of cysteines. Light and heavy versions of ICAT are added to sets of non-reduced samples, which are expected to have differences in the thiol state. Samples are subsequently mixed, digested, purified, and analyzed by mass spectrometry. Because the ICAT reagent only modifies reduced cysteines, a decrease in intensity in a given sample (light or heavy) indicates a decrease in available free cysteines and in reverse an increase in oxidation. Using this approach, Sethuraman et al. identified and quantified oxidized cysteines in a hydrogen peroxide-treated particulate membrane fraction of rabbit heart (Sethuraman et al. 2004). However, as two different samples are compared, differential protein expression and protein stability needs to be taken into account as they will also change signal intensities. A modified technique, which uses a differential thiol trapping approach as it labels reduced cysteines with light ICAT and oxidized cysteines with heavy ICAT avoids this problem, and was termed OxICAT (Leichert et al. 2008). In this setup, the ratio of reduced and oxidized cysteines in a peptide sample can be quantified and these ratios can be compared across samples to identify redox-sensitive proteins. This technique has successfully been used in *E. coli*, *Caenorhabditis elegans* and *S. cerevisiae*, identifying a number of putative redox-regulated proteins affecting a variety of cellular pathways (Brandes et al. 2011; Kumsta et al. 2011; Leichert et al. 2008).

6.2.4.6 Global Approaches for the Identification of Specific Thiol Modifications

It is often desirable to distinguish between different reversible thiol modifications, such as disulfide bonds, glutathionylation, sulfenic acids, or nitrosothiols. This can be accomplished by the use of specific reductants, such as for example ascorbate, which selectively reduces nitrosothiols with some specificity. Another approach is the use of probes that react specifically with the modified cysteine, such as dimedone, which exclusively reacts with sulfenic acids. However, these approaches can be sometimes complicated by the instability of the modification in question or the lack of specificity of the reductant.

The biotin-switch technique, which detects S-nitrosylated proteins in a three step procedure is based on the use of ascorbate as a specific reductant (Jaffrey and Snyder 2001). Free thiol groups are blocked by a thiolating agent, such as MMTS (methyl methanethiosulfonate) or an alkylating agent, such as NEM. Proteins are then treated with ascorbic acid, which reduces S-nitrosylated cysteines. These cysteines are then biotinylated by biotin-HPDP. Biotinylated proteins can be detected either by electrophoresis and subsequent western blotting using streptavidin-conjugated fluorophores or anti-biotin antibodies or by mass spectrometry after tryptic digest and affinity purification (Huang et al. 2009). Instead of biotin-HPDP, probes such as the cysTMT sixplex reagent may be used for protein labeling. The cysTMT is a mass spectrometric probe that has six different isobars, which release different reporter ions when fragmented in an MS/MS experiment. These reporter ions can then be used to quantify up to six different samples in a single experiment (Murray et al. 2012).

Although ascorbic acid is considered to be a specific reductant of S-nitrosothiols, there is evidence that it may also reduce disulfide bonds in the colorimetric thiol probe 5,5′-dithiobis-2-nitrobenzoic acid (DTNB), cystine, mixed disulfides, and even in biotin-HPDH in a pH and concentration-dependent manner (Giustarini et al. 2008; Landino et al. 2006). To circumvent this problem, triphenylphosphate ester derivatives such as PE PEG or PE SO3, which have been shown to be even more specific reductants of S-nitrosothiols, can be used (Li et al. 2011).

Sulfenic acid is a very unstable intermediate and often reacts with free thiols to disulfides or is further oxidized to sulfinic and sulfonic acid. This instability complicates the specific detection of sulfenic acid. Different dimedone-based probes including DAz-1 and DCP-Bio1 have been developed that react specifically with sulfenic acid. These reagents contain affinity or fluorescent tags, which enable enrichment or visual detection, respectively (Klomsiri et al. 2010; Poole et al. 2005; Reddie et al. 2008). DAz-1 was successfully used to demonstrate that the thioredoxin-related protein DsbG protects single cysteine residues of periplasmic proteins from oxidation to sulfenic acid (Depuydt et al. 2009).

Specific detection of S-glutathionylation can be achieved by reduction of S-glutathionylated cysteines with glutaredoxin (Grx) and subsequent labeling with N-ethylmaleimide–biotin (Lind et al. 2002). Other methods that have been used for detection of S-glutathionylated proteins include the detection of proteins modified by ^{35}S-labeled or biotinylated glutathione (Brennan et al. 2006; Lind et al. 1998).

6.3 Summary

Redox-proteomics methods can provide us with global insights into protein damage caused by oxidation, as well as into the molecular mechanisms that underlie redox-regulation. But the identification of target proteins of oxidative stress is only a first step in understanding the connection between oxidative stress and its pathological consequences. The availability of precise mass spectrometers and highly specific derivatization techniques already facilitated the identification of numerous proteins that are prone to oxidation and the role of their oxidative modification in redox-regulation, oxidative protein folding, and protein damage. The ongoing development of probes that show high specificity for defined oxidative modifications has the potential to further expand our insights, especially when these efforts are targeted at unstable and transient oxidative modifications. Global data sets gathered from redox-proteomics experiments will continue to provide a useful starting point for concise analyses of the role of protein oxidation within global cellular networks and will help us understanding the role of oxidative stress and redox signaling in health and disease.

References

Abello N, Kerstjens HAM, Postma DS, Bischoff R (2009) Protein tyrosine nitration: selectivity, physicochemical and biological consequences, denitration, and proteomics methods for the identification of tyrosine-nitrated proteins. J Proteome Res 8:3222–3238

Abulimiti A, Qiu X, Chen J et al (2003) Reversible methionine sulfoxidation of *Mycobacterium tuberculosis* small heat shock protein Hsp16.3 and its possible role in scavenging oxidants. Biochem Biophys Res Commun 305:87–93

Agbas A, Moskovitz J (2009) The role of methionine oxidation/reduction in the regulation of immune response. Curr Signal Transduct Ther 4:46–50

Aksenov M, Aksenova M, Butterfield DA, Markesbery WR (2000) Oxidative modification of creatine kinase BB in Alzheimer's disease brain. J Neurochem 74:2520–2527

Alamuri P, Maier RJ (2006) Methionine sulfoxide reductase in *Helicobacter pylori*: interaction with methionine-rich proteins and stress-induced expression. J Bacteriol 188:5839–5850
Alexandrova ML, Bochev PG (2005) Oxidative stress during the chronic phase after stroke. Free Radic Biol Med 39:297–316
Anderson LB, Maderia M, Ouellette AJA et al (2002) Posttranslational modifications in the CP43 subunit of photosystem II. Proc Natl Acad Sci USA 99:14676–14681
Beal MF, Ferrante RJ, Browne SE et al (1997) Increased 3-nitrotyrosine in both sporadic and familial amyotrophic lateral sclerosis. Ann Neurol 42:644–654
Bigelow DJ, Squier TC (2005) Redox modulation of cellular signaling and metabolism through reversible oxidation of methionine sensors in calcium regulatory proteins. Biochim Biophys Acta 1703:121–134
Bindoli A, Fukuto JM, Forman HJ (2008) Thiol chemistry in peroxidase catalysis and redox signaling. Antioxid Redox Signal 10:1549–1564
Biteau B, Labarre J, Toledano MB (2003) ATP-dependent reduction of cysteine-sulphinic acid by *S. cerevisiae* sulphiredoxin. Nature 425:980–984
Bochner BR, Lee PC, Wilson SW et al (1984) AppppA and related adenylylated nucleotides are synthesized as a consequence of oxidation stress. Cell 37:225–232
Boileau C, Eme L, Brochier-Armanet C et al (2011) A eukaryotic-like sulfiredoxin involved in oxidative stress responses and in the reduction of the sulfinic form of 2-Cys peroxiredoxin in the cyanobacterium *Anabaena* PCC 7120. New Phytol 191:1108–1118
Bollineni RC, Hoffmann R, Fedorova M (2011) Identification of protein carbonylation sites by two-dimensional liquid chromatography in combination with MALDI- and ESI-MS. J Proteomics 74:2338–2350
Brandes N, Reichmann D, Tienson H et al (2011) Using quantitative redox proteomics to dissect the yeast redoxome. J Biol Chem 286:41893–41903
Brennan JP, Miller JIA, Fuller W et al (2006) The utility of N, N-biotinyl glutathione disulfide in the study of protein S-glutathiolation. Mol Cell Proteomics 5:215–225
Bridgewater JD, Srikanth R, Lim J, Vachet RW (2007) The effect of histidine oxidation on the dissociation patterns of peptide ions. J Am Soc Mass Spectrom 18:553–562
Buick R (2008) When did oxygenic photosynthesis evolve? Philos Trans R Soc Lond B Biol Sci 363:2731–2743
Calabrese V, Cornelius C, Rizzarelli E et al (2009) Nitric oxide in cell survival: a janus molecule. Antioxid Redox Signal 11:2717–2739
Chae HZ, Uhm TB, Rhee SG (1994) Dimerization of thiol-specific antioxidant and the essential role of cysteine 47. Proc Natl Acad Sci USA 91:7022–7026
Chai YC, Jung CH, Lii CK et al (1991) Identification of an abundant S-thiolated rat liver protein as carbonic anhydrase III; characterization of S-thiolation and dethiolation reactions. Arch Biochem Biophys 284:270–278
Chambers DE, Parks DA, Patterson G et al (1985) Xanthine oxidase as a source of free radical damage in myocardial ischemia. J Mol Cell Cardiol 17:145–152
Chan H-L, Gharbi S, Gaffney PR et al (2005) Proteomic analysis of redox- and ErbB2-dependent changes in mammary luminal epithelial cells using cysteine- and lysine-labelling two-dimensional difference gel electrophoresis. Proteomics 5:2908–2926
Chance B, Sies H, Boveris A (1979) Hydroperoxide metabolism in mammalian organs. Physiol Rev 59:527–605
Chang T-S, Jeong W, Woo HA et al (2004) Characterization of mammalian sulfiredoxin and its reactivation of hyperoxidized peroxiredoxin through reduction of cysteine sulfinic acid in the active site to cysteine. J Biol Chem 279:50994–51001
Chang Y-C, Huang C-N, Lin C-H et al (2010) Mapping protein cysteine sulfonic acid modifications with specific enrichment and mass spectrometry: an integrated approach to explore the cysteine oxidation. Proteomics 10:2961–2971
Cheal SM, Ng M, Barrios B et al (2009) Mapping protein-protein interactions by localized oxidation: consequences of the reach of hydroxyl radical. Biochemistry 48:4577–4586

Chiappetta G, Corbo C, Palmese A et al (2009) Quantitative identification of protein nitration sites. Proteomics 9:1524–1537
Ciorba MA, Heinemann SH, Weissbach H et al (1997) Modulation of potassium channel function by methionine oxidation and reduction. Proc Natl Acad Sci USA 94:9932–9937
Craven PA, DeRubertis FR (1978) Restoration of the responsiveness of purified guanylate cyclase to nitrosoguanidine, nitric oxide, and related activators by heme and hemeproteins. Evidence for involvement of the paramagnetic nitrosyl-heme complex in enzyme activation. J Biol Chem 253:8433–8443
D'Autréaux B, Toledano MB (2007) ROS as signalling molecules: mechanisms that generate specificity in ROS homeostasis. Nat Rev Mol Cell Biol 8:813–824
Dai D-F, Johnson SC, Villarin JJ et al (2011) Mitochondrial oxidative stress mediates angiotensin II-induced cardiac hypertrophy and Galphaq overexpression-induced heart failure. Circ Res 108:837–846
Dalle-Donne I, Giustarini D, Colombo R et al (2003a) Protein carbonylation in human diseases. Trends Mol Med 9:169–176
Dalle-Donne I, Rossi R, Giustarini D et al (2003b) Protein carbonyl groups as biomarkers of oxidative stress. Clin Chim Acta 329:23–38
Dalle-Donne I, Aldini G, Carini M et al (2006) Protein carbonylation, cellular dysfunction, and disease progression. J Cell Mol Med 10:389–406
Dalle-Donne I, Rossi R, Giustarini D et al (2007) S-glutathionylation in protein redox regulation. Free Radic Biol Med 43:883–898
Darby N, Creighton TE (1995) Disulfide bonds in protein folding and stability. Methods Mol Biol 40:219–252
Depuydt M, Leonard SE, Vertommen D et al (2009) A periplasmic reducing system protects single cysteine residues from oxidation. Science 326:1109–1111
Depuydt M, Messens J, Collet J-F (2011) How proteins form disulfide bonds. Antioxid Redox Signal 15:49–66
Dickinson DA, Forman HJ (2002) Glutathione in defense and signaling: lessons from a small thiol. Ann N Y Acad Sci 973:488–504
Donoghue PCJ, Antcliffe JB (2010) Early life: origins of multicellularity. Nature 466:41–42
Dremina ES, Li X, Galeva NA et al (2011) A methodology for simultaneous fluorogenic derivatization and boronate affinity enrichment of 3-nitrotyrosine-containing peptides. Anal Biochem 418:184–196
Fedorova M, Kuleva N, Hoffmann R (2010) Identification of cysteine, methionine and tryptophan residues of actin oxidized in vivo during oxidative stress. J Proteome Res 9:1598–1609
Fridovich I (1983) Superoxide radical: an endogenous toxicant. Annu Rev Pharmacol Toxicol 23:239–257
Fu S, Dean R, Southan M, Truscott R (1998) The hydroxyl radical in lens nuclear cataractogenesis. J Biol Chem 273:28603–28609
Ghesquière B, Jonckheere V, Colaert N et al (2011) Redox proteomics of protein-bound methionine oxidation. Mol Cell Proteomics 10:M110.006866
Ghezzi P, Bonetto V, Fratelli M (2005) Thiol-disulfide balance: from the concept of oxidative stress to that of redox regulation. Antioxid Redox Signal 7:964–972
Giles NM, Watts AB, Giles GI et al (2003) Metal and redox modulation of cysteine protein function. Chem Biol 10:677–693
Giustarini D, Dalle-Donne I, Colombo R et al (2008) Is ascorbate able to reduce disulfide bridges? A cautionary note. Nitric Oxide 19:252–258
Godon C, Lagniel G, Lee J et al (1998) The H_2O_2 stimulon in *Saccharomyces cerevisiae*. J Biol Chem 273:22480–22489
Gross AJ, Sizer IW (1959) The oxidation of tyramine, tyrosine, and related compounds by peroxidase. J Biol Chem 234:1611–1614
Gross E, Sevier CS, Heldman N et al (2006) Generating disulfides enzymatically: reaction products and electron acceptors of the endoplasmic reticulum thiol oxidase Ero1p. Proc Natl Acad Sci USA 103:299–304

Grosvenor AJ, Morton JD, Dyer JM (2010) Profiling of residue-level photo-oxidative damage in peptides. Amino Acids 39:285–296
Grune T, Davies KJ (1997) Breakdown of oxidized proteins as a part of secondary antioxidant defenses in mammalian cells. Biofactors 6:165–172
Grune T, Merker K, Sandig G, Davies KJA (2003) Selective degradation of oxidatively modified protein substrates by the proteasome. Biochem Biophys Res Commun 305:709–718
Guan J-Q, Chance MR (2005) Structural proteomics of macromolecular assemblies using oxidative footprinting and mass spectrometry. Trends Biochem Sci 30:583–592
Gygi SP, Rist B, Gerber SA et al (1999) Quantitative analysis of complex protein mixtures using isotope-coded affinity tags. Nat Biotechnol 17:994–999
Halliwell B, Gutteridge J (1989) Free radicals in biology and medicine. Clarendon, Oxford
Harman D (1956) Aging: a theory based on free radical and radiation chemistry. J Gerontol 11: 298–300
Harris I, Meriwether BP, Park JH (1963) Chemical nature of the catalytic sites in glyceraldehyde-3-phosphate dehydrogenase. Nature 198:154–157
Herbig AF, Helmann JD (2001) Roles of metal ions and hydrogen peroxide in modulating the interaction of the *Bacillus subtilis* PerR peroxide regulon repressor with operator DNA. Mol Microbiol 41:849–859
Hochgräfe F, Mostertz J, Albrecht D, Hecker M (2005) Fluorescence thiol modification assay: oxidatively modified proteins in *Bacillus subtilis*. Mol Microbiol 58:409–425
Hochgräfe F, Mostertz J, Pöther D-C et al (2007) S-cysteinylation is a general mechanism for thiol protection of *Bacillus subtilis* proteins after oxidative stress. J Biol Chem 282:25981–25985
Hofmann B, Hecht H-J, Flohé L (2002) Peroxiredoxins. Biol Chem 383:347–364
Holmgren A (1989) Thioredoxin and glutaredoxin systems. J Biol Chem 264:13963–13966
Hondorp ER, Matthews RG (2004) Oxidative stress inactivates cobalamin-independent methionine synthase (MetE) in *Escherichia coli*. PLOS Biol 2:e336
Huang B, Liao CL, Lin YP et al (2009) S-nitrosoproteome in endothelial cells revealed by a modified biotin switch approach coupled with Western blot-based two-dimensional gel electrophoresis. J Proteome Res 8:4835–4843
Hurd TR, Prime TA, Harbour ME et al (2007) Detection of reactive oxygen species-sensitive thiol proteins by redox difference gel electrophoresis: implications for mitochondrial redox signaling. J Biol Chem 282:22040–22051
Ischiropoulos H, al-Mehdi AB (1995) Peroxynitrite-mediated oxidative protein modifications. FEBS Lett 364:279–282
Ischiropoulos H, Zhu L, Chen J et al (1992) Peroxynitrite-mediated tyrosine nitration catalyzed by superoxide dismutase. Arch Biochem Biophys 298:431–437
Ito S, Kato T, Shinpo K, Fujita K (1984) Oxidation of tyrosine residues in proteins by tyrosinase. Formation of protein-bonded 3,4-dihydroxyphenylalanine and 5-S-cysteinyl-3,4-dihydroxyphenylalanine. Biochem J 222:407–411
Ivanov B, Khorobrykh S (2003) Participation of photosynthetic electron transport in production and scavenging of reactive oxygen species. Antioxid Redox Signal 5:43–53
Jaffrey SR, Snyder SH (2001) The biotin switch method for the detection of S-nitrosylated proteins. Sci STKE 2001:pl1
Jakob U, Muse W, Eser M, Bardwell JC (1999) Chaperone activity with a redox switch. Cell 96:341–352
Janssen-Heininger YMW, Mossman BT, Heintz NH et al (2008) Redox-based regulation of signal transduction: principles, pitfalls, and promises. Free Radic Biol Med 45:1–17
Katsuki S, Arnold W, Mittal C, Murad F (1977) Stimulation of guanylate cyclase by sodium nitroprusside, nitroglycerin and nitric oxide in various tissue preparations and comparison to the effects of sodium azide and hydroxylamine. J Cyclic Nucleotide Res 3:23–35
Klomsiri C, Nelson KJ, Bechtold E et al (2010) Use of dimedone-based chemical probes for sulfenic acid detection evaluation of conditions affecting probe incorporation into redox-sensitive proteins. Methods Enzymol 473:77–94

Kumsta C, Thamsen M, Jakob U (2011) Effects of oxidative stress on behavior, physiology, and the redox thiol proteome of *Caenorhabditis elegans*. Antioxid Redox Signal 14:1023–1037
Kuthan H, Ullrich V (1982) Oxidase and oxygenase function of the microsomal cytochrome P450 monooxygenase system. Eur J Biochem 126:583–588
Landino LM, Koumas MT, Mason CE, Alston JA (2006) Ascorbic acid reduction of microtubule protein disulfides and its relevance to protein S-nitrosylation assays. Biochem Biophys Res Commun 340:347–352
Lee J-W, Helmann JD (2006) The PerR transcription factor senses H_2O_2 by metal-catalysed histidine oxidation. Nature 440:363–367
Lee S, Young NL, Whetstone PA et al (2006) Method to site-specifically identify and quantitate carbonyl end products of protein oxidation using oxidation-dependent element coded affinity tags (O-ECAT) and nanoliquid chromatography Fourier transform mass spectrometry. J Proteome Res 5:539–547
Lee J-W, Soonsanga S, Helmann JD (2007) A complex thiolate switch regulates the *Bacillus subtilis* organic peroxide sensor OhrR. Proc Natl Acad Sci USA 104:8743–8748
Lee S, Chen Y, Luo H et al (2010) The first global screening of protein substrates bearing protein-bound 3,4-Dihydroxyphenylalanine in *Escherichia coli* and human mitochondria. J Proteome Res 9:5705–5714
Leeuwenburgh C, Rasmussen JE, Hsu FF et al (1997) Mass spectrometric quantification of markers for protein oxidation by tyrosyl radical, copper, and hydroxyl radical in low density lipoprotein isolated from human atherosclerotic plaques. J Biol Chem 272:3520–3526
Leichert LI, Jakob U (2004) Protein thiol modifications visualized in vivo. PLOS Biol 2:e333
Leichert LI, Gehrke F, Gudiseva HV et al (2008) Quantifying changes in the thiol redox proteome upon oxidative stress in vivo. Proc Natl Acad Sci USA 105:8197–8202
Levine RL (2002) Carbonyl modified proteins in cellular regulation, aging, and disease. Free Radic Biol Med 32:790–796
Levine RL, Mosoni L, Berlett BS, Stadtman ER (1996) Methionine residues as endogenous antioxidants in proteins. Proc Natl Acad Sci USA 93:15036–15040
Levine RL, Moskovitz J, Stadtman ER (2000) Oxidation of methionine in proteins: roles in antioxidant defense and cellular regulation. IUBMB Life 50:301–307
Li S, Wang H, Xian M, Whorton AR (2011) Identification of protein nitrosothiols using phosphine-mediated selective reduction. Nitric Oxide 26:20–26
Lind C, Gerdes R, Schuppe-Koistinen I, Cotgreave IA (1998) Studies on the mechanism of oxidative modification of human glyceraldehyde-3-phosphate dehydrogenase by glutathione: catalysis by glutaredoxin. Biochem Biophys Res Commun 247:481–486
Lind C, Gerdes R, Hamnell Y et al (2002) Identification of S-glutathionylated cellular proteins during oxidative stress and constitutive metabolism by affinity purification and proteomic analysis. Arch Biochem Biophys 406:229–240
Lindahl M, Mata-Cabana A, Kieselbach T (2011) The disulfide proteome and other reactive cysteine proteomes: analysis and functional significance. Antioxid Redox Signal 14:2581–2642
Loschen G, Flohé L, Chance B (1971) Respiratory chain linked H_2O_2 production in pigeon heart mitochondria. FEBS Lett 18:261–264
Madian AG, Regnier FE (2010) Proteomic identification of carbonylated proteins and their oxidation sites. J Proteome Res 9:3766–3780
Maria CS, Revilla E, Ayala A et al (1995) Changes in the histidine residues of Cu/Zn superoxide dismutase during aging. FEBS Lett 374:85–88
Maskos Z, Rush JD, Koppenol WH (1992) The hydroxylation of phenylalanine and tyrosine: a comparison with salicylate and tryptophan. Arch Biochem Biophys 296:521–529
Matayatsuk C, Poljak A, Bustamante S et al (2007) quantitative determination of *ortho-* and *meta-*tyrosine as biomarkers of protein oxidative damage in β-thalassemia. Redox Rep 12:219–228
Meng T-C, Fukada T, Tonks NK (2002) Reversible oxidation and inactivation of protein tyrosine phosphatases in vivo. Mol Cell 9:387–399
Mirzaei H, Regnier F (2005) Affinity chromatographic selection of carbonylated proteins followed by identification of oxidation sites using tandem mass spectrometry. Anal Chem 77:2386–2392

Mirzaei H, Regnier F (2006) Enrichment of carbonylated peptides using Girard P reagent and strong cation exchange chromatography. Anal Chem 78:770–778
Miseta A, Csutora P (2000) Relationship between the occurrence of cysteine in proteins and the complexity of organisms. Mol Biol Evol 17:1232–1239
Møller IM, Kristensen BK (2006) Protein oxidation in plant mitochondria detected as oxidized tryptophan. Free Radic Biol Med 40:430–435
Moskovitz J, Rahman MA, Strassman J et al (1995) *Escherichia coli* peptide methionine sulfoxide reductase gene: regulation of expression and role in protecting against oxidative damage. J Bacteriol 177:502–507
Moskovitz J, Berlett BS, Poston JM, Stadtman ER (1997) The yeast peptide-methionine sulfoxide reductase functions as an antioxidant in vivo. Proc Natl Acad Sci USA 94:9585–9589
Moskovitz J, Flescher E, Berlett BS et al (1998) Overexpression of peptide-methionine sulfoxide reductase in *Saccharomyces cerevisiae* and human T cells provides them with high resistance to oxidative stress. Proc Natl Acad Sci USA 95:14071–14075
Murray CI, Uhrigshardt H, O'Meally RN et al (2012) Identification and quantification of S-nitrosylation by cysteine reactive tandem mass Tag switch assay. Mol Cell Proteomics 11(2):M111.013441
Nakagawa M, Watanabe H, Kodato S et al (1977) A valid model for the mechanism of oxidation of tryptophan to formylkynurenine-25 years later. Proc Natl Acad Sci USA 74:4730–4733
Nakamura A, Goto S (1996) Analysis of protein carbonyls with 2,4-dinitrophenyl hydrazine and its antibodies by immunoblot in two-dimensional gel electrophoresis. J Biochem 119:768–774
Nyström T (2005) Role of oxidative carbonylation in protein quality control and senescence. EMBO J 24:1311–1317
Ogawa O, Zhu X, Perry G, Smith MA (2002) Mitochondrial abnormalities and oxidative imbalance in neurodegenerative disease. Sci Aging Knowledge Environ 2002:pe16
Oien DB, Osterhaus GL, Latif SA et al (2008) MsrA knockout mouse exhibits abnormal behavior and brain dopamine levels. Free Radic Biol Med 45:193–200
Perdivara I, Deterding LJ, Przybylski M, Tomer KB (2010) Mass spectrometric identification of oxidative modifications of tryptophan residues in proteins: chemical artifact or post-translational modification? J Am Soc Mass Spectrom 21:1114–1117
Poole LB, Zeng B-B, Knaggs SA et al (2005) Synthesis of chemical probes to map sulfenic acid modifications on proteins. Bioconjug Chem 16:1624–1628
Proctor P, McGinness JE (1970) Levodopa side-effects and the Lesch-Nyhan syndrome. Lancet 2:1367
Pryor WA (1986) Oxy-radicals and related species: their formation, lifetimes, and reactions. Annu Rev Physiol 48:657–667
Ralser M, Wamelink MM, Kowald A et al (2007) Dynamic rerouting of the carbohydrate flux is key to counteracting oxidative stress. J Biol 6:10
Ravichandran V, Seres T, Moriguchi T et al (1994) S-thiolation of glyceraldehyde-3-phosphate dehydrogenase induced by the phagocytosis-associated respiratory burst in blood monocytes. J Biol Chem 269:25010–25015
Reddie KG, Seo YH, Muse Iii WB et al (2008) A chemical approach for detecting sulfenic acid-modified proteins in living cells. Mol Biosyst 4:521–531
Reichard P, Ehrenberg A (1983) Ribonucleotide reductase–a radical enzyme. Science 221:514–519
Reinheckel T, Körn S, Möhring S et al (2000) Adaptation of protein carbonyl detection to the requirements of proteome analysis demonstrated for hypoxia/reoxygenation in isolated rat liver mitochondria. Arch Biochem Biophys 376:59–65
Rinalducci S, Murgiano L, Zolla L (2008) Redox proteomics: basic principles and future perspectives for the detection of protein oxidation in plants. J Exp Bot 59:3781–3801
Rinna A, Torres M, Forman HJ (2006) Stimulation of the alveolar macrophage respiratory burst by ADP causes selective glutathionylation of protein tyrosine phosphatase 1B. Free Radic Biol Med 41:86–91
Ruan H, Tang XD, Chen M-L et al (2002) High-quality life extension by the enzyme peptide methionine sulfoxide reductase. Proc Natl Acad Sci USA 99:2748–2753

Sabens Liedhegner EA, Gao XH, Mieyal JJ (2012) Mechanisms of altered redox regulation in neurodegenerative diseases–focus on S-glutathionylation. Antioxid Redox Signal 16(6): 543–566

Salsbury FR, Knutson ST, Poole LB, Fetrow JS (2008) Functional site profiling and electrostatic analysis of cysteines modifiable to cysteine sulfenic acid. Protein Sci 17:299–312

Sato S, Shimoda Y, Muraki A et al (2007) A large-scale protein protein interaction analysis in *Synechocystis sp*. PCC6803. DNA Res 14:207–216

Sethuraman M, McComb ME, Huang H et al (2004) Isotope-coded affinity tag (ICAT) approach to redox proteomics: identification and quantitation of oxidant-sensitive cysteine thiols in complex protein mixtures. J Proteome Res 3:1228–1233

Sharov V, Dremina E, Pennington J (2008) Selective fluorogenic derivatization of 3-nitrotyrosine and 3,4-dihydroxyphenylalanine in peptides: a method designed for quantitative proteomic analysis. Methods Enzymol 441:19–32

Sharov VS, Dremina ES, Galeva NA et al (2010) Fluorogenic tagging of peptide and protein 3-nitrotyrosine with 4-(aminomethyl)-benzenesulfonic acid for quantitative analysis of protein tyrosine nitration. Chroma 71:37–53

Sheehan D, McDonagh B, Bárcena JA (2010) Redox proteomics. Expert Rev Proteomics 7:1–4

Sies H (1993) Strategies of antioxidant defense. Eur J Biochem 215:213–219

Sies H, Cadenas E (1985) Oxidative stress: damage to intact cells and organs. Philos Trans R Soc Lond B Biol Sci 311:617–631

Simpson JA, Narita S, Gieseg S et al (1992) Long-lived reactive species on free-radical-damaged proteins. Biochem J 282:621–624

Simpson JA, Gieseg SP, Dean RT (1993) Free radical and enzymatic mechanisms for the generation of protein bound reducing moieties. Biochim Biophys Acta 1156:190–196

Staal FJ, Roederer M, Herzenberg LA, Herzenberg LA (1990) Intracellular thiols regulate activation of nuclear factor kappa B and transcription of human immunodeficiency virus. Proc Natl Acad Sci USA 87:9943–9947

Stadtman ER (1990) Metal ion-catalyzed oxidation of proteins: biochemical mechanism and biological consequences. Free Radic Biol Med 9:315–325

Stamler JS, Jaraki O, Osborne J et al (1992) Nitric oxide circulates in mammalian plasma primarily as an S-nitroso adduct of serum albumin. Proc Natl Acad Sci USA 89:7674–7677

Storz G, Tartaglia LA, Ames BN (1990) Transcriptional regulator of oxidative stress-inducible genes: direct activation by oxidation. Science 248:189–194

Sun H, Gao J, Ferrington DA et al (1999) Repair of oxidized calmodulin by methionine sulfoxide reductase restores ability to activate the plasma membrane Ca-ATPase. Biochemistry 38:105–112

Tarrago L, Kieffer-Jaquinod S, Lamant T et al (2011) Affinity chromatography: a valuable strategy to isolate substrates of methionine sulfoxide reductases? Antioxid Redox Signal 16:79–84

Taylor SW, Fahy E, Murray J et al (2003a) Oxidative post-translational modification of tryptophan residues in cardiac mitochondrial proteins. J Biol Chem 278:19587–19590

Taylor SW, Fahy E, Zhang B et al (2003b) Characterization of the human heart mitochondrial proteome. Nat Biotechnol 21:281–286

Tsikas D (2010) Analytical methods for 3-nitrotyrosine quantification in biological samples: the unique role of tandem mass spectrometry. Amino Acids 42:45–63

Turrens JF (2003) Mitochondrial formation of reactive oxygen species. J Physiol (Lond) 552: 335–344

Uchida K, Kawakishi S (1993) 2-Oxo-histidine as a novel biological marker for oxidatively modified proteins. FEBS Lett 332:208–210

van der Vliet A, O'Neill CA, Halliwell B et al (1994) Aromatic hydroxylation and nitration of phenylalanine and tyrosine by peroxynitrite. Evidence for hydroxyl radical production from peroxynitrite. FEBS Lett 339:89–92

van Dooren SHJ, Raijmakers R, Pluk H et al (2011) Oxidative stress-induced modifications of histidyl-tRNA synthetase affect its tRNA aminoacylation activity but not its immunoreactivity. Biochem Cell Biol 89:545–553

Vogt W (1995) Oxidation of methionyl residues in proteins: tools, targets, and reversal. Free Radic Biol Med 18:93–105

Weissbach H, Resnick L, Brot N (2005) Methionine sulfoxide reductases: history and cellular role in protecting against oxidative damage. Biochim Biophys Acta 1703:203–212

Wolff J, Cook GH, Goldhammer AR, Berkowitz SA (1980) Calmodulin activates prokaryotic adenylate cyclase. Proc Natl Acad Sci USA 77:3841–3844

Wong CM, Cheema AK, Zhang L, Suzuki YJ (2008) Protein carbonylation as a novel mechanism in redox signaling. Circ Res 102:310–318

Woo HA, Jeong W, Chang T-S et al (2005) Reduction of cysteine sulfinic acid by sulfiredoxin is specific to 2-cys peroxiredoxins. J Biol Chem 280:3125–3128

Wood ZA, Schröder E, Robin Harris J, Poole LB (2003) Structure, mechanism and regulation of peroxiredoxins. Trends Biochem Sci 28:32–40

Woods AA, Linton SM, Davies MJ (2003) Detection of HOCl-mediated protein oxidation products in the extracellular matrix of human atherosclerotic plaques. Biochem J 370:729–735

Yao Y, Yin D, Jas GS et al (1996) Oxidative modification of a carboxyl-terminal vicinal methionine in calmodulin by hydrogen peroxide inhibits calmodulin-dependent activation of the plasma membrane Ca-ATPase. Biochemistry 35:2767–2787

Zaffagnini M, Michelet L, Marchand C et al (2007) The thioredoxin-independent isoform of chloroplastic glyceraldehyde-3-phosphate dehydrogenase is selectively regulated by glutathionylation. FEBS J 274:212–226

Zheng M, Aslund F, Storz G (1998) Activation of the OxyR transcription factor by reversible disulfide bond formation. Science 279:1718–1721

Chapter 7
Computational Redox Biology: Methods and Applications

Stefano M. Marino, Goedele Roos, and Vadim N. Gladyshev

Abstract Cysteine (Cys) is the most important amino acid in redox biology: it is the premier residue used by proteins to maintain redox homeostasis, sense redox changes in the environment, and counteract oxidative stress. Cys is often used as a catalytic redox-active residue and plays a key role in protein structure stabilization via disulfides and metal binding. Cys is much different from other common amino acids in proteins: its unique chemical and physical properties provide high affinity for metal ions, support formation of covalent bonds with other Cys, and confer response to changes in the environment. These features are largely responsible for the broad variety of its biological functions. Thus, a better understanding of basic properties of Cys is essential for understanding the fundamental roles Cys plays in redox biology, as well as for prediction and classification of functional Cys residues in proteins. In this chapter, we provide an overview of theoretical and computational tools that have been developed in the area of thiol regulation and redox biology. In particular, we introduce and discuss methods to investigate basic properties of Cys, such as exposure and pKa, and a variety of algorithms for functional prediction of different types of Cys in proteins.

S.M. Marino • V.N. Gladyshev (✉)
Division of Genetics, Department of Medicine, Brigham and Women's Hospital, Harvard Medical School, Boston, MA, USA
e-mail: vgladyshev@rics.bwh.harvard.edu

G. Roos
General Chemistry, Vrije Universiteit Brussel, VUB, B-1050 Brussels, Belgium

Department of Structural Biology, VIB, B-1050 Brussels, Belgium

U. Jakob and D. Reichmann (eds.), *Oxidative Stress and Redox Regulation*, DOI 10.1007/978-94-007-5787-5_7,

7.1 Introduction

One of the major mechanisms of regulating protein function involves posttranslational modifications of specific amino acid residues. The modifications can occur at certain regulatable positions (residues not directly involved in protein function) or functional sites in proteins (e.g., catalytic residues). Through these modifications, enzymes are able to sense and respond to changes in the cellular environment. In the case of the response to redox perturbations, the major sites involved typically include the amino acid cysteine (Cys). This residue is the predominant target of redox control of protein function and is key to the redox regulation of cellular processes. This chapter provides a general overview of computational methods used in redox biology. Because of their importance for the subject of this chapter, we will initially focus on Cys properties and theoretical methods available to investigate these properties.

Among the 20 common amino acid in proteins, Cys is often an outlier with respect to descriptors used to classify and systematize properties and features of amino acid residues. For a start, it is one of the least abundant amino acids in organisms, but in spite of this, it is frequently observed in functional sites of proteins, where it serves catalytic, regulatory, structure stabilizing, cofactor binding, or other functions. Furthermore, Cys is thought to be a relatively recent addition to the genetic code (Trifonov 2004), but it appears to accumulate more than any other amino acid in present day organisms (Jordan et al. 2005). Other distinctive features of Cys, highlighting its importance in biology, include the following: (i) in humans, Cys mutations lead to genetic diseases more often than expected on the basis of its abundance (Wu et al. 2007); (ii) the content of this amino acid in mitochondrially encoded proteins is particularly low and inversely correlates with lifespan of animals (Moosmann and Behl 2008). The biological interpretation of this relationship is debated (Jobson et al. 2010; Moosmann 2011; Schindeldecker et al. 2011), but it highlights the fact that Cys residues in proteins appear to be under strict evolutionary control. This feature has to be associated with several unique biological functions of Cys, as detected by genome-wide analyses of its tendency to form functional clusters, such as structural disulfides and metal-binding sites (Beeby et al. 2005; Marino and Gladyshev 2010a). Indeed, in the protein realm, a very unique property of Cys is its ability to react with another Cys forming disulfide bonds. The only other naturally occurring amino acid, albeit much less frequent in proteins, capable of a similar reactivity is selenocysteine (Sec). Sec is a selenium-containing amino acid that differs from Cys by a single atom (i.e., Se in place of S). In all characterized selenoproteins, Sec is a redox active functional residue (e.g. directly involved in catalytic function), albeit recently an exception to this rule has been reported (Lee et al. 2011). Its function can be partially preserved when Cys replaces Sec, although the replacement often leads to a considerable decrease in catalytic efficiency (Bock et al. 1991; Kim and Gladyshev 2005). A recent study made the relation between Cys and Sec even more intriguing. It has been found that Cys can be inserted in proteins in place of Sec (Xu et al. 2010).

What are the relevant physico-chemical features of Cys? What makes this amino acid so special for protein structure, function and regulation? The main features appear to be the high reactivity and chemical capabilities of its sulfur-based functional group. As mentioned above, the nature of its side chain allows for a unique reactivity in the protein world, i.e. the covalent interactions with other thiols create intra- and intermolecular disulfide bonds. In addition, its functional group can coordinate a variety of metals and metalloids. Therefore, Cys residues hold a special role in terms of protein structure stabilization (e.g., through metal binding and covalent bonds that help hold the structure together).

As incorporated into proteins, the side chain of Cys contains a sulfhydryl group, representing the most reduced state of sulfur in proteins. This moiety is very reactive when it comes in contact with various oxidants and other electrophiles under both physiological and pathophysiological conditions. Reversible oxidation of Cys thiols is known to play a role in redox regulation of proteins via the formation of sulfenic acid intermediates (Cys-SOH), overoxidation to sulfinic acids (Cys-SO_2H) (Leonard et al. 2009; Poole et al. 2004; Shenton and Grant 2003; Wood et al. 2003), intra- and intermolecular disulfide bonds (Paget and Buttner 2003), and mixed disulfides with glutathione (Cabiscol and Levine 1996). Additionally, Cys is the main target of nitrosative stress, leading to the formation of reversible S-nitrosothiols (Hess et al. 2005). The susceptibility of Cys to these modifications is largely dependent on the reactivity of each specific thiol: Cys thiolates are good nucleophiles and more prone to oxidation than Cys thiols (Winterbourn and Metodiewa 1999). Cys residues are also very polarizable (i.e., the thiol dipole can be easily perturbed by interaction with other residues of the protein). Documented examples include the effects of N-terminal helix dipole (Iqbalsyah et al. 2006), proximity to other titrable residues (Marino and Gladyshev 2009; Salsbury et al. 2008) and hydrogen bond partners (Foloppe et al. 2001). Therefore, Cys reactivity ultimately depends on local environmental features (e.g., secondary structure composition, proximity with charged residues, H-bond donors, etc.). A relevant functional consequence is that its sensitivity to oxidation and over-oxidation greatly varies with changes in the environment: an exposed Cys residue can be turned into a very reactive residue and thus more likely to be either functional or dysfunctional.

7.2 Computational Methods to Investigate Thiol Reactivity

7.2.1 General Aspects of Cys Reactivity in Proteins

The general features of Cys in proteins are difficult to define. Their chemico-physical properties are greatly influenced by the environment: deeply buried Cys residues tend to behave as hydrophobic residues (due to hydrophobic packing inside the protein body) while solvent accessible Cys residues have an opportunity to interact with many H-bond partners and titratable groups of polar residues, which

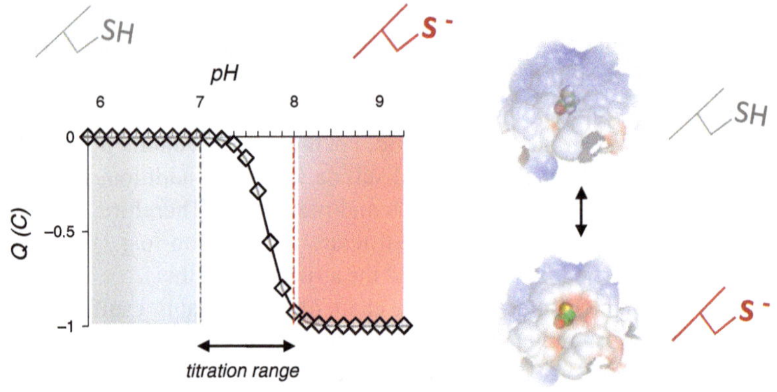

Fig. 7.1 Effect of pH perturbation on Cys located on molecular surfaces. For many exposed and polar Cys, pKa is close to the physiological pH (shown in *white*, between pH values of 7 and 8, in the figure). In this scenario, assuming Henderson-Hasselbach (HH) behavior for a monoprotic acid (in a physiological solution), sudden net charge switches can happen in response to even small local pH shifts, e.g. small pH increase, favoring deprotonation. Cys is in the anionic form, as shown above the *red shaded* portion on the graph. Conversely, relatively small pH changes may promote protonation and loss of the net negative charge. This can translate into sudden charge switches, which can affect considerably physico-chemical properties of molecular surfaces in proximity to the "switching" Cys. The latter event can affect protein-protein interaction patterns

are abundant on protein surfaces. These interactions can polarize considerably the exposed Cys. Indeed, Cys can be easily perturbed by closely positioned titratable residues (and also by interaction with the solvent), significantly affecting both its pKa and its titration range (Marino and Gladyshev 2010a).

According to computer-based evaluations, exposed Cys are the naturally occurring titratable residues with the closest pKa (i.e., average pKa in proteins) to the physiological pH (Marino and Gladyshev 2010a). This observation is significant *per se*: it implies that even for small variations of local pH (within the physiological range), exposed Cys residues may experience sudden charge shifts and significant electrostatic changes, which may extend to proximal portions of the molecular surface (e.g., Cys can act as a charge switch, potentially affecting local electrostatic properties and protein-protein interactions) (Fig. 7.1). For the same reason (i.e., pKa of many exposed Cys being close to physiological pH), Cys residues on molecular surface can experience sudden variation in their ability to serve as nucleophiles.

These observations highlight the intrinsically high responsiveness of exposed Cys to changes in physiological states and environmental conditions, with considerable implications in terms of protein function. Consequently, Cys aptitude may provide a biological explanation for why Cys residues are found much less frequently (than expected) on molecular surfaces. Unless employed for a specific function, exposed Cys residues tend to be removed from protein surfaces (Marino and Gladyshev 2010a).

7.2.2 Exposure and pKa

It follows from the discussion above that, in computational biology, two particularly important descriptors of Cys reactivity are its exposure to solvent and the protonation status of its functional group. In both cases, structural information is needed to evaluate Cys properties. To estimate protein exposure, a common approach consists of rolling a molecular probe (e.g., a sphere of 1.4 Å radius to mimic dimensions of the water molecule) over the protein body; the latter, usually, is treated as rigid body: the probe can just touch its residues, but cannot penetrate the surface. Commonly used (and free for download) programs include Naccess (v2.1.1, http://www.bioinf.manchester.ac.uk/naccess/) and Surface Racer© (http://apps.phar.umich.edu/tsodikovlab/index_files/Page756.htm). These methods exist in stand-alone versions and, therefore, are suitable for automated large-scale analysis. Relevant to exposure calculations are algorithms searching for residues, which, albeit accessible to the solvent, are fostered in small pockets of the protein body (Laurie and Jackson 2005). These methods are usually called pocket predictors (e.g. http://www.modelling.leeds.ac.uk/pocketfinder/) and can be useful for investigation of Cys modifications that occur in particularly controlled environments (e.g., some cases of S-nitrosylation) (Marino and Gladyshev 2010b). Additionally, some programs use simplified docking procedures to discover regions of proteins, which, alongside being exposed, are also suitable to accommodate larger substrates (Laurie and Jackson 2005). One such program is Q-site finder (http://www.modelling.leeds.ac.uk/qsitefinder/). It can be useful for detection of particularly reactive types of residues, such as catalytic Cys.

In the case of pKa predictions, no definitive computational protocols exist. As hinted before, thiolates are better nucleophiles than their protonated counterparts. Also, thiolates can react more rapidly with natural oxidants, such as peroxide (Winterbourn and Metodiewa 1999), even though this depends on the local environment (Winterbourn and Hampton 2008). For these reasons, the ability to correctly predict Cys pKa is a much sought after feature that would prove to be extremely valuable in redox biology.

Different approaches have been used to study reactive Cys. One method makes use of DFT (density functional theory) calculations to calculate pKa through Natural Population Analysis (NPA) charge on Cys sulfur atoms (Roos et al. 2009a, b). The method worked well when tested with some known cases (e.g. thioredoxins, and other thioredoxin fold thiol oxidoreductases). Considering the intrinsic complexity of this analysis, DFT methods for pKa prediction can be computationally demanding: usually, reduced protein models of the active site (i.e. the functional site under investigation) are studied instead. Another method, which has been applied to the investigation of reactive Cys is the empirical pKa predictor PROPKA (Sanchez et al. 2008). For a titratable residue, a pKa shift is evaluated as a function of the sum of energy contributions provided by surrounding residues. Although the approach's theory is relatively simple, it has been praised for its balance of speed and performance (Marino and Gladyshev 2011a, b).

A third category of methods, which found applications in the analysis of redox Cys (Tosatto et al. 2008; Foloppe et al. 2001), is based on the numerical solution of the Poisson-Boltzmann (PB) equation. The electrostatic calculations provide (free) energies of each of the protonation microstates in the system. This allows calculation of partition functions and, ultimately, for each titrable residue, the probability of protonation over a range of pH values. The latter is a bonus of PB-based pKa predictors: along with the pKa value, they can produce information about the titration curve for pH values close to the $pK_{1/2}$, a feature which can be very informative (Ondrechen et al. 2001; Marino and Gladyshev 2009). PB-based approaches for pKa prediction have been applied successfully to various redox related problems, such as glutathione peroxidases (Tosatto et al. 2008).

Each approach has its unique features, e.g. PROPKA is ultrafast while still providing acceptable results. If properly set-up, the quantum mechanics (QM)-based methods can be very accurate, and PB methods can compute, besides the pKa, the whole titration curve in proximity to the $pK_{1/2}$. So, rather than in competition, these methods can be considered complementary, as they can provide insights from different perspectives and ultimately help establish a more complete picture of Cys reactivity.

Related to Cys reactivity, a new proteomics approach, called isoTOP-ABPP (Weerapana et al. 2010), allows high-throughput identification of all reactive Cys in proteins, while at the same time quantifying their reactivity (through a reactivity score, R). As R-values reflect Cys nucleophilicity, and the latter depends (among other factors) on the acid dissociation constant of its functional group, methods like isoTOP-ABPP could be very useful for large-scale comparison between theoretical prediction of reactivity and experimental data. It would be interesting to compare theoretically based ranking of Cys reactivity such as pKa and exposure with R-values, experimentally obtained for all reactive Cys.

Besides pKa prediction, QM investigations can be useful in unraveling other aspects of Cys properties and reactivity, as detailed in the following section.

7.2.3 DFT in the Study of Redox Proteins

Cys reactivity can be assessed using reactivity descriptors founded in conceptual density functional theory, DFT (Parr and Yang 1995; Geerlings and De Proft 2008), describing the preferred reaction energetics and thus the kinetics in terms of the properties of the reagents in the ground state. Conceptual DFT looks at energy perturbation (Parr and Yang 1995; Geerlings and De Proft 2008) upon changing the number of electrons or the external potential, i.e. the potential felt by the electrons due to the nuclei. Sulfur is an electron rich element having d-electrons not tightly bound to the nucleus making it polarizable. Therefore, sulfur can be classified as a 'soft' element. The softness (S) is the reciprocal of the hardness, which is defined as the second derivative of the energy to the number of electrons. It measures the resistance of the system to changes in the number of electrons (Pearson and Parr 1983; Geerlings and De Proft 2008). The *S* can be approximated via the ionization energy (IE) and the electron affinity (EA), i.e. $S \approx 1/(IE - EA)$. When the reactive part of a molecule (e.g., the functional group, the thiol/thiolate for Cys) is considered, the Fukui function f(r) can be used as a local descriptor (Pearson and Parr 1983; Geerlings and De Proft 2008). The condensed Fukui functions on A can be approximated as:

1. $f_A^- = q_A(N_0 - 1) - q_A(N_0)$, for the nucleophilic attack (performed by atom A)
2. $f_A^+ = q_A(N_0) - q_A(N_0 + 1)$, for the electrophilic attack (received by atom A)

where $q_A(N_0)$, $q_A(N_0 + 1)$ and $q_A(N_0 - 1)$ are the atomic populations for atom A in the neutral molecule (N_0 electrons) and the corresponding anion ($N_0 + 1$) or cation ($N_0 - 1$), as evaluated at the geometry of the N_0 electron system. Both, the Fukui function and the softness have been successfully applied to reveal details of enzymatic reaction mechanisms, as well as in the interpretation of biochemical experimental observations (Roos et al. 2009a, b). For example, the detailed mechanism of the reaction for thioredoxin mediated reduction of target disulfides (a process involving formation and dissociation of mixed disulfides) was studied via these descriptors (Roos et al. 2009a, b). Using the Fukui function and the softness it was possible to identify and rationalize the succession of events involved in the reaction. In the catalytic C_1xxC_2 motif (where x denotes any amino acid, and C stands for Cys) of thioredoxin, the first Cys (C_1) is not only the attacking nucleophile (i.e., the one which initiates the reaction) but also the site receiving the final nucleophilic attack from the second Cys (C_2) of thioredoxin (i.e., C_1 attacks the substrate, then C_2 attacks C_1, releasing the reduced substrate). It was found that C_2 activation, after the mixed disulfide bond complex is formed, is due to hydrogen bonds (Roos et al. 2009a, b). This example shows that the Fukui function is able to identify the preferential sites for nucleophilic attack in disulfide bonds, a useful feature that can provide valuable insights in the investigation of reaction mechanisms involving this type of covalent modifications (e.g., the Dsb family).

As discussed above, sulfur is very prone to oxidation (Winterbourn and Metodiewa 1999; Winterbourn and Hampton 2008) and can adopt oxidation states ranging from −2 (the thiol group in Cys residues), to +4 (the sulfonic acid, $R\text{-}SO_3H$). In most biologically relevant forms, Cys oxidation states include −2 (thiol), −1 (disulfide), and 0 (sulfenic acid, R-SOH). Cys oxidation thermodynamics have not been experimentally studied, due to the high reaction rates between thiols and oxidants (Roos and Messens 2011). From a computational perspective, reduction potentials are real challenges: for small and medium sized systems, high level free energy calculations can be performed, albeit being computationally expensive, in a thermodynamic cycle linking the process in the gas phase with that in solvent (Baik and Friesner 2002; Schmidt Am Busch and Knapp 2005). For large protein systems, the free energies can be calculated in a quantum mechanics-molecular mechanics framework (Hu and Yang 2008; Mark et al. 2002; Kamerlin et al. 2009). However, to the best of our knowledge, these techniques have never been applied in redox biology, e.g. to study Cys oxidation in proteins. Recently, an alternative approach, called the *R*eduction potentials from *E*lectronic *E*nergies

(REE) method, was developed (Billiet et al. 2012). The REE approach is based on the correlation between reaction path, independent reaction energies and free energies. To assess its applicability in redox biology, the oxidation of thiol to sulfenic acid in human peroxiredoxin was studied (Billiet et al. 2012). It was found that a polar environments favor the sulfenylation of the deprotonated thiolate RS^-. Furthermore, when redox couples at different protonation states were compared, it was calculated that RSH/RSOH oxidation is thermodynamically favored over deprotonated RS^-/RSO^- oxidation. This thermodynamic preference for the protonation state can, however, be reversed in the active site of human Prx by a conserved arginine: the hydrogen bond between this arginine and the active site Cys favors the oxidation of the deprotonated couple (RS^-/RSO^-). With this interaction, in the active site of peroxiredoxin, not only kinetics but also thermodynamics favor the oxidation of thiolates (i.e., the reaction between oxidants and RS^- groups is favored over the reaction with RSH groups, both energetically and kinetically).

The REE method is not limited to the study of sulfenic acid/thiol redox couple and can also apply to other oxidized Cys couples, such as sulfinic acid or sulfonic acid derivatives. Furthermore, methionine oxidation and Sec oxidation can, in principle, be studied this way too. Therefore REE is one of few existing tools for detailed mechanistic studies of redox control. Once available to the community, it would offer a significant addition to the current arsenal of computational tools in the redox biology area (e.g., to provide *in silico* estimation of redox potentials for particular protein residues).

7.3 Bioinformatics Approaches for Prediction of Reactive Cys in Proteins

As introduced above, Cys may serve different functions in proteins. These residues can act as stabilizing elements, e.g. by binding metals or forming stable structural disulfides, or serve as posttranslational modification sites. Thus, a common classification of Cys residues in proteins is based on their function: different functional classes of Cys have different salient features determining their ability to complete the specific function. For example, Cys residues serving catalytic functions have to be accessible to their substrates, and they often show features of high reactivity, such as low pKa or high electrostatic perturbation of their functional groups (Marino and Gladyshev 2009).

Different functional categories of Cys include: (i) catalytic Cys residues, (ii) structural cystine residues (i.e., stable disulfide-bonded Cys), (iii) metal-coordinating Cys residues, and (iv) Cys residues, which serve as sites of posttranslational modifications (often called regulatory Cys). However, it has to be noted that while this classification is certainly useful, not all Cys residues can be unambiguously assigned to one of the above functions (e.g., redox switchable

Cys). In the following paragraphs, we briefly introduce relevant biological aspects of each Cys functional category. Then, for each category, a brief discussion of how bioinformatics approaches can be used to investigate the subject, and with which tools, is provided (for an overview of most relevant methods discussed in this chapter, see Table 7.1). Finally, we comment on the challenges to bioinformatics analyses posed by redox switches.

7.3.1 Catalytic Cys Residues

In many enzymes, Cys plays a critical role as a nucleophile in enzyme-catalyzed reactions. Such Cys residues belong to the functional category of catalytic Cys residues. Depending on whether or not Cys residues change redox states during catalysis, a further division can be made between redox and non-redox catalytic Cys functions. Examples of enzymes with non-redox catalytic Cys are protein tyrosine phosphatases, glyceraldehyde-3-phosphate dehydrogenase (GAPDH), Cys peptidases, various members of the deubiquitination system, and dCMP hydroxymethylases. It has to be noted that, while in their normal catalytic cycle the active site Cys of these enzymes do not change redox state, some of them (e.g., GAPDH) are redox sensitive (i.e. they are easily targeted by various oxidant, such as hydrogen peroxide) and experience redox regulation. Other enzymes, called thiol oxidoreductases possess an active site nucleophilic Cys that, during the catalytic function, changes redox state. Here, the catalytic Cys function involves substrate oxidation or reduction, disulfide bond isomerization and detoxification of various compounds. To our knowledge, no computational approaches for the detection of catalytic non-redox Cys residues have been developed. So, we further focus on thiol oxidoreductases, for which better progress has been made.

Thiol oxidoreductases are the only known enzymes that make use of Sec as the catalytic residue (Fomenko et al. 2007). In cases where Sec is not employed, Cys is used in its place. Corollary considerations follow: (i) Sec in active sites of enzymes is always redox-active, (ii) in thiol oxidoreductases, homologous Sec/Cys pairs of residues (i.e., Cys substituting Sec in orthologous proteins) indicate that both residues ought to be considered redox-active. The latter observation was used as a foundation for designing an algorithm (Fig. 7.2) for high-throughput identification of catalytic redox Cys in protein sequences. The method searches for sporadic Cys/Sec pairs in homologous sequences (Fomenko et al. 2007). First, it identifies unique Cys/Sec pairs flanked by homologous sequences within a pool of translated nucleotide sequences. These pairs then serve as seeds for sequence analysis at the level of protein families and subfamilies. In other words, because Sec is exclusively used in redox catalysis, an alignment between two protein sequences where Sec and a conserved Cys are paired points to the catalytic role for the Cys. A useful characteristic of the method (Fig. 7.2a) consists of it ability to identify thiol oxidoreductases through their catalytic residues (e.g. not only thiol oxidoreductases, but also the exact identity of their redox active Cys). Together with

Table 7.1 Methods for prediction and analysis of functional Cys

Method	Input	Methodology	Usage[a]/type of prediction	Availability
PROSITE	Protein sequence	Regular expression search	Cys in MBS or in DB[b]	Web accessible[d]
METAL DETECTOR	Protein sequence	SVM prediction	Cys in MBS or in DB	Web accessible[e], Standalone Program
FOLDX	Protein structure	Search for energetically avorable MBS geometries	Cys in MBS	Standalone Program
DISULFIND	Protein sequence	SVM prediction	Cys in DB	Web accessible[f]
DCPB	Protein sequence	Automated homology modeling, SVM prediction	Cys in DB or MBS[c]	Web accessible[g]
FINDSITE-metal	Protein structure	Automated protein threading, SVM prediction	Cys in MBS	Web accessible[h], Standalone Program
PROPKA	Protein structure	Heuristic pKa calculation	pKa	Web accessible[i], Standalone Program
Naccess	Protein structure	Probe based sampling	atomic exposure	Standalone Program
Q-SiteFinder	Protein structure	Probe based sampling	Active/functional site	Web accessible[j]

[a]Some of the listed programs are not limited to Cys evaluation (but can process all other suitable type of residues)
[b]*DB* disulfide bond, *MBS* metal binding site
[c]The SVM is not trained specifically for MBS, but only for DS. For MBS predicted by DCPB, additional analysis (e.g., with METALDETECTOR) maybe required
[d]http://prosite.expasy.org/ and http://prosite.expasy.org/scanprosite/
[e]http://metaldetector.dsi.unifi.it/
[f]http://disulfind.dsi.unifi.it/
[g]http://biomedical.ctust.edu.tw/edbcp/
[h]http://cssb.biology.gatech.edu/findsite-metal
[i]http://propka.ki.ku.dk/
[j]http://www.modelling.leeds.ac.uk/qsitefinder/

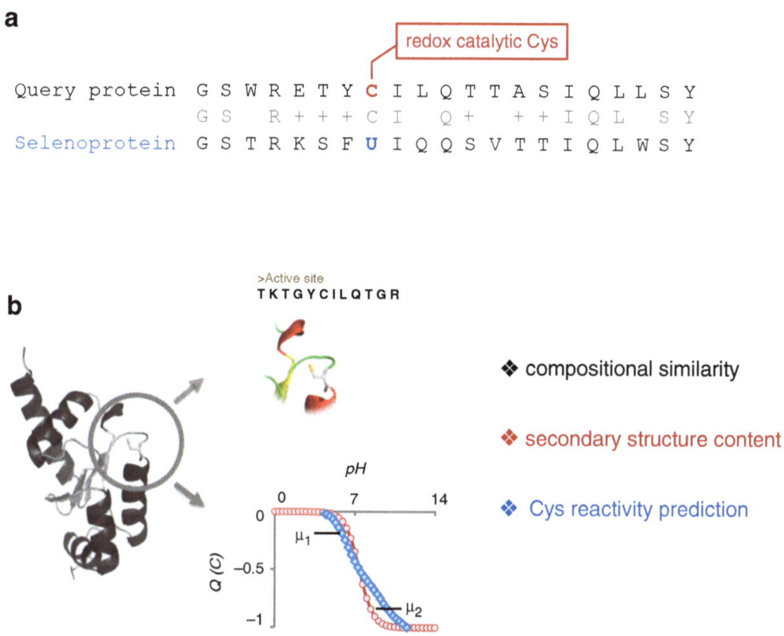

Fig. 7.2 Methods for thiol oxidoreductase prediction. (**a**) Sec2Cys method. A query protein is analyzed with tBlastn against a database of nucleotide sequences. Sec pairing with Cys in the sequence directly points to the catalytic redox function of the Cys. (**b**) Structure-based prediction of thiol oxidoreductases. By analyzing sequence and structural homology with known thiol oxidoreductases, and by theoretical estimation of Cys reactivity, a query protein is evaluated. Important parameters include amino acid and secondary structure composition within 8 Å from the catalytic Cys, and the deviation from the Henderson–Hasselbach (HH) behavior. A typical representation of the latter is shown in the *bottom part* of panel B, where the deviation ($\mu_1 + \mu_2$) reflects the difference between the theoretical titration curve calculated from the structure (*blue curve*), and the corresponding curve (*red curve*) for a standard HH behavior

sensitivity, a key advantage of this approach is speed. High-throughput analyses are possible in reasonable amount of time, allowing genome-wide analyses of thiol oxidoreductases. When tested, the method was capable of correctly recognizing nearly all known thiol oxidoreductases and predicted several new families, such as the one within a superfamily of AdoMet-dependent methyltransferases, as having thiol-based redox functions. This specific family (arsenic methyltransferases) was then experimentally verified to contain the catalytic redox-reactive Cys (Fomenko et al. 2007), further validating the approach. More recently, this approach was applied to assess the general use of thiol-based oxidoreduction in biology (Fomenko and Gladyshev 2012). These analyses showed that approximately 1% of all proteins are thiol oxidoreductases and that these proteins are used in all living organisms. Thus, thiol-based redox control is an essential feature of all cellular life.

A different computational approach, developed principally to investigate common structural features of catalytic residues in thiol oxidoreductases, was also employed for predictive purposes (Fig. 7.2b) (Marino and Gladyshev 2009). Features describing the majority of catalytic residues in thiol oxidoreductases included: (i) weak, yet detectable sequence similarities between different families of thiol oxidoreductases, (ii) preference for an α-helix or helix/coil secondary structure, (iii) theoretical titration spectra deviating from the corresponding Henderson-Hasselbalch (HH) behavior (this feature described catalytic redox Cys better than the pKa, particularly when compared to non-redox catalytic Cys), and (iv) docking affinity for small and uncharged molecular probes, used as mimics for generic substrates (Laurie and Jackson 2005). These features were combined into a multi-parameter scoring function and implemented in the form of a predictive algorithm. The method was examined with different test cases, correctly predicting known thiol oxidoreductases. Additionally, the method predicted some new candidates, among which particularly interesting was 6-O-methylguanine-DNA methylase (MGMT). This protein showed high scores, higher than those for many known thiol oxidoreductases. Because some MGMTs also have sporadic Cys/Sec pairs in homologous sequences, these enzymes can be predicted as interesting thiol oxidoreductase candidates.

The main advantages of the structure-based approach described above reside in its ability to specifically trace back the contribution of each component to the overall prediction. The contribution of each sub-part of the algorithm can be separately analyzed, and therefore the weight of each to the final output value can be immediately retrieved. As each sub-part corresponds to specific physical, chemical or biological aspects of Cys reactivity (e.g., exposure and accessibility, titration curve and its deviation from HH, sequence homology with known thiol oxidoreductases), the user can easily extract biological information from the final output of scoring function. In addition, it showed an ability to detect new thiol oxidoreductases. On the other hand, the main disadvantage of the method is its speed: several different structure-based calculations are needed, some of which are computationally demanding. Therefore, the method is not well suited for extensive high-throughput analyses.

Bioinformatics approaches applied to the study of thiol oxidoreductases are certainly not limited to their prediction. Indeed, computational investigations examined a variety of aspects of thiol oxidoreductase evolution, distribution and functional classification. We further discuss several specific cases. Babbitt and co-workers developed computational strategies to analyze the structure and sequence landscape of thiol oxidoreductases and examined how, in particular within the thioredoxin fold, they evolved into different protein families and functionalities (Atkinson and Babbitt 2009a). For example, the authors inferred the evolution of a protein from a thioredoxin-like common ancestor of peroxiredoxins (Copley et al. 2004); the two protein families shared the same fold, but evolved different catalytic mechanisms (with peroxiredoxin forming sulfenic acid intermediates on the catalytic Cys, a feature not present in thioredoxins) and rescue modality (i.e., reductants, which in the case of peroxiredoxin are type- and cell-specific, while a single enzyme, thioredoxin reductase, evolved in the case of thioredoxin) (Copley et al. 2004). The effect of evolutionary divergence in shaping functional specificities can be assessed with Blast-based evaluation of sequence distance, and the functional consequences of such variation can be traced to the protein fold through structural superimpositions.

In another study, evolution of glutathione transferases (GSTs) was assessed though evaluation of both structure and sequence similarity networks (Atkinson and Babbitt 2009b). The study showed the effects of sequence variability (e.g., the loss, in most members, of the nucleophilic catalytic Cys, typically found in CxxC like motifs in thiol oxidoreductases with the thioredoxin fold) in term of structure and function, a process that ultimately produces proteins with different functions. Other bioinformatics approaches to redox biology include the use of functional active site profiling to classify different subfamilies of peroxiredoxins (Nelson et al. 2011; Soito et al. 2011). By using active site signatures, this method allowed researchers to define six peroxiredoxin subfamilies, each of them with specific functionalities. A carefully curated database for peroxiredoxin sub-families classification (PREX) has been developed and is publicly available (http://csb.wfu.edu/prex/).

7.3.2 Structural Disulfides

As discussed above, Cys is one of only two naturally occurring amino acids (the other one being Sec) capable of covalent interaction with other residues of the same type. This property provides fundamental advantages, including structural stabilization and folding guidance. Prior to discussing the relevant computational methods, an important note is due: all the approaches described address (and, thus, apply to) the case of stable, structurally relevant disulfides. In fact, the prediction of transient disulfides (e.g., those present in reducing compartments of the cell) represents a very challenging task because of protein movement associated with disulfide formation or reduction. Therefore, to date, their identification still relies on the case by case in depth analysis, either through theoretical analysis of protein mobility (e.g., molecular dynamics approaches), or by direct solution of protein structure at different oxidation states (e.g., crystal structures of both reduced and oxidized forms).

Computational approaches to predict stable structural disulfides can be divided into sequence-based and structure-based (Fig. 7.3). The former strategy has been, and still is, the elective choice for large-scale analysis of proteomes, where structural information is not available. In turn, structure-based algorithms, albeit slower and inapplicable to proteins, where no structural information is available, can provide distinct advantage in terms of accuracy and precision. The approach to structure-based computation of disulfides can be very straightforward. Starting from the atomic coordinates, the distance between each Cys sulfur is computed. As a common practice, the pairs found to lie within 2.5 Å are considered safe candidate for residues engaged in covalent bonds (Fig. 7.3b). Disulfide bonds detected by analyzing the S-S distance often can be found already annotated in the PDB repository (they can be found by directly accessing the PDB file header).

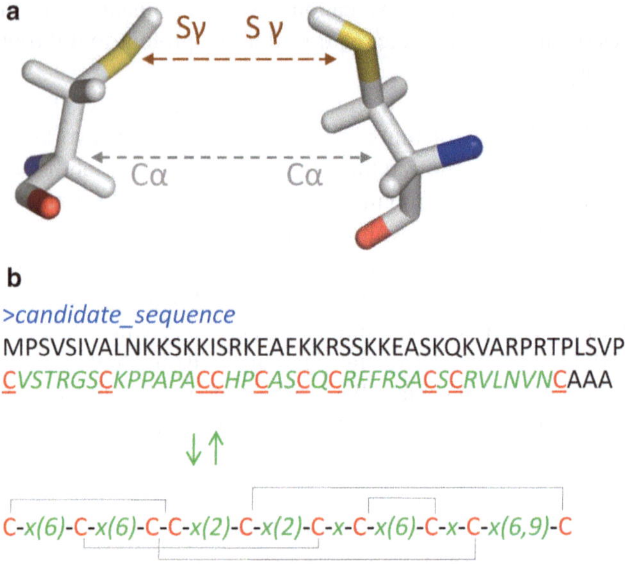

Fig. 7.3 Sequence- and structure-based approaches for disulfide detection. (**a**) A simple structure-based approach for disulfide identification. Two distances are routinely considered: (i) sulfur (Sγ) to sulfur distance, with the usual cut-off distance of 2.5 Å; and carbon alpha (Cα) to carbon alpha distance, with the usual cut-off of 8 Å. (**b**) A typical PROSITE based approach is shown. A candidate sequence under investigation is parsed with PROSITE patterns describing known disulfides. A match is found for the pattern PDOC60024 (also called, Agouti domain signature). For explanation of the regular expression vocabulary for PDOC60024, see the main text. Conserved Cys are shown in *red*, which form disulfides in a sequence specific arrangement (shown as *dotted lines*, connecting Cys in the figure)

A powerful modification of this method consists of the analysis of distances between carbon alpha atoms. In this case, a common cut-off distance is 8 Å (Fig. 7.3b). The main advantage in doing this sort of analysis is the fact that the prediction of "only" the alpha-trace represents a considerable computational advantage. Disregarding positioning side chains and focusing only on the main chain folding, it allows (i) prediction for very difficult homology model cases (e.g., very low homology with a template, using threading approaches), and (ii) reduction of computational time needed to obtain the model. This approach is well suited for large-scale comparative analyses (Beeby et al. 2005; Marino and Gladyshev 2011a), which benefit from a reasonable speed of calculations, along with the ability to analyze scarcely refined structural models.

In some cases, however, it is desirable to work without structural information at all: the structural coverage of natural proteins is still largely incomplete. In this area, various computational approaches have been developed. As a common theme, they search protein sequences for recurring motifs, as found in the primary structure. Usually, these features are discovered by manual curation (i.e., investigating known cases), and are then compiled in the form of rigorous patterns. An example of one such pattern is C-x(6)-C-x(6)-C-C-x(2)-C-x(2)-C-x-C-x(6)-C-x-C-x(6,9)-C

(Fig. 7.3a), where C is Cys, x denotes any amino acid, and the number in parenthesis indicates the number of such amino acids. This is an example of a pattern, which includes several disulfides, compiled, curated and maintained by PROSITE (Sigrist et al. 2002). Indeed, perhaps the simplest approach for sequence-based prediction of disulfides is the direct implementation of sequence patterns and profiles found in PROSITE. These patterns are annotated regular expressions that describe relatively short portions of protein sequence that may have biological meaning or function. They can be directly implementable in simple programming scripts for relatively quick searches of entire proteomes. Alternatively, for non-specialist users, the PROSITE web server provides a simple access interface (ScanProsite, de Castro et al. 2006) to browse for S-S patterns in any input sequence. However, although many S-S patterns with perfect specificity have been compiled, PROSITE profiles and regular expressions can detect only a minority of disulfide-bonded Cys. It has been estimated that only one out of four of them are matched by a pattern with high specificity (Passerini and Frasconi 2004).

Improved sequence-based approaches have been developed in recent years that could overcome this limitation. To enhance performance, machine learning approaches were implemented (Chen et al. 2004; Cheng et al. 2006; Ceroni et al. 2006), which employ various levels of sequence-based information, such as nature of adjacent amino acids, conservation of flanking residues, etc. Very generally, the scheme includes the definition of (i) a set of true positive cases (i.e., known disulfides), and (ii) a set of true negatives (known non-covalently bonded Cys residues). Then, the algorithm is trained to distinguish the two groups. Rules and parameters allowing for a better resolution are kept and become part of the search program. In other words, starting from the training set of manually collected known cases, the algorithm can learn a classification function, which can then be used for prediction purposes. One such approach is called DISULFIND (Ceroni et al. 2006): it uses support vector machine (SVM) and neural networks to classify and rank different Cys in protein sequence. It is available as a user-friendly web service, at http://disulfind.dsi.unifi.it/. The algorithm is fast and performs, overall, better than PROSITE (Ceroni et al. 2006). A viable alternative is represented by the machine learning approach called EDBCP (Lin and Tseng 2010). Its main feature is that it employs structural information through automatic modeling (starting from an input consisting of a protein sequence) via the Modeller package (Lin and Tseng 2010). After the model is created, an SVM-based algorithm uses the information to score and rank all Cys in proteins. These values are then compared with reference scores for true positive candidates. Therefore, only Cys pairs that show structural resemblance to known disulfides are predicted as cystine residues. Similarly to DISULFIND, EDBCP is publicly available as a web accessible service (http://biomedical.ctust.edu.tw/edbcp/).

7.3.3 Metal-Binding Cys

Metal-binding Cys residues are found in structurally and evolutionary distinct groups of proteins, present in all branches of life. Together with histidine (His), Cys is the most frequently employed amino acids in metal coordination (Dokmanić et al. 2008), especially for binding zinc, cadmium, copper, iron, and nickel. A particularly common occurrence is the coordination of Zn: biologically relevant assets include tetra-Cys coordination ($Zn\text{-}Cys_4$) and mixed Cys and His binding (e.g. $His_2\text{-}Zn\text{-}Cys_2$). These structures are present in zinc fingers motifs, where they exert a crucial role in scaffolding the local structure of the broad variety of proteins that use this motif to bind DNA (Klug 2010). In the case of $Zn\text{-}Cys_4$ complexes, four thiolate-Zn^{2+} bridges act as stabilizing elements for protein structure (Kröncke and Klotz 2009). This structure can be thought of as a sort of "substitutive" for structural disulfide in cellular compartments characterized by overall reducing conditions, such as the cytosol, where the formation of disulfides is disfavored while $Zn\text{-}Cys_4$ complexes can form and be stable.

Besides structural roles, metal binding sites carry out many other important functions, such as serving as catalysts and regulatory sites. In this regard, Cys properties make this amino acid an excellent site for redox-dependent regulation of metal binding. For example, Zn^{2+}-Cys complexes allow for tight binding, but at the same time the involved Cys are potentially amenable to oxidation, which would destabilize the complex and release the metal. From a computational perspective, metal-Cys sites can be studied with various approaches, somewhat mirroring the case of disulfides. Thus, the simplest approach uses manually curated sequence patterns/profiles recurrently found in different classes of metal-binding sites. These motifs can be found, meticulously curated, at the PROSITE website. In spite of their speed and simplicity, pattern-based approaches can be affected by low performance. Furthermore, while several patterns provide perfect specificity (e.g., PS00198, PS00190, PS00463, at the PROSITE website), other patterns are much less specific, and some Cys residues can match more than one pattern (i.e., circa 12% of metal bound Cys can be detected by multiple patterns, Passerini et al. 2006). Obviously, another limitation includes their inapplicability for detecting new types of metal-binding sites.

Thus, more sophisticated approaches have been developed based on machine learning (Passerini and Frasconi 2004; Passerini et al. 2006) and non-linear statistical methods (Lin et al. 2005). A generic (and simplified) presentation of similar approaches could be as follows: for each candidate Cys, feature vectors are derived that include sequence information in the form of (i) conservation, for example, through a position specific substitution matrix (PSSM), and (ii) distance, in primary structure, between pairs of candidates (e.g.$[C]_1\text{-}x(d)\text{-}[C]_2$, where $[C]_i$) denotes a candidate site, x any amino acid, and d the number of amino acids separating the two Cys. Various additional layers can be added to the analysis, such as conservation of amino acids within a certain distance from the candidate site. A particularly interesting approach of this type is MetalDetector (Passerini et al. 2006). It considers only a subset of binding sites, where Cys and/or His are involved. Here, feature vectors include PSSM based information and global descriptors, such as protein length and amino acid composition. The feature vector is first classified by an SVM, and then a bi-directional recurrent neural network is used to distinguish metal and non-metal binding sites. These steps are meant to help separate the two types of structural Cys, i.e. those present in the disulfide form from those in metal-binding sites. The latter is a prominent feature of MetalDetector: for any candidate site, a likelihood score is provided that reflects the probability that this Cys is metal bound, as opposed to disulfide bonded (or vice versa). The method is freely available both as a web accessible service and as a standalone program (http://metaldetector.dsi.unifi.it/). Various other SVM ranking methods have been developed for predicting metal-binding Cys (Shu et al. 2008). These approaches usually outperform simpler sequence based descriptors (e.g. pattern-based, such as PROSITE, or homology-based, such as Blast-derived comparative analyses) while maintaining many of the advantages, such as speed (Shu et al. 2008).

Like in the case of disulfide prediction, structure-based approaches can be a valuable alternative for the prediction of metal-binding Cys. One interesting method involves the use of empirical force field FoldX (Schymkowitz et al. 2005). The search algorithm uses geometric information typically found in metal-binding sites as a starting point for prediction of new sites. It first analyzes typical (e.g. recurrent) arrangements of Cys ligands around zinc coordination sites. The method can recognize similar geometrical and compositional patterns, such as the nature of ligands clustered in space and their relative geometries. The information gained for each type of metal binding site is then stored, and in its final form, the procedure is capable of scanning a candidate protein structure recognizing potential metal-binding sites and identifying residues and types of metal involved (Schymkowitz et al. 2005). To be noted, FoldX, available free of charge for academic users at http://foldx.crg.es/, is not specific for Cys, or His or zinc. In turn, it can work with almost all common metal-binding site types. However, due to the use of structure-based information and employment of energy-based calculations, it has its limitations: it heavily depends on the quality of structural data (highly resolved experimental structure is the ideal situation for such analysis). Thus, it may be not the best choice for large scale analyses of protein datasets, or for the analysis of models with low refinement.

Alternative structure-based strategies that could overcome this issue are now available, such as the algorithm implemented in FINDSITE-metal (Brylinski and Skolnick 2011). This method allows prediction of metal-binding sites in weakly homologous proteins. This is achieved by combining the power of threading techniques for structural modeling based on distantly related templates with evolutionary information (e.g. through comparative analysis of conservation patterns) and, finally, machine learning evaluation and classification of feature vectors. This program is capable of working with entire proteomes in a reasonable time (Brylinski and Skolnick 2011) and is freely available to the community at http://cssb.biology.gatech.edu/findsite-metal.

7.3.4 Regulatory Cys

As discussed above, Cys residues are subject to different types of redox posttranslational modifications (PTMs), with sulfenic acid (Cys-SOH), disulfide bonds (both intramolecular and intermolecular), S-nitrosylation (NO-Cys) and glutathionylation being the most common. Additionally, Cys can react with endogenous hydrogen sulfide (H_2S), a modification that can lead to significant physiological (Yang et al. 2004; Johansen et al. 2006) and structural effects (Jiang et al. 2010). All of the above can be classified as redox-based PTMs and are reversible. However, other important Cys modifications, which are stable and do not involve a change in the redox state, also occur. They include, for example, the formation of thioether bonds with farnesyl or geranylgeranyl groups, leading to protein lipidation and membrane anchoring (Zhang and Casey 1996) or covalent binding of protein cofactors, such as heme. These Cys modifications could be classified into a separate category of functional Cys residues. In the following text, we discuss the role of bioinformatics in the study of reversible Cys modifications that can affect protein properties, function and interaction networks. For these reasons, they are often referred to as regulatory Cys (Marino and Gladyshev 2011a, b).

Large-scale computational investigation of regulatory Cys was hindered in the past by scarcity of experimental data. However, in recent years several proteomic approaches have been developed, which provided substantial improvements with regard to experimental data (Fratelli et al. 2004; Hao et al. 2006; Leonard et al. 2009; Michelet et al. 2008). Benefitting from this large influx of information, bioinformaticians could start addressing the existence of recurrent patterns defining the specificity of Cys to regulatory modifications. To the day, no sequence-based predictive pattern has been found, for any type of Cys PTM sites. Instead, computational studies revealed heterogeneity of sequence features around different Cys PTMs. In some cases, enriched motifs were detected (Greco et al. 2006; Wu et al. 2010), like adjacent negatively charged residue in the case of some NO-Cys sites, but these could not describe the majority of regulatory sites of this type (Marino and Gladyshev 2010b).

Better results were obtained with structure-based approaches, particularly in the case of NO-Cys (Greco et al. 2006; Marino and Gladyshev 2010b). First, a quantum mechanics (QM)-based study demonstrated that NO modification induces a significant charge redistribution, which largely affects its side chain atoms in a manner inversely proportional to the distance from the modification sites (e.g., carbon alpha is the least affected). At the same time, the effect is negligible for main chain atoms of the same residue (Han 2008). NO-Cys specific force field parameters and charge schemes for NO-Cys were developed, by using a restrained electrostatic potential (RESP) approach (Cieplak et al. 1995). Structural restrains for dihedral angles for NO-Cys sites were derived from the analysis of crystal structures, and geometrical optimization was conducted at the Hartree-Fock (HF) level of theory (HF/6-31G*). With these modifications, molecular dynamics (MD) simulation tests were performed and found to be in good agreement with previously available data (e.g., analysis of human thioredoxin 1, for which the structures of both unmodified and modified protein forms are available in PDB) (Han 2008).

These theoretical findings provide the basis for docking simulation studies with NO-Cys containing proteins or substrates. A first computational approach, using the new and *ad hoc* developed parameters for NO-Cys to investigate a database of potential Cys targets for GSNO mediated trans-nitrosylation, was recently reported

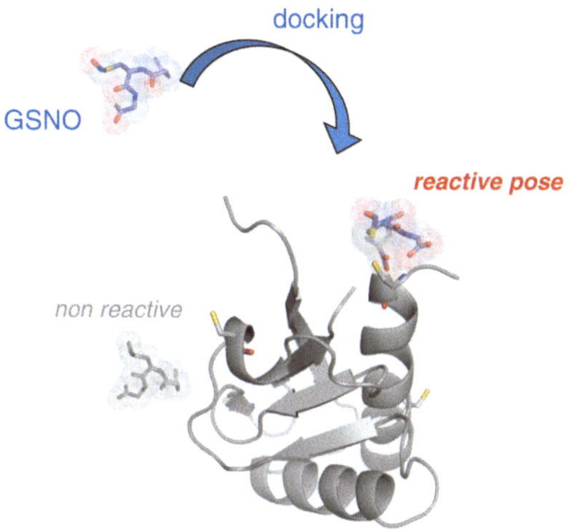

Fig. 7.4 Docking-based approaches for identification of S-nitrosylation sites. Availability of *ad hoc* parameters for S-nitrosylation of Cys residues in proteins (see main text) allowed their implementation for NO-modified proteins. In the example in the figure, S-nitrosylated glutathione (GSNO) is shown. As a putative trans-nitrosylating agent, GSNO is docked to a protein. Among the best ranked docking complexes, one (or more) poses might be consistent with reactivity (e.g. GSNO approaches an exposed Cys with a favorable interaction, and a sulfur to sulfur distance is less than 3 Å with no steric clashes). In the example, a reactive pose is found (labeled "reactive pose"). The Cys is, in this docking based assay, a positive target for GSNO mediated trans-nitrosylation. In addition to GSNO, similar approaches can be applied to many other agents (e.g. Cys-NO, or even small S-nitrosylated proteins)

(Marino and Gladyshev 2010b). This docking approach was able to address the issue of prediction of a specific subset of NO-Cys sites, i.e. Cys modified via trans-nitrosylation with GSNO. Similarly, other trans-nitrosylating agents could be explored with this strategy (Fig. 7.4). Indeed, docking calculations could be a valid computational alternative for detection of specific Cys amenable to modification with different NO-sylated substrates (NO-Cys, S-nitrosylated small peptides, etc.). Particularly challenging, but certainly feasible, is the investigation of the role of protein-protein interactions in the transfer of NO groups from one protein to another, the so-called protein interaction-based trans-nitrosylation. This process has so far escaped detailed computational studies. However, the steady development of suitable docking software (e.g., Rosetta dock) and the information gained from previous studies (Han 2008) may make it soon possible to investigate protein interaction-based trans- nitrosylation.

7.3.5 Cys as Redox Switches

Classification of Cys residues based on their biological roles is certainly reasonable in the context of a general overview. However, in some known cases (and, perhaps, in many still unknown) the situation is more promiscuous, with the division between different Cys types being much less clear cut, i.e. Cys residues could belong to two or more of these categories. For instance, catalytic non-redox Cys in GAPDH also undergoes regulation by reactive oxygen species and thiols (Fermani et al. 2007; Hara et al. 2005; Hook and Harding 1997). Indeed, thiol redox regulation of enzymes may act on different Cys types, such as catalytic and metal binding (Ilbert et al. 2007; Jakob et al. 1999). Whenever the modification leads to a change in protein structure and/or function, the Cys can be viewed as a sort of switch. In cases where the modification is redox, it is often called the redox switch (Klomsiri et al. 2011). Redox switches are widely employed in a variety of processes and proteins, e.g. transcription factors OxyR and Yap1, kinases, phosphatases, chaperones, mitochondrial branched chain aminotransferase and other proteins. For a detailed discussion on the subject, we refer the reader to representative and updated reviews (Maret 2006; Brandes et al. 2009; Kumsta and Jakob 2009; Nagahara 2010; Klomsiri et al. 2011).

A very interesting case is represented by Hsp33 (Jakob et al. 1999, 2000). This protein protects bacteria from oxidative stress that results in protein unfolding and aggregation. In unstressed conditions, the protein is monomeric and inactive. It has four Cys residues (Cys232, Cys234, Cys265, and Cys268) involved in binding a zinc atom. After exposure to increasing levels of oxidants, e.g. hydrogen peroxide, one of the four Cys is oxidized to sulfenic acid. This transient state leads to the formation of a disulfide, and the loss of the zinc atom. Ultimately, the process triggers a large change in structure (tertiary and quaternary) and activates the chaperone function of Hsp33.

In humans, up to 10% of proteins are believed to possess Zn-binding sites (Kröncke and Klotz 2009; Maret 2006). The majority of these sites contain Cys residues. As discussed in various sections of this chapter, when Cys binds zinc, a very relevant property is the tight binding, but also susceptibility to oxidative modification. This property, together with reversibility of binding as a function of the intracellular redox state, open up the possibility that many Cys-coordinated Zn sites are candidate redox switch sites. The distribution, function and general physiological relevance (i.e. beyond the few known cases) of metal-based redox switches have yet to be thoroughly explored.

Currently, no computational methods can properly deal with Cys redox switches. Sequence-based computational predictors for metal-binding Cys are often unable to distinguish them from Cys capable of disulfide bonding, and vice versa. For instance, when *E. coli* Hsp33 sequence is scanned with a state-of-the-art machine learning-based metal-binding site prediction program (Passerini et al. 2006), Metal detector (http://metaldetector.dsi.unifi.it/), two of its Cys (Cys265, Cys268) are classified as metal binding, whereas the other two Cys are not. Additionally, the program predicts that all Hsp33 Cys residues have a negligible tendency to form disulfides. But we know that Cys265 and Cys268 form the first Hsp33 redox switch, and thus are indeed capable of disulfide bonding, even if only after one of them is oxidized. This is not surprising as the algorithm has not been trained to distinguish redox switches, and training is the most essential part of machine learning approaches.

Therefore, it is perhaps possible that similar approaches, properly devised and implemented, could provide a starting point to initiate bioinformatics to this very important, intriguing and still largely unexplored area of redox biology.

7.4 Concluding Remarks

The main subject of this chapter is Cys, a relatively scarce, but often functional, residue in proteins. Cys residues are leading characters in the play of redox regulation, and their many unique properties make them a very interesting, albeit challenging, subject to study. Believing that many aspects of the biology of Cys are tightly linked to its unique chemical and structural features, we dedicated the first part of this review for discussion of fundamental properties of this amino acid, as well as theoretical descriptors employed by researchers to describe Cys reactivity. In our opinion, a better understanding of basic chemical and physical features of Cys is crucial in order to improve currently available tools for recognition and subsequent functional annotation of reactive thiols in proteins. In the second part of the chapter, we shifted the discussion to the various functional roles played by reactive Cys residues in proteins. Moreover, we reviewed the current state of computational methods for investigation and prediction of Cys functions. In some cases, bioinformatics provided important insights and tools, especially for catalytic redox Cys, metal-binding Cys and disulfide bonds. In other cases, progress has been limited, e.g. for regulatory Cys, sites of stable posttranslational modifications and catalytic non-redox Cys.

On the other hand, experimental advances, especially in proteomics and structural and post-translational datasets, provide researchers with new opportunities to address critical issues in thiol-based redox control. We expect a steady growth of the use of bioinformatics in redox biology, a boost that will help filling the gap between different types of Cys reactivity, ultimately enabling the community to have a more extensive array of tools for large scale investigation (and prediction) of all main types of Cys function, including the ones which have been so far more elusive (e.g. regulatory Cys and redox switches).

References

Atkinson HJ, Babbitt PC (2009a) An atlas of the thioredoxin fold class reveals the complexity of function-enabling adaptations. PLOS Comput Biol 5:e1000541

Atkinson HJ, Babbitt PC (2009b) Divergence of function in the thioredoxin fold suprafamily: evidence for evolution of peroxiredoxins from a thioredoxin-like ancestor. Biochemistry 48:11108–11116

Baik MH, Friesner RA (2002) Computing redox potentials in solution: density functional theory as a tool for rational design of redox agents. J Phys Chem 106:7407–7412

Beeby M, O'Connor BD, Ryttersgaard C, Boutz DR, Perry LJ, Yeates TO (2005) The genomics of disulfide bonding and protein stabilization in thermophiles. PLOS Biol 3:e309

Billiet L, Messens J, Geerlings P, Roos G (2012) The thermodynamics of thiol sulfenylation. Free Radic Biol Med 52:1473–1485

Bock A, Forchhammer K, Heider J, Baron C (1991) Selenoprotein synthesis: an expansion of the genetic code. Trends Biochem Sci 16:463–467

Brandes N, Schmitt S, Jakob U (2009) Thiol-based redox switches in eukaryotic proteins. Antioxid Redox Signal 11:997–1014

Brylinski M, Skolnick J (2011) FINDSITE-metal: integrating evolutionary information and machine learning for structure-based metal-binding site prediction at the proteome level. Proteins 79:735–751

Cabiscol E, Levine RL (1996) The phosphatase activity of carbonic anhydrase III is reversibly regulated by glutathiolation. Proc Natl Acad Sci USA 93:4170–4174

Ceroni A, Passerini A, Vullo A, Frasconi P (2006) DISULFIND: a disulfide bonding state and cysteine connectivity prediction server. Nucleic Acids Res 34:W177–W181

Chen YC, Lin YS, Lin CJ, Hwang JK (2004) Prediction of the bonding states of cysteines using the support vector machines based on multiple feature vectors and cysteine state sequences. Proteins 55:1036–1042

Cheng J, Saigo H, Baldi P (2006) Large-scale prediction of disulphide bridges using kernel methods, two-dimensional recursive neural networks, and weighted graph matching. Proteins 62:617–629

Cieplak P, Cornell WD, Bayly CI, Kollman PA (1995) Application of the multimolecule and multiconformational RESP methodology to biopolymers: charge derivation for DNA, RNA, and proteins. J Comput Chem 16:1357–1376

Copley SD, Novak WR, Babbitt PC (2004) Divergence of function in the thioredoxin fold suprafamily: evidence for evolution of peroxiredoxins from a thioredoxin-like ancestor. Biochemistry 43:13981–13995

de Castro E, Sigrist CJA, Gattiker A, Bulliard V, Langendijk-Genevaux PS, Gasteiger E, Bairoch A, Hulo N (2006) ScanProsite: detection of PROSITE signature matches and ProRule-associated functional and structural residues in proteins. Nucleic Acids Res 34:W362–W365

Dokmanić I, Sikić M, Tomić S (2008) Metals in proteins: correlation between the metal-ion type, coordination number and the amino-acid residues involved in the coordination. Acta Crystallogr D Biol Crystallogr 64:257–263

Fermani S, Sparla F, Falini G, Martelli PL, Casadio R, Pupillo P, Ripamonti A, Trost P (2007) Molecular mechanism of thioredoxin regulation in photosynthetic A2B2-glyceraldehyde-3-phosphate dehydrogenase. Proc Natl Acad Sci USA 104:11109–11114

Foloppe N, Sagemark J, Nordstrand K, Berndt KD, Nilsson L (2001) Structure, dynamics and electrostatics of the active site of glutaredoxin 3 from *Escherichia coli*: comparison with functionally related proteins. J Mol Biol 310:449–470

Fomenko DE, Gladyshev VN (2012) Comparative genomics of thiol oxidoreductases reveals widespread and essential functions of thiol-based redox control of cellular processes. Antioxid Redox Signal 16:193–201

Fomenko DE, Xing W, Adair BM, Thomas DJ, Gladyshev VN (2007) High-throughput identification of catalytic redox-active cysteine residues. Science 315:387–389

Fratelli M, Gianazza E, Ghezzi P (2004) Redox proteomics: identification and functional role of glutathionylated proteins. Expert Rev Proteomics 1:365–376

Geerlings P, De Proft F (2008) Conceptual DFT: the chemical relevance of higher response functions. Phys Chem Chem Phys 10:3028–3042

Greco TM, Hodara R, Parastatidis I, Heijnen HF, Dennehy MK, Liebler DC, Ischiropoulos H (2006) Identification of S-nitrosylation motifs by site-specific mapping of the S-nitrosocysteine proteome in human vascular smooth muscle cells. Proc Natl Acad Sci USA 103:7420–7425

Han S (2008) Force field parameters for S-nitrosocysteine and molecular dynamics simulations of S-nitrosated thioredoxin. Biochem Biophys Res Commun 377:612–616

Hao G, Derakhshan B, Shi L, Campagne F, Gross SS (2006) SNOSID, a proteomic method for identification of cysteine S-nitrosylation sites in complex protein mixtures. Proc Natl Acad Sci USA 103:1012–1017

Hara MR, Agrawal N, Kim SF, Cascio MB, Fujimuro M, Ozeki Y, Takahashi M, Cheah JH, Tankou SK, Hester LD, Ferris CD, Hayward SD, Snyder SH, Sawa A (2005) S-nitrosylated GAPDH initiates apoptotic cell death by nuclear translocation following Siah1 binding. Nat Cell Biol 7:665–674

Hess DT, Matsumoto A, Kim SO, Marshall HE, Stamler JS (2005) Protein S-nitrosylation: purview and parameters. Nat Rev Mol Cell Biol 62:150–166

Hook DW, Harding JJ (1997) Inactivation of glyceraldehyde 3-phosphate dehydrogenase by sugars, prednisolone-21-hemisuccinate, cyanate and other small molecules. Biochim Biophys Acta 1362:232–242

Hu H, Yang WT (2008) Free energies of chemical reactions in solution and in enzymes with ab initio quantum mechanics/molecular mechanics methods. Annu Rev Phys Chem 59:573–601

Ilbert M, Horst J, Ahrens S, Winter J, Graf PC, Lilie H, Jakob U (2007) The redox-switch domain of Hsp33 functions as dual stress sensor. Nat Struct Mol Biol 14:556–563

Iqbalsyah TM, Moutevelis E, Warwicker J, Errington N, Doig AJ (2006) The CXXC motif at the N terminus of an alpha-helical peptide. Protein Sci 15:1945–1950

Jakob U, Muse W, Eser M, Bardwell JC (1999) Chaperone activity with a redox switch. Cell 96:341–352

Jakob U, Eser M, Bardwell JC (2000) Redox switch of hsp33 has a novel zinc-binding motif. J Biol Chem 275:38302–38310

Jiang B, Tang G, Cao K, Wu L, Wang R (2010) Molecular mechanism for H(2)S-induced activation of K(ATP) channels. Antioxid Redox Signal 12:1167–11678

Jobson RW, Dehne-Garcia A, Galtier N (2010) Apparent longevity-related adaptation of mitochondrial amino acid content is due to nucleotide compositional shifts. Mitochondrion 10:540–547

Johansen D, Ytrehus K, Baxter GF (2006) Exogenous hydrogen sulfide (H2S) protects against regional myocardial ischemiareperfusion injury: evidence for a role of KATP channels. Basic Res Cardiol 101:53–60

Jordan IK, Kondrashov FA, Adzhubei IA, Wolf YI, Koonin EV, Kondrashov AS, Sunyaev S (2005) A universal trend of amino acid gain and loss in protein evolution. Nature 433:633–638

Kamerlin SC, Haranczyk M, Warshel A (2009) Progress in ab initio QM/MM free-energy simulations of electrostatic energies in proteins: accelerated QM/MM studies of pKa, redox reactions and solvation free energies. J Phys Chem B 113:1253–1272

Kim HY, Gladyshev VN (2005) Different catalytic mechanisms in mammalian selenocysteine- and cysteine-containing methionine-R-sulfoxide reductases. PLOS Biol 3:e375

Klomsiri C, Karplus PA, Poole LB (2011) Cysteine-based redox switches in enzymes. Antioxid Redox Signal 14:1065–1077

Klug A (2010) The discovery of zinc fingers and their development for practical applications in gene regulation and genome manipulation. Q Rev Biophys 43:1–21

Kröncke KD, Klotz LO (2009) Zinc fingers as biologic redox switches? Antioxid Redox Signal 11:1015–1027

Kumsta C, Jakob U (2009) Redox-regulated chaperones. Biochemistry 48:4666–4676

Laurie AT, Jackson RM (2005) Q-SiteFinder: an energy-based method for the prediction of protein-ligand binding sites. Bioinformatics 21:1908–1916

Lee BC, Lobanov AV, Marino SM, Kaya A, Seravalli J, Hatfield DL, Gladyshev VN (2011) A 4-selenocysteine, 2-selenocysteine insertion sequence (SECIS) element methionine sulfoxide reductase from Metridium senile reveals a non-catalytic function of selenocysteines. J Biol Chem 286:18747–18755

Leonard SE, Reddie KG, Carroll KS (2009) Mining the thiol proteome for sulfenic acid modifications reveals new targets for oxidation in cells. ACS Chem Biol 4:783–799

Lin HH, Tseng LY (2010) DBCP: a web server for disulfide bonding connectivity pattern prediction without the prior knowledge of the bonding state of cysteines. Nucleic Acids Res 38: W503–W507

Lin CT, Lin KL, Yang CH, Chung IF, Huang CD, Yang YS (2005) Protein metal binding residue prediction based on neural networks. Int J Neural Syst 15:71–84

Maret W (2006) Zinc coordination environments in proteins as redox sensors and signal transducers. Antioxid Redox Signal 8:1419–1441

Marino SM, Gladyshev VN (2009) A structure-based approach for detection of thiol oxidoreductases and their catalytic redox-active cysteine residues. PLOS Comput Biol 5:e1000383

Marino SM, Gladyshev VN (2010a) Cysteine function governs its conservation and degeneration and restricts its utilization on protein surfaces. J Mol Biol 404:902–916

Marino SM, Gladyshev VN (2010b) Structural analysis of cysteine S-nitrosylation: a modified acid-based motif and the emerging role of trans-nitrosylation. J Mol Biol 395:844–859

Marino SM, Gladyshev VN (2011a) Redox biology: computational approaches to the investigation of functional cysteine residues. Antioxid Redox Signal 15:135–146

Marino SM, Gladyshev VN (2011b) Analysis and functional prediction of reactive cysteine residues. J Biol Chem 287:4419–4425

Mark S, Formaneck GL, Zhang X, Cui Q (2002) Calculating accurate redox potentials in enzymes with a combined QM/MM free energy perturbation approach. J Theor Comput Chem 1:53–67

Michelet L, Zaffagnini M, Vanacker H, Le Maréchal P, Marchand C, Schroda M, Lemaire SD, Decottignies P (2008) In vivo targets of S-thiolation in Chlamydomonas reinhardtii. J Biol Chem 283:21571–21578

Moosmann B (2011) Respiratory chain cysteine and methionine usage indicate a causal role for thiyl radicals in aging. Exp Gerontol 46:164–169

Moosmann B, Behl C (2008) Mitochondrially encoded cysteine predicts animal lifespan. Aging Cell 7:32–46

Nagahara N (2010) Intermolecular disulfide bond to modulate protein function as a redox-sensing switch. Amino Acids 41:59–72

Nelson KJ, Knutson ST, Soito L, Klomsiri C, Poole LB, Fetrow JS (2011) Analysis of the peroxiredoxin family: using active-site structure and sequence information for global classification and residue analysis. Proteins 79:947–964

Ondrechen MJ, Clifton JG, Ringe D (2001) THEMATICS: a simple computational predictor of enzyme function from structure. Proc Natl Acad Sci USA 98:12473–12478

Paget MS, Buttner MJ (2003) Thiol-based regulatory switches. Annu Rev Genet 37:91–121

Parr RG, Yang W (1995) Density-functional theory of the electronic structure of molecules. Annu Rev Phys Chem 46:701–728

Passerini A, Frasconi P (2004) Learning to discriminate between ligand-bound and disulfide-bound cysteines. Protein Eng Des Sel 17:367–373

Passerini A, Punta M, Ceroni A, Rost B, Frasconi P (2006) Identifying cysteines and histidines in transition-metal-binding sites using support vector machines and neural networks. Proteins 65:305–316

Pearson RG, Parr RG (1983) Absolute hardness: companion parameter to absolute electronegativity. J Am Chem Soc 105:7512–7516

Poole LB, Karplus PA, Claiborne A (2004) Protein sulfenic acids in redox signaling. Annu Rev Pharmacol Toxicol 44:325–347

Roos G, Messens J (2011) Protein sulfenic acid formation: from cellular damage to redox regulation. Free Radic Biol Med 51:314–326

Roos G, Foloppe N, Van Laer K, Wyns L, Nilsson L, Geerlings P, Messens J (2009a) How thioredoxin dissociates its mixed disulfide. PLOS Comput Biol 5:e1000461

Roos G, Geerlings P, Messens J (2009b) Enzymatic catalysis: the emerging role of conceptual density functional theory. J Phys Chem B 113:13465–13475

Salsbury FR Jr, Knutson ST, Poole LB, Fetrow JS (2008) Functional site profiling and electrostatic analysis of cysteines modifiable to cysteine sulfenic acid. Protein Sci 17:299–312

Sanchez R, Riddle M, Woo J, Momand J (2008) Prediction of reversibly oxidized protein cysteine thiols using protein structure properties. Protein Sci 17:473–481

Schindeldecker M, Stark M, Behl C, Moosmann B (2011) Differential cysteine depletion in respiratory chain complexes enables the distinction of longevity from aerobicity. Mech Ageing Dev 132:171–179
Schmidt am Busch M, Knapp EW (2005) One-electron reduction potential for oxygen- and sulfur-centered organic radicals in protic and aprotic solvents. J Am Chem Soc 127:15730–15737
Schymkowitz J, Borg J, Stricher F, Nys R, Rousseau F, Serrano L (2005) The FoldX web server: an online force field. Nucleic Acids Res 33:W382–W388
Shenton D, Grant CM (2003) Protein S-thiolation targets glycolysis and protein synthesis in response to oxidative stress in the yeast Saccharomyces cerevisiae. Biochem J 374:513–519
Shu N, Zhou T, Hovmöller S (2008) Prediction of zinc-binding sites in proteins from sequence. Bioinformatics 24:775–782
Sigrist CJA, Cerutti L, Hulo N, Gattiker A, Falquet L, Pagni M, Bairoch A, Bucher P (2002) PROSITE: a documented database using patterns and profiles as motif descriptors. Brief Bioinform 3:265–274
Soito L, Williamson C, Knutson ST, Fetrow JS, Poole LB, Nelson KJ (2011) PREX: PeroxiRedoxin classification indEX, a database of subfamily assignments across the diverse peroxiredoxin family. Nucleic Acids Res 39:D332–D337
Tosatto SC, Bosello V, Fogolari F, Mauri P, Roveri A, Toppo S, Flohé L, Ursini F, Maiorino M (2008) The catalytic site of glutathione peroxidases. Antioxid Redox Signal 10:1515–1526
Trifonov EN (2004) The triplet code from first principles. J Biomol Struct Dyn 22:1–11
Weerapana E, Wang C, Simon GM, Richter F, Khare S, Dillon MB, Bachovchin DA, Mowen K, Baker D, Cravatt BF (2010) Quantitative reactivity profiling predicts functional cysteines in proteomes. Nature 468:790–795
Winterbourn CC, Hampton MB (2008) Thiol chemistry and specificity in redox signaling. Free Radic Biol Med 45:549–561
Winterbourn CC, Metodiewa D (1999) Reactivity of biologically important thiol compounds with superoxide and hydrogen peroxide. Free Radic Biol Med 27:322–328
Wood ZA, Schroder E, Robin Harris J, Poole LB (2003) Structure, mechanism and regulation of peroxiredoxins. Trends Biochem Sci 28:32–40
Wu H, Ma BG, Zhao JT, Zhang HY (2007) How similar are amino acid mutations in human genetic diseases and evolution. Biochem Biophys Res Commun 362:233–237
Wu C, Liu T, Chen W, Oka S, Fu C, Jain MR, Parrott AM, Baykal AT, Sadoshima J, Li H (2010) Redox regulatory mechanism of transnitrosylation by thioredoxin. Mol Cell Proteomics 9:2262–2275
Xu XM, Turanov AA, Carlson BA, Yoo MH, Everley RA, Nandakumar R, Sorokina I, Gygi SP, Gladyshev VN, Hatfield DL (2010) Targeted insertion of cysteine by decoding UGA codons with mammalian selenocysteine machinery. Proc Natl Acad Sci USA 107:21430–21434
Yang G, Sun X, Wang R (2004) Hydrogen sulfide-induced apoptosis of human aorta smooth muscle cells via the activation of mitogen-activated protein kinases and caspase-3. FASEB J 18: 1782–1784
Zhang FL, Casey PJ (1996) Protein prenylation: molecular mechanisms and functional consequences. Annu Rev Biochem 65:241–269

Chapter 8
Redox Regulation in Plants: Glutathione and "Redoxin" Related Families

J.P. Jacquot, K.J. Dietz, N. Rouhier, E. Meux, P.A. Lallement, B. Selles, and A. Hecker

Abstract Photosynthetic organisms and especially terrestrial plants contain expanded redox regulatory pathways. Proteins involved in these redox chains and detoxification reactions often evolved on a thioredoxin fold structural basis. We discuss in this review the proteins associated with this superfamily, their modes of reduction/regeneration and their known protein targets, focusing in particular on the protein disulfide isomerase and glutathione transferase families. We also discuss the evolution of the redoxin and target families along the plant kingdom and provide information on the systems biology of the redox circuitry.

8.1 General Background

In general, redox reactions involve changes in the oxidoreduction state of participating compounds. Single redox reactions are combined to redox chains or pathways. In this respect, the electron transfer chain components are redox actors, with transitions at the oxidation state of metals like copper and iron for example. In this review, however, we will consider only the oxidation/reduction reactions that involve cysteine or methionine residues. In this context, the most documented redox reactions in plants concern the dithiol-disulfide exchange cascades involving components of the thioredoxin (TRX) and glutaredoxin (GRX) systems. Some

J.P. Jacquot (✉) • N. Rouhier • E. Meux • P.A. Lallement • B. Selles • A. Hecker
Interactions Arbres Microorganismes, IFR 110 EFABA, Faculté des Sciences, Unité Mixte de Recherches 1136 INRA, Université de Lorraine, Vandoeuvre Cedex, France
e-mail: j2p@univ-lorraine.fr; nrouhier@univ-lorraine.fr; edgar.meux@univ-lorraine.fr; pierre-alexandre.lallement@univ-lorraine.fr; benjamin.selles@univ-lorraine.fr; arnaud.hecker@univ-lorraine.fr

K.J. Dietz
Plant Biochemistry and Physiology, Bielefeld University, 33501 Bielefeld, Germany
e-mail: karl-josef.dietz@uni-bielefeld.de

U. Jakob and D. Reichmann (eds.), *Oxidative Stress and Redox Regulation*, DOI 10.1007/978-94-007-5787-5_8,

glutathione S-transferases (GST) have the capacity to reduce disulfides in the presence of glutathione. Likewise, protein disulfide isomerases (PDI) belong to that category being able to oxidize, reduce or isomerize disulfide bonds. Besides these dithiol-disulfide exchange reactions, we will also consider those reactions, which are catalyzed by peroxiredoxins (PRX) or methionine sulfoxide reductases (MSR) and in which the redox state of catalytic cysteines varies. Overall, the changes in cysteine oxidoreduction state are essential for the catalysis of certain enzymes (e.g., formation of sulfenic acid on the catalytic cysteine of methionine sulfoxide reductases and peroxiredoxins) but they can also induce structural transitions needed to control the catalytic rates of enzymes (e.g., dithiol-disulfide control of fructose-1,6-bisphosphatase or NADP malate dehydrogenase in the chloroplasts) and of many other target proteins with diverse functions in transcription, translation, signalling and regulation.

8.2 Why Is There Redox Regulation in Plants?

It has become apparent in the last decades that redox reactions, as defined above, are particularly exacerbated in plants. For example, there are between 3 and 6 thioredoxin and glutaredoxin genes in non-photosynthetic prokaryotes and mammalian systems but about 40 genes of each family in photosynthetic eukaryotes. The PDI, MSR and PRX families are also extended but to a minor extent. The reasons for this expansion of redoxins are probably multiple. Part of it could be related to the existence of specialized tissues in plants, like the phloem and xylem for transport, epidermal and stomatal cells for leaf protection and gas exchange, stamen and pistil for reproduction etc. A more likely explanation is related to the physiology of plants. Unlike animal cells, plants contain plastids/chloroplasts, which play an essential energetic role by providing ATP and NADPH needed for CO_2 fixation. In animal cells, the energetic demands are mostly met by mitochondria, which are also present in plant cells. Chloroplasts not only fix CO_2 but they also release O_2 in a process called water photolysis at the level of photosystem II. As a result, in oxygenic photosynthetic organisms, in addition to mitochondria, NAD(P)H oxidases and peroxisomes, the chloroplast is an additional site of reactive oxygen species (ROS) generation. In plant mitochondria, ROS are generated at the level of complex I, ubiquinone and complex III (Navrot et al. 2007). In the chloroplasts, ROS are essentially generated at the level of photosystem II and photosystem I (Asada 2006). One plausible reason for the need of elevated redoxin systems in oxygenic photosynthetic organisms might therefore be related to the increased production of ROS in the light. Another specificity of plants is that unlike animal systems, they rely on light as an energy source, and this energy is converted into utilisable chemical energy by chlorophyll, a pigment specifically present in the chloroplasts. The presence of this specific additional compartment in plants has several consequences. First, for a number of reasons, it is essential to switch on selected enzymes in the light and turn them off in the dark. This could prevent futile cycles, like the one potentially occurring between fructose-1,6-bisphosphatase (FBPase) and phosphofructokinase. Would such a futile cycle occur, it would induce the consumption of ATP with no net gain for the plant. Another observation is that plants lack in general efficient alternative splicing that allows in animal cells for the multiplication of isoforms from a single gene. Thus, possible explanations for the expansion in the number of genes would be first to keep "security copies" of a given gene, second to generate variants with improved selectivity or efficiency without having to rely on alternative splicing. Another relevant observation is that there is much duplication in plant genomes, which are often polyploïd. Finally the sessile lifestyle of plants requires integration of many environmental input parameters to optimize plant performance and fitness. It appears that redox cues are sensed and extensively exploited to modulate or switch developmental and acclimation processes.

8.3 Glutathione and Redoxin-Linked Pathways in Plants

Several regulatory or detoxification pathways depend on glutathione (GSH) or TRX and other thiol-linked systems in plants. In general, the proteins that have the capacity to mediate dithiol-disulfide exchange reactions (redox transmitters) possess a so-called TRX fold, namely a central pleated β sheet with 5β-strands surrounded by three or more α-helices. In TRX itself (classically about 110 amino acids), the active site with the conserved sequence WC[G/P]PC is rather surface exposed which favors the reactivity of the catalytic cysteine. Its reactivity is further enhanced by an adjacent helix dipole effect, which lowers the pKa of the thiol. In plants, TRXs are reduced either by light and the ferredoxin-thioredoxin system in plastids, or by NADPH-dependent thioredoxin reductases (NTR) in the cytosol and mitochondria. An alternative pathway in plastids consists of a fused enzyme (NtrC), composed of a NTR and a Trx module (Spínola et al. 2008). Ferredoxin thioredoxin reductase (FTR) is essentially restricted to photosynthetic organisms and is absent in mammals and non-photosynthetic organisms. This heterodimeric enzyme is uniquely shaped to accommodate on one side the substrate ferredoxin and on the other side TRX. At the interface between the two subunits of FTR lies a critical [4Fe4S] cluster and a catalytic disulfide in close proximity (Dai et al. 2007). The convergent structural evolution between FTR and NTR has been discussed in a recent review together with an exhaustive listing of the thioredoxin reductases in photosynthetic organisms (Jacquot et al. 2009). The thioredoxin structural model has given rise to a number of variants, which differ in their active site sequences, their modular organization, and/or their regeneration mechanisms. GRXs are proteins with a size similar to TRXs but with active sites that have been modified into C[P/G/S][Y/F][C/S] (class 1), CGFS (class 2), CCx[C/S] (class 3) or Cxx[C/S] (class 4) (Couturier et al. 2009). Most GRXs are reduced by the tripeptide glutathione, which itself is maintained in its reduced form via NADPH-dependent glutathione reductase. As mentioned previously, in land plants, the TRX and GRX repertoires have been greatly enhanced with about 40 genes of each category identified. TRXs, GRXs and their reducing

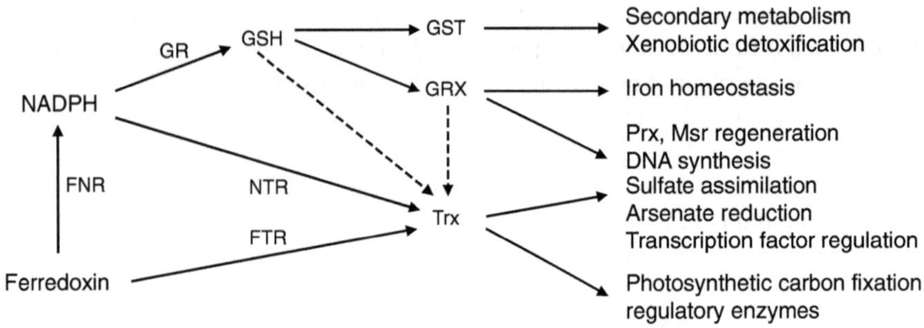

Fig. 8.1 Major functions associated with TRX and GSH/GRX reducing pathways. TRXs are usually reduced by the NADPH/NADPH-thioredoxin reductases (*NTR*) or the Ferredoxin/ferredoxin thioredoxin reductase (*FTR*) systems but a few isoforms have been shown to rely on GSH and/or GRX for their regeneration (Koh et al. 2008). GSH, maintained in its reduced form by the NADPH/glutathione reductase (*GR*) system, provides electrons to both GSTs and GRXs. FNR: ferredoxin NADP reductase

pathways have been discussed extensively in a number of recent reviews (Dietz 2003; Gelhaye et al. 2005; Rouhier et al. 2008; Chibani et al. 2009; Couturier et al. 2009; Meyer et al. 2009) and we will not detail them any further in this section. Figure 8.1 depicts the reducing systems of TRXs and GRXs in plants and some of the known functions associated to each system. The reader is invited to consult these papers and the chapter by Collet and Messens for more information on the biochemical and structural properties of these two protein families.

At least two other types of enzymes, glutathione S-transferases and protein disulfide isomerases, contain thioredoxin domains. The protein disulfide isomerases are located predominantly in the ER and the most common forms contain four thioredoxin-like modules, with an overall organization of the a-b-b′-a′ type. The a and a′ domains display CGHC active sites whereas the b and b′ modules present a TRX fold but lack the active sites. However, they are essential for some aspects of PDI activities, particularly the b′ module, which is involved in substrate recognition (Byrne et al. 2009). In the ER, the functions of PDIs are to create and isomerise disulfide bonds to ensure correct protein folding. Consequently, PDIs have both an oxidizing and reducing activity. After PDI oxidizes its protein substrates, its own disulfide bonds become reduced and need to be reoxidized. Based on results published on human and yeast proteins, different pathways apparently lead to PDI oxidation. The first candidates are the sulfhydryl oxidases Ero1 α and β flavoproteins, which catalyze the formation of disulfide bonds on PDIs with concomitant reduction of oxygen to hydrogen peroxide (Sevier and Kaiser 2008). The produced H_2O_2 can in turn be reduced by ER-located thiol-peroxidases, such as 2-Cys peroxiredoxins or glutathione peroxidases, which, once oxidized, also contribute to PDI re-oxidation (Tavender et al. 2010; Nguyen et al. 2011). Hence, the consumption of one oxygen molecule can serve to form two disulfides, allowing a tight control of ROS accumulation into the ER. Finally, it was also proposed that the proportion of oxidized glutathione, which is elevated in the ER compared to other subcellular compartments, could be thermodynamically sufficient to reoxidize PDI active sites. It should be mentioned that these different modes of reoxidation might not equally target all PDI isoforms and that there might be specificities linked to the variety of PDI representatives, especially in photosynthetic organisms. A recent genomic analysis of PDIs in photosynthetic organisms has demonstrated that apart from the classical multimodular L-type PDI (that contains the a-b-b′-a′ modules), 8 additional PDI classes exist, which differ in their modular composition, some being monomodular, some trimodular and some others containing thioredoxin unrelated modules. Six of the PDI classes are present in terrestrial plants but three of them are restricted to algae. The detailed phylogenetic analysis can be found in Selles et al. (2011).

Another example of enzymes, which contain a TRX-like module, is glutathione-S-transferase. These enzymatic systems are often ignored although some of these enzymes possess GRX-like and peroxidase activities. For this reason, we will give a comprehensive overview of these enzymes in plants in this review. Glutathione transferases, better known as glutathione S-transferases, have been identified more than 40 years ago in rat liver (Booth et al. 1961) and in plants through their roles in the detoxification of herbicides in annual species (Lamoureux et al. 1970; Shimabukuro et al. 1971). GSTs are present in both eukaryotes and prokaryotes and constitute a multigenic superfamily that is widely distributed in nature. GSTs are the main players involved in the metabolism of electrophilic compounds within the cell. Through the addition or the substitution of a glutathione molecule (GSH, a non-ribosomally synthesized γ-Glu–Cys–Gly tripeptide) to an electrophilic center present in a small acceptor molecule, these enzymes are able to metabolize a broad range of endogenous and exogenous substrates for their detoxification, subsequent transport, export or secretion. In addition, they are also able to reduce peroxides and to recycle antioxidant molecules (ascorbate in particular) thereby providing protection against endogenous or xenobiotic substances with genotoxic and carcinogenic effects. GSTs are also involved in several other cell functions through their capacity to bind non-catalytically to a large number of endogenous and exogenous compounds.

Based on their primary sequence, 3D structure, function and cellular localization, GSTs have been initially classified into three main subfamilies: (i) the cytosolic (or soluble) GSTs, which form the largest family, (ii) the mitochondrial and peroxisomal GSTs also known as kappa GSTs, and (iii) the microsomal GSTs belonging to the superfamily of the trimeric Membrane Associated Proteins involved in Eicosanoids and Glutathione metabolism (MAPEG) (Pearson 2005). A further distinct family is represented by the plasmid-encoded bacterial Fosfomycin-resistance GSTs (FosA), a manganese-containing metalloglutathione transferase conferring resistance to the antibiotic fosfomycin (Bernat et al. 1997). The ambiguity here is that kappa GSTs are also soluble enzymes and that many GSTs included in the cytosolic/soluble class are located in organelles (nuclei, mitochondria, peroxisomes and plastids in the case of plants). Hence, although cytosolic GSTs, kappa GSTs and MAPEGs have indeed evolved separately and have to be distinguished, this classification is clearly no longer fitting (Sheehan et al. 2001).

Kappa GSTs have been discovered and characterized in both prokaryotes and eukaryotes and fulfill, in the latter, specific functions associated with mitochondria and peroxisomes, particularly in energy production and lipid metabolism, respectively (Morel and Aninat 2011). Microsomal MAPEG GSTs are enzymes, which differ substantially in size and structure from other GST families, having an average length of 150 amino acids comprising four possible membrane spanning domains and a less than 10% identity with other GST classes (Jakobsson et al. 1999, 2000; Bresell et al. 2005). In mammals, these membrane-bound trimers are involved in eicosanoid, prostaglandin and leukotriene synthesis by catalyzing GSH-dependent transferase and peroxidase reactions (Jakobsson et al. 1999).

Other GST classes initially classified as members of the cytosolic GST subfamily are present in mammals, plants, insects, parasites, fungi and bacteria. They are notably involved in primary and secondary metabolisms, and in stress response through their herbicide and xenobiotic detoxification capacities. Whereas some GST classes are found only in certain kingdoms or phyla (lambda (L), phi (F) and tau (U) in plants; delta (D), epsilon (E) in insects; beta (B) in prokaryotes, rho (R) in fishes; nu (N) in nematodes; gamma (G), etherases (GTE) and glutathione transferase (GTT) in fungi), some others are present in several kingdoms (mu (M), alpha (A), pi (P), theta (T), sigma (S), zeta (Z), omega (O), Ure2p-like as well as tetrachlorohydroquinone dehalogenase (TCHQD) and S-glutathionyl hydroquinone reductase (GHR) classes) (Table 8.1).

Although GSTs share less than 30% identity among different classes, almost all GSTs are homo- or heterodimeric enzymes with subunits of 25–35 kDa (i.e., 200–250 amino acids). Structural analyses have shown that GSTs can be divided into two domains: a highly conserved N-terminal GSH-binding domain (G-site), which is structurally similar to, and has evolved from, the TRX-fold (Atkinson and Babbitt 2009), and a less conserved C-terminal alpha-helical domain (H-site), which allows binding of a wide range of hydrophobic co-substrates (Sheehan et al. 2001). Amino acids from both domains form the catalytic site of the protein. Most typical GSTs have serine, tyrosine or cysteine as catalytic residue, but some atypical GSTs seem to use arginine or asparagine. In dehydroascorbate reductase (DHAR) and lambda GSTs, for example, the residue responsible for the enzymatic activity is a cysteine, included in a CPxA motif reminiscent of the CPxC motif found in GRXs. As a consequence, the presence of this cysteine completely changes the biochemical properties of these GSTs. Instead of promoting the formation of the reactive thiolate anion of GSH necessary for its reaction with electrophilic compounds, the presence of the cysteine provides the capability to reduce glutathionylated substrates in a manner similar to GRXs.

In plants, GSTs (encoded by 83 genes in *Populus trichocarpa* and 54 in *Arabidopsis thaliana*) can be divided into at least seven classes: theta, zeta, phi, tau, lambda, DHAR and tetrachlorohydroquinone dehalogenase (TCHQD) (Smith et al. 2004; Basantani and Srivastava 2007; Lan et al. 2009). Genes encoding GSTs are generally close to each other and grouped by class in genomes forming gene clusters. In *A. thaliana*, GSTs are divided into 34 class-specific clusters, with isolated genes representing less than 30% of the GST gene panel (Dixon et al. 2002b).

Table 8.1 Classification of GSTs

GST class	Symbol	Origin	Subunit structure	Catalytic residue
Alpha	A	Mammals	Dimer	Tyr
Beta	B	Bacteria	Dimer	Cys
Gamma	G	Fungi	Dimer	Tyr
Delta	D	Insects	Dimer	Ser
Epsilon	E	Insects	Dimer	Ser
Zeta	Z	Animals, plants and bacteria	Dimer	Ser
Theta	T	Animals, plants and bacteria	Dimer	Ser
Kappa	K	Eukaryotes and prokaryotes	Dimer	Ser
Lambda	L	Plants	Monomer	Cys
Mu	M	Mammals	Dimer	Tyr
Nu	N	Nematodes	?	Tyr
Pi	P	Mammals	Dimer	Tyr
Rho	R	Fishes	Dimer	?
Sigma	S	Animals, red and brown algae	Dimer	Tyr
Tau	U	Plants	Dimer	Ser
Phi	F	Plants	Dimer	Ser
Chi	C	Bacteria	Dimer	?
Omega	O	Mammals, fungi, insects	Dimer	Cys
DHAR	–	Plants	Monomer	Cys
GHR	–	Fungi and prokaryotes	Dimer	Cys
GTE	–	Fungi	Dimer	?
GTT	–	Fungi	Dimer	?
MAPEG	–	Eukaryotes and bacteria	Trimer	Arg
TCHQD	–	Bacteria, plants	Dimer	?
Ure2p-like	–	Fungi and bacteria	Dimer	Asn

GSTs are classified according to several criteria: (i) two GSTs belong to the same class if they share more than 40% identity, (ii) less than 25% identity separate classes in mammals, and less than 20% separate classes of GST in plants. When the percentages of identity are between 25 and 40%, other techniques are used to refine the classification such as gene structure and their number in a given species, size and position of introns and exons, immunological properties and finally enzyme activity. A few other protein families are usually classified as GSTs (EF1Bγ, MAK16, CLIC), but in general this is related to structural similarities, not to the existence of a glutathione-dependent activity. This table has been essentially constructed from the references, Pearson (2005) and Morel et al. (2009)

As an example, seven Tau GST genes are located on a fragment of 14 kb on chromosome 2 (Lin et al. 1999). In rice, less than 25% of GST genes (15 genes) are present in isolated loci in the genome. The most important cluster includes 23 Tau GSTs on a 239 kb genomic fragment of chromosome 10 (Soranzo et al. 2004). In poplar, 37 of the full-length tau GSTs and 20 tau-type fragments are arranged in six clusters (clusters I to V and VII) on five chromosomes (1, 8, 10, 11 and 19) and 4 phi GST genes constitute one cluster (cluster VI) on chromosome 14. Members of minor GST classes are sparsely distributed at single loci on different chromosomes (Lan et al. 2009). Interestingly in mammals, human alpha, mu and theta (Morel et al. 2002) and murine pi GST genes (Henderson et al. 1998) are located on specific sections of the genome whereas delta GSTs form a single compact block of genes in insect genomes (Ding et al. 2003). The almost systematic presence of GST gene clusters in plant and animal genomes reveals common evolutionary mechanisms allowing the generation of gene redundancies.

Phi and tau GSTs are the most abundant in plants and are plant-specific. Tau GSTs catalyze conjugation reactions not only on conventional substrates but also on herbicides, such as atrazine, metolachlor, fluorodifen and thiocarbamates (Edwards and Dixon 2005). These GSTs seem to be involved in cell protection by playing a role in tolerance to environmental stress by enhancing the detoxification of herbicides (Kawahigashi 2009). Tau GSTs are also able to transfer GSH on anthocyanins, allowing their import into the vacuole by ABC transporters (Edwards et al. 2000). Members of the phi class have peroxidase and transferase activity, notably on herbicides (Marrs 1996). In addition, expression of phi GSTs is induced in response to auxin. Interestingly, AtGSTF2, the best-studied Phi GST of *A. thaliana*, is strongly induced by oxidative stress and phytohormone treatment and it is able to bind auxin non-enzymatically. Thus, this protein is involved not only in response to stress, but also in growth and development (Gong et al. 2005; Smith et al. 2003). A GST class present both in plants and animal systems is the theta class. Theta GSTs function as GSH-dependent peroxidases using glutathione to reduce organic hydroperoxides into the corresponding alcohols during oxidative stress (Frova 2003). Zeta GSTs, highly conserved GSTs in eukaryotes, act as GSH-dependent isomerases in the catabolism of tyrosine by catalyzing the isomerization of maleylacetoacetate to fumarylacetoacetate (Dixon et al. 2000). In contrast, lambda and DHAR classes show neither transferase nor peroxidase activity but exhibit GRX-like activities through their capacity to reduce disulfides (hydroxyethyldisulfide) and dehydroascorbate using, as described below, a conserved reactive cysteinyl residue. Thus, like GRXs, they might be involved in the maintenance of the cellular redox balance (Dixon et al. 2002a). The ability of GSTs to bind a large range of small molecules led to hypothesize that these enzymes could be involved in various metabolic pathways. Supporting this view, *Arabidopsis* lambda GSTs were recently shown to bind S-glutathionylated flavonols, porphyrins or tocopherols (Dixon and Edwards 2010; Mueller et al. 2000; Dixon et al. 2011). However, their exact functions *in planta* remain unknown, mostly because the study of knock-out mutants is complex considering the existence of multigenic families. TCHQD, initially discovered in the soil bacterium *Sphingobium chlorophenolicum*, which can grow on pentachlorophenol as carbon source, are found in animals, terrestrial plants and bacteria. In *S. chlorophenolicum*, TCHQD catalyses the reductive dehalogenation of tetrachlorohydroquinone to trichlorohydroquinone and then to 2,6-dichlorohydroquinone during the biodegradation of pentachlorophenol (Xun et al. 1992), an environmental xenobiotic used as fungicide for wood preservation. Members of the TCHQD class, which are closely related to zeta class, present an unusual ability to metabolize chlorinated xenobiotics. Finally, plants also possess members with similarities to microsomal MAPEG or prostaglandin E synthase families, but their function and properties have not been explored to date.

8.4 Thioredoxin Targets

A number of plant thioredoxin targets are extremely well described both in terms of physiology and structure/function. Most of these redox-regulated target enzymes are involved in carbon fixation and metabolism, and are located in plastids (Jacquot et al. 1997). Interestingly, many of these enzymes (e.g., NADP-malate dehydrogenase, fructose-1,6-bisphosphatase, CF1 ATPase) have non-regulatory counterparts in other subcellular compartments. Sequence comparisons have led to the identification of redox regulatory elements that contain critical cysteine residues (Schürmann and Jacquot 2000). These elements vary from one enzyme model to another, both in their position (sometimes as N- or C-terminal extensions, sometimes as insertions), their length (generally between 5 and 30 amino acids) and their amino acid sequence. The multiplication of known regulatory sequences and structures of redox-regulated enzymes led to the following concept: there is no "magic regulatory module" for redox control in enzymes. In fact, some of these enzymes do not even contain sequence insertions or additions but simply harbor cysteine residues that have been introduced at critical positions in their sequence. It appears that the amino acid sequence evolution of each enzyme responds to structural constraints specific for each individual case. 3D-structures have revealed at least for two of these enzymes (NADP-MDH and FBPase) that the regulatory disulfides are distant from the active site and that large molecular movements are needed to open the active sites in order for catalysis to proceed (Ruelland and Miginiac-Maslow 1999). As a consequence, these enzymes are reduced quickly via the FTR system but their catalytic activity appears on a much slower time scale. This property, linked to the additional observation that the rate of reduction is much slower than the rate of catalysis in these enzymes, classifies them as hysteretic enzymes (Hertig and Wolosiuk 1983). Of significant interest is the observation that redox regulation has been acquired along with the evolution of the green lineage and especially together with the onset of oxygenic photosynthesis. One very interesting example is NADP-MDH. In terrestrial plants, each of the two subunits of the homodimeric enzyme has amino acid extensions at both N- and C-terminus, each containing two regulatory cysteines (overall the dimeric enzyme contains four disulfide bonds in its oxidized, inactive state). In the green alga *Chlamydomonas reinhardtii*, however, the enzyme lacks the N-terminal disulfide but contains the C-terminal disulfide and is hence less tightly redox controlled. The *Chlamydomonas* NADP-MDH has been proposed to represent an evolutionary step leading to the terrestrial plant model, where full redox control has been achieved with the subsequent addition of the N-terminal disulfide (Lemaire et al. 2005). Other well-recognized targets of thioredoxins in plants participate in the detoxification of peroxides or repair of their molecular damages. PRXs are thiol peroxidases that are present in various plant cell compartments and detoxify a broad range of peroxides into alcohol, whereas methionine sulfoxide reductases are enzymes catalyzing the reduction of methionine sulfoxide into methionine. Atypical type II PRX, 2 Cys PRX, PRX Q and MSRA are all regenerated by TRXs (Rouhier et al. 2002, 2004,

2007). Interestingly, so-called glutathione peroxidases lack selenocysteines in plants and they are in fact thioredoxin-dependent enzymes (Navrot et al. 2006). A large number of additional thioredoxin targets have been identified by various methods, but principally by fishing them using TRX variants, in which the second cysteine of the active site has been substituted by site-directed mutagenesis into serine. These TRX engineered variants attack proteins with disulfides but cannot release their substrates. Client proteins are then dissociated with dithiothreitol and analyzed with proteomics techniques. This approach has led to the discovery of hundreds of putative targets involved in many different physiological pathways (Balmer et al. 2003, 2004; Montrichard et al. 2009). Some of these target enzymes have been confirmed biochemically but for many of them experimental confirmation is still lacking. A comforting observation is that a large number of these sequences contain conserved cysteines, which could potentially be involved in regulatory sites.

8.5 Glutaredoxin Targets

The first glutaredoxin-linked enzymatic system recorded is probably the ribonucleotide reductase in *E. coli*, and in fact GRX was discovered in a *trx* mutant strain still able to reduce ribonucleotides into deoxyribonucleotides (Holmgren 1976). It is likely that GRXs play a similar role in plants but there is not much evidence about it yet. A second enzymatic system linked to GRX is involved in sulphur assimilation. In terrestrial plants, the step generating reduced sulfide from adenosine phosphosulfate is catalysed by APS reductase. Higher plant APS reductases contain a built-in GRX module located at the C-terminus of the polypeptide (Bick et al. 1998). Glutaredoxin was then identified as a good electron donor to atypical type II PRXs (Rouhier et al. 2001, 2002). Interestingly, it was later found that *prx* and *grx* genes are fused in several pathogenic bacteria and cyanobacteria lending further support to this proposal (Rouhier and Jacquot 2003). GRX was also demonstrated to be a preferred electron donor to a specific subclass of B-type methionine sulfoxide reductases (Tarrago et al. 2009). Besides those "direct" biochemical observations, the need for GRX was also evaluated using a strategy similar to the one employed with TRXs, i.e. generation of GRX variants containing a single active site cysteine, formation of a stable disulfide between GRX and its targets, release of the target proteins with a reductant and identification of target proteins by mass spectrometry. This has led to the identification of nearly 100 potential targets, some of them being redundant with TRX but some others specific to GRX. This work confirmed PRXs as possible targets and also that β-carbonic anhydrase was indeed redox-regulated via GRX *in vitro* (Rouhier et al. 2005). Recent advances have indicated that in addition to their redox role, plant glutaredoxins especially of the CGFS type are involved in iron sulphur cluster assembly and iron homeostasis. We have not detailed these exciting developments in this chapter as they are included in the paper by Messens et al. in this volume. More details can also be obtained in recent reviews published by Rouhier et al. (2008) and Rouhier et al. (2010).

8.6 Evolution of the Redox Systems in Photosynthetic Organisms

Genome comparisons allow the identification of orthologous and paralogous genes and to construct robust phylogenetic relationships. This approach has been used extensively to identify and categorize components of plant redox systems. It became clear that the number of gene family members of redox proteins in plants usually exceeds the one found in non-photosynthetic multicellular organisms as impressively demonstrated when comparing TRX and GRX diversification in plants and humans. Thus, mammalian genomes code for two dithiol GRXs (Grx1 and 2) and two monothiol GRXs (Grx3 and Grx5) while 33 gene loci coding for GRXs and GRX domain-containing proteins have been identified in *A. thaliana* (Couturier et al. 2009). The largest expansion of GRX occurred in the CC-type GRX subfamily, which took over specific functions in developmental control and interact with transcription factors (Ziemann et al. 2009). Likewise 44 open reading frames for TRXs and TRX-like proteins in *A. thaliana* compare with two human TRXs. Regarding peroxiredoxins, a detailed comparison of the fully annotated genomes of *Arabidopsis thaliana*, *Oryza sativa*, *Populus trichocarpa*, *Chlamydomonas reinhardtii* and the cyanobacterium *Synechocystis PPC6803* allows us to define the minimum PRX inventory in plants and green alga (Table 8.2). Thus, plastids contain at least one 2-Cys PRX, one PRX Q and one type II PRX named PRX IIE. One 1-Cys PRX localizes to the higher plant nucleus and cytosol, the PRX IIF to the mitochondrion and another type II PRX to the cytosol (Dietz 2011). Thus 7–10 *prx* genes in higher plant genomes compare with 6 *prx* genes in humans. In contrast to some animal glutathione peroxidases (GPXs), plant GPXs do not employ a selenocysteine in their catalytic peroxidatic site but, like PRXs, use a Cys residue instead. Plant GPXs are localized in plastids, mitochondria, cytosol and the secretory pathway (Navrot et al. 2006). After each peroxide reduction reaction, PRXs and GPXs are oxidized and need to be regenerated by thiol-dependent electron donors such as GRX or TRX. Usually there is specificity or at least strong preference for specific electron donors, although type II PRXs generally accept electrons from both GRXs and TRXs (Rouhier et al. 2001). In conclusion, the present knowledge suggests that redox systems in plants have evolved to particular complexity, which is linked to the multiplicity of stress factors that act on sessile land plants during their life cycle.

8.7 Systems Biology of the Redox Regulatory Circuits

The thiol/disulfide redox system of the cell can be considered as a network of components linked by electron flow. In its central part, thiol/disulfide exchange reactions take place in a cascade like arrangement (Fig. 8.2). The thiol/disulfide network is connected to other redox reactions in metabolism, such as to photosynthetic

Table 8.2 Compilation of gene families encoding GRX, TRX, PRX, GPX and SRX in *Synechocystis pcc6803*, *Chlamydomonas reinhardtii*, *Arabidopsis thaliana*, *Populus trichocarpa* and *Oryza sativa*

Gene families		*S. pcc6803*		*C. reinhardtii*		*A. thaliana*		*P. trichocarpa*		*O. sativa*	
PRX	2-CysPrx	5	1	7	2+1[a]	9	2	9	2	7	1
	1-CysPrx		1		–	10[b]	1		1		1
	PrxIIB		–		1		3		2		1
	PrxIIE		1		1		(+1)		1		2
	PrxIIF		–		1		1		1		1
	PrxQ		2		1		1		2		1
GPX	Cys	2	2	5	3	8	8	6	6	5	5
	SeCys		–		2		—		—		
SRX				1		1		1		1	
TRX	Trx-f	4		10	2	44	2	44[c]	1	30[d]	
	Trx-m		1		1		4		8		
	Trx-x		1		1		1		1		
	Trx-y		1		1		2		2		
	Trx-z				1		1		1		
	Trx-h				2		8		7		
	Trx-o				1		2		1		
	Others		1		2		24		23		
GRX[e]	Class I (CPY/FC)	3[f]	2	7	2	33	6	38	6	29	5
	Class II (CGFS)		1		4		4		5		5
	Class III (CCxx)				1		21		24		17
	Class IV						2		3		2

[a]The *C. reinhardtii* genome codes for a fusion protein consisting of 2-CysPRX and TRX domains (Dayer et al. 2008)
[b]PrxIIA appears to be a non functional gene (Dietz 2011)
[c]Chibani et al. (2009)
[d]Nuruzzaman et al. (2008); the TRX assignment in *O. saliva* to subfamilies is still incomplete
[e]Ziemann et al. (2009)
[f]Marteyn et al. (2009). The numbers in the first column assigned to each organism give the total sum of family members, and the numbers in the second column depict the numbers of each isoform within the family

electron transport by means of electron input elements (Dietz and Pfannschmidt 2011). Redox transmitters, as described before, mediate electron transfer to redox target proteins. Ultimately the electrons are passed on to electron acceptors such as reactive oxygen species and reactive nitrogen species. During oxidative stress when ROS and RNS accumulate to high levels, they may directly react with some regulatory targets converting them into oxidized disulfide, sulfenylated, sulfinylated, sulfonylated, nitrosylated, nitrated or glutathionylated forms (Jacob and Anwar 2008). However, during regular metabolism, most ROS and RNS are safely detoxified minimizing uncontrolled redox reaction. This role is achieved

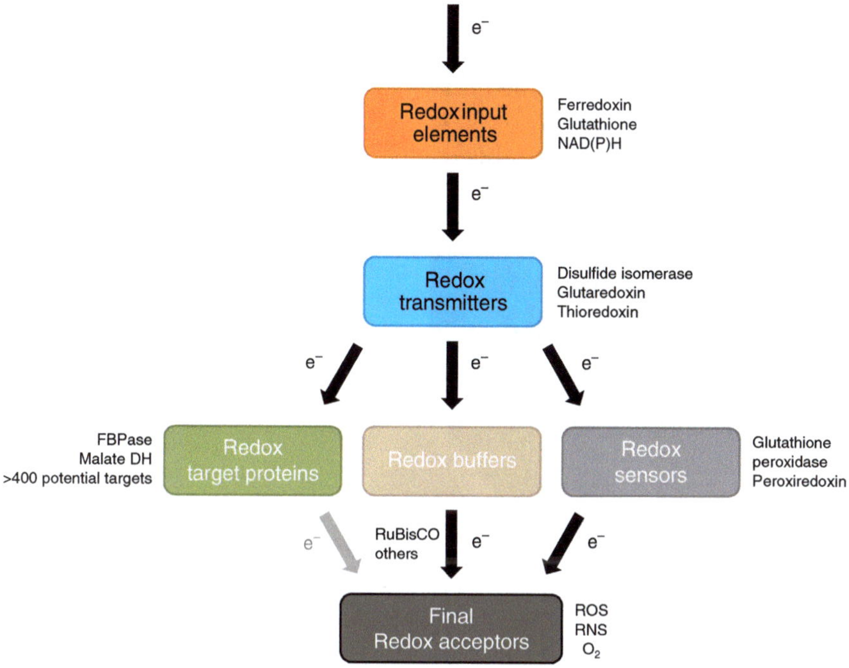

Fig. 8.2 Simplified schematics of the principle structure of the redox regulatory network of the cell. Redox input elements, transmitters, target proteins, sensors, buffers and ROS and RNS constitute a thiol/disulfide redox regulatory network that controls essentially all aspects of metabolism and development. Examples for each type of component are indicated

by the antioxidant systems of the cell. PRXs and GPXs are both antioxidants and elements of the thiol/disulfide network. Therefore, they function as redox sensors, reducing peroxides and withdrawing electrons from the thiol-based redox transmitters, thereby assisting the system to adjust to the prevailing redox condition of the cell. During more severe oxidative conditions, ROS and RNS not only react with protein thiols of these sensors or of less preferred regulated target proteins but also with other protein thiols that likely buffer ROS bursts. Thus, the concept of redox input elements, transmitters, targets, sensors and final electron acceptors, needs to be extended to redox buffer proteins (Fig. 8.2). Proteomics approaches have identified more than 400 potential thiol-oxidation targets in plants (Lindahl and Kieselbach 2009). The distinction between regulatory targets and redox buffer proteins is often difficult, as illustrated by ribulose-1,5-bisphosphate carboxylase/oxygenase (RubisCO). With its stromal concentration of 0.5 mmol/L and 8 large and 8 small subunits with 9 + 4 Cys residues in *A. thaliana*, RubisCO represents the largest thiol pool of any compound in the chloroplast. Moreno et al. (2008) titrated RubisCO with redox reagents and observed two effects; moderate oxidation inhibited RubisCO carboxylation activity while strong oxidation triggered proteolysis. The increasing wealth of information on plant redox regulation and

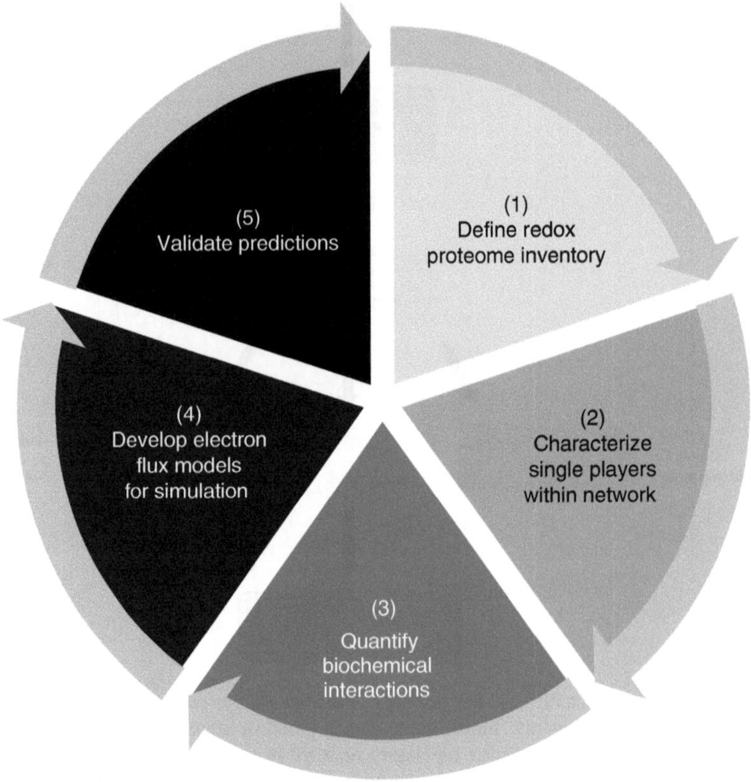

Fig. 8.3 Systems biology approach needed for complete understanding of the thiol/disulfide redox regulatory network (see text for details)

its unique complexity, calls for the next step in analytical pervasion. Figure 8.3 depicts the work flow of a typical systems biology approach, applied to redox biology (in this respect, theoretical and practical concepts concerning hubs and bottlenecks in signalling networks can be found in a recent review by Dietz et al. in 2010). As pointed out above, the basic inventory of participating proteins but also first sets of qualitative and quantitative interactions are available as well as genetic approaches for functional validation. In each work module, additional experimental efforts are still needed, for instance to identify less abundant thiol/dithiol transition proteins and to deepen our quantitative understanding on interactions. In addition, the development of descriptive qualitative and quantitative models for simulation and prediction are needed to understand the competition for electrons and the functionality of feedback and feed forward regulation achieved through the redox network. First models on H_2O_2 reduction by the ascorbate-dependent water-water cycle have been developed by Polle (2001) and were combined with photosynthesis by Laisk et al. (2006). However these models do not include the thiol-disulfide network as central controlling mechanism. Adimora et al. (2010) constructed a predictive model of the rat liver cell thiol system and analyzed steady-state fluxes. The simulation indicates an unequal flux through the thioredoxin-dependent and glutathione-dependent antioxidant pathways. In human cells, TRX is predicted to bear the major antioxidant defense load due to the function of peroxiredoxins and protein disulfide oxidation and their regeneration (Adimora et al. 2010). Similar attempts are needed for photosynthesizing cells.

8.8 Concluding Remarks

A nagging question remains in photosynthetic organisms: why are there so many TRX and GRX isoforms and what about their possible redundancy? We have given a number of arguments in this paper that provide possible answers to this question but many are awaiting experimental confirmation. The questions are of course difficult to answer because of the complexity of the redox network but also because genetic tools for gene knockout are lagging behind in land plants compared to yeast. In this respect, photosynthetic bacterial models such as *Synechocystis* are of interest. Another possibility would be to use *Physcomitrella patens*, a moss in which gene knockout is much more feasible. Of course high throughput gene sequencing will also help us compare the redoxin equipment along the evolution of the green lineage and we will certainly obtain useful information by comparing the photosynthetic genomes in the future. Finally, comparing redox regulation networks in perennial photosynthetic organisms (*e.g.*, trees) *versus* annual species will also be of interest, and comparing the poplar and *Arabidopsis* models is certainly a first step in this direction.

References

Adimora NJ, Jones DP, Kemp ML (2010) A model of redox kinetics implicates the thiol proteome in cellular hydrogen peroxide responses. Antioxid Redox Signal 13:731–743

Asada K (2006) Production and scavenging of reactive oxygen species in chloroplasts and their functions. Plant Physiol 141:391–396

Atkinson HJ, Babbitt PC (2009) Glutathione transferases are structural and functional outliers in the thioredoxin fold. Biochemistry 48:11108–11116

Balmer Y, Koller A, del Val G, Manieri W, Schürmann P, Buchanan BB (2003) Proteomics gives insight into the regulatory function of chloroplast thioredoxins. Proc Natl Acad Sci USA 100:370–375

Balmer Y, Vensel WH, Tanaka CK, Hurkman WJ, Gelhaye E, Rouhier N, Jacquot JP, Manieri W, Schürmann P, Droux M, Buchanan BB (2004) Thioredoxin links redox to the regulation of fundamental processes of plant mitochondria. Proc Natl Acad Sci USA 101:2642–2647

Basantani M, Srivastava A (2007) Plant glutathione transferases – a decade falls short. Can J Bot 85:443–456

Bernat BA, Laughlin LT, Armstrong RN (1997) Fosfomycin resistance protein (FosA) is a manganese metalloglutathione transferase related to glyoxalase I and the extradiol dioxygenases. Biochemistry 36:3050–3055

Bick JA, Aslund F, Chen Y, Leustek T (1998) Glutaredoxin function for the carboxyl-terminal domain of the plant-type 5′-adenylylsulfate reductase. Proc Natl Acad Sci USA 95:8404–8409

Booth J, Boyland E, Sims P (1961) An enzyme from rat liver catalysing conjugations with glutathione. Biochem J 79:516–524

Bresell A, Weinander R, Lundqvist G, Raza H, Shimoji M, Sun TH, Balk L, Wiklund R, Eriksson J, Jansson C, Persson B, Jakobsson PJ, Morgenstern R (2005) Bioinformatic and enzymatic characterization of the MAPEG superfamily. FEBS J 272:1688–1703

Byrne LJ, Sidhu A, Wallis AK, Ruddock LW, Freedman RB, Howard MJ, Williamson RA (2009) Mapping of the ligand-binding site on the b′ domain of human PDI: interaction with peptide ligands and the x-linker region. Biochem J 423:209–217

Chibani K, Wingsle G, Jacquot JP, Gelhaye E, Rouhier N (2009) Comparative genomic study of the thioredoxin family in photosynthetic organisms with emphasis on *Populus trichocarpa*. Mol Plant 2:308–322

Couturier C, Jacquot JP, Rouhier N (2009) Evolution and diversity of glutaredoxins in photosynthetic organisms. Cell Mol Life Sci 66:2539–2545

Dai S, Friemann R, Glauser DA, Bourquin F, Manieri W, Schürmann P, Eklund H (2007) Structural snapshots along the reaction pathway of ferredoxin-thioredoxin reductase. Nature 448:92–96

Dayer R, Fischert BB, Eggen RIL, Lemaire SD (2008) The peroxiredoxin and glutathione peroxidase families in *Chlamydomonas reinhardtii*. Genetics 179:41–57

Dietz KJ (2003) Plant peroxiredoxins. Annu Rev Plant Biol 54:93–107

Dietz KJ (2011) Peroxiredoxins in plants and cyanobacteria. Antioxid Redox Signal 15:1129–1159

Dietz KJ, Pfannschmidt T (2011) Novel regulators in photosynthetic redox control of plant metabolism and gene expression. Plant Physiol 155:1477–1485

Dietz KJ, Jacquot JP, Harris GC (2010) Tansley review: hubs and bottlenecks in plant molecular signalling networks. New Phytol 188:919–938

Ding YC, Ortelli F, Rossiter LC, Hemingway J, Ranson H (2003) The *Anopheles gambiae* glutathione transferase supergene family: annotation, phylogeny and expression profiles. BMC Genomics 4:35

Dixon DP, Edwards R (2010) Roles for stress-inducible lambda glutathione transferases in flavonoid metabolism in plants as identified by ligand fishing. J Biol Chem 285:36322–36329

Dixon DP, Cole DJ, Edwards R (2000) Characterisation of a zeta class glutathione transferase from *Arabidopsis thaliana* with a putative role in tyrosine catabolism. Arch Biochem Biophys 384:407–412

Dixon DP, Davis BG, Edwards R (2002a) Functional divergence in the glutathione transferase superfamily in plants. Identification of two classes with putative functions in redox homeostasis in *Arabidopsis thaliana*. J Biol Chem 277:30859–30869

Dixon DP, Lapthorn A, Edwards R (2002b) Plant glutathione transferases. Genome Biol 3:1–10, REVIEWS 3004.1–3004.10

Dixon DP, Steel PG, Edwards R (2011) Roles for glutathione transferases in antioxidant recycling. Plant Signal Behav 6:1223–1227

Edwards R, Dixon DP (2005) Plant glutathione transferases. Methods Enzymol 401:169–186

Edwards R, Dixon DP, Walbot V (2000) Plant glutathione S-transferases: enzymes with multiple functions in sickness and in health. Trends in Plant Science 5:193–198

Frova C (2003) The plant glutathione transferase gene family: genomic structure, functions, expression and evolution. Physiol Plant 119:469–479

Gelhaye E, Rouhier N, Navrot N, Jacquot JP (2005) The plant thioredoxin system. Cell Mol Life Sci 62:24–35

Gong H, Jiao Y, Hu WW, Pua EC (2005) Expression of glutathione-S-transferase and its role in plant growth and development in vivo and shoot morphogenesis *in vitro*. Plant Mol Biol 57: 53–66

Henderson CJ, Smith AG, Ure J, Brown K, Bacon EJ, Wolf CR (1998) Increased skin tumorigenesis in mice lacking pi class glutathione S-transferases. Proc Natl Acad Sci USA 95:5275–5280

Hertig CM, Wolosiuk RA (1983) Studies on the hysteretic properties of chloroplast fructose-1,6-bisphosphatase. J Biol Chem 258:984–989

Holmgren A (1976) Hydrogen donor system for *Escherichia coli* ribonucleoside-diphosphate reductase dependent upon glutathione. Proc Natl Acad Sci USA 73:2275–2279

Jacob C, Anwar A (2008) The chemistry behind redox regulation with a focus on sulphur redox systems. Physiol Plant 133:469–480

Jacquot JP, Lancelin JM, Meyer Y (1997) Tansley review n° 94: thioredoxins: structure and function in plant cells. New Phytol 136:543–570

Jacquot JP, Eklund H, Rouhier N, Schürmann P (2009) Structural and evolutionary aspects of thioredoxin reductases in photosynthetic organisms. Trends Plant Sci 14:336–3343

Jakobsson PJ, Morgenstern R, Mancini J, Ford-Hutchinson A, Persson B (1999) Common structural features of MAPEG – a widespread superfamily of membrane associated proteins with highly divergent functions in eicosanoid and glutathione metabolism. Protein Sci 8:689–692

Jakobsson PJ, Morgenstern R, Mancini J, Ford-Hutchinson A, Persson B (2000) Membrane-associated proteins in eicosanoid and glutathione metabolism (MAPEG). A widespread protein superfamily. Am J Respir Crit Care Med 161:S20–S24

Kawahigashi H (2009) Transgenic plants for phytoremediation of herbicides. Curr Opin Biotechnol 20:225–230

Koh CS, Navrot N, Didierjean C, Rouhier N, Hirasawa M, Knaff DB, Wingsle G, Samian R, Jacquot JP, Corbier C, Gelhaye E (2008) An atypical catalytic mechanism involving three cysteines of thioredoxin. J Biol Chem 283:23062–23072

Laisk A, Eichelmann H, Oja V (2006) C_3 photosynthesis *in silico*. Photosynth Res 90:45–66

Lamoureux GL, Shimabukuro RH, Swanson HR, Frear DS (1970) Metabolism of 2-chloro-4-ethylamino-6-isopropylamino-s-triazine (atrazine) in excised sorghum leaf sections. J Agric Food Chem 18:81–86

Lan T, Yang ZL, Yang X, Liu YJ, Wang XR, Zeng QY (2009) Extensive functional diversification of the populus glutathione S-transferase supergene family. Plant Cell 21:3749–3766

Lemaire SD, Quesada A, Merchan F, Corral JM, Igeno MI, Keryer E, Issakidis-Bourguet E, Hirasawa M, Knaff DB, Miginiac-Maslow M (2005) NADP-malate dehydrogenase from unicellular green alga *Chlamydomonas reinhardtii*. A first step toward redox regulation? Plant Physiol 137:514–521

Lin X, Kaul S, Rounsley S, Shea TP, Benito MI, Town CD, Fujii CY, Mason T, Bowman CL, Barnstead M, Feldblyum TV, Buell CR, Ketchum KA, Lee J, Ronning CM, Koo HL, Moffat KS, Cronin LA, Shen M, Pai G, Van Aken S, Umayam L, Tallon LJ, Gill JE, Adams MD, Carrera AJ, Creasy TH, Goodman HM, Somerville CR, Copenhaver GP, Preuss D, Nierman WC, White O, Eisen JA, Salzberg S, Fraser CM, Venter JC (1999) Sequence and analysis of chromosome 2 of the plant *Arabidopsis thaliana*. Nature 402:761–768

Lindahl M, Kieselbach T (2009) Disulphide proteomes and interactions with thioredoxin on the track towards understanding redox regulation in chloroplasts and cyanobacteria. J Proteomics 72:416–438

Marrs KA (1996) The functions and regulation of glutathione s-transferases in plants. Annu Rev Plant Physiol Plant Mol Biol 47:127–158

Marteyn B, Domain F, Legrain P, Chauvat F, Cassier-Chauvat C (2009) The thioredoxin reductase-glutaredoxins-ferredoxin crossroad pathway for selenate tolerance in *Synechocystis* PCC6803. Mol Microbiol 71:520–532

Meyer Y, Buchanan BB, Vignols F, Reichheld JP (2009) Thioredoxins and glutaredoxins: unifying elements in redox biology. Annu Rev Genet 43:335–367

Montrichard F, Alkhalfioui F, Yano H, Vensel WH, Hurkman WJ, Buchanan BB (2009) Thioredoxin targets in plants: the first 30 years. J Proteomics 72:452–474

Morel F, Aninat C (2011) The glutathione transferase kappa family. Drug Metab Rev 43:281–291

Morel F, Rauch C, Coles B, Le Ferrec E, Guillouzo A (2002) The human glutathione transferase alpha locus: genomic organization of the gene cluster and functional characterization of the genetic polymorphism in the hGSTA1 promoter. Pharmacogenetics 12:277–286

Morel M, Ngadin AA, Droux M, Jacquot JP, Gelhaye E (2009) The fungal glutathione S-transferase system. Evidence of new classes in the wood-degrading basidiomycete *Phanerochaete chrysosporium*. Cell Mol Life Sci 66:3711–3725

Moreno J, García-Murria MJ, Marín-Navarro J (2008) Redox modulation of Rubisco conformation and activity through its cysteine residues. J Exp Bot 59:1605–1614

Mueller LA, Goodman CD, Silady RA, Walbot V (2000) AN9, a petunia glutathione S-transferase required for anthocyanin sequestration, is a flavonoid-binding protein. Plant Physiol 123: 1561–1570

Navrot N, Collin V, Gualberto J, Gelhaye E, Hirasawa M, Rey P, Knaff DB, Issakidis E, Jacquot JP, Routhier N (2006) Plant glutathione peroxidases are functional peroxiredoxins distributed in several subcellular compartments and regulated during biotic and abiotic stresses. Plant Physiol 142:1364–1379

Navrot N, Rouhier N, Gelhaye E, Jacquot JP (2007) ROS generation and antioxidant systems in plant mitochondria. Physiol Plant 129:185–195

Nguyen VD, Saaranen MJ, Karala AR, Lappi AK, Wang L, Raykhel IB, Alanen HI, Salo KE, Wang CC, Ruddock LW (2011) Two endoplasmic reticulum PDI peroxidases increase the efficiency of the use of peroxide during disulfide bond formation. J Mol Biol 406:503–515

Nuruzzaman M, Gupta M, Zhang CJ, Wang L, Xie WB, Xiong LH, Zhang QF, Lian XM (2008) Sequence and expression analysis of the thioredoxin protein gene family in rice. Mol Genet Genom 280:139–151

Pearson WR (2005) Phylogenies of glutathione transferase families. Methods Enzymol 401: 186–204

Polle A (2001) Dissecting the superoxide dismutase-ascorbate-glutathione-pathway in chloroplasts by metabolic modeling. Computer simulations as a step towards flux analysis. Plant Physiol 126:445–462

Rouhier N, Jacquot JP (2003) Molecular and catalytic properties of a peroxiredoxin-glutaredoxin hybrid from *Neisseria meningitidis*. FEBS Lett 554:149–153

Rouhier N, Gelhaye E, Sautiere PE, Brun A, Laurent P, Tagu D, Gerard J, de Faÿ E, Meyer Y, Jacquot JP (2001) Isolation and characterization of a new peroxiredoxin from poplar sieve tubes that uses either glutaredoxin or thioredoxin as a proton donor. Plant Physiol 127:1299–1309

Rouhier N, Gelhaye E, Jacquot JP (2002) Glutaredoxin dependent peroxiredoxin from poplar: protein-protein interaction and catalytic mechanism. J Biol Chem 277:13609–13614

Rouhier N, Gelhaye E, Gualberto JM, Jordy MN, De Fay E, Hirasawa M, Duplessis S, Lemaire SD, Frey P, Martin F, Manieri W, Knaff DB, Jacquot JP (2004) Poplar peroxiredoxin Q. A thioredoxin-linked chloroplast antioxidant functional in pathogen defense. Plant Physiol 134:1027–1038

Rouhier N, Villarejo A, Srivastava M, Gelhaye E, Keech O, Droux M, Finkemeier I, Samuelsson G, Dietz KJ, Jacquot JP, Wingsle G (2005) Identification of plant glutaredoxin targets. Antioxid Redox Signal 7:919–929

Rouhier N, Kauffmann B, Tete-Favier F, Paladino P, Gans P, Branlant G, Jacquot JP, Boschi-Muller S (2007) Functional and structural aspects of poplar cytosolic and plastidial type A methionine sulfoxide reductases. J Biol Chem 282:3367–3378

Rouhier N, Lemaire S, Jacquot JP (2008) The role of glutathione in photosynthetic organisms: emerging functions for glutaredoxins and glutathionylation. Annu Rev Plant Biol 59:143–166

Rouhier N, Couturier J, Jacquot JP, Johnson MK (2010) Glutaredoxins and iron homeostasis. Trends Biochem Sci 35:43–52

Ruelland E, Miginiac-Maslow M (1999) Regulation of chloroplast enzyme activities by thioredoxins: activation or relief from inhibition? Trends Plant Sci 4:136–141

Schürmann P, Jacquot JP (2000) Plant Thioredoxin systems revisited. Annu Rev Plant Physiol and Plant Mol Biol 51:371–400

Selles B, Jacquot JP, Rouhier N (2011) Comparative genomic study of protein disulfide isomerases from photosynthetic organisms. Genomics 97:37–50

Sevier CS, Kaiser CA (2008) Ero1 and redox homeostasis in the endoplasmic reticulum. Biochim Biophys Acta 1783:549–556

Sheehan D, Meade G, Foley VM, Dowd CA (2001) Structure, function and evolution of glutathione transferases: implications for classification of non-mammalian members of an ancient enzyme superfamily. Biochem J 360:1–16

Shimabukuro RH, Frear DS, Swanson HR, Walsh WC (1971) Glutathione conjugation. An enzymatic basis for atrazine resistance in corn. Plant Physiol 47:10–14

Smith AP, Nourizadeh SD, Peer WA, Xu J, Bandyopadhyay A, Murphy AS, Goldsbrough PB (2003) Arabidopsis AtGSTF2 is regulated by ethylene and auxin, and encodes a glutathione S-transferase that interacts with flavonoids. Plant J 36:433–442

Smith AP, DeRidder BP, Guo WJ, Seeley EH, Regnier FE, Goldsbrough PB (2004) Proteomic analysis of Arabidopsis glutathione S-transferases from benoxacor- and copper-treated seedlings. J Biol Chem 279:26098–26104

Soranzo N, Gorla MS, Mizzi L, De Toma G, Frova C (2004) Organisation and structural evolution of the rice glutathione S-transferase gene family. Mol Genet Genomics 271:511–521

Spínola MC, Pérez-Ruiz JM, Pulido P, Kirchsteiger K, Guinea M, González M, Cejudo FJ (2008) NTRC new ways of using NADPH in the chloroplast. Physiol Plant 133:516–524

Tarrago L, Laugier E, Zaffagnini M, Marchand C, Le Maréchal P, Rouhier N, Lemaire SD, Rey P (2009) Regeneration mechanisms of *Arabidopsis thaliana* methionine sulfoxide reductases B by glutaredoxins and thioredoxins. J Biol Chem 284:18963–18971

Tavender TJ, Springate JJ, Bulleid NJ (2010) Recycling of peroxiredoxin IV provides a novel pathway for disulphide formation in the endoplasmic reticulum. EMBO J 15:4185–4197

Xun L, Topp E, Orser CS (1992) Purification and characterization of a tetrachloro-p-hydroquinone reductive dehalogenase from a *Flavobacterium sp.* J Bacteriol 174:8003–8007

Ziemann M, Bhave M, Zachgo S (2009) Origin and diversification of land plant CC-type glutaredoxins. Genome Biol Evol 1:265–277

Sheehan D, Meade G, Foley VM, Dowd CA (2001) Structure, function and evolution of glutathione transferases: implications for classification of non-mammalian members of an ancient enzyme superfamily. Biochem J 360:1–16

[illegible]

Chapter 9
Prokaryotic Redox Switches

John D. Helmann

Abstract Reduction-oxidation (redox) reactions are ubiquitous in metabolism. In microbial systems, individual cells monitor the internal and external environment with a diverse array of regulatory redox sensors that control the expression of appropriate adaptive responses. These sensors detect the presence of redox active molecules in the environment as well as those produced by metabolic processes within the cell. Molecular oxygen is the prototypic oxidant, and sensors of O_2 help coordinate the switch between anaerobic and aerobic lifestyles. The overall redox balance of the cell can be sensed by proteins monitoring the ratio of oxidized and reduced cofactors, such as low molecular weight thiols, the NADH pool, or membrane-associated quinones. Reactive oxygen and nitrogen species, produced endogenously or present in the environment, are also important signal molecules that can be readily sensed by their redox activity. This suite of sensor proteins is as diverse as their activating signals. Some sensors contain redox active metal centers that may include mononuclear iron, iron-sulfur clusters, and heme cofactors. Other redox switches rely on the facile oxidation and reduction of bound flavin cofactors or cysteine thiolates to effect changes in protein activity. Finally, some regulators assess redox-related signals by reversibly binding to molecules that signal redox status. A survey of these diverse mechanisms provides insights into the manifold pathways that enable cells to adapt to a range of environmental conditions that collectively comprise redox stress.

Keywords Flavoprotein • Heme sensor • Iron-sulfur cluster • Metal-catalyzed oxidation • Thiol

---

J.D. Helmann (✉)
Department of Microbiology, Cornell University, Wing Hall, Ithaca, NY 14853-8101, USA
e-mail: jdh9@cornell.edu

U. Jakob and D. Reichmann (eds.), *Oxidative Stress and Redox Regulation*,
DOI 10.1007/978-94-007-5787-5_9,

9.1 Electron Transfer Reactions Within the Cell: The Flavors of Biological Electron Transfer

Life processes are inextricably linked to the transfer of electrons. Photosynthesis relies on the generation of high potential electrons that serve to generate NAPDH cofactors for the reduction of CO_2 to organic molecules. Conversely, catabolism often relies on the oxidation of organic molecules to yield both energy (in the form of ATP) and reducing equivalents (in the form of NADH). These two currencies (or their equivalents) fuel anabolic reactions. Redox reactions therefore help drive many of the central processes that underlie metabolism.

Most electron transfers within the cell involve either a metal center or an organic electron carrier as cofactor. Iron is particularly well suited for cellular redox chemistry, although both manganese and copper also play key roles (Frausto da Silva and Williams 2001). Redox active iron centers may include a single ion (a mononuclear iron site), multiple ions, iron-sulfur clusters, or heme iron (Outten and Theil 2009). The local protein environment serves to tune both the redox midpoint of the metal center and the specificity of interaction with redox partners. In addition to metal centers, redox active proteins may use protein thiols (or more accurately, the thiolate ion) or a bound flavin cofactor (Antelmann and Helmann 2011; Becker et al. 2011). These protein-bound redox centers equilibrate with diffusible electron carriers in the cell including low molecular weight thiols and thioredoxins, the NAD^+/NADH couple and, within membranes, the quinones (Green and Paget 2004). While most of the time electrons are transferred to their intended recipients, aberrant reactions also occur and can give rise to reactive oxygen and nitrogen species (RNOS). These species can, in turn, react with protein thiols, cofactors, or metal centers (Imlay 2003).

9.2 Bacterial Redox Sensors and Switches

Redox reactions provide the molecular mechanism for a wide variety of redox sensors and switches. Here, we provide a broad survey of the best-characterized examples within prokaryotes (Bacteria and Archaea), many of which are paradigms for large classes of redox sensors. As defined here, a *redox sensor* has a regulatory role: it detects redox active compounds (e.g., O_2, RNOS) or compounds that report on a change in cellular redox status (e.g., NADH, quinones) and regulates the expression or activity of other proteins in response. Sensing may result from the reversible binding of redox active compounds or by reactions leading to redox changes in the sensor protein or its bound cofactors. Classic examples include the OxyR and Fnr-based redox sensors in *Escherichia coli*. A *redox switch* can be defined as a reversible modification (switch) that turns the activity of a protein on or off. Thus, redox sensors often contain an embedded redox switch. In addition to their role in redox-based regulators, redox switches also occur in

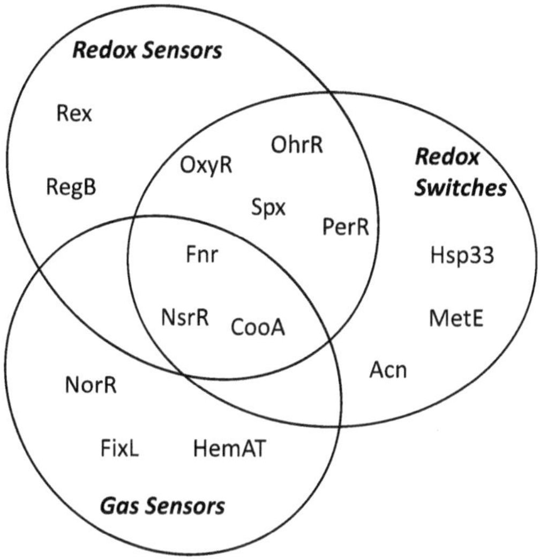

Fig. 9.1 A Venn diagram illustrating a conceptual framework for thinking about redox sensors and switches and their relationship to gas sensors. *Redox sensors* are those proteins that both sense redox changes and in response control gene expression. *Redox switch* refers to a reversible modification of a protein or protein-bound cofactor to control protein activity. Many, but not all, redox sensors contain an embedded redox switch as indicated by the representative examples and discussed in the text. However, the term redox switch is more often associated with non-regulatory proteins (e.g. Hsp33). Some proteins that sense redox active cofactors (e.g. sensing of NADH/NAD^+ ratios by Rex) do not function as redox switches. Bacterial gas sensors use some of the same cofactors as redox sensors, although many do not undergo a redox change upon ligand-binding

redox-regulated enzymes. Classic examples include the redox regulation of the *E. coli* Hsp33 chaperone (Kumsta and Jakob 2009) and the reversible inactivation of *E. coli* methionine synthase by S-glutathionylation (Hondorp and Matthews 2004). Bacteria also contain gas sensors that, at least in some cases, also rely on redox chemistry to sense their ligands (Fig. 9.1).

Redox sensors and switches play a variety of roles within the cell and, as a group, they take advantage of nearly the full range of redox chemistry that is used in metabolic pathways (Fig. 9.2). First, they can sense the redox balance within the cell by monitoring the levels of reduced NADH and quinones within the electron transfer chain. Second, they can sense the oxidation of normally reduced species, including protein and low molecular thiols (disulfide stress), that results from the depletion of reducing equivalents within the cytosol or exposure to oxidizing agents. Third, redox sensors can sense molecular oxygen, RNOS, and non-specific oxidants using a variety of metal centers, flavin cofactors, and redox active thiols (reviewed in Green and Paget 2004).

Of the various types of redox reactions and cofactors in the cell, there are relatively few that have not yet been described as also serving a regulatory role

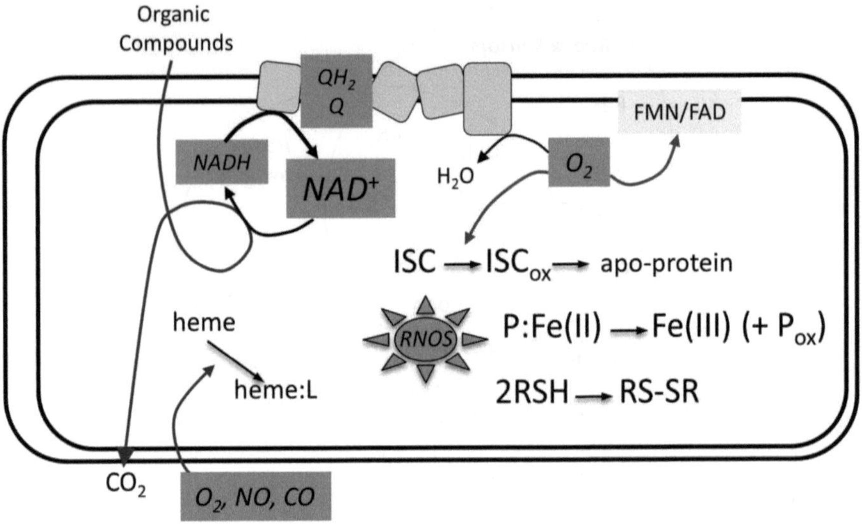

Fig. 9.2 Conceptual overview of redox sensing within a typical bacterial cell. The major classes of redox signals in the cell are in each case shown against a shaded background. Catabolism of organic compounds generates reduced electron carriers (e.g., NADH) which delivers electrons to the electron transport system. With the help of membrane soluble organic carriers (the quinones, Q), electrons are transferred to a terminal electron acceptor (here, O_2). Oxidative metabolism and redox-cycling compounds give rise to ROS, and RNS such as NO are generated during denitrification or by NO synthases. Collectively, these RNOS species are sensed by heme-containing gas sensors (which sense primarily NO and O_2 by formation of a heme:ligand complex; heme:L), ISC-containing proteins (NO, O_2, and RNOS), and thiol-containing redox proteins (RNOS). For non-heme Fe(II)-containing proteins, MCO may lead to protein oxidation (P_{ox}). Proteins containing FMN/FAD cofactors can also sense O_2 and other redox signals

within a prokaryotic redox switch or sensor protein. For example, there are few if any regulatory switches that utilize the redox activity of either copper or manganese. Another notable example is oxidation of methionine (Met) to methionine sulfoxide, a reversible protein modification that would seem well suited to provide a redox switch. Met oxidation is widespread in response to strong oxidants like hypochlorite and it has been suggested that this is a significant contributor to cell killing by activated neutrophils (Rosen et al. 2009). Since oxidized Met can be repaired by two distinct and diasterospecific methionine sulfoxide reductases, this could regulate protein activity. Indeed, such a mechanism has been described for calmodulin in mammals (Bigelow and Squier 2011), but has not yet been documented in the Bacteria.

Here we provide a survey of redox sensors and switches organized according to the chemical nature of the primary redox switch. For each class of sensors, more focused reviews are cited that provide much greater detail on both the mechanistic and structural basis of the switch mechanism and the resulting conformational changes. For a broad overview, the reader is referred specifically to an excellent

2004 review (Green and Paget 2004). The present work takes inspiration from this earlier review and includes an update on some of the same regulatory systems as well as additional systems that have come to prominence in the interim.

9.3 Redox Regulation by Mononuclear Metal Centers

The redox activity of metal ions, and iron in particular, has been co-opted by numerous regulators as a facile mechanism to signal the presence of RNOS (Helmann 2010; Outten and Theil 2009; Green and Paget 2004). Four representative examples include the PerR, NorR, IrrR, and HbpS sensor proteins (Table 9.1). In each of these cases, a redox active species is sensed by coordination and often chemical reaction with an iron atom leading to a conformational change in the regulator that leads to changes in gene expression.

9.3.1 Bacillus subtilis PerR

Bacillus subtilis PerR is a dimeric, metal-dependent repressor of the Fur (ferric uptake repressor) family (Herbig and Helmann 2001; Lee and Helmann 2007). In response to sub-micromolar levels of H_2O_2, PerR is functionally inactivated leading to the induction of protective functions including catalase, alkylhydroperoxide reductase, and an iron sequestration protein (Lee and Helmann 2006b; Helmann et al. 2003). PerR has evolved to be highly sensitive to oxidative inactivation to ensure that protective functions are induced before damage to other macromolecules is widespread. PerR orthologs are widespread in the Bacteria and, together with OxyR, represent one of the most prevalent mechanisms for regulating adaptation to H_2O_2 (Duarte and Latour 2010).

PerR senses H_2O_2 by metal-catalyzed oxidation (MCO) of the protein at one of two distinct metal-binding sites (Lee and Helmann 2006b). The first site plays a structural role and contains Zn(II) held with four Cys ligands. This thiolate rich site is not reactive with even high concentrations of H_2O_2 (Lee and Helmann 2006a), but can be oxidized by the strong oxidant, hypochlorite (Chi et al. 2011). This contrasts with some other Zn(II) sites in which coordinating thiolates are highly reactive with ROS (e.g., RsrA; see below). The second metal-binding site serves to regulate DNA-binding activity and can coordinate either Mn(II) or Fe(II). The iron-loaded form of PerR (PerR:Fe) is highly sensitive to H_2O_2 with rates of inactivation in excess of 10^5 M^{-1} s^{-1}. In contrast, the Mn(II) form (PerR:Mn) is non-reactive (Lee and Helmann 2006b). Thus, the PerR regulon can be derepressed by very low levels of H_2O_2 unless the cytosol has very low Fe(II) and comparatively high Mn(II) levels (Fuangthong et al. 2002). This provides a mechanism whereby PerR can tune the induction of peroxide stress genes to the bioavailability of Fe(II) in the cell. This is likely adaptive since the major toxic effects of H_2O_2 arise from

Table 9.1 Representative iron-containing redox sensors (Fe, ISC, heme)

Organism	Sensor	Targets and/or function	Class	Signal(s)	Mechanism
B. subtilis	PerR	Peroxide stress response	Fe(II)	H_2O_2	MCO (2-oxo-His)
E. coli	Fur	Iron homeostasis	Fe(II)	H_2O_2	Unknown
E. coli	Fur	Iron homeostasis	Fe(II)	NO	DNIC
E. coli	NorR	NO detoxification	GAF:Fe(II)	NO	Mononitrosyl iron complex
S. reticuli	HbpS	Catalase regulation	Fe(II)?	ROS?	MCO (protein crosslinking)
E. coli	Fnr	Anaerobic metabolism	$[4Fe\text{-}4S]^{2+}$	O_2	Cluster degradation
E. coli	SoxR	Chemical protection	$[2Fe\text{-}2S]^{+}$	Oxidants	$[2Fe\text{-}2S]^{+}$ » $[2Fe\text{-}2S]^{2+}$
E. coli	IscR	FeS cluster homeostasis	$[2Fe\text{-}2S]^{+}$	Oxidants	ISC disassembly
E. coli	NsrR	NO detoxification	$[2Fe\text{-}2S]^{+}$	NO	DNIC
E. coli	AcnB	mRNA stability/translation	$[4Fe\text{-}4S]^{2+}$	Fe, ROS	Cluster degradation
B. japonicum/ S. meliloti	FixL	N_2 fixation	PAS:heme	O_2	Heme:O_2-binding
E. coli	DOS-RNC	mRNA turnover (DOS-RNC = DosP + DosC + RNases)	PAS:heme (DosP) GCS:heme (DosC)	O_2	Heme:O_2-binding
M. tuberculosis	DosS	Dormancy functions	GAF:heme	O_2, MQ, NO?	Heme:ligand-binding
M. tuberculosis	DosT	Dormancy functions	GAF:heme	O_2	Heme:O_2-binding
V. fisheri	H-NOX	Symbiosis functions	H-NOX:heme	NO	Heme:NO-binding
B. japonicum	Irr	Iron homeostasis	Heme	heme, ROS	MCO
R. rubrum	CooA	CO oxidation	CRP:heme	CO ($-O_2$)	Heme:CO-binding

Fenton chemistry in which H_2O_2 reacts with Fe(II) generating hydroxyl radicals that damage macromolecules including DNA, proteins, and potentially the cell membrane (Imlay 2003).

Both biochemical and structural studies support a model in which H_2O_2 is sensed by reaction with Fe(II) in PerR:Fe leading to the generation of a reactive hydroxyl radical that modifies one of three His residues coordinated to the Fe(II) atom (Lee and Helmann 2006b; Traore et al. 2009). The most frequent reaction is at His37 in *B. subtilis* PerR, with a lesser degree of oxidation at His91. His93, which is also an Fe(II) ligand, is refractory to modification apparently due to its own position opposite the site of approach of the incoming H_2O_2 (Jacquamet et al. 2009). Oxidation of His to 2-oxo-histidine and the concomitant oxidation of Fe(II) to Fe(III), which binds much less tightly to PerR, leads to derepression of the peroxide stress response (Lee and Helmann 2006b). Although multiple cycles of oxidation may be possible *in vitro*, in the cell the predominant product is PerR oxidized on a single amino acid. Since His37 is located in the DNA-binding domain, whereas the bulk of the Fe(II) coordinating residues are in the dimerization domain, loss of Fe(II) leads to a reorientation of the DNA-binding domains and PerR adopts a conformation no longer suited for DNA-binding (Jacquamet et al. 2009; Duarte and Latour 2010; Giedroc 2009).

The PerR-regulated peroxide stress response is known to be transient (Helmann et al. 2003), yet the mechanisms enabling the cell to re-establish repression are not well defined. There is no known repair pathway for 2-oxo-His residues, and it is most likely that oxidized PerR is degraded in the cell. Repression is presumably re-established by newly synthesized PerR, which is autoregulated (Fuangthong et al. 2002). PerR thereby functions as a sacrificial regulator. While unusual, this is not unique to PerR: the LexA repressor of the SOS response is regulated by its self-catalyzed degradation (Pruteanu and Baker 2009) and the activity of Ada as a transcription factor for the adaptive response to alkylation stress is regulated by irreversible protein modification (He et al. 2005).

9.3.2 Sensing of H_2O_2 and NO by Fur Family Proteins

PerR regulators have evolved to be exquisitely sensitive to H_2O_2-mediated protein oxidation. In contrast with *B. subtilis* PerR, the Fur protein in this organism functions physiologically as an Fe(II) sensor and is less sensitive to inactivation by H_2O_2 (Lee and Helmann 2006b; Helmann et al. 2003). Since both PerR and Fur are homologs with similar Fe(II)-binding sites, it will be of interest to determine how these proteins have evolved to enhance or suppress, respectively, their reactivity with H_2O_2. These proteins bind and sense Fe(II) by reversible binding to a solvent exposed site. In PerR, Fe(II) is five-coordinate with an open coordination site for the incoming H_2O_2. The details of Fe(II) coordination by *B. subtilis* Fur are unknown, but Fe(II) may be six-coordinate, which could account for the lower reactivity with oxidants. Managing the reactivity of this potentially exposed ferrous ion to endogenous oxidants therefore presents a challenge.

In contrast with *B. subtilis* Fur, *E. coli* Fur is inactivated by sub-micromolar levels of H_2O_2 and, in this sense, resembles PerR (Varghese et al. 2007). The chemical mechanism of this inactivation has not been reported, but is suggested to be due to oxidation of the bound ferrous ion, the protein, or both. Paradoxically, inactivation of Fur leads to derepression of iron uptake under precisely those conditions of oxidative stress where this would be disadvantageous. To counter this effect, the OxyR transcription factor upregulates Fur protein levels to re-establish effective repression of the Fur regulon (Zheng et al. 1999; Varghese et al. 2007). This represents one evolutionary solution to the problem of the intrinsic reactivity of protein-bound Fe(II).

At first glance, *B. subtilis* appears to have a similar regulatory circuitry since PerR is a repressor of the *fur* gene (Fuangthong et al. 2002). However, the *fur* gene itself is not peroxide inducible even though it is regulated by PerR (Fuangthong et al. 2002). This is due to the fact that PerR:Mn is the favored repressor for the *fur* gene and PerR:Fe, the form that responds to H_2O_2, is ineffective. The significance of the PerR regulation of the *fur* gene is not entirely clear, but it may be related to modulating the expression of iron uptake pathways in response to the relative levels of Mn(II) and Fe(II) in the cell.

PerR and Fur proteins also respond to nitric oxide (NO). In the presence of NO, the Fe(II) atom in *E. coli* Fur is dinitrosylated (generating a dinitrosyl iron complex; DNIC) and this results in a conformational change that leads to derepression (D'Autreaux et al. 2004). In *E. coli*, Fur and NorR together regulate most genes that are strongly induced by NO stress (Mukhopadhyay et al. 2004). Similarly, in *B. subtilis* NO is known to derepress both the PerR and Fur regulons, presumably by a similar mechanism of iron nitrosylation (Moore et al. 2004). This is apparently a gratuitous induction since there is no apparent benefit to the cell of inducing the Fur and PerR regulons in response to NO stress. Indeed, *B. subtilis fur* mutants are more sensitive to NO than wild-type under at least some growth conditions (Moore et al. 2004). Although nitrosylation of the non-heme iron of Fur family proteins does not confer any obvious benefit to the cell, a related mechanism is used by the NO-specific sensor, NorR.

9.3.3 Sensing of NO by NorR

E. coli has a complex adaptive response to RNOS that involves multiple regulatory proteins (Mukhopadhyay et al. 2004; Pullan et al. 2007). These include OxyR, a sensor of H_2O_2 (and perhaps other RNOS species) that utilizes a highly reactive thiolate, and two sensors of NO: the non-heme iron-dependent NorR and the iron-sulfur containing NO sensor, NsrR. In addition to these dedicated sensors, RNOS perturb the regulation of numerous other regulons by affecting the integrity or activity of regulatory proteins (Flatley et al. 2005; Pullan et al. 2007).

The ability of NorR to sense NO relies on the mononitrosylation of a non-heme iron atom (Bush et al. 2011) and is, in this respect, similar to the described derepression of the Fur and PerR regulons. Nitrosylation of NorR leads to up-regulation of the *norVW* operon encoding a flavorubredoxin and associated oxidoreductase that detoxify NO by reduction to N_2O (nitrous oxide) under anaerobic conditions.

NorR is an activator protein that functions with the σ^{54} RNAP to enhance transcription (reviewed by Bush et al. 2011). Activators functioning as partners for the σ^{54} RNAP are distinct from most other bacterial transcription factors. Much like some eukaryotic transcription factors, these ATP-dependent enhancer binding proteins (bEBP) can function from a variety of positions relative to the core promoter elements (Shingler 2011). At most promoters, σ^{54} RNAP requires a bEBP in order to effectively engage the promoter as an open complex competent for transcription initiation (Wigneshweraraj et al. 2008). Like other bEBPs, NorR is an ATPase that couples the energy of ATP hydrolysis to the remodeling of the σ^{54} RNAP pre-initiation complex.

NorR is a complex protein with three distinct functional domains (Bush et al. 2011). An amino-terminal GAF domain binds the non-heme iron atom and senses NO. In the absence of NO, this GAF domain inhibits the activity of the AAA^+ ATPase domain and thereby prevents promoter activation. A carboxy-terminal domain contains a DNA-binding helix-turn-helix motif. As described in detail elsewhere (Bush et al. 2011), the current model for NorR activation at the *norVW* promoter region posits that an oligomer (likely a hexamer) of NorR is bound in a higher order complex that is functionally inactive. Sensing of NO relieves the GAF-domain mediated repression of the ATPase domain, and enables the hexameric protein to engage productively with RNAP. It is presently unknown whether binding of NO to a single subunit of the oligomer is sufficient for activation, whether activation is modulated by the extent of NO modification of the oligomer, or whether there is cooperativity in ligand binding.

9.3.4 Sensing of ROS by HbpS

HbpS is an extracellular, oligomeric protein that functions in transducing redox signals to the SenS-SenR two-component system (TCS) in the cellulose-degrading actinomycete *Streptomyces reticuli* (Siedenburg et al. 2011). HbpS is a heme-binding protein that is exported through the twin-arginine transport (TAT) system. Structural studies reveal both a hemin-binding site and a bound iron. It is suggested that iron would normally be provided by the non-enzymatic oxidation of heme (Ortiz de Orue Lucana et al. 2010). In the absence of ROS, extracellular HbpS interacts with the SenS sensor kinase to inhibit autophosphorylation. In the presence of compounds likely to lead to metal-catalyzed protein oxidation (including heme, iron, or H_2O_2), conformational changes within the HbpS octamer activate the SenS sensor kinase (Ortiz de Orue Lucana et al. 2010). These findings support a model in which HbpS alters the activity of the SenS-SenR TCS in response to redox stress.

Studies conducted *in vitro* reveal that HbpS oxidation leads to protein-protein crosslinks, likely due to dityrosine formation (between Tyr77 in two protomers) as well as protein carbonylation (Ortiz de Orue Lucana et al. 2010). The nature of the oxidized products of HbpS in cells is not yet clear. The relevant oxidation event may be the decomposition of the bound heme, the formation of protein-protein crosslinks, or other modifications of the HbpS redox sensor.

Once activated by exposure to ROS in the presence of free (or hemin-associated) iron, the HbpS-SenS-SenR signaling cascade activates transcription of the *senS, senR,* and *hbpS* genes themselves, the *cpeB* catalase-peroxidase, and *furS* encoding a Fur family redox sensor (Ortiz de Orue Lucana and Groves 2009). Although the details are not yet clear, the current model suggests that oxidation of HbpS relieves the inhibition of the SenS-SenR TCS. Newly synthesized HbpS then serves to downregulate SenS activity if, upon secretion, it is not also oxidized. The CpeB catalase is proposed to protect newly synthesized HbpS from being oxidized and to enable the system to reestablish equilibrium. This feedback mechanism may help ensure that the cell produces sufficient catalase-peroxidase to protect against ROS. If correct, this mechanism is similar in some ways to the sacrificial oxidation of PerR leading to the induction of catalase and peroxidases in *B. subtilis.*

FurS is itself a Fur family repressor, which is redox regulated by H_2O_2-mediated disulfide bond formation between two Zn(II) ligands, Cys93 and Cys96, leading to derepression of the *furS-cpeB* operon (Ortiz de Orue Lucana et al. 2003, 2012). Thus, expression of CpeB catalase is proposed to respond to both extracellular redox signals, transduced by the HbpS-SenS-SenR cascase, and intracellular RNOS sensed by FurS.

9.3.5 Mononuclear Iron Centers Function as Redox Switches for Regulating Enzyme Function

The susceptibility of protein-associated Fe(II) sites to oxidation by H_2O_2 and/or nitrosylation by NO varies enormously. As noted above, PerR and *E. coli* Fur react rapidly with H_2O_2 leading to changes in gene expression, whereas *B. subtilis* Fur is less reactive. Bacterial cells contain many enzymes that are known or likely to use Fe(II) as a cofactor for catalysis, although the full extent of Fe(II) metallation of enzymes is poorly understood (Anjem and Imlay 2012). This is due, in part, to the fact that many enzymes will function (albeit at variable levels) with more than one divalent metal cofactor and *in vitro* studies conducted aerobically may not include Fe(II) when testing metal specificity (Anjem et al. 2009).

Enzymes that have a mononuclear Fe(II) center are potentially susceptible to inactivation by RNOS (Sobota and Imlay 2011). However, there are no clear examples where oxidation of a mononuclear iron center is used as a redox switch for directly regulating enzyme function. Indeed, Fe(II) oxidation seems ill-suited for monitoring H_2O_2 levels since oxidation leads to the production of highly reactive hydroxyl radicals, which can covalently modify and irreversibly inactivate the protein (as seen with PerR). Some mononuclear Fe(II) enzymes appear to use a Cys residue as a sacrificial ligand that, in response to H_2O_2, is reversibly oxidized to the cysteine sulfenate (Anjem and Imlay 2012) and this also has the potential to regulate enzyme activity. Fe(II) centers are also adept at sensing other oxidants (e.g., NO), leading to the loss or inactivity of the catalytically essential metal center and thereby altering metabolism. One solution to the problem of redox sensitivity of Fe(II) centers is the substitution with a different metal ion. In *E. coli*, low levels of H_2O_2 activate OxyR-dependent transcription of the MntH Mn(II) uptake system (Kehres et al. 2002). The resulting increase in cytosolic Mn(II) levels metallates normally Fe(II) dependent enzymes (Anjem et al. 2009; Anjem and Imlay 2012). In some cases this may allow the enzyme to function as a mangano-enzyme, whereas in others this may simply be a way of excluding redox-active Fe(II) from the active site and thereby protecting the protein against destruction by MCO reactions (Sobota and Imlay 2011).

9.4 Sensing of Redox with Iron-Sulfur Clusters

Iron-sulfur clusters (ISC) play a variety of roles in protein structure and function (Johnson et al. 2005). In some proteins, an ISC seems to play a primarily structural role analogous to the widespread utilization of Zn(II) coordination in stabilizing zinc finger domains. In other enzymes, the ISC functions in coordination or activation of substrates but without redox changes in the cluster (analogous to the use of Zn(II) as a Lewis acid). Finally, ISCs may serve as a conduit for electrons, often as part of a multistep electron transport process (Johnson et al. 2005). Here, we focus on those examples where the ISC serves as a redox switch to control the activity of regulatory proteins. ISCs are particularly well suited for the detection of molecular oxygen (FNR), NO (NsrR), and other one-electron oxidants (SoxR) (Outten and Theil 2009; Crack et al. 2012).

9.4.1 FNR as a Sensor of Molecular Oxygen

The fumarate and nitrate reduction regulator (FNR) senses molecular oxygen and functions as a global regulator to effect large changes in cellular metabolism including the activation of alternative respiratory pathways. The *E. coli* FNR protein is the best-understood example of this class of sensor, which is widespread in the Bacteria (Dufour et al. 2010). Molecular oxygen is sensed by the oxidation of a $[4Fe\text{-}4S]^{2+}$ cluster within FNR as discussed in detail elsewhere (Crack et al. 2008). Many of our insights into the biochemical mechanisms of oxygen sensing by FNR were sparked by the development of techniques for the purification of active FNR under strictly anaerobic conditions (Yan and Kiley 2009) and the subsequent application of a variety of spectroscopic and biochemical techniques for monitoring iron-sulfur cluster chemistry.

In the current model for FNR function, the dimeric FNR protein contains a single $[4Fe\text{-}4S]^{2+}$ cluster per protomer and binds tightly to the promoter regions of target operons to activate transcription. In the presence of molecular oxygen, FNR is oxidized to an intermediate $[3Fe\text{-}4S]^{1+}$ cluster with release of Fe^{2+} and superoxide anion ($O_2^{-\cdot}$). This initial oxidation is sensitive to relatively low levels of oxygen (a second order rate constant of $k_1 \sim 250\ M^{-1}\ s^{-1}$). In a second step ($k_2 \sim 0.008\ s^{-1}$), the cluster decays further to form a relatively stable $[2Fe\text{-}2S]^{2+}$ with the release of 2 sulfide ions and Fe^{3+} (Crack et al. 2008). Further decomposition of the $[2Fe\text{-}2S]^{2+}$ to generate apo-FNR is likely to occur *in vivo* since the major form of FNR in aerobic *E. coli* is the monomeric apo-protein. The inactivation of FNR by oxygen is reversible, since the $[4Fe\text{-}4S]^{2+}$-cluster can be reassembled into inactive apo-protein in the cell (Mettert et al. 2008; Dibden and Green 2005). ISC assembly is thought to be essential for dimerization by helping neutralize charge repulsion between monomers (Moore et al. 2006).

While the broad outlines of O_2-sensing by FNR are now established, some features of this regulatory switch are still poorly understood. It has been noted, for example, that a product of O_2-sensing is superoxide anion, which can dismutate to yield H_2O_2 and O_2. Dismutation of superoxide occurs spontaneously but is greatly accelerated by superoxide dismutase (SOD) and, *in vitro*, FNR has also been reported to have SOD activity. These observations have been interpreted to indicate that a single O_2 molecule can potentially oxidize up to four $[4Fe\text{-}4S]^{2+}$-clusters (Green et al. 2009). An additional complexity derives from the observation that the presence of iron chelators affects the rate of cluster decomposition. Since iron chelators are ubiquitous in the cell, this could affect the kinetics of the cluster conversion processes *in vivo* (Crack et al. 2008). It is also not clear whether the mechanisms established for *E. coli* FNR will apply, in whole or in part, to other FNR orthologs. For example, *B. subtilis* FNR has a $[4Fe\text{-}4S]^{2+}$-cluster with three (rather than four) Cys ligands together with an Asp (Gruner et al. 2011), suggesting that it may be functionally divergent (Reents et al. 2006). Indeed, unlike *E. coli* FNR, the *B. subtilis* ortholog does not require cluster assembly for dimerization.

9.4.2 SoxR as a Sensor of One-Electron Oxidants

The *E. coli* SoxR protein is a $[2Fe\text{-}2S]^{+}$-containing transcription factor originally described as a sensor of superoxide stress (Storz and Imlay 1999). SoxR is a dimeric activator protein and a member of the MerR family of transcription factors. In response to redox-cycling compounds, oxidized SoxR activates the expression of SoxS, a MarR family protein that regulates the expression of genes that protect *E. coli* against redox-cycling compounds. The SoxRS regulon includes genes encoding the SodA superoxide dismutase, ISC assembly proteins, and proteins implicated in modification and efflux of redox-cycling compounds.

The activity of SoxR is regulated by the reversible, one electron oxidation of a solvent-exposed $[2Fe\text{-}2S]^{+}$ cluster (Watanabe et al. 2008). Since SoxR was initially described as a regulator of SodA, and SoxR is activated by various, structurally diverse redox-active compounds, it had long been assumed that superoxide itself was sensed directly by SoxR. However, more recent studies suggest that SoxR reacts directly with redox active compounds and that superoxide is not a physiologically relevant mediator of SoxR activation (Gu and Imlay 2011). This revised model of SoxR redox sensing is supported by several observations. First, cells lacking SOD have significantly elevated fluxes of superoxide, sufficient to damage iron-sulfur containing enzymes, but do not activate the SoxRS regulon. Second, the rate of reaction of SoxR with superoxide anion is insufficient to be relevant at the concentrations of superoxide likely to be present in the cell. Third, induction of the SoxRS regulon is observed even under anaerobic conditions when superoxide cannot be formed (Krapp et al. 2011; Gu and Imlay 2011). The notion that SoxR has a much broader role in sensing the presence of redox active compounds by their direct oxidation of the exposed $[2Fe\text{-}2S]^+$ cluster, also fits with the observation that SoxR does not even regulate SOD in several non-enteric bacteria (Dietrich and Kiley 2011).

9.4.3 IscR as a Sensor of Redox Stress Affecting Iron-Sulfur Cluster Proteins

E. coli IscR is a $[2Fe\text{-}2S]^+$-containing repressor that regulates the expression of the *isc* operon encoding one of two major pathways for ISC assembly. The requirement of a $[2Fe\text{-}2S]^+$ cluster for IscR-mediated repression allows IscR to repress the *isc* operon in response to the catalytic potential for ISC assembly. The IscR $[2Fe\text{-}2S]^+$ cluster is coordinated by three conserved Cys residues in the C-terminal domain of the protein. IscR also has a regulatory role when the $[2Fe\text{-}2S]^+$ cluster is not present: apo-IscR acts as a transcription activator of the second, stress-induced pathway for ISC assembly encoded by the *suf* operon (Yeo et al. 2006). This system is essential under conditions of even low level oxidative stress due to the intrinsic redox sensitivity of the Isc-mediated pathway (Jang and Imlay 2010). Remarkably, the apo-form of IscR binds to a distinct set of target operons. Thus, the IscR regulon includes two sets of genes: those regulated by metallated IscR and those regulated by apo-IscR (Nesbit et al. 2009).

In addition to the feedback regulation of the *isc*-encoded pathway, IscR also contributes to regulation of the alternative, stress-induced *suf* pathway for ISC assembly. Induction of the *suf* operon upon oxidative stress reflects the coordinate action of several regulatory proteins. Fur binds as a repressor, and inactivation of Fur by ROS is a prerequisite for activation of *suf* transcription (Lee et al. 2008). Activation of *suf* transcription can be mediated by OxyR (Outten et al. 2004; Lee et al. 2004b), which appears to play the major role and is essential for achieving sufficient levels of the SUF machinery to support robust growth (Jang and Imlay 2010), and by IscR (Yeo et al. 2006). The integration of signals sensed by these various transcription factors ensures that *suf* operon induction is enabled when ISC assembly pathways are limited in the cell or when oxidants are present that destroy existing ISCs.

9.4.4 NsrR: Sensing of NO with an Iron-Sulfur Cluster

E. coli NsrR is, together with IscR, a member of the Rfr2 family of regulatory proteins. NsrR shares with NorR the property of being a direct NO-sensor that controls NO detoxification functions. NsrR was discovered in *E. coli* in an attempt to understand the NO-specific induction of *hmp* encoding an NO-detoxifying flavohemoglobin (Bodenmiller and Spiro 2006). Genomic-based analyses indicate that the *E. coli* NsrR regulon is comprised of ~60 genes and includes functions for NO-detoxification, for nitrogen respiration, as well as for the other dedicated NO-sensor, NorR (Partridge et al. 2009). NsrR orthologs are found in many bacteria and representatives have been purified from *E. coli, Neisseria gonorrhoeae, B. subtilis,* and *S. coelicolor* (Tucker et al. 2010).

Studies of the detailed mechanism of NO-sensing by NsrR proteins have yielded conflicting results. Each purified *E. coli* NsrR protomer contains a $[2Fe\text{-}2S]^{+}$ cluster which, like IscR, is coordinated by three Cys residues. The fourth ligand in the case of NsrR is predicted to be a His residue (Tucker et al. 2010). The NsrR $[2Fe\text{-}2S]^{+}$ cluster is dinitrosylated (to a dinitrosyl iron complex; DNIC) in the presence of NO leading to derepression. Similarly, both the *N. gonorrhoeae* and *S. coelicolor* NsrR proteins were purified with a $[2Fe\text{-}2S]^{+}$ cluster, which responds to NO with formation of a DNIC and loss of DNA-binding activity (Isabella et al. 2009; Tucker et al. 2008). In contrast, *B. subtilis* NsrR purified anaerobically contains a $[4Fe\text{-}4S]^{2+}$ cluster (Yukl et al. 2008). Since the *N. gonorrhoeae, B. subtilis,* and *S. coelicolor* proteins were all purified after overexpression in *E. coli*, and only the *B. subtilis* protein was purified under anaerobic conditions, this raises the possibility that the observed $[2Fe\text{-}2S]^{+}$ clusters may be an artifact of heterologous expression or a product of the aerobic decomposition of an originally present $[4Fe\text{-}4S]^{2+}$ cluster. Indeed, anaerobic purification of *S. coelicolor* NsrR was recently reported to yield protein with a $[4Fe\text{-}4S]^{2+}$ cluster although it is unclear whether this form of the protein can sense NO (Tucker et al. 2010). It is possible that some NsrR proteins use a $[2Fe\text{-}2S]^{+}$ cluster and others use a $[4Fe\text{-}4S]^{2+}$ cluster to sense NO, but this would be unusual given the overall similarity in protein sequence and function.

9.4.5 Aconitase: A Metabolic Enzyme with a Redox-Sensitive Switch

Aconitases catalyze the interconversion of citrate and isocitrate in the tricarboxylic acid (TCA) cycle. This reaction occurs when the substrate binds to a $[4Fe\text{-}4S]^{2+}$ cluster coordinated by three Cys residues with one exposed iron atom. Like other, related $[4Fe\text{-}4S]^{2+}$ dehydratases, aconitase functions as a dehydratase, converting citrate to cis-aconitate, followed by rehydration to generate the isomer, iso-citrate. As a class, $[4Fe\text{-}4S]^{2+}$-containing dehydratases are sensitive to redox stress and, in some cases, to iron depletion due to the presence of a weakly bound iron atom

(Varghese et al. 2003). In the presence of superoxide, oxidation of the $[4Fe\text{-}4S]^{2+}$ cluster leads to an unstable $[4Fe\text{-}4S]^{3+}$ cluster, which decomposes with loss of Fe^{2+} to a $[3Fe\text{-}4S]^{+}$ cluster (Flint et al. 1993). Thus, oxidation of this ISC constitutes a switch that, minimally, converts aconitase from an active to an inactive form. What makes this switch particularly noteworthy is that, in at least some systems, the oxidized form of aconitase has other roles in the cell.

In mammalian cells, a cytosolic aconitase functions as a redox- and iron-regulated RNA-binding protein to control both translation initiation and mRNA stability of target mRNAs (Rouault 2006). This Iron-Responsive Element (IRE)-binding protein (IRP1), together with a homolog IRP2, responds to iron status, oxidative stress, and NO to control the expression of iron homeostasis functions. Detailed structural and functional characterization of IRP1 demonstrate that this cytosolic aconitase has a $[4Fe\text{-}4S]^{2+}$ cluster and, upon loss of the ISC, functions as an mRNA binding protein (reviewed in Wang and Pantopoulos 2011).

There is suggestive evidence that at least some bacterial aconitases may also have an additional regulatory role. *E. coli* encodes two, differentially regulated aconitases. The major enzyme in non-stressed cells is aconitase B (AcnB), which is sensitive to inactivation by one-electron oxidants and under conditions of iron-limitation (Varghese et al. 2003). The inactive form of AcnB is not well characterized *in vivo* but is presumed to be the $[3Fe\text{-}4S]^{+}$ form. This conversion not only inactivates the enzymatic activity of AcnB, but the demetallated protein binds to the 3′-untranslated region of its own mRNA thereby stabilizing the message against degradation (Tang and Guest 1999).

A second aconitase, AcnA, is expressed at a lower level but further induced under iron-replete conditions (as sensed by Fur) and as part of the SoxRS-regulon. Unlike AcnB, this aconitase isozyme is relatively resistant to both oxidative stress and iron fluctuations. These observations have led to a model in which one function of AcnB is to sense iron availability in the cell (Varghese et al. 2003). Under iron-restricted conditions, AcnB will be inactive leading to the accumulation and excretion of citrate, which can serve as a siderophore. Conversely, if iron levels are sufficient but AcnB is inactivated by redox stress, induction of AcnA can provide aconitase activity (Varghese et al. 2003). AcnB also reacts rapidly with nitric oxide to generate the DNIC. This reaction occurs faster than the competing reactions of NO with either molecular oxygen (which generates peroxynitrite) or glutathione. This suggests that aconitase, and related $[4Fe\text{-}4S]^{2+}$ dehydratases (e.g., IlvD) may be primary targets of NO toxicity in *E. coli* (Duan et al. 2009).

The ability of demetallated aconitases to bind to their own mRNA suggests that aconitases may comprise a molecular switch (Tang and Guest 1999). Initial studies suggest that this switch may be mediated by $[4Fe\text{-}4S]^{2+}$ cluster oxidation, much like mammalian IRPs. Both *E. coli* aconitases affect the stability of their own mRNAs. They also tune the translation of *sodA*, encoding manganese superoxide dismutase, with AcnA playing a positive role and AcnB a negative role (Tang et al. 2002). AcnB also affects translation of *ftsH* and thereby affects motility (Tang et al. 2004). Although the $[4Fe\text{-}4S]^{2+}$ cluster is implicated in this regulatory switch, structural studies of AcnB revealed that this protein contains an additional HEAT domain that may mediate iron-responsive protein dimerization and be part of the functional switch (Williams et al. 2002; Tang et al. 2005).

Further studies are required to determine whether aconitases routinely play a dual role in the Bacteria, serving both as a central metabolic enzyme and regulatory switch. To date, mRNA binding activity has been reported for aconitases from *B. subtilis* (Alen and Sonenshein 1999) and *Mycobacterium tuberculosis* (Banerjee et al. 2007). However, it is not yet clear whether these switches respond physiologically to iron depletion, RNOS, or both and the scope and significance of the changes elicited by the RNA-binding activity have yet to be determined. Ultimately, deciphering the regulatory role of aconitases will require a concerted effort to define their global impact on gene expression under various stress conditions.

9.4.6 WhiB-Like Regulators: A Family of Redox Sensors in Actinobacteria

In addition to the relatively well-characterized model systems described above, there is suggestive evidence that iron-sulfur clusters may play a regulatory role in several other contexts. One emerging area of research is the role of the WhiB protein and, more generally, the WhiB-like (Wbl) family of regulators in the actinobacteria (reviewed in den Hengst and Buttner 2008). The *whiB* gene was originally described in *S. coelicolor* as the locus of a mutation affecting development, antibiotic production, and pigmentation. The extended WhiB family of proteins includes 14 members in *S. coelicolor* and it is generally assumed, based largely on their pleiotropic phenotypes, that *wbl* genes encode transcription factors although detailed biochemical characterization of their DNA-binding activity and regulatory potential is just beginning to emerge.

The first Wbl protein found to have an associated $[4Fe\text{-}4S]^{2+}$ cluster was *S. coelicolor* WhiD (Jakimowicz et al. 2005). Subsequently, several Wbl proteins from another member of the actinobacteria, *M. tuberculosis*, were also found to have associated ISCs although, in this case, several were found to contain $[2Fe\text{-}2S]^{+}$ clusters (Alam et al. 2009). The *M. tuberculosis* WhiB4 protein appears to purify with a bound $[2Fe\text{-}2S]^{+}$ cluster, but can be reconstituted with a $[4Fe\text{-}4S]^{2+}$ cluster. In this case, WhiB4 was proposed to function as a protein disulfide reductase in which the active site is occluded, in non-stress conditions, by an assembled ISC (Alam et al. 2007). If confirmed, this will be the first example of a redox-gated switch in a bacterial protein disulfide reductase. However, this activity was not shared with the demetallated *S. coelicolor* WhiD (Crack et al. 2009).

We are still far from understanding the presumed regulatory functions of the iron sulfur clusters in the extended Wbl family of proteins. Current evidence suggests a role for Wbl proteins in sensing RNOS. *M. tuberculosis* WhiB3 appears to respond to NO by oxidation of its $[4Fe\text{-}4S]^{2+}$ cluster (Singh et al. 2007) and has been demonstrated to bind directly to the promoter regions of target operons (Singh et al. 2009). Recent biochemical results indicate that both *S. coelicolor* WhiD and

M. tuberculosis WhiB1 react extremely rapidly with NO leading to complete cluster dissolution. Together with genetic evidence linking Wbl proteins to nitrosative stress, this rapid reaction suggests a likely regulatory role in NO signaling (Crack et al. 2011).

9.4.7 Iron-Sulfur Clusters as Redox Switches to Regulate Protein Function

As the exploration of the aconitase and Wbl proteins illustrates, the assembly, modification, and ultimate dissolution of ISCs may serve as regulatory switches to control enzyme activity. In the case of *E. coli* AcnB, the loss of the labile iron atom from the $[4Fe\text{-}4S]^{2+}$ cluster inactivates enzyme function potentially leading to the accumulation of citrate, a known siderophore (Varghese et al. 2003). This would thus appear to be a biologically relevant switch, albeit not necessarily a redox switch.

Bacterial metabolism relies on a large number of enzymes that utilize ISCs. It has been estimated that perhaps 5% of all proteins encoded by *E. coli* contain ISCs (predominantly $[4Fe\text{-}4S]^{2+}$) (Fontecave 2006). Although in many cases, the role of the ISC is apparent from the mechanism of the enzyme, in other cases its presence is seemingly unrelated to catalysis. It seems likely that some ISCs will play a role in redox-sensing instead of, or in addition to, any role in catalysis. One example of such a mechanism is *B. subtilis* PurF (glutamine phosphoribosylpyrophosphate amidotransferase) encoding the first enzyme of *de novo* purine biosynthesis. This enzyme contains an oxygen-labile $[4Fe\text{-}4S]^{2+}$ cluster proposed to function as a redox switch controlling protein stability (Grandoni et al. 1989; Switzer 1989). Another example is *B. subtilis* CysH1 (adenylyl sulfate/phosphoadenylyl sulfate reductase). This enzyme has a $[4Fe\text{-}4S]^{2+}$ cluster also proposed to function as an oxygen-sensitive redox switch (Berndt et al. 2004). It is widely appreciated that ISCs are very sensitive to oxidative degradation and this can have profound impacts on cellular physiology (Py et al. 2011; Imlay 2006). The more challenging question is to ascertain those specific examples where this sensitivity has been exploited by the cell to an adaptive advantage.

9.5 Redox Regulation by Heme Proteins

The third class of iron-containing redox sensors are those containing heme as a prosthetic group (Gilles-Gonzalez and Gonzalez 2005). Heme is bound non-covalently to a hydrophobic pocket in many heme sensors with the heme iron coordinated to four nitrogen atoms from the porphyrin macrocycle and one or two protein ligands. Typically, the fifth coordination site of the heme iron is a histidine imidazole nitrogen (designated the proximal ligand) with the sixth coordination site interacting, often weakly, with an additional (distal) ligand and available for the binding of a regulatory ligand. The distal ligand to the heme iron can be an amino acid, or even a low molecular weight thiol (Gusarov and Nudler 2005).

Heme proteins are particularly well suited for sensing of small diatomic gases such as O_2, NO, and CO (reviewed in Gilles-Gonzalez and Gonzalez 2005; Aono 2012). In a subset of these sensors, there is a redox change in the heme iron that accompanies ligand binding, but in many cases heme proteins function as gas sensors but not, strictly speaking, as redox sensors. However, the detailed mechanism of sensing is not, in all cases, established and, in any event, we have here extended the idea of redox sensors to include those proteins that sense redox active molecules. Therefore, we will consider several of the best-characterized examples of heme-based gas sensors.

As a class, most heme-based gas sensors have a proximal histidine that binds the heme ferrous ion and a relatively apolar pocket where a gas molecule can bind as a distal ligand. It has been noted that, for a wide-range of heme-based sensors, the ratio of binding affinities are similar with highest affinity for NO, a 1000-fold decreased affinity for CO, and a roughly 10^6-fold decreased affinity for O_2 (Tsai et al. 2012). However, individual sensor proteins can increase or decrease their ability to discriminate amongst these diatomic gases by a variety of chemical mechanisms including variation of the proximal ligand or altering the structure and properties of the ligand-binding pocket.

9.5.1 *Oxygen Sensing by FixL*

FixL was the first described representative of a class of heme-based oxygen sensors that use a PAS domain-associated heme (Gilles-Gonzalez et al. 1994). FixL is the sensor kinase of the FixL-FixJ TCS found in the alpha-proteobacteria. FixL proteins have been best characterized in *Sinorhizobium meliloti* and *Bradyrhizobium japonicum*, two organisms that establish a symbiotic relationship with the roots of specific host plants and are responsible for nitrogen fixation. In both systems, O_2 inhibits the activity of FixL, which in turn regulates a phosphorylation signaling cascade. The kinase activity of FixL leads to phosphotransfer to FixJ which then activates transcription of nitrogen fixation and microaerobic respiration genes (Green et al. 2009). Inhibition of FixL by O_2 ensures that the highly O_2 sensitive nitrogenase is not synthesized under conditions where it would be oxidatively destroyed.

The specificity of FixL for O_2 derives from the presence of a key Arg residue in the ligand-binding pocket. In the absence of O_2, this key Arg residue interacts with a proprionate side chain on the edge of the porphyrin macrocycle and the kinase is active. When bound to O_2 ($K_d \sim 50$–$140\ \mu M$) the heme iron adopts a low spin state (Green et al. 2009) and the electronegative oxygen attracts the Arg guanidinium group and leads to inhibition of kinase activity (Dunham et al. 2003). Although FixL can bind to CO and NO, these ligands do not inhibit kinase activity.

9.5.2 Oxygen Sensing by the *E. coli* Dos Ribonucleoprotein Complex (DOS-RNC)

The *E. coli* direct oxygen sensor (Ec DosP; formerly EcDos) protein is, like FixL, an oxygen sensor that utilizes a heme group bound to a PAS domain (reviewed in Sasakura et al. 2006). In DosP, the heme is bound to one of two PAS domains (PAS-A) and regulates the activity of a carboxyl-terminal region with similarity to GGDEF/EAL domains related to cyclic-di-GMP signaling (c-di-GMP). In this case, the relevant activity is presumed to be c-di-GMP hydrolysis, which is enhanced in the presence of oxygen (Kobayashi et al. 2010). Thus, DosP likely serves to help modulate levels of the key second messenger, c-di-GMP (Mills et al. 2011).

Recent work suggests that DosP functions in *E. coli* as part of a multiprotein complex (Tuckerman et al. 2011). This complex includes the DosP c-di-GMP phosphodiesterase and a co-transcribed, heme-regulated c-di-GMP cyclase (DosC). Together, DosC and DosP co-purify with polynucleotide phosphorylase (PNPase), which is part (together with enolase and RNase E) of an RNA degradosome. Collectively, these proteins have been described as the Dos ribonucleoprotein complex (DOS-RNC). Both DosC and DosP are regulated by binding to oxygen. Under anaerobic conditions, the DosC cyclase is active and c-di-GMP activates PNPase in the complex. Conversely, when O_2 is present, the DosP phosphodiesterase hydrolyzes c-di-GMP (and the cyclase is now inactive) thereby stabilizing target mRNAs against degradation (Tuckerman et al. 2011; Mills et al. 2011). The mechanisms by which mRNAs might be targeted for O_2-dependent regulation are presently unknown.

Although both DosP and DosC are heme-based O_2 sensors, they are structurally distinct. The DosP sensor domain contains a heme-binding PAS domain but is unusual in that the heme iron is coordinated by both a proximal His residue and a distal Met. Binding to O_2 therefore requires displacement of Met. DosP binds O_2 with an affinity (K_d) of $\sim 74\ \mu M$ (Tuckerman et al. 2009). In contrast, DosC contains heme bound to a globin-like domain (a globin-coupled sensor; Hou et al. 2001), which binds oxygen with a K_d of 20 μM (Tuckerman et al. 2009).

9.5.3 *Mycobacterium* DosS and DosT: Sensing by Heme-GAF Domain Proteins

The mycobacterial DosS and DosT proteins are paralogous sensor kinase proteins containing heme bound within a GAF domain. When activated, these kinases phosphorylate the response regulator DosR, which controls the expression of genes required for the entry of *M. tuberculosis* into the dormant state that allows for long term persistence within the host (Gerasimova et al. 2011). These three proteins (DosRST; also known as DevRST) are therefore thought to be key regulators for coordinating gene expression with environmental signals during the process of infection. These signals include O_2, and perhaps also CO and NO.

The DosS and DosT sensor kinases are thought to respond differently to O_2 and both are necessary for the full activation of the DosR regulon. DosT is more sensitive to decreasing oxygen levels and is the first activated during the transition to anaerobiosis (Kim et al. 2010; Honaker et al. 2009). The sequential activation of these two kinases leads to a gradual increase in active DosR, which may account for the sequential induction of various dormancy regulon target genes depending on the number and affinity of DosR binding sites (Chauhan et al. 2011). DosT functions as a direct gas sensor: the kinase is active in the absence of bound oxygen and binding of O_2 to the heme inactivates the kinase. In contrast, DosS is easily oxidized and binds H_2O at the distal position in the oxidized (Fe^{3+}) state. The DosS heme is reduced to the ferrous form under anoxic conditions, and this serves as a redox sensor (the heme iron changes valence). These differences may be due to the structure of the channel leading to the heme-iron within the GAF domain: point mutations that affect the size of the channel affect the access of O_2 to the heme iron and control whether the protein functions as a direct gas sensor or a redox sensor (Cho et al. 2011; Kim et al. 2010).

The nature of the signals that elicit changes in DosS and DosT activity *in vivo* are not yet fully understood. It is likely that O_2 is the relevant signal for DosT. Recent evidence indicates that a reduced menaquinone pool may also activate DosS, thus providing a mechanism to sense changes in activity of the electron transport chain and thereby regulating induction of the dormancy regulon (Honaker et al. 2010). Further, when coordinated with NO, the DosS Fe(III)-heme is much more easily reduced to the active, Fe(II) state suggesting the NO may be a physiologically relevant signal in cells (Yukl et al. 2011). Finally, it has been noted that the DosRST system is responsible for up-regulation of dormancy-associated genes in response to CO, which can be produced from the host heme-oxygenase system of macrophages (Shiloh et al. 2008). Defining the full range of signals that are active *in vivo* during the various stages of infection, and understanding how these signals work with the DosRST system and other convergent signaling pathways (Chao et al. 2010), presents a formidable challenge for future work.

9.5.4 Sensing by H-NOX Proteins

NO came to prominence as a second messenger in mammalian systems, where it exerts its effects, in part, by interactions with a soluble guanylate cyclase (sGC). A combination of genomics and structural studies led to the appreciation that related, heme-containing sensors are widespread in the Bacteria. These domains can bind either NO or oxygen as ligands and were christened as H-NOX (heme-NO and oxygen binding) domains (Pellicena et al. 2004).

Structural studies of bacterial H-NOX proteins have focused on the representatives from *Thermoanaerobacter tengcongensis* (Tt H-NOX), an anaerobic thermophile, and one from the cyanobacterium *Nostoc punctiforme* (NpH-NOX). Detailed biophysical studies of these proteins have led to a model in which distortion of the porphyrin ring by the H-NOX protein modulates the electron density on the heme iron and alters the iron redox potential (Olea et al. 2010; Spiro 2008).

While several biophysical studies have focused on the properties of bacterial H-NOX proteins, the output of these sensors, and their physiological roles are largely unknown. Recent results suggest that one H-NOX protein, from the squid symbiont *Vibrio fischeri*, functions to detect NO during colonization of the squid light organ (Wang et al. 2010). In this system the H-NOX protein is encoded adjacent to a gene encoding a histidine protein kinase suggesting that the output of NO-sensing might be an alteration in kinase activity. In the presence of NO and the H-NOX protein, *V. fisheri* downregulates the expression of genes involved in the utilization of heme iron. Since, as noted above, NO leads to nitrosylation of Fur it can lead to derepression of iron uptake functions. The pathway controlled by the H-NOX protein in this system may serve, in part, to counter this derepression (Wang et al. 2010).

9.5.5 Sensing of Heme Iron by Irr

In the alpha-proteobacteria, Fur proteins play a relatively minor role in regulating gene expression in response to iron status (Johnston et al. 2007; Rudolph et al. 2006). Indeed, in several species within the Rhizobiales and the Rhodobacterales the Fur-like protein functions instead as a manganese responsive regulator (Mur). In *Bradyrhizobium japonicum*, Fur responds to both manganese (Hohle and O'Brian 2009, 2010) and iron (Yang et al. 2006b). However, since iron and manganese regulation are closely intertwined (Puri et al. 2010), the apparent Fe(II) responsiveness in cells may be indirect.

Instead of Fur, many alpha-proteobacteria sense iron in the form of heme by the iron-responsive repressor (Irr). This is likely expedient since it is estimated that one-half of all iron in *Bradyrhizobium japonicum* is in the form of heme (Small et al. 2009). Irr contains two heme-binding sites and obtains heme directly from the site of synthesis on ferrochelatase (Qi and O'Brian 2002). Binding of heme not only signals iron sufficiency to Irr, but also primes the protein to function as a sensor or oxidative stress. Irr binds a ferric heme with an N-terminal heme-recognition motif (HRM) and a ferrous heme in an internal His-rich domain. Irr normally represses gene expression when iron is limited (the opposite of a typical Fur protein) and in response to iron (heme) sufficiency, Irr is inactivated. The inactivation of Irr is associated with heme-dependent protein oxidation, which targets the protein for rapid degradation (Small et al. 2009; Yang et al. 2006a). This MCO reaction is reminiscent of the inactivation of *B. subtilis* PerR by reaction of peroxide at a non-heme iron center, although the molecular details of this heme-mediated event are not yet clear. The inactivation of Irr by heme is not obligately coupled to protein degradation since this regulated degradation is not observed with the *Rhizobium leguminosarum* Irr protein (Singleton et al. 2010).

9.5.6 Sensing of CO and Redox by CooA

The heme-based *Rhodospirillum rubrum* CO-specific sensor, CooA, is a remarkable example of a protein with dual roles as both a redox and gas sensor (Roberts et al. 2005). Unlike other examples discussed, CooA is a member of the CRP family of DNA-binding activators. In response to CO-binding, CooA activates the transcription of genes involved in CO oxidation. Analysis of the CooA sensing mechanism has been aided by the availability of a crystal structure of the sensor protein (Lanzilotta et al. 2000). Upon binding of CO, a distal iron ligand is displaced resulting in a conformational change of CooA. Structural studies revealed that, in the off state, the heme iron is coordinated by both a proximal His ligand (H77) with the amino-terminal Pro residue (P2) from the other protomer serving as distal ligand. Remarkably, the proximal ligand is regulated by the oxidation state of the heme iron with H77 being replaced by Cys75 in Fe(III) CooA. It has been suggested that this redox switch is biologically relevant: under aerobic conditions CooA would be in the inactive Fe(III)-heme state, which would thereby prevent activation of the CO reduction system which is known to be oxygen labile. CooA would activate transcription only in response to CO and in the absence of O_2 (Bonam et al. 1989). *Pseudomonas aeruginosa* DNR is another member of the CRP superfamily that utilizes a heme moiety to sense a diatomic gas, NO (Giardina et al. 2011). In response to NO, DNR activates the expression of denitrification genes, but whether this is strictly a gas sensor, or additionally involves a redox change, is not yet resolved (Rinaldo et al. 2012).

9.5.7 Other Heme-Based Gas Sensors

We have here reviewed some of the best-characterized examples of heme-based sensor proteins. Recent work has led to the identification of numerous other examples that further emphasize the utility of heme as a sensor of environmental changes. Two additional families of heme-based sensors worth noting are represented by the *Burkholderia xenovorans* RcoM proteins (Marvin et al. 2008) and the globin-coupled sensor (GCS) proteins HemAT (Hou et al. 2000).

The *B. xenovorans* RcoM-1 and RcoM-2 proteins contain heme in a PAS domain and regulate CO-dependent transcription. In the absence of bound ligand, the heme is in the Fe(III) oxidation state with six ligands (including a proximal His and a distal Cys thiolate). Upon reduction to Fe(II), the Cys ligand is displaced and the Fe(II) is coordinated instead by a Met residue (Marvin et al. 2008). Coordination of either CO or NO can displace the weakly bound Met residue resulting in activation of transcription. This is analogous to the O_2-sensing mechanism employed by DosP, as noted above.

The best characterized O_2-binding heme proteins are, of course, myoglobin and hemoglobin. A widely distributed family of sensor proteins, found in both Bacteria and Archaea, use a related heme-containing domain and define the globin-coupled sensor (GCS) family (Hou et al. 2001). GCS family proteins include *E. coli* DosC (see above) and the aerotaxis sensor (HemAT) involved in coordinating motility in response to O_2 availability. The HemAT proteins were originally described in the archaeon *Halobacterium salinarum* (HemAT-HS) and in *B. subtilis* (HemAT-BS) (Hou et al. 2000). HemAT proteins contain an amino-terminal domain related to methyl-accepting chemotaxis proteins (MCPs) and a carboxy-terminal GCS domain. In response to increasing cytosolic O_2, binding to the heme iron leads to movement of the iron into the heme plane, which leads to movement of the proximal histidine. Selectivity for O_2 is derived, in part, from its ability to form a hydrogen bond with Thr95, which is coupled to bonding between His86 and a heme propionate side chain (Yoshimura et al. 2006). The dimeric HemAT protein is thought to bind O_2 with negative cooperativity. Activation of HemAT-coupled CheA protein initiates the chemotaxis signal transduction cascade that ultimately couples changes in O_2 levels to alterations in the direction of rotation of the flagellum and hence motility.

9.6 Thiol-Based Redox Regulators

The cysteine side-chain contains a thiol (–SH) group that provides a redox-switch for a variety of sensor proteins and enzymes. Ionization of the thiol (to form the thiolate anion) greatly enhances the reactivity of proteins with RNOS. Oxidation of cysteine (Cys) leads in many proteins to a disulfide in which two Cys residues are covalently linked (the oxidation of two Cys residues to form the disulfide generates cystine). Disulfide bond formation is initiated when one partner reacts with an oxidant to generate a sulfenic acid (R-SOH). Condensation of the initially formed sulfenate with another thiol (such as another Cys residue) generates a disulfide. Protein disulfides may form between two Cys residues in a single polypeptide chain (an intramolecular disulfide), between two subunits of a multimeric protein (an intersubunit disulfide), or between a reactive Cys thiolate and a low molecular weight thiol in the cell (a mixed disulfide).

There are a large and ever-growing number of thiol-containing redox sensors in both prokaryotes and eukaryotes and this area has been the focus of numerous recent reviews. These include reviews focused on cysteine chemistry and efforts to monitor redox status *in vivo* (Klomsiri et al. 2011; Leonard and Carroll 2011), thiol-based sensors (Antelmann and Helmann 2011), and thiol-based switches in enzymes (Klomsiri et al. 2011). Related topics are also addressed elsewhere in this Volume. Here, we will briefly review the key features of some of the best-characterized prokaryotic thiol-based redox sensors with a specific emphasis on the unique features of each system (Table 9.2).

Table 9.2 Thiol-based redox sensors and switches

Organism	Sensor (class)	Targets and/or function	Signal(s)	Mechanism
E. coli	OxyR	Peroxide stress response	H_2O_2	Intrasubunit disulfide
B. subtilis	OhrR (1-Cys)	Peroxidase regulation	ROOH, NaOCl	S-bacillithiolation
X. campestris	OhrR (2-Cys)	Peroxidase regulation	ROOH,	Intersubunit disulfide
S. aureus	SarZ (1-Cys)	Virulence gene expression	ROS	S-thiolation
B. subtilis	HypR	Hypochlorite resistance	NaOCl	Intersubunit disulfide
S. coelicolor	RsrA (Cys_4:Zn)	Anti-σ for σ^R thiol homeostasis regulon	ROS	Intrasubunit disulfide
B. subtilis	Spx	Thiol homeostasis regulon	ROS	Intrasubunit disulfide
S. enterica	SsrB	SPI-2 regulation (virulence)	RNOS	Cys S-nitrosylation
E. coli	Hsp33 (Cys_4:Zn)	Protein chaperone (holdase)	NaOCl	Intrasubunit disulfide
E. coli	MetE	Methionine synthase	NaOCl	S-glutathionylation
B. subtilis	MetE	Methionine synthase	NaOCl	S-bacillithiolation

9.6.1 E. coli OxyR: An Intramolecular Disulfide Switch in a Transcription Activator

OxyR was the first regulatory protein shown to be controlled by reversible disulfide-bond formation (Storz et al. 1990). In the presence of sub-micromolar levels of H_2O_2, OxyR is converted from a reduced form to an oxidized form in which two Cys residues (C199 and C208) are disulfide-linked. OxyR is a tetrameric DNA-binding protein of the LysR family and binds to its target DNA in both the reduced and oxidized forms. However, the oxidized form is altered in conformation and often activates transcription from adjacent promoter sites (Toledano et al. 1994). The targets of OxyR include enzymes for degradation of peroxides (catalase and alkylhydroperoxide reductase), maintenance of thiol homeostasis (glutathione reductase and a glutaredoxin), and for metal homeostasis (the Fur regulator, the MntH manganese uptake channel, and the iron-sequestration protein Dps).

The molecular mechanism of peroxide-sensing by OxyR has been characterized both structurally and through detailed biochemical analysis (Lee et al. 2004a; Choi et al. 2001). The initial oxidation occurs on C199 (the peroxidatic Cys) to yield the Cys-sulfenate (Cys-SOH), which then reacts with the C208 thiol to generate the disulfide. This is accompanied by a significant conformational change in the protein since these two thiols are normally 17 Å apart. The C199 thiolate is ionized at neutral pH due to an abnormally low pKa and this accounts, in part, for its high reactivity. This thiolate can also react with other RNOS including nitrosating agents or by trapping of the initially produced C199-sulfenate by reaction with glutathione (S-glutathionylation). The physiological relevance of these other types of modifications is a matter of debate. It has been shown, however, that the OxyR regulon can be induced by depletion of reduced thiols in the cytosol (disulfide stress, Aslund et al. 1999).

9.6.2 *OhrR Proteins as Oxidation-Sensitive Repressors*

The OhrR family of repressors bind DNA in their reduced state, typically as transcription repressors, and oxidation of the protein leads to derepression (Duarte and Latour 2010; Zuber 2009). OhrR family repressors represent a subset of the MarR family of regulators and are characterized by a conserved Cys residue in the amino-terminal, DNA-binding domain of the protein and often one or more Cys residues in a carboxyl-terminal domain. OhrR proteins were first characterized in *B. subtilis* (Fuangthong et al. 2001) and in *Xanthomonas campestris* (Sukchawalit et al. 2001) where they serve as repressor of a peroxidase (OhrA).

B. subtilis OhrR ($OhrR_{BS}$) is the prototype for the 1-Cys family of OhrR redox sensors, so-named since each protein subunit contains only a single redox active Cys residue. The structure of the complex of $OhrR_{BS}$ bound to operator DNA reveals a typical winged-helix, helix-turn-helix architecture (Hong et al. 2005). $OhrR_{BS}$ has a single Cys residue per protomer (Cys15) that reacts with organic peroxides to generate a sulfenate (Fuangthong and Helmann 2002). However, this reaction leads to only a small conformational change (Hong et al. 2005) and is not sufficient to inactivate the repressor (Lee et al. 2007). Since there is no nearby Cys residue (from either subunit of the dimer), protein disulfide bond formation does not occur. Instead, the initial sulfenate reacts rapidly with low molecular weight (LMW) thiols in the cytosol to generate a mixed disulfide. Isolation of oxidized OhrR from cells treated with organic hydroperoxide led to the identification of mixed disulfides between OhrR-Cys15 and both free cysteine and a novel thiol of 398 Da (Lee et al. 2007). This latter molecule was subsequently characterized as bacillithiol and is now appreciated as the major LMW thiol in many low GC Gram positive bacteria (reviewed in Helmann 2011).

Oxidation of $OhrR_{BS}$ can lead to a variety of products. When thiols are not available, the initial sulfenate slowly condenses with a backbone amide to generate a cyclic sulfenamide and this also inactivates DNA-binding activity. There is evidence that this reaction may also occur *in vivo* under some conditions. Both the mixed disulfide and sulfenamide form of OhrR can be reduced by thiol reducing agents to regenerate active repressor (Lee et al. 2007). Oxidation of OhrR to the mixed disulfide form has been studied with cumene hydroperoxide (a model organic peroxide) and also occurs in cells treated with the strong oxidant sodium hypochlorite (NaOCl) (Chi et al. 2011). When linoleic acid hydroperoxide was used as oxidant, the protein was instead rapidly overoxidized to the sulfinic and sulfonic acid forms (Soonsanga et al. 2008b). Unlike the mixed disulfide form, these cannot be repaired in the cell and, in this case, $OhrR_{BS}$ appears to function as a sacrificial regulator. Studies using single-chain variants demonstrate that oxidation of a single active site per dimer is likely sufficient to mediate induction (Eiamphungporn et al. 2009), which may be important to allow the cell to respond to even very low levels of ROS.

Xanthomonas campestris OhrR ($OhrR_{XC}$) is homologous to $OhrR_{BS}$ but has additional Cys residues in the C-terminal domain of the protein. In this case, initial oxidation of the N-terminal Cys residue leads to the rapid formation of an intersubunit disulfide (Newberry et al. 2007). This 2-Cys mechanism differs from the 1-Cys mechanism in that the regulator is much more resistant to overoxidation and does not rely on the LMW thiol pool for modification of the peroxidatic Cys (Soonsanga et al. 2008a). In $OhrR_{XC}$ the reactive Cys (C22) is separated by > 15 Å from its redox partner (C127) on the opposing subunit of the dimer. The redox active thiolate is held in a hydrogen-bonding network with two Tyr residues from the opposing subunit (as also seen with $OhrR_{BS}$) and disulfide bond formation is accompanied by disruption of these contacts leading to a reconfiguration of the dimer interface. As a result, the two DNA-binding domains undergo a rigid body rotation of 28° relative to the core of the protein dimer. A similar conformational change is likely to lead to DNA dissociation of other redox-active MarR regulators (Poor et al. 2009).

Redox-regulated MarR family regulators related to OhrR are present in many bacteria and there are often multiple paralogs within a single cell (Chen et al. 2011). Many of these repressors appear to be redox-regulated and this may involve either formation of an intersubunit disulfide (2-Cys mechanism) or a mixed disulfide (1-Cys mechanism). Regulators of this family include *S. aureus* MgrA and SarZ and *P. aeruginosa* OspR and MexR (reviewed in Chen et al. 2011). These regulators likely play important roles in coordinating the expression of virulence genes in response to RNOS. While some are closely related to OhrR, others are members of the structurally related MarR/DUF24 family of proteins (Antelmann and Helmann 2011). Of particular note, the redox-sensing mechanism of the *S. aureus* SarZ protein has been visualized by structure determination of the protein in the reduced, sulfenic acid, and mixed-disulfide forms (Poor et al. 2009). The structural basis for redox-sensing by the *B. subtilis* DUF24 family sensor, HypR, has also been determined. HypR is oxidized by hypochlorite to an intersubunit disulfide linking Cys14 of one subunit with Cys49 of the other and the resulting crosslinked protein functions as a transcription activator both *in vitro* and *in vivo* (Palm et al. 2012).

9.6.3 *Streptomyces coelicolor* RsrA: A Redox-Regulated Anti-sigma Factor

The *S. coelicolor* alternative σ factor, σ^R, regulates gene expression in response to oxidative stress (Paget et al. 2001; Kallifidas et al. 2010). In this system, oxidative stress is sensed by an anti-σ factor (RsrA) that binds and sequesters σ^R in an inactive state (Kang et al. 1999; Paget et al. 1998). When redox-active compounds oxidize cytosolic thiols, RsrA is inactivated and σ^R is released to activate genes involved in thiol homeostasis, synthesis of LMW thiols (Newton and Fahey 2008), and repair or degradation of oxidatively damaged proteins (Kallifidas et al. 2010).

RsrA is inactivated by reversible disulfide bond formation between 2 Cys residues in a Zn(II)-binding domain (Zdanowski et al. 2006; Bae et al. 2004; Li et al. 2003). In contrast with the other examples described above, the RsrA Zn(II) finger motif serves to activate the reactivity of (at least one) of the thiolate ligands. Zn(II) is bound by Cys11 together with three ligands from a conserved HCC motif (H37, C41, and C44). The primary disulfide in oxidized RsrA joins Cys11 and Cys44 (Zdanowski et al. 2006; Bae et al. 2004).

9.6.4 *B. subtilis* Spx: An Intramolecular Disulfide Switch

The *B. subtilis* Spx protein regulates a large group of perhaps 120 genes in response to disulfide stress (reviewed in Zuber 2009). Spx is unusual amongst prokaryotic transcription factors in that it lacks an obvious DNA-binding motif and interacts weakly if at all with target operons in the absence of bound RNAP. Indeed, Spx can interact with the C-terminal domain (CTD) of the alpha subunit of RNAP and interfere with activation by some more classical activator proteins (Nakano et al. 2003b). Spx is related in structure to the ArsC family of As(III) reductases and has a CxxC motif near the N-terminus. Oxidation to the disulfide activates Spx to engage with RNAP, which leads to elevated transcription of thiol homeostasis genes, methionine sulfoxide reductase, and various detoxification functions (Nakano et al. 2003a). When functioning as a transcription activator, Spx binds to DNA together with the alpha-CTD of RNA polymerase: a complex that has been characterized by X-ray crystallography (Newberry et al. 2005; Nakano et al. 2010).

The activity of Spx is tightly regulated in the cell at multiple levels. Transcription of Spx is controlled by at least three promoters including one repressed by PerR and the redox-sensitive MarR homolog YodB and one activated by the antibiotic stress inducible alternative σ factor, σ^M (Zuber 2009). The latter contributes to increasing oxidative stress resistance in cells exposed to antibiotics, many of which trigger cell killing pathways dependent on oxidative stress (Kohanski et al. 2007). Spx activity is also regulated by reversible disulfide bond formation. Finally, the levels of Spx are regulated by proteolysis by the ClpXP protease. Indeed, activation of Spx leads to upregulation of YjbH, which serves as an adaptor protein for delivery of Spx to the ClpXP protease (Kommineni et al. 2011). This therefore constitutes a negative feedback loop. Intriguingly, YjbH itself has a redox-sensitive Zn(II)-finger domain, suggesting that redox also regulates its activity. Moreover, YjbH activity is also regulated by an anti-adaptor protein, YirB (Kommineni et al. 2011). Thus, the Spx regulon is regulated by numerous known or potential redox-sensing proteins including Spx itself, PerR, YodB, and YjbH. The mechanisms by which all these signals are integrated to control the Spx stress response is presently unknown.

9.6.5 *Salmonella SsrB: A Reactive Cys Residue as a Sensor of Nitrosative Stress*

Salmonella enterica serovar Typhimurium is an intracellular pathogen that uses a type III secretion system to deliver effector proteins into the host cell (Fass and Groisman 2009). The genes encoding this secretion system are encoded on Salmonella Pathogenicity Island-2 (SPI-2) and are regulated by the two-component system SsrAB. The presence of RNOS is one of the signals that regulate virulence gene expression during infection. Remarkably, SsrB is a direct sensor of RNOS and is S-nitrosylated on Cys203 in the C-terminal dimerization domain. Modification of C203 inhibits the DNA-binding activity of SsrB. *In vivo* studies indicate that this modification is physiologically relevant: a C203S mutant protein retains DNA-binding activity but is insensitive to RNOS. Further, the ability of SsrB to respond to RNOS is correlated with attenuated growth in a mouse disease model (Husain et al. 2010).

9.6.6 *Hsp33: A Redox Regulated Chaperone*

The *E. coli* heat shock protein, Hsp33, functions as a protein chaperone, whose activity is regulated by a reversible thiol-dependent redox switch. This system, recently reviewed in detail (Kumsta and Jakob 2009), is an excellent example of a redox switch devoted to control of protein activity rather than effecting changes in gene expression. Hsp33 is an abundant protein "holdase" that helps prevent protein aggregation under stress conditions. Hsp33 activity is strongly upregulated by ROS (Jakob et al. 1999) and can thereby compensate for the inactivity of DnaK under severe stress conditions (Winter et al. 2005). Although it was initially noted that Hsp33 activity was activated by the simultaneous presence of high levels of H_2O_2 and high temperatures (Ilbert et al. 2007), the physiological relevance of this was unclear. The realization that Hsp33 is strongly and selectively activated by comparatively low levels of bleach suggests that this is likely to be a physiologically relevant oxidant for Hsp33 activation (Winter et al. 2008). Indeed, Hsp33 protects proteins against hypochlorite-induced aggregation and thereby confers resistance to this strong oxidant.

The mechanism of Hsp33 activation has been defined through detailed biochemical and structural studies (Kumsta and Jakob 2009). In non-stressed conditions, Hsp33 is a monomer with a C-terminal Cys_4-Zn regulatory domain. Oxidation leads to release of Zn(II), formation of two intramolecular disulfide bonds, and dimerization of Hsp33 to an active chaperone. Once the stress conditions have been alleviated, Hsp33 can be reduced by the thioredoxin or glutaredoxin systems. The reduced Hsp33, together with any bound client proteins, interacts with the ATP-dependent foldase, DnaK, to release refolded proteins and allow release of the inactive Hsp33 chaperone.

9.6.7 Redox Regulated Enzymes

Disulfide stress can lead to oxidation of thiols in numerous proteins as monitored by a variety of redox proteomics technologies (Leonard and Carroll 2011; Lindahl et al. 2011). In some cases, it is likely that these oxidation events play a regulatory role. For example, oxidation of a Cys in or near a protein active site can potentially provide a redox switch for reversible regulation of enzyme function, analogous to the regulation of aconitase activity in response to Fe(II) availability noted above. One example of such a mechanism is methionine synthase.

Exposure of *E. coli* to H_2O_2 leads to the oxidation of Cys residues in numerous proteins including MetE, encoding cobalamin-independent methionine synthase (Hondorp and Matthews 2004). Oxidation of MetE leads to methionine deficiency (Hondorp and Matthews 2009). Proteomic analyses indicated highly specific S-glutathionylation of MetE on Cys645, which is near the entrance to the active site. This oxidation event, which can be triggered by H_2O_2 or the thiol-specific oxidant diamide, can quantitatively inactivate MetE (Hondorp and Matthews 2009). The reduction of MetE to an active form is correlated with resumption of growth. Mutation of C645 to Ala eliminated the Met auxotrophy imposed by diamide, demonstrating that this modification is necessary for the observed physiological effects. Remarkably, this same mechanism is conserved in *B. subtilis* (Chi et al. 2011). In this case, exposure to bleach leads to the highly specific S-bacillithiolation of MetE (and a paralog, YxjG). In addition to MetE, several other enzymes are S-bacillithiolated in response to NaOCl stress in *B. subtilis* (Chi et al. 2011). It is presently unclear whether this is a general protective mechanism or whether this serves a regulatory role by acting as a redox switch to control enzyme function, as shown for MetE. It can be anticipated that the increasing power of redox proteomics will uncover additional examples of post-translational modifications of enzymes, some of which will be manifestations of redox processes. Deciphering which of these are adventitious events, and which have regulatory consequences, will require careful physiological and genetic studies of each modification event.

9.7 Redox Regulation Mediated by Flavin Cofactors

Flavin cofactors, including both flavin mononucleotide (FMN) and flavin adenine dinucleotide (FAD) are involved in a wide variety of one- and two-electron transfer reactions in the cell. There are now numerous examples where the flavin cofactor serves a regulatory function in addition to, or instead of, a catalytic function. The roles of flavins in the redox switching of protein functions has recently been reviewed in depth (Becker et al. 2011), and we will here summarize some of the highlights from this field. Unlike many of the other redox switches described, flavoproteins do not typically function as DNA-binding regulatory proteins, although they can control transcription indirectly.

9.7.1 Flavin-Containing PAS Domains: The LOV Subfamily

The PAS domain is one of the most versatile ligand-binding motifs in bacterial sensor proteins and is associated with binding of flavins in several different classes of sensor proteins (Henry and Crosson 2011). Flavin-based sensors utilizing a PAS domain include the LOV domain photosensors, the *E. coli* Aer aerotaxis receptor, and the *Azotobacter* NifL protein.

The most thoroughly characterized FAD/FMN-based light sensors are the Vivid (VVD) protein from the fungus *Neurospora* and the *B. subtilis* YtvA protein that ultimately contributes to the regulation of the σ^B general stress response (Losi and Gartner 2012; Herrou and Crosson 2011). Here, we focus on the latter example. YtvA is thought to be a dimeric protein that associates as part of a multiprotein "stressosome" complex that ultimately helps to regulate activity of σ^B. YtvA itself has two domains: a LOV domain with an FMN chromophore for signal perception and a STAS (sulfate transporter and anti-sigma factor) domain that binds GTP and is involved in signal transmission. YtvA senses blue light as a signal and this sensing results in the formation of a covalent adduct between a protein Cys residue and the bound FMN cofactor. Absorption of blue light excites the bound FMN cofactor to triplet state, which decays with the generation of a FMN-cysteine C(4a)-thiol adduct (Losi and Gartner 2012; Herrou and Crosson 2011). This covalently bound (light) state (referred to as LOV390 to indicate the peak of absorbance at 390 nm) reverts, on a time-scale that varies dramatically from one photosensor to another, to regenerate the (dark) state (LOV450). For YtvA, this process takes about one hour, making it one of the slower LOV-based photoswitching events. Detailed studies of the YtvA protein in the dark and light states have provided insights into signal propagation in this system as described elsewhere (Losi and Gartner 2012; Herrou and Crosson 2011). The physiological significance of light-mediated regulation of the σ^B general stress response is not entirely clear.

Genomic studies have uncovered numerous examples of LOV-domain associated signaling systems, including several associated with pathogens that await further study. One notable example, from the anaerobic bacterium *Erythrobacter litoralis*, is EL222. This protein, notable as the first example of a full length LOV domain signaling protein to have a high resolution structure (Nash et al. 2011), has an N-terminal LOV domain and C-terminal DNA-binding domain related the LuxR family or repressors. In this system, light is proposed to directly activate the DNA-binding function of the EL222 protein.

9.7.2 Flavin Containing PAS Domains Involved in Redox Signaling

PAS domains play an integral role in numerous bacterial signal transduction systems (reviewed in Taylor and Zhulin 1999; Becker et al. 2011). In the nitrogen fixing bacterium *Azotobacter vinelandii*, NifL provides a redox sensor that controls the activity of the NifA transcription factor. When oxygen is present, and NifL is in the oxidized conformation, it binds to NifA preventing the expression of the *nif* genes, which encode the O_2-sensitive nitrogenase enzyme. In the absence of O_2, FAD bound to one of two PAS domains in NifL is in the reduced state and NifA is released to activate gene expression (Martinez-Argudo et al. 2004). Oxidation of FAD in one PAS domain propagates a signal through the second domain and ultimately affects the NifA sequestration function as defined by recent genetic and physiological studies (Little et al. 2011). The pathway by which NifL is reduced is less well understood.

A second example of redox-sensing role for a PAS-domain associated FAD is the *E. coli* aerotaxis receptor, Aer (Taylor 2007). This transmembrane chemotaxis receptor forms a four-helical bundle, stabilized by adjacent HAMP domains, in which the cytosolic FAD-containing domains monitor cellular redox status (Taylor 2007). The redox state of the PAS-FAD domain alters the conformation of the HAMP helical bundle, as monitored in detail using structure-based protein-protein crosslinking studies, and ultimately leads to activation of the CheA kinase and a chemotactic response (Watts et al. 2011). This FAD domain coordinates aerotaxis in response to changes in the reduction potential of electron carriers, and stands in contrast the HemAT protein of *B. subtilis*, which senses O_2-binding directly.

9.7.3 Redox Regulation of Flavin-Containing Enzymes

The oxidation and reduction of a bound flavin cofactor is necessary for catalysis by many enzymes, but may also serve a regulatory function. Two well characterized examples, of what is perhaps a much broader phenomenon, are the *E. coli* pyruvate:quinone oxidoreductase (PQO) and proline dehydrogenase (PutA) involved in proline utilization. In both of these systems, changes in FAD redox status alter protein localization and function within the cell.

PQO localizes to the inner surface of the cytosolic membrane when the flavin is reduced. This form of the enzyme is activated for catalysis and transfers electrons from pyruvate, via the flavin, to ubiquinone in the membrane. In contrast, when PQO is oxidized the enzyme is cytosolic rather than membrane-associated (Becker et al. 2011). The cytosolic PQO is not known to have a separate function, so this redistribution can be viewed as a simple on:off switch for enzyme activity.

A more dramatic example of a FAD-mediated redox switch is the PutA enzyme needed for proline utilization. PutA has an N-terminal DNA-binding domain and when the FAD is oxidized PutA localizes to five binding sites in the regulatory region controlling transcription of the divergent *putP* and *putA* genes. In contrast, when proline is present PutA relocalizes to the inner surface of the membrane where it catalyzes electron transfer from proline to ubiquinone (Becker et al. 2011).

Table 9.3 Redox sensors using or sensing organic cofactors (FMN, FAD, NADH/NAD^+, quinones)

Organism	Sensor (cofactors)	Targets and/or function	Signal(s)	Mechanism
B. subtilis	YtvA (LOV:FMN)	Regulation of σ^B general stress response	Blue light	FMN-Cys adduct
A. vinelandii	NifL (PAS:FAD)	Regulation of NifA and nitrogen fixation	O_2	FAD redox
S. coelicolor	Rex	Regulation of respiratory pathways	NADH/NAD^+	Cofactor binding
E. coli	ArcB	ArcA response regulator; anaerobic respiration	Q, MK	Cofactor binding; thiol redox?
Rhodobacter spp.	RegB	RegA response regulator; photosynthesis	Q/QH_2 ratio	Cofactor binding; thiol redox

9.8 Direct Sensing of Electron Carriers (NADH/NAD^+ and Quinone Pools)

The final group of prokaryotic redox sensors we will consider are those proteins that sense cellular redox status by the direct binding to, or reaction with, electron carriers in the cell (Table 9.3). The Rex protein, initially described in *S. coelicolor*, is the prototype for a family of redox sensors that monitor the ratio of reduced to oxidized NADH in the cytosol. The *E. coli* ArcB transmembrane sensor kinase is an example of a family of sensors that monitor the status of the membrane quinone pools. The ability of regulatory proteins to sense electron carriers has been recently reviewed (Delgado et al. 2012).

9.8.1 Sensing of the NADH/NAD^+ Ratio by Rex

S. coelicolor Rex regulates respiratory processes by binding to operator sites controlling transcription of the cytochrome bd oxidase (*cydABCD*), the proton translocating NADH dehdyrogenase (*nuoA-N*), and the *rex-hemACD* operon encoding both Rex and heme biosynthetic enzymes (Brekasis and Paget 2003). Rex proteins from several organisms (including *Thermus aquaticus, S. coelicolor, B. subtilis,* and *S. aureus*) have been interrogated using a variety of structural, genetic and biochemical approaches. These analyses support a mechanism of Rex regulation in which the protein DNA-binding activity is controlled by the relative ratio of NADH and NAD^+ in the cytosol.

Rex binds DNA in the presence of NAD^+ and binding is inhibited by NADH. Indeed, *B. subtilis* Rex (Rex_{BS}) binds NAD^+ 30 times tighter when bound to DNA, implying that the converse is also true: NAD^+ enhances DNA-binding affinity (Wang et al. 2008). Normally, the rapid oxidation of NADH by transfer of electrons to the respiratory chain results in a much greater level of NAD^+ than NADH in the cell and Rex is tuned to sense changes in this ratio. Indeed, since Rex represses the NADH dehydrogenase which is the major pathway by which NADH is oxidized by ongoing respiration, this serves as a feedback loop (Gyan et al. 2006). Consistent with this notion, Rex_{BS} binds NADH 20,000 tighter than NAD^+ (K_d's of 24 nM for NADH vs. 0.5 mM for NAD^+) (Wang et al. 2008). A similar difference is noted for *S. aureus* Rex with a K_d of 95 nM for NADH and 150 μM for NAD^+ (Pagels et al. 2010). For *Thermus aquaticus* Rex (T-Rex), the first protein of this family to be structurally characterized (Sickmier et al. 2005), binding of NADH leads to a significant 40° rotation between the two subunits of the dimer leading to a conformation incompatible with DNA-binding (McLaughlin et al. 2010). The regulons controlled by Rex proteins have been at least partially defined and efforts are underway to better understand how the action of this regulator is integrated with other sensors monitoring the availability of electron acceptors (including oxygen) and other redox signals.

9.8.2 Two-Component Systems That Monitor Quinone Pools

After electrons are transferred to the electron transport system (ETS) from the soluble NADH pool, the major non-protein redox carriers are the quinones. *E. coli* synthesizes ubiquinone (Q), which is the major carrier for respiration with O_2 and NO_3 as electron acceptors, and menaquinones (MK) that are more commonly used for other types of anaerobic respiration. In *E. coli*, the oxidation status of the Q pool is thought to be directly sensed by the ArcB membrane kinase, which is inhibited in response to oxidized Q (Georgellis et al. 2001). Conversely, when Q is largely in the reduced form (QH_2) the ArcB kinase is active and phosphorylates the ArcA response regulator. ArcA, in turn, activates expression of an estimated 30 operons involved in respiration. The precise mechanism of regulation of ArcB by changes in redox status is still incompletely understood. One model posits that ArcB binds directly to oxidized Q and this allosterically regulates kinase activity. This binding may also lead to protein oxidation leading to disulfide bond formation between two ArcB monomers. The implicated Cys residues are in a linker domain between the transmembrane domains (where Q presumably interacts) and the C-terminal transmitter and kinase domains. It has also been noted that regulation of ArcB is not limited to the sensing of Q; the redox state of the MK pool also appears to play an important role (Bekker et al. 2010). Finally, transcriptional analyses combined with modeling approaches have led Poole and colleagues to question whether Q is a relevant signal for ArcA. They propose instead that the relevant signal, at least under their experimental conditions, is inhibition of the ArcB phosphatase activity by fermentation products (Rolfe et al. 2011). This system is further complicated by the ability of the ArcBA system to activate transcription of a small RNA (sRNA), ArcZ, that is encoded convergent with and overlapping the *arcB* gene. This sRNA is proposed to function, in part, as a negative feedback loop on the expression of the ArcB (Mandin and Gottesman 2010).

A conceptually related regulatory system is present in several photosynthetic bacteria and we here focus on the RegBA two-component system from *Rhodobacter* spp. In this system, RegB is the membrane-spanning sensor kinase and phosphorylates the RegA response regulator (Elsen et al. 2004). Together, RegBA regulate numerous metabolic processes including photosynthesis, carbon and nitrogen fixation, and respiration. Current models for RegB regulation propose that the transmembrane domain senses the ratio of Q/QH_2 in the membrane. In this system, both Q and QH_2 bind to RegB, but only Q inhibits the kinase activity. Although RegB also contains potentially redox active Cys in the linker region, studies with a Cys to Ser mutant demonstrate that this site is not required for regulation (Wu and Bauer 2010). Two-component systems related to RegBA are widespread in the Bacteria and are likely involved in redox regulation of central metabolic processes, although the biochemical details of these regulatory switches are still under emerging.

9.9 Concluding Remarks

Bacteria encode a remarkable diversity of redox-sensing regulators as evidenced by the broad survey presented in this chapter. In light of the central role that redox reactions in metabolism, it should perhaps not be surprising that redox-sensing regulators are tightly interwoven into the regulatory network of the cell. In many cases the biochemical basis of sensing is now well understood and this is increasingly supported by high resolution structural studies. An equally formidable challenge is to define the physiological role of each sensor and how multiple sensor proteins work together to integrate information about cellular redox status. Genetic and physiological studies, often buttressed by genomics-enabled approaches, are allowing rapid progress in this area as well. Redox-sensing regulators play important roles in maintaining homeostasis during major metabolic transitions, such as from aerobic to anaerobic growth. Moreover, many redox active gases (O_2, NO) and secondary metabolites (including several antibiotics) are involved in cell-cell signaling. As a result, redox regulators play key roles both in pathogenesis and the establishment of symbiotic interactions in a variety of systems. Here, the reader has been introduced to many of the best known types of regulators which will hopefully provide a useful framework for ongoing studies of these and yet to be discovered redox regulators.

Acknowledgements I thank Dr. J.P. Shapleigh for helpful comments. Work in the Helmann laboratory on metal ion homeostasis is supported by a grant from the National Institutes of Health (GM-059323) and studies of redox biology are supported by a grant from the National Science Foundation (MCB-1020481).

References

Alam MS, Garg SK, Agrawal P (2007) Molecular function of WhiB4/Rv3681c of *Mycobacterium tuberculosis* H37Rv: a [4Fe-4S] cluster co-ordinating protein disulphide reductase. Mol Microbiol 63:1414–1431

Alam MS, Garg SK, Agrawal P (2009) Studies on structural and functional divergence among seven WhiB proteins of *Mycobacterium tuberculosis* H37Rv. FEBS J 276:76–93

Alen C, Sonenshein AL (1999) *Bacillus subtilis* aconitase is an RNA-binding protein. Proc Natl Acad Sci USA 96:10412–10417

Anjem A, Imlay JA (2012) Mononuclear iron enzymes are primary targets of hydrogen peroxide stress. J Biol Chem 287(19):15544–15556

Anjem A, Varghese S, Imlay JA (2009) Manganese import is a key element of the OxyR response to hydrogen peroxide in *Escherichia coli*. Mol Microbiol 72:844–858

Antelmann H, Helmann JD (2011) Thiol-based redox switches and gene regulation. Antioxid Redox Signal 14:1049–1063

Aono S (2012) Novel bacterial gas sensor proteins with transition metal-containing prosthetic groups as active sites. Antioxid Redox Signal 16(7):678–686

Aslund F, Zheng M, Beckwith J, Storz G (1999) Regulation of the OxyR transcription factor by hydrogen peroxide and the cellular thiol-disulfide status. Proc Natl Acad Sci USA 96: 6161–6165

Bae JB, Park JH, Hahn MY, Kim MS, Roe JH (2004) Redox-dependent changes in RsrA, an anti-sigma factor in *Streptomyces coelicolor*: zinc release and disulfide bond formation. J Mol Biol 335:425–435

Banerjee S, Nandyala AK, Raviprasad P, Ahmed N, Hasnain SE (2007) Iron-dependent RNA-binding activity of *Mycobacterium tuberculosis* aconitase. J Bacteriol 189:4046–4052

Becker DF, Zhu W, Moxley MA (2011) Flavin redox switching of protein functions. Antioxid Redox Signal 14:1079–1091

Bekker M, Alexeeva S, Laan W, Sawers G, De Mattos JT, Hellingwerf K (2010) The ArcBA two-component system of *Escherichia coli* is regulated by the redox state of both the ubiquinone and the menaquinone pool. J Bacteriol 192:746–754

Berndt C, Lillig CH, Wollenberg M, Bill E, Mansilla MC, de Mendoza D, Seidler A, Schwenn JD (2004) Characterization and reconstitution of a 4Fe-4S adenylyl sulfate/phosphoadenylyl sulfate reductase from *Bacillus subtilis*. J Biol Chem 279:7850–7855

Bigelow DJ, Squier TC (2011) Thioredoxin-dependent redox regulation of cellular signaling and stress response through reversible oxidation of methionines. Mol Biosyst 7:2101–2109

Bodenmiller DM, Spiro S (2006) The *yjeB (nsrR)* gene of *Escherichia coli* encodes a nitric oxide-sensitive transcriptional regulator. J Bacteriol 188:874–881

Bonam D, Lehman L, Roberts GP, Ludden PW (1989) Regulation of carbon monoxide dehydrogenase and hydrogenase in *Rhodospirillum rubrum*: effects of CO and oxygen on synthesis and activity. J Bacteriol 171:3102–3107

Brekasis D, Paget MS (2003) A novel sensor of NADH/NAD + redox poise in *Streptomyces coelicolor* A3(2). EMBO J 22:4856–4865

Bush M, Ghosh T, Tucker N, Zhang X, Dixon R (2011) Transcriptional regulation by the dedicated nitric oxide sensor, NorR: a route towards NO detoxification. Biochem Soc Trans 39:289–293

Chao JD, Papavinasasundaram KG, Zheng X, Chavez-Steenbock A, Wang X, Lee GQ, Av-Gay Y (2010) Convergence of Ser/Thr and two-component signaling to coordinate expression of the dormancy regulon in *Mycobacterium tuberculosis*. J Biol Chem 285:29239–29246

Chauhan S, Sharma D, Singh A, Surolia A, Tyagi JS (2011) Comprehensive insights into *Mycobacterium tuberculosis* DevR (DosR) regulon activation switch. Nucleic Acids Res 39:7400–7414

Chen PR, Brugarolas P, He C (2011) Redox signaling in human pathogens. Antioxid Redox Signal 14:1107–1118

Chi BK, Gronau K, Mader U, Hessling B, Becher D, Antelmann H (2011) S-Bacillithiolation protects against hypochlorite stress in *Bacillus subtilis* as revealed by transcriptomics and redox proteomics. Mol Cell Proteomics 10:M111.009506

Cho HY, Cho HJ, Kim MH, Kang BS (2011) Blockage of the channel to heme by the E87 side chain in the GAF domain of *Mycobacterium tuberculosis* DosS confers the unique sensitivity of DosS to oxygen. FEBS Lett 585:1873–1878

Choi H, Kim S, Mukhopadhyay P, Cho S, Woo J, Storz G, Ryu SE (2001) Structural basis of the redox switch in the OxyR transcription factor. Cell 105:103–113

Crack JC, Jervis AJ, Gaskell AA, White GF, Green J, Thomson AJ, le Brun NE (2008) Signal perception by FNR: the role of the iron-sulfur cluster. Biochem Soc Trans 36:1144–1148

Crack JC, den Hengst CD, Jakimowicz P, Subramanian S, Johnson MK, Buttner MJ, Thomson AJ, le Brun NE (2009) Characterization of [4Fe-4S]-containing and cluster-free forms of *Streptomyces* WhiD. Biochemistry 48:12252–12264

Crack JC, Smith LJ, Stapleton MR, Peck J, Watmough NJ, Buttner MJ, Buxton RS, Green J, Oganesyan VS, Thomson AJ, le Brun NE (2011) Mechanistic insight into the nitrosylation of the [4Fe-4S] cluster of WhiB-like proteins. J Am Chem Soc 133:1112–1121

Crack JC, Green J, Hutchings MI, Thomson AJ, Le Brun NE (2012) Bacterial iron-sulfur regulatory proteins as biological sensor-switches. Antioxid Redox Signal 17(9):1215–1231

D'Autreaux B, Horner O, Oddou JL, Jeandey C, Gambarelli S, Berthomieu C, Latour JM, Michaud-Soret I (2004) Spectroscopic description of the two nitrosyl-iron complexes responsible for Fur inhibition by nitric oxide. J Am Chem Soc 126:6005–6016

Delgado MJ, Bueno E, Mesa S, Bedmar EJ, Richardson DJ (2012) Bacterial adaptation of respiration from oxic to microoxic and anoxic conditions: redox control. Antioxid Redox Signal 16(8):819–852

den Hengst CD, Buttner MJ (2008) Redox control in actinobacteria. Biochim Biophys Acta 1780:1201–1216

Dibden DP, Green J (2005) In vivo cycling of the *Escherichia coli* transcription factor FNR between active and inactive states. Microbiology 151:4063–4070

Dietrich LE, Kiley PJ (2011) A shared mechanism of SoxR activation by redox-cycling compounds. Mol Microbiol 79:1119–1122

Duan X, Yang J, Ren B, Tan G, Ding H (2009) Reactivity of nitric oxide with the [4Fe-4S] cluster of dihydroxyacid dehydratase from *Escherichia coli*. Biochem J 417:783–789

Duarte V, Latour JM (2010) PerR vs OhrR: selective peroxide sensing in *Bacillus subtilis*. Mol Biosyst 6:316–323

Dufour YS, Kiley PJ, Donohue TJ (2010) Reconstruction of the core and extended regulons of global transcription factors. PLOS Genet 6:e1001027

Dunham CM, Dioum EM, Tuckerman JR, Gonzalez G, Scott WG, Gilles-Gonzalez MA (2003) A distal arginine in oxygen-sensing heme-PAS domains is essential to ligand binding, signal transduction, and structure. Biochemistry 42:7701–7708

Eiamphungporn W, Soonsanga S, Lee JW, Helmann JD (2009) Oxidation of a single active site suffices for the functional inactivation of the dimeric *Bacillus subtilis* OhrR repressor in vitro. Nucleic Acids Res 37:1174–1181

Elsen S, Swem LR, Swem DL, Bauer CE (2004) RegB/RegA, a highly conserved redox-responding global two-component regulatory system. Microbiol Mol Biol Rev 68:263–279

Fass E, Groisman EA (2009) Control of *Salmonella* pathogenicity island-2 gene expression. Curr Opin Microbiol 12:199–204

Flatley J, Barrett J, Pullan ST, Hughes MN, Green J, Poole RK (2005) Transcriptional responses of *Escherichia coli* to S-nitrosoglutathione under defined chemostat conditions reveal major changes in methionine biosynthesis. J Biol Chem 280:10065–10072

Flint DH, Tuminello JF, Emptage MH (1993) The inactivation of Fe-S cluster containing hydrolyases by superoxide. J Biol Chem 268:22369–22376

Fontecave M (2006) Iron-sulfur clusters: ever-expanding roles. Nat Chem Biol 2:171–174

Frausto Da Silva J, Williams R (2001) The biological chemistry of the elements: the inorganic chemistry of life. Oxford University Press, Oxford

Fuangthong M, Helmann JD (2002) The OhrR repressor senses organic hydroperoxides by reversible formation of a cysteine-sulfenic acid derivative. Proc Natl Acad Sci USA 99(10): 6690–6695

Fuangthong M, Atichartpongkul S, Mongkolsuk S, Helmann JD (2001) OhrR is a repressor of ohrA, a key organic hydroperoxide resistance determinant in *Bacillus subtilis*. J Bacteriol 183:4134–4141

Fuangthong M, Herbig AF, Bsat N, Helmann JD (2002) Regulation of the *Bacillus subtilis fur* and *perR* genes by PerR: not all members of the PerR regulon are peroxide inducible. J Bacteriol 184:3276–3286

Georgellis D, Kwon O, Lin EC (2001) Quinones as the redox signal for the arc two-component system of bacteria. Science 292:2314–2316

Gerasimova A, Kazakov AE, Arkin AP, Dubchak I, Gelfand MS (2011) Comparative genomics of the dormancy regulons in mycobacteria. J Bacteriol 193:3446–3452

Giardina G, Castiglione N, Caruso M, Cutruzzola F, Rinaldo S (2011) The *Pseudomonas aeruginosa* DNR transcription factor: light and shade of nitric oxide-sensing mechanisms. Biochem Soc Trans 39:294–298

Giedroc DP (2009) Hydrogen peroxide sensing in *Bacillus subtilis*: it is all about the (metallo)regulator. Mol Microbiol 73:1–4

Gilles-Gonzalez MA, Gonzalez G (2005) Heme-based sensors: defining characteristics, recent developments, and regulatory hypotheses. J Inorg Biochem 99:1–22

Gilles-Gonzalez MA, Gonzalez G, Perutz MF, Kiger L, Marden MC, Poyart C (1994) Heme-based sensors, exemplified by the kinase FixL, are a new class of heme protein with distinctive ligand binding and autoxidation. Biochemistry 33:8067–8073

Grandoni JA, Switzer RL, Makaroff CA, Zalkin H (1989) Evidence that the iron-sulfur cluster of *Bacillus subtilis* glutamine phosphoribosylpyrophosphate amidotransferase determines stability of the enzyme to degradation in vivo. J Biol Chem 264:6058–6064

Green J, Paget MS (2004) Bacterial redox sensors. Nat Rev Microbiol 2:954–966

Green J, Crack JC, Thomson AJ, Lebrun NE (2009) Bacterial sensors of oxygen. Curr Opin Microbiol 12:145–151

Gruner I, Fradrich C, Bottger LH, Trautwein AX, Jahn D, Hartig E (2011) Aspartate 141 is the fourth ligand of the oxygen-sensing [4Fe-4S]2+ cluster of *Bacillus subtilis* transcriptional regulator Fnr. J Biol Chem 286:2017–2021

Gu M, Imlay JA (2011) The SoxRS response of *Escherichia coli* is directly activated by redox-cycling drugs rather than by superoxide. Mol Microbiol 79:1136–1150

Gusarov I, Nudler E (2005) NO-mediated cytoprotection: instant adaptation to oxidative stress in bacteria. Proc Natl Acad Sci USA 102:13855–13860

Gyan S, Shiohira Y, Sato I, Takeuchi M, Sato T (2006) Regulatory loop between redox sensing of the NADH/NAD(+) ratio by Rex (YdiH) and oxidation of NADH by NADH dehydrogenase Ndh in *Bacillus subtilis*. J Bacteriol 188:7062–7071

He C, Hus JC, Sun LJ, Zhou P, Norman DP, Dotsch V, Wei H, Gross JD, Lane WS, Wagner G, Verdine GL (2005) A methylation-dependent electrostatic switch controls DNA repair and transcriptional activation by *E. coli* Ada. Mol Cell 20:117–129

Helmann JD (2010) Metal-dependent and metal-responsive regulatory systems. In: Spiro S, Dixon R (eds) Sensory mechanisms in bacteria: molecular aspects of signal recognition. Caister Academic Press, Portland

Helmann JD (2011) Bacillithiol, a new player in bacterial redox homeostasis. Antioxid Redox Signal 15:123–133

Helmann JD, Wu MF, Gaballa A, Kobel PA, Morshedi MM, Fawcett P, Paddon C (2003) The global transcriptional response of *Bacillus subtilis* to peroxide stress is coordinated by three transcription factors. J Bacteriol 185:243–253

Henry JT, Crosson S (2011) Ligand-binding PAS domains in a genomic, cellular, and structural context. Annu Rev Microbiol 65:261–286

Herbig AF, Helmann JD (2001) Roles of metal ions and hydrogen peroxide in modulating the interaction of the *Bacillus subtilis* PerR peroxide regulon repressor with operator DNA. Mol Microbiol 41:849–859

Herrou J, Crosson S (2011) Function, structure and mechanism of bacterial photosensory LOV proteins. Nat Rev Microbiol 9:713–723

Hohle TH, O'Brian MR (2009) The *mntH* gene encodes the major Mn(2+) transporter in *Bradyrhizobium japonicum* and is regulated by manganese via the Fur protein. Mol Microbiol 72:399–409

Hohle TH, O'Brian MR (2010) Transcriptional control of the *Bradyrhizobium japonicum irr* gene requires repression by fur and antirepression by Irr. J Biol Chem 285:26074–26080

Honaker RW, Leistikow RL, Bartek IL, Voskuil MI (2009) Unique roles of DosT and DosS in DosR regulon induction and *Mycobacterium tuberculosis* dormancy. Infect Immun 77:3258–3263

Honaker RW, Dhiman RK, Narayanasamy P, Crick DC, Voskuil MI (2010) DosS responds to a reduced electron transport system to induce the *Mycobacterium tuberculosis* DosR regulon. J Bacteriol 192:6447–6455

Hondorp ER, Matthews RG (2004) Oxidative stress inactivates cobalamin-independent methionine synthase (MetE) in *Escherichia coli*. PLOS Biol 2:e336

Hondorp ER, Matthews RG (2009) Oxidation of cysteine 645 of cobalamin-independent methionine synthase causes a methionine limitation in *Escherichia coli*. J Bacteriol 191:3407–3410

Hong M, Fuangthong M, Helmann JD, Brennan RG (2005) Structure of an OhrR-*ohrA* operator complex reveals the DNA binding mechanism of the MarR family. Mol Cell 20:131–141

Hou S, Larsen RW, Boudko D, Riley CW, Karatan E, Zimmer M, Ordal GW, Alam M (2000) Myoglobin-like aerotaxis transducers in Archaea and Bacteria. Nature 403:540–544

Hou S, Freitas T, Larsen RW, Piatibratov M, Sivozhelezov V, Yamamoto A, Meleshkevitch EA, Zimmer M, Ordal GW, Alam M (2001) Globin-coupled sensors: a class of heme-containing sensors in Archaea and Bacteria. Proc Natl Acad Sci USA 98:9353–9358

Husain M, Jones-Carson J, Song M, McCollister BD, Bourret TJ, Vazquez-Torres A (2010) Redox sensor SsrB Cys203 enhances Salmonella fitness against nitric oxide generated in the host immune response to oral infection. Proc Natl Acad Sci USA 107:14396–14401

Ilbert M, Horst J, Ahrens S, Winter J, Graf PC, Lilie H, Jakob U (2007) The redox-switch domain of Hsp33 functions as dual stress sensor. Nat Struct Mol Biol 14:556–563

Imlay JA (2003) Pathways of oxidative damage. Annu Rev Microbiol 57:395–418

Imlay JA (2006) Iron-sulphur clusters and the problem with oxygen. Mol Microbiol 59:1073–1082

Isabella VM, Lapek JD Jr, Kennedy EM, Clark VL (2009) Functional analysis of NsrR, a nitric oxide-sensing Rrf2 repressor in *Neisseria gonorrhoeae*. Mol Microbiol 71:227–239

Jacquamet L, Traore DA, Ferrer JL, Proux O, Testemale D, Hazemann JL, Nazarenko E, El Ghazouani A, Caux-Thang C, Duarte V, Latour JM (2009) Structural characterization of the active form of PerR: insights into the metal-induced activation of PerR and Fur proteins for DNA binding. Mol Microbiol 73:20–31

Jakimowicz P, Cheesman MR, Bishai WR, Chater KF, Thomson AJ, Buttner MJ (2005) Evidence that the *Streptomyces* developmental protein WhiD, a member of the WhiB family, binds a [4Fe-4S] cluster. J Biol Chem 280:8309–8315

Jakob U, Muse W, Eser M, Bardwell JC (1999) Chaperone activity with a redox switch. Cell 96:341–352

Jang S, Imlay JA (2010) Hydrogen peroxide inactivates the *Escherichia coli* Isc iron-sulphur assembly system, and OxyR induces the Suf system to compensate. Mol Microbiol 78:1448–1467

Johnson DC, Dean DR, Smith AD, Johnson MK (2005) Structure, function, and formation of biological iron-sulfur clusters. Annu Rev Biochem 74:247–281

Johnston AW, Todd JD, Curson AR, Lei S, Nikolaidou-Katsaridou N, Gelfand MS, Rodionov DA (2007) Living without Fur: the subtlety and complexity of iron-responsive gene regulation in the symbiotic bacterium *Rhizobium* and other alpha-proteobacteria. Biometals 20:501–511

Kallifidas D, Thomas D, Doughty P, Paget MS (2010) The sigmaR regulon of *Streptomyces coelicolor* A32 reveals a key role in protein quality control during disulphide stress. Microbiology 156:1661–1672

Kang JG, Paget MS, Seok YJ, Hahn MY, Bae JB, Hahn JS, Kleanthous C, Buttner MJ, Roe JH (1999) RsrA, an anti-sigma factor regulated by redox change. EMBO J 18:4292–4298

Kehres DG, Janakiraman A, Slauch JM, Maguire ME (2002) Regulation of *Salmonella enterica* serovar Typhimurium mntH transcription by H(2)O(2), Fe(2+), and Mn(2+). J Bacteriol 184:3151–3158

Kim MJ, Park KJ, Ko IJ, Kim YM, Oh JI (2010) Different roles of DosS and DosT in the hypoxic adaptation of Mycobacteria. J Bacteriol 192:4868–4875

Klomsiri C, Karplus PA, Poole LB (2011) Cysteine-based redox switches in enzymes. Antioxid Redox Signal 14:1065–1077

Kobayashi K, Tanaka A, Takahashi H, Igarashi J, Ishitsuka Y, Yokota N, Shimizu T (2010) Catalysis and oxygen binding of Ec DOS: a haem-based oxygen-sensor enzyme from *Escherichia coli*. J Biochem 148:693–703

Kohanski MA, Dwyer DJ, Hayete B, Lawrence CA, Collins JJ (2007) A common mechanism of cellular death induced by bactericidal antibiotics. Cell 130:797–810

Kommineni S, Garg SK, Chan CM, Zuber P (2011) YjbH-enhanced proteolysis of Spx by ClpXP in *Bacillus subtilis* is inhibited by the small protein YirB (YuzO). J Bacteriol 193:2133–2140

Krapp AR, Humbert MV, Carrillo N (2011) The soxRS response of *Escherichia coli* can be induced in the absence of oxidative stress and oxygen by modulation of NADPH content. Microbiology 157:957–965

Kumsta C, Jakob U (2009) Redox-regulated chaperones. Biochemistry 48:4666–4676

Lanzilotta WN, Schuller DJ, Thorsteinsson MV, Kerby RL, Roberts GP, Poulos TL (2000) Structure of the CO sensing transcription activator CooA. Nat Struct Biol 7:876–880

Lee JW, Helmann JD (2006a) Biochemical characterization of the structural Zn2+ site in the *Bacillus subtilis* peroxide sensor PerR. J Biol Chem 281(33):23567–23578

Lee JW, Helmann JD (2006b) The PerR transcription factor senses H_2O_2 by metal-catalysed histidine oxidation. Nature 440:363–367

Lee JW, Helmann JD (2007) Functional specialization within the fur family of metalloregulators. Biometals 20:485–499

Lee C, Lee SM, Mukhopadhyay P, Kim SJ, Lee SC, Ahn WS, Yu MH, Storz G, Ryu SE (2004a) Redox regulation of OxyR requires specific disulfide bond formation involving a rapid kinetic reaction path. Nat Struct Mol Biol 11:1179–1185

Lee JH, Yeo WS, Roe JH (2004b) Induction of the *sufA* operon encoding Fe-S assembly proteins by superoxide generators and hydrogen peroxide: involvement of OxyR, IHF and an unidentified oxidant-responsive factor. Mol Microbiol 51:1745–1755

Lee JW, Soonsanga S, Helmann JD (2007) A complex thiolate switch regulates the *Bacillus subtilis* organic peroxide sensor OhrR. Proc Natl Acad Sci USA 104:8743–8748

Lee KC, Yeo WS, Roe JH (2008) Oxidant-responsive induction of the *suf* operon, encoding a Fe-S assembly system, through Fur and IscR in *Escherichia coli*. J Bacteriol 190:8244–8247

Leonard SE, Carroll KS (2011) Chemical 'omics' approaches for understanding protein cysteine oxidation in biology. Curr Opin Chem Biol 15:88–102

Li W, Bottrill AR, Bibb MJ, Buttner MJ, Paget MS, Kleanthous C (2003) The role of zinc in the disulphide stress-regulated anti-sigma factor RsrA from *Streptomyces coelicolor*. J Mol Biol 333:461–472

Lindahl M, Mata-Cabana A, Kieselbach T (2011) The disulfide proteome and other reactive cysteine proteomes: analysis and functional significance. Antioxid Redox Signal 14:2581–2642

Little R, Salinas P, Slavny P, Clarke TA, Dixon R (2011) Substitutions in the redox-sensing PAS domain of the NifL regulatory protein define an inter-subunit pathway for redox signal transmission. Mol Microbiol 82:222–235

Losi A, Gartner W (2012) The evolution of flavin-binding photoreceptors: an ancient chromophore serving trendy blue-light sensors. Annu Rev Plant Biol 63:49–72

Mandin P, Gottesman S (2010) Integrating anaerobic/aerobic sensing and the general stress response through the ArcZ small RNA. EMBO J 29:3094–3107

Martinez-Argudo I, Little R, Shearer N, Johnson P, Dixon R (2004) The NifL-NifA System: a multidomain transcriptional regulatory complex that integrates environmental signals. J Bacteriol 186:601–610

Marvin KA, Kerby RL, Youn H, Roberts GP, Burstyn JN (2008) The transcription regulator RcoM-2 from *Burkholderia xenovorans* is a cysteine-ligated hemoprotein that undergoes a redox-mediated ligand switch. Biochemistry 47:9016–9028

McLaughlin KJ, Strain-Damerell CM, Xie K, Brekasis D, Soares AS, Paget MS, Kielkopf CL (2010) Structural basis for NADH/NAD^+ redox sensing by a Rex family repressor. Mol Cell 38:563–575

Mettert EL, Outten FW, Wanta B, Kiley PJ (2008) The impact of O_2 on the Fe-S cluster biogenesis requirements of *Escherichia coli* FNR. J Mol Biol 384:798–811

Mills E, Pultz IS, Kulasekara HD, Miller SI (2011) The bacterial second messenger c-di-GMP: mechanisms of signalling. Cell Microbiol 13:1122–1129

Moore CM, Nakano MM, Wang T, Ye RW, Helmann JD (2004) Response of *Bacillus subtilis* to nitric oxide and the nitrosating agent sodium nitroprusside. J Bacteriol 186:4655–4664

Moore LJ, Mettert EL, Kiley PJ (2006) Regulation of FNR dimerization by subunit charge repulsion. J Biol Chem 281:33268–33275

Mukhopadhyay P, Zheng M, Bedzyk LA, Larossa RA, Storz G (2004) Prominent roles of the NorR and Fur regulators in the *Escherichia coli* transcriptional response to reactive nitrogen species. Proc Natl Acad Sci USA 101:745–750

Nakano S, Kuster-Schock E, Grossman AD, Zuber P (2003a) Spx-dependent global transcriptional control is induced by thiol-specific oxidative stress in *Bacillus subtilis*. Proc Natl Acad Sci USA 100:13603–13608

Nakano S, Nakano MM, Zhang Y, Leelakriangsak M, Zuber P (2003b) A regulatory protein that interferes with activator-stimulated transcription in bacteria. Proc Natl Acad Sci USA 100:4233–4238

Nakano MM, Lin A, Zuber CS, Newberry KJ, Brennan RG, Zuber P (2010) Promoter recognition by a complex of Spx and the C-terminal domain of the RNA polymerase alpha subunit. PLOS One 5:e8664

Nash AI, McNulty R, Shillito ME, Swartz TE, Bogomolni RA, Luecke H, Gardner KH (2011) Structural basis of photosensitivity in a bacterial light-oxygen-voltage/helix-turn-helix (LOV-HTH) DNA-binding protein. Proc Natl Acad Sci USA 108:9449–9454

Nesbit AD, Giel JL, Rose JC, Kiley PJ (2009) Sequence-specific binding to a subset of IscR-regulated promoters does not require IscR Fe-S cluster ligation. J Mol Biol 387:28–41

Newberry KJ, Nakano S, Zuber P, Brennan RG (2005) Crystal structure of the *Bacillus subtilis* anti-alpha, global transcriptional regulator, Spx, in complex with the alpha C-terminal domain of RNA polymerase. Proc Natl Acad Sci USA 102:15839–15844

Newberry KJ, Fuangthong M, Panmanee W, Mongkolsuk S, Brennan RG (2007) Structural mechanism of organic hydroperoxide induction of the transcription regulator OhrR. Mol Cell 28:652–664

Newton GL, Fahey RC (2008) Regulation of mycothiol metabolism by sigma(R) and the thiol redox sensor anti-sigma factor RsrA. Mol Microbiol 68:805–809

Olea C Jr, Kuriyan J, Marletta MA (2010) Modulating heme redox potential through protein-induced porphyrin distortion. J Am Chem Soc 132:12794–12795

Ortiz De Orue Lucana D, Groves MR (2009) The three-component signalling system HbpS-SenS-SenR as an example of a redox sensing pathway in bacteria. Amino Acids 37:479–486

Ortiz De Orue Lucana D, Troller M, Schrempf H (2003) Amino acid residues involved in reversible thiol formation and zinc ion binding in the *Streptomyces reticuli* redox regulator FurS. Mol Genet Genomics 268:618–627

Ortiz De Orue Lucana D, Roscher M, Honigmann A, Schwarz J (2010) Iron-mediated oxidation induces conformational changes within the redox-sensing protein HbpS. J Biol Chem 285:28086–28096

Ortiz De Orue Lucana D, Wedderhoff I, Groves MR (2012) ROS-mediated signalling in bacteria: zinc-containing Cys-X-X-Cys Redox centres and iron-based oxidative stress. J Signal Transduct 2012:605905

Outten FW, Theil EC (2009) Iron-based redox switches in biology. Antioxid Redox Signal 11:1029–1046

Outten FW, Djaman O, Storz G (2004) A *suf* operon requirement for Fe-S cluster assembly during iron starvation in *Escherichia coli*. Mol Microbiol 52:861–872

Pagels M, Fuchs S, Pane-Farre J, Kohler C, Menschner L, Hecker M, McNamarra PJ, Bauer MC, von Wachenfeldt C, Liebeke M, Lalk M, Sander G, von Eiff C, Proctor RA, Engelmann S (2010) Redox sensing by a Rex-family repressor is involved in the regulation of anaerobic gene expression in *Staphylococcus aureus*. Mol Microbiol 76:1142–1161

Paget MS, Kang JG, Roe JH, Buttner MJ (1998) sigmaR, an RNA polymerase sigma factor that modulates expression of the thioredoxin system in response to oxidative stress in *Streptomyces coelicolor* A3(2). EMBO J 17:5776–5782

Paget MS, Molle V, Cohen G, Aharonowitz Y, Buttner MJ (2001) Defining the disulphide stress response in *Streptomyces coelicolor* A3(2): identification of the sigmaR regulon. Mol Microbiol 42:1007–1020

Palm G, Chi BK, Waack P, Gronau K, Becher DR, Albrecht D, Hinrichs W, Read RJ, Antelmann H (2012) Structural insights into the redox-switch mechanism of the MarR/DUF24-type regulator HypR. Nucleic Acids Res 40(9):4178–4192

Partridge JD, Bodenmiller DM, Humphrys MS, Spiro S (2009) NsrR targets in the *Escherichia coli* genome: new insights into DNA sequence requirements for binding and a role for NsrR in the regulation of motility. Mol Microbiol 73:680–694

Pellicena P, Karow DS, Boon EM, Marletta MA, Kuriyan J (2004) Crystal structure of an oxygen-binding heme domain related to soluble guanylate cyclases. Proc Natl Acad Sci USA 101:12854–12859

Poor CB, Chen PR, Duguid E, Rice PA, He C (2009) Crystal structures of the reduced, sulfenic acid, and mixed disulfide forms of SarZ, a redox active global regulator in *Staphylococcus aureus*. J Biol Chem 284:23517–23524

Pruteanu M, Baker TA (2009) Proteolysis in the SOS response and metal homeostasis in *Escherichia coli*. Res Microbiol 160:677–683

Pullan ST, Gidley MD, Jones RA, Barrett J, Stevanin TM, Read RC, Green J, Poole RK (2007) Nitric oxide in chemostat-cultured *Escherichia coli* is sensed by Fnr and other global regulators: unaltered methionine biosynthesis indicates lack of S nitrosation. J Bacteriol 189:1845–1855

Puri S, Hohle TH, O'Brian MR (2010) Control of bacterial iron homeostasis by manganese. Proc Natl Acad Sci USA 107:10691–10695

Py B, Moreau PL, Barras F (2011) Fe-S clusters, fragile sentinels of the cell. Curr Opin Microbiol 14:218–223

Qi Z, O'Brian MR (2002) Interaction between the bacterial iron response regulator and ferrochelatase mediates genetic control of heme biosynthesis. Mol Cell 9:155–162

Reents H, Gruner I, Harmening U, Bottger LH, Layer G, Heathcote P, Trautwein AX, Jahn D, Hartig E (2006) *Bacillus subtilis* Fnr senses oxygen via a [4Fe-4S] cluster coordinated by three cysteine residues without change in the oligomeric state. Mol Microbiol 60:1432–1445

Rinaldo S, Castiglione N, Giardina G, Caruso M, Arcovito A, Della Longa S, D'Angelo P, Cutruzzola F (2012) Unusual heme binding properties of the dissimilative nitrate respiration regulator (Dnr), a bacterial nitric oxide sensor. Antioxid Redox Signal 17(9):1178–1189

Roberts GP, Kerby RL, Youn H, Conrad M (2005) CooA, a paradigm for gas sensing regulatory proteins. J Inorg Biochem 99:280–292

Rolfe MD, ter Beek A, Graham AI, Trotter EW, Asif HM, Sanguinetti G, de Mattos JT, Poole RK, Green J (2011) Transcript profiling and inference of *Escherichia coli* K-12 ArcA activity across the range of physiologically relevant oxygen concentrations. J Biol Chem 286:10147–10154

Rosen H, Klebanoff SJ, Wang Y, Brot N, Heinecke JW, Fu X (2009) Methionine oxidation contributes to bacterial killing by the myeloperoxidase system of neutrophils. Proc Natl Acad Sci USA 106:18686–18691

Rouault TA (2006) The role of iron regulatory proteins in mammalian iron homeostasis and disease. Nat Chem Biol 2:406–414

Rudolph G, Hennecke H, Fischer HM (2006) Beyond the Fur paradigm: iron-controlled gene expression in rhizobia. FEMS Microbiol Rev 30:631–648

Sasakura Y, Yoshimura-Suzuki T, Kurokawa H, Shimizu T (2006) Structure-function relationships of EcDOS, a heme-regulated phosphodiesterase from *Escherichia coli.* Acc Chem Res 39: 37–43

Shiloh MU, Manzanillo P, Cox JS (2008) *Mycobacterium tuberculosis* senses host-derived carbon monoxide during macrophage infection. Cell Host Microbe 3:323–330

Shingler V (2011) Signal sensory systems that impact sigma -dependent transcription. FEMS Microbiol Rev 35:425–440

Sickmier EA, Brekasis D, Paranawithana S, Bonanno JB, Paget MS, Burley SK, Kielkopf CL (2005) X-ray structure of a Rex-family repressor/NADH complex insights into the mechanism of redox sensing. Structure 13:43–54

Siedenburg G, Groves MR, Ortiz De Orue Lucana D (2011) Novel redox-sensing modules: accessory proteins- and nucleic acids-mediated signalling. Antioxid Redox Signal 16(7): 668–677

Singh A, Guidry L, Narasimhulu KV, Mai D, Trombley J, Redding KE, Giles GI, Lancaster JR Jr, Steyn AJ (2007) Mycobacterium tuberculosis WhiB3 responds to O2 and nitric oxide via its [4Fe-4S] cluster and is essential for nutrient starvation survival. Proc Natl Acad Sci USA 104:11562–11567

Singh A, Crossman DK, Mai D, Guidry L, Voskuil MI, Renfrow MB, Steyn AJ (2009) *Mycobacterium tuberculosis* WhiB3 maintains redox homeostasis by regulating virulence lipid anabolism to modulate macrophage response. PLOS Pathog 5:e1000545

Singleton C, White GF, Todd JD, Marritt SJ, Cheesman MR, Johnston AW, le Brun NE (2010) Heme-responsive DNA binding by the global iron regulator Irr from *Rhizobium leguminosarum.* J Biol Chem 285:16023–16031

Small SK, Puri S, O'Brian MR (2009) Heme-dependent metalloregulation by the iron response regulator (Irr) protein in *Rhizobium* and other Alpha-proteobacteria. Biometals 22:89–97

Sobota JM, Imlay JA (2011) Iron enzyme ribulose-5-phosphate 3-epimerase in *Escherichia coli* is rapidly damaged by hydrogen peroxide but can be protected by manganese. Proc Natl Acad Sci USA 108:5402–5407

Soonsanga S, Lee JW, Helmann JD (2008a) Conversion of *Bacillus subtilis* OhrR from a 1-Cys to a 2-Cys peroxide sensor. J Bacteriol 190:5738–5745

Soonsanga S, Lee JW, Helmann JD (2008b) Oxidant-dependent switching between reversible and sacrificial oxidation pathways for *Bacillus subtilis* OhrR. Mol Microbiol 68:978–986

Spiro T (2008) A twist on heme signaling. ACS Chem Biol 3:673–675

Storz G, Imlay JA (1999) Oxidative stress. Curr Opin Microbiol 2:188–194

Storz G, Tartaglia LA, Ames BN (1990) Transcriptional regulator of oxidative stress-inducible genes: direct activation by oxidation. Science 248:189–194

Sukchawalit R, Loprasert S, Atichartpongkul S, Mongkolsuk S (2001) Complex regulation of the organic hydroperoxide resistance gene (ohr) from Xanthomonas involves OhrR, a novel organic peroxide-inducible negative regulator, and posttranscriptional modifications. J Bacteriol 183:4405–4412

Switzer RL (1989) Non-redox roles for iron-sulfur clusters in enzymes. Biofactors 2:77–86

Tang Y, Guest JR (1999) Direct evidence for mRNA binding and post-transcriptional regulation by *Escherichia coli* aconitases. Microbiology 145:3069–3079

Tang Y, Quail MA, Artymiuk PJ, Guest JR, Green J (2002) *Escherichia coli* aconitases and oxidative stress: post-transcriptional regulation of *sodA* expression. Microbiology 148: 1027–1037

Tang Y, Guest JR, Artymiuk PJ, Read RC, Green J (2004) Post-transcriptional regulation of bacterial motility by aconitase proteins. Mol Microbiol 51:1817–1826

Tang Y, Guest JR, Artymiuk PJ, Green J (2005) Switching aconitase B between catalytic and regulatory modes involves iron-dependent dimer formation. Mol Microbiol 56:1149–1158

Taylor BL (2007) Aer on the inside looking out: paradigm for a PAS-HAMP role in sensing oxygen, redox and energy. Mol Microbiol 65:1415–1424

Taylor BL, Zhulin IB (1999) PAS domains: internal sensors of oxygen, redox potential, and light. Microbiol Mol Biol Rev 63:479–506

Toledano MB, Kullik I, Trinh F, Baird PT, Schneider TD, Storz G (1994) Redox-dependent shift of OxyR-DNA contacts along an extended DNA-binding site: a mechanism for differential promoter selection. Cell 78:897–909

Traore DA, El Ghazouani A, Jacquamet L, Borel F, Ferrer JL, Lascoux D, Ravanat JL, Jaquinod M, Blondin G, Caux-Thang C, Duarte V, Latour JM (2009) Structural and functional characterization of 2-oxo-histidine in oxidized PerR protein. Nat Chem Biol 5:53–59

Tsai AL, Berka V, Martin E, Olson JS (2012) A "sliding-scale rule" for selectivity between NO, CO and O2 by heme protein sensors. Biochemistry 51:172–186

Tucker NP, Hicks MG, Clarke TA, Crack JC, Chandra G, le Brun NE, Dixon R, Hutchings MI (2008) The transcriptional repressor protein NsrR senses nitric oxide directly via a [2Fe-2S] cluster. PLOS One 3:e3623

Tucker NP, le Brun NE, Dixon R, Hutchings MI (2010) There's NO stopping NsrR, a global regulator of the bacterial NO stress response. Trends Microbiol 18:149–156

Tuckerman JR, Gonzalez G, Sousa EH, Wan X, Saito JA, Alam M, Gilles-Gonzalez MA (2009) An oxygen-sensing diguanylate cyclase and phosphodiesterase couple for c-di-GMP control. Biochemistry 48:9764–9774

Tuckerman JR, Gonzalez G, Gilles-Gonzalez MA (2011) Cyclic di-GMP activation of polynucleotide phosphorylase signal-dependent RNA processing. J Mol Biol 407:633–639

Varghese S, Tang Y, Imlay JA (2003) Contrasting sensitivities of *Escherichia coli* aconitases A and B to oxidation and iron depletion. J Bacteriol 185:221–230

Varghese S, Wu A, Park S, Imlay KR, Imlay JA (2007) Submicromolar hydrogen peroxide disrupts the ability of Fur protein to control free-iron levels in *Escherichia coli*. Mol Microbiol 64: 822–830

Wang J, Pantopoulos K (2011) Regulation of cellular iron metabolism. Biochem J 434:365–381

Wang E, Bauer MC, Rogstam A, Linse S, Logan DT, von Wachenfeldt C (2008) Structure and functional properties of the *Bacillus subtilis* transcriptional repressor Rex. Mol Microbiol 69:466–478

Wang Y, Dufour YS, Carlson HK, Donohue TJ, Marletta MA, Ruby EG (2010) H-NOX-mediated nitric oxide sensing modulates symbiotic colonization by *Vibrio fischeri*. Proc Natl Acad Sci USA 107:8375–8380

Watanabe S, Kita A, Kobayashi K, Miki K (2008) Crystal structure of the [2Fe-2S] oxidative-stress sensor SoxR bound to DNA. Proc Natl Acad Sci USA 105:4121–4126

Watts KJ, Johnson MS, Taylor BL (2011) Different conformations of the kinase-on and kinase-off signaling states in the Aer HAMP domain. J Bacteriol 193:4095–4103

Wigneshweraraj S, Bose D, Burrows PC, Joly N, Schumacher J, Rappas M, Pape T, Zhang X, Stockley P, Severinov K, Buck M (2008) Modus operandi of the bacterial RNA polymerase containing the sigma54 promoter-specificity factor. Mol Microbiol 68:538–546

Williams CH, Stillman TJ, Barynin VV, Sedelnikova SE, Tang Y, Green J, Guest JR, Artymiuk PJ (2002) *E. coli* aconitase B structure reveals a HEAT-like domain with implications for protein-protein recognition. Nat Struct Biol 9:447–452

Winter J, Linke K, Jatzek A, Jakob U (2005) Severe oxidative stress causes inactivation of DnaK and activation of the redox-regulated chaperone Hsp33. Mol Cell 17:381–392

Winter J, Ilbert M, Graf PC, Ozcelik D, Jakob U (2008) Bleach activates a redox-regulated chaperone by oxidative protein unfolding. Cell 135:691–701

Wu J, Bauer CE (2010) RegB kinase activity is controlled in part by monitoring the ratio of oxidized to reduced ubiquinones in the ubiquinone pool. MBio 1(5):e00272-10–e00272-18

Yan A, Kiley PJ (2009) Techniques to isolate O_2-sensitive proteins: [4Fe-4S]-FNR as an example. Methods Enzymol 463:787–805

Yang J, Panek HR, O'Brian MR (2006a) Oxidative stress promotes degradation of the Irr protein to regulate haem biosynthesis in *Bradyrhizobium japonicum*. Mol Microbiol 60:209–218

Yang J, Sangwan I, O'Brian MR (2006b) The *Bradyrhizobium japonicum* Fur protein is an iron-responsive regulator in vivo. Mol Genet Genomics 276:555–564

Yeo WS, Lee JH, Lee KC, Roe JH (2006) IscR acts as an activator in response to oxidative stress for the suf operon encoding Fe-S assembly proteins. Mol Microbiol 61:206–218

Yoshimura H, Yoshioka S, Kobayashi K, Ohta T, Uchida T, Kubo M, Kitagawa T, Aono S (2006) Specific hydrogen-bonding networks responsible for selective O_2 sensing of the oxygen sensor protein HemAT from *Bacillus subtilis*. Biochemistry 45:8301–8307

Yukl ET, Elbaz MA, Nakano MM, Moenne-Loccoz P (2008) Transcription factor NsrR from *Bacillus subtilis* senses nitric oxide with a 4Fe-4S cluster (dagger). Biochemistry 47: 13084–13092

Yukl ET, Ioanoviciu A, Sivaramakrishnan S, Nakano MM, Ortiz De Montellano PR, Moenne-Loccoz P (2011) Nitric oxide dioxygenation reaction in DevS and the initial response to nitric oxide in *Mycobacterium tuberculosis*. Biochemistry 50:1023–1028

Zdanowski K, Doughty P, Jakimowicz P, O'Hara L, Buttner MJ, Paget MS, Kleanthous C (2006) Assignment of the zinc ligands in RsrA, a redox-sensing ZAS protein from *Streptomyces coelicolor*. Biochemistry 45:8294–8300

Zheng M, Doan B, Schneider TD, Storz G (1999) OxyR and SoxRS regulation of fur. J Bacteriol 181:4639–4643

Zuber P (2009) Management of oxidative stress in *Bacillus*. Annu Rev Microbiol 63:575–597

Chapter 10
Combating Oxidative/Nitrosative Stress with Electrophilic Counterattack Strategies

Takumi Satoh, Mohd Waseem Akhtar, and Stuart A. Lipton

Abstract Redox stress is thought to contribute to neurodegenerative disorders, including Alzheimer's and Parkinson's diseases. Cysteine-based redox regulation, via glutathione- and thioredoxin-mediated pathways, represents an acute defense system. Additionally, with prolonged oxidative stress, cells mount a 'counterattack' to activate transcription-dependent pathways, including the Keap1/Nrf2 and HSP90/HSF-1 pathways, which induce phase 2 enzymes and heat-shock proteins, respectively. Oxidative/nitrosative stress itself is an activator of the Keap1/Nrf2 pathway via cysteine thiol oxidation. Moreover, stress-induced oxidation of endogenous compounds can generate electrophiles, including active aldehydes, nitroso-compounds, fatty acids, nitro-guanosine, and quinone-based dopamine derivatives. These electrophilic compounds were considered toxic, but recently have been shown to be neuroprotective under certain conditions. These endogenously-produced electrophiles signal an "electrophilic counterattack," binding to specific cysteines of Keap1 and HSP90 to activate these pathways. Finally, we describe novel pro-electrophilic drugs (PEDs) that are activated by the very oxidative/nitrosative stress that they subsequently counteract. One example is carnosic acid (CA), found in the herb rosemary. CA itself is not electrophilic, but in response to oxidation becomes electrophilic, and then activates the Keap1/Nrf2 pathway. PEDs appear to have minimal side effects, in part because they are generated preferentially in cells

T. Satoh (✉)
Del E. Webb Center for Neuroscience, Aging, and Stem Cell Research, Sanford-Burnham Medical Research Institute, 10901 North Torrey Pines Road, La Jolla, CA, USA

Department of Welfare Engineering, Faculty of Engineering, Iwate University, Morioka, Iwate, 020-8551, Japan
e-mail: tsatoh@iwate-u.ac.jp

M.W. Akhtar • S.A. Lipton (✉)
Del E. Webb Center for Neuroscience, Aging, and Stem Cell Research, Sanford-Burnham Medical Research Institute, 10901 North Torrey Pines Road, La Jolla, CA, USA
e-mail: slipton@sanfordburnham.org

U. Jakob and D. Reichmann (eds.), *Oxidative Stress and Redox Regulation*,
DOI 10.1007/978-94-007-5787-5_10,

experiencing oxidative stress. In contrast, in the absence of oxidative stress, true electrophiles, unlike PEDs, react with and thus deplete glutathione, paradoxically rendering these normal cells susceptible to damage.

Keywords Electrophile • Nrf2 • HSF-1 • Keap1 • HSP90 • Pro-electrophilic drug • Pathologically-activated therapeutic

Abbreviations

ABCC	ATP-binding cassette, sub-family C
AD	Alzheimer's disease
ARE	antioxidant response element
BDNF	brain-derived neurotrophic factor
CA	carnosic acid
CAB	biotinylated carnosic acid
CNTF	ciliary neurotrophic factor
DA	dopamine
EP	electrophile
ER	endoplasmic reticulum
GCLM	glutamyl cysteine ligase modifier subunit
GCLC	glutamyl cysteine ligase catalytic subunit
GSSG	disulfide form of glutathione
GPX	glutathione peroxidase
GR	NADPH-dependent GSSG reductase
GST	glutathione-S-transferase
HHE	*trans*-4-hydroxy-2-hexenal
HNE	*trans*-4-hydroxy-2-nonenal
H_2O_2	hydrogen peroxide
HO-1	hemeoxygenase-1
HSE	heat-shock factor response element
HSF-1	heat-shock factor-1
HSP	heat-shock protein
HO	hydroxyl radical
NEPP	neurite outgrowth-promoting prostaglandin
NMDA	*N*-methyl-D-aspartate
NO_2	nitric dioxide
NO	nitric oxide
NQO1	NADPH quinone oxidoreductase 1
NGF	nerve growth factor
O_2^-	superoxide radical anion
$ONOO^-$	peroxynitrite
PAT	pathologically activated therapeutic
PUFA	poly-unsaturated fatty acids
PRX	peroxyredoxin

PED	pro-electrophilic drug
RNS	reactive nitrogen species
ROS	reactive oxygen species
SRXN	ATP-dependent reductase, sulfiredoxin
TBHQ	*tert-butyl* hydroquinone
TRX	thioredoxin
TRXR	thioredoxin reductase
xCT	Na^{+}-independent cystine-glutamate exchanger

10.1 Direct Effects of ROS/RNS and Electrophilic Counterreaction

10.1.1 Oxidative and Nitrosative Stress

The brain is sensitive to oxidative and nitrosative damage (Coyle and Puttfarcken 1993; Halliwell and Gutteridge 1999). In neurodegenerative diseases, increased levels of reactive oxygen species (ROS) and reactive nitrogen species (RNS) within specific brain regions lead to selective neurodegeneration (Fig. 10.1), thereby suggesting that ROS/RNS can contribute to the development of disease through primary or secondary pathological mechanisms (Coyle and Puttfarcken 1993; Pacher et al. 2007; Bredesen 2008; Hardingham and Lipton 2011). Many ROS and RNS contain one or more unpaired electrons in their outer pi molecular orbitals, and are thus characterized as free radicals with high reactivity. The insults promulgated by ROS and RNS are termed oxidative stress and nitrosative stress, respectively. One mechanism for such damage in biological systems involves dysregulation of cysteine thiol-based redox regulation (Fig. 10.2) due to overproduction or altered spatiotemporal distribution of ROS/RNS (Coyle and Puttfarcken 1993; Pacher et al. 2007; Bredesen 2008).

10.1.2 Direct Effects of ROS/RNS and Electrophilic Counterreactions

There are at least two modes of counterreaction against excessive ROS/RNS; the first is mediated directly by ROS/RNS themselves, and the other occurs via endogenous electrophiles (EPs; Talalay 2000; Satoh and Lipton 2007). The broad definition of an EP is a molecule having one or more electron-poor atoms that can accept electrons from electron-rich donor molecules (nucleophiles). ROS/RNS can directly oxidize critical amino acids, such as cysteine thiols, and this oxidation can result, for example, in reaction with critical thiols to activate Keap1/Nrf2 and HSP90/HSF-1 transcriptional pathways. These reactions are termed "direct counterreaction," as is shown in Fig. 10.1B (Fourquet et al. 2010; Um et al.

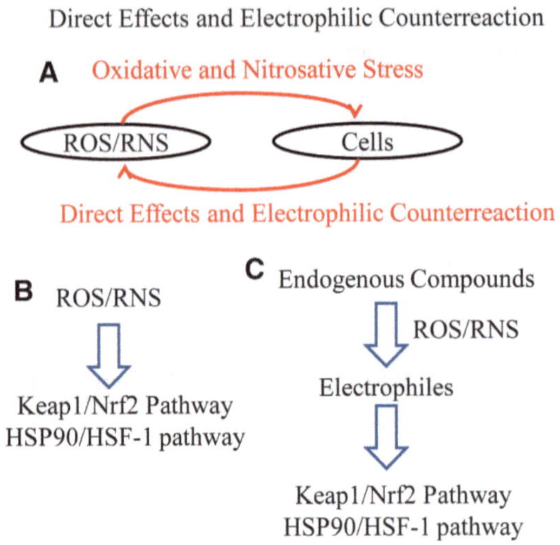

Fig. 10.1 Direct effects of ROS/RNS and electrophilic counterreaction. *A* Interaction between ROS/RNS and cells. Although ROS/RNS exert toxic effects by oxidation of cellular components, cells have two counterreactions to remove ROS/RNS. ROS/RNS and EPs cause direct effects and electrophilic counterreactions, respectively. *B* Direct counterreactions. ROS/RNS can directly modify critical cysteines of modulatory proteins such as Keap1 and HSP90, and thus induce activation of the Keap1/Nrf2 and HSP90/HSF-1 pathways. *C* Electrophilic counterreactions. ROS/RNS oxidize endogenous compounds to produce bioactive EPs, which then activate the Keap1/Nrf2 and HSP90/HSF-1 pathways

2011). In some cases, this type of counterreaction provides cells full activation of pathways that afford efficient protection against oxidative and nitrosative stress. From studies conducted over the past 20 years, we know that ROS/RNS oxidize certain endogenous compounds to produce bioactive EPs (Groeger and Freeman 2010; Akaike et al. 2010). These ROS/RNS-produced EPs can oxidize specific cysteine thiols much more effectively than ROS/RNS themselves, and also activate the same signaling pathways activated by the direct counterreaction. Hence, this mode of activation is termed "electrophilic counterreaction," as shown in Fig. 10.1C. Biologically relevant EPs can be generated during tightly controlled metabolic processes or during dysregulated pathological processes as byproducts of oxidation (Groeger and Freeman 2010; Akaike et al. 2010). Some EPs are toxic in and of themselves, e.g., due to their indiscriminate reaction with thiols such as glutathione, thus depleting cellular defensive reserves against oxidative stress. In response to such EP stress, some cells mount a second kind of 'electrophilic counterreaction,' characterized by a system that detoxifies EPs and removes them immediately (Satoh and Lipton 2007). This electrophilic counterreaction usually lies relatively dormant, with only mild baseline activity, but becomes activated by EPs themselves (Satoh and Lipton 2007). This type of counterreaction can prevent neurodegeneration and tumor growth because it eliminates not only the EPs themselves but also ROS (Mattson and Cheng 2006; Satoh and Lipton 2007). Thus, EPs could possibly be

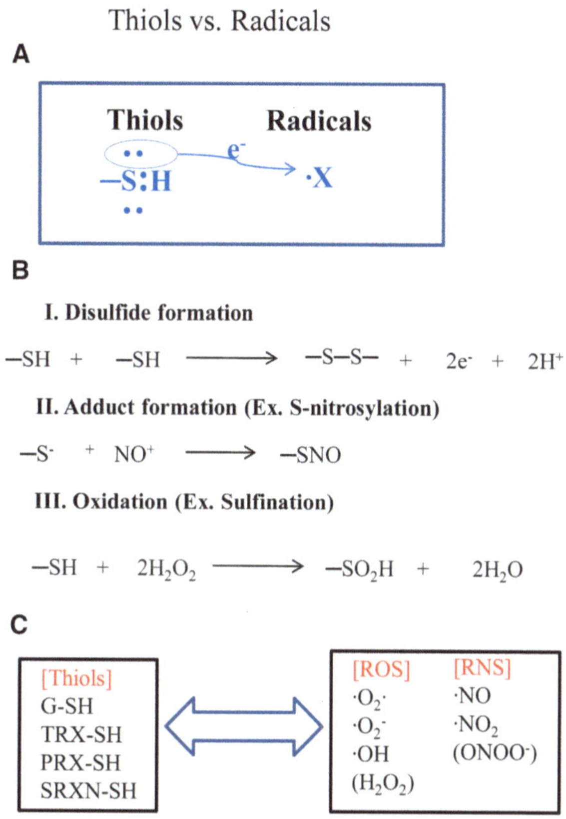

Fig. 10.2 Thiols *vs.* ROS/RNS. *A* Basic reaction. Thiols have two lone pairs of electrons and donate electrons to radicals, which have an unpaired electron(s), although the reaction mechanism(s) and other participating molecules are still under contention. *B* Three modes of reaction. There are three enzymatic reactions (I. Disulfide formation, II. Adduct formation, and III. Oxidation) that allow the basic reaction in panel *A*. *C* Thiols and ROS/RNS. There are various molecules representing thiols, ROS, and RNS

used as neuroprotective agents if depletion of glutathione and other critical thiols could be avoided. In this review, we will discuss four molecules that are typical endogenous EPs produced by ROS/RNS:

1. Active aldehydes including *trans*-4-hydroxy-2-nonenal (HNE) and *trans*-4-hydroxy-2-hexenal (HHE) produced by a series of lipid peroxidations of $n=3$ and $n=6$ poly-unsaturated fatty acids (PUFA), respectively (Uchida 2003; Catalá 2009; Long and Picklo 2010);
2. Nitrated fatty acids, including nitrated derivatives of oleic and linoleic acids produced by nitration of fatty acids in phospholipids (Groeger and Freeman 2010; Kansanen et al. 2011);
3. Nitrated-guanosines, including 8-nitro-cGMP produced by nitration of guanosine-containing compounds (Akaike et al. 2010; Fujii et al. 2010);
4. Quinone-formation of catecholamines, including dopamine (DA), in the presence of Cu^{2+} and O_2 (Shih et al. 2007).

Free radical formation is involved in the production of endogenous EPs, and is initiated by reactive radicals or related molecules ($\cdot OH$, NO_2, and O_2^-). Hence, until recently, these compounds were assumed to produce only toxic effects so they were studied in neurodegenerative conditions (Coyle and Puttfarcken 1993; Halliwell and Gutteridge 1999). However, an emerging concept in cellular stress responses is that under certain conditions these EPs can be neuroprotective. The pivotal event in developing this concept was the discovery that EPs could influence the activity of the Keap1/Nrf2 and HSP90/HSF-1 pathways (Talalay 2000; Itoh et al. 2004; Bukau et al. 2006; Morimoto 2008; Satoh et al. 2011). In the context of the Keap1/Nrf2 pathway, EPs are now considered physiological activators (Mattson and Cheng 2006; Satoh and Lipton 2007). Furthermore, these "electrophilic counterreactions" can be used to develop neuroprotective drugs (Satoh and Lipton 2007). Along these lines, we have been developing Pro-Electrophilic Drugs (PEDs) that are activated to their electrophilic form by oxidative/nitrosative stress, the very insult that they then counteract. Since activation of PEDs occurs preferentially in cells undergoing such stress (and therefore that are already relatively depleted of glutathione), PEDs are innocuous in normal, unstressed cells. We have found that such drugs can afford neuroprotection in animal models of chronic neurodegenerative diseases such as Alzheimer's and Parkinson's (Satoh et al. 2008a, 2011).

10.2 Thiols *vs*. Radicals

10.2.1 Basic Reaction

As shown in Fig. 10.2A, the basic reaction for cysteine-based redox regulation against oxidative/nitrosative stress is the electron donation from cysteine thiols to free radicals (Halliwell and Gutteridge 1999; Arteel and Sies 2001). Thiols have two lone pairs of electrons and tend to donate electrons. In contrast, free radicals have an unpaired electron and thus accept electrons from donors. Thus, an electron may potentially be donated from thiol to radical, as shown in Fig. 10.2A. Cells contain high concentrations of reduced thiols, such as reduced glutathione (GSH), thioredoxin (TRX), peroxiredoxin (PRX), and sulfiredoxin (SRXN; Immenschuh and Baumgart-Vogt 2005; Arteel and Sies 2001; Holmgren and Lu 2010), which serve to combat free radicals such as molecular oxygen (O_2), superoxide anion (O_2^-), hydroxyl radical (·OH), nitric oxide (·NO) and nitric dioxide (·NO_2) radicals. Critical thiols on a number of proteins can also undergo reaction with GSH to form glutathionylated cysteine residues (Dalle-Donne et al. 2009). H_2O_2 and $ONOO^-$ do not have unpaired electrons and thus are not free radicals, but they may produce related free radicals (Radi 2004; Pacher et al. 2007; Butterfield et al. 2011), ·OH and ·NO_2. Therefore, they can be included as reactive oxygen species (ROS) and reactive nitrogen species (RNS). Importantly, however, the reaction mechanisms are not well characterized and are most probably not direct. For example, reaction of cysteine with an NO group may possibly proceed via thiolate anion (RS^-) reacting with a nitrosonium intermediate (NO^+, with one less electron than NO) in the presence of an electron acceptor such as a transition metal to balance the electron equation (Lipton et al. 1993; Lipton and Stamler 1994).

Cells continuously have to combat ROS and RNS, and thus cysteine-based redox regulation, as shown in Fig 10.2B, is an effective frontline defense for cells. Protein and glutathione cysteine thiols (–SH) are some of the most powerful nucleophilic compounds, having two lone pairs of electrons; thus, they can donate electrons to oxygen, ROS, and RNS. There are basically three types of reduction pathways used for redox regulation, as shown in Fig. 10.2B, i.e., disulfide formation, adduct formation, and oxidation.

10.2.2 ROS

Molecular oxygen requires a total of four electrons for reduction to water, and reacts slowly with many biomolecules (Halliwell and Gutteridge 1999). If reduced, however, O_2 becomes more reactive (hence, a ROS) (Fig. 10.2C). One-electron reduction results in O_2^-; two-electron reduction, in hydrogen peroxide (H_2O_2); and three-electron reduction, in the highly destructive hydroxyl radical (·OH). ROS are frequently interconverted, for example, superoxide into peroxide (by dismutase enzymes) and peroxide into ·OH (the so-called Fe^{2+}-catalyzed Fenton reaction).

10.2.3 RNS

Nitric oxide (NO) is a gaseous free radical with a relatively short half-life, and the NO group also exists, at least at very low pH, as a nitrosonium cation (NO^+), with one less electron, or nitroxyl anion (NO^-), with an additional electron in its outer pi molecular orbital. In the case of NO^-, the electron pair can be in a high-energy singlet state or a lower energy triplet state, with distinctive biological reactivity (Kim et al. 1999). These various NO species regulate an ever-growing list of biological processes in the brain (Hess et al. 2005; Hara and Snyder 2007; Nakamura and Lipton 2009, 2011). NO represents a pleiotropic signaling molecule, regulating diverse cellular processes and is produced endogenously from L-arginine by NO synthases. The classical NO signaling pathway involves reaction with heme in soluble guanylate cyclase to activate the enzyme and generate cGMP, which in turn stimulates protein kinase G (Hess et al. 2005; Hara and Snyder 2007). However, in the past several years, protein S-nitrosylation, which is the covalent attachment of an – NO group to a cysteine thiol (or more properly thiolate anion), has been recognized as a reversible posttranslational modification by which NO regulates the function of many target enzymes, transcription factors, and ion channels. These targets include protein disulfide isomerase (Uehara et al. 2006), matrix metalloproteinases (Gu et al. 2002), parkin (Chung et al. 2004; Yao et al. 2004), cyclin-dependent kinase 5 (Cdk5, Qu et al. 2011), peroxiredoxin 2 (PRX2, Fang et al. 2007), dynamin related protein 1 (Drp1, Cho et al. 2009), X-linked inhibitor of apoptosis protein (Nakamura et al. 2010), phosphatase with sequence homology to tensin (PTEN, Numajiri et al. 2011), and many other molecules (Hess et al. 2005; Hara and Snyder 2007; Nakamura and Lipton 2009, 2011; Nakamura et al. 2011). Interestingly, while some proteins are S-nitrosylated by physiological levels of endogenous NO as part of a normal signaling pathway, other targets appear to be nitrosylated only under pathological conditions with high levels of NO, as seen in a number of neurodegenerative disorders. These aberrant nitrosylation reactions can lead to abnormal protein activity and thus contribute to the pathogenesis of the disease (Nakamura and Lipton 2009, 2011).

Oxidative metabolism leads to the formation of numerous RNS, including NO_2, peroxynitrite ($ONOO^-$), and *S*-nitrosoglutathione. RNS can react directly or indirectly with proteins and other molecules in the cell, inducing chemical modifications that lead to changes in their structure and function. For example, $ONOO^-$ can act as an effector of NO-dependent signals as a strong oxidizing agent by targeting cysteine thiols, in some cases facilitating disulfide formation in proteins. By targeting cysteine thiols, $ONOO^-$ inhibits tyrosine phosphatases, antioxidant enzymes, and cysteine proteases (Radi 2004; Pacher et al. 2007). $ONOO^-$ also modifies proteins by nitrating several amino acids, including tyrosine residues (Pacher et al. 2007; Butterfield et al. 2011). Much research has focused on peroxynitrite-mediated modification of tyrosine residues, which forms 3-nitrotyrosine, considered a key player in $ONOO^-$-mediated toxicity in animals (Radi 2004). This nitration of tyrosine residues, in counterdistinction to S-nitrosylation of cysteine thiols, is mediated by reaction of $ONOO^-$ with tyrosine not cysteine residues as is a common misconception. Recently, $ONOO^-$ has been reported to also induce nitration (by adding -NO_2) of unsaturated fatty acids and guanosine, which function as intracellular signaling molecules (Niles et al. 2006). Interestingly, these nitrated molecules (nitro-fatty acid and nitro-guanosine) are thought to be physiological activators of the Keap1/Nrf2 pathway and thus function in electrophilic counterreaction (Groeger and Freeman 2010; Kansanen et al. 2011). $ONOO^-$ itself is highly reactive and thus unstable, but may be stabilized when complexed with carbonate radical, which can produce NO_2 radicals (Radi 2004; Pacher et al. 2007). These radicals mediate nitration of fatty acid and guanosine under physiological conditions, suggesting a potential normal signaling function of these molecules in addition to their known toxic effects (Groeger and Freeman 2010; Kansanen et al. 2011).

10.3 Cysteine-Based Redox Regulation

Cysteine-based redox regulation is an effective frontline of defense in cells that is used to combat oxidative and nitrosative stress, as shown in Fig. 10.3 (Winyard et al. 2005; Immenschuh and Baumgart-Vogt 2005; Satoh and Lipton 2007). Importantly, virtually all of the molecules and proteins involved in this defense mechanism are regulated by the Keap1/Nrf2 and/or the HSP90/HSF-1 transcriptional pathways (Takahashi et al. 2009; Tamaki et al. 2010; Satoh et al. 2011). For cysteine-based regulation to work effectively, transcriptional regulation is essential, and EPs are effective activators of these transcriptional pathways. Three of the cysteine-based redox regulatory cascades that are activated by these transcriptional pathways are summarized below, and they include (i) the glutathione peroxidase (GPX) system, (ii) the PRX/TRX system, and (iii) the PRX/SRXN system.

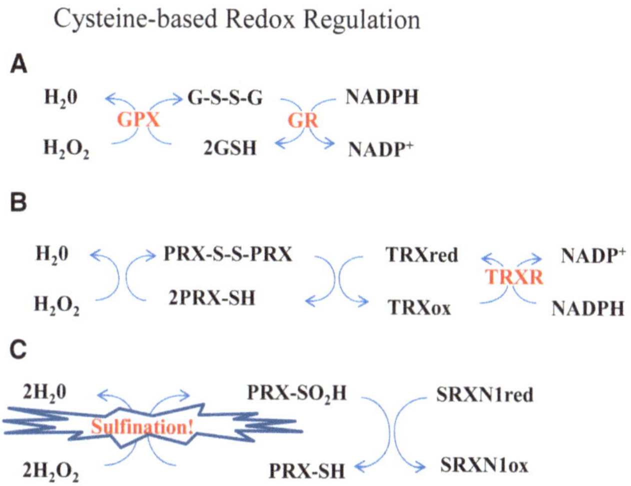

Fig. 10.3 Cysteine-based Redox regulation. Cells manifest three cascades for cysteine-based redox regulation: the GPX/GSH (A), PRX/TRX (B), and PRX/SRXN (C) systems

10.3.1 GPX/GSH System

Glutathione is amongst the most important intracellular antioxidants and is present at millimolar concentrations within human cells (Arteel and Sies 2001). As shown in Fig. 10.3A, reduced glutathione (GSH) exists in equilibrium with its disulfide form (GSSG), and the ratio of GSH to GSSG can be used as an indicator of the redox status of the cell. In addition to catalase and PRX, the antioxidant system composed of the selenium-dependent enzyme GPX and GSH constitute the major cellular enzymatic mechanisms to reduce hydrogen peroxide to water (Arteel and Sies 2001). GSH is also used by glutathione-S-transferase (GST), which catalyzes the conjugation of GSH to various electrophiles, thereby removing these toxic products of free radical reactions. GSH-dependent defenses rely, of course, on the availability of GSH. Therefore, systems involved in the reduction of GSSG are essential, including NADPH-dependent GSSG reductase (GR), as well as GSSG transporters. GR achieves reduction of GSSG by a combination of a thiol–disulfide exchange reaction, two-electron transfer (from $FADH_2$ to reduce the GR disulfide) and hydride transfer from NADPH to FAD (Arteel and Sies 2001).

10.3.2 PRX/TRX System

TRX has the capacity to repair oxidized proteins by reducing protein disulfides at the expense of the oxidation of its own redox-active cysteine residues, as shown in

Fig. 10.3B (Chang et al. 2004; Immenschuh and Baumgart-Vogt 2005; Holmgren and Lu 2010). The oxidized TRX cysteines are then reduced back to the free thiol form by thioredoxin reductase (TRXR). TRXR contains redox-active cysteine and selenocysteine residues, and regenerates the reduced form of TRX by a series of thiol–disulfide and seleno–disulfide exchange reactions and electron transfer from NADPH. PRX reacts with peroxides to detoxify them to water, forming disulfides on PRX, which are then reduced by TRX to regenerate active PRX with free thiols. The TRX-PRX pathway is arguably the most important antioxidant system to limit accumulation of intracellular peroxides via redox reactions at critical cysteine residues. Expression of TRX-PRX is controlled by the Keap1/Nrf2 system (Talalay 2000; Itoh et al. 2004). In particular, the TRX-PRX system is extremely abundant in the brain. In summary, the TRX-PRX system detoxifies peroxides by transferring reducing equivalents from NADPH to peroxides via TRX reductase, TRX, and finally PRX (Chang et al. 2004; Immenschuh and Baumgart-Vogt 2005; Holmgren and Lu 2010). Recently, we showed that excessive NO can react with the free thiols of PRX to prevent its reaction with peroxides, thus increasing susceptibility to neuronal cell death and injury (Fang et al. 2007).

10.3.3 PRX/SRXN System

In response to fulminant oxidative stress, PRX forms PRX-SOH via reaction with peroxide, and can be further oxidized to a sulfinic ($-SO_2H$) or sulfonic ($-SO_3H$) acid derivative (Rhee et al. 2007; Soriano et al. 2008, 2009). This hyperoxidation of PRX causes inactivation of its peroxidase/neuroprotective activity. Although PRX-SO_2H cannot be reduced by TRX, it can be reduced back to the catalytically active free thiol form in eukaryotic cells including neurons by an ATP-dependent reductase, sulfiredoxin (SRXN). The activity of SRXN1 restores inactive PRX-SO_2H back to the TRX cycle and prevents permanent oxidative inactivation of PRX by strong oxidative insults, as shown in Fig. 10.3C (Rhee et al. 2007; Soriano et al. 2008, 2009). However, SRXN1 must be generated by transcriptional activation in cells in order for it to be produced in sufficient quantities to reverse hyperoxidation of PRX, and this transcriptional activation requires activation of the Keap1/Nrf2 pathway (Soriano et al. 2008).

10.4 GSH Metabolism as a Key to Protection

The level of GSH is important for protection against oxidative/nitrosative stress. Through the study of various EPs, we found that the intracellular GSH level is a key in determining whether an EP will be protective or toxic to neurons (Fig. 10.4). EPs can exert opposite effects on the level of GSH: they can directly react with GSH to deplete it (action #1) as well as induce phase 2 enzymes that synthesize GSH (action

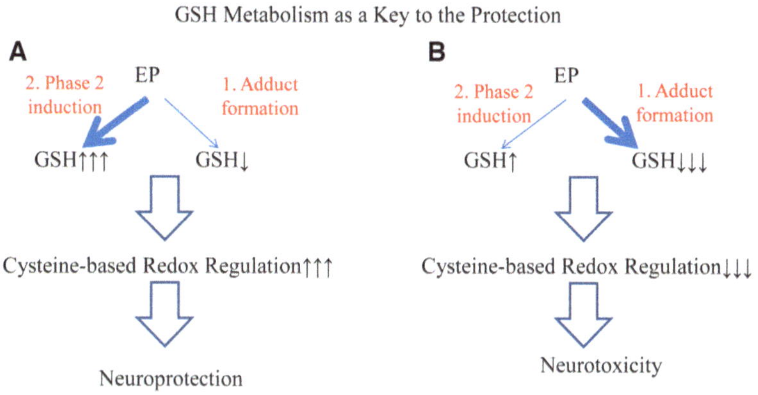

Fig. 10.4 GSH metabolism as a key to redox protection. EPs do not have unpaired electrons like ROS/RNS, but have atoms (carbon, nitrogen, and sulfur) with electron deficiency. EPs react in facile fashion with GSH. In the case of neurotoxic EPs, the depletion of GSH can proceed by adduct formation. In contrast, an increase in GSH is dominant in neuroprotection via transcriptional activation by EPs. Note that this increase in GSH is due to the induction of phase 2 enzymes, which is involved in the cysteine-based redox regulation described in Fig. 10.3

#2). If action #1 overwhelms action #2, then the EPs will be toxic and actually render the cells more susceptible to oxidative and nitrosative stress. In contrast, if phase 2 induction overcomes adduct formation, then the EP is neuroprotective (Fig. 10.4A). When adduct formation overcomes phase 2 induction, the EP will be neurotoxic (Fig. 10.4B). Cysteine-based redox regulation is a powerful defense against acute oxidative and nitrosative stress only if the various forms of cysteines (GSH, TRX, PRX, SRXN and other proteins) are constitutively supplied, i.e., they are available at the time of the insult. If they have to be transcriptionally induced, then the cell may not survive an acute fulminant episode of oxidative/nitrosative stress since the insult may overwhelm the number of reduced cysteines that are present; the same caveat exists if a milder stress is present for a prolonged time. However, in the latter case, cells may have time to generate additional cysteine-based redox compounds by activating various phase 2 enzymes such as the glutamyl cysteine ligase modifier subunit (GCLM), glutamyl cysteine ligase catalytic subunit (GCLC), or Na^+-independent cystine-glutamate exchanger (xCT).

As a master transcriptional regulator of redox activity, Nrf2 can regulate the expression of many proteins relevant to the cell's reducing capacity, including phase 2 enzymes. The Keap1/Nrf2 pathway can be activated by various agents, including ROS/RNS themselves as well as various forms of electrophilic compounds. In the context of continued oxidative/nitrosative stress, the continued supply of reduced cysteine-containing compounds via the Keap1/Nrf2 pathway is a key event for the regulation of cysteine-based redox regulation.

The importance of GSH metabolism for the neuroprotective effects of EPs has also been implicated in many experimental paradigms (Ahlgren-Beckendorf et al. 1999; Lee et al. 2003; Kraft et al. 2004; Sun et al. 2005; Satoh et al. 2008a, b,

2009a, b; Takahashi et al. 2009; Tamaki et al. 2010). Here we discuss carnosic acid (CA) as an example of a compound that can be converted to a neuroprotective EP, and the protective mechanism can be understood at least in part in terms of GSH metabolism. We recently found that CA induces transcription of four phase 2 genes encoding GST (A2 subtype), GCLC, and two ATP-binding cassette family members, represented by ATP-binding cassette sub-family C member 4 (ABCC4) and multidrug resistance-associated protein 1 (ABCC1). All of these are involved in GSH regulation. GSTA2 catalyzes the conjugation of GSH with electrophilic compounds, GCLC is involved in the synthesis of GSH, and ABCC1 and ABCC4 actively transport GSH-conjugated electrophilic compounds. The induction of these genes is known to up-regulate GSH metabolism and hence increases the redox capacity of the cell against oxidative stress.

10.5 Two Transcriptional Pathways for Electrophilic Counterattack

As shown in Fig. 10.5, the Keap1/Nrf2 and HSP90/HSF-1 transcriptional pathways provide an effective 'backup system' for cysteine-based redox regulation by activating endogenous gene networks that are involved in antioxidant defense. The work of Satoh et al. (2011) and Zhang et al. (2011) showed that several activators of Nrf2 via reaction with Keap1 also covalently bind to cysteines of HSP90 to activate HSF-1, thereby inducing molecular chaperones, including additional HSPs. These results suggest that S-alkylation of critical cysteine residues by EPs can activate both of these transcriptional pathways, representing electrophilic counterreactions. Since neurons have little capacity for redox regulation, these backup systems become very important in the nervous system. After electrophilic activation, the transcription factors Nrf2 and HSF-1 bind to the antioxidant response element (ARE)- and heat-shock factor response element (HSE), respectively, to initiate transcription of genes whose products provide neuroprotection from oxidative and nitrosative insults (Satoh et al. 2011).

10.5.1 The Keap1/Nrf2 Pathway

Nrf2 is a member of a small family of transcription factors that share a conserved bZIP dimerization/DNA-binding domain and the ability to bind ARE-like DNA sequence motifs. Characterization of mice that lack Nrf2 has revealed an essential role for this transcription factor in both basal and inducible expression of classical phase 2 genes (Talalay 2000; Itoh et al. 2004). The ability of Nrf2 to activate transcription of its target genes is regulated, in large part, through association with a cytoplasmic protein termed Keap1 (Talalay 2000; Itoh et al. 2004). Keap1 was first identified by two-hybrid screening that used the N-terminal regulatory domain of Nrf2 as bait (Talalay 2000; Itoh et al. 2004). Keap1 is a member of a large family of proteins that contain an N-terminal Broad complex, Tramtrack, and Bric a brac

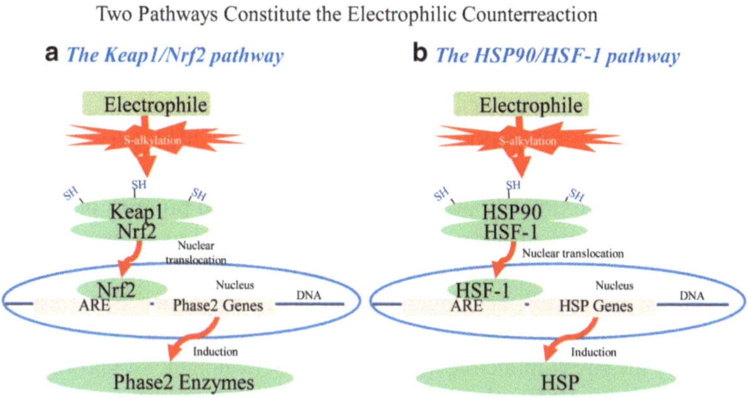

Fig. 10.5 Two pathways for electrophilic counterattack. Two essential pathways (*A* Keap1/Nrf2 and *B* HSP90/HSF-1) constitute electrophilic counterreactions. Some EPs activate both pathways and others activate one or the other. Note that EPs activate these pathways by triggering S-alkylation of the regulatory thiol proteins, Keap1 and HSP90

(BTB) domain, and a C-terminal Kelch repeat domain (Padmanabhan et al. 2006). Previous reports have demonstrated that Keap1 retains Nrf2 in the cytoplasm via a direct protein-protein interaction between the C-terminal Kelch repeat domain of Keap1 and the N-terminal Neh2 regulatory domain of Nrf2. The role of Keap1 as a physiological regulator of Nrf2 is further supported by the observation that macrophages from mice that lack Keap1 have constitutive nuclear accumulation of Nrf2 (Padmanabhan et al. 2006; Satoh et al. 2009b).

Because EPs are effective activators of the Keap1/Nrf2 pathway, they can induce the expression of a set of metabolic antioxidant enzymes, called 'phase 2 enzymes,' which include hemeoxygenase-1 (HO-1), NADPH quinone oxidoreductase 1 (NQO1), sulfiredoxin (SRXN), glutamyl cysteine ligase modifier subunit (GCLM) and the Na^+-independent-cystine/glutamate exchanger (xCT), all of which provide efficient cytoprotection by regulating the intracellular redox state (Kraft et al. 2004; Satoh et al. 2006, 2008a). Mechanistically, when EPs react with critical cysteine residues on Keap1 protein to form a covalent adduct, they perturb this system, thereby releasing Nrf2 and allowing it to be translocated from the cytoplasm into the nucleus, where it binds to AREs and stimulates the transcription of phase 2 genes (Kraft et al. 2004; Satoh et al. 2006, 2008a). Thus, Nrf2 has emerged as a potential therapeutic target for the treatment of neurodegenerative diseases (Satoh and Lipton 2007; Alfieri et al. 2011).

10.5.2 The HSP90/HSF-1 Pathway

The HSP90/HSF-1 pathway is the principal regulatory pathway leading to expression of heat-shock proteins (HSPs), which rescue cells from endoplasmic reticulum

(ER) stress and related insults (Bukau et al. 2006; Morimoto 2008). Several reports have indicated that ROS/RNS regulate HSP expression by activating this pathway. We and others have recently reported that EPs, which activate the Keap1/Nrf2 pathway through S-alkylation of a critical cysteine thiol on Keap1, also induce HSPs via activation of the HSP90/HSF-1 pathway (Satoh et al. 2011; Zhang et al. 2011). The ER is a reticulated organelle in which proteins are synthesized and modified for proper folding. Approximately 30% of newly synthesized proteins in normal cells are misfolded, but some of these proteins can be refolded to achieve their correct structures. This process of refolding is facilitated by ER chaperones. However, other proteins remain misfolded, accumulate in the ER, and induce ER stress (Bukau et al. 2006; Morimoto 2008). Transcriptional induction of molecular chaperones is governed by the stress-inducible heat-shock transcription factor known as HSF-1, which when bound to HSEs, plays a key regulatory role in activating transcription in response to environmental stress (Bukau et al. 2006; Morimoto 2008). Compounds that modulate HSF-1 to increase HSP expression can exert protective effects against various types of stress including ER stress (Bukau et al. 2006; Morimoto 2008). The HSF-1/HSE system can protect against oxidative stress in addition to ER stress (Kim et al. 2008; Nakamura and Lipton 2009, 2011).

In this background, we recently found that induction of HSPs mediated by D1, a cyclic EP, plays a role in neuroprotection against ER stress as well as against oxidative stress (Satoh et al. 2011). Oxidative stress can also lead to ER stress, which may contribute to several chronic neurodegenerative diseases. Accumulation of misfolded proteins in the ER induces ER dysfunction (Kim et al. 2008; Nakamura and Lipton 2009). To combat such stress, there is a powerful endogenous protective mechanism for induction of molecular chaperones, including heat-shock protein 70 (HSP70) and DnaJ, as well as heat-shock factor-binding protein (HSBP) and heat-shock 105-kDa/110-kDa protein 1 (HSPH1). These molecular chaperones are known to suppress protein misfolding (Bukau et al. 2006; Morimoto 2008). The expression of these chaperones after exposure to various types of cell stress is known to be regulated by HSF-1. Under unstressed conditions, HSF-1 is localized in the cytosol and is inactivated in a protein complex that includes HSP90. As mentioned above, upon exposure to stress-inducing compounds including EPs, HSF-1 dissociates from the HSP90 protein complex, is translocated into the nucleus, and binds there to the HSE in the promoter region of various molecular chaperone genes to induce their expression. For this reason, HSF-1 has been considered as a potential target for the treatment of neurodegenerative diseases (Morimoto 2008).

10.6 Keap1(Cys151) as a Hyperactive Thiol

For EPs to be effective in electrophilic counterreaction, an initial event is the S-alkylation of essential cysteine residues of regulatory proteins such as Keap1 and HSP90. Several publications have studied the critical cysteines of Keap1 that can be S-alkylated, but we will limit our discussion to the critical cysteines of

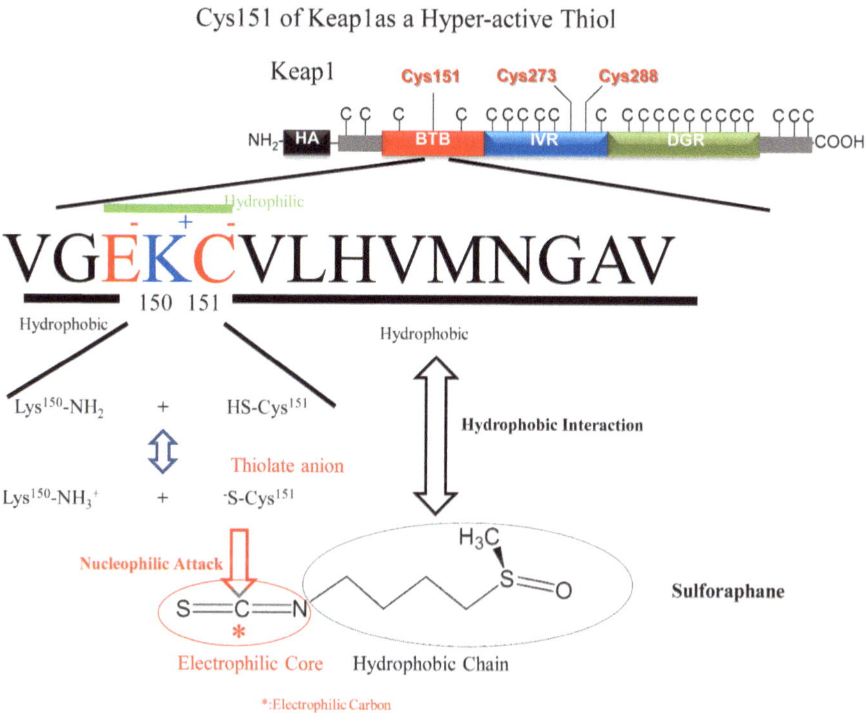

Fig. 10.6 Keap1(Cys151) as a hyperactive thiol. Keap1 has 25 cysteines, and the amino acid sequence surrounding the critical residue at Cys151 is shown in the middle column. The amino acids Glu149 and Lys150 are the only hydrophilic residues in a stretch of hydrophobic amino acids. The NH_2 of Lys150 is thought to remove a proton from the -SH of Cys151, thus forming a thiolate anion ($-S^-$). This hyperactive thiol can bind to the electrophilic core of sulforaphane, represented by isocyanate

Keap1 whose modification leads to activation of Nrf2 (Talalay 2000; Itoh et al. 2004; Padmanabhan et al. 2006). The studies pertaining to Keap1 activation by EPs have concentrated on Cys273, Cys288, Cys151 and Cys434, while activation by ROS/RNS has centered around Cys226 and Cys613. Keap1 actively targets the N-terminal Neh2 domain of Nrf2 for ubiquitination and proteosome-mediated degradation. Two cysteine residues located in the linker domain of Keap1, Cys273 and Cys288, are critically required for both Keap1-dependent ubiquitination and Keap1-mediated repression of Nrf2-dependent transcription under basal conditions (Dinkova-Kostova et al. 2002). A third cysteine residue located in the BTB domain of Keap1, Cys151, is required for stabilization of Nrf2 and for activation of Nrf2-dependent transcription by the EPs tBHQ and sulforaphane, as shown in Fig. 10.6 (Zhang et al. 2004; Hong et al. 2005; Kobayashi et al. 2009). Cys151 provides a link between the ability of chemical inducers to block Keap1-mediated repression of Nrf2 and to induce posttranslational modification(s) of Keap1.

Why is Cys151 of Keap1 distinct from its other cysteines? One possible answer is the formation of a thiolate anion, $-S^-$ (Kobayashi et al. 2009). Cys151 is the direct molecular target of a number of chemical inducers of Nrf2-dependent transcription (Zhang et al. 2004). Kobayashi et al. (2009) reported the importance of Lys150 in addition to Cys151, suggesting that the –SH group of Cys151 donates H^+ to the NH_2 of Lys150, thus forming a thiolate anion ($-S^-$) and an ammonium cation ($-NH_3{}^+$). Thiolate anion, as opposed to a thiol of the sulfhydryl group itself (–SH), displays an increased capacity to donate electrons to ROS, RNS, and EPs. This proposed chemical reaction scheme makes Keap1(Cys151) distinct from other cysteines in Keap1. Furthermore, Lys150 and Cys151 are surrounded by hydrophobic amino-acid residues, suggesting that the reaction of EPs with Cys151 requires an interaction between the nearby hydrophobic amino acids in Keap1 and the hydrophobic chain of EPs. For example, sulphoraphane, which is reported to bind to Cys151, is a typical example of this chemical reaction (Fig. 10.6). The electrophilic core (isocyanate) of sulphoraphane binds to Cys151, and its hydrophobic tail binds to the hydrophobic chain of Keap1, resulting in the release of Nrf2 and explaining the effective activation of the Keap1/Nrf2 pathway by sulphoraphane (Zhang et al. 2004; Hong et al. 2005; Kobayashi et al. 2009). In contrast, ROS and RNS themselves, including hydrogen peroxide (H_2O_2) and NO, have an electrophilic core but lack a hydrophobic chain; thus, they require a much higher concentration to facilitate their reaction with Cys151 (Fourquet et al. 2010; Um et al. 2011).

Although H_2O_2 does not apparently directly oxidize Cys 151 to form a stable sulfenic, sulfinic, or sulfonic acid derivative, it does facilitate disulfide bond formation with a second Cys151. Fourquet et al. (2010) found that in untreated cells, a fraction of Keap1 carries a long-range disulfide linking Cys226 and Cys613. Exposing cells to H_2O_2 or certain NO donors increases the incidence of this disulfide and also promotes formation of a disulfide linking two Keap1 molecules via their Cys151 residues. The involvement of other cysteines was also reported, for example, Cys434 (Fujii et al. 2010). In contrast to Cys151, Cys434 is not surrounded by many hydrophobic amino acids and thus hydrophilic compounds, such as 8-nitroguanosine, can approach it for reaction. In fact, Keap1(Cys434) is one of the cysteine residues that is most sensitive to adduct formation with 8-nitroguanosine. Additionally, nitro-fatty acids can reportedly make an adduct with Keap1 in a Cys151-independent manner (Kansanen et al. 2011).

10.7 Electrophilic Counterreactions by Endogenous Electrophiles

In addition to direct reaction of ROS and RNS with Keap1 or HSP90, endogenous EPs produced by ROS/RNS can activate these pathways, a process termed "electrophilic counterreaction" or "electrophilic counterattack" against oxidative and nitrosative stress (Groeger and Freeman 2010; Akaike et al. 2010; Satoh and Lipton 2007). Examples of the reactions that generate EPs are illustrated below.

10.7.1 Active Aldehyde

Lipid peroxidation yields multiple aldehyde species. Of these, *trans*-4-hydroxy-2-nonenal (HNE), derived from *n*-6 poly-unsaturated fatty acids (PUFA), is one of the most studied products of lipid peroxidation (Fig. 10.7A) (Uchida 2003; Catalá 2009; Long and Picklo 2010). Additionally, oxidative damage to *n*-3 PUFAs, e.g., docosahexaenoic acid (22:6, *n*-3) and eicosapentaenoic acid, is now recognized as an important effector of oxidative stress and is of particular interest in *n*-3-rich tissues such as the brain and retina (Uchida 2003; Catalá 2009; Long and Picklo 2010). *Trans*-4-hydroxy-2-hexenal (HHE) is a major-unsaturated aldehyde product of *n*-3 PUFA oxidation and, like HNE, is an active biochemical mediator resulting from lipid peroxidation. HHE adducts are elevated in disease states, in some cases to higher levels than the corresponding HNE adduct. HHE has properties in common with HNE, but there are also important differences, particularly with respect to adduct targets and detoxification pathways (Uchida 2003; Catalá 2009; Long and Picklo 2010). During the generation of HHE and HNE, the withdrawal of the π electron from the conjugated double bond by OH is thought to initiate a series of lipid peroxidation steps shown in Fig. 10.7A. Although HNE and HHE activate the Keap1/Nrf2 and HSP90/HSF-1 pathways, this action is also accompanied by toxic effects (Uchida 2003; Mangialasche et al. 2009). The reason for this is that the reaction produces various toxic radicals and lipid hydroperoxides, and thus this pathway is considered toxic rather than beneficial against oxidative stress.

10.7.2 Nitro-Fatty Acid and 8-Nitro-cGMP

As shown in Fig. 10.7B and C, the chemical reaction for the generation of nitro-fatty acid and 8-nitroguanosine is basically via the same pathway, although they are achieved under hydrophobic and hydrophilic conditions, respectively (Groeger and Freeman 2010; Kansanen et al. 2011; Akaike et al. 2010; Fujii et al. 2010). NO_2, as an initiator of the chemical reaction, withdraws a π electron from the double bond and makes an adduct (Radi 2004; Groeger and Freeman 2010). The resultant alkyl radical is attacked by another NO_2, and finally HNO_2 is released from the molecule, and nitro-fatty acid and 8-nitroguanosine are produced. Surprisingly, the production of NO_2 seems to proceed continuously at very low levels. Free and esterified polyunsaturated fatty acids (PUFAs), such as linoleate, are in particular predisposed to formation of both reactive intermediates and bioactive products (Groeger and Freeman 2010; Kansanen et al. 2011). Nitrated derivatives of oleic, linoleic, linolenic, arachidonic, and eicosapentaenoic acids have also been identified

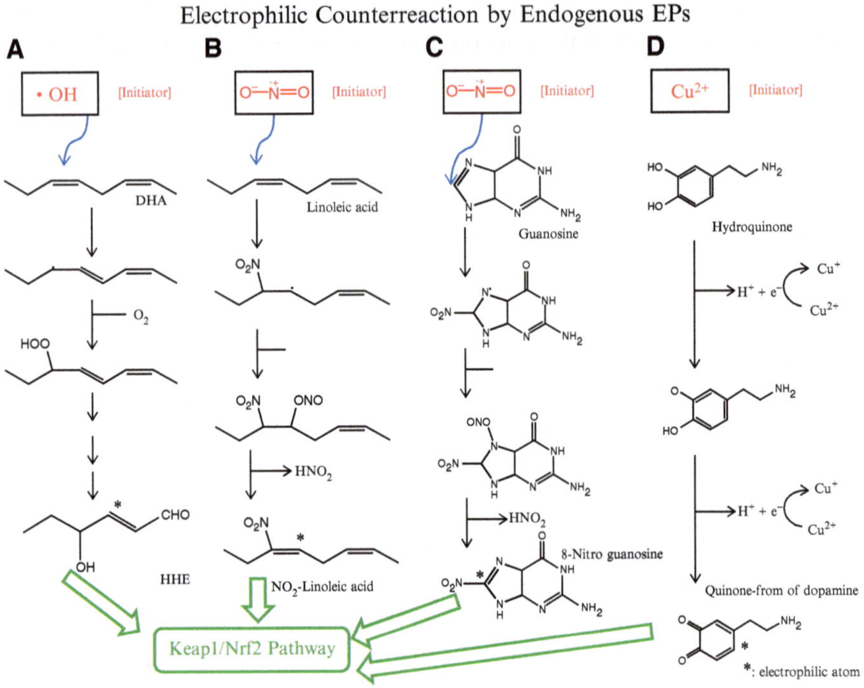

Fig. 10.7 Electrophilic counterreactions triggered by endogenous EPs. Chemical reactions of ROS to generate endogenous EPs are illustrated. The precise chemical reactions 1–3 are not fully understood but are thought to involve free radical-initiated lipid peroxidation. The EPs thus produced activate the regulatory Keap1/Nrf2 and HSP90/HSF-1 pathways

in mammals. These molecules might exist under physiological conditions because nitrated derivatives of unsaturated fatty acids are present in healthy human blood, and nitrohydroxy derivatives of linoleate have been detected in healthy human plasma and lipoproteins (Groeger and Freeman 2010; Kansanen et al. 2011). Acting as an EP, 8-nitro-cGMP reacts with particular protein thiols to generate a unique posttranslational modification (Akaike et al. 2010; Fujii et al. 2010).

10.7.3 Dopamine Quinone

Dopamine (DA) is synthesized from tyrosine via the enzymatic activities of tyrosine hydroxylase and aromatic amino acid decarboxylase (Winyard et al. 2005; Hastings 2009). Once formed, DA is safely stored in high millimolar concentrations in synaptic vesicles. If vesicular storage is disrupted, DA levels increase in the cytoplasm (Winyard et al. 2005; Hastings 2009). The catechol ring of DA can be oxidized by ROS to form electron-deficient DA quinone (Fig. 10.7D). DA quinone will react readily with cellular nucleophiles, such as reduced thiol groups of small peptides and protein cysteinyl residues (Winyard et al. 2005; Hastings 2009). There appear to be two opposite opinions as to whether DA quinone is protective or toxic to neurons (Hastings 2009; Shih et al. 2007). The classical opinion is that the process produces peroxides and electrophilic quinones, and these events can lead to neurodegeneration via increased oxidative stress (Hastings 2009). A more recent opinion is that this reaction of DA and other catecholamines produces electrophilic quinones that activate the Keap1/Nrf2 pathway and can thus protect neurons (Shih et al. 2007; Takahata et al. 2009). Both protective and toxic effects of DA quinone require high concentrations ($> 100\ \mu M$), and these processes therefore occur under pathological conditions (Shih et al. 2007; Hastings 2009). The critical difference between the opinions is whether DA quinone is to be regarded as toxic waste or as a mediator of electrophilic counterreaction. In fact, both lines of work may be correct. One possible interpretation is that DA quinone is toxic to neurons but protective to glial cells. Because neurons are highly sensitive to oxidative stress, DA quinone appears to be directly toxic to neurons, but, in at least some cases, appears to be tolerated by glia, activating the Keap1/Nrf2 pathway. Thus, the glia may in turn protect themselves as well as nearby neurons in a non-cell autonomous fashion, for example, by generating protective factors such as GSH and neurotrophic factors (Shih et al. 2007; Hastings 2009). Nonetheless, as discussed further below, an EP that is only protective would, in our opinion, be preferred so that potential consequences of toxic side reactions could be avoided entirely.

10.8 Two Distinct Groups of Electrophiles

In our opinion, it is possible to take advantage of the intracellular mechanism of electrophilic counterreaction by endogenous EPs, as described above, in order to develop a novel strategy for drug development against neurodegenerative diseases (Fig. 10.8). The core concept of this drug development strategy is based on the following principles (Satoh et al. 2008a, 2011; Satoh and Lipton 2007):

1. We are developing pro-electrophilic compounds rather than electrophilic compounds, and thus call these new drugs "Pro-Electrophilic Drugs (PEDs)."
2. PEDs become electrophilic (i.e., are converted to the active form to trigger electrophilic counterreaction) in a Cu^{2+}/O_2-recycling system-dependent manner, and thus PEDs preferentially become active in the face of oxidative stress rather than under physiological conditions. Hence, only cells undergoing oxidative stress will generate active EPs, making the PEDs relatively innocuous for normal cells not facing free radical insult.

This approach still takes advantage of an important property of EPs, namely, their superiority to antioxidant molecules because of the sustained action and amplification of the effect of EPs via transcription-mediated signaling pathways

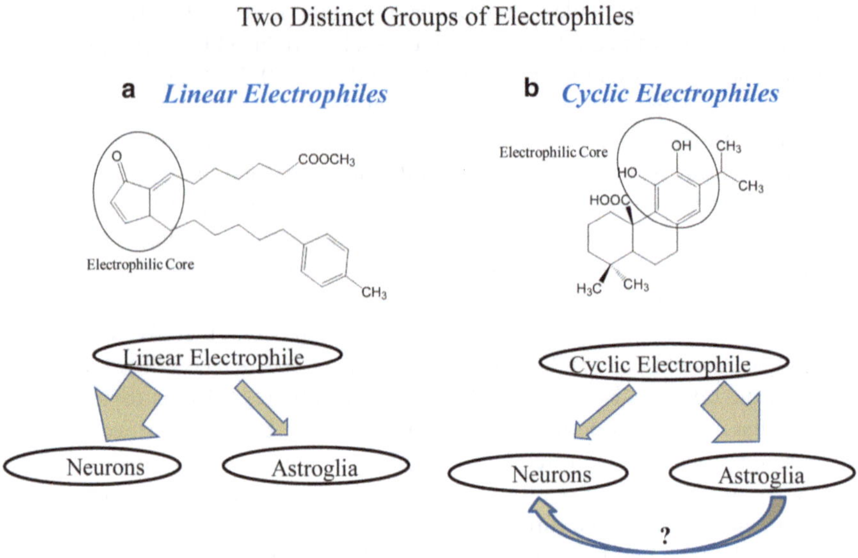

Fig. 10.8 Two distinct groups of electrophiles. There are two types of EPs, one of which has a linear electrophilic core, such as NEPP11, whereas the other has cyclic structures with hydroquinone as the electrophilic core, e.g., CA. After treatment of cells, linear EPs appear to concentrate in neurons and cyclic EPs, in astroglia. Thus, paracrine effects may be involved in the neuroprotective effect of cyclic EPs

(Talalay 2000; Itoh et al. 2004). The neuroprotective EPs reported previously can be divided into two major groups, linear EPs and cyclic EPs (Satoh et al. 2008a). These two types of compounds manifest distinct features. One difference lies in their degree of electrophilicity and thus their ability to activate the Nrf2/ARE pathway. Linear EPs, including neurite outgrowth-promoting prostaglandin (NEPP11) (Satoh et al. 2000, 2001, 2003, 2006), curcumin (Mandel et al. 2009) and sulforaphane (Kraft et al. 2004), are electrophilic in nature; i.e., they readily react with electron donors. In contrast, cyclic EPs are more properly termed PEDs since they are not in and of themselves truly electrophilic, but they become electrophilic via oxidative conversion to their *quinone* form, which occurs when they encounter free radical oxidative stress (Satoh and Lipton 2007). Thus, these cyclic EPs function as pro-drugs, requiring conversion from *hydroquinone* to *quinone* in order to exert their neuroprotective effect (Satoh et al. 2009a; Wang et al. 2010). The cyclic EPs or PEDs include compounds such as carnosic acid (CA), *tert*-butyl hydroquinone (TBHQ), strongylophorine 8 (Sasaki et al. 2011), and D1 (Satoh et al. 2011).

We found that the cellular distribution of linear EPs and cyclic EPs in the brain treated with these compounds appears to be different (Satoh et al. 2006, 2008a). Linear EPs, including NEPP11, appear to act preferentially on neurons, based on our earlier findings that NEPP11 accumulates in neurons as opposed to astrocytes, and consequently induces phase 2 enzymes such as neuronal HO-1 (Satoh et al. 2006).

In contrast, in general cyclic EPs, including CA and TBHQ, appear to preferentially act on astrocytes, as evidenced by the fact that TBHQ activates the ARE in astrocytes to induce phase 2 enzyme synthesis rather than in neurons (Lee et al. 2003: Kraft et al. 2004; Satoh et al. 2008a; Ahlgren-Beckendorf et al. 1999; Vargas and Johnson 2009). In view of these findings, we propose that linear EPs such NEPP11 exert direct protective effects in neurons while cyclic EPs/PEDs such as CA and TBHQ exert paracrine effects to protect neurons (Kraft et al. 2004; Satoh et al. 2008a). In all likelihood, therefore, paracrine or non-cell autonomous effects are the most plausible mechanism for neuroprotection of PEDs, and some candidate mediators of neuroprotection that are released from astrocytes have been identified by microarray analysis (Lee et al. 2003). These non-cell autonomous neuroprotective molecules may include neurotrophins, such as nerve growth factor (NGF), brain-derived neurotrophic factor (BDNF) or ciliary neurotrophic factor (CNTF), since we have found that CA potently releases NGF protein from glioblastoma cells. Another possible candidate is GSH since TBHQ potently induces GSH formation in astrocytes (Satoh et al. 2008a; Ahlgren-Beckendorf et al. 1999; Vargas and Johnson 2009). Importantly, tissues undergoing oxidative stress generally involve both neurons and glia, so glia contiguous to stressed neurons may also be expected to have undergone insult. In a number of neurodegenerative diseases, in recent years it has become increasingly clear that it is the nonneuronal cells or glia that mediate the toxic effect on neurons (Clement et al. 2003; Lioy et al. 2011; Garden and La Spada 2012), so in fact it may be the glia that are injured initially by the disease. Hence, mitigating oxidative stress in the glia with PEDs may in fact be more effective than activating the transcriptional pathways in neurons. In any event, after treatment with higher concentrations of PEDs such as CA, drugs may also enter neurons and may therefore exert direct effects on the Nrf2/ARE pathway in neurons as well as in glia (Satoh et al. 2008a, b). Another important point concerning the potential "drugability" of PEDs is that in many cases their catechol chemical formula renders them permeable to the blood–brain-barrier, thus producing excellent levels in the brain (this property of catechols is one of the reasons that L-dopa is used to treat Parkinson's disease rather than dopamine itself).

10.9 Concept of Pro-electrophilic Drugs

Linear EPs have been shown to protect neurons through an electrophilic chemical reaction. However, linear EPs have a serious flaw as potential therapeutic agents because systemic administration can result in reaction with thiol substrates prior to reaching the intended target in the brain (Satoh et al. 2006, 2008a; Satoh and Lipton 2007). It would be far better to have a pro-electrophilic compound that remains non-reactive until it is converted to an electrophile by the oxidative insult at or near the pathological site of its intended action (Satoh and Lipton 2007; Lipton 2007; Satoh et al. 2008a). Such a drug would represent what we have termed a "Pro-Electrophilic Drug," and we have shown that CA constitutes this type of a compound (Fig. 10.9). We have demonstrated that cyclic EPs/PEDs appear to have

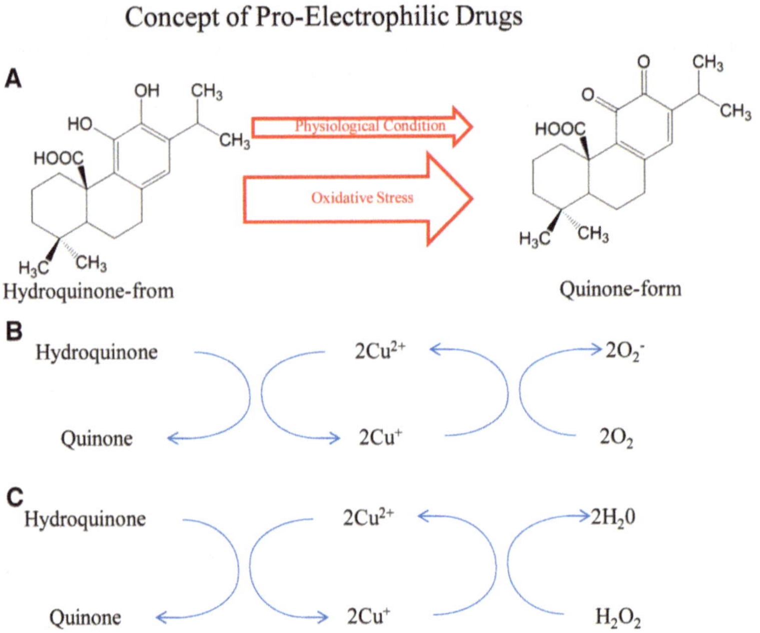

Fig. 10.9 Concept of pro-electrophilic drugs. The *catechol-type* CA that is oxidized to a *quinone-type* form, with the carbon at position 14 [C(14)] becoming electrophilic. This CA quinone is subject to nucleophilic attack by the cysteine thiol of Keap1 to form an adduct. Importantly, via this chemical reaction, the cysteine thiol of Keap1 forms a Keap1-CA adduct, resulting in release of Nrf2 protein from the Keap1/Nrf2 complex. Nrf2 can then be translocated into the nucleus, where it activates transcription of phase 2 enzymes via ARE transcriptional elements of the corresponding genes. These phase 2 enzymes improve the redox state of neurons, contributing to an endogenous anti-oxidant defense system. Note that quinone formation is enhanced under oxidative stress, as described in the text

two clear-cut advantages over linear EPs and similar compounds: *(i)* low potential toxicity due to conversion of the drug to an EP by the oxidative insult at the pathological site, and *(ii)* effective penetrance into brain tissue. These characteristics stem from their chemical structure. Moreover, when the hydroquinone form is oxidized to a quinone, it becomes more hydrophobic and will tend to stay in the injured tissue.

The time course of conversion from *hydroquinone* to *quinone* is rather slow in cell-free systems. How, then, can CA be neuroprotective? One possible answer may lie in the difference between *cell-free* and *cell-based* experiments. Since normal cells have many thiols, including GSH and cysteine-bearing proteins, a quinone would rapidly react with these thiols to form an adduct. Via this mechanism, removal of free quinone would shift the equilibrium between hydroquinone and quinone towards formation of more quinone (Satoh et al. 2008a, b). Thus, the conversion to quinone proceeds much faster in cells than in cell-free systems. Another possible explanation is the difference in reactivity of quinones for the thiols of GSH versus Keap1. Some proteins have active cysteines, which can be easily converted to thiolate anion if surrounded by basic amino acids, as shown in Fig. 10.6. Keap1(Cys151) is a typical example of such an active cysteine (Zhang et al. 2004; Kobayashi et al. 2009). This cysteine is thus a sensor for electrophilic compounds, and in fact EPs appear to be more reactive with Keap1(Cys151) than GSH. In other words, low concentrations of CA quinone preferentially react with Keap1 over GSH and thus can contribute to activation of cell defense systems while sparing GSH (Satoh et al. 2008a, 2009b, 2011). Other researchers have reported similar results for TBHQ and other simple benzene-type hydroquinones (Nakamura et al. 2003; Wang et al. 2010). Additionally, in cells undergoing oxidative stress, GSH is generally depleted, as it is a first line of defense against free radical insult. Thus, in cells stressed cells where PEDs are converted to the active quinone electrophilic form, the resulting EP would also be more likely to react with Keap1(Cys151) since GSH is already depleted.

As stated above, cyclic EPs are converted from an electrophilic precursor (or pro-electrophilic) compound to an electrophilic form, thereby activating the neuroprotective Keap1/Nrf2 pathway (Fig. 10.9b, C). There are additional factors that influence the rapidity of quinone formation from PEDs, i.e., Cu^{2+} and molecular oxygen (Nakamura et al. 2003; Satoh et al. 2009a, b; Wang et al. 2010). Copper is present in many tissues at micromolar concentrations, and in some tissues, including kidney and liver its concentration exceeds 100 μmol/l (Linder and Hazegh-Azam 1996). In the body, copper exists as either oxidized Cu^{2+} or reduced Cu^{+}. Moreover, it is a redox-active transition metal and can catalyze oxidative activation of a number of phenolic compounds via a Cu^{2+}/Cu^{+} redox cycling mechanism that produces ROS. Under physiological conditions, an electron acceptor from Cu^{+} is O_2 (Fig. 10.9B); whereas under pathological conditions, the electron acceptor is likely be a ROS, for example, H_2O_2, as shown in Fig. 10.9C (Nakamura et al. 2003; Satoh et al. 2009a, b; Wang et al. 2010). It is important to note that ROS are much better electron acceptors from Cu^{+} than from O_2, and thus oxidation of Cu^{+} to Cu^{2+} should proceed more easily under pathological conditions. This also means that conversion of a PED to the quinone form will proceed more rapidly under these conditions. In conclusion, we have developed a platform of drug development using PEDs, which provides us with an important framework for producing neuroprotective drugs that become electrophilic only when they encounter oxidative damage, the very condition that they are designed to combat via transcriptional activation of phase 2 enzyme counterreaction (Satoh et al. 2008a; Satoh and Lipton 2007; Lipton 2007).

Importantly, these PED compounds remain relatively innocuous in their pro-drug form and only become active at the site of oxidative stress when converted from their hydroquinone to quinone form. These compounds therefore represent a type of pathologically-activated therapeutic or PAT drug, representing a more gentle ‘tap’ than previous attempts to develop neuroprotective drugs that failed because of their lack of clinical tolerability (Lipton 2004, 2006, 2007). As reviewed here, we believe that this chemical conversion is the key to understanding the molecular mechanism of the protective yet well tolerated effects afforded by cyclic EPs (Satoh and Lipton

2007; Satoh et al. 2008a). In our work, we have focused on hydroquinone-type pro-electrophiles because they have the distinct property of being "pro-drug"-like compounds. During neurodegenerative insults, oxidative stress plays a critical role in disease progression. Our published studies have demonstrated that this pathological insult can be used to activate pro-electrophilic compounds via their oxidation, which occurs in the target tissue to provide neuroprotection where it is needed. Thus, this approach represents a novel strategy against neurodegenerative disorders that could activate electrophilic drugs via pathological activity.

We have thus delineated a strategy for the development of novel neuroprotective drugs that are clinically well tolerated (Lipton 2004, 2007; Satoh and Lipton 2007). This strategy is based on the principle that drugs should interact with their target only during states of pathological activation but not interfere with the target if it functions normally. Such drugs should therefore exhibit little inhibition of normal physiological function. We have had previous success developing such a PAT drug, resulting in FDA approval of the NMDA-type glutamate receptor antagonist memantine for moderate-to-severe Alzheimer's disease (Lipton 2004, 2006, 2007). PEDs represent another possible candidate for a PAT drug because conversion to the active quinone is redox-sensitive and thus enhanced in the presence of oxidative insults. For this reason, we have studied the neuroprotective effects afforded by cyclic EPs/PEDs, including CA and D1 (Satoh et al. 2008a, 2011).

10.10 List of Potential Pro-electrophilic Drugs

Based on their chemical structures and biological actions, we list in Fig. 10.10 potential PEDs from among published compounds. The quinone form of each of these cyclic EPs can activate electrophilic counterreactions, including the Keap1/Nrf2 and/or HSP90/HSF-1 pathways. Each drug has been reported to induce phase 2 enzymes via activation of the Nrf2/ARE pathway. They have para- or ortho-hydroquinones as their electrophilic cores and thus have PED properties. These compounds have been reported to protect neuronal cells against oxidative and nitrosative stress, and some of them (e.g., CA, quercetin, fisetin, and TBHQ) can protect the rodent brain in various models of neurodegenerative diseases. In particular, neuroprotective effects of diterpene-type PEDs, including CA (Satoh et al. 2008a, b, 2009a, b; Tamaki et al. 2010), CS (Satoh et al. 2008b; Tamaki et al. 2010), strongylophorine 8 (Sasaki et al. 2011), TBHQ (Murphy et al. 1991; Lee et al. 2003; Kraft et al. 2004; Sun et al. 2005; Shih et al. 2005; Jakel et al. 2007), and D1 (Satoh et al. 2011), are quite striking in their ability to trigger the Keap1/Nrf2 pathway. Flavonoid-type PEDs, such as quercetin (Ishige et al. 2001; Milioli et al. 2007; Bureau et al. 2008; Ahmad et al. 2011; Suematsu et al. 2011), eriodictyol (Hanneken et al. 2006; Johnson et al. 2009), and fisetin (Burdo et al. 2008; Maher 2008; Lewerenz et al. 2010; Maher et al. 2011), also manifest neuroprotective effects, although their action may occur via other pathways, including direct anti-oxidant effects.

List of Pro-Electrophilic Drugs (PEDs)

Name	Structure	Origin	Type	Examined cells	Examined models	Ref
Carnosic acid	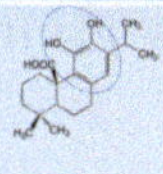	Rosemary (Rosmarinus officinalis)	Ortho-hydroquinone Diterpene	HT22 PC12 Cortical Neurons	MCAO	Satoh T et al., 2008a Satoh T et al., 2008b Satoh T et al., 2009a Saqtoh T et al., 2009b Takahashi et al., 2010 Tamaki Y et al., 2010
Carnosol	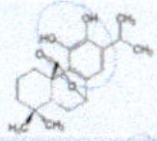	Rosemary (Rosmarinus officinalis)	Ortho-hydroquinone Diterpene	HT22	N.D.	Satoh T et al., 2008a Satoh T et al., 2008b
Strongylophorine 8	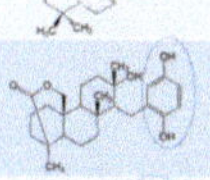	Sponge Petrosia (Strongylophora) corticata	Para-hydroquinone Diterpene	HT22	N.D.	Sasaki et al., 2011
Quercetin	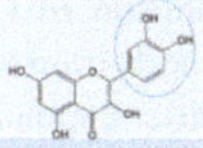	Green plant	Ortho-hydroquinone Flavonoid	Cortical Neurons PC12 HT22 SH-SY5Y	MCAO MPTP	Bureau et al., 2008 Milioli EM et al., 2007 Ahmad A et al., 2011 Ishige K et al., 2001 Suematsu N et al., 2011
Fisetin	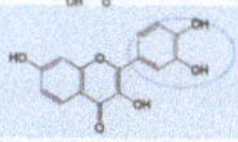	Ggreen plant	Ortho-hydroquinone Flavonoid	PC12 ARPE-19 Cortical Neurons	HD	Maher P et al., 2011 Lewerntz J et al., 2010 Burdo J et al., 2008 Mahere P 2008
Eriodictyol	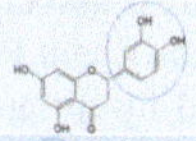	Green plant	Ortho-hydroquinone Flavonoid	PC12 ARPE-19	N.D.	Johnson J et al., 2009 Hanneken et al., 2009
TBHQ	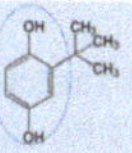	Synthetic Compound	Para-hydroquinone	Cortical Neurons HT22 PC12 IMR-32	MCAO MPTP 6-OHDA	Murphy TH et al., 1991 Lee et al., 2001 Kraft AD et al., 2004 Sun X et al., 2005 Shih AY et al., 2005 Jakel RJ et al., 2007
D1	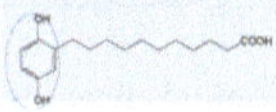	Synthetic Compound	Para-hydroquinone	ARPE-19 HT22	N.D.	Satoh T et al., 2011

Fig. 10.10 List of possible pro-electrophilic drugs. We list possible PEDs based on published data and chemical structures. Structures: The ovals on the chemical structures highlight the electrophilic core. Origins: Origins are given in the case of natural compounds. Type: Isomer type of the electrophilic core on the hydroquinone is given. Examined cells: All of the EPs shown exerted protective effects in neuronal HT22 cells (mouse hippocampal neurons), PC12 (mouse pheochromecytoma cells), SH-SY5Y (mouse neuroblastoma cells), IMR-32 (human neuroblastoma cells), or ARPE (human retinal pigment epithelial cells). Examined models: Some of the compounds (CA, quercetin, fisetin and TBHQ) have been shown to be neuroprotective in rodent neurodegenerative models such as MCAO (Middle Cerebral Artery Occlusion, a mouse model of brain ischemia), MPTP (1-methyl-4-phenyl-1,2,3,6-tetrahydropyridine toxicity, a mouse model of Parkinson's disease), HD (mouse Huntington's disease model), or 6-OHDA (6-hydroxydopamine toxicity, a mouse model of Parkinson's disease). N.D., not determined

Interestingly, we have recently shown that *para*-hydroquinones (rather than the *ortho* or *meta* positions for the hydroxyl groups) form more electrophilic compounds and thus display maximal activation of the Nrf2/ARE pathway (Satoh et al. 2009a, 2011); thus, we feel that these drugs may have the greatest clinical potential as neuroprotectants. Importantly, however, the ultimate success of PEDs in the clinic will depend on their balance between anti-oxidant effects and clinical tolerability. Only clinical studies in humans will be able to determine this outcome, but the approach does appear to be a promising one.

Acknowledgments We thank Drs. Larry D. Frye (Biomdcom) and Scott McKercher for help with manuscript preparation and Dr. Tomohiro Nakamura (Sanford-Burnham Medical Research Institute) for valuable discussions concerning chemical reactions. Our studies described in this chapter were supported in part by a grant from the JSPS, Joint Project of Japan–U.S., from the MEXT Japan, from Grants-in-Aid for Scientific Research (No.19500261; No. 22500282), and from Grants-in-Aid for Scientific Research on Innovative Areas (No. 2011701) to T.S. Other support for our studies came from NIH grants R01 EY05477, P01 ES016738, P01 HD29587, and P30 NS057096/P30 NS076411 to S.A.L.

References

Ahlgren-Beckendorf JA, Reising AM, Schander MA, Herdler JW, Johnson JA (1999) Coordinate regulation of NAD(P)H:quinone oxidoreductase and glutathione-S-transferase in primary cultures of rat neurons and glias: role of the antioxidant/electrophile responsive element. Glia 15:131–142

Ahmad A, Khan MM, Hoda MN, Raza SS, Khan MB, Javed H, Ishrat T, Ashafaq M, Ahmad ME, Safhi MM, Islam F (2011) Quercetin protects against oxidative stress associated damages in a rat model of transient focal cerebral ischemia and reperfusion. Neurochem Res 36:1360–1371

Akaike T, Fujii S, Sawa T, Ihara H (2010) Cell signaling mediated by nitrated cyclic guanine nucleotide. Nitric Oxide 23:166–174

Alfieri A, Srivastava S, Siow RC, Modo M, Fraser PA, Mann GE (2011) Targeting the Nrf2-Keap1 antioxidant defence pathway for neurovascular protection in stroke. J Physiol 589:4125–4136

Arteel GE, Sies H (2001) The biochemistry of selenium and the glutathione system. Environ Toxicol Pharmacol 10:153–158

Bredesen DE (2008) Programmed cell death mechanism in neurological diseases. Curr Mol Med 8:173–186

Bukau B, Weisman J, Horwich A (2006) Molecular chaperones and protein quality control. Cell 125:443–451

Burdo J, Schubert D, Maher P (2008) Glutathione production is regulated via distinct pathways in stressed and non-stressed cortical neurons. Brain Res 1189:12–22

Bureau G, Longpré F, Martinoli MG (2008) Resveratrol and quercetin, two natural polyphenols, reduce apoptotic neuronal cell death induced by neuroinflammation. J Neurosci Res 86: 403–410

Butterfield DA, Reed T, Sultana R (2011) Roles of 3-nitrotyrosine- and 4-hydroxynonenal-modified brain proteins in the progression and pathogenesis of Alzheimer's disease. Free Radic Res 45:59–72

Catalá A (2009) Lipid peroxidation of membrane phospholipids generates hydroxy-alkenals and oxidized phospholipids active in physiological and/or pathological conditions. Chem Phys Lipids 157:1–11

Chang TS, Jeong W, Woo HA, Lee SM, Park S, Rhee SG (2004) Characterization of mammalian sulfiredoxin and its reactivation of hyperoxidized peroxiredoxin through reduction of cysteine sulfinic acid in the active site to cysteine. J Biol Chem 279:50994–51001

Cho DH, Nakamura T, Fang J, Cieplak P, Godzik A, Gu Z, Lipton SA (2009) S-Nitrosylation of Drp1 mediates beta-amyloid-related mitochondrial fission and neuronal injury. Science 324:102–105

Chung KK, Thomas B, Li X, Pletnikova O, Troncoso JC, Marsh L, Dawson VL, Dawson TM (2004) S-Nitrosylation of parkin regulates ubiquitination and compromises parkin's protective function. Science 304:1328–1331

Clement AM, Nguyen MD, Roberts EA, Garcia ML, Boillée S, Rule M, McMahon AP, Doucette W, Siwek D, Ferrante RJ, Brown RH Jr, Julien JP, Goldstein LS, Cleveland DW (2003) Wild-type nonneuronal cells extend survival of SOD1 mutant motor neurons in ALS mice. Science 302:113–117

Coyle JT, Puttfarcken P (1993) Oxidative stress, glutamate and neurodegenerative disorders. Science 262:689–695

Dalle-Donne I, Rossi R, Colombo G, Giustarini D, Milzani A (2009) Protein S-glutathionylation: a regulatory device from bacteria to humans. Trends Biochem Sci 34:85–96

Dinkova-Kostova AT, Holtzclaw WD, Cole RN, Itoh K, Wakabayashi N, Katoh Y, Yamamoto M, Talalay P (2002) Direct evidence that sulfhydryl groups of Keap1 are the sensors regulating induction of phase 2 enzymes that protect against carcinogens and oxidants. Proc Natl Acad Sci USA 99:11908–11913

Fang J, Nakamura T, Cho DH, Gu Z, Lipton SA (2007) S-Nitrosylation of peroxiredoxin 2 promotes oxidative stress-induced neuronal cell death in Parkinson's disease. Proc Natl Acad Sci USA 104:18742–18747

Fourquet S, Guerois R, Biard D, Toledano MB (2010) Activation of NRF2 by nitrosative agents and H2O2 involves KEAP1 disulfide formation. J Biol Chem 285:8463–8471

Fujii S, Sawa T, Ihara H, Tong KI, Ida T, Okamoto T, Ahtesham AK, Ishima Y, Motohashi H, Yamamoto M, Akaike T (2010) The critical role of nitric oxide signaling, via protein S-guanylation and nitrated cyclic GMP, in the antioxidant adaptive response. J Biol Chem 285:23970–23984

Garden GA, La Spada AR (2012) Intercellular (Mis)communication in neurodegenerative disease. Neuron 73:886–901

Groeger AL, Freeman BA (2010) Signaling actions of electrophiles: anti-inflammatory therapeutic candidates. Mol Interv 10:39–50

Gu Z, Kaul M, Yan B, Kridel SJ, Cui J, Strongin A, Smith JW, Liddington RC, Lipton SA (2002) S-nitrosylation of matrix metalloproteinases: signaling pathway to neuronal cell death. Science 297:1186–1190

Halliwell B, Gutteridge JMC (1999) Free radicals in biology and medicine. Oxford University Press, Oxford

Hanneken A, Lin FF, Johnson J, Maher P (2006) Flavonoids protect human retinal pigment epithelial cells from oxidative-stress-induced death. Invest Ophthalmol Vis Sci 47:3164–3177

Hara MR, Snyder SH (2007) Cell signaling and neuronal death. Annu Rev Pharmacol Toxicol 47:117–141

Hardingham GE, Lipton SA (2011) Regulation of neuronal oxidative and nitrosative stress by endogenous protective pathways and disease processes. Antioxid Redox Signal 14:1421–1424

Hastings TG (2009) The role of dopamine oxidation in mitochondrial dysfunction: implications for Parkinson's disease. J Bioenerg Biomembr 41:469–472

Hess DT, Matsumoto A, Kim SO, Marshall HE, Stamler JS (2005) Protein S-nitrosylation: purview and parameters. Nat Rev Mol Cell Biol 6:150–166

Holmgren A, Lu J (2010) Thioredoxin and thioredoxin reductase: current research with special reference to human disease. Biochem Biophys Res Commun 396:120–124

Hong F, Freeman ML, Lieber DC (2005) Identification of sensor cysteines in human Keap1 modified by the cancer chemopreventive agent sulforaphane. Chem Res Toxicol 18:1917–1926

Immenschuh S, Baumgart-Vogt E (2005) Peroxiredoxin, oxidative stress and cell proliferation. Antioxid Redox Signal 7:768–777

Ishige K, Schubert D, Sagara Y (2001) Flavonoids protect neuronal cells from oxidative stress by three distinct mechanisms. Free Radic Biol Med 30:433–446

Itoh K, Tong KI, Yamamoto M (2004) Molecular mechanism activating Nrf2-Keap1 pathway in regulation of adaptive response to electrophiles. Free Radic Biol Med 36:1208–1213

Jakel RJ, Townsend JA, Kraft AD, Johnson JA (2007) Nrf2-mediated protection against 6-hydroxydopamine. Brain Res 1144:192–201

Johnson J, Maher P, Hanneken A (2009) The flavonoid, eriodictyol, induces long-term protection in ARPE-19 cells through its effects on Nrf2 activation and phase 2 gene expression. Invest Ophthalmol Vis Sci 50:2398–2406

Kansanen E, Bonacci G, Schopfer FJ, Kuosmanen SM, Tong KI, Leinonen H, Woodcock SR, Yamamoto M, Carlberg C, Ylä-Herttuala S, Freeman BA, Levonen AL (2011) Electrophilic nitro-fatty acids activate NRF2 by a KEAP1 cysteine 151-independent mechanism. J Biol Chem 286:14019–14027

Kim W-K, Choi Y-B, Rayudu PV, Das P, Asaad W, Arnelle DR, Stamler JS, Lipton SA (1999) Attenuation of NMDA receptor activity and neurotoxicity by nitroxyl (NO^-). Neuron 24: 461–469
Kim I, Xu W, Reed JC (2008) Cell death and endoplasmic reticulum stress: disease relevance and therapeutic opportunities. Nat Rev Drug Discov 7:1013–1030
Kobayashi M, Li L, Iwamoto N, Nakajima-Takagi Y, Kaneko H, Nakayama Y, Eguchi M, Wada Y, Kumagai Y, Yamamoto M (2009) The antioxidant defense system Keap1-Nrf2 comprises a multiple sensing mechanism for responding to a wide range of chemical compounds. Mol Cell Biol 29:493–502
Kraft AD, Johnson DA, Johnson JA (2004) Nuclear factor E2-related factor 2-dependent antioxidant response element activation by tert-butylhydroquinone and sulforaphane occurring preferentially in astrocytes conditions neurons against oxidative insult. J Neurosci 24: 1101–1112
Lee JM, Calkins MJ, Chan K, Kan YW, Johonson JA (2003) Identification of the NF-E2-related factor-2-dependent genes conferring protection against oxidative stress in primary cortical astrocytes using oligonucleotide microarray analysis. J Biol Chem 278:12029–12038
Lewerenz J, Dargusch R, Maher P (2010) Lactacidosis modulates glutathione metabolism and oxidative glutamate toxicity. J Neurochem 113:502–514
Linder MC, Hazegh-Azam M (1996) Copper biochemistry and molecular biology. Am J Clin Nutr 63:797S–811S
Lioy DT, Garg SK, Monaghan CE, Raber J, Foust KD, Kaspar BK, Hirrlinger PG, Kirchhoff F, Bissonnette JM, Ballas N, Mandel G (2011) A role for glia in the progression of Rett's syndrome. Nature 475:497–500
Lipton SA (2004) Concepts: turning down but not off–neuroprotection requires a paradigm shift in drug development. Nature 428:473
Lipton SA (2006) Paradigm shift in neuroprotection by NMDA receptor blockade: memantine and beyond. Nat Rev Drug Discov 5:160–170
Lipton SA (2007) Pathologically-activated therapeutics. Nat Rev Neurosci 8:803–808
Lipton SA, Stamler JS (1994) Actions of redox-related congeners of nitric oxide at the NMDA receptor. Neuropharmacology 33:1229–1233
Lipton SA, Choi Y-B, Pan Z-H, Lei SZ, Chen H-SV, Sucher NJ, Singel DJ, Loscalzo J, Stamler JS (1993) A redox-based mechanism for the neuroprotective and neurodestructive effects of nitric oxide and related nitroso-compounds. Nature 364:626–632
Long EK, Picklo MJ Sr (2010) Trans-4-hydroxy-2-hexenal, a product of n-3 fatty acid peroxidation: make some room HNE. Free Radic Biol Med 49:1–8
Maher P (2008) The flavonoid fisetin promotes nerve cell survival from trophic factor withdrawal by enhancement of proteasome activity. Arch Biochem Biophys 476:139–144
Maher P, Dargusch R, Bodai L, Gerard PE, Purcell JM, Marsh JL (2011) ERK activation by the polyphenols fisetin and resveratrol provides neuroprotection in multiple models of Huntington's disease. Hum Mol Genet 20:261–270
Mandel MN, Patlolla JM, Zheng L, Agbaga MP, Tran JT, Wicker L, Kasus-Jacobi A, Elliott MH, Rao CV, Anderson RE (2009) Curcumin protects retinal cells from light-and oxidant stress-induced cell death. Free Radic Biol Med 46:672–679
Mangialasche F, Polidori MC, Monastero R, Ercolani S, Camarda C, Cecchetti R, Mecocci P (2009) Biomarkers of oxidative and nitrosative damage in Alzheimer's disease and mild cognitive impairment. Ageing Res Rev 8:285–305
Mattson MP, Cheng A (2006) Neurohormetic phytochemicals: low-dose toxins that induce adaptive neuronal stress response. Trends Neurosci 29:632–639
Milioli EM, Cologni P, Santos CC, Marcos TD, Yunes VM, Fernandes MS, Schoenfelder T, Costa-Campos L (2007) Effect of acute administration of hydroalcohol extract of Ilex paraguariensis St Hilaire (Aquifoliaceae) in animal models of Parkinson's disease. Phytother Res 21:771–776
Morimoto RI (2008) Proteotoxic stress and inducible chaperone networks in neurodegenerative disease and aging. Genes Dev 22:1427–1438

Murphy TH, De Long MJ, Coyle JT (1991) Enhanced NAD(P)H:quinone reductase activity prevents glutamate toxicity produced by oxidative stress. J Neurochem 56:990–995

Nakamura Y, Kumagai T, Yoshida C, Naito Y, Miyamoto M, Ohigashi H, Osawa T, Uchida K (2003) Pivotal role of electrophilicity in glutathione S-transferase induction by tert-butylhydroquinone. Biochemistry 15:4300–4309

Nakamuara T, Lipton SA (2009) Cell death: protein misfolding and neurodegenerative diseases. Apoptosis 14:455–468

Nakamura T, Wang L, Wong CC, Scott FL, Eckelman BP, Han X, Tzitzilonis C, Meng F, Gu Z, Holland EA, Clemente AT, Okamoto S, Salvesen GS, Riek R, Yates JR 3rd, Lipton SA (2010) Transnitrosylation of XIAP regulates caspase-dependent neuronal cell death. Mol Cell 39:184–195

Nakamura T, Lipton SA (2011) Redox modulation by S-nitrosylation contributes to protein misfolding, mitochondrial dynamics, and neuronal synaptic damage in neurodegenerative diseases. Cell Death Differ 18:1478–1486

Nakamura T, Wang L, Wong CC, Scott FL, Eckelman BP, Han X, Tzitzilonis C, Meng F, Gu Z, Holland EA, Clemente AT, Okamoto S, Salvesen GS, Riek R, Yates JR III, Lipton SA (2011) Transnitrosylation of XIAP regulates caspase-dependent neuronal cell death. Mol Cell 39:184–195

Niles JC, Wishnok JS, Tannenbaum SR (2006) Peroxynitrite-induced oxidation and nitration products of guanine and 8-oxoguanine: structures and mechanisms of product formation. Nitric Oxide 14:109–121

Numajiri N, Takasawa K, Nishiya T, Hayakawa W, Asada M, Matsuda H, Azumi K, Tanaka H, Hyakkoku K, Kamata H, Nakamura T, Hara H, Minami M, Lipton SA, Uehara T (2011) On-off system for PI3-kinase-Akt signaling through S-nitrosylation of phosphatase with sequence homology to tensin (PTEN). Proc Natl Acad Sci USA 108:10349–10354

Pacher P, Beckman JS, Liaudet L (2007) Nitric oxide and peroxynitrite in health and disease. Physiol Rev 87:315–424

Padmanabhan B, Tong KI, Ohta T, Nakamura Y, Scharlock M, Ohtsuji M, Kang MI, Kobayashi A, Yokoyama S, Yamamoto M (2006) Structural basis for defects of Keap1 activity provoked by its point mutations in lung cancer. Mol Cell 21:689–700

Qu J, Nakamura T, Cao G, Holland EA, McKercher SR, Lipton SA (2011) S-Nitrosylation activates Cdk5 and contributes to synaptic spine loss induced by β-amyloid peptide. Proc Natl Acad Sci USA 108:14330–14335

Radi R (2004) Nitric oxide, oxidants, and protein tyrosine nitration. Proc Natl Acad Sci USA 101:4003–4008

Rhee SG, Jeong W, Chang TS, Woo HA (2007) Sulfiredoxin, the cysteine sulfinic acid reductase specific to 2-Cys peroxiredoxin: its discovery, mechanism of action, and biological significance. Kidney Int Suppl 106:S3–S8

Sasaki S, Tozawa T, Van Wagoner RM, Ireland CM, Harper MK, Satoh T (2011) Strongylophorine-8, a pro-electrophilic compound from the marine sponge Petrosia (Strongylophora) corticata, provides neuroprotection through Nrf2/ARE pathway. Biochem Biophys Res Commun 415: 6–10

Satoh T, Lipton SA (2007) Redox regulation of neuronal survival by electrophilic compounds. Trends Neurosci 30:38–45

Satoh T, Furuta K, Tomokiyo K, Nakatsuka D, Tanikawa M, Nakanishi M, Miura M, Tanaka S, Koike T, Hatanaka H, Ikuta K, Suzuki M, Watanabe Y (2000) Facilitatory roles of novel compounds designed from cyclopentenone prostaglandins on neurite outgrowth-promoting activities of nerve growth factor. J Neurochem 75:1092–1102

Satoh T, Furuta K, Tomokiyo K, Namura S, Nakatsuka D, Sugie Y, Ishikawa Y, Hatanaka H, Suzuki M, Watanabe Y (2001) Neurotrophic actions of novel compounds designed from cyclopentenone prostaglandins. J Neurochem 77:50–62

Satoh T, Baba M, Nakatsuka D, Ishikawa Y, Aburatani H, Furuta K, Ishikawa T, Hatanaka H, Suzuki M, Watanabe Y (2003) Role of heme oxygenase-1 protein in the neuroprotective effects by cyclopentenone prostaglandin derivatives as a sustained phase of neuronal survival promoting mechanism under oxidative stress. Eur J Neurosci 17:2249–2255

Satoh T, Okamoto S, Cui J, Watanabe Y, Furuta K, Suzuki M, Tohyama K, Lipton SA (2006) Activation of the Keap1/Nrf2 pathway for neuroprotection by electrophilic phase II inducers. Proc Natl Acad Sci USA 103:768–773

Satoh T, Kosaka K, Itoh K, Kobayashi A, Yamamoto M, Shimojo Y, Kitajima C, Cui J, Kamins J, Okamoto S, Shirasawa T, Lipton SA (2008a) Carnosic acid, a catechol-type electrophilic compound, protects neurons both in vitro and in vivo through activation of the Keap1/Nrf2 pathway via S-alkylation of specific cysteines. J Neurochem 104:1116–1131

Satoh T, Izumi M, Inukai Y, Tsutumi Y, Nakayama N, Kosaka K, Kitajima C, Itoh K, Yokoi T, Shirasawa T (2008b) Carnosic acid protects neuronal HT22 cells through activation of the antioxidant-responsive element in free carboxylic acid- and catechol hydroxyl moieties-dependent manners. Neurosci Lett 434:260–265

Satoh T, Saitoh S, Hosaka H, Kosaka K (2009a) Simple ortho- and para-hydroquinones as neuroprotective compounds against oxidative stress associated with a specific transcriptional activation. Biochem Biophys Res Commun 379:537–541

Satoh T, Harada N, Hosoya T, Tohyama K, Yamamoto M, Itoh K (2009b) Keap1/Nrf2 system regulates neuronal survival as revealed through study of keap1 gene knockout mice. Biochem Biophys Res Commun 380:298–302

Satoh T, Rezaie T, Seki M, Sunico CR, Tabuchi T, Kitagawa T, Yanagitai M, Senzaki M, Kosegawa C, Taira H, McKercher SR, Hoffman JK, Roth GP, Lipton SA (2011) Dual neuroprotective pathways of a pro-electrophilic compound via HSF-1-activated heat-shock proteins and Nrf2-activated phase 2 antioxidant response enzymes. J Neurochem 119:569–578

Shih AY, Li P, Murphy TH (2005) A small-molecule-inducible Nrf2-mediated antioxidant response provides effective prophylaxis against cerebral ischemia in vivo. J Neurosci 25:10321–10335

Shih AY, Erb H, Murphy TH (2007) Dopamine activates Nrf2-regulated neuroprotective pathways in astrocytes and meningeal cells. J Neurochem 101:109–119

Soriano FX, Leveille F, Papadia S, Higgins LG, Varley J, Baxter P, Hayes JD, Hardingham GE (2008) Induction of sulfiredoxin expression and reduction of peroxiredoxin hyperoxidation by neuroprotective Nrf2 activator 3H-1,2-dithiol-3-thione. J Neurochem 107:533–543

Soriano FX, Baxter P, Murray LM, Sporn MB, Gillingwater TH, Hardingham GE (2009) Transcriptional regulation of the AP-1 and Nrf2 target gene sulfiredoxin. Mol Cells 27: 279–282

Suematsu N, Hosoda M, Fujimori K (2011) Protective effects of quercetin against hydrogen peroxide-induced apoptosis in human neuronal SH-SY5Y cells. Neurosci Lett 504:223–227

Sun X, Erb H, Murphy TH (2005) Coordinate regulation of glutathione metabolism in astrocytes by Nrf2. Biochem Biophys Res Commun 326:371–377

Takahashi T, Tabuchi T, Tamaki Y, Kosaka K, Takikawa Y, Satoh T (2009) Carnosic acid and carnosol inhibit adipocyte differentiation in mouse 3 T3-L1 cells through induction of phase 2 enzymes and activation of glutathione metabolism. Biochem Biophys Res Commun 382: 549–554

Takahata Y, Takarada T, Iemata M, Yamamoto T, Nakamura Y, Kodama A, Yoneda Y (2009) Functional expression of beta2 adrenergic receptors responsible for protection against oxidative stress through promotion of glutathione synthesis after Nrf2 upregulation in undifferentiated mesenchymal C3H10T1/2 stem cells. J Cell Physiol 218:268–275

Talalay P (2000) Chemoprotection against cancer by induction of phase 2 enzymes. Biofactors 12:5–11

Tamaki Y, Tabuchi T, Takahashi T, Kosaka K, Satoh T (2010) Activated glutathione metabolism participates in protective effects of carnosic acid against oxidative stress in neuronal HT22 cells. Planta Med 76:683–688

Uchida K (2003) 4-Hydroxy-2-nonenal: a product and mediator of oxidative stress. Prog Lipid Res 42:318–343

Uehara T, Nakamura T, Yao D, Shi ZQ, Gu Z, Ma Y, Masliah E, Nomura Y, Lipton SA (2006) S-nitrosylated protein-disulphide isomerase links protein misfolding to neurodegeneration. Nature 441:513–517

Um HC, Jang JH, Kim DH, Lee C, Surh YJ (2011) Nitric oxide activates Nrf2 through S-nitrosylation of Keap1 in PC12 cells. Nitric Oxide 25:161–168

Vargas MR, Johnson JA (2009) The Nrf2-ARE cytoprotective pathway in astrocytes. Expert Rev Mol Med 11:e17

Wang XJ, Hayes JD, Higgins LJ, Wolf CR, Dinkova-Kostova AT (2010) Activation of the NRF2 signaling pathway by copper-mediated redox cycling of para- and ortho-hydroquinones. Chem Biol 17:75–85

Winyard PG, Moody CJ, Jacob C (2005) Oxidative activation of antioxidant defence. Trends Biochem 30:453–461

Yao D, Gu Z, Nakamura T, Shi Z-Q, Ma Y, Gaston B, Palmer LA, Rockenstein EM, Zhang Z, Masliah E, Uehara T, Lipton SA (2004) Nitrosative stress linked to sporadic Parkinson's disease: S-Nitrosylation of parkin regulates it E3 ligase activity. Proc Natl Acad Sci USA 101:10810–10814

Zhang DD, Lo SC, Cross JV, Templeton DJ, Hannink M (2004) Keap1 is a redox-regulated substrate adaptor protein for a Cul3-dependent ubiquitin ligase complex. Mol Cell Biol 24:10941–10953

Zhang Y, Ahn YH, Benjamin IJ, Honda T, Hicks RJ, Calabrese V, Cole PA, Dinkova-Kostova AT (2011) HSF1-dependent upregulation of Hsp70 bysulfhydryl-reactive inducers of the KEAP1/NRF2/ARE pathway. Chem Biol 18:1355–1361

Chapter 11
Reactive Oxygen Species, Kinase Signaling, and Redox Regulation of Epigenetics

Isaac K. Sundar and Irfan Rahman

Abstract Reactive oxygen species (ROS) cause alterations in the cellular redox status (GSH/GSSG ratio) and hence activate various kinase signaling pathways, redox sensitive transcription factors (NF-κB and AP-1), and chromatin modification enzymes leading to increased pro-inflammatory and immunological responses. Chromatin modification enzymes include histone acetyltransferases (HATs), histone and non-histone deacetylases (HDACs, sirtuins), histone methyltransferases (HMTs) and histone demethylases (HDMs). Activation of these enzymes results in transcriptional gene activation/repression. Histone methyltransferases are enzymes that dynamically cause histone mono-, di- or tri-methylation at lysine residues, which either result in gene activation (H3K4, H3K36, and H3K79) or repression (H3K9, H3K27, and H3K20). Histone demethylases catalyze the removal of methyl groups from lysine or arginine residue of histones, hence regulating gene expression. Recent evidences have indicated that oxidative stress and environmental agents can alter nuclear histone acetylation/deacetylation/methylation, allowing access for transcription factor DNA binding and leading to enhanced pro-inflammatory gene expression. Cross-talks between histone modifications and DNA methylation also occur during inflammation. Understanding the mechanisms of ROS and redox epigenetic regulation via stress signaling kinases, redox sensitive transcription factors, the balance between histone acetylation/deacetylation and histone methylation/demethylation may lead to the development of novel therapies based on epigenetics against chronic inflammatory diseases, and cancer.

Keywords Redox • Epigenetics • Oxidative stress • Histone acetylation • Gene expression

---

I.K. Sundar • I. Rahman, Ph.D. (✉)
Department of Environmental Medicine, Lung Biology and Disease Program, University of Rochester Medical Center, Box 850, 601 Elmwood Avenue, Rochester, NY 14642, USA
e-mail: irfan_rahman@urmc.rochester.edu

U. Jakob and D. Reichmann (eds.), *Oxidative Stress and Redox Regulation*,
DOI 10.1007/978-94-007-5787-5_11,

Abbreviations

AP-1	activator protein-1
CBP	CREB-binding protein
COPD	chronic obstructive pulmonary disease
CREB	cAMP-response-element-binding protein
CS	cigarette smoke
ECSOD	extracellular superoxide dismutase
ERK	extracellular signal-related kinase
FOXO	forkhead box O
GR	glucocorticoid receptor
GSH	glutathione
GSSG	glutathione disulfide
HAT	histone acetyltransferase
HDAC	histone deacetylase
HDM	histone demethylase
HMT	histone methytransferase
4-HNE	4-hydroxy-2-nonenal
H_2O_2	hydrogen peroxide
$^{\bullet}OH$	hydroxyl radical
IKK	IkappaB kinase
IL-8	interleukin-8
JNK	c-Jun N-terminal protein kinase
LPS	lipopolysaccharide
MMP-9	matrix metalloproteinase-9
MnSOD	manganese SOD
MSK1	mitogen- and stress-activated kinase 1
NAD^+	nicotinamide adenine dinucleotide
NF-κB	nuclear factor κB
NIK	NF-κB inducing kinase
NO	nitric oxide
NSCLC	non-small cell lung cancer
PCAF	p300-CBP associated factor
$ONOO^-$	peroxynitrite anion
PGC-1α	peroxisome-proliferator-activated receptor (PPAR) gamma co-activator-1 alpha
PI3K	phosphoinositide 3-kinase
PKC	Protein kinase C
RNS	reactive nitrogen species
ROS	reactive oxygen species
SIRT1	sirtuin 1
STAT3	signal transducer and activator of transcription 3
$O_2^{\bullet -}$	superoxide anion
SOD	superoxide dismutase

TGF-β1	transforming growth factor-beta 1
TNF-α	tumor necrosis factor-alpha
TSA	trichostatin A

11.1 Introduction

Lung tissue is a direct target for oxidative injury from reactive oxygen species (ROS) and free radicals derived by cigarette smoke (CS) or environmental agents, which results in increased cellular kinase signaling associated with post-translational modifications (acetylation, deacetylation, methylation, phosphorylation, carbonylation, ubiquitination, sumoylation, and poly-ADP-ribosylation) of histones and non-histone proteins, and redox modification of proteins implicated in chromatin remodeling (Gilmour et al. 2003; Tomita et al. 2003; Yang et al. 2008, 2009; Bast et al. 1991; Rahman et al. 2004; Repine et al. 1997; Sundar et al. 2012).

Lung cells have developed antioxidant defense mechanisms that can efficiently counteract the oxidative challenge, which is continuously imposed by the environment. The most important antioxidant is the non-protein sulfhydryl glutathione (GSH), which plays an important role in maintaining the cellular redox status in lung cells (Forman and Dickinson 2003). The ratio of reduced GSH to glutathione disulphide GSSG in the cytoplasm is a key determinant in the regulation of different intracellular processes, such as signal transduction, gene expression, proliferation and apoptosis (Rahman et al. 2005). The ratio between GSH/GSSG serves as an indicator of the cellular redox state (Ghezzi et al. 2005). This ratio is determined by the rates of H_2O_2 reduction by glutathione peroxidase and GSSG reduction by glutathione reductase. The intracellular redox status is shown to play an important role in the regulation of epigenetic events (Cyr and Domann 2011).

Chromatin remodeling by histone modifications play a key role in the regulation of cellular processes including gene transcription/repression, DNA repair, and proliferation (Sundar et al. 2012). Posttranslational histone modifications predominantly occur via acetylation/deacetylation and methylation/demethylation reactions (Fukuda et al. 2006; Roth et al. 2001; Sundar et al. 2012) (Fig. 11.1). The amino-terminal region of histones is a hotspot for such post-translational modifications, which affect the interaction of histones with DNA leading to chromatin remodeling events on gene promoters (Fischle et al. 2003).

Histone acetyltransferases, HATs (PCAF, p300/CBP and GCN5) and histone deacetylases, HDACs (e.g. HDAC2 and SIRT1) maintain the balance in histone acetylation/deacetylation events (Sundar et al. 2012). HATs affect the binding of DNA sequence-specific transcription factors and subsequently recruit coactivators or corepressors on gene specific regions to form either coactivator or corepressor complexes (Wang et al. 2001). Several reports support the fact that transcriptional regulators possess intrinsic HAT (CBP/p300) and HDAC activities, suggesting that histone acetylation and deacetylation play a causal role in regulating gene transcription in various organisms (Kuo and Allis 1998; Wolffe 1997; Workman and

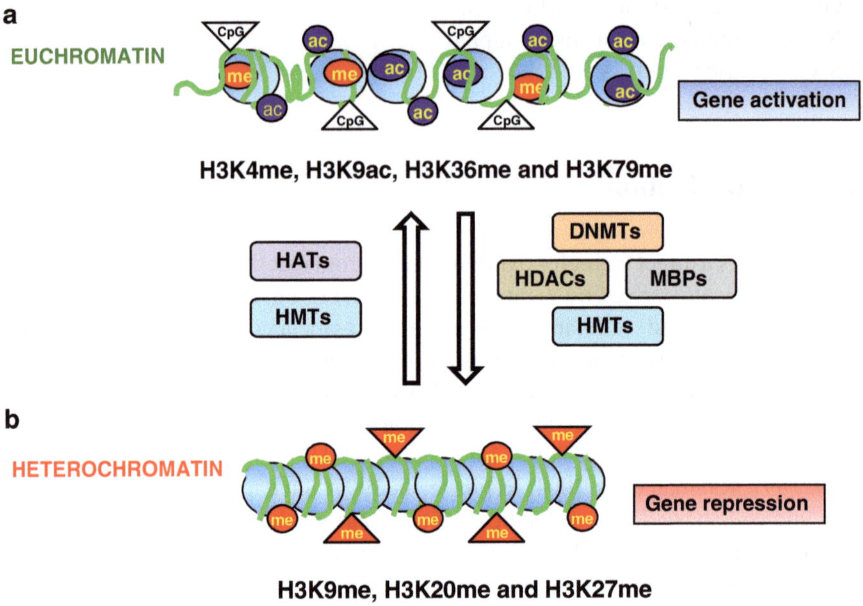

Fig. 11.1 Schematic representation of epigenetic regulation of gene expression by alterations in DNA/histone modifications. (**a**) Open chromatin confirmation with widely spaced nucleosomes (euchromatin) represents active gene transcription, associated with unmethylated CpG residues on DNA and post-translational histone modifications, such as increased acetylation and methylation. The euchromatin structure is readily accessible to redox sensitive transcription factors NF-κB and AP-1, HATs and cofactors/coactivators to regulate gene expression. Acetylation of histone H3K9 and methylation of histone H3K4, H3K36, and H3K79 are the hallmarks for gene activation. (**b**) Closed chromatin confirmation (heterochromatin) represents transcriptional repression characterized by tightly packed nucleosomes with DNA methylation at CpG islands, deacetylation of histones and post-translational histone modifications, such as methylation of specific histone H3 residues. The heterochromatin structure renders the DNA inaccessible to transcription factors but provides access to gene repressive proteins, including histone deacetylases (HDACs), DNA methyltransferases (DNMTs), and methyl-CpG binding proteins (MBPs). Methylation of histone H3K9, H3K20, and H3K27 are associated with gene repression

Kingston 1998). There are several evidences showing that an increase in histone acetylation leads to increased gene transcription while histone hypoacetylation is linked to decreased gene transcription (Kuo and Allis1998; Wolffe 1997; Workman and Kingston 1998). Histone acetylation by HATs and deacetylation by HDACs are also linked to cell-cycle progression, proliferation, senescence, and DNA damage/repair, as well as to inflammatory gene transcription, which can be affected by intracellular redox signaling (Kouzarides 1999; Taplick et al. 1998; Biswas and Rahman 2009). A recent report suggests a novel role for histone carbonylation on specific histone proteins. This may have implications in redox epigenetics, which occurs in proliferating cells (particularly in fibroblasts) during the active phase of DNA synthesis (Garcia-Gimenez et al. 2012). Evidence from recent studies also suggest that epigenetic events, such as DNA methylation, histone modifications, and non-coding RNAs or microRNAs (miRNAs) affect gene expression/repression in several chronic inflammatory lung diseases, such as asthma, COPD and interstitial lung disease, which are influenced by environment (tobacco smoke and inhaled pollutants/gases), diet, and aging (Yang and Schwartz 2011). Further, the role of these epigenetic events in prenatal or early childhood exposure to environmental stress conditions, including tobacco smoke, pollutants/oxidants, and dietary agents rendering children susceptible to asthma and allergic airways disease later in life has been suggested (Martino and Prescott 2011). Thus, post-translational modifications of histone proteins by oxidants and environmental stresses can trigger genes that are involved in various cellular processes and chronic inflammatory events. This chapter focuses on ROS and redox epigenetic regulators, such as stress signaling kinases, transcription factor (NF-κB), HATs, HDACs, histone methytransferases (HMTs), and histone demethylases (HDMs), and their role in epigenetic regulation of inflammatory response in chronic lung diseases.

11.2 ROS and Intracellular Redox Regulation in Epigenetics

ROS, such as superoxide anions ($O_2^{\cdot -}$) and hydroxyl radicals ($^{\cdot}OH$) are highly unstable oxygen species with unpaired electrons, capable of initiating oxidation. ROS cause oxidation of proteins, DNA and lipids, which may directly affect the lung or induce several other cellular responses through the generation of secondary metabolic reactive species. Similarly, the reactive nitrogen species (RNS) nitric oxide (NO) is generated by nitric oxide synthase (iNOS) and forms potent and damaging peroxynitrite molecules in the presence of superoxide anion. ROS cause alterations in extracellular matrix (ECM) and blood vessels, stimulate mucus secretion, lead to apoptosis, and regulate cell proliferation (Rahman 2005). Ozone, cigarette smoke and inflammatory cells are major sources of ROS/RNS generation. NADPH oxidase is an important ROS-generating enzyme found in inflammatory cells. Additional mechanisms include mitochondrial respiration via the release of reactive oxygen intermediates. Other enzyme systems include xanthine or the xanthine oxidase system and the heme peroxidases, which are also involved in the pathogenesis of COPD (Rahman and MacNee 1996; Pinamonti et al. 1998). Cigarette smoke is the major source for environmentally derived ROS that are involved in pathogenesis of COPD. There are some striking features found in the COPD phenotype, characterized by an inflammatory-immune response, activation of epithelial cells, resident macrophages, and the recruitment and activation of neutrophils, monocytes and B and T lymphocytes into the lung. Inflammatory cells are activated in response to cytokines/chemokines/chemoattractants after recruitment into the airspace, thereby generating various ROS. ROS are highly reactive and, when generated close to cell membranes, can induce lipid peroxidation (oxidation of membrane phospholipids) and the accumulation of malondialdehyde, 4-hydroxy-2-nonenal, acrolein, and F_2-isoprostanes (Rahman and Adcock 2006; Rahman et al.

2002b). Alveolar repair responses and immune modulation in the lung may also be influenced by ROS (Rahman and MacNee 1998), suggesting the role of ROS/RNS in epigenetic regulation of chronic lung diseases where oxidant/antioxidant imbalance is a common phenomenon (Rahman and MacNee 1996).

11.2.1 Kinases and Redox Signaling

ROS generated from cellular metabolic response (inflammatory cells) and environmental sources (air pollutants or cigarette smoke) can cause activation of a number of signal transducing molecules via oxidation-prone cysteine rich domains, resulting in active gene transcription (Rahman and MacNee 1998; Adler et al. 1999). Among these are members of the MAPK family, extracellular signal regulated kinase (ERK), c-Jun N-terminal kinase (JNK), p38 kinase and PI3K/Akt. Based on the oxidative/redox status of the cells, members of the MAPK family are activated, leading to a multifaceted transactivation of redox sensitive transcription factors (c-Jun, ATF-2, CREB-binding protein (CBP) and Elk-1) (Adler et al. 1999; Thannickal and Fanburg 2000). Furthermore, activation of upstream kinases, such as NF-κB-inducing kinase (NIK), mitogen-and stress-activated kinase 1 (MSK1), and IκB kinase-α (IKKα) results in downstream chromatin remodeling events (Chung et al. 2010, 2011; Yang et al. 2008, 2009; Sundar et al. 2012), which in turn modulates a series of specific pro-inflammatory gene transcription events that mediate cellular functions, such as apoptosis, proliferation, transformation, and differentiation. Lipid peroxidation by cigarette smoke generates aldehydes, which have been shown to trigger the inflammatory response-mediated signal transduction and gene activation (Uchida et al. 1999). Cigarette smoke-mediated oxidative stress specifically induces the JNK pathway, leading to endothelial cell injury (Hoshino et al. 2005). Cigarette smoke also induces activation of protein kinase C (PKC), mediating regulation of intracellular signaling pathways (Wyatt et al. 1999). PKCζ-knockout mice exposed to cigarette smoke or LPS showed a reduction in lung inflammatory response via chromatin modifications suggesting that PKCζ is an important modifier of lung inflammatory response (Yao et al. 2010b). Hence, activation of kinases by ROS and aldehyde/lipid peroxidation products can lead to a pro-inflammatory response by chromatin remodeling in eukaryotic cells (Sundar et al. 2012).

Stress signaling kinases, such as phosphoinositide 3-kinase (PI3K) and p38 are shown to be involved in the pathogenesis of COPD and lung cancer (Doukas et al. 2009; Engelman and Janne 2008; Gustafson et al. 2010). Inhibition of PI3K/p38 lowers tobacco/cigarette smoke-induced lung inflammatory response (Doukas et al. 2009). Interestingly, PI3K inhibition significantly increased the steroid efficacy to inhibit an inflammatory response under CS-induced oxidative stress in the lung (Marwick et al. 2009; To et al. 2010; Bhavsar et al. 2010). Overexpression of p110, a subunit of PI3K, in type II alveolar epithelial cells is involved in pathogenesis of lung adenocarcinomas, which can be repressed by dual pan-PI3K and mTOR

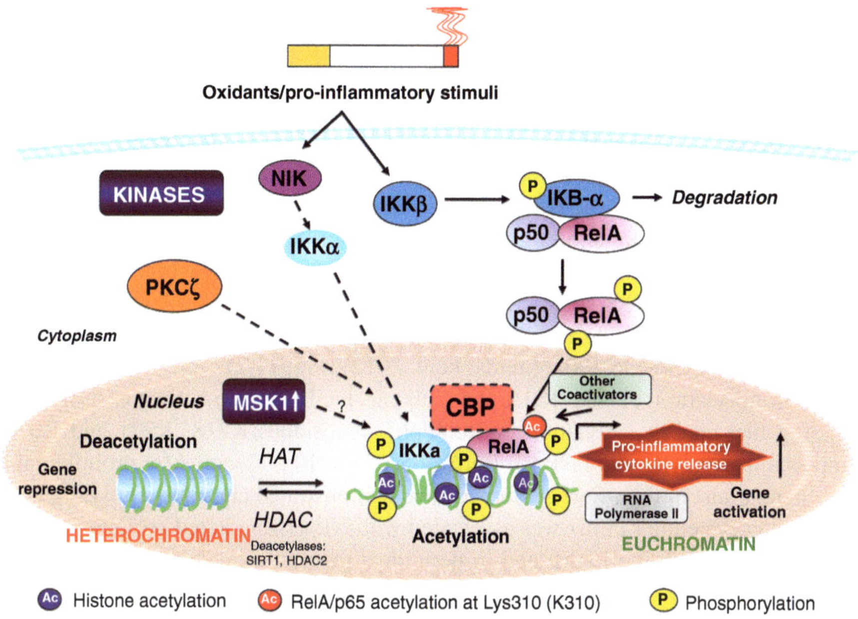

Fig. 11.2 Role of redox kinase signaling in chromatin modifications. Cigarette smoke-derived oxidants/aldehydes activate kinase signaling mechanisms, such as NIK, IKKα, MSK1, and PKCζ by redox modulation, and induce pro-inflammatory gene transcription via chromatin modifications

pharmacological inhibitors (Engelman and Janne 2008). However, treatment of mice bearing lung adenocarcinomas with rapamycin, an inhibitor of mTOR, was unable to shrink the tumors. Furthermore, PI3K is required for Kras-induced lung tumorigenesis (Engelman and Janne 2008). Therefore, in addition to p38 MAPK and PKCζ, oxidants/cigarette smoke-mediated activation of PI3K signaling pathway may be a potential target for the intervention of COPD and other chronic inflammatory lung diseases. These oxidant and redox sensitive pathways activate various kinases such as NIK, MSK1, PI3K, PKCzeta and IKKα resulting in histone acetylation events (Yang et al. 2008; Chung et al. 2011; Sundar et al. 2012) (Fig. 11.2).

11.2.2 NF-κB and Redox Regulation of Inflammation

Among transcription factors, NF-κB remains to be of particular interest due to its role in regulating inflammatory mediators (Christman et al. 2000). The important NF-κB-dependent proinflammatory mediators include IL-1, IL-6, IL-8, MCP-1, and TNFα, which are known to be involved in lung airway inflammation in COPD. Nuclear accumulation of NF-κB RelA/p65 was increased in smokers and patients with COPD as well as in lung epithelial cells and macrophages exposed to cigarette smoke (Di Stefano et al. 2002; Yagi et al. 2006; Rajendrasozhan et al. 2008b). Furthermore, the number of RelA/p65-positive epithelial cells and macrophages correlated with the severity of airflow limitation in COPD patients (Di Stefano et al. 2002), suggesting a role of NF-κB activation in chronic inflammatory lung diseases. NF-κB activating upstream kinases are redox-sensitive, thus oxidative stress is associated with NF-κB activation in the lungs of patients with COPD (Bowie and O'Neill 2000; Pantano et al. 2006). Phosphorylation of NF-κB RelA/p65 is regulated by several redox-regulating kinases, such as protein kinase A (PKA), MSK1 at Ser276, IKKβ at Ser536 and protein kinase C (PKC) ζ at Ser311, and p38 further leading to increased RelA/p65-mediated cytokine release (Yang et al. 2003; Chen and Greene 2004; Sundar et al. 2012) (Fig. 11.2).

Phosphorylation of the RelA/p65 subunit facilitates co-activator CBP/p300 binding that causes acetylation of RelA/p65 at residues K218, K221 and K310. Importantly, acetylation of NF-κB RelA/p65 at Lys310 is essential for proinflammatory gene transcription. Earlier studies have demonstrated cigarette smoke-induced inflammatory cellular influx was associated with NF-κB-mediated proinflammatory cytokine release in the lungs, and in peritoneal macrophages and monocytes (i.e., MonoMac6 cells). However, IKKβ inhibitors attenuated cigarette smoke extract-induced NF-κB-dependent proinflammatory mediators (Yang et al. 2006; Yao et al. 2008a). Specific knockdown of NF-κB subunit in epithelial cells or IKKβ deficiency in macrophages significantly attenuated lung tumorigenesis in mice (Stathopoulos et al. 2007; Takahashi et al. 2010). Alternative or non-canonical pathways require NIK, along with IKKα to induce the processing of p100 C-terminus (termed IκBδ), resulting in nuclear translocation of p52/RelB (Senftleben et al. 2001; Yin et al. 2001; Xiao et al. 2004). Cigarette smoke-mediated increase in the levels of p52, RelB, IKKα and NIK, as well as interaction of RelB with p52 and NIK occur in mouse lung (Yang et al. 2008, 2009). RelB was recruited to the promoters of proinflammatory genes signifying the role of NF-κB alternative pathway in cigarette smoke-mediated lung inflammatory response (Yang et al. 2008). In addition, cigarette smoke-mediated RelA/p65 NF-κB phospho-acetylation (ser276 and lys310) increased nuclear levels of RelB, and augmented IL-8 release in MonoMac6 cells transfected with IKKα (Yang et al. 2008) implicating the functional role of IKKα in NF-κB activation by cigarette smoke. These findings are in agreement with earlier reports demonstrating a marked decrease in proinflammatory cytokine expression in IKKα deficient fibroblasts, though IκBα degradation and *in vitro* NF-κB DNA binding activity were normal in these cells (Sizemore et al. 2002; Anest et al. 2003; Yamamoto et al. 2003). A recent study demonstrated the protective role of lung-targeted overexpression of the NF-κB member RelB in attenuating cigarette smoke-mediated lung neutrophil infiltration, inflammatory cytokine and chemokine production, cyclooxygenase-2 (COX-2) and prostaglandin E2 (PGE2) in mice (McMillan et al. 2011). The mechanism of this protection observed in acute cigarette smoke-induced pulmonary inflammation was due to decreased expression of redox sensitive NF-κB-dependent intercellular adhesion molecule 1, through which RelB regulates inflammatory cell migration into the vasculature

(McMillan et al. 2011). Therefore, activation or increased expression of RelB could be a novel therapeutic strategy against cigarette smoke-mediated acute lung inflammation and other chronic lung inflammatory diseases including COPD (Yang et al. 2009). Another recent report showed a novel role of NIK in cigarette smoke and TNFα-induced RelA/p65 activation and histone H3K9 acetylation. Human bronchial epithelial (H292) cells treated with cigarette smoke extract and TNFα showed increased nuclear accumulation of NIK, and recruitment of NIK onto proinflammatory gene promoters associated with sustained NF-κB-dependent gene transcription (Chung et al. 2011).

Pharmacological administration of NF-κB inhibitors attenuated cigarette smoke-induced acute lung inflammation in rodents (Rajendrasozhan et al. 2010b). Further studies are required to assess whether NF-κB inhibitors are effective in controlling cigarette smoke-induced chronic inflammation, as well as overcoming steroid resistance, and subsequent pathological changes in patients with COPD. Interestingly, deficiency of p50, another subunit of NF-κB, enhances cigarette smoke-induced lung inflammation and emphysema, suggesting a differential role of p50 and RelA/p65 in inflammation (Rajendrasozhan et al. 2010a; Cao et al. 2006). Furthermore, pharmacological administration of NF-κB inhibitor (e.g. pyrrolidine dithiocarbamate) and IκB protease inhibitor (tosylphenylalanylchloromethane) reduced TGF-β1-induced human lung cancer cell migration (Fong et al. 2009). Therefore, down-regulation of the redox sensitive IKKβ-NF-κB pathway may provide the therapeutic targets in inflammation that occurs in lung cancer. Activated STAT3 also induces tumor-promoting local inflammation by activating NF-κB, as shown by a study demonstrating the occurrence of lung adenocarcinomas in alveolar epithelial cells of mice overexpressing STAT3 (Li et al. 2007). Indeed, STAT3 is activated in the lungs of patients with lung carcinomas and COPD (Qu et al. 2009). Hence, the multiple redox-sensitive signaling pathways can regulate NF-κB in response to oxidative stress in chronic inflammatory lung diseases.

11.2.3 HATs and Redox Epigenetics

Histone acetyltransferease (HATs) transfer acetyl groups from acetyl-CoA to the N-terminal lysine residues of histones (H3 and H4), thereby uncoiling the DNA resulting in transcription factor binding followed by gene transcription (Imhof and Wolffe 1998; Tan 2001; Rahman et al. 2004; Ito et al. 2007; Sundar et al. 2012). HATs and HDACs play opposing roles in regulating acetylation of core nucleosomal histones. HATs are classified into five distinct families (Torchia et al. 1998; Tan 2001). Each of these HAT families has diverse roles in regulating chromatin structure (via histone acetylation/deacetylation) as they share a common enzymatic activity (Racey and Byvoet 1971).

Cigarette smoke/oxidant-mediated acetylation of histone H3 in macrophages and epithelial cells, and in lungs of humans and rodents suggests that histone acetylation plays an important role in chromatin remodeling (Sundar et al. 2012).

Cigarette smoke-mediated alterations in chromatin remodeling are shown to be involved in sustained lung inflammatory responses which occurs in patients with COPD (Marwick et al. 2004; Yang et al. 2008; Szulakowski et al. 2006; Yao et al. 2010b). CREB-binding protein (CBP) and p300 are the known coactivators that possess intrinsic HAT activity. They are regulated by the p38 MAP and other kinase pathway, which activates redox sensitive transcription factors, such as NF-κB and AP-1 (Thomson et al. 1999). Therefore, histone acetylation by CBP/p300 has an impact on the activation of specific proinflammatory mediators linked with NF-κB/AP-1-mediated gene expression (Kamei et al. 1996; Carrero et al. 2000; Rahman et al. 2002a). Transcription factor NF-κB is acetylated by other HATs (CB/p300) but also cause histone acetylation in a temporal manner by recruiting other cofactors/coactivators and chromatin remodeling complexes to activate proinflammatory gene transcription (Ito et al. 2000; Ghosh and Karin 2002; Lee et al. 2006). Ito et al. have reported that (CBP/p300) induced acetylation of histone H3, but not histone H2A, H2B or H3, occurs in epithelial cells on specific lysine residues (Lys8 and Lys12) at NF-κB-responsive regulatory elements on proinflammatory genes in human cells (Ito et al. 2000). Mutations and deletions in CBP gene has been detected in patients with lung cancer, suggesting the role of CBP in the tumorigenesis and/or progression of human lung cancer (Kishimoto et al. 2005). Oxidative stress and cellular GSH/GSSG levels regulate histone acetylation/deacetylation and thereby chromatin remodeling. For example, H_2O_2 and cigarette smoke activate NF-κB, HAT, CBP/p300 leading to specific histone acetylation in human macrophages and lung cells (Adenuga et al. 2009; Ito et al. 2005; Moodie et al. 2004; Yang et al. 2006; Yao and Rahman 2011). Glutathione depletion is associated with increased histone acetylation (Rahman et al. 2002b, 2004).

Corticosteroid resistance is also associated with increased oxidative stress imposed by cigarette smoke. Corticosteroids do not block inflammatory response when there is an increased oxidative stress, such as in severe asthmatics and patients with COPD. This inefficacy is not associated with alterations in nuclear HAT activity, (Hew et al. 2006). The corticosteroid insensitivity to inhibit inflammatory response is due to the reduction in HDAC2 levels/activity (a redox sensitive protein), which is lost in lungs of patients with COPD and severe asthmatics (see below under HDAC2 and steroid resistance section). Therefore, development of small molecule inhibitors (HAT inhibitors) against coactivators, such as CBP/p300, PCAF, and GCN, or activators of HDAC2 may be the key targets for therapeutic intervention of COPD and asthma as well as other chronic inflammatory lung diseases where oxidative stress occurs.

11.2.4 HDACs and Redox Epigenetics

Acetylation of histones by HATs is not an irreversible post-translational modification (Sundar et al. 2012). Histone deacetylases play a crucial role in removing the acetyl groups from histones (Hay and Candido 1983). Presence of HDACs within the gene regulatory regions leads to formation of closed chromatin complexes resulting in epigenetic gene silencing (chromatin condensation). So far 18 mammalian HDAC enzymes are known, which are classified into four classes based on their homology to a prototypical HDAC found in yeast (de Ruijter et al. 2003). Class I HDACs (HDACs 1, 2, 3, and 8) are ubiquitously expressed with the possible exception of HDAC3 and HDAC10, which are predominantly localized in the nucleus (Taplick et al. 2001). Class II HDACs (HDACs 4, 5, 6, 7, 9, and 10) are expressed in a tissue specific manner, and they have the ability to shuttle between the nucleus and cytoplasm during specific cellular signals (Bertos et al. 2001; Grozinger et al. 1999). The shuttling of class II HDACs from the nucleus is mainly regulated by nuclear export signaling 14-3-3 proteins (Bertos et al. 2001; Johnstone 2002). The class III HDACs or sirtuins are named based on their homology to the yeast Sir2 gene, which is a highly conserved gene family. In humans, sirtuins comprise seven members, sirtuin (SIRT) 1–7 (Michan and Sinclair 2007). Class IV deacetylases may undergo redox modifications and carbonylation due to the presence of cysteine, lysine, and histidine moieties in their catalytic sites (Caito et al. 2010; Yang et al. 2006, 2007).

HDACs are also reported to deacetylate non-histone proteins, such as the redox sensitive transcription factor NF-κB (RelA/p65) and FOXO3, thereby regulating transcription of NF-κB-dependent proinflammatory genes (Sengupta and Seto 2004; Sundar et al. 2012). The levels and activities of HDACs are significantly decreased in macrophages, lung cells *in vitro*, and in lungs of COPD patients in response to oxidative/carbonyl stress and via alterations in intracellular GSH/GSSG redox ratio (caused by H_2O_2, PM_{10}, and cigarette smoke-derived oxidants/free radicals), and are associated with COPD severity, asthma, and steroid resistance in asthmatics, who smoke tobacco (Adenuga et al. 2009; Ito et al. 2005; Moodie et al. 2004; Yang et al. 2006; Yao and Rahman 2011). Intracellular levels of GSH retain HDAC2 in a reducing environment since an increase in intracellular reduced GSH levels can inhibit the oxidative post-translational modifications of HDAC2 (Adenuga et al. 2009; Yang et al. 2006; Meja et al. 2008). Furthermore, a decrease in levels of HDACs is due to post-translational modifications, such as oxidation/carbonylation, nitrosylation, acetylation, and phosphorylation in response to oxidants and aldehydes derived from cigarette smoke (Adenuga et al. 2009; Ito et al. 2004; Moodie et al. 2004; Yang et al. 2006; Adenuga and Rahman 2010). Hence, HDAC2 levels/activities and the status of redox/oxidative post-translational modifications dictate the effects of coactivators on the chromatin status whether to drive the gene activation or repression in chronic inflammatory diseases and cancer (Sundar et al. 2012).

11.2.4.1 HDAC2 and Steroid Resistance

Cigarette smoke/oxidants/carbonyl-mediated reduction in HDAC2 is associated with activation of NF-κB RelA/p65 subunit (increased levels of total and acetylated RelA/p65). RelA/p65 interacts with HDAC2 (possibly HDAC1 and HDAC3) and nuclear accumulation of RelA/p65 results in transcription of proinflammatory genes

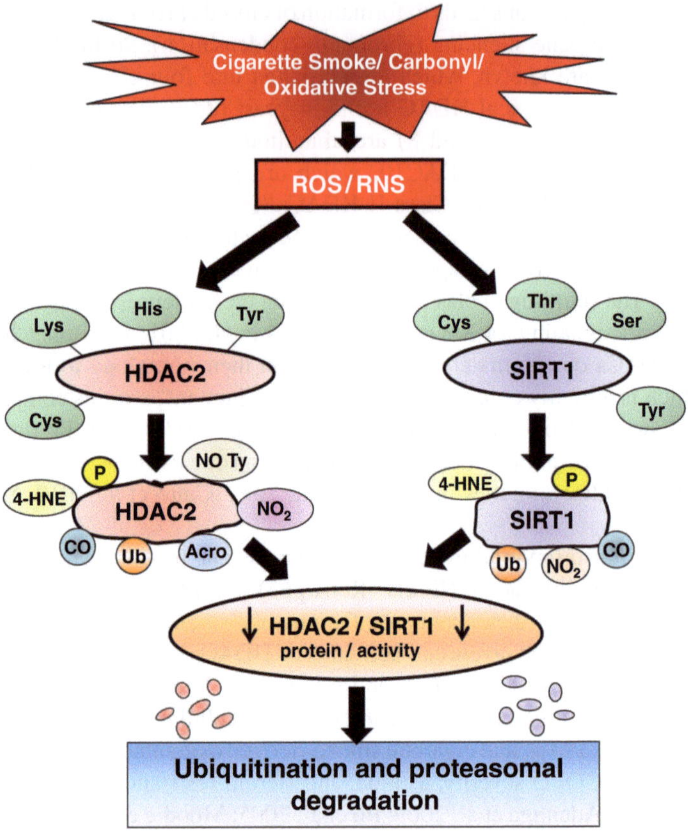

Fig. 11.3 Oxidative post-translational modifications of deacetylases (HDAC2 and SIRT1). Cigarette smoke or carbonyl/oxidant stress-mediated reactive oxygen species (ROS)/reactive nitrogen species (RNS) and/or alterations of the GSH/GSSG ratio cause various post-translational modifications of HDACs (HDAC2 and SIRT1), including phosphorylation (P), nitration (NO_2), carbonylation (CO), adduct formation (4-HNE: 4-hydroxy-2-nonenal, Acro: Acrolein), nitrotyrosine (NO-Ty), and ubiquitination (Ub) leading to ubiquitination-proteosomal degradation of HDACs. Post-translational modifications of HDAC2 and SIRT1 can include (i) direct nitration/nitrosylation on cysteine and tyrosine residues, (ii) covalent adduct formation on cysteine, lysine, and histidine residues, or (iii) phosphorylation of serine and threonine residues. Covalent modifications of the deacetylases on cysteine, histidine, or lysine residues can be irreversible, whereas other modifications can be reversible. The oxidative covalent modifications can lead to protein ubiquitination and degradation

when HDAC2 is oxidatively modified, ubiquitinated, and degraded (Yang et al. 2006; Yao et al. 2008b; Londhe et al. 2011; Yao and Rahman 2011) (Fig. 11.3). HDAC2 modification is dependent on the intra cellular thiol redox status (Yang et al. 2006; Meja et al. 2008). *In vitro* studies using trichostatin A (HDAC inhibitor) in different cell lines showed enhanced NF-κB-driven inflammatory gene transcription (Ito et al. 2000; Chen et al. 2001). Hence, alteration of HDACs by cigarette smoke/oxidative stress leads to increased acetylation of histones (histone H3 and H4) and activation of transcription factor (NF-κB), thus enhancing transcription of proinflammatory genes (Sundar et al. 2012; Yang et al. 2006; Adenuga et al. 2008a, b; Rajendrasozhan et al. 2009). In humans, a marked reduction of HDAC2 expression/activity in lung parenchyma, bronchial biopsies, and alveolar macrophages of patients with COPD is correlated with the severity of inflammation and the disease state (Ito et al. 2005). The underlying mechanism for reduced HDAC2 levels/activities includes post-translational modifications (i.e., nitrosylation, phosphorylation and ubiquitination) further leading to proteasome-dependent degradation particularly in response to cigarette smoke/oxidants/reactive aldehydes (Galasinski et al. 2002; Adenuga et al. 2009; Rajendrasozhan et al. 2009; Sundar et al. 2012), and/or oxidative/carbonyl modifications of HDAC2 (Yang et al. 2006). HDAC2 is also required for the anti-inflammatory effects of glucocorticoid efficacy as reduced levels/activity of HDAC2 are reported in lungs of patients with COPD that are resistant to corticosteroids to inhibit proinflammatory mediators (Ito et al. 2005). Increasing the activity of HDACs by curcumin and theophylline (HDAC2 activators) significantly enhanced the suppression of IL-8 release by steroids in monocytes and alveolar macrophage from patients with COPD, and was blocked by HDAC inhibitor (Cosio et al. 2004; Meja et al. 2008). Furthermore, deacetylation of glucocorticoid receptor (GR) by HDAC2 enhanced the association of GR and RelA/p65, which resulted in the attenuation of proinflammatory gene transcription (Ito et al. 2006). Consequently, restoration or attenuation of decreased HDAC2 levels/activities will further enhance the glucocorticoid sensitivity by deacetylating RelA/p65 and GR to inhibit transcription of proinflammatory genes (Sundar et al. 2012). This can possibly be achieved by reversing the post-translational modifications of HDAC2, such as decarbonylation or dephosphorylation, via inducing Nrf2-dependent aldehyde dehydrogenases/reductases, thioredoxin reductase, and phosphatases, or inducing the antioxidant buffer systems by using Nrf2 activators and ECSOD mimetics (Adenuga et al. 2010; Yao et al. 2010a). Hence, HDAC2 is a redox sensitive protein, whose levels/activities are decreased under oxidative stress/carbonyl conditions, resulting in chromatin modifications associated with steroid resistance leading to lung inflammatory response (Sundar et al. 2012) (Fig. 11.4).

11.2.4.2 SIRT1 Deacetylase, Inflammation, and Senescence

Sirtuin 1 (SIRT1) belongs to class III HDACs, which possess anti-inflammatory, anti-aging/senescence, and anti-apoptotic properties due to their ability to deacetylate both histones and non-histone proteins, including transcription factors and other signaling molecules (NF-κB, FOXO3, p53, Ku70, and PGC-1α) (Sundar et al. 2012; Yang and Sauve 2006) (Fig. 11.5). Earlier reports showed that SIRT1 levels and activities were decreased in monocytes (MonoMac6 cells), bronchial epithelial cells (Beas-2B), mouse lungs exposed to cigarette smoke, as well as

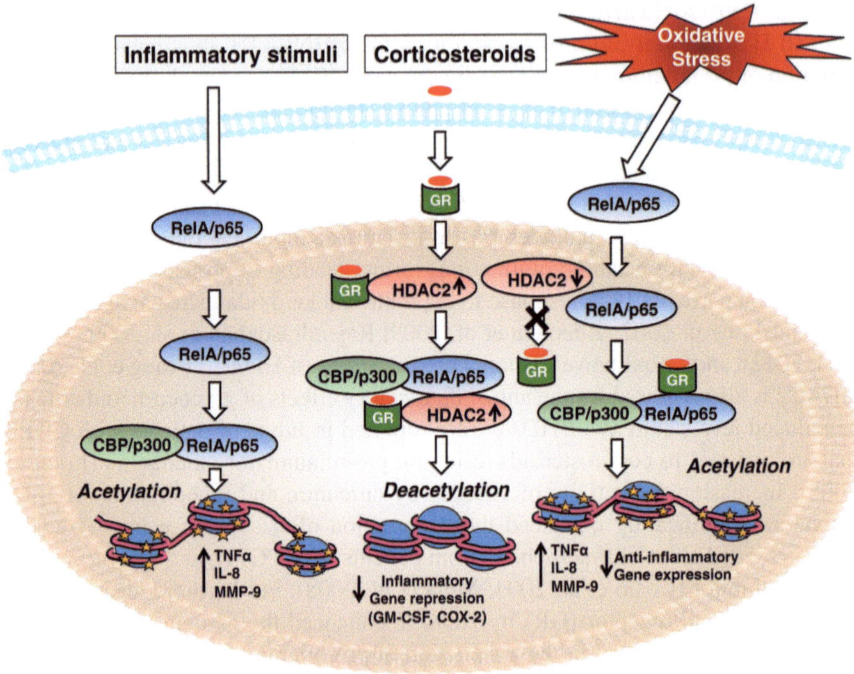

Fig. 11.4 Oxidative depletion of HDAC2 and redox-mediated chromatin modifications of pro-inflammatory genes in chronic inflammation and steroid resistance. Stimulation of alveolar macrophages and epithelial cells activates RelA/p65 (NF-κB) leading to recruitment of CBP/p300-RelA/p65 onto the promoters of proinflammatory genes (TNFα, IL-8, and MMP9), thereby resulting in proinflammatory gene transcription. Corticosteroids reverse this process by binding to gluococorticoid receptors (GR) and recruiting HDAC2 into the transcriptome complex, thereby reversing histone acetylation and RelA/p65 activation and causing transcriptional repression. Cigarette smoke-mediated degradation HDAC2 leads to the disruption of GR and NF-κB complex, and results in NF-κB activation and histone acetylation on promoters of pro-inflammatory genes. This leads to augmented gene expression of pro-inflammatory mediators

in the lungs of COPD patients and smokers (Rajendrasozhan et al. 2008a, b; Yang et al. 2007; Hwang et al. 2010), implicating the involvement of SIRT1 in the pathogenesis of COPD (Sundar et al. 2012). The mechanism for SIRT1 reduction involves post-translational modifications, such as phosphorylation, nitrosylation, and oxidation/carbonylation by ROS/carbonyls in response to cigarette smoke/oxidative/carbonyl stress, ultimately leading to SIRT1 oxidative modifications and degradation (Caito et al. 2010; Rajendrasozhan et al. 2008a, b) (Fig. 11.3). SIRT1 is also regulated by the intracellular redox thiol (GSH/GSSG) pool (Yang et al. 2007; Caito et al. 2010). SIRT1 activators or endogenous SIRT1 regulators (Milne and Denu 2008; Milne et al. 2007) are implicated in the therapeutic intervention of COPD (Rajendrasozhan et al. 2008b). Nakamaru et al. showed that administration of the SIRT1 activator SRT2172 attenuated cigarette smoke-induced lung inflammation in mice (Nakamaru et al. 2009). Furthermore,

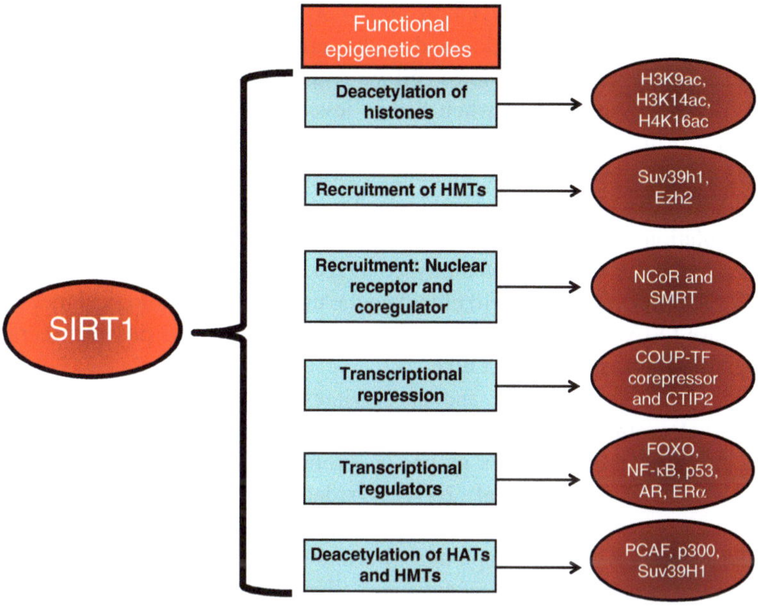

Fig. 11.5 Functional epigenetic roles of Sirtuin 1. SIRT1 regulates chromatin modifications and transcriptional activation/repression through different mechanisms. SIRT1 interacts with several factors and regulatory molecules, particularly those which are closely associated with chromatin regulation events. SIRT1 deacetylates histones H3K9, H3K14, and H4K16. Deacetylation of histones by SIRT1 synergizes with increased tri-methylation of H3K9 (gene repressive modification), thus resulting in transcriptional repression. SIRT1-mediated deacetylation of histone H3K9 activates histone methyltransferase Suv39h1, Ezh2 to further repress transcription. Similarly, SIRT1 interacts with nuclear receptor NCoR and coregulator SMRT, and transcriptional repression by COUP-TF corepressor, CTIP2, is also enhanced during deacetylation of histone H3 and H4 by SIRT1. SIRT1-mediated deacetylation of transcription regulators (FOXO, NF-κB, p53, AR, ERα), HATs (PCAF, p300) and HMT (Suv39h1) also cause transcriptional repression. Suv39h1 suppressor of variegation 3–9 homolog 1, Ezh2 Enhancer of Zeste 2 gene, NCoR nuclear corepressor, SMRT silencing mediator for retinoid or thyroid-hormone receptors, COUP-TF chicken ovalbumin upstream promoter transcription factor, CTIP2 COUP-TF-interacting protein 2, FOXO Forkhead Box class O, NF-κB nuclear factor-kappaB, AR androgen receptor, ERα estrogen receptor alpha, PCAF p300/CBP associated factor

SIRT1-mediated protection in response to cigarette smoke/oxidative stress against lung inflammatory response and injury are linked to the deacetylation of NF-κB RelA/p65 and negative regulation of MMP-9 (Chen et al. 2002; Yang et al. 2007; Nakamaru et al. 2009). Further studies are required to investigate the role of SIRT1 regulation and progression of COPD/emphysema using genetic/pharmacological approaches.

SIRT1 also deacetylates other transcription factors, such as forkhead box class (FOXO3), p53, and NF-κB, thereby regulating oxidative stress-induced cell cycle arrest, apoptosis and cellular senescence, which play important roles in the pathogenesis of COPD (Yao and Rahman 2011; Sundar et al. 2012) (Fig. 11.5). FOXO3 is regulated by its phosphorylation and acetylation status. SIRT1 deacetylates FOXO3, leading to its activation and to the induction of cell cycle arrest. Hence, increased SIRT1 activity promotes cell survival (Motta et al. 2004; You and Mak 2005). FOXO3 is acetylated when SIRT1 levels/activity are reduced in response to cigarette smoke in mouse lung (Rajendrasozhan et al. 2008a). Furthermore, FOXO3 has a regulatory role in lung inflammatory response and regulation of antioxidant genes (MnSOD and catalase). Targeted disruption of FOXO3 results in downregulation of antioxidant genes in mouse lungs and increases the susceptibility for the development of COPD/emphysema (Hwang et al. 2011). Further studies on the SIRT1-FOXO3 pathway will elucidate the underlying pathological mechanisms in response to oxidative/carbonyl stress, and will provide the possible therapeutic interventions in treatment of COPD.

SIRT1 interacts with p53, and deacetylates lysine residue at the C-terminal regulatory domain (Vaziri et al. 2001), whereas SIRT1 reduction leads to increased p53 acetylation, thus increases its pro-apoptotic function and cellular senescence (Vaziri et al. 2001; Luo et al. 2004). Oxidative stress is known to accelerate cellular senescence, due to increased p53 acetylation via decreasing the function of SIRT1 through NAD^+ depletion (Furukawa et al. 2007; Ota et al. 2007). Furthermore, blockade of p53 by antisense oligonucleotides reversed the inhibitory effect of SIRT1 on cellular senescence (Ota et al. 2007). Earlier studies from the authors laboratory have shown that the nuclear levels of SIRT1 were decreased in response to CS both *in vitro* and *in vivo* (Yang et al. 2007), but still it remains unclear whether SIRT1-mediated regulation of p53 (acetylation/deacetylation) and possibly p21 plays an important role in oxidants/CS-mediated apoptosis, and cellular senescence/lung aging (Sundar et al. 2012). Other members of sirtuins (eg. SIRT6 have also shown to modify cellular functions under various oxidative stress conditions) (Mao et al. 2011; Michishita et al. 2008; Kawahara et al. 2009; Van Gool et al. 2009). Therefore, other sirtuin members possibly gain similar credence as SIRT1 in understanding the pathogenesis of COPD and other chronic lung diseases (Sundar et al. 2012).

SIRT1 is reported as a key regulator of vascular endothelial homeostasis controlling angiogenesis, vascular tone and endothelial dysfunction by regulating endothelial nitric oxide synthase (eNOS) activity (Wright and Churg 2008; Potente and Dimmeler 2008a; Potente and Dimmeler 2008b). Endothelial dysfunction plays an important role in pathogenesis of emphysema, which is associated with decreased expression of eNOS (Edirisinghe et al. 2008; Ferrer et al. 2009). SIRT1 interacts with eNOS and deacetylates lysines at position 496 and 506 in the calmodulin-binding domain of eNOS, leading to enhanced nitric oxide (NO) production (Mattagajasingh et al. 2007; Arunachalam et al. 2010). Hence, the activation of SIRT1 by small molecules may help in resetting the eNOS activity during endothelial dysfunction, e.g. in susceptible smokers, where NO availability is limited (Michaud et al. 2006). Another study demonstrated that SIRT1 overexpression attenuated cigarette smoke-induced apoptosis and inflammatory response in cultured coronary arterial endothelial cells (Csiszar et al. 2008). Therefore, SIRT1 acts as a possible redox regulator in the treatment and prevention of chronic lung and cardiovascular diseases including COPD (emphysema) and atherosclerosis by protecting endothelial cells from stress-induced premature senescence, apoptosis and inflammatory response (Sundar et al. 2012).

11.2.5 Histone Acetylation/Deacetylation and Its Cross-Talk with Histone Methylation in Redox Epigenetics

Unlike histone acetylation, which drives gene activation, histone methylation can either serve as signal for activation or repression of gene transcription, depending on the sites of histone methylation (Zhang and Reinberg 2001; Bannister and Kouzarides 2005). HMTs utilize S-adenosylmethionine (SAM) as a cofactor during transmethylation and produce S-adenosylhomocysteine (SAH) as by-product (Sundar et al. 2012). The substrate specificity and activity of HMT depends on SET domains and associated motifs (Zhang et al. 2003). HMTs are deregulated in several chronic lung diseases, thereby affecting the global methylation status. Histone H3K4, H3K9, and H3K27, and H4K20 are frequently and preferentially methylated as mono-, di- or tri- methylated histone H3 and histone H4 (Seligson et al. 2009). Methylation at H3K4, H3K36, and H3K79 is linked to gene activation, whereas H3K9, H3K20, and H3K27 methylation is associated with gene repression (Gosden and Feinberg 2007). Histone modifications are governed by specific chromatin modification enzymes, which have a residue-specific role in regulating gene expression with respect to histone modifications in epigenetic events (Table 11.1). Only lysine and serine modifications are shown in this table due to the focus of this review. Furthermore, the role of these enzymes in regulation of the histone code is nebulous. Barlési et al. demonstrated two key epigenetic histone modifications, H3K4me2 ($\geq$85% tumor cells) and H3K9ac ($\geq$68% tumor cells) that may contribute to the development and progression of Non-Small Cell Lung Cancer (NSCLC) (Barlesi et al. 2007). Cancer cells have a very different epigenotype compared to the cells from normal tissues from which they arise. This epigenetic switch in cancer cells is due to changes in DNA methylation and histone modifications caused by increased oxidant burden. Cancer cells have increased expression of epigenetic modification enzymes, yet their reaction products do not correlate, suggesting that there may be other factors affecting their activity (Ehrlich et al. 2006). A recent study has demonstrated a distinct mechanism by which histone methyltransferase G9a domains and different molecular mechanisms (ligand-dependent ERα binding function) regulate G9a-mediated coactivator versus corepressor function (Purcell et al. 2011). Transient reporter gene assays using MCF-7 breast cancer cells showed an involvement of endogenous G9a in contribution of estrogen estradiol (E_2)-dependent activation of estrogen receptor (ER)α target genes while concurrently limiting the expression of E_2-induced expression of other ERα target genes (Purcell et al. 2011). *In vitro* exposures to cigarette smoke condensate or extract and oxidants using airway epithelial cells and immortalized bronchial epithelial cells showed reduced H4K16ac and H4K20me3 but increased H3K27me3 levels, which may

Table 11.1 Post-translational histone modification sites (serine and lysine residues), chromatin modification enzymes, and their proposed functional role in gene regulation

Histone/ modification type	Post-translational modification sites	Chromatin modification enzymes	Proposed function
Histone H3 phosphorylation	H3S10	Aurora-B kinase, MSK1, MSK2, IKKα, Snf1	Mitosis, meiosis, immediate-early gene activation, transcriptional activation (proinflammatory genes)
	H3S28	Aurora-B kinase, MSK1, MSK2	Mitosis, immediate-early gene activation
Histone H3 acetylation	H3K4 (*S. cerevisiae*)	Esa1	Transcriptional activation
		Hpa2	Unknown
	H3K9	Gcn5, SRC-1	Transcriptional activation and histone deposition
	H3K14	Gcn5, PCAF, Esal, Tip60, SRC-1, Elp3, Hpa2, hTFIIIC90, TAF1, Sas2, Sas3, p300	DNA repair, transcriptional activation, RNA polymerase II & III transcription, euchromatin
	H3K18	Gcn5, CBP/p300	Transcriptional activation, DNA replication and repair
	H3K23	Gcn5, Sas3	Transcriptional activation (elongation) and DNA repair
	H3K27	Gcn5	Transcriptional activation
	H3K56 (*S. cerevisiae*)	Spt10	Transcriptional activation and DNA repair
Histone H3 methylation	H3K4	SET1 (*S. cerevisiae*), SET7/9 (vertebrates), MLL, ALL01,	Permissive euchromatin (di-me), transcriptional activation (tri-me)
	H3K9	SUV39h, Clr4, G9a, SETDB1	Transcriptional repression (tri-me)
	H3R17	CARM1	Transcriptional activation
	H3K27	Ezh2, G9a	Transcriptional repression, X-inactivation (tri-me)
	H3K36	Set2	Transcriptional activation (elongation)
	H3K79	Dot1	Euchromatin, transcriptional activation (elongation), checkpoint response
Histone H4 phosphorylation	H4S1	CK2	Mitosis, chromatin assembly and DNA repair

(continued)

Table 11.1 (continued)

Histone/ modification type	Post-translational modification sites	Chromatin modification enzymes	Proposed function
Histone H4 acetylation	H4K5	Hat1, Esal, Tip60, ATF2, Hpa2, p300	Histone deposition, transcriptional activation and DNA repair
	H4K8	Gcn5, PCAF, Esal, Tip60, ATF2, Elp3, p300	Transcriptional activation (elongation) and DNA repair
	H4K12	Hat1, Esal, Tip60, Hpa2, p300	Histone deposition, telomeric silencing, transcriptional activation and DNA repair
	H4K16	Gcn5, Esal, Tip60, ATF2, Sas2	Transcriptional activation, DNA repair, euchromatin
Histone H4 methylation	H4R3	PRMT1, PRMT5	Transcriptional activation/repression
	H4K20	PR-SET7, SUV4-20h	Transcriptional repression (mono-me), heterochromatin (tri-me)

have implications in tumor development in a wide array of cancer types (Liu et al. 2010). However, it remains to be seen whether the regulation of these enzymes and specific histone modifications via modulation by oxidative stress and/or redox status of the cells also occur in redox epigenetics regulation of pro-inflammatory genes particularly in chronic inflammatory diseases (Sundar et al. 2012).

The cross-talks between different histone methylation sites also epigenetically control the transcriptional activation of target genes (Cheung and Lau 2005; Wang et al. 2008). The positive and negative cross-talks ultimately generate the complex pattern of gene- or locus-specific histone marks, which are linked with distinct chromatin states (euchromatin or heterochromatin), leading to transcriptional repression or activation (Sundar et al. 2012). Cross-talk exists between various epigenetic markers on histones and DNA methylation (Fig. 11.6). For example, the methyl-cytosine binding protein, MeCP2, recruits HDACs to methylated DNA as well as to methylated histones (Gilliland et al. 2002; Ng et al. 1999). It has been shown that inhibition of HDACs and histone acetylation/methyation, and/or DNA methylases leads to upregulation of pro-inflammatory mediators, such as GM-CSF by IL-1β in lung epithelial cells (Kagoshima et al. 2001). Histone H3K4 tri-methylation has been reported to play a role in tumorigenesis and X-chromatin inactivation in human cells (Lachner and Jenuwein 2002). Furthermore, hypermethylation of histone H3K4 was found to be unrelated to acute TNF-α-induced gene expression (Saccani and Natoli 2002). In contrast, the involvement of H3K4 tri-methylation in IL-1β-induced gene expression has been reported (Wada et al. 2005). Furthermore, H3K4 tri-methylation plays a critical role in IL-1β stimulated secretory leukocyte protease inhibitor (SLPI) expression (Wada et al. 2005). However, it remains to be

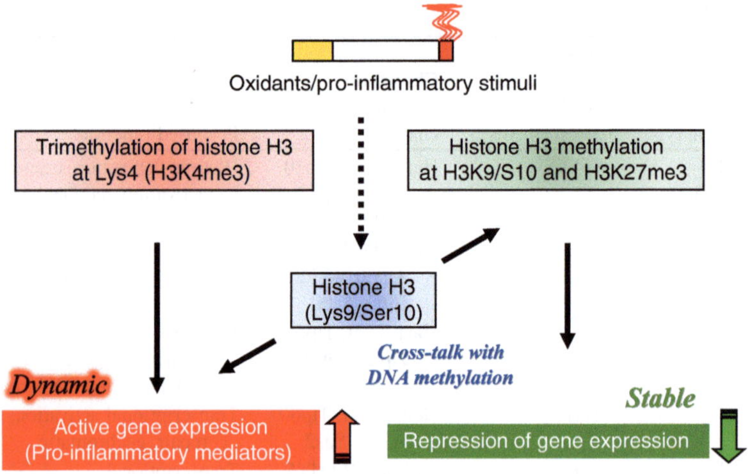

Fig. 11.6 Cross-talk between histone acetylation and methylation. Oxidative/carbonyl stress can activate histone acetylation on specific lysine residues. Trimethylation of histone H3 at K4 (H3K4me3) can lead to dynamic gene transcription, whereas histone H3 methylation at H3K9 and H3K27me3 can result in stable gene repression. Histone H3 acetylation at K9 and serine 10 (S10) phosphorylation can culminate in gene transcription. Histone acetylation/deacetylation and histone methylation can cross-talk with DNA methylation on CpG islands to activate or repress gene transcription

unknown whether the cross-talk occurs between histone acetylation and methylation in response to cigarette smoke/oxidant during lung inflammatory response. Studies on sequential events of histone acetylation/histone methylation will provide the understanding on the new epigenetics-based biomarkers and/or treatment for chronic inflammatory diseases including COPD/emphysema and cancer (Sundar et al. 2012) (Fig. 11.7).

11.2.6 HDMs and Redox Epigenetics

Histone demethylases catalyze the removal of methyl groups from lysine or arginine residue of histones (Sundar et al. 2012). They are classified into two types, the lysine specific demethylase 1 (LSD1) and Jumonji C (JmjC) domain family proteins involved in the regulation of gene expression (Tian and Fang 2007). LSD1 specifically demethylates histone H3K4me2, and is an important member of HDMs among other transcriptional repression complexes (Shi et al. 2004). JmjC demethylates mono-, di- or trimethyl lysine residues by a different mechanism that requires cofactors, such as molecular oxygen, α-ketoglutarate, Fe^{2+}, and ascorbate via a redox modulating process (Klose et al. 2006a, b). Aberrant expression of HDMs occurs during the course of tumor initiation and progression (Lim et al. 2010). However, the role of ROS/redox GSH status in modulation of HDMs in COPD

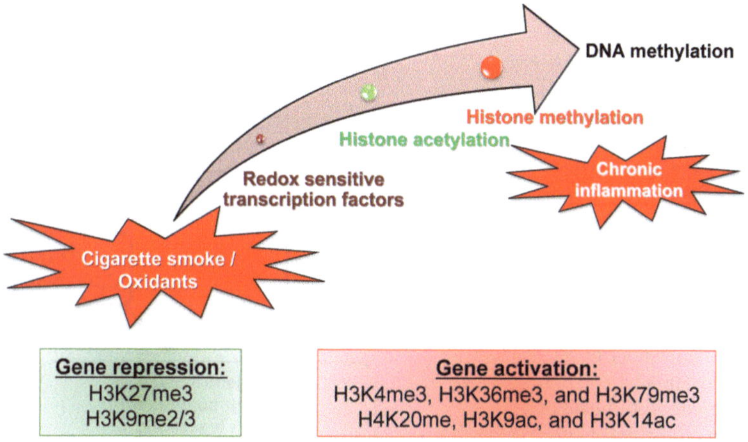

Fig. 11.7 Redox chromatin modifications in pathogenesis of chronic inflammatory diseases. Specific histone modifications can cross-talk with CpG methylation to activate or repress gene transcription in inflammation involved in chronic inflammatory diseases

and other chronic lung diseases remains unknown (Sundar et al. 2012). Hypoxia is known to occur in tumor microenvironments, as well as in lungs of patients with COPD and other chronic inflammatory lung diseases. The level of JMJD2B is upregulated in response to hypoxia, which depends on hypoxia-inducible factor 1 alpha (HIF-1α) (Yang et al. 2010). Inhibition of H3K4 demethylase JMJD1A sufficiently reduces tumor growth *in vivo* demonstrating its role in regulating histone methylation in hypoxia (Krieg et al. 2010). In contrast, hypoxia increases global levels of H3K4me3 in alveolar A549 and bronchial Beas-2B cell lines, due to the inhibition of demethylation process, particularly demethylase (JARID1A) (Zhou et al. 2010). The involvement of JARID1B (KDM5B) in growth of cancer cells has been shown through the E2F/RB1 cell cycle regulation pathway in different cancer cell lines. Lung tumor samples from NSCLC and SCLC showed elevated expression of JARID1B compared to non-neoplastic tissues confirmed by microarray analysis and immunohistochemistry suggesting JARID1B activation plays an important role in lung carcinogenesis (Hayami et al. 2010). Thus, identification of target specific histone demethylase inhibitors may be a viable tool in epigenetic therapeutics or via potentiating the activity of hypomethylating agents in cancer (Grant 2009).

11.2.7 DNA Methylation in Redox Epigenetics

DNA methylation is a nonhistone epigenetic event that plays an important role in transcriptional regulation. Mostly in all eukaryotes, methylation occurs on cytosine residues, producing 5-meC (5-methylcytosine), but the local context of the methylation event differs from one organism to the other (Suzuki and Bird

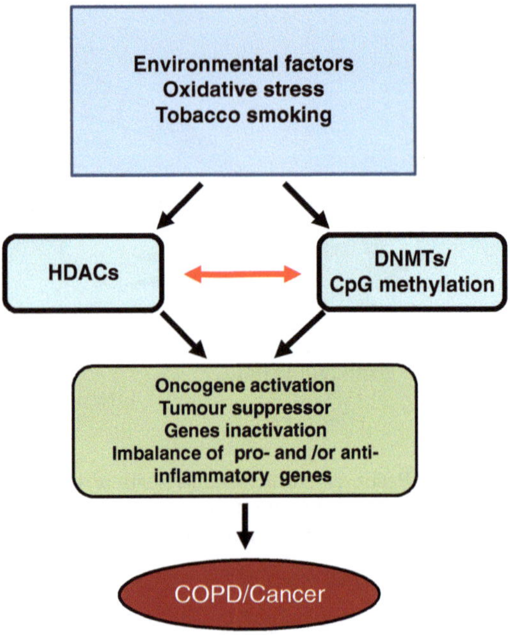

Fig. 11.8 Epigenetic regulation of gene expression in COPD and cancer. Environmental factors, oxidative stress, diet, genetic predisposition factors, and aging can increase promoter CpG methylation by recruitment of MeCP2 and DNMTs on various promoters along with HDACs leading to alterations in tumor suppressor, onocogenes, pro- and anti-inflammatory genes. DNA methylation is associated with recruitment of HDACs on promoters of various genes. These processes are intertwined in pathogenesis of COPD and lung cancer

2008). DNA methylation is associated with epigenetic silencing, and this effect is in part mediated by recruitment of HDACs through the methyl-DNA binding motifs, including components of several HDAC-containing complexes (Nan et al. 1998). Environmental factors/agents that trigger oxidative stress, such as diet, genetic predisposition factors, and aging can gradually affect the promoter CpG methylation by recruiting MeCP2 and DNMTs to various promoters along with HDACs (Sundar et al. 2012). This leads to alterations in the expression of tumor suppressor, onocogenes, and pro- and anti-inflammatory genes (Fig. 11.8). Promoter methylation in multiple genes has been demonstrated in adenocarcinomas and NSCLC, and they are linked with tumor progression (Zochbauer-Muller et al. 2001). Hence, identifying gene specific DNA methylation will provide further insights in the development of useful markers as tools for early detection and treatment of different cancers. There are several reports available that describe the promoter hypermethylation and associated gene-silencing of various genes in lung cancer (Belinsky 2004; Anglim et al. 2008). However, the relationship between gene-specific DNA methylation and smoking history is not known (Tsou et al. 2002). Georgiou et al. reported methylation of p16 promoter in sputum of patients with COPD that significantly correlated with heavy cigarette smoking (Georgiou et al. 2007). Gene ontology analysis suggested genes involved in immune and inflammatory system pathways, stress response, as well as wound healing and coagulation cascades. Significant association was shown between SERPINA1 hypomethylation and pathogenesis of COPD with lower lung function (Qiu et al.

2012). This suggests that DNA methylation/hypomethylation, which is caused by oxidative stress in specific genes may be a epigenetic mark in the pathogenesis of COPD (Qiu et al. 2012).

Cigarette smoke downregulates DNMT3B in adenocarcinoma alveolar A549 epithelial cells resulting in demethylation of the CpG island in the oncogene SNCG and leading to high expression of this prometastatic oncogene in A549 cells (Liu et al. 2007). Promoter specific hypermethylation in genes, such as *APC1* (Brabender et al. 2001), *DAPK* (Kim et al. 2001) and *p16* (Sanchez-Cespedes et al. 2001) from lung tissue of NSCLC patients showed no significant association between smoking history. On the contrary, significant association between promoter hypermethylation and smoking history has been observed in NSCLC patients in *CDKN2A* (Kim et al. 2001) *HIC1* (Eguchi et al. 1997), *HtrA3* (Beleford et al. 2010), and *CHFR* (Takeshita et al. 2008). Promoter hypermethylation leads to gene-silencing of various genes in lung cancer (Belinsky 2004; Anglim et al. 2008). Sood et al. identified the promoter methylation from sputum samples of patients demonstrating wood smoke exposure associated with COPD phenotype (decline in lung function, airflow obstruction and chronic bronchitis), and smokers with aberrant *p16* and *GATA4* methylation (Sood et al. 2010). Epigenetics in human asthma is not well understood, particularly in older adults. Investigating the methylation profiles in sputum DNA among older adults with asthmatics, who are smokers revealed methylation of selective genes such as Protocadherin-20. Hence, the use of novel demethylating agents may possibly help to prevent and treat asthma among older smokers (Sood et al. 2012). Suzuki et al. demonstrated methylation of *IL-12Rb2* and *Wif-1* in COPD patients compared to non-COPD group, suggesting the molecular epigenetic events that influence COPD-related NSCLC (Suzuki et al. 2010). An interesting report described the association of prenatal exposure to tobacco smoke with significant changes in two types of DNA methylation (global methylation and promoter CpG island methylation) (Breton et al. 2009). Hence, the findings from this study supports that epigenetic effects of *in utero* exposures may be due to alterations in DNA methylation patterns (Breton et al. 2009).

A recent study demonstrated the influence of cigarette smoke extract on WW domain containing oxidoreductase (WWOX) expression and methylation status in human bladder cancer cells (T-24). Cigarette smoke extract treatment significantly increased the levels of DNMT1 but not DNMT3A and DNMT3B. The expression levels of DBMT1 corroborate with hypermethylation of WWOX induced by cigarette smoke which may be the underlying mechanism for the loss of WWOX expression in bladder cancer (Yang et al. 2012). Another study on gender-specific methylation differences in relation to prenatal exposure to cigarette smoke demonstrates two differentially methylated regions (DMRs) regulating insulin-like growth factor 2 (IGF2) and H19. There was a positive correlation between smoking-related increased DNA methylation at the IGF2 DMR in male offsprings, who had smoking-related low birth weight (approximately 20%). Hence, these observations suggest that IGF2 DMR plasticity is an important mechanism that regulates *in utero* adaptation to environmental toxicants (Murphy et al. 2012). Coordinated changes in DNA methylation particularly at the aryl hydrocarbon receptor repressor

(AHRr) was observed in lymphoblasts and alveolar macrophages from smokers suggesting the implications of epigenetic effects mediated by cigarette smoke on carcinogenesis and other related co-morbidities (Monick et al. 2012). The DNA methylation status on susceptibility genes may have implications in smokers, who are at high risk for chronic lung diseases including COPD. The role of various histone modifications versus DNA methylation, particularly in response to oxidative stress in inflammatory conditions is an emerging field of further research (Sundar et al. 2012).

11.3 Conclusions

The role of ROS and redox epigenetic regulation in pro-inflammatory/anti-inflammatory gene transcription is discussed in this chapter. ROS produced by extracellular or by intracellular processes have the potential to induce considerable post-translational modifications of intracellular proteins. Oxidative stress by cigarette smoke/oxidants is critical for chronic lung inflammatory response via activation of stress signaling kinases, redox-sensitive transcription factors, and modulation of epigenetic chromatin modifications, which result in gene transcription (Sundar et al. 2012). Epigenetic events, such as histone acetylation, deacetylation, phosphorylation, and histone/DNA methylation play an important role in oxidative stress-induced chronic inflammatory diseases, such as asthma, COPD, and cancer. HDAC2 is required for glucocorticoids to function as anti-inflammatory agents. Thus, activation of HDAC2 or reversal of oxidative post-translational modifications of HDAC2 is another avenue to devise epigenetic based therapy in treatment of severe asthma and COPD (Sundar et al. 2012). Further studies are required to unravel the mechanisms of redox epigenetics for the development of epigenetics-based therapies in chronic inflammatory diseases and cancer.

Acknowledgements This study was supported by the NIH 1R01HL085613, 1R01HL097751, 1R01HL092842, and NIEHS Environmental Health Science Center grant P30-ES01247.

References

Adenuga D, Rahman I (2010) Protein kinase CK2-mediated phosphorylation of HDAC2 regulates co-repressor formation, deacetylase activity and acetylation of HDAC2 by cigarette smoke and aldehydes. Arch Biochem Biophys 498:62–73

Adenuga D, March TH, Seagrave J, Rahman I (2008a) Cigarette smoke-induced loss of nuclear HDAC2 is associated with chronic inflammation and emphysema in A/J mice. Am J Respir Crit Care Med 177:A866

Adenuga D, Yang SR, Rajendrasozhan S, Rahman I (2008b) HDAC2 degradation is associated with increased hyperphosphorylation and a proteasome-dependent mechanism in response to cigarette smoke in macrophages. Am J Respir Crit Care Med 177:A866

Adenuga D, Yao H, March TH, Seagrave J, Rahman I (2009) Histone deacetylase 2 is phosphorylated, ubiquitinated, and degraded by cigarette smoke. Am J Respir Cell Mol Biol 40:464–473

Adenuga D, Caito S, Yao H, Sundar IK, Hwang JW, Chung S, Rahman I (2010) Nrf2 deficiency influences susceptibility to steroid resistance via HDAC2 reduction. Biochem Biophys Res Commun 403:452–456

Adler V, Yin Z, Tew KD, Ronai Z (1999) Role of redox potential and reactive oxygen species in stress signaling. Oncogene 18:6104–6111

Anest V, Hanson JL, Cogswell PC, Steinbrecher KA, Strahl BD, Baldwin AS (2003) A nucleosomal function for IkappaB kinase-alpha in NF-kappaB-dependent gene expression. Nature 423:659–663

Anglim PP, Galler JS, Koss MN, Hagen JA, Turla S, Campan M, Weisenberger DJ, Laird PW, Siegmund KD, Laird-Offringa IA (2008) Identification of a panel of sensitive and specific DNA methylation markers for squamous cell lung cancer. Mol Cancer 7:62

Arunachalam G, Yao H, Sundar IK, Caito S, Rahman I (2010) SIRT1 regulates oxidant- and cigarette smoke-induced eNOS acetylation in endothelial cells: role of resveratrol. Biochem Biophys Res Commun 393:66–72

Bannister AJ, Kouzarides T (2005) Reversing histone methylation. Nature 436:1103–1106

Barlesi F, Giaccone G, Gallegos-Ruiz MI, Loundou A, Span SW, Lefesvre P, Kruyt FA, Rodriguez JA (2007) Global histone modifications predict prognosis of resected non small-cell lung cancer. J Clin Oncol 25:4358–4364

Bast A, Haenen GR, Doelman CJ (1991) Oxidants and antioxidants: state of the art. Am J Med 91:2S–13S

Beleford D, Liu Z, Rattan R, Quagliuolo L, Boccellino M, Baldi A, Maguire J, Staub J, Molina J, Shridhar V (2010) Methylation induced gene silencing of HtrA3 in smoking-related lung cancer. Clin Cancer Res 16:398–409

Belinsky SA (2004) Gene-promoter hypermethylation as a biomarker in lung cancer. Nat Rev Cancer 4:707–717

Bertos NR, Wang AH, Yang XJ (2001) Class II histone deacetylases: structure, function, and regulation. Biochem Cell Biol 79:243–252

Bhavsar P, Khorasani N, Hew M, Johnson M, Chung KF (2010) Effect of p38 MAPK inhibition on corticosteroid suppression of cytokine release in severe asthma. Eur Respir J 35:750–756

Biswas SK, Rahman I (2009) Environmental toxicity, redox signaling and lung inflammation: the role of glutathione. Mol Asp Med 30:60–76

Bowie A, O'Neill LA (2000) Oxidative stress and nuclear factor-kappaB activation: a reassessment of the evidence in the light of recent discoveries. Biochem Pharmacol 59:13–23

Brabender J, Usadel H, Danenberg KD, Metzger R, Schneider PM, Lord RV, Wickramasinghe K, Lum CE, Park J, Salonga D, Singer J, Sidransky D, Holscher AH, Meltzer SJ, Danenberg PV (2001) Adenomatous polyposis coli gene promoter hypermethylation in non-small cell lung cancer is associated with survival. Oncogene 20:3528–3532

Breton CV, Byun HM, Wenten M, Pan F, Yang A, Gilliland FD (2009) Prenatal tobacco smoke exposure affects global and gene-specific DNA methylation. Am J Respir Crit Care Med 180:462–467

Caito S, Rajendrasozhan S, Cook S, Chung S, Yao H, Friedman AE, Brookes PS, Rahman I (2010) SIRT1 is a redox-sensitive deacetylase that is post-translationally modified by oxidants and carbonyl stress. FASEB J 24:3145–3159

Cao S, Zhang X, Edwards JP, Mosser DM (2006) NF-kappaB1 (p50) homodimers differentially regulate pro- and anti-inflammatory cytokines in macrophages. J Biol Chem 281:26041–26050

Carrero P, Okamoto K, Coumailleau P, O'Brien S, Tanaka H, Poellinger L (2000) Redox-regulated recruitment of the transcriptional coactivators CREB-binding protein and SRC-1 to hypoxia-inducible factor 1alpha. Mol Cell Biol 20:402–415

Chen LF, Greene WC (2004) Shaping the nuclear action of NF-kappaB. Nat Rev Mol Cell Biol 5:392–401

Chen L, Fischle W, Verdin E, Greene WC (2001) Duration of nuclear NF-kappaB action regulated by reversible acetylation. Science 293:1653–1657

Chen LF, Mu Y, Greene WC (2002) Acetylation of RelA at discrete sites regulates distinct nuclear functions of NF-kappaB. EMBO J 21:6539–6548

Cheung P, Lau P (2005) Epigenetic regulation by histone methylation and histone variants. Mol Endocrinol 19:563–573

Christman JW, Sadikot RT, Blackwell TS (2000) The role of nuclear factor-kappa B in pulmonary diseases. Chest 117:1482–1487

Chung S, Sundar IK, Yao H, Ho YS, Rahman I (2010) Glutaredoxin 1 regulates cigarette smoke-mediated lung inflammation through differential modulation of I{kappa}B kinases in mice: impact on histone acetylation. Am J Physiol Lung Cell Mol Physiol 299:L192–L203

Chung S, Sundar IK, Hwang JW, Yull FE, Blackwell TS, Kinnula VL, Bulger M, Yao H, Rahman I (2011) NF-kappaB inducing kinase, NIK mediates cigarette smoke/TNFalpha-induced histone acetylation and inflammation through differential activation of IKKs. PLOS One 6:e23488

Cosio BG, Tsaprouni L, Ito K, Jazrawi E, Adcock IM, Barnes PJ (2004) Theophylline restores histone deacetylase activity and steroid responses in COPD macrophages. J Exp Med 200: 689–695

Csiszar A, Labinskyy N, Jo H, Ballabh P, Ungvari Z (2008) Differential proinflammatory and prooxidant effects of bone morphogenetic protein-4 in coronary and pulmonary arterial endothelial cells. Am J Physiol Heart Circ Physiol 295:H569–H577

Cyr AR, Domann FE (2011) The redox basis of epigenetic modifications: from mechanisms to functional consequences. Antioxid Redox Signal 15:551–589

de Ruijter AJ, van Gennip AH, Caron HN, Kemp S, van Kuilenburg AB (2003) Histone deacetylases (HDACs): characterization of the classical HDAC family. Biochem J 370: 737–749

di Stefano A, Caramori G, Oates T, Capelli A, Lusuardi M, Gnemmi I, Ioli F, Chung KF, Donner CF, Barnes PJ, Adcock IM (2002) Increased expression of nuclear factor-kappaB in bronchial biopsies from smokers and patients with COPD. Eur Respir J 20:556–563

Doukas J, Eide L, Stebbins K, Racanelli-Layton A, Dellamary L, Martin M, Dneprovskaia E, Noronha G, Soll R, Wrasidlo W, Acevedo LM, Cheresh DA (2009) Aerosolized phosphoinositide 3-kinase gamma/delta inhibitor TG100-115 [3-[2,4-diamino-6-(3-hydroxyphenyl)pteridin-7-yl]phenol] as a therapeutic candidate for asthma and chronic obstructive pulmonary disease. J Pharmacol Exp Ther 328:758–765

Edirisinghe I, Yang SR, Yao H, Rajendrasozhan S, Caito S, Adenuga D, Wong C, Rahman A, Phipps RP, Jin ZG, Rahman I (2008) VEGFR-2 inhibition augments cigarette smoke-induced oxidative stress and inflammatory responses leading to endothelial dysfunction. FASEB J 22:2297–2310

Eguchi K, Kanai Y, Kobayashi K, Hirohashi S (1997) DNA hypermethylation at the D17S5 locus in non-small cell lung cancers: its association with smoking history. Cancer Res 57:4913–4915

Ehrlich M, Woods CB, Yu MC, Dubeau L, Yang F, Campan M, Weisenberger DJ, Long T, Youn B, Fiala ES, Laird PW (2006) Quantitative analysis of associations between DNA hypermethylation, hypomethylation, and DNMT RNA levels in ovarian tumors. Oncogene 25:2636–2645

Engelman JA, Janne PA (2008) Mechanisms of acquired resistance to epidermal growth factor receptor tyrosine kinase inhibitors in non-small cell lung cancer. Clin Cancer Res 14: 2895–2899

Ferrer E, Peinado VI, Diez M, Carrasco JL, Musri MM, Martinez A, Rodriguez-Roisin R, Barbera JA (2009) Effects of cigarette smoke on endothelial function of pulmonary arteries in the guinea pig. Respir Res 10:76–87

Fischle W, Wang Y, Allis CD (2003) Histone and chromatin cross-talk. Curr Opin Cell Biol 15:172–183

Fong YC, Hsu SF, Wu CL, Li TM, Kao ST, Tsai FJ, Chen WC, Liu SC, Wu CM, Tang CH (2009) Transforming growth factor-beta1 increases cell migration and beta1 integrin up-regulation in human lung cancer cells. Lung Cancer 64:13–21

Forman HJ, Dickinson DA (2003) Oxidative signaling and glutathione synthesis. Biofactors 17: 1–12

Fukuda H, Sano N, Muto S, Horikoshi M (2006) Simple histone acetylation plays a complex role in the regulation of gene expression. Brief Funct Genomic Proteomic 5:190–208

Furukawa A, Tada-Oikawa S, Kawanishi S, Oikawa S (2007) H2O2 accelerates cellular senescence by accumulation of acetylated p53 via decrease in the function of SIRT1 by NAD + depletion. Cell Physiol Biochem 20:45–54

Galasinski SC, Resing KA, Goodrich JA, Ahn NG (2002) Phosphatase inhibition leads to histone deacetylases 1 and 2 phosphorylation and disruption of corepressor interactions. J Biol Chem 277:19618–19626

Garcia-Gimenez JL, Ledesma AM, Esmoris I, Roma-Mateo C, Sanz P, Vina J, Pallardo FV (2012) Histone carbonylation occurs in proliferating cells. Free Radic Biol Med 52(8):1453–1464

Georgiou E, Valeri R, Tzimagiorgis G, Anzel J, Krikelis D, Tsilikas C, Sarikos G, Destouni C, Dimitriadou A, Kouidou S (2007) Aberrant p16 promoter methylation among Greek lung cancer patients and smokers: correlation with smoking. Eur J Cancer Prev 16:396–402

Ghezzi P, Bonetto V, Fratelli M (2005) Thiol-disulfide balance: from the concept of oxidative stress to that of redox regulation. Antioxid Redox Signal 7:964–972

Ghosh S, Karin M (2002) Missing pieces in the NF-kappaB puzzle. Cell 109(Suppl):S81–S96

Gilliland FD, Harms HJ, Crowell RE, Li YF, Willink R, Belinsky SA (2002) Glutathione S-transferase P1 and NADPH quinone oxidoreductase polymorphisms are associated with aberrant promoter methylation of P16(INK4a) and O(6)-methylguanine-DNA methyltransferase in sputum. Cancer Res 62:2248–2252

Gilmour PS, Rahman I, Donaldson K, Macnee W (2003) Histone acetylation regulates epithelial IL-8 release mediated by oxidative stress from environmental particles. Am J Physiol Lung Cell Mol Physiol 284:L533–L540

Gosden RG, Feinberg AP (2007) Genetics and epigenetics–nature's pen-and-pencil set. N Engl J Med 356:731–733

Grant S (2009) Targeting histone demethylases in cancer therapy. Clin Cancer Res 15:7111–7113

Grozinger CM, Hassig CA, Schreiber SL (1999) Three proteins define a class of human histone deacetylases related to yeast Hda1p. Proc Natl Acad Sci USA 96:4868–4873

Gustafson AM, Soldi R, Anderlind C, Scholand MB, Qian J, Zhang X, Cooper K, Walker D, Mcwilliams A, Liu G, Szabo E, Brody J, Massion PP, Lenburg ME, Lam S, Bild AH, Spira A (2010) Airway PI3K pathway activation is an early and reversible event in lung cancer development. Sci Transl Med 2:26ra25

Hay CW, Candido EP (1983) Histone deacetylase. Association with a nuclease resistant, high molecular weight fraction of HeLa cell chromatin. J Biol Chem 258:3726–3734

Hayami S, Yoshimatsu M, Veerakumarasivam A, Unoki M, Iwai Y, Tsunoda T, Field HI, Kelly JD, Neal DE, Yamaue H, Ponder BA, Nakamura Y, Hamamoto R (2010) Overexpression of the JmjC histone demethylase KDM5B in human carcinogenesis: involvement in the proliferation of cancer cells through the E2F/RB pathway. Mol Cancer 9:59

Hew M, Bhavsar P, Torrego A, Meah S, Khorasani N, Barnes PJ, Adcock I, Chung KF (2006) Relative corticosteroid insensitivity of peripheral blood mononuclear cells in severe asthma. Am J Respir Crit Care Med 174:134–141

Hoshino S, Yoshida M, Inoue K, Yano Y, Yanagita M, Mawatari H, Yamane H, Kijima T, Kumagai T, Osaki T, Tachiba I, Kawase I (2005) Cigarette smoke extract induces endothelial cell injury via JNK pathway. Biochem Biophys Res Commun 329:58–63

Hwang JW, Chung S, Sundar IK, Yao H, Arunachalam G, McBurney MW, Rahman I (2010) Cigarette smoke-induced autophagy is regulated by SIRT1-PARP-1-dependent mechanism: implication in pathogenesis of COPD. Arch Biochem Biophys 500:203–209

Hwang JW, Rajendrasozhan S, Yao H, Chung S, Sundar IK, Huyck HL, Pryhuber GS, Kinnula VL, Rahman I (2011) FOXO3 deficiency leads to increased susceptibility to cigarette smoke-induced inflammation, airspace enlargement, and chronic obstructive pulmonary disease. J Immunol 187:987–998

Imhof A, Wolffe AP (1998) Transcription: gene control by targeted histone acetylation. Curr Biol 8:R422–R424

Ito K, Barnes PJ, Adcock IM (2000) Glucocorticoid receptor recruitment of histone deacetylase 2 inhibits interleukin-1beta-induced histone H4 acetylation on lysines 8 and 12. Mol Cell Biol 20:6891–6903

Ito K, Hanazawa T, Tomita K, Barnes PJ, Adcock IM (2004) Oxidative stress reduces histone deacetylase 2 activity and enhances IL-8 gene expression: role of tyrosine nitration. Biochem Biophys Res Commun 315:240–245

Ito K, Ito M, Elliott WM, Cosio B, Caramori G, Kon OM, Barczyk A, Hayashi S, Adcock IM, Hogg JC, Barnes PJ (2005) Decreased histone deacetylase activity in chronic obstructive pulmonary disease. N Engl J Med 352:1967–1976

Ito K, Yamamura S, Essilfie-Quaye S, Cosio B, Ito M, Barnes PJ, Adcock IM (2006) Histone deacetylase 2-mediated deacetylation of the glucocorticoid receptor enables NF-kappaB suppression. J Exp Med 203:7–13

Ito K, Charron CE, Adcock IM (2007) Impact of protein acetylation in inflammatory lung diseases. Pharmacol Ther 116:249–265

Johnstone RW (2002) Histone-deacetylase inhibitors: novel drugs for the treatment of cancer. Nat Rev Drug Discov 1:287–299

Kagoshima M, Wilcke T, Ito K, Tsaprouni L, Barnes PJ, Punchard N, Adcock IM (2001) Glucocorticoid-mediated transrepression is regulated by histone acetylation and DNA methylation. Eur J Pharmacol 429:327–334

Kamei Y, Xu L, Heinzel T, Torchia J, Kurokawa R, Gloss B, Lin SC, Heyman RA, Rose DW, Glass CK, Rosenfeld MG (1996) A CBP integrator complex mediates transcriptional activation and AP-1 inhibition by nuclear receptors. Cell 85:403–414

Kawahara TL, Michishita E, Adler AS, Damian M, Berber E, Lin M, McCord RA, Ongaigui KC, Boxer LD, Chang HY, Chua KF (2009) SIRT6 links histone H3 lysine 9 deacetylation to NF-kappaB-dependent gene expression and organismal life span. Cell 136:62–74

Kim DH, Nelson HH, Wiencke JK, Zheng S, Christiani DC, Wain JC, Mark EJ, Kelsey KT (2001) p16(INK4a) and histology-specific methylation of CpG islands by exposure to tobacco smoke in non-small cell lung cancer. Cancer Res 61:3419–3424

Kishimoto M, Kohno T, Okudela K, Otsuka A, Sasaki H, Tanabe C, Sakiyama T, Hirama C, Kitabayashi I, Minna JD, Takenoshita S, Yokota J (2005) Mutations and deletions of the CBP gene in human lung cancer. Clin Cancer Res 11:512–519

Klose RJ, Kallin EM, Zhang Y (2006a) JmjC-domain-containing proteins and histone demethylation. Nat Rev Genet 7:715–727

Klose RJ, Yamane K, Bae Y, Zhang D, Erdjument-Bromage H, Tempst P, Wong J, Zhang Y (2006b) The transcriptional repressor JHDM3A demethylates trimethyl histone H3 lysine 9 and lysine 36. Nature 442:312–316

Kouzarides T (1999) Histone acetylases and deacetylases in cell proliferation. Curr Opin Genet Dev 9:40–48

Krieg AJ, Rankin EB, Chan D, Razorenova O, Fernandez S, Giaccia AJ (2010) Regulation of the histone demethylase JMJD1A by hypoxia-inducible factor 1 alpha enhances hypoxic gene expression and tumor growth. Mol Cell Biol 30:344–353

Kuo MH, Allis CD (1998) Roles of histone acetyltransferases and deacetylases in gene regulation. Bioessays 20:615–626

Lachner M, Jenuwein T (2002) The many faces of histone lysine methylation. Curr Opin Cell Biol 14:286–298

Lee KY, Ito K, Hayashi R, Jazrawi EP, Barnes PJ, Adcock IM (2006) NF-kappaB and activator protein 1 response elements and the role of histone modifications in IL-1beta-induced TGF-beta1 gene transcription. J Immunol 176:603–615

Li Y, Du H, Qin Y, Roberts J, Cummings OW, Yan C (2007) Activation of the signal transducers and activators of the transcription 3 pathway in alveolar epithelial cells induces inflammation and adenocarcinomas in mouse lung. Cancer Res 67:8494–8503

Lim S, Metzger E, Schule R, Kirfel J, Buettner R (2010) Epigenetic regulation of cancer growth by histone demethylases. Int J Cancer 127:1991–1998

Liu H, Zhou Y, Boggs SE, Belinsky SA, Liu J (2007) Cigarette smoke induces demethylation of prometastatic oncogene synuclein-gamma in lung cancer cells by downregulation of DNMT3B. Oncogene 26:5900–5910
Liu F, Killian JK, Yang M, Walker RL, Hong JA, Zhang M, Davis S, Zhang Y, Hussain M, Xi S, Rao M, Meltzer PA, Schrump DS (2010) Epigenomic alterations and gene expression profiles in respiratory epithelia exposed to cigarette smoke condensate. Oncogene 29:3650–3664
Londhe VA, Sundar IK, Lopez B, Maisonet TM, Yu Y, Aghai ZH, Rahman I (2011) Hyperoxia impairs alveolar formation and induces senescence through decreased histone deacetylase activity and up-regulation of p21 in neonatal mouse lung. Pediatr Res 69:371–377
Luo J, Li M, Tang Y, Laszkowska M, Roeder RG, Gu W (2004) Acetylation of p53 augments its site-specific DNA binding both in vitro and in vivo. Proc Natl Acad Sci USA 101:2259–2264
Mao Z, Hine C, Tian X, van Meter M, Au M, Vaidya A, Seluanov A, Gorbunova V (2011) SIRT6 promotes DNA repair under stress by activating PARP1. Science 332:1443–1446
Martino D, Prescott S (2011) Epigenetics and prenatal influences on asthma and allergic airways disease. Chest 139:640–647
Marwick JA, Kirkham PA, Stevenson CS, Danahay H, Giddings J, Butler K, Donaldson K, Macnee W, Rahman I (2004) Cigarette smoke alters chromatin remodeling and induces proinflammatory genes in rat lungs. Am J Respir Cell Mol Biol 31:633–642
Marwick JA, Caramori G, Stevenson CS, Casolari P, Jazrawi E, Barnes PJ, Ito K, Adcock IM, Kirkham PA, Papi A (2009) Inhibition of PI3Kdelta restores glucocorticoid function in smoking-induced airway inflammation in mice. Am J Respir Crit Care Med 179:542–548
Mattagajasingh I, Kim CS, Naqvi A, Yamamori T, Hoffman TA, Jung SB, Dericco J, Kasuno K, Irani K (2007) SIRT1 promotes endothelium-dependent vascular relaxation by activating endothelial nitric oxide synthase. Proc Natl Acad Sci USA 104:14855–14860
McMillan DH, Baglole CJ, Thatcher TH, Maggirwar S, Sime PJ, Phipps RP (2011) Lung-targeted overexpression of the NF-kappaB member RelB inhibits cigarette smoke-induced inflammation. Am J Pathol 179:125–133
Meja KK, Rajendrasozhan S, Adenuga D, Biswas SK, Sundar IK, Spooner G, Marwick JA, Chakravarty P, Fletcher D, Whittaker P, Megson IL, Kirkham PA, Rahman I (2008) Curcumin restores corticosteroid function in monocytes exposed to oxidants by maintaining HDAC2. Am J Respir Cell Mol Biol 39:312–323
Michan S, Sinclair D (2007) Sirtuins in mammals: insights into their biological function. Biochem J 404:1–13
Michaud SE, Dussault S, Haddad P, Groleau J, Rivard A (2006) Circulating endothelial progenitor cells from healthy smokers exhibit impaired functional activities. Atherosclerosis 187:423–432
Michishita E, McCord RA, Berber E, Kioi M, Padilla-Nash H, Damian M, Cheung P, Kusumoto R, Kawahara TL, Barrett JC, Chang HY, Bohr VA, Ried T, Gozani O, Chua KF (2008) SIRT6 is a histone H3 lysine 9 deacetylase that modulates telomeric chromatin. Nature 452:492–496
Milne JC, Denu JM (2008) The Sirtuin family: therapeutic targets to treat diseases of aging. Curr Opin Chem Biol 12:11–17
Milne JC, Lambert PD, Schenk S, Carney DP, Smith JJ, Gagne DJ, Jin L, Boss O, Perni RB, Vu CB, Bemis JE, Xie R, Disch JS, Ng PY, Nunes JJ, Lynch AV, Yang H, Galonek H, Israelian K, Choy W, Iffland A, Lavu S, Medvedik O, Sinclair DA, Olefsky JM, Jirousek MR, Elliott PJ, Westphal CH (2007) Small molecule activators of SIRT1 as therapeutics for the treatment of type 2 diabetes. Nature 450:712–716
Monick MM, Beach SR, Plume J, Sears R, Gerrard M, Brody GH, Philibert RA (2012) Coordinated changes in AHRR methylation in lymphoblasts and pulmonary macrophages from smokers. Am J Med Genet B Neuropsychiatr Genet 159B:141–151
Moodie FM, Marwick JA, Anderson CS, Szulakowski P, Biswas SK, Bauter MR, Kilty I, Rahman I (2004) Oxidative stress and cigarette smoke alter chromatin remodeling but differentially regulate NF-kappaB activation and proinflammatory cytokine release in alveolar epithelial cells. FASEB J 18:1897–1899
Motta MC, Divecha N, Lemieux M, Kamel C, Chen D, Gu W, Bultsma Y, McBurney M, Guarente L (2004) Mammalian SIRT1 represses forkhead transcription factors. Cell 116:551–563

Murphy SK, Adigun A, Huang Z, Overcash F, Wang F, Jirtle RL, Schildkraut JM, Murtha AP, Iversen ES, Hoyo C (2012) Gender-specific methylation differences in relation to prenatal exposure to cigarette smoke. Gene 494:36–43

Nakamaru Y, Vuppusetty C, Wada H, Milne JC, Ito M, Rossios C, Elliot M, Hogg J, Kharitonov S, Goto H, Bemis JE, Elliott P, Barnes PJ, Ito K (2009) A protein deacetylase SIRT1 is a negative regulator of metalloproteinase-9. FASEB J 23:2810–2819

Nan X, Ng HH, Johnson CA, Laherty CD, Turner BM, Eisenman RN, Bird A (1998) Transcriptional repression by the methyl-CpG-binding protein MeCP2 involves a histone deacetylase complex. Nature 393:386–389

Ng HH, Zhang Y, Hendrich B, Johnson CA, Turner BM, Erdjument-Bromage H, Tempst P, Reinberg D, Bird A (1999) MBD2 is a transcriptional repressor belonging to the MeCP1 histone deacetylase complex. Nat Genet 23:58–61

Ota H, Akishita M, Eto M, Iijima K, Kaneki M, Ouchi Y (2007) Sirt1 modulates premature senescence-like phenotype in human endothelial cells. J Mol Cell Cardiol 43:571–579

Pantano C, Reynaert NL, van der Vliet A, Janssen-Heininger YM (2006) Redox-sensitive kinases of the nuclear factor-kappaB signaling pathway. Antioxid Redox Signal 8:1791–1806

Pinamonti S, Leis M, Barbieri A, Leoni D, Muzzoli M, Sostero S, Chicca MC, Carrieri A, Ravenna F, Fabbri LM, Ciaccia A (1998) Detection of xanthine oxidase activity products by EPR and HPLC in bronchoalveolar lavage fluid from patients with chronic obstructive pulmonary disease. Free Radic Biol Med 25:771–779

Potente M, Dimmeler S (2008a) Emerging roles of SIRT1 in vascular endothelial homeostasis. Cell Cycle 7:2117–2122

Potente M, Dimmeler S (2008b) NO targets SIRT1: a novel signaling network in endothelial senescence. Arterioscler Thromb Vasc Biol 28:1577–1579

Purcell DJ, Jeong KW, Bittencourt D, Gerke DS, Stallcup MR (2011) A distinct mechanism for coactivator versus corepressor function by histone methyltransferase G9a in transcriptional regulation. J Biol Chem 286:41963–41971

Qiu W, Baccarelli A, Carey VJ, Boutaoui N, Bacherman H, Klanderman B, Rennard S, Agusti A, Anderson W, Lomas DA, Demeo DL (2012) Variable DNA methylation is associated with chronic obstructive pulmonary disease and lung function. Am J Respir Crit Care Med 185: 373–381

Qu P, Roberts J, Li Y, Albrecht M, Cummings OW, Eble JN, Du H, Yan C (2009) Stat3 downstream genes serve as biomarkers in human lung carcinomas and chronic obstructive pulmonary disease. Lung Cancer 63:341–347

Racey LA, Byvoet P (1971) Histone acetyltransferase in chromatin. Evidence for in vitro enzymatic transfer of acetate from acetyl-coenzyme A to histones. Exp Cell Res 64:366–370

Rahman I (2005) The role of oxidative stress in the pathogenesis of COPD: implications for therapy. Treat Respir Med 4:175–200

Rahman I, Adcock IM (2006) Oxidative stress and redox regulation of lung inflammation in COPD. Eur Respir J 28:219–242

Rahman I, Macnee W (1996) Oxidant/antioxidant imbalance in smokers and chronic obstructive pulmonary disease. Thorax 51:348–350

Rahman I, Macnee W (1998) Role of transcription factors in inflammatory lung diseases. Thorax 53:601–612

Rahman I, Gilmour PS, Jimenez LA, Macnee W (2002a) Oxidative stress and TNF-alpha induce histone acetylation and NF-kappaB/AP-1 activation in alveolar epithelial cells: potential mechanism in gene transcription in lung inflammation. Mol Cell Biochem 234–235:239–248

Rahman I, van Schadewijk AA, Crowther AJ, Hiemstra PS, Stolk J, Macnee W, de Boer WI (2002b) 4-Hydroxy-2-nonenal, a specific lipid peroxidation product, is elevated in lungs of patients with chronic obstructive pulmonary disease. Am J Respir Crit Care Med 166:490–495

Rahman I, Marwick J, Kirkham P (2004) Redox modulation of chromatin remodeling: impact on histone acetylation and deacetylation, NF-kappaB and pro-inflammatory gene expression. Biochem Pharmacol 68:1255–1267

Rahman I, Biswas SK, Jimenez LA, Torres M, Forman HJ (2005) Glutathione, stress responses, and redox signaling in lung inflammation. Antioxid Redox Signal 7:42–59
Rajendrasozhan S, Yang SR, Caito S, Rahman I (2008a) Nucleocytoplasmic Shuttling and Post-Translational Modifications of Sirtuin in Response to Cigarette Smoke Lead to Increased Acetylation of NF-kappaB and FOXO3. Am J Respir Crit Care Med 177:A266
Rajendrasozhan S, Yang SR, Kinnula VL, Rahman I (2008b) SIRT1, an antiinflammatory and antiaging protein, is decreased in lungs of patients with chronic obstructive pulmonary disease. Am J Respir Crit Care Med 177:861–870
Rajendrasozhan S, Yao H, Rahman I (2009) Current perspectives on role of chromatin modifications and deacetylases in lung inflammation in COPD. COPD 6:291–297
Rajendrasozhan S, Chung S, Sundar IK, Yao H, Rahman I (2010a) Targeted disruption of NF-{kappa}B1 (p50) augments cigarette smoke-induced lung inflammation and emphysema in mice: a critical role of p50 in chromatin remodeling. Am J Physiol Lung Cell Mol Physiol 298:L197–L209
Rajendrasozhan S, Hwang JW, Yao H, Kishore N, Rahman I (2010b) Anti-inflammatory effect of a selective IkappaB kinase-beta inhibitor in rat lung in response to LPS and cigarette smoke. Pulm Pharmacol Ther 23:172–181
Repine JE, Bast A, Lankhorst I (1997) Oxidative stress in chronic obstructive pulmonary disease. Oxidative Stress Study Group. Am J Respir Crit Care Med 156:341–357
Roth SY, Denu JM, Allis CD (2001) Histone acetyltransferases. Annu Rev Biochem 70:81–120
Saccani S, Natoli G (2002) Dynamic changes in histone H3 Lys 9 methylation occurring at tightly regulated inducible inflammatory genes. Genes Dev 16:2219–2224
Sanchez-Cespedes M, Decker PA, Doffek KM, Esteller M, Westra WH, Alawi EA, Herman JG, Demeure MJ, Sidransky D, Ahrendt SA (2001) Increased loss of chromosome 9p21 but not p16 inactivation in primary non-small cell lung cancer from smokers. Cancer Res 61:2092–2096
Seligson DB, Horvath S, McBrian MA, Mah V, Yu H, Tze S, Wang Q, Chia D, Goodglick L, Kurdistani SK (2009) Global levels of histone modifications predict prognosis in different cancers. Am J Pathol 174:1619–1628
Senftleben U, Cao Y, Xiao G, Greten FR, Krahn G, Bonizzi G, Chen Y, Hu Y, Fong A, Sun SC, Karin M (2001) Activation by IKKalpha of a second, evolutionary conserved, NF-kappa B signaling pathway. Science 293:1495–1499
Sengupta N, Seto E (2004) Regulation of histone deacetylase activities. J Cell Biochem 93:57–67
Shi Y, Lan F, Matson C, Mulligan P, Whetstine JR, Cole PA, Casero RA (2004) Histone demethylation mediated by the nuclear amine oxidase homolog LSD1. Cell 119:941–953
Sizemore N, Lerner N, Dombrowski N, Sakurai H, Stark GR (2002) Distinct roles of the Ikappa B kinase alpha and beta subunits in liberating nuclear factor kappa B (NF-kappa B) from Ikappa B and in phosphorylating the p65 subunit of NF-kappa B. J Biol Chem 277:3863–3869
Sood A, Petersen H, Blanchette CM, Meek P, Picchi MA, Belinsky SA, Tesfaigzi Y (2010) Wood smoke exposure and gene promoter methylation are associated with increased risk for COPD in smokers. Am J Respir Crit Care Med 182:1098–1104
Sood A, Petersen H, Blanchette CM, Meek P, Picchi MA, Belinsky SA, Tesfaigzi Y (2012) Methylated genes in sputum among older smokers with asthma. Chest 142(2):425–431
Stathopoulos GT, Sherrill TP, Cheng DS, Scoggins RM, Han W, Polosukhin VV, Connelly L, Yull FE, Fingleton B, Blackwell TS (2007) Epithelial NF-kappaB activation promotes urethane-induced lung carcinogenesis. Proc Natl Acad Sci USA 104:18514–18519
Sundar IK, Chung S, Hwang JW, Lapek JD Jr, Bulger M, Friedman AE, Yao H, Davie JR, Rahman I (2012) Mitogen- and stress-activated kinase 1 (MSK1) regulates cigarette smoke-induced histone modifications on NF-kappaB-dependent genes. PLOS One 7:e31378
Sundar IK, Yao H, Rahman I (2012) Oxidative stress and chromatin remodeling in chronic obstructive pulmonary disease and smoking-related diseases. Antioxid Redox Signal. doi:10.1089/ars.2012.4863
Suzuki MM, Bird A (2008) DNA methylation landscapes: provocative insights from epigenomics. Nat Rev Genet 9:465–476

Suzuki M, Wada H, Yoshino M, Tian L, Shigematsu H, Suzuki H, Alaa M, Tamura H, Fujiwara T, Nagato K, Motohashi S, Moriya Y, Hoshino H, Yoshida S, Shibuya K, Hiroshima K, Nakatani Y, Yoshino I (2010) Molecular characterization of chronic obstructive pulmonary disease-related non-small cell lung cancer through aberrant methylation and alterations of EGFR signaling. Ann Surg Oncol 17:878–888

Szulakowski P, Crowther AJ, Jimenez LA, Donaldson K, Mayer R, Leonard TB, Macnee W, Drost EM (2006) The effect of smoking on the transcriptional regulation of lung inflammation in patients with chronic obstructive pulmonary disease. Am J Respir Crit Care Med 174:41–50

Takahashi H, Ogata H, Nishigaki R, Broide DH, Karin M (2010) Tobacco smoke promotes lung tumorigenesis by triggering IKKbeta- and JNK1-dependent inflammation. Cancer Cell 17: 89–97

Takeshita M, Koga T, Takayama K, Kouso H, Nishimura-Ikeda Y, Yoshino I, Maehara Y, Nakanishi Y, Sueishi K (2008) CHFR expression is preferentially impaired in smoking-related squamous cell carcinoma of the lung, and the diminished expression significantly harms outcomes. Int J Cancer 123:1623–1630

Tan S (2001) One HAT size fits all? Nat Struct Biol 8:8–10

Taplick J, Kurtev V, Lagger G, Seiser C (1998) Histone H4 acetylation during interleukin-2 stimulation of mouse T cells. FEBS Lett 436:349–352

Taplick J, Kurtev V, Kroboth K, Posch M, Lechner T, Seiser C (2001) Homo-oligomerisation and nuclear localisation of mouse histone deacetylase 1. J Mol Biol 308:27–38

Thannickal VJ, Fanburg BL (2000) Reactive oxygen species in cell signaling. Am J Physiol Lung Cell Mol Physiol 279:L1005–L1028

Thomson S, Mahadevan LC, Clayton AL (1999) MAP kinase-mediated signalling to nucleosomes and immediate-early gene induction. Semin Cell Dev Biol 10:205–214

Tian X, Fang J (2007) Current perspectives on histone demethylases. Acta Biochim Biophys Sin (Shanghai) 39:81–88

To Y, Ito K, Kizawa Y, Failla M, Ito M, Kusama T, Elliott WM, Hogg JC, Adcock IM, Barnes PJ (2010) Targeting phosphoinositide-3-kinase-delta with theophylline reverses corticosteroid insensitivity in chronic obstructive pulmonary disease. Am J Respir Crit Care Med 182: 897–904

Tomita K, Barnes PJ, Adcock IM (2003) The effect of oxidative stress on histone acetylation and IL-8 release. Biochem Biophys Res Commun 301:572–577

Torchia J, Glass C, Rosenfeld MG (1998) Co-activators and co-repressors in the integration of transcriptional responses. Curr Opin Cell Biol 10:373–383

Tsou JA, Hagen JA, Carpenter CL, Laird-Offringa IA (2002) DNA methylation analysis: a powerful new tool for lung cancer diagnosis. Oncogene 21:5450–5461

Uchida K, Shiraishi M, Naito Y, Torii Y, Nakamura Y, Osawa T (1999) Activation of stress signaling pathways by the end product of lipid peroxidation. 4-hydroxy-2-nonenal is a potential inducer of intracellular peroxide production. J Biol Chem 274:2234–2242

van Gool F, Galli M, Gueydan C, Kruys V, Prevot PP, Bedalov A, Mostoslavsky R, Alt FW, de Smedt T, Leo O (2009) Intracellular NAD levels regulate tumor necrosis factor protein synthesis in a sirtuin-dependent manner. Nat Med 15:206–210

Vaziri H, Dessain SK, Ng Eaton E, Imai SI, Frye RA, Pandita TK, Guarente L, Weinberg RA (2001) hSIR2(SIRT1) functions as an NAD-dependent p53 deacetylase. Cell 107:149–159

Wada H, Kagoshima M, Ito K, Barnes PJ, Adcock IM (2005) 5-Azacytidine suppresses RNA polymerase II recruitment to the SLPI gene. Biochem Biophys Res Commun 331:93–99

Wang H, Cao R, Xia L, Erdjument-Bromage H, Borchers C, Tempst P, Zhang Y (2001) Purification and functional characterization of a histone H3-lysine 4-specific methyltransferase. Mol Cell 8:1207–1217

Wang Z, Zang C, Rosenfeld JA, Schones DE, Barski A, Cuddapah S, Cui K, Roh TY, Peng W, Zhang MQ, Zhao K (2008) Combinatorial patterns of histone acetylations and methylations in the human genome. Nat Genet 40:897–903

Wolffe AP (1997) Transcriptional control. Sinful repression. Nature 387:16–17

Workman JL, Kingston RE (1998) Alteration of nucleosome structure as a mechanism of transcriptional regulation. Annu Rev Biochem 67:545–579

Wright JL, Churg A (2008) Short-term exposure to cigarette smoke induces endothelial dysfunction in small intrapulmonary arteries: analysis using guinea pig precision cut lung slices. J Appl Physiol 104:1462–1469

Wyatt TA, Heires AJ, Sanderson SD, Floreani AA (1999) Protein kinase C activation is required for cigarette smoke-enhanced C5a-mediated release of interleukin-8 in human bronchial epithelial cells. Am J Respir Cell Mol Biol 21:283–288

Xiao G, Fong A, Sun SC (2004) Induction of p100 processing by NF-kappaB-inducing kinase involves docking IkappaB kinase alpha (IKKalpha) to p100 and IKKalpha-mediated phosphorylation. J Biol Chem 279:30099–30105

Yagi O, Aoshiba K, Nagai A (2006) Activation of nuclear factor-kappaB in airway epithelial cells in patients with chronic obstructive pulmonary disease. Respiration 73:610–616

Yamamoto Y, Verma UN, Prajapati S, Kwak YT, Gaynor RB (2003) Histone H3 phosphorylation by IKK-alpha is critical for cytokine-induced gene expression. Nature 423:655–659

Yang T, Sauve AA (2006) NAD metabolism and sirtuins: metabolic regulation of protein deacetylation in stress and toxicity. AAPS J 8:E632–E643

Yang IV, Schwartz DA (2011) Epigenetic control of gene expression in the lung. Am J Respir Crit Care Med 183:1295–1301

Yang F, Tang E, Guan K, Wang CY (2003) IKK beta plays an essential role in the phosphorylation of RelA/p65 on serine 536 induced by lipopolysaccharide. J Immunol 170:5630–5635

Yang SR, Chida AS, Bauter MR, Shafiq N, Seweryniak K, Maggirwar SB, Kilty I, Rahman I (2006) Cigarette smoke induces proinflammatory cytokine release by activation of NF-kappaB and posttranslational modifications of histone deacetylase in macrophages. Am J Physiol Lung Cell Mol Physiol 291:L46–L57

Yang SR, Wright J, Bauter M, Seweryniak K, Kode A, Rahman I (2007) Sirtuin regulates cigarette smoke-induced proinflammatory mediator release via RelA/p65 NF-kappaB in macrophages in vitro and in rat lungs in vivo: implications for chronic inflammation and aging. Am J Physiol Lung Cell Mol Physiol 292:L567–L576

Yang SR, Valvo S, Yao H, Kode A, Rajendrasozhan S, Edirisinghe I, Caito S, Adenuga D, Henry R, Fromm G, Maggirwar S, Li JD, Bulger M, Rahman I (2008) IKK alpha causes chromatin modification on pro-inflammatory genes by cigarette smoke in mouse lung. Am J Respir Cell Mol Biol 38:689–698

Yang SR, Yao H, Rajendrasozhan S, Chung S, Edirisinghe I, Valvo S, Fromm G, McCabe MJ Jr, Sime PJ, Phipps RP, Li JD, Bulger M, Rahman I (2009) RelB is differentially regulated by IkappaB Kinase-alpha in B cells and mouse lung by cigarette smoke. Am J Respir Cell Mol Biol 40:147–158

Yang J, Jubb AM, Pike L, Buffa FM, Turley H, Baban D, Leek R, Gatter KC, Ragoussis J, Harris AL (2010) The histone demethylase JMJD2B is regulated by estrogen receptor alpha and hypoxia, and is a key mediator of estrogen induced growth. Cancer Res 70:6456–6466

Yang W, Cui S, Ma J, Lu Q, Kong C, Liu T, Sun Z (2012) Cigarette smoking extract causes hypermethylation and inactivation of WWOX gene in T-24 human bladder cancer cells. Neoplasma 59:216–223

Yao H, Rahman I (2011) Current concepts on oxidative/carbonyl stress, inflammation and epigenetics in pathogenesis of chronic obstructive pulmonary disease. Toxicol Appl Pharmacol 254:72–85

Yao H, Edirisinghe I, Rajendrasozhan S, Yang SR, Caito S, Adenuga D, Rahman I (2008a) Cigarette smoke-mediated inflammatory and oxidative responses are strain dependent in mice. Am J Physiol Lung Cell Mol Physiol 294:L1174–L1186

Yao H, Edirisinghe I, Yang SR, Rajendrasozhan S, Kode A, Caito S, Adenuga D, Rahman I (2008b) Genetic ablation of NADPH oxidase enhances susceptibility to cigarette smoke-induced lung inflammation and emphysema in mice. Am J Pathol 172:1222–1237

Yao H, Arunachalam G, Hwang JW, Chung S, Sundar IK, Kinnula VL, Crapo JD, Rahman I (2010a) Extracellular superoxide dismutase protects against pulmonary emphysema by attenuating oxidative fragmentation of ECM. Proc Natl Acad Sci USA 107:15571–15576

Yao H, Hwang JW, Moscat J, Diaz-Meco MT, Leitges M, Kishore N, Li X, Rahman I (2010b) Protein kinase C zeta mediates cigarette smoke/aldehyde- and lipopolysaccharide-induced lung inflammation and histone modifications. J Biol Chem 285:5405–5416

Yin L, Wu L, Wesche H, Arthur CD, White JM, Goeddel DV, Schreiber RD (2001) Defective lymphotoxin-beta receptor-induced NF-kappaB transcriptional activity in NIK-deficient mice. Science 291:2162–2165

You H, Mak TW (2005) Crosstalk between p53 and FOXO transcription factors. Cell Cycle 4: 37–38

Zhang Y, Reinberg D (2001) Transcription regulation by histone methylation: interplay between different covalent modifications of the core histone tails. Genes Dev 15:2343–2360

Zhang X, Yang Z, Khan SI, Horton JR, Tamaru H, Selker EU, Cheng X (2003) Structural basis for the product specificity of histone lysine methyltransferases. Mol Cell 12:177–185

Zhou X, Sun H, Chen H, Zavadil J, Kluz T, Arita A, Costa M (2010) Hypoxia induces trimethylated H3 lysine 4 by inhibition of JARID1A demethylase. Cancer Res 70:4214–4221

Zochbauer-Muller S, Fong KM, Virmani AK, Geradts J, Gazdar AF, Minna JD (2001) Aberrant promoter methylation of multiple genes in non-small cell lung cancers. Cancer Res 61:249–255

Chapter 12
Redox Regulation of Stem Cell Function

Heinrich Jasper and Dirk Bohmann

Abstract Stem cells are critical for tissue regeneration and homeostasis, and their regenerative potential has to be carefully managed to prevent degenerative diseases and cancer. In recent years, it has emerged that stem cells are uniquely sensitive to changes in intracellular reactive oxygen species (ROS) (Hochmuth et al., Cell Stem Cell 8:1–12, 2011a; Ito et al., Nature 431:997–1002, 2004; Owusu-Ansah and Banerjee, Nature 461:537–541, 2009; Smith et al., Proc Natl Acad Sci USA 97:10032–10037, 2000; Tothova et al., Cell 128:325–339, 2007; Tsatmali et al., Brain Res 1040:137–150, 2005). Dynamic regulation of intracellular ROS seems to be critical to control the regenerative potential and to avoid loss of self-renewal capabilities and pluripotency of stem cells. Here, we review these findings, highlighting recent observations in genetic model organisms that clarify the role of ROS in stem cell regulation.

12.1 Introduction

Two major stem cell types can be found in most metazoans: germline stem cells (GSCs) and somatic stem cells (SSCs). GSCs are the precursors for egg and sperm cells, which, after fertilization, can give rise to all cell types that constitute the organism. They are therefore truly totipotent. SSCs, on the other hand, are more restricted in their developmental potential, i.e. they are normally

H. Jasper (✉)
Biology Department, University of Rochester, Rochester, NY 14627, USA

The Buck Institute for Research on Aging, Novato, CA 94945, USA
e-mail: HJasper@buckinstitute.org

D. Bohmann (✉)
Department of Biomedical Genetics, University of Rochester Medical Center, Rochester, NY 14642, USA
e-mail: Dirk_Bohmann@urmc.rochester.edu

U. Jakob and D. Reichmann (eds.), *Oxidative Stress and Redox Regulation*,
DOI 10.1007/978-94-007-5787-5_12,

not totipotent, but rather pluripotent with a limited set of cell types that can arise from the differentiation of SSCs. Accordingly, multiple distinct SSC types, such as hematopoetic, neuronal or epithelial stem cells have been identified. SSCs serve as a reservoir to replenish somatic cells that have been lost or rendered dysfunctional, and thus contribute critically to tissue homeostasis.

SCs are characterized by the ability to undergo asymmetric cell division, whereby they give rise to one daughter cell that recapitulates the stem cell characteristic of the mother cell (self-renewal), and one daughter that initiates a process towards terminal differentiation. This latter cell may already express markers and phenotypes of a terminally differentiated cell, or represent an intermediary, mitotically active, but specified state. It is currently being debated, if SC division is invariably asymmetric, or whether the stem cell pool in a tissue can be maintained through a process called "population asymmetry", where symmetric and asymmetric divisions are in balance to ensure proper self renewal and differentiation rates (Biteau et al. 2011; Simons and Clevers 2011; Snippert et al. 2010; Tian et al. 2011).

A thorough understanding of SC molecular and cell biology is a paramount priority for biomedical research. The integrity of adult stem cell reservoirs is critical for health and homeostasis, and preserving this integrity requires a well-regulated balance between stem cell self renewal and differentiation. Declining stem cell numbers and function cause loss of regenerative potential and lead to degenerative diseases. Conversely, unchecked stem cell amplification can disrupt stem cell homeostasis and cause hyperplasia and/or aberrant differentiation, a condition that may favor the initiation and progression of malignant diseases and cancer (Biteau et al. 2011; Radtke and Clevers 2005; Radtke et al. 2006; Rando 2006).

Self-renewal and multipotency set stem cells apart from the "mortal" differentiated cells of the soma. Reflecting these special properties, stem cells have unique molecular and signaling characteristics that distinguish them from their fellow cells of the organism. One such unique aspect of stem cell physiology and molecular biology is the relationship of stem cells to oxidative stress and redox homeostasis. This chapter will review stem cell specific functions of redox signaling and stress response that have emerged as general features shared among many types of stem cells, including adult as well as germ line stem cells. To illustrate the research that led up to the current understanding of redox signals in the regulation of stem cell differentiation and maintenance, we will describe studies in specific models of stem cell biology in greater detail. We will concentrate this discussion on research that was conducted in *Drosophila melanogaster*, which is emerging as a powerful, genetically tractable model system for the study of germline and adult stem cell biology (Biteau et al. 2011; Jasper and Jones 2010; Jiang and Edgar 2011; Wang and Jones 2010). For readers interested in research related to redox regulation of vertebrate stem cells, we refer to an excellent recent review by T. Finkel and colleagues (Liu et al. 2011).

12.2 REDOX and Stem Cells: A Special Relationship

Several observations highlight the special role of redox regulation in stem cells.

1. *Stress resistance.* Undifferentiated, pluripotent stem cells and multipotent progenitor cells are generally resilient against oxidative damage.
2. *Reducing conditions favor stemness.* Research on a number of specific stem cell systems has revealed that a reducing intracellular environment characterizes the undifferentiated state of stem and progenitor cells. Conversely, an increasing oxidized state correlates with, and in some cases has been shown to be required for, differentiation and loss of stemness, the state in which stem cells retain the potential to self renew and give rise to differentiated descendants.
3. *Stem cell specific oxidative stress responses.* Specific signaling pathways permit eukaryotic cells to sense and respond to increased oxidative stress caused either by changes in metabolism or by extracellular conditions. Stem cells also employ these signaling pathways, but, interestingly, they differ from many other cell types in the way they respond to them. Such stem cell specific signal responses may explain some of the unique biological properties of stem cells.

Below we will elaborate on these three points:

12.2.1 Stress Resistance

Germ line cells have unlimited replicative potential and therefore are, at least in principle, immortal. Adult stem cells are also long-lived. Many adult stem cells divide rarely or remain quiescent for months or years. Over such protracted periods of time the genetic and physical integrity of stem cells has to be safeguarded. A major threat to any cell, including stem cells, are reactive oxygen species (ROS), which can emerge as byproducts of mitochondrial respiration or originate from external sources, for example due to exposure to xenobiotic insults or as a consequence of microbial infection. ROS can damage proteins, lipids and nucleic acids. Importantly, they can cause DNA damage, which, if not repaired, can result in mutations. Loss or genetic damage of stem cells would have fatal consequences for the organism as damage to stem cells might comprise their regenerative and developmental potential or might cause the descendants of these cells to inherit deleterious mutations. It appears, therefore, that stem cells should be more "precious" to the organism than expendable somatic cells. Consistent with this notion, they are generally especially well protected. Stem cells frequently reside in special "niches" (Jones and Wagers 2008; Voog and Jones 2010; Walker et al. 2009). Such stem cell niches not only provide a reservoir for regeneration and release the cues that control stem cell identity and function, they are also a protective environment. Niches such as those for hematopoetic stem cells in the bone marrow are located in areas of low oxygen tension, reducing the probability of oxidative damage (Jang and Sharkis 2007;

Parmar et al. 2007). In addition to their residence in such sheltered niches, stem cells are themselves well equipped to handle extraneous ROS that might cause damage. An effective mechanism to avoid oxidative damage to vital cell components is to keep the intracellular milieu reducing. Indeed many stem and progenitor cells have been found to have more reducing intracellular conditions than the differentiated cells that they generate (Hochmuth et al. 2011b; Noble et al. 2003; Smith et al. 2000).

12.2.2 Reducing Milieu = Stemness

A reducing intracellular environment that was introduced in the preceding paragraph as a characteristic of stem cells that may confer oxidative stress resistance has also been suggested as a cue for regulating stem cell differentiation. Multiple examples, some of which are discussed in more detail below, have shown that the transition from a reduced to an oxidized state occurs during stem cell differentiation, the departure from the state of stemness. In many cases, this change of redox state is not only a consequence but a condition for stem cell differentiation (see schematic in Fig. 12.1). It is now a broadly accepted view that regulated changes of redox state are central parts of the mechanisms that control stem cell maintenance and differentiation (Hochmuth et al. 2011b; Ito et al. 2004; Liu et al. 2011; Owusu-Ansah and Banerjee 2009; Ray et al. 2012; Smith et al. 2000; Tothova and Gilliland 2007; Tsatmali et al. 2005). The molecular mechanisms that control stem cell fate via changes in redox state are intensively investigated in the stem cell field.

12.2.3 Stem Cell Specific Signal Responses

Stem cell are distinct from other cells in terms of their developmental potential and their unlimited self-renewal. They are different in other respects too. One example is their response to stress signaling pathways. Multiple conserved signaling systems function to inform eukaryotic cells of changes in oxidative load. Well-known components of these signaling pathways include protein kinases such as JNK, p38 and ATM (Bhatti et al. 2011; Biteau et al. 2011; Cuadrado and Nebreda 2010; Weston and Davis 2007) as well as transcription factors like Nrf2, p53 and Fos (Biteau et al. 2010; Borras et al. 2011; Maher and Yamamoto 2010).

Interestingly, cells differ in the biological responses that they mount upon signaling through such pathways. When a typical body cell is exposed to acute oxidative stress, it will arrest the cell cycle attempting to limit damage and avoiding damage to newly replicated DNA. Stem cells are different. They often respond to oxidative stress by leaving quiescence and embarking on an expansion and

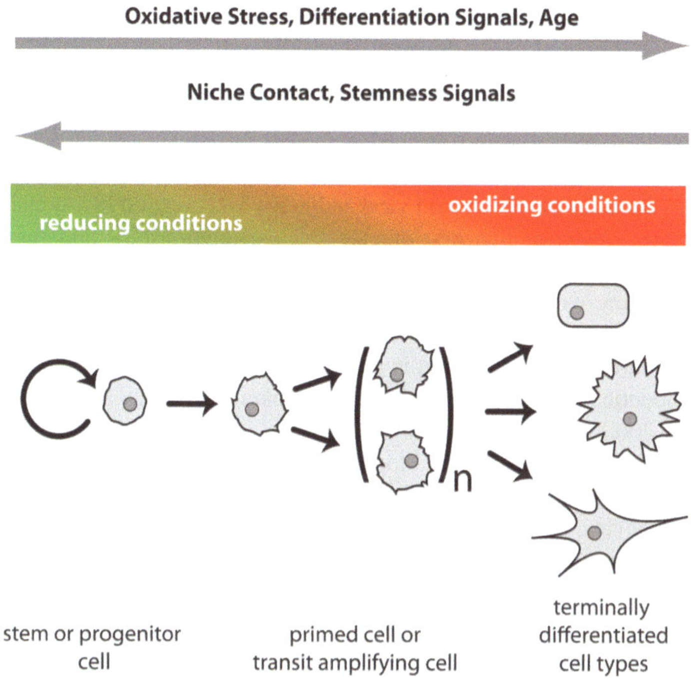

Fig. 12.1 Cartoon depicting the correlation between redox state and stem cell function. Undifferentiated, pluripotent or multipotent stem and progenitor cells keep their intracellular milieu reducing. In this state they are less vulnerable to oxidative damage, and stay quiescent or divide very infrequently. Regenerative responses to stress or tissue injury invoke increased stem cell differentiation and/or expansion. The transition from a state of stemness to terminal differentiation can involve intermediary steps such cell types that are primed for differentiation or transit amplifying cells that expand the population of differentiated descents of a limited number of stem cells. The progression along the path from a stem cell to a terminally differentiated cell coincides with shift in redox balance from a reduced to an oxidized state

differentiation program (Abou-Khalil and Brack 2010; Biteau et al. 2008, 2011; Hochmuth et al. 2011a; Tothova et al. 2007). This response can be viewed as the onset of a regenerative response. The wiring and regulatory logic of a stem cell appears in many aspects to be fundamentally different from a non-stem cell. A molecular and genetic description of such differences is now beginning to emerge. It is important to understand the unique rules by which stem cells interpret oxidative inputs and other signals from their environment.

Next, we will discuss several specific examples that have illustrated this concept and advanced our understanding of redox mechanisms that control in the maintenance and differentiation of stem and progenitor cells.

12.3 Redox Mechanisms in Stem Cells

12.3.1 EXAMPLE 1: Hematopoetic Stem Cell in Drosophila

The ancestors of red and white blood cells are arguably the best understood of all stem cells. Decades of research on viral and non-virally caused blood malignancies have yielded important insights into the lineage relationships and regulators of hematopoesis (Orkin and Zon 2008). Blood cell lineages are comparatively easy to monitor using well-established cytological criteria and markers. Hematopoetic stem cells are also the only stem cells that are currently in widespread therapeutic use. Bone marrow transplantation is an established treatment to replenish the loss of hematopoetic stem cells in myeloma and leukemia patients after chemotherapy (Burt et al. 2008; Gratwohl et al. 2010).

While *Drosophila* has no red blood cells, the fly's hemolymph contains three types of mature white blood cell types that resemble mammalian myeloid cells: the plasmatocytes, crystal cells and lamellocytes (Evans et al. 2003). These cells mediate innate immune responses to various types of infectious agents and parasites. All of these cells originate in the lymph gland, the fly's hematopoietic organ. The lymph gland forms during larval development from mesodermal cell lineages (Evans et al. 2003). As the organ matures, multipotent stem-like cells emerge that can differentiate into the three terminally differentiated immune cells mentioned above. The differentiation process can be stimulated by microbial infection or by parasite invasion. The generation of the immune cells is accompanied by a shift to an oxidized cell state that can be visualized by redox dyes (Owusu-Ansah and Banerjee 2009). This increase in redox levels is critical for myeloid differentiation. Experiments in which the intracellular milieu of the hematopoetic progenitor cells is forced towards a more reducing level, for example by overexpression of antioxidant scavenger proteins or catalase, prevent immune cell formation, proving that the oxidative shift is necessary for the response. Conversely, increasing the levels of ROS in the stem cells by mutations in electron transport chain components causes ectopic myeloid differentiation, even in the absence of immune challenges (Owusu-Ansah and Banerjee 2009). Thus, oxidative stimuli are sufficient to trigger stem cell differentiation. Interestingly, *Drosophila* hematopoesis follows a two-step redox mechanism to control differentiation. As the lymph gland matures, the myeloid progenitors become increasingly oxidized, which confers to them a "primed state" in which they are receptive to immune signals that push ROS levels across a threshold and stimulates the differentiation and release of mature myeloid cells.

Molecular analysis has shown that a key signaling pathway that triggers myeloid differentiation in response to increased ROS levels in the Drosophila hematopoetic lineage is comprised of the JNK MAP kinase cascade and its downstream target, the FoxO transcription factor (Owusu-Ansah and Banerjee 2009). Oxidative stress will increase JNK's enzymatic activity causing it to phosphorylate FoxO. FoxO, when phosphorylated by JNK, is transcriptionally competent and can activate a gene expression program that drives myeloid cell differentiation. Active JNK will also cause the repression of polycomb-like transcriptional repressors, and this regulation contributes to progenitor cell differentiation independently of Foxo.

The mechanism that causes stimulation of JNK activity after oxidative stress exposure is complex and still not completely worked out. A heterogeneous group of JNKKKs can connect JNK activity to a range of upstream signals and stresses. The JNKKK Ask1, for example can respond to changes in thioredoxin levels and/or oxidation state. Tak1 has been implicated in responses to inflammatory signals and microbial challenges. Which one of these, if any, are operative in the *Drosophila* hematopoietic system is still an open question. The activation of FoxO by oxidative stress and JNK mediated phosphorylation is not unique to stem cells and has been described as a ubiquitous mechanism in many cell types (Essers et al. 2004; Salih and Brunet 2008; Wang et al. 2005). Interestingly, however, the biological readout to FoxO activation seems to be markedly different in stem cells and non-stem cells. Rather than inducing differentiation, FoxO activation in the developing eye causes growth arrest and, eventually, apoptosis (Owusu-Ansah et al. 2008; Biteau et al. 2010). The comparison between the genetically well-described *Drosophila* lymph gland and eye systems might thus provide insights into the unique biology and the idiosyncrasies of stem cells.

The impact of redox changes on hematopoietic progenitor function is likely to be conserved in vertebrates: Interestingly, FoxO seems to play a critical role in managing reactive oxygen species in mouse HSCs. Loss of FoxO in these cells results in elevated ROS levels, triggering exit from quiescence, and in consequence depletion of the HSC pool and reduced long-term regenerative potential (Miyamoto et al. 2007; Miyamoto et al. 2008; Tothova and Gilliland 2007; Tothova et al. 2007). This exit from quiescence of HSCs is accompanied by increased myeloid differentiation, suggesting that Foxo acts in HSCs primarily to prevent excessive ROS accumulation, while Foxo-independent redox responses govern hematopoietic progenitor cell differentiation in mice. While the general response of stem and progenitor cells to increased ROS levels may thus be conserved between flies and mice, it is possible that the molecular regulation of these responses has diverged. The role of Foxo in vertebrate HSCs, however, needs to be further examined. It remains unclear, for example, whether activation of FoxO in mammalian HSCs or restricted progenitors is also sufficient to promote myeloid cell expansion. Interestingly, a recent study suggests that this might indeed be the case, as FoxO and/or JNK/Jun activity is found to be high in patient samples of acute myeloid leukemia (AML). In mouse models for AML, it was found that FoxO and JNK/Jun signaling maintain a differentiation blockade that is critical for the progression of AML. While FoxO proteins are expressed at the transition from HSCs to myeloid progenitors, it remains to be established precisely at which stage in the HSC to myeloid progenitor lineage FoxO activity responds to ROS or controls ROS levels to direct cell cycle arrest, cell expansion, or differentiation (Sykes et al. 2011; Tothova and Gilliland 2007).

12.3.2 EXAMPLE 2: Redox-Dependent Control of Germline Stem Cell Function

The *Drosophila* male and female germlines have served as productive model systems to characterize the molecular mechanisms governing the interaction between stem cells and their niches (reviews Fuller and Spradling 2007; Ohlstein et al. 2004; Spradling et al. 1997, 2001; Voog and Jones 2010). A particularly active field of study in this context is the characterization of age-related changes in GSC function, which are caused by both stem cell intrinsic and extrinsic factors, including oxidative stress (Cheng et al. 2008; Jasper and Jones 2010; Pan et al. 2007; Wang and Jones 2010). In the female germline, GSCs are attached to niche cells through adherens junctions, in which E-cadherin plays a major role. Proper niche attachment is critical for self-renewal, and prevents differentiation of GSCs. Interestingly, the E-cadherin-mediated interaction between GSCs and their niche declines with age, causing an age-related decrease of GSC numbers (Pan et al. 2007). This age-related decline can be rescued by over-expression of superoxide dismutase, indicating that the accumulation of ROS in GSCs plays a major role in this loss of function.

A role for reactive oxygen species in influencing progenitor cell attachment to other cells has recently been established in primordial germ cells (PGCs) of the fly. In these cells, the peroxiredoxin Jafrac1 plays a critical role in promoting E-cadherin protein stability and thus PGC attachment and motility (DeGennaro et al. 2011). The exact mechanism by which E-cadherin stabilization is achieved remains unclear. Nevertheless, the common role of reactive oxygen species in modulating stem and progenitor cell adhesiveness through this particular molecular interaction suggests that ROS-mediated changes in E-cadherin function play a critical and general role in stem cell regulation.

12.3.3 EXAMPLE 3: Intestinal Stem Cells

A particularly illustrative example for the intersection between regulatory, physiological and deleterious functions of redox signaling is observed in intestinal stem cells (ISCs) of Drosophila. ISCs are critical for tissue regeneration along the gastrointestinal tract of flies, and their regulation and function have been subject of intense investigation in recent years. Best understood so far are ISCs of the posterior midgut epithelium. These cells regenerate the two functional cell types of the posterior midgut, enterocytes (ECs) and enteroendocrine cells (EEs) (Micchelli and Perrimon 2006; Ohlstein and Spradling 2006). ISCs commonly divide asymmetrically, producing a new stem cell and an undifferentiated progenitor cell, an enteroblast (EB). ISCs are the only proliferating cells in the intestinal epithelium and direct EB differentiation by activating the Notch signaling pathway in these cells.

ISC proliferation is regulated dynamically according to the needs of the intestinal epithelium. In conditions in which the intestine is healthy and not exposed to excessive stress, most ISCs are found in a quiescent state, as characterized by the fact that few of these cells incorporate BrdU in a window of 24 h (Hochmuth et al. 2011a). When challenged by a variety of stressors (including infection, cell death, tissue damage, DNA damage, and oxidative stress), ISCs embark on a regenerative program that is characterized by a high rate of proliferation in the intestine.

The signaling pathways responding to these challenges and inducing ISC proliferation include the Jun-N-terminal Kinase (JNK) and Jak/Stat signaling pathways. These inductive pathways control ISC proliferation in coordination with signals that are essential for proliferative competence of ISCs, including insulin, Epidermal Growth Factor Receptor (EGFR), and Wnt signaling (Amcheslavsky et al. 2009; Biteau et al. 2008; Biteau and Jasper 2011; Buchon et al. 2009; Choi et al. 2008; Lee et al. 2009; Lin et al. 2008). The proliferative response to JNK signaling is an illustration of the unique signaling characteristics of stem cells. The more common response to JNK activation in non stem cells is growth arrest or apoptosis (Liu and Lin 2005; McEwen and Peifer 2005; Weston and Davis 2007).

Stimulation of ISC proliferation represents the first step in a regenerative response that is required to recover intestinal integrity after a challenge, and as such can be viewed as a mechanism that should benefit the organism. However, this process needs to be tightly regulated to ensure intestinal homeostasis (Biteau et al. 2008; Jiang et al. 2009). Accordingly, the stress-induced increase in ISC proliferation rates declines rapidly after the stress subsides. When excessive proliferation is induced in the gut, either artificially, or through chronic of excessive damage, disruption of epithelial homeostasis ensues. This is caused by the accumulation of misdifferentiated ISC daughter cells (Amcheslavsky et al. 2009; Biteau et al. 2008; Choi et al. 2008).

Multiple recent studies have established that modulation of the redox balance in the intestinal epithelium significantly influences ISC proliferation rates and epithelial homeostasis. Increased proliferation is observed in response to treatment with the ROS-inducing compound paraquat (Biteau et al. 2008), as well as in mutants for the ROS detoxifying enzyme catalase (Choi et al. 2008), while treating flies with N-acetyl-cysteine (NAC) or glutathione is sufficient to restrain ISC proliferation (Buchon et al. 2009). Interestingly, loss of proliferative homeostasis in the intestine of aging flies limits Drosophila lifespan (Biteau et al. 2008, 2010; Hochmuth et al. 2011a). Excessive ISC proliferation in aging intestines seems to be caused by dramatically increased oxidative load due to a chronic inflammatory state. This state is triggered by age-related changes in the intestinal commensal flora that induce dual oxidase (DuOX)-mediated innate immune responses, generating large amounts of reactive oxygen species (Biteau et al. 2008; Buchon et al. 2009; Choi et al. 2008; Ha et al. 2005, 2009b; Hochmuth et al. 2011b).

Given the potential exposure of ISCs to an adverse oxidative environment, and the critical regenerative role of ISC in such an environment, it is not too surprising that proliferative activity of these cells is regulated by a well-known master regulator of the antioxidant response, the transcription factor Nrf2.

12.4 Nrf2: Stem Cell Control by the Master Regulator of the Antioxidant Response

Nrf2 is a central regulator of the intracellular redox state in vertebrates and invertebrates. This transcription factor is a member of the 'cap-and-collar' (Cnc) family that counteracts excessive ROS accumulation in cells by inducing genes encoding key antioxidant molecules, such as enzymes involved in glutathione (e.g., γ-glutamyl-cysteine-ligase catalytic subunit, Gclc) and thioredoxin metabolism (Hayes and McMahon 2009; Lee et al. 2005; McMahon et al. 2001; Motohashi et al. 2002), and antioxidant enzymes such as peroxiredoxins (Lee et al. 2005; Maher and Yamamoto 2010; Motohashi et al. 2002).

This function of Nrf2 significantly influences overall stress sensitivity and lifespan of several tested model organisms: gain-of-function conditions for the *Drosophila* homologue of Nrf2, CncC, result in significantly increased stress tolerance and longevity, and activation of the *C. elegans* homologue SKN1 has similar consequences in worms (Inoue et al. 2005; Sykiotis and Bohmann 2008; Tullet et al. 2008). Interestingly, gene expression analysis has suggested that SKN1 increases tissue homeostasis and extends lifespan by promoting a germcell – like environment in somatic cells (Curran et al. 2009).

In vertebrates and in *Drosophila*, Nrf2 is negatively regulated by the cytoplasmic repressor Kelch-like ECH-associated protein 1 (Keap1) (Hayes and McMahon 2009; Nguyen et al. 2009; Sykiotis and Bohmann 2008, 2010; Toledano 2009). Keap1 acts as an adaptor of a Cul3-ubiquitin ligase complex that promotes Nrf2 degradation. Accordingly, loss of Keap1 results in lifespan extension and increased oxidative stress tolerance in flies (Sykiotis and Bohmann 2008).

In somatic cells, oxidative stress results in disruption of the interaction between Keap1 and Nrf2, causing nuclear accumulation of Nrf2, and activation of gene expression. In contrast, recent studies in *Drosophila* suggest that the *Drosophila* homologue of Nrf2, CncC, is constitutively active in ISCs and that oxidative stress results in a Keap1-mediated inhibition of Nrf2 in these cells. Constitutive Nrf2 activity is required to maintain low ROS concentrations in ISCs. This activity is therefore required to prevent spontaneous activation of ISC proliferation, promoting epithelial homeostasis. Inhibition of Nrf2 activity, on the other hand, is critical to engage in regenerative responses, as artificially sustaining Nrf2 activity in ISCs exposed to stress inhibits such a response (Hochmuth et al. 2011b). These results suggest that constitutive Nrf2/CncC activity in ISCs prevents excessive ROS accumulation in response to challenges, setting a threshold for ROS levels at which a regenerative response is warranted. The Keap1/CncC regulatory module thus emerges as central in the control of ISC proliferation and intestinal regeneration.

The origin of redox regulation in the control of ISC proliferation remains unclear. It can be speculated, however, that repression of CncC activity by Keap1 is required to induce a permissive state in ISCs that ensures sufficient ROS accumulation in response to a mitogenic signal engaging receptor tyrosine kinases (RTKs). ROS accumulation is critical for signal transduction and signal stabilization in growth factor signaling pathways (Bashan et al. 2009), as RTK signal transduction requires transient inactivation of peroxiredoxins to promote local ROS accumulation at the membrane (Woo et al. 2010). The increased local ROS concentration is believed to cause oxidation and inactivation of redox-sensitive protein tyrosine phosphatases, thus strengthening the signal downstream of the receptor (Lambeth 2004; Toledano et al. 2010; Tonks 2006). Supporting this view, CncC over-expression inhibits insulin receptor-mediated proliferation of ISCs (Hochmuth et al. 2011b).

This model predicts that appropriate regulation of ROS production and detoxification is critical to ensure proper cellular responses to mitogenic and survival signals (Bashan et al. 2009). At the same time, this regulatory system would be sensitive to external oxidative challenges, promoting regenerative stress responses, but also promoting excessive proliferation under chronic stress conditions. This is particularly prevalent in barrier epithelia like the intestinal epithelium, where a vigorous oxidative burst response is employed as a defense against pathogenic bacteria (Ha et al. 2005, 2009a, b).

Interestingly, Keap1 mutant mice show significant hyperkeratosis of the esophageal epithelium, suggesting that Keap1/Nrf2 also influences cell differentiation and proliferation in intestinal epithelia of higher organisms (Wakabayashi et al. 2003). This function is reminiscent of the requirement for SKN1 in intestinal development in *C. elegans* (An and Blackwell 2003), highlighting a potential evolutionary conservation of this function, but the exact role of Nrf2 in the regulation of stem cell activity in vertebrates remains to be established.

12.5 Perspectives and Outlook, Therapeutics

The examples listed above highlight the critical role of redox signaling in the regulation of stem cell function. It emerges that regulated changes in redox state affect stem cell function at different levels in addition to simply providing a defense against ROS inflicted damage to this precious cell type. Redox changes also modulate the specificity of signal response and even the adhesive properties of stem cells and therefore their interaction with specific niches. The observation that stem and progenitor cells are controlled by regulatory networks that distinguish the response of these cells to redox signals from the response of differentiated cells has been unexpected. Stem cell-specific signaling and regulatory networks are therefore attractive targets for therapy, both in efforts to boost stem cell maintenance and self-renewal, as well as in efforts to limit the expansion of aberrant stem-like cells. Interestingly, a role for redox changes in the biology of cancer stem cells has emerged in recent years (Diehn et al. 2009; Sykes et al. 2011), supporting this idea. A particularly interesting target for such therapies is Nrf2, as it is a central regulator of the intracellular redox state and has been the target of numerous clinical trials with a wide range of pharmacological activators and inhibitors (Glintborg et al. 2006; Yu and Kensler 2005). For a productive approach to Nrf2-based therapies, it will be critical to clearly establish the role of Nrf2 in the regulation of mammalian stem cell lineages, following the conceptual insight generated by the *Drosophila* model.

Acknowledgements This work was supported by grants two joint grants to DB and HJ: N08G-048 from New York State stem cell science and R01 GM100196 from NIH/NIGMS.

References

Abou-Khalil R, Brack AS (2010) Muscle stem cells and reversible quiescence: the role of sprouty. Cell Cycle 9:2575–2580

Amcheslavsky A, Jiang J, Ip YT (2009) Tissue damage-induced intestinal stem cell division in Drosophila. Cell Stem Cell 4:49–61

An JH, Blackwell TK (2003) SKN-1 links C. elegans mesendodermal specification to a conserved oxidative stress response. Genes Dev 17:1882–1893

Bashan N, Kovsan J, Kachko I, Ovadia H, Rudich A (2009) Positive and negative regulation of insulin signaling by reactive oxygen and nitrogen species. Physiol Rev 89:27–71

Bhatti S, Kozlov S, Farooqi AA, Naqi A, Lavin M, Khanna KK (2011) ATM protein kinase: the linchpin of cellular defenses to stress. Cell Mol Life Sci 68:2977–3006

Biteau B, Jasper H (2011) EGF signaling regulates the proliferation of intestinal stem cells in Drosophila. Development 138:1045–1055

Biteau B, Hochmuth CE, Jasper H (2008) JNK activity in somatic stem cells causes loss of tissue homeostasis in the aging Drosophila gut. Cell Stem Cell 3:442–455

Biteau B, Karpac J, Hwangbo D, Jasper H (2010) Regulation of Drosophila lifespan by JNK signaling. Exp Gerontol 46:349–354

Biteau B, Hochmuth CE, Jasper H (2011) Maintaining tissue homeostasis: dynamic control of somatic stem cell activity. Cell Stem Cell 9:402–411

Borras C, Gomez-Cabrera MC, Vina J (2011) The dual role of p53: DNA protection and antioxidant. Free Radic Res 45:643–652

Buchon N, Broderick NA, Chakrabarti S, Lemaitre B (2009) Invasive and indigenous microbiota impact intestinal stem cell activity through multiple pathways in Drosophila. Genes Dev 23:2333–2344

Burt RK, Loh Y, Pearce W, Beohar N, Barr WG, Craig R, Wen Y, Rapp JA, Kessler J (2008) Clinical applications of blood-derived and marrow-derived stem cells for nonmalignant diseases. JAMA 299:925–936

Cheng J, Turkel N, Hemati N, Fuller MT, Hunt AJ, Yamashita YM (2008) Centrosome misorientation reduces stem cell division during ageing. Nature 456:599–604

Choi NH, Kim JG, Yang DJ, Kim YS, Yoo MA (2008) Age-related changes in Drosophila midgut are associated with PVF2, a PDGF/VEGF-like growth factor. Aging Cell 7:318–334

Cuadrado A, Nebreda AR (2010) Mechanisms and functions of p38 MAPK signalling. Biochem J 429:403–417

Curran SP, Wu X, Riedel CG, Ruvkun G (2009) A soma-to-germline transformation in long-lived Caenorhabditis elegans mutants. Nature 459:1079–1084

DeGennaro M, Hurd TR, Siekhaus DE, Biteau B, Jasper H, Lehmann R (2011) Peroxiredoxin stabilization of DE-Cadherin promotes primordial germ cell adhesion. Dev Cell 20(2):233–243

Diehn M, Cho RW, Lobo NA, Kalisky T, Dorie MJ, Kulp AN, Qian D, Lam JS, Ailles LE, Wong M et al (2009) Association of reactive oxygen species levels and radioresistance in cancer stem cells. Nature 458:780–783

Essers MA, Weijzen S, de Vries-Smits AM, Saarloos I, de Ruiter ND, Bos JL, Burgering BM (2004) FOXO transcription factor activation by oxidative stress mediated by the small GTPase Ral and JNK. EMBO J 23:4802–4812

Evans CJ, Hartenstein V, Banerjee U (2003) Thicker than blood: conserved mechanisms in Drosophila and vertebrate hematopoiesis. Dev Cell 5:673–690

Fuller MT, Spradling AC (2007) Male and female Drosophila germline stem cells: two versions of immortality. Science 316:402–404

Glintborg B, Weimann A, Kensler TW, Poulsen HE (2006) Oltipraz chemoprevention trial in Qidong, People's Republic of China: unaltered oxidative biomarkers. Free Radic Biol Med 41:1010–1014

Gratwohl A, Baldomero H, Aljurf M, Pasquini MC, Bouzas LF, Yoshimi A, Szer J, Lipton J, Schwendener A, Gratwohl M et al (2010) Hematopoietic stem cell transplantation: a global perspective. JAMA 303:1617–1624

Ha EM, Oh CT, Bae YS, Lee WJ (2005) A direct role for dual oxidase in Drosophila gut immunity. Science 310:847–850

Ha EM, Lee KA, Park SH, Kim SH, Nam HJ, Lee HY, Kang D, Lee WJ (2009a) Regulation of DUOX by the Galphaq-phospholipase Cbeta-Ca2+ pathway in Drosophila gut immunity. Dev Cell 16:386–397

Ha EM, Lee KA, Seo YY, Kim SH, Lim JH, Oh BH, Kim J, Lee WJ (2009b) Coordination of multiple dual oxidase-regulatory pathways in responses to commensal and infectious microbes in drosophila gut. Nat Immunol 10:949–957

Hayes JD, McMahon M (2009) NRF2 and KEAP1 mutations: permanent activation of an adaptive response in cancer. Trends Biochem Sci 34:176–188

Hochmuth CE, Biteau B, Bohmann D, Jasper H (2011a) Redox regulation by Keap1 and Nrf2 controls intestinal stem cell proliferation in Drosophila. Cell Stem Cell 8:1–12

Hochmuth CE, Biteau B, Bohmann D, Jasper H (2011b) Redox regulation by Keap1 and Nrf2 controls intestinal stem cell proliferation in Drosophila. Cell Stem Cell 8:188–199

Inoue H, Hisamoto N, An JH, Oliveira RP, Nishida E, Blackwell TK, Matsumoto K (2005) The C. elegans p38 MAPK pathway regulates nuclear localization of the transcription factor SKN-1 in oxidative stress response. Genes Dev 19:2278–2283

Ito K, Hirao A, Arai F, Matsuoka S, Takubo K, Hamaguchi I, Nomiyama K, Hosokawa K, Sakurada K, Nakagata N et al (2004) Regulation of oxidative stress by ATM is required for self-renewal of haematopoietic stem cells. Nature 431:997–1002

Jang YY, Sharkis SJ (2007) A low level of reactive oxygen species selects for primitive hematopoietic stem cells that may reside in the low-oxygenic niche. Blood 110:3056–3063

Jasper H, Jones DL (2010) Metabolic regulation of stem cell behavior and implications for aging. Cell Metab 12:561–565

Jiang H, Edgar BA (2011) Intestinal stem cells in the adult Drosophila midgut. Exp Cell Res 317:2780–2788

Jiang H, Patel PH, Kohlmaier A, Grenley MO, McEwen DG, Edgar BA (2009) Cytokine/Jak/Stat signaling mediates regeneration and homeostasis in the Drosophila midgut. Cell 137: 1343–1355

Jones DL, Wagers AJ (2008) No place like home: anatomy and function of the stem cell niche. Nat Rev Mol Cell Biol 9:11–21

Lambeth JD (2004) NOX enzymes and the biology of reactive oxygen. Nat Rev Immunol 4: 181–189

Lee JM, Li J, Johnson DA, Stein TD, Kraft AD, Calkins MJ, Jakel RJ, Johnson JA (2005) Nrf2, a multi-organ protector? FASEB J 19:1061–1066

Lee WC, Beebe K, Sudmeier L, Micchelli CA (2009) Adenomatous polyposis coli regulates Drosophila intestinal stem cell proliferation. Development 136:2255–2264

Lin G, Xu N, Xi R (2008) Paracrine Wingless signalling controls self-renewal of Drosophila intestinal stem cells. Nature 455:1119–1123

Liu J, Lin A (2005) Role of JNK activation in apoptosis: a double-edged sword. Cell Res 15:36–42

Liu J, Cao L, Finkel T (2011) Oxidants, metabolism, and stem cell biology. Free Radic Biol Med 51:2158–2162
Maher J, Yamamoto M (2010) The rise of antioxidant signaling–the evolution and hormetic actions of Nrf2. Toxicol Appl Pharmacol 244:4–15
McEwen DG, Peifer M (2005) Puckered, a Drosophila MAPK phosphatase, ensures cell viability by antagonizing JNK-induced apoptosis. Development 132:3935–3946
McMahon M, Itoh K, Yamamoto M, Chanas SA, Henderson CJ, McLellan LI, Wolf CR, Cavin C, Hayes JD (2001) The Cap'n'Collar basic leucine zipper transcription factor Nrf2 (NF-E2 p45-related factor 2) controls both constitutive and inducible expression of intestinal detoxification and glutathione biosynthetic enzymes. Cancer Res 61:3299–3307
Micchelli CA, Perrimon N (2006) Evidence that stem cells reside in the adult Drosophila midgut epithelium. Nature 439:475–479
Miyamoto K, Araki KY, Naka K, Arai F, Takubo K, Yamazaki S, Matsuoka S, Miyamoto T, Ito K, Ohmura M et al (2007) Foxo3a is essential for maintenance of the hematopoietic stem cell pool. Cell Stem Cell 1:101–112
Miyamoto K, Miyamoto T, Kato R, Yoshimura A, Motoyama N, Suda T (2008) FoxO3a regulates hematopoietic homeostasis through a negative feedback pathway in conditions of stress or aging. Blood 112:4485–4493
Motohashi H, O'Connor T, Katsuoka F, Engel JD, Yamamoto M (2002) Integration and diversity of the regulatory network composed of Maf and CNC families of transcription factors. Gene 294:1–12
Nguyen T, Nioi P, Pickett CB (2009) The Nrf2-antioxidant response element signaling pathway and its activation by oxidative stress. J Biol Chem 284:13291–13295
Noble M, Smith J, Power J, Mayer-Proschel M (2003) Redox state as a central modulator of precursor cell function. Ann N Y Acad Sci 991:251–271
Ohlstein B, Spradling A (2006) The adult Drosophila posterior midgut is maintained by pluripotent stem cells. Nature 439:470–474
Ohlstein B, Kai T, Decotto E, Spradling A (2004) The stem cell niche: theme and variations. Curr Opin Cell Biol 16:693–699
Orkin SH, Zon LI (2008) Hematopoiesis: an evolving paradigm for stem cell biology. Cell 132:631–644
Owusu-Ansah E, Banerjee U (2009) Reactive oxygen species prime Drosophila haematopoietic progenitors for differentiation. Nature 461:537–541
Owusu-Ansah E, Yavari A, Mandal S, Banerjee U (2008) Distinct mitochondrial retrograde signals control the G1-S cell cycle checkpoint. Nat Genet 40:356–361
Pan L, Chen S, Weng C, Call G, Zhu D, Tang H, Zhang N, Xie T (2007) Stem cell aging is controlled both intrinsically and extrinsically in the Drosophila ovary. Cell Stem Cell 1: 458–469
Parmar K, Mauch P, Vergilio JA, Sackstein R, Down JD (2007) Distribution of hematopoietic stem cells in the bone marrow according to regional hypoxia. Proc Natl Acad Sci USA 104: 5431–5436
Radtke F, Clevers H (2005) Self-renewal and cancer of the gut: two sides of a coin. Science 307:1904–1909
Radtke F, Clevers H, Riccio O (2006) From gut homeostasis to cancer. Curr Mol Med 6:275–289
Rando TA (2006) Stem cells, ageing and the quest for immortality. Nature 441:1080–1086
Ray PD, Huang BW, Tsuji Y (2012) Reactive oxygen species (ROS) homeostasis and redox regulation in cellular signaling. Cell Signal 24(5):981–990
Salih DA, Brunet A (2008) FoxO transcription factors in the maintenance of cellular homeostasis during aging. Curr Opin Cell Biol 20:126–136
Simons BD, Clevers H (2011) Stem cell self-renewal in intestinal crypt. Exp Cell Res 317: 2719–2724
Smith J, Ladi E, Mayer-Proschel M, Noble M (2000) Redox state is a central modulator of the balance between self-renewal and differentiation in a dividing glial precursor cell. Proc Natl Acad Sci USA 97:10032–10037

Snippert HJ, van der Flier LG, Sato T, van Es JH, van den Born M, Kroon-Veenboer C, Barker N, Klein AM, van Rheenen J, Simons BD et al (2010) Intestinal crypt homeostasis results from neutral competition between symmetrically dividing Lgr5 stem cells. Cell 143:134–144
Spradling AC, de Cuevas M, Drummond-Barbosa D, Keyes L, Lilly M, Pepling M, Xie T (1997) The Drosophila germarium: stem cells, germ line cysts, and oocytes. Cold Spring Harb Symp Quant Biol 62:25–34
Spradling A, Drummond-Barbosa D, Kai T (2001) Stem cells find their niche. Nature 414:98–104
Sykes SM, Lane SW, Bullinger L, Kalaitzidis D, Yusuf R, Saez B, Ferraro F, Mercier F, Singh H, Brumme KM et al (2011) AKT/FOXO signaling enforces reversible differentiation blockade in myeloid leukemias. Cell 146:697–708
Sykiotis GP, Bohmann D (2008) Keap1/Nrf2 signaling regulates oxidative stress tolerance and lifespan in Drosophila. Dev Cell 14:76–85
Sykiotis GP, Bohmann D (2010) Stress-activated Cap'n'collar transcription factors in aging and human disease. Sci Signal 3:re3
Tian H, Biehs B, Warming S, Leong KG, Rangell L, Klein OD, de Sauvage FJ (2011) A reserve stem cell population in small intestine renders Lgr5-positive cells dispensable. Nature 478: 255–259
Toledano MB (2009) The guardian recruits cops: the p53-p21 axis delegates prosurvival duties to the Keap1-Nrf2 stress pathway. Mol Cell 34:637–639
Toledano MB, Planson AG, Delaunay-Moisan A (2010) Reining in H(2)O(2) for safe signaling. Cell 140:454–456
Tonks NK (2006) Protein tyrosine phosphatases: from genes, to function, to disease. Nat Rev Mol Cell Biol 7:833–846
Tothova Z, Gilliland DG (2007) FoxO transcription factors and stem cell homeostasis: insights from the hematopoietic system. Cell Stem Cell 1:140–152
Tothova Z, Kollipara R, Huntly BJ, Lee BH, Castrillon DH, Cullen DE, McDowell EP, Lazo-Kallanian S, Williams IR, Sears C et al (2007) FoxOs are critical mediators of hematopoietic stem cell resistance to physiologic oxidative stress. Cell 128:325–339
Tsatmali M, Walcott EC, Crossin KL (2005) Newborn neurons acquire high levels of reactive oxygen species and increased mitochondrial proteins upon differentiation from progenitors. Brain Res 1040:137–150
Tullet JMA, Hertweck M, An JH, Baker J, Oliviera RP, Baumeister R, Blackwell TK (2008) Direct inhibition of the longevity promoting factor SKN-1 by Insulin like signaling in *C. elegans*. Cell 132:1025–1038
Voog J, Jones DL (2010) Stem cells and the niche: a dynamic duo. Cell Stem Cell 6:103–115
Wakabayashi N, Itoh K, Wakabayashi J, Motohashi H, Noda S, Takahashi S, Imakado S, Kotsuji T, Otsuka F, Roop DR et al (2003) Keap1-null mutation leads to postnatal lethality due to constitutive Nrf2 activation. Nat Genet 35:238–245
Walker MR, Patel KK, Stappenbeck TS (2009) The stem cell niche. J Pathol 217:169–180
Wang L, Jones DL (2010) The effects of aging on stem cell behavior in Drosophila. Exp Gerontol 46:340–344
Wang MC, Bohmann D, Jasper H (2005) JNK extends life span and limits growth by antagonizing cellular and organism-wide responses to insulin signaling. Cell 121:115–125
Weston CR, Davis RJ (2007) The JNK signal transduction pathway. Curr Opin Cell Biol 19: 142–149
Woo HA, Yim SH, Shin DH, Kang D, Yu DY, Rhee SG (2010) Inactivation of peroxiredoxin I by phosphorylation allows localized H(2)O(2) accumulation for cell signaling. Cell 140:517–528
Yu X, Kensler T (2005) Nrf2 as a target for cancer chemoprevention. Mutat Res 591:93–102

[illegible]

Chapter 13
Oxidative Stress in Infectious Diseases

Esther Jortzik and Katja Becker

Abstract The "big three" infectious diseases HIV/AIDS, tuberculosis, and malaria were collectively responsible for nearly 260 million infected people in 2010. HIV, *Mycobacterium tuberculosis*, and *Plasmodium falciparum*, the causative agents of AIDS, tuberculosis, and malaria, are continuously exposed to reactive oxygen and nitrogen species endogenously produced or derived from the host immune system in response to infection. Oxidative stress has a key function in the pathogenesis of many infectious diseases, and represents moreover a promising strategy for chemotherapeutic development. Understanding the redox interactions and redox signaling mechanisms of pathogens and their hosts is crucial for developing (1) drugs that support the host antioxidant defense in order to protect cells from oxidative damage, (2) drugs that enhance specific reactive oxygen or nitrogen species to improve the host defense against pathogens, and (3) drugs that interfere with the redox system of the pathogen in order to block its growth and survival.

Keywords Antioxidants • HIV/AIDS • Malaria • Tuberculosis • Redox system

Abbreviations

AhpC	alkyl hydroperoxide reductase
AIDS	acquired immunodeficiency syndrome
AOP	antioxidant protein
ASK	apoptosis signal-regulating kinase
G6PD	glucose 6-phosphate dehydrogenase
GR	glutathione reductase

E. Jortzik • K. Becker (✉)
Biochemistry and Molecular Biology, Interdisciplinary Research Center,
Justus Liebig University, Heinrich-Buff-Ring 26-32, 35392 Giessen, Germany
e-mail: katja.becker@uni-giessen.de

U. Jakob and D. Reichmann (eds.), *Oxidative Stress and Redox Regulation*,
DOI 10.1007/978-94-007-5787-5_13,

GST	glutathione S-transferase
gp120	glycoprotein 120
Grx	glutaredoxin
GSH	reduced glutathione
GSSG	oxidized glutathione
HAART	highly active anti-retroviral therapy
HIV	human immunodeficiency virus
KatG	catalase-peroxidase
iNOS	nitric oxide synthase
LTR	long terminal repeat
Mca	mycothiol *S*-conjugate amidase
MscR	mycothiol-*S*-nitrosoreductase/-formaldehyde reductase
MSH	mycothiol
MSNO	*S*-nitrosomycothiol
MSSM	mycothiol disulfide
Msr	methionine sulfoxide reductases
Mtr	mycothiol reductase
NAC	N-acetylcysteine
NFκB	nuclear factor κB
NO	nitric oxide
NOX	NADPH oxidase
nPrx	nuclear peroxiredoxin
Plrx	plasmoredoxin
Prx	peroxiredoxin
RNS	reactive nitrogen species
ROS	reactive oxygen species
SOD	superoxide dismutase
Tat	transactivator protein
TNF	tumor necrosis factor
TPx	thioredoxin peroxidase
Trx	thioredoxin
TrxR	thioredoxin reductase

13.1 Introduction

Oxidative stress is not a disease, but an unspecific pathological state that can be worsened by depletion of antioxidants, immunosuppressants, zinc and selenium deficiency, xenobiotics, and infections, thus being involved in the pathogenesis of a variety of diseases (Stehbens 2004). Mammalian cells of the immune system use the production of reactive oxygen and nitrogen species (ROS, RNS) to control and defend themselves against infections with bacterial pathogens, parasites, or viruses. Infectious agents counteract oxidative stress derived from the host immune system by highly efficient antioxidant defense systems. Inhibiting the antioxidant defense of parasites, bacteria, and viruses is an intensely studied strategy in order to develop effective anti-infectious drugs. AIDS (HIV), tuberculosis (*Mycobacterium tuberculosis*), and malaria (*Plasmodium falciparum*) are the three leading infectious diseases worldwide. In their pathogenesis, oxidative stress plays a major role, which will be comprehensively discussed in this review.

13.2 Oxidative Stress in HIV/AIDS

According to the WHO, 2.7 million people were newly infected and 34 million people were living with the human immunodeficiency virus (HIV) in 2010 (WHO 2011a). HIV infection leads to a complex disease with broad clinical symptoms and complex pathogenic mechanisms. Dysregulation of the immune system is caused by a constant decline of $CD4^+$ T cells (Pantaleo et al. 1993), which has been intensely studied. ROS and oxidative stress are critically involved in the pathogenesis of HIV infection, with the most important aspects being discussed in the following paragraphs.

13.2.1 Mechanisms of Oxidative Stress in the Pathogenesis of HIV Infection

A role of ROS in viral infections had already been described in 1970 (Belding et al. 1970) and has been intensely investigated since that time. Oxidative stress, ROS, and antioxidant defense systems appear to be implicated in many aspects of HIV infection and AIDS such as immune function, inflammatory response, virus replication, and apoptosis. Humans infected with HIV show chronically increased levels of oxidative stress, decreased concentrations of antioxidants, and perturbations in the antioxidant systems (Gil et al. 2010). Furthermore, high plasma concentrations of hydroperoxides and malondialdehyde were found in HIV patients, indicating increased lipid peroxidation (Suresh et al. 2009). The activation of $CD4^+$ T cells and monocytes/macrophages upon HIV infection is triggered by hydrogen peroxide, superoxide anion, and peroxynitrite. Moreover, HIV infection of macrophages results in enhanced production of peroxynitrite and superoxide (Elbim et al. 1999). Oxidative stress upon HIV infection can be induced by the HIV transactivator protein (Tat) and the envelope glycoprotein gp120 (Price et al. 2005). However, the mechanisms of oxidative stress induction have not been completely unraveled.

13.2.1.1 Viral Replication

Redox signaling has an important function in the regulation of viral gene transcription and replication. Oxidative stress has been found to increase HIV replication *in vitro*, while antioxidants show the opposite effect and reduce virus replication (Schreck et al. 1991; Roederer et al. 1990). The promoter of HIV-1 is located in the long terminal repeat (LTR) at the 5′-end and contains binding sites for different transcription factors, including NFκB that mediate the activation of viral transcription in complex interplay with both viral and host factors (for review please see Hiscott et al. 2001). Reactive oxygen intermediates such as H_2O_2 and oxygen radicals disturb the redox balance towards oxidizing conditions and thereby activate the translocation of NFκB into the nucleus, where it binds to the HIV LTR in order to increase the expression of viral genes and replication (Schreck et al. 1991). Moreover, peroxynitrite was shown to play a major role in virus replication, since a synthetic peroxynitrite decomposition catalyst (MnTBAP) had a strong inhibitory effect on virus replication. Peroxynitrite is suggested to also mediate its effects via the NFκB pathway (Aquaro et al. 2007). Overexpression of peroxiredoxin IV in T cells deactivated HIV LTR and thus inhibited HIV transcription by regulating hydrogen peroxide-mediated activation of NFκB (Jin et al. 1997). Both degradation of IκB and binding of NFκB to the LTR are suggested to be redox-controlled. Furthermore, the HIV LTR contains binding sites for the transcription factors p53 and AP-1, which are also sensitive to redox regulation (Pereira et al. 2000).

An HIV-encoded protein required for efficient viral transcription is Tat, a transactivator protein activated upon oxidizing conditions and inhibited upon reduction, thus functioning as a redox sensor (Koken et al. 1994; Washington et al. 2010). Dysfunction of Tat leads to disturbed transcription with the formation of prematurely terminated short transcripts (Kao et al. 1987). HIV Tat protein contains a cysteine-rich domain with several CxxC motifs that form intramolecular disulfide bonds required for the transactivation activity (Koken et al. 1994). *In vitro* inhibition of HIV expression by N-acetylcysteine (NAC) might at least in part be mediated by reduction and deactivation of Tat (Washington et al. 2010). Similarly, thioredoxin reductase has been shown to act as a negative regulator of Tat transactivation activity by reducing critical disulfide bonds (Fig. 13.1) (Kalantari et al. 2008). Moreover, Tat itself enhances oxidative stress by increasing the expression of tumor necrosis factor (TNF α/β) and by repressing the expression of antioxidant defense enzymes such as manganese superoxide dismutase, glutathione peroxidase, and γ-glutamylcysteine synthetase (Romani et al. 2010; Buonaguro et al. 1994; Flores et al. 1993; Richard et al. 2001; Choi et al. 2000). Therefore, Tat contributes to the oxidizing environment during HIV infection by decreasing the levels of several antioxidant enzymes.

13.2.1.2 Apoptosis

HIV manipulates apoptotic pathways of host cells by multiple factors in order to kill non-infected inflammatory or immune cells and infected $CD4^+$ T cells by apoptosis (reviewed in (Gougeon 2005)). Balancing of ROS and redox signaling are suggested to be critically involved in HIV-induced apoptosis (Agrawal et al. 2007; Romero-Alvira and Roche 1998; Buccigrossi et al. 2011; Banki et al. 1998). HIV infection

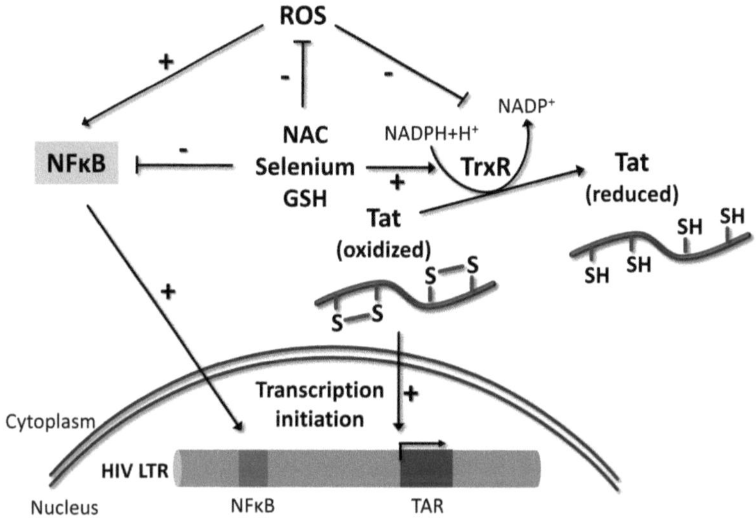

Fig. 13.1 Scheme of the influence of reactive oxygen species on HIV transcription and replication. Increased cellular concentrations of reduced glutathione (GSH) and supplementation with selenium or N-acetylcysteine (NAC) decrease the concentrations of reactive oxygen species (ROS) and activate antioxidant enzymes such as thioredoxin reductase (TrxR). ROS activate NFκB-dependent gene expression, which is decreased under reducing conditions. HIV transactivator protein (Tat) is inhibited by reduction of critical disulfide bonds. TrxR has been shown to reduce Tat and thereby inhibits Tat-mediated HIV transcription (modified after Kalantari et al. 2008)

leads to an increase in mitochondrial ROS that subsequently trigger apoptosis together with other agents such as viral proteins (Tat, gp120) and cytokines (TNFα). Moreover, caspases are cysteine-dependent enzymes and are thus sensitive towards the redox status of the cell (Hampton and Orrenius 1998).

Members of the thioredoxin (Trx) family including Trx itself, glutaredoxin (Grx), and peroxiredoxins have been discussed in the context of apoptotic signaling during HIV infection (Masutani et al. 2005). Trx is upregulated in the plasma of HIV-infected patients (Nakamura et al. 1996), and is known to inhibit the activity of apoptosis signal-regulating kinase 1 (ASK1) under reducing conditions, and thereby blocks induction of apoptosis, an effect that is reversed under oxidative stress and subsequent oxidation of Trx (Saitoh et al. 1998). In contrast, thioredoxin-dependent peroxidases, peroxiredoxins, are able to inhibit apoptosis by scavenging hydrogen peroxide (Kim et al. 2000).

Recently, it has been reported that HIV Tat protein can directly mediate apoptosis in enterocytes via a redox-dependent mechanism, leading to damage of the intestinal mucosa (Buccigrossi et al. 2011). Similarly, ROS have been shown to be crucial for HIV Tat-induced apoptosis in neurons (Agrawal et al. 2007). Tat-induced apoptotic signaling is suggested to be based on an increase in intracellular ROS concentrations and an imbalance in the GSH/GSSG ratio (Buccigrossi et al. 2011; Agrawal et al. 2007; Choi et al. 2000). Moreover, NAC not only prevents oxidative stress by balancing ROS and GSSG concentrations, but also appears to inhibit Tat-induced apoptosis (Buccigrossi et al. 2011).

13.2.2 Antioxidant Systems During HIV Infection

Several clinical studies showed that the redox balance in HIV-infected patients without highly active anti-retroviral therapy (HAART) is severely disturbed, with perturbations affecting almost all components of the antioxidant defense system including glutathione, tocopherol, ascorbate, selenium, the thioredoxin system, superoxide dismutase, and glutathione peroxidase (Gil et al. 2010), as summarized in the following paragraphs.

13.2.2.1 Thioredoxin Family

In order to enter host cells, HIV requires the viral glycoprotein gp120, which undergoes conformational changes depending on the reduction of disulfides. Reduction of gp120 can be catalyzed by Trx1, Grx1, and protein disulfide isomerase, thus demonstrating that the host redox system is required for virus entry into host cells (Reiser et al. 2012; Azimi et al. 2010; Auwerx et al. 2009).

Dysregulation of the Trx system appears to be involved in apoptosis in HIV-infected cells as outlined above. *In vitro*, HIV infection of T cell lines decreased the expression of Trx3 shortly after infection (Masutani et al. 2005). Similarly, Trx levels were decreased in monocytes from asymptomatic untreated patients, while Trx expression in cells from AIDS patients was shown to be at a higher level when compared to uninfected cells, which has been suggested to limit ROS levels and apoptosis at later disease stages (Elbim et al. 1999). Thioredoxin reductase 1 (TrxR1) negatively regulates the activity of Tat and thereby Tat-dependent transcription in human macrophages by reducing two disulfide bonds within the cysteine-rich motif of Tat (Kalantari et al. 2008). Inhibition of Grx1 activity by anti-Grx antibodies blocks HIV replication *in vitro*, most likely by inhibiting HIV entry into $CD4^+$ T cells (Auwerx et al. 2009).

Peroxiredoxins reduce hydrogen peroxide by using Trx as an electron donor. In HIV-infected T cell lines, peroxiredoxin IV is downregulated, while T cells overexpressing peroxiredoxin IV show a decreased HIV transcription (Jin et al. 1997). Altogether, these results show that the Trx system is intriguingly involved in HIV infection.

13.2.2.2 Glutathione

In HIV-infected patients, reduced levels of glutathione were found in the plasma, lymphocytes, monocytes, as well as in $CD4^+$ and $CD8^+$ T cells (Eck et al. 1989;

de Quay et al. 1992; Roederer et al. 1991; Buhl et al. 1989). Low levels of reduced glutathione (GSH) are associated with a poor survival of HIV-infected patients (Herzenberg et al. 1997). The mechanisms responsible for impaired GSH production and GSSG reduction upon HIV infection are unclear. Depletion in GSH is accompanied by increased concentrations of oxidized glutathione, indicating a shift in the GSH:GSSG ratio (Aukrust et al. 1995), which has been confirmed in infected macrophages (Morris et al. 2012). Progressively depleted intracellular GSH levels are at least partially mediated by reduced GSH synthesis, since HIV Tat protein decreases transcription and protein levels of γ-glutamylcysteinyl synthetase (Choi et al. 2000). This was supported by a recent study showing that expression of GSH biosynthesis enzymes is decreased in HIV patients, which was attributed to chronically enhanced production of inflammatory cytokines (IL-1, IL-17, and TNF-α) that might interfere with GSH biosynthesis (Morris et al. 2012).

13.2.2.3 Superoxide Dismutase and Catalase

The first enzyme in the defense against superoxide anions is superoxide dismutase (SOD), which converts superoxide anions into hydrogen peroxide, which is subsequently reduced by catalase. The expression of mitochondrial manganese superoxide dismutase (MnSOD) is decreased in HIV-infected patients, with the effect being mediated by HIV Tat protein followed by increased oxidative stress with enhanced protein carbonylation and lipid peroxidation (Flores et al. 1993). In contrast, the expression of cytosolic CuZnSOD in macrophages is increased during HIV infection *in vitro*, most likely to counteract elevated superoxide anion concentrations (Delmas-Beauvieux et al. 1996). Hydrogen peroxide produced by SOD is scavenged by catalase, which shows an increasing activity with progressing HIV infection, and most likely compensates GSH deficiency in HIV-infected cells (Leff et al. 1992).

13.2.3 Antioxidants in the Treatment of AIDS

Decreased concentrations of antioxidants observed in HIV-positive patients are associated with deficiencies of micronutrients with antioxidant properties such as vitamins C and E, thiamine, selenium, and zinc. There are many indications that supplementation with antioxidants can have a beneficial therapeutic effect for HIV-infected patients (reviewed e.g. in Lanzillotti and Tang 2005; Singhal and Austin 2002).

13.2.3.1 N-Acetylcysteine

Enhancing cysteine bioavailability in order to increase GSH concentrations is most likely the major effect of the antioxidant NAC. Oral supplementation with NAC was reported to increase the glutathione pool in plasma and muscle (Atkuri et al. 2007).

NAC has been found to efficiently inhibit HIV replication. NAC is a potent inhibitor of NFκB by counteracting the effect of ROS (Staal et al. 1990), but has also been shown to block the TNFα-stimulated replication of HIV-1 (Roederer et al. 1990). Moreover, treatment with NAC can prevent Tat-induced apoptosis by restoring the GSH:GSSG ratio *in vitro* and *ex vivo*, and thereby protects the intestinal mucosa (Buccigrossi et al. 2011). An increase in GSH can be achieved by both NAC and glutamine supplementation, with NAC supplying cysteine and glutamine by delivering glycine (Borges-Santos et al. 2012). Similar to NAC, glutamine supplementation enhances plasma GSH levels and significantly increases lean body mass in patients with HIV infection (Borges-Santos et al. 2012).

13.2.3.2 Selenium

Selenium deficiency is strongly associated with disease progression and mortality in HIV-infected patients. Selenium supplementation of HIV-infected individuals improves the CD4 count, suppresses the virus load, decreases anxiety, and reduces the need for hospitalization (Hurwitz et al. 2007; Shor-Posner et al. 2003; McDermid et al. 2002). However, information on how selenium influences HIV infection is limited. The effect is most likely mediated via selenoproteins such as TrxR and glutathione peroxidase that shift the redox balance towards a reducing environment and thereby inhibit Tat and thus HIV transcription as explained above (Kalantari et al. 2008).

13.3 Oxidative Stress in Tuberculosis

Tuberculosis infected 8.8 million and killed 1.4 million people in 2010 and is after HIV the second leading cause of death from an infectious disease worldwide (WHO 2011b). *Mycobacterium tuberculosis*, its causative agent in humans, infects and multiplies in lung macrophages of the host. In immunodeficient individuals (e.g., HIV-positive individuals), *M. tuberculosis* can multiply to high numbers and induce an active disease. In contrast, the immune system of immunocompetent individuals is in most cases able to control the infection, and more than 90% of the infected remain asymptomatic (Lawn and Zumla 2011). Despite the bactericidal environment, *M. tuberculosis* is able to survive in macrophages, a fact that demonstrates the adaptation of the pathogen to the oxidant burden in the host.

13.3.1 Mechanisms of Oxidative Stress in the Pathogenesis of Tuberculosis

The host immune system is able to influence the virulence of *M. tuberculosis* by producing ROS and RNS. As a reaction towards phagocytosis of *M. tuberculosis*, macrophages can induce the expression of NADPH oxidase (NOX) and nitric oxide synthase 2 (iNOS, NOS2) and thereby increase the oxidative burden. NOX generates superoxide by catalyzing the one-electron reduction of O_2 (Leto and Geiszt 2006). iNOS produces $NO^{\bullet}$, which can react with superoxide to form highly reactive peroxynitrite and can kill mycobacteria (Shiloh and Nathan 2000). iNOS expression and release of RNS are suggested to be important for controlling the virulence of *M. tuberculosis* in humans. A range of studies based on murine models of tuberculosis using iNOS inhibitors and iNOS-deficient mice show that iNOS and RNS are required in order to control tuberculosis infection (MacMicking et al. 1997; Chan et al. 1995; Scanga et al. 2001). Moreover, NO appears to be important for latent tuberculosis infection in mice and prevents reactivation together with RNS-independent mechanisms (Flynn et al. 1998). Contrarily, another study reported no influence of a lack of iNOS on tuberculosis infection (Jung et al. 2002). Similarly, the function of ROS produced by NADPH oxidase in controlling *M. tuberculosis* is controversially discussed. Experiments using NOX-deficient mice models showed that the absence of NOX-produced superoxide increased bacterial growth during an early infection stage (Cooper et al. 2000; Adams et al. 1997). In contrast, another study did not show any differences between a murine knockout model lacking NOX and wildtype mice in ability to control *M. tuberculosis* infection (Jung et al. 2002). However, clear experimental evidence on the ROS/RNS-based mechanisms of infection control in humans is rare. Alveolar macrophages from *M. tuberculosis*-infected patients show increased concentrations of iNOS compared to non-infected individuals (Nicholson et al. 1996). Several single nucleotide polymorphisms of the *nos2a* gene in African Americans show associations with susceptibility to tuberculosis, thus supporting the role of iNOS for pathogenesis of tuberculosis in humans (Velez et al. 2009; Gomez et al. 2007).

13.3.2 Antioxidant Defense Systems of Mycobacterium tuberculosis

Several studies reported a remarkable resistance of *M. tuberculosis* towards oxidative stress both *in vitro* and *in vivo*; the bacterium can tolerate H_2O_2 concentrations up to 10 mM (Voskuil et al. 2011). In contrast to enteric bacteria, *M. tuberculosis* shows very low but differential transcriptional responses after exposure to a range of H_2O_2 and NO concentrations, and is resistant to DNA damage-mediated killing by H_2O_2 (Voskuil et al. 2011; Garbe et al. 1996). Low H_2O_2 or NO concentrations lead to an induction of few H_2O_2-/NO-responsive genes such as those encoding proteins involved in H_2O_2 scavenging, repair mechanisms, and iron acquisition (Voskuil et al. 2011). Similarly, treatment with cumene hydroperoxide also did not have a major impact on gene transcription (Garbe et al. 1996). Interestingly, the expression of many genes involved in antioxidant defense did not show major changes after H_2O_2 or NO exposition indicating that they are constitutively expressed at a high level (Voskuil et al. 2011). Several antimycobacterial drugs, such as ethionamide, isoniazid, and the nitroimidazopyran PA-824, have to be reduced for activation and thus depend on the intracellular redox state. Moreover, an increased NADH/NAD$^+$ ratio, mutations in NADH dehydrogenase, and mutations in mycothiol biosynthesis have been related to resistance against isoniazid (Vilcheze et al. 2005; Xu et al. 2011; Miesel et al. 1998). Therefore, understanding the redox homeostasis of *M. tuberculosis* is important for drug identification strategies. As an intracellular pathogen, *M. tuberculosis* employs different antioxidant systems to counteract oxidative stress of the host defense system, as discussed in the following paragraphs.

13.3.2.1 Response Towards Oxidative Stress

M. tuberculosis shows an unusually low transcriptional response towards oxidative and nitrosative stress when compared to other enteric bacteria (Garbe et al. 1996). In different bacteria, transcriptional response towards oxidative stress is regulated by the transcription factors OxyR and SoxR, which modulate transcription in response to peroxide and superoxide, respectively. Both regulate the expression of a variety of genes, including a range of redox-active proteins (summarized in (Zahrt and Deretic 2002)). OxyR from *M. tuberculosis* is nonfunctional and is not able to sense peroxide stress. Several mutations (deletions and frameshifts) in the *oxyR* gene render the transcription factor inactive, which impairs the oxidative and nitrosative stress response in *M. tuberculosis* (Deretic et al. 1995; Sherman et al. 1995). Moreover, the *soxRS* regulon is missing in the genome of mycobacteria (Cole et al. 1998).

In *M. tuberculosis*, oxidative stress response can be mediated by the regulator *furA* and catalase-peroxidase KatG, which are mutated and dysfunctional in human pathogenic *M. leprae*. *furA* is located upstream of the catalase-peroxidase *katG*, and is supposed to act as a negative regulator of *katG* expression (Zahrt et al. 2001; Pym et al. 2001). This regulation was suggested to be important for virulence of *M. tuberculosis*, since *M. tuberculosis* mutants lacking *katG* showed a severely impaired persistence in mice and guinea pigs (Li et al. 1998). Moreover, KatG is required for the conversion of the prodrug isoniazid into its active form (Zhang et al. 1992).

The expression of *katG* can also be negatively regulated in a FurA-independent manner by the transcriptional regulator OxyS. Via a cysteine residue in its DNA-binding domain, OxyS binds directly to the katG promoter region, with the DNA-binding capacity being diminished upon oxidation of OxyS. Mycobacteria overexpressing OxyS show an increased susceptibility towards oxidative stress, thereby indicating that OxyS functions as a redox sensor (Li and He 2012).

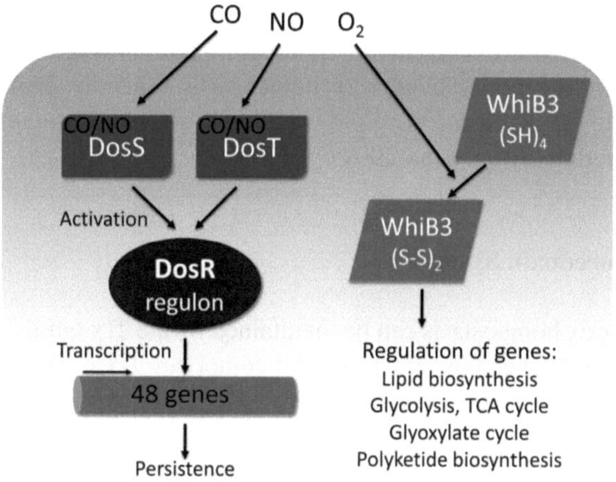

Fig. 13.2 Regulation of oxidative stress response in *Mycobacterium tuberculosis*. The DosR regulon is modulated by the histidine kinases DosS and DosT that sense NO, CO, and O_2. The DosR regulon comprises 48 genes with functions for persistence and stress adaptation. On the basis of a redox switch mechanism, WhiB3 regulates lipid biosynthesis and other central metabolic pathways involved in virulence, persistence, and redox homeostasis in response to oxidative stress (modified after Kumar et al. 2011 and Singh et al. 2009)

The DosR/S/T system (also called DevR system) is a redox-sensing system involved in virulence and persistence of *M. tuberculosis* (Leistikow et al. 2010; Shiloh et al. 2008; Park et al. 2003). The DosR regulon is controlled by DosR (DevR), a response regulator, and DosS and DosT, two sensor histidine kinases. The DosR regulon is induced by conditions that inhibit aerobic respiration, since the heme proteins DosS and DosT can sense NO, O_2, and CO by binding them to iron in their heme group (Shiloh et al. 2008). Subsequently, the DosR dormancy regulon is induced, which includes at least 48 genes suggested to play a role in persistence, latent infection, and stress adaptation (Fig. 13.2) (Park et al. 2003). The DosR regulon enables *M. tuberculosis* to follow and respond to conditions that do not allow aerobic respiration and is required for maintaining the redox balance and energy levels under anaerobic conditions (Leistikow et al. 2010).

Another redox signaling pathway in *M. tuberculosis* is controlled by WhiB3, a redox regulator that senses NO and O_2 from the host and is involved in virulence and pathogenesis of mycobacteria (Fig. 13.2) (Steyn et al. 2002). WhiB3 from *M. tuberculosis* contains four cysteine residues coordinating an iron-sulfur cluster that can specifically bind NO and O_2 (Singh et al. 2007). DNA binding of WhiB3 occurs independently of its iron-sulfur cluster but depends on the oxidation state of its four cysteines: oxidation stimulates DNA binding, while reduction diminishes it (Singh et al. 2009). Oxidation of WhiB3 stimulates DNA binding to several lipid biosynthetic genes and directly regulates production of lipids including poly- and diacyltrehaloses, sulfolipids, and triacylglycerol. WhiB3 is regarded as a redox sensor that connects host redox signals with its intermediary metabolism, since it induces a metabolic shift to fatty acids by regulating lipid anabolism in response to oxidative stress during tuberculosis infection (Singh et al. 2009). Thus, WhiB3 maintains the intracellular redox homeostasis in part by channeling reducing equivalents into *M. tuberculosis* lipid synthesis, which modulate inflammatory cytokine production (Singh et al. 2009).

13.3.2.2 Thioredoxin System

Intracellular redox homeostasis can be maintained by the Trx family, which can catalyze thiol/disulfide exchange reactions and comprises TrxR, Trx, and thioredoxin peroxidases (TPx). *M. tuberculosis* encodes one TrxR and three Trx (A, B, and C) (Akif et al. 2008). However, no mRNA transcripts of *trxA* have been observed. This is supported by the finding that recombinant TrxA cannot be reduced by TrxR. Thus, *trxA* was suggested to be a cryptic gene in *M. tuberculosis* and is most likely not involved in antioxidant defense (Akif et al. 2008). The TrxR/Trx couple from *M. tuberculosis* is able to reduce peroxides and dinitrobenzenes, while cumene hydroperoxide can only be reduced by TrxR and not by Trx *in vitro* (Zhang et al. 1999). Moreover, TrxC is able to reduce mycothiol, GSSG, and *S*-nitrosoglutathione *in vitro*, which was suggested to be important for defense against host-derived oxidative stress (Attarian et al. 2009). The crystal structure of *M. tuberculosis* TrxC, both oxidized and in complex with an inhibitor, has been solved and can be exploited for structure-based inhibitor development (Hall et al. 2006, 2011)

Thioredoxin peroxidase (TPx) from *M. tuberculosis* shows homology to atypical 2-cys peroxiredoxins but has been characterized as a 1-cys peroxiredoxin with Cys60 being the peroxidatic cysteine (Trujillo et al. 2006). TPx can use TrxB and C as electron donors to efficiently reduce hydrogen peroxide, *t*-butyl hydroperoxide, cumene hydroperoxide, and peroxynitrite (Trujillo et al. 2006; Jaeger et al. 2004). The crystal structure of dimeric TPx revealed a Trx fold similar to that of other peroxiredoxin family members (Rho et al. 2006). *M. tuberculosis* mutants lacking TPx show an increased sensitivity towards hydrogen peroxide and nitric oxide. Moreover, they are unable to survive and grow in macrophages, cannot establish acute infections or maintain persistent infections in murine tuberculosis models, and show an attenuated virulence. This indicates that TPx is an essential component of the antioxidant defense of *M. tuberculosis* (Hu and Coates 2009).

13.3.2.3 Mycothiol

Like other actinomycetes, *M. tuberculosis* lacks detectable levels of glutathione and glutathione peroxidases and employs mycothiol (MSH, AcCys-GlcN-Ins) as its major low molecular weight thiol. MSH is a cysteinyl pseudo-disaccharide consisting of N-acetylcysteine, glucosamine, and *myo*-inositol, and was discovered in 1993 (Newton et al. 1993). In *M. smegmatis* and *M. tuberculosis*, MSH concentrations are in a low millimolar range, comparable to the concentrations of glutathione in eukaryotes (Newton et al. 1996). MSH is required for the survival of *M. tuberculosis* (Sareen et al. 2003), but not for *M. smegmatis* (Rawat et al. 2002).

The biosynthetic pathway of MSH consists of five steps catalyzed by four enzymes, which cannot be found in eukaryotes or eubacteria and are thus discussed as highly feasible targets for chemotherapeutic interventions against mycobacteria. MshA is an N-acetylglucosamine transferase that forms 3-phospho-GlcNAc-Ins, which is converted into GlcNAc-Ins by dephosphorylation catalyzed by a phosphatase termed MshA2. The deacetylase MshB deacetylates GlcNAc-Ins to form Gln-Ins, which is then ligated to cysteine by MshC, a mycothiol ligase. Cys-Gln-Ins is subsequently acetylated by mycothiol synthase MshD generating MSH (for detailed reviews of MSH biosynthesis please see (Jothivasan and Hamilton 2008; Fan et al. 2009)).

The functions of MshA-D have been intensely investigated by gene knockout studies in *M. smegmatis* and *M. tuberculosis*. MshA and C are indeed essential for growth and survival of *M. tuberculosis* (Buchmeier and Fahey 2006; Sareen et al. 2003), while mutants lacking MshB and D are still viable (Buchmeier et al. 2006). These studies show that only the absence of MshA or C leads to complete depletion of MSH, while still 10 and 2% of normal MSH levels could be detected in the mutants lacking MshC or MshD, respectively (Sareen et al. 2003; Buchmeier and Fahey 2006; Buchmeier et al. 2003, 2006). Moreover, disruption of the MshD gene in *M. tuberculosis* results in high levels of the MshD substrate Cys-GlcN-Ins and N-formyl-Cys-GlcN-Ins, with the latter being maintained in a reduced state. Thus, N-formyl-Cys-GlcN-Ins might function as a surrogate for MSH under normal conditions, but is not sufficient under increased oxidative stress (Buchmeier et al. 2006). These studies indicate that MshA and C might be more feasible as targets for anti-mycobacterial drug development (Fan et al. 2009).

MSH exerts a range of protective functions in mycobacteria by serving as an intracellular redox buffer as a functional analog of GSH and by maintaining the redox balance. Low MSH levels in *M. smegmatis* result in a remarkably increased sensitivity towards free radicals, oxidants, alkylating agents, and antibiotics including erythromycin, rifamycin, and penicillin G (Rawat et al. 2002; Buchmeier et al. 2006). Thus, MSH is involved in protection against oxidative stress and antibiotics. In contrast, MSH-depleted mutants show a 200-fold increased resistance towards isoniazid, demonstrating a function of MSH in the drug mechanism of isoniazid (Rawat et al. 2002). Furthermore, MSH protects against electrophilic xenobiotics by forming MSH-toxin conjugates (Rawat et al. 2004).

MSH-dependent enzymes are involved in several cellular processes, including defense against ROS and RNS and detoxification of electrophilic xenobiotics (Fig. 13.3). Mycothiol disulfide (MSSM) can be reduced by the NADPH-dependent flavoenzyme mycothiol disulfide reductase (Mtr, mycothione reductase), which maintains a high MSH:MSSM ratio by the same disulfide reducing mechanism found in glutathione reductase (Argyrou et al. 2004; Argyrou and Blanchard 2004). MSH-toxin conjugates can be hydrolyzed by mycothiol *S*-conjugate amidase (Mca) forming a mercapturic acid and GlcN-Ins. While the mercapturic acid derivative can be exported from the cell, GlcN-Ins is recycled for re-synthesis of MSH (Newton

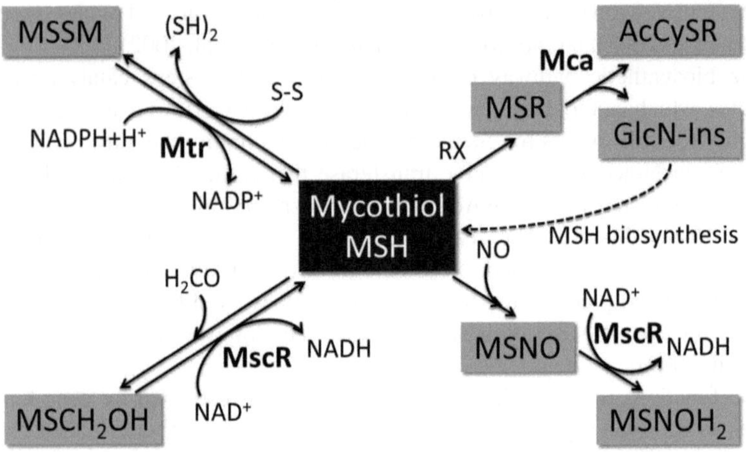

Fig. 13.3 Functions of mycothiol, the major low molecular weight thiol in *Mycobacterium tuberculosis*. Mycothiol (MSH) can reduce proteins and oxidants and is thereby oxidized to mycothiol disulfide (MSSM). MSSM is recycled by mycothiol reductase (Mtr) under consumption of NADPH + H^+. *S*-nitrosomycothiol (MSNO) can be reduced by mycothiol-*S*-nitrosoreductase/-formaldehyde reductase (MscR) to MSH sulfonamide ($MSNOH_2$). MSH can react with formaldehyde (HCOOH) to form the hemothioacetal formylmycothiol ($MSCH_2OH$), which is reduced by MscR. Thiol-reactive reagents (RX) form MSH-toxin conjugates that are hydrolyzed by mycothiol *S*-conjugate amidase (Mca) yielding a mercapturic acid and GlcN-Ins. The latter can enter the MSH biosynthesis pathway to be recycled to MSH (modified after Newton et al. 2008)

et al. 2000). *M. tuberculosis* Mca has a crucial function in drug resistance, a fact that directed interest towards the development of inhibitors against Mca (Nicholas et al. 2003). MSH can rapidly react with formaldehyde forming a hemithioacetal, which is a substrate of mycothiol-*S*-nitrosoreductase/-formaldehyde reductase (MscR) (Vogt et al. 2003). Moreover, MscR can reduce *S*-nitrosomycothiol (Vogt et al. 2003) and is required for growth of *M. tuberculosis* (Sassetti et al. 2003).

13.3.2.4 Further Enzymes of Antioxidant Defense

M. tuberculosis encodes a catalase peroxidase (KatG) that exhibits catalase, peroxidase, and peroxynitritase activity and can therefore detoxify ROS and RNS (Rouse et al. 1996; Manca et al. 1999; Wengenack et al. 1999). Experiments using transgenic murine models indicated that the major role of KatG is the defense against the oxidative burden in the host by metabolizing peroxides generated by the host phagocyte NADPH oxidase (Ng et al. 2004). In order to exhibit its antimycobacterial activity, the most effective and specific antimycobacterial drug isoniazid has to be converted from a prodrug form into its active form by KatG (Zhang et al. 1992). Mutations in the mycobacterial *katG* gene are associated with resistance to isoniazid. At least 130 known mutations of *katG* are characterized by a decreased or diminished activity of KatG and thus reduced formation of the adduct INH-NAD (reviewed in Vilcheze and Jacobs 2007).

The peroxiredoxin-type alkyl hydroperoxide reductase (AhpC) can directly detoxify hydroperoxides and peroxynitrite (Bryk et al. 2000; Master et al. 2002). Although AhpC is considered to be a 2-cys peroxiredoxin, it involves a third cysteine in catalysis (Koshkin et al. 2004). AhpC employs the Trx-like protein AhpD, a protein with a low alkylhydroperoxidase activity of its own, as an electron donor but cannot be reduced by Trx (Hillas et al. 2000; Bryk et al. 2002). Moreover, *M. tuberculosis* codes for a 1-cys peroxiredoxin named alkyl hydroperoxide reductase E (AhpE), which is involved in peroxide and peroxynitrite detoxification (Hugo et al. 2009). In order to detoxify superoxide radicals, *M. tuberculosis* encodes two SODs, an iron-dependent SOD, and a Cu and Zn-dependent one (Zhang et al. 1991; D'Orazio et al. 2009). Two methionine sulfoxide reductases (MsrA and B) are supposed to reduce methionine sulfoxide to methionine in *M. tuberculosis*. MsrA and B appear to have redundant functions, since only a mutant lacking both genes would be more sensitive to nitrite and hypochlorite compare to wildtype (Lee et al. 2009).

13.4 Oxidative Stress in Malaria

Besides HIV/AIDS and tuberculosis, malaria is one of the world's most devastating diseases. Although the number of reported malaria cases was reduced by more than 50% between 2000 and 2010, the estimated number of malaria-related deaths is with 665,000–1,133,000 people in 2010 still high (WHO 2011c; Murray et al. 2012). Approximately 86% of malaria deaths globally were of children under 5 years of age, and 91% of the deaths were in Africa (WHO 2011c). Tropical malaria is caused by *Plasmodium falciparum*, a unicellular eukaryotic parasite that lives and multiplies both in *Anopheles* mosquitoes and humans. An infected *Anopheles* mosquito injects sporozoites into the subcutaneous tissue of the human host. The sporozoites migrate to the liver, where they develop into merozoites, which subsequently can invade red blood cells. There, the parasites undergo asexual replication until merozoites are formed and released by rupture of the red blood cell membrane. Most merozoites infect new red blood cells and thus restart the intraerythrocytic cycle leading to the typical symptoms of tropical malaria. Some parasites differentiate into the sexual forms required for transmission into the mosquito vector (for review please see Tuteja 2007; Kappe et al. 2010).

13.4.1 Sources of Oxidative Stress in Malaria Parasites

During their intraerythrocytic development, malaria parasites face high concentrations of oxygen and iron and are exposed to intense oxidative stress derived from different sources. *Plasmodium* parasites degrade host hemoglobin as their major source of nutrients for protein synthesis, which is at the same time also a major source of oxidative stress. Hemoglobin is taken up from the erythrocyte cytoplasm into the acidic food vacuole and is systematically degraded by proteases into free heme and amino acids (Francis et al. 1997). Most of the highly toxic free heme (ferriprotoporphyrin IX) aggregates to crystalline hemozoin, the so-called malaria pigment (Stiebler et al. 2011). Alternative detoxification pathways of heme comprise heme degradation and binding to glutathione or heme-binding proteins (Loria et al. 1999; Ginsburg et al. 1998). However, small amounts of heme escape from the detoxification mechanisms and cause major oxidative damage including lipid peroxidation and DNA damage (reviewed in Kumar and Bandyopadhyay 2005). In the acidic food vacuole, free heme is oxidized from Fe(II) to Fe(III) with concomitant production of superoxide and H_2O_2 (Atamna and Ginsburg 1993). Additionally, malaria parasites have to counteract ROS released by the host immune system in order to fight the infection (Becker et al. 2004).

13.4.2 Oxidative Stress in the Pathogenesis of Malaria

Oxidative stress has been implicated in different aspects of malaria pathogenesis (reviewed in Becker et al. 2004; Hunt and Stocker 1990). Human glucose-6-phosphate dehydrogenase (G6PD) deficiency is the most common enzymopathy and confers resistance to infection and/or the development of severe clinical symptoms of malaria (Ruwende and Hill 1998). Protection from malaria is thought to be based on a lack of reducing equivalents in the form of NADPH, leading to oxidation of hemoglobin to membrane-associated hemochromes, which finally lead to early IgG-based detection and degradation of G6PD-deficient parasitized erythrocytes (Cappadoro et al. 1998). Moreover, a reduced multiplication rate was attributed to an intracellular accumulation of toxic oxidized molecules such as oxidized glutathione and hemozoin. Similar protection mechanisms were proposed for glutathione reductase-deficient erythrocytes (Gallo et al. 2009).

During disease manifestation, the redox balance of the patient is disturbed. The levels of glutathione, tocopherol, catalase, and superoxide dismutase in the erythrocyte and the concentrations of ascorbate and albumin in the plasma are significantly decreased in patients with malaria when compared to uninfected erythrocytes (Das and Nanda 1999; Pabon et al. 2003; Narsaria et al. 2011). Reduced α-tocopherol content in the erythrocyte membrane was associated with malaria and may contribute to erythrocyte loss and anemia in severe malaria (Griffiths et al. 2001). Moreover, increased lipid peroxidation has been reported in patients infected with *falciparum* or *vivax* malaria (Polat et al. 2002; Das et al. 1990; Pabon et al. 2003; Narsaria et al. 2011). Micronutrients including vitamins A, E, and zinc may have a beneficial effect on the severity of malaria by modulating immune response and decreasing oxidative stress (Nussenblatt and Semba 2002; Zeba et al. 2008).

The role of macrophage-derived ROS in controlling malaria infection has been controversially discussed. It has been hypothesized that phagocyte-derived ROS are involved in host immunity against malaria infections (Hunt and Stocker 1990). Superoxide produced by NADPH oxidase was reported to be involved in malaria transmission and gametocyte development but not in parasitemia patterns, as shown in murine malaria models lacking NADPH oxidase (Harada et al. 2001). Similarly, another study did not show any effect of the absence of functional NADPH oxidase and thus of phagocyte-derived ROS on parasitemia and parasite burden in murine malaria models. This indicates that ROS generated by phagocytes are not involved in controlling malaria infection (Potter et al. 2005). Furthermore, phagocyte-derived ROS are not associated with the pathogenesis of cerebral malaria in mice malaria models (Sanni et al. 1999) but have been implicated in neuronal damage in humans (Becker et al. 2004).

In uncomplicated and severe malaria, low levels of nitric oxide (NO) were detected and are associated with increased disease severity and mortality. Plasma levels of IL-10, a cytokine that suppresses NO synthesis, were increased with severity, while levels of NOS2 were decreased in cerebral malaria (Anstey et al. 1996). Moreover, severe malaria is associated with depletion of arginine, the precursor of NO, and elevated levels of plasma arginase, which catabolizes arginine (Weinberg et al. 2008). Therefore, suppression of NO was suggested to protect against severe disease rather than to contribute to malaria pathogenesis (Anstey et al. 1996).

A beneficial role of antioxidants as an adjunctive treatment of severe malaria has been discussed. Treatment of adults with NAC increased the rate of normalization of plasma lactate by a TNF-independent mechanism that was attributed to improved red cell deformability or increased GSH levels (Watt et al. 2002). However, NAC as an adjunctive treatment with artesunate did not influence the disease outcome in patients with severe *falciparum* malaria (Charunwatthana et al. 2009). The effect of NAC appears to be concentration-dependent: low doses of NAC decrease H_2O_2 levels and lipid peroxidation by increasing GSH concentrations, while supplementation with high doses of NAC had the opposite effect *in vitro* (Fitri et al. 2011). Thus, putative beneficial effects of NAC treatment remain controversial.

As explained above, NO is supposed to have a significant function in controlling malaria infection (Weinberg et al. 2008). Supplementation with L-arginine, the substrate of NO synthase and precursor of NO, leads to recovery of arginine levels and improvement of endothelial function in patients with severe malaria as shown in a Phase I trial (Yeo et al. 2008).

13.4.3 Antioxidant Defense System

In order to avoid oxidative damage and maintain a redox balance, malaria parasites employ an efficient combination of antioxidant systems based on GSH and Trx as outlined in the following paragraphs. Moreover, *Plasmodium* codes for two SODs, which convert superoxide radicals to hydrogen peroxide (Gratepanche et al. 2002; Sienkiewicz et al. 2004). However, enzymes with major antioxidant functions in other organisms such as catalase, glutathione peroxidase, and methionine sulfoxide reductase are missing in *Plasmodium* (Sztajer et al. 2001; Clarebout et al. 1998).

13.4.3.1 The Glutathione System

Glutathione is the major low molecular weight thiol in malaria parasites and functions as a thiol redox buffer by cycling between a reduced and an oxidized form (Becker et al. 2003b). GSH is a cofactor for detoxification of electrophilic compounds and methylglyoxal (Harwaldt et al. 2002; Iozef et al. 2003), reduces the dithiol glutaredoxin (Rahlfs et al. 2001), and directly reduces a range of ROS, RNS, protein disulfides, and sulfenates (Becker et al. 2003b). GSH is synthesized by the consecutive activity of γ-glutamyl-cysteine synthetase and glutathione synthetase. A lack of the glutathione biosynthesis pathway leads to decreased GSH levels and a significant growth delay during the intraerythrocytic stage but completely blocks oocyte development in the *Anopheles* vector (Vega-Rodriguez et al. 2009). During the intraerythrocytic stage, the parasites depend on either GSH *de novo* synthesis or efficient reduction of GSSG by glutathione reductase, since a knockout of both enzymes is lethal for the parasites (Pastrana-Mena et al. 2010).

High intracellular concentrations of GSH in malaria parasites are maintained by an NADPH-dependent reaction catalyzed by the flavoenzyme glutathione reductase (GR). A range of studies examined the kinetic mechanism of GR in detail and developed selective inhibitors of the parasite enzymes (e.g. Bohme et al. 2000; Krauth-Siegel et al. 1996; Sarma et al. 2003). As a consequence of its central position in antioxidant defense, *P. falciparum* GR was regarded as a highly attractive drug target (Becker et al. 2003b, 2004). However, recent studies demonstrated that *P. berghei* GR is not essential during the intraerythrocytic stage but is required for sporogony in the *Anopheles* vector (Buchholz et al. 2010; Pastrana-Mena et al. 2010). High functional redundancy in the antioxidant network of *P. falciparum* allows reduction of GSSG by several GR-independent pathways such as direct reduction by Trx, reduction by protein *S*-glutathionylation, and by dihydrolipoamide-dependent reactions catalyzed by Grx (Becker et al. 2003b, 2004; Kanzok et al. 2000). In order to maintain low intracellular GSSG concentrations, the parasite exports GSSG into the erythrocyte cytosol (Atamna and Ginsburg 1997).

A major function of glutathione is the non-enzymatic reduction of Grx, a dithiol protein that catalyzes the reduction of a variety of proteins on the basis of a dithiol exchange mechanism (Holmgren 1989). *Plasmodium* encodes one classic dithiol Grx (i.e., PfGrx) and three monothiol Grx (Rahlfs et al. 2001; Deponte et al. 2005). PfGrx can serve as a hydrogen donor for ribonucleotide reductase (Rahlfs et al. 2001), interacts with several proteins from different metabolic pathways in *Plasmodium* (Sturm et al. 2009), and catalyzes the deglutathionylation of proteins *in vitro* (Kehr et al. 2011).

GSH is a cofactor of glutathione *S*-transferase, an enzyme that catalyzes the conjugation of electrophiles with GSH for detoxification of xenobiotics (reviewed in Eaton and Bammler 1999). *P. falciparum* GST cannot be assigned to any of the described GST classes and has been investigated in detail (Harwaldt et al. 2002; Fritz-Wolf et al. 2003).

13.4.3.2 The Thioredoxin System

Thioredoxins are central proteins in redox homeostasis and reduce a large variety of protein and non-protein substrates. Trx are directly involved in antioxidant defense and furthermore contribute to redox signaling and regulation by modulating the structure and/or activity of many proteins in response to the intracellular redox state (Arner and Holmgren 2000; Holmgren 1989). *Plasmodium* employs three Trx and two Trx-like proteins with distinct subcellular localization (Nickel et al. 2006). Cytosolic Trx1, the most intensely studied *P. falciparum* Trx, detoxifies hydroperoxides and glutathione disulfide directly, functions as an electron donor for peroxiredoxins and plasmoredoxin (see below), and regulates a variety of proteins (Nickel et al. 2006; Sturm et al. 2009; Jortzik et al. 2010).

Thioredoxin reductase (TrxR) maintains Trx in a reduced state in an NADPH-dependent reaction. TrxR regulates antioxidant defense and cell growth either indirectly by reducing Trx or directly by reducing other substrates including hydrogen and lipid peroxides, dehydroascorbate, selenium compounds, Grx, protein disulfide isomerase, and ubiquinone (reviewed in Becker et al. 2000). *P. falciparum* TrxR contains two catalytic centers each consisting of two cysteine residues, while mammalian TrxR contains a selenocysteine-cysteine in its C-terminal active site (Gladyshev et al. 1996; Williams et al. 2000). By using alternative translation initiation, *P. falciparum* expresses two isoforms of TrxR, which are located in the cytosol and in the mitochondrion (Kehr et al. 2010). *P. falciparum* TrxR was discussed as an excellent drug target (Becker et al. 2000; Nickel et al. 2006). Knockout studies demonstrated that TrxR is indeed essential for *P. falciparum* (Krnajski et al. 2002) but not for the rodent malaria parasite *P. berghei* (Buchholz et al. 2010). Whether the discrepancy is due to *in vitro* culture conditions, technical aspects, or indeed major differences in the redox metabolism of *P. falciparum* and *P. berghei* remains to be studied. However, high redundancies in the Trx and GSH systems might allow compensation for the loss of single components. Simultaneous inhibition of the two *Plasmodium* disulfide reductases GR and TrxR provides a good strategy for the development of antimalarial chemotherapeutic interventions (Buchholz et al. 2010).

Trx protects from oxidative damage by serving as an electron donor for Trx-dependent peroxidases, which catalyze the reduction of different peroxides including hydrogen peroxide, peroxynitrite, cumene hydroperoxide, and *tert*-butylhydroperoxide. According to the number of catalytic cysteines, the enzymes are clustered into 1-cys and 2-cys peroxiredoxins (Wood et al. 2003). *Plasmodium* harbors six peroxidases (recently reviewed in Gretes et al. 2012): Prx1a (TPx1) and

Prx1m (TPx2) (Nickel et al. 2005; Komaki-Yasuda et al. 2003), Prx5 (antioxidant protein, AOP) (Sarma et al. 2005), and Prx6 (1-cys peroxiredoxin) (Nickel et al. 2006) are classic peroxiredoxins. Additionally, a glutathione peroxidase-like TPx and a nuclear peroxiredoxin (PrxQ, nPrx) associated with chromatin have been described (Sztajer et al. 2001; Richard et al. 2011). Except for nPrx (Richard et al. 2011), all *P. falciparum* peroxidases prefer Trx as a reducing substrate. Additionally, malaria parasites import human peroxiredoxin 2 as an enzymatic scavenger of hydroperoxides into their cytosol (Koncarevic et al. 2009).

A *Plasmodium*-specific oxidoreductase is plasmoredoxin (Plrx), a dithiol protein belonging to the thioredoxin superfamily. Plrx can be reduced by glutathione but more effectively by Trx and Grx *in vitro*; it transfers electrons to ribonucleotide reductase and glutathione disulfide and interacts with several enzymes involved in different cellular pathways (Becker et al. 2003a; Nickel et al. 2005; Sturm et al. 2009). In *P. berghei*, Plrx is dispensable under unstressed conditions, indicating a high functional redundancy with other redox-active proteins (Buchholz et al. 2008).

References

Adams LB, Dinauer MC, Morgenstern DE, Krahenbuhl JL (1997) Comparison of the roles of reactive oxygen and nitrogen intermediates in the host response to Mycobacterium tuberculosis using transgenic mice. Tuber Lung Dis 78:237–246

Agrawal L, Louboutin JP, Strayer DS (2007) Preventing HIV-1 Tat-induced neuronal apoptosis using antioxidant enzymes: mechanistic and therapeutic implications. Virology 363:462–472

Akif M, Khare G, Tyagi AK, Mande SC, Sardesai AA (2008) Functional studies of multiple thioredoxins from Mycobacterium tuberculosis. J Bacteriol 190:7087–7095

Anstey NM, Weinberg JB, Hassanali MY, Mwaikambo ED, Manyenga D, Misukonis MA, Arnelle DR, Hollis D, McDonald MI, Granger DL (1996) Nitric oxide in Tanzanian children with malaria: inverse relationship between malaria severity and nitric oxide production/nitric oxide synthase type 2 expression. J Exp Med 184:557–567

Aquaro S, Muscoli C, Ranazzi A, Pollicita M, Granato T, Masuelli L, Modesti A, Perno CF, Mollace V (2007) The contribution of peroxynitrite generation in HIV replication in human primary macrophages. Retrovirology 4:76

Argyrou A, Blanchard JS (2004) Flavoprotein disulfide reductases: advances in chemistry and function. Prog Nucleic Acid Res Mol Biol 78:89–142

Argyrou A, Vetting MW, Blanchard JS (2004) Characterization of a new member of the flavoprotein disulfide reductase family of enzymes from Mycobacterium tuberculosis. J Biol Chem 279:52694–52702

Arner ES, Holmgren A (2000) Physiological functions of thioredoxin and thioredoxin reductase. Eur J Biochem 267:6102–6109

Atamna H, Ginsburg H (1993) Origin of reactive oxygen species in erythrocytes infected with Plasmodium falciparum. Mol Biochem Parasitol 61:231–241

Atamna H, Ginsburg H (1997) The malaria parasite supplies glutathione to its host cell–investigation of glutathione transport and metabolism in human erythrocytes infected with Plasmodium falciparum. Eur J Biochem 250:670–679

Atkuri KR, Mantovani JJ, Herzenberg LA, Herzenberg LA (2007) N-Acetylcysteine – a safe antidote for cysteine/glutathione deficiency. Curr Opin Pharmacol 7:355–359

Attarian R, Bennie C, Bach H, Av-Gay Y (2009) Glutathione disulfide and S-nitrosoglutathione detoxification by Mycobacterium tuberculosis thioredoxin system. FEBS Lett 583:3215–3220

Aukrust P, Svardal AM, Muller F, Lunden B, Berge RK, Ueland PM, Froland SS (1995) Increased levels of oxidized glutathione in CD4+ lymphocytes associated with disturbed intracellular redox balance in human immunodeficiency virus type 1 infection. Blood 86:258–267

Auwerx J, Isacsson O, Soderlund J, Balzarini J, Johansson M, Lundberg M (2009) Human glutaredoxin-1 catalyzes the reduction of HIV-1 gp120 and CD4 disulfides and its inhibition reduces HIV-1 replication. Int J Biochem Cell Biol 41:1269–1275

Azimi I, Matthias LJ, Center RJ, Wong JW, Hogg PJ (2010) Disulfide bond that constrains the HIV-1 gp120 V3 domain is cleaved by thioredoxin. J Biol Chem 285:40072–40080

Banki K, Hutter E, Gonchoroff NJ, Perl A (1998) Molecular ordering in HIV-induced apoptosis. Oxidative stress, activation of caspases, and cell survival are regulated by transaldolase. J Biol Chem 273:11944–11953

Becker K, Gromer S, Schirmer RH, Muller S (2000) Thioredoxin reductase as a pathophysiological factor and drug target. Eur J Biochem 267:6118–6125

Becker K, Kanzok SM, Iozef R, Fischer M, Schirmer RH, Rahlfs S (2003a) Plasmoredoxin, a novel redox-active protein unique for malarial parasites. Eur J Biochem 270:1057–1064

Becker K, Rahlfs S, Nickel C, Schirmer RH (2003b) Glutathione – functions and metabolism in the malarial parasite Plasmodium falciparum. Biol Chem 384:551–566

Becker K, Tilley L, Vennerstrom JL, Roberts D, Rogerson S, Ginsburg H (2004) Oxidative stress in malaria parasite-infected erythrocytes: host-parasite interactions. Int J Parasitol 34:163–189

Belding ME, Klebanoff SJ, Ray CG (1970) Peroxidase-mediated virucidal systems. Science 167:195–196

Bohme CC, Arscott LD, Becker K, Schirmer RH, Williams CH Jr (2000) Kinetic characterization of glutathione reductase from the malarial parasite Plasmodium falciparum. Comparison with the human enzyme. J Biol Chem 275:37317–37323

Borges-Santos MD, Moreto F, Pereira PC, Ming-Yu Y, Burini RC (2012) Plasma glutathione of HIV(+) patients responded positively and differently to dietary supplementation with cysteine or glutamine. Nutrition 28(7–8):753–756

Bryk R, Griffin P, Nathan C (2000) Peroxynitrite reductase activity of bacterial peroxiredoxins. Nature 407:211–215

Bryk R, Lima CD, Erdjument-Bromage H, Tempst P, Nathan C (2002) Metabolic enzymes of mycobacteria linked to antioxidant defense by a thioredoxin-like protein. Science 295:1073–1077

Buccigrossi V, Laudiero G, Nicastro E, Miele E, Esposito F, Guarino A (2011) The HIV-1 transactivator factor (Tat) induces enterocyte apoptosis through a redox-mediated mechanism. PLOS One 6:e29436

Buchholz K, Rahlfs S, Schirmer RH, Becker K, Matuschewski K (2008) Depletion of Plasmodium berghei plasmoredoxin reveals a non-essential role for life cycle progression of the malaria parasite. PLOS One 3:e2474

Buchholz K, Putrianti ED, Rahlfs S, Schirmer RH, Becker K, Matuschewski K (2010) Molecular genetics evidence for the in vivo roles of the two major NADPH-dependent disulfide reductases in the malaria parasite. J Biol Chem 285:37388–37395

Buchmeier N, Fahey RC (2006) The mshA gene encoding the glycosyltransferase of mycothiol biosynthesis is essential in Mycobacterium tuberculosis Erdman. FEMS Microbiol Lett 264:74–79

Buchmeier NA, Newton GL, Koledin T, Fahey RC (2003) Association of mycothiol with protection of Mycobacterium tuberculosis from toxic oxidants and antibiotics. Mol Microbiol 47: 1723–1732

Buchmeier NA, Newton GL, Fahey RC (2006) A mycothiol synthase mutant of Mycobacterium tuberculosis has an altered thiol-disulfide content and limited tolerance to stress. J Bacteriol 188:6245–6252

Buhl R, Jaffe HA, Holroyd KJ, Wells FB, Mastrangeli A, Saltini C, Cantin AM, Crystal RG (1989) Systemic glutathione deficiency in symptom-free HIV-seropositive individuals. Lancet 2: 1294–1298

Buonaguro L, Buonaguro FM, Giraldo G, Ensoli B (1994) The human immunodeficiency virus type 1 Tat protein transactivates tumor necrosis factor beta gene expression through a TAR-like structure. J Virol 68:2677–2682

Cappadoro M, Giribaldi G, O'Brien E, Turrini F, Mannu F, Ulliers D, Simula G, Luzzatto L, Arese P (1998) Early phagocytosis of glucose-6-phosphate dehydrogenase (G6PD)-deficient erythrocytes parasitized by Plasmodium falciparum may explain malaria protection in G6PD deficiency. Blood 92:2527–2534

Chan J, Tanaka K, Carroll D, Flynn J, Bloom BR (1995) Effects of nitric oxide synthase inhibitors on murine infection with Mycobacterium tuberculosis. Infect Immun 63:736–740

Charunwatthana P, Abul Faiz M, Ruangveerayut R, Maude RJ, Rahman MR, Roberts LJ 2nd, Moore K, Bin Yunus E, Hoque MG, Hasan MU, Lee SJ, Pukrittayakamee S, Newton PN, White NJ, Day NP, Dondorp AM (2009) N-acetylcysteine as adjunctive treatment in severe malaria: a randomized, double-blinded placebo-controlled clinical trial. Crit Care Med 37:516–522

Choi J, Liu RM, Kundu RK, Sangiorgi F, Wu W, Maxson R, Forman HJ (2000) Molecular mechanism of decreased glutathione content in human immunodeficiency virus type 1 Tat-transgenic mice. J Biol Chem 275:3693–3698

Clarebout G, Slomianny C, Delcourt P, Leu B, Masset A, Camus D, Dive D (1998) Status of Plasmodium falciparum towards catalase. Br J Haematol 103:52–59

Cole ST, Brosch R, Parkhill J, Garnier T, Churcher C, Harris D, Gordon SV, Eiglmeier K, Gas S, Barry CE 3rd, Tekaia F, Badcock K, Basham D, Brown D, Chillingworth T, Connor R, Davies R, Devlin K, Feltwell T, Gentles S, Hamlin N, Holroyd S, Hornsby T, Jagels K, Krogh A, McLean J, Moule S, Murphy L, Oliver K, Osborne J, Quail MA, Rajandream MA, Rogers J, Rutter S, Seeger K, Skelton J, Squares R, Squares S, Sulston JE, Taylor K, Whitehead S, Barrell BG (1998) Deciphering the biology of Mycobacterium tuberculosis from the complete genome sequence. Nature 393:537–544

Cooper AM, Segal BH, Frank AA, Holland SM, Orme IM (2000) Transient loss of resistance to pulmonary tuberculosis in p47(phox−/−) mice. Infect Immun 68:1231–1234

D'Orazio M, Cervoni L, Giartosio A, Rotilio G, Battistoni A (2009) Thermal stability and redox properties of M. tuberculosis CuSOD. Arch Biochem Biophys 486:119–124

Das BS, Nanda NK (1999) Evidence for erythrocyte lipid peroxidation in acute falciparum malaria. Trans R Soc Trop Med Hyg 93:58–62

Das BS, Thurnham DI, Patnaik JK, Das DB, Satpathy R, Bose TK (1990) Increased plasma lipid peroxidation in riboflavin-deficient, malaria-infected children. Am J Clin Nutr 51:859–863

de Quay B, Malinverni R, Lauterburg BH (1992) Glutathione depletion in HIV-infected patients: role of cysteine deficiency and effect of oral N-acetylcysteine. AIDS 6:815–819

Delmas-Beauvieux MC, Peuchant E, Couchouron A, Constans J, Sergeant C, Simonoff M, Pellegrin JL, Leng B, Conri C, Clerc M (1996) The enzymatic antioxidant system in blood and glutathione status in human immunodeficiency virus (HIV)-infected patients: effects of supplementation with selenium or beta-carotene. Am J Clin Nutr 64:101–107

Deponte M, Becker K, Rahlfs S (2005) Plasmodium falciparum glutaredoxin-like proteins. Biol Chem 386:33–40

Deretic V, Philipp W, Dhandayuthapani S, Mudd MH, Curcic R, Garbe T, Heym B, Via LE, Cole ST (1995) Mycobacterium tuberculosis is a natural mutant with an inactivated oxidative-stress regulatory gene: implications for sensitivity to isoniazid. Mol Microbiol 17:889–900

Eaton DL, Bammler TK (1999) Concise review of the glutathione S-transferases and their significance to toxicology. Toxicol Sci 49:156–164

Eck HP, Gmunder H, Hartmann M, Petzoldt D, Daniel V, Droge W (1989) Low concentrations of acid-soluble thiol (cysteine) in the blood plasma of HIV-1-infected patients. Biol Chem Hoppe Seyler 370:101–108

Elbim C, Pillet S, Prevost MH, Preira A, Girard PM, Rogine N, Matusani H, Hakim J, Israel N, Gougerot-Pocidalo MA (1999) Redox and activation status of monocytes from human immunodeficiency virus-infected patients: relationship with viral load. J Virol 73:4561–4566

Fan F, Vetting MW, Frantom PA, Blanchard JS (2009) Structures and mechanisms of the mycothiol biosynthetic enzymes. Curr Opin Chem Biol 13:451–459

Fitri LE, Sardjono TW, Simamora D, Sumarno RP, Setyawati SK (2011) High dose of N-acetylcysteine increase HO and MDA levels and decrease GSH level of HUVECs exposed with malaria serum. Trop Biomed 28:7–15

Flores SC, Marecki JC, Harper KP, Bose SK, Nelson SK, McCord JM (1993) Tat protein of human immunodeficiency virus type 1 represses expression of manganese superoxide dismutase in HeLa cells. Proc Natl Acad Sci USA 90:7632–7636

Flynn JL, Scanga CA, Tanaka KE, Chan J (1998) Effects of aminoguanidine on latent murine tuberculosis. J Immunol 160:1796–1803

Francis SE, Sullivan DJ Jr, Goldberg DE (1997) Hemoglobin metabolism in the malaria parasite Plasmodium falciparum. Annu Rev Microbiol 51:97–123

Fritz-Wolf K, Becker A, Rahlfs S, Harwaldt P, Schirmer RH, Kabsch W, Becker K (2003) X-ray structure of glutathione S-transferase from the malarial parasite Plasmodium falciparum. Proc Natl Acad Sci USA 100:13821–13826

Gallo V, Schwarzer E, Rahlfs S, Schirmer RH, van Zwieten R, Roos D, Arese P, Becker K (2009) Inherited glutathione reductase deficiency and Plasmodium falciparum malaria–a case study. PLOS One 4:e7303

Garbe TR, Hibler NS, Deretic V (1996) Response of Mycobacterium tuberculosis to reactive oxygen and nitrogen intermediates. Mol Med 2:134–142

Gil L, Tarinas A, Hernandez D, Riveron BV, Perez D, Tapanes R, Capo V, Perez J (2010) Altered oxidative stress indexes related to disease progression marker in human immunodeficiency virus infected patients with antiretroviral therapy. Biomed Pharmacother 1(1):8–15

Ginsburg H, Famin O, Zhang J, Krugliak M (1998) Inhibition of glutathione-dependent degradation of heme by chloroquine and amodiaquine as a possible basis for their antimalarial mode of action. Biochem Pharmacol 56:1305–1313

Gladyshev VN, Jeang KT, Stadtman TC (1996) Selenocysteine, identified as the penultimate C-terminal residue in human T-cell thioredoxin reductase, corresponds to TGA in the human placental gene. Proc Natl Acad Sci USA 93:6146–6151

Gomez LM, Anaya JM, Vilchez JR, Cadena J, Hinojosa R, Velez L, Lopez-Nevot MA, Martin J (2007) A polymorphism in the inducible nitric oxide synthase gene is associated with tuberculosis. Tuberculosis (Edinb) 87:288–294

Gougeon ML (2005) To kill or be killed: how HIV exhausts the immune system. Cell Death Differ 12(Suppl 1):845–854

Gratepanche S, Menage S, Touati D, Wintjens R, Delplace P, Fontecave M, Masset A, Camus D, Dive D (2002) Biochemical and electron paramagnetic resonance study of the iron superoxide dismutase from Plasmodium falciparum. Mol Biochem Parasitol 120:237–246

Gretes MC, Poole LB, Karplus PA (2012) Peroxiredoxins in parasites. Antioxid Redox Signal 17(4):608–633

Griffiths MJ, Ndungu F, Baird KL, Muller DP, Marsh K, Newton CR (2001) Oxidative stress and erythrocyte damage in Kenyan children with severe Plasmodium falciparum malaria. Br J Haematol 113:486–491

Hall G, Shah M, McEwan PA, Laughton C, Stevens M, Westwell A, Emsley J (2006) Structure of Mycobacterium tuberculosis thioredoxin C. Acta Crystallogr D Biol Crystallogr 62:1453–1457

Hall G, Bradshaw TD, Laughton CA, Stevens MF, Emsley J (2011) Structure of Mycobacterium tuberculosis thioredoxin in complex with quinol inhibitor PMX464. Protein Sci 20:210–215

Hampton MB, Orrenius S (1998) Redox regulation of apoptotic cell death. Biofactors 8:1–5

Harada M, Owhashi M, Suguri S, Kumatori A, Nakamura M, Kanbara H, Matsuoka H, Ishii A (2001) Superoxide-dependent and -independent pathways are involved in the transmission blocking of malaria. Parasitol Res 87:605–608

Harwaldt P, Rahlfs S, Becker K (2002) Glutathione S-transferase of the malarial parasite Plasmodium falciparum: characterization of a potential drug target. Biol Chem 383:821–830

Herzenberg LA, de Rosa SC, Dubs JG, Roederer M, Anderson MT, Ela SW, Deresinski SC, Herzenberg LA (1997) Glutathione deficiency is associated with impaired survival in HIV disease. Proc Natl Acad Sci USA 94:1967–1972

Hillas PJ, del Alba FS, Oyarzabal J, Wilks A, Ortiz De Montellano PR (2000) The AhpC and AhpD antioxidant defense system of Mycobacterium tuberculosis. J Biol Chem 275:18801–18809
Hiscott J, Kwon H, Genin P (2001) Hostile takeovers: viral appropriation of the NF-kappaB pathway. J Clin Invest 107:143–151
Holmgren A (1989) Thioredoxin and glutaredoxin systems. J Biol Chem 264:13963–13966
Hu Y, Coates AR (2009) Acute and persistent Mycobacterium tuberculosis infections depend on the thiol peroxidase TpX. PLOS One 4:e5150
Hugo M, Turell L, Manta B, Botti H, Monteiro G, Netto LE, Alvarez B, Radi R, Trujillo M (2009) Thiol and sulfenic acid oxidation of AhpE, the one-cysteine peroxiredoxin from Mycobacterium tuberculosis: kinetics, acidity constants, and conformational dynamics. Biochemistry 48:9416–9426
Hunt NH, Stocker R (1990) Oxidative stress and the redox status of malaria-infected erythrocytes. Blood Cells 16:499–526; discussion 527–530
Hurwitz BE, Klaus JR, Llabre MM, Gonzalez A, Lawrence PJ, Maher KJ, Greeson JM, Baum MK, Shor-Posner G, Skyler JS, Schneiderman N (2007) Suppression of human immunodeficiency virus type 1 viral load with selenium supplementation: a randomized controlled trial. Arch Intern Med 167:148–154
Iozef R, Rahlfs S, Chang T, Schirmer H, Becker K (2003) Glyoxalase I of the malarial parasite Plasmodium falciparum: evidence for subunit fusion. FEBS Lett 554:284–288
Jaeger T, Budde H, Flohe L, Menge U, Singh M, Trujillo M, Radi R (2004) Multiple thioredoxin-mediated routes to detoxify hydroperoxides in Mycobacterium tuberculosis. Arch Biochem Biophys 423:182–191
Jin DY, Chae HZ, Rhee SG, Jeang KT (1997) Regulatory role for a novel human thioredoxin peroxidase in NF-kappaB activation. J Biol Chem 272:30952–30961
Jortzik E, Fritz-Wolf K, Sturm N, Hipp M, Rahlfs S, Becker K (2010) Redox regulation of Plasmodium falciparum ornithine delta-aminotransferase. J Mol Biol 402:445–459
Jothivasan VK, Hamilton CJ (2008) Mycothiol: synthesis, biosynthesis and biological functions of the major low molecular weight thiol in actinomycetes. Nat Prod Rep 25:1091–1117
Jung YJ, Lacourse R, Ryan L, North RJ (2002) Virulent but not avirulent Mycobacterium tuberculosis can evade the growth inhibitory action of a T helper 1-dependent, nitric oxide Synthase 2-independent defense in mice. J Exp Med 196:991–998
Kalantari P, Narayan V, Natarajan SK, Muralidhar K, Gandhi UH, Vunta H, Henderson AJ, Prabhu KS (2008) Thioredoxin reductase-1 negatively regulates HIV-1 transactivating protein Tat-dependent transcription in human macrophages. J Biol Chem 283:33183–33190
Kanzok SM, Schirmer RH, Turbachova I, Iozef R, Becker K (2000) The thioredoxin system of the malaria parasite Plasmodium falciparum. Glutathione reduction revisited. J Biol Chem 275:40180–40186
Kao SY, Calman AF, Luciw PA, Peterlin BM (1987) Anti-termination of transcription within the long terminal repeat of HIV-1 by tat gene product. Nature 330:489–493
Kappe SH, Vaughan AM, Boddey JA, Cowman AF (2010) That was then but this is now: malaria research in the time of an eradication agenda. Science 328:862–866
Kehr S, Sturm N, Rahlfs S, Przyborski JM, Becker K (2010) Compartmentation of redox metabolism in malaria parasites. PLOS Pathog 6:e1001242
Kehr S, Jortzik E, Delahunty C, Yates JR 3rd, Rahlfs S, Becker K (2011) Protein s-glutathionylation in malaria parasites. Antioxid Redox Signal 15:2855–2865
Kim H, Lee TH, Park ES, Suh JM, Park SJ, Chung HK, Kwon OY, Kim YK, Ro HK, Shong M (2000) Role of peroxiredoxins in regulating intracellular hydrogen peroxide and hydrogen peroxide-induced apoptosis in thyroid cells. J Biol Chem 275:18266–18270
Koken SE, Greijer AE, Verhoef K, van Wamel J, Bukrinskaya AG, Berkhout B (1994) Intracellular analysis of in vitro modified HIV Tat protein. J Biol Chem 269:8366–8375
Komaki-Yasuda K, Kawazu S, Kano S (2003) Disruption of the Plasmodium falciparum 2-Cys peroxiredoxin gene renders parasites hypersensitive to reactive oxygen and nitrogen species. FEBS Lett 547:140–144

Koncarevic S, Rohrbach P, Deponte M, Krohne G, Prieto JH, Yates J 3rd, Rahlfs S, Becker K (2009) The malarial parasite Plasmodium falciparum imports the human protein peroxiredoxin 2 for peroxide detoxification. Proc Natl Acad Sci USA 106:13323–13328

Koshkin A, Knudsen GM, Ortiz De Montellano PR (2004) Intermolecular interactions in the AhpC/AhpD antioxidant defense system of Mycobacterium tuberculosis. Arch Biochem Biophys 427:41–47

Krauth-Siegel RL, Muller JG, Lottspeich F, Schirmer RH (1996) Glutathione reductase and glutamate dehydrogenase of Plasmodium falciparum, the causative agent of tropical malaria. Eur J Biochem 235:345–350

Krnajski Z, Gilberger TW, Walter RD, Cowman AF, Muller S (2002) Thioredoxin reductase is essential for the survival of Plasmodium falciparum erythrocytic stages. J Biol Chem 277:25970–25975

Kumar S, Bandyopadhyay U (2005) Free heme toxicity and its detoxification systems in human. Toxicol Lett 157:175–188

Kumar A, Farhana A, Guidry L, Saini V, Hondalus M, Steyn AJ (2011) Redox homeostasis in mycobacteria: the key to tuberculosis control? Expert Rev Mol Med 13:e39

Lanzillotti JS, Tang AM (2005) Micronutrients and HIV disease: a review pre- and post-HAART. Nutr Clin Care 8:16–23

Lawn SD, Zumla AI (2011) Tuberculosis. Lancet 378:57–72

Lee WL, Gold B, Darby C, Brot N, Jiang X, de Carvalho LP, Wellner D, St John G, Jacobs WR Jr, Nathan C (2009) Mycobacterium tuberculosis expresses methionine sulphoxide reductases A and B that protect from killing by nitrite and hypochlorite. Mol Microbiol 71:583–593

Leff JA, Oppegard MA, Curiel TJ, Brown KS, Schooley RT, Repine JE (1992) Progressive increases in serum catalase activity in advancing human immunodeficiency virus infection. Free Radic Biol Med 13:143–149

Leistikow RL, Morton RA, Bartek IL, Frimpong I, Wagner K, Voskuil MI (2010) The Mycobacterium tuberculosis DosR regulon assists in metabolic homeostasis and enables rapid recovery from nonrespiring dormancy. J Bacteriol 192:1662–1670

Leto TL, Geiszt M (2006) Role of Nox family NADPH oxidases in host defense. Antioxid Redox Signal 8:1549–1561

Li Y, He ZG (2012) The mycobacterial LysR-type regulator OxyS responds to oxidative stress and negatively regulates expression of the catalase-peroxidase gene. PLOS One 7:e30186

Li Z, Kelley C, Collins F, Rouse D, Morris S (1998) Expression of katG in Mycobacterium tuberculosis is associated with its growth and persistence in mice and guinea pigs. J Infect Dis 177:1030–1035

Loria P, Miller S, Foley M, Tilley L (1999) Inhibition of the peroxidative degradation of haem as the basis of action of chloroquine and other quinoline antimalarials. Biochem J 339(Pt 2): 363–370

Macmicking JD, North RJ, Lacourse R, Mudgett JS, Shah SK, Nathan CF (1997) Identification of nitric oxide synthase as a protective locus against tuberculosis. Proc Natl Acad Sci USA 94:5243–5248

Manca C, Paul S, Barry CE 3rd, Freedman VH, Kaplan G (1999) Mycobacterium tuberculosis catalase and peroxidase activities and resistance to oxidative killing in human monocytes in vitro. Infect Immun 67:74–79

Master SS, Springer B, Sander P, Boettger EC, Deretic V, Timmins GS (2002) Oxidative stress response genes in Mycobacterium tuberculosis: role of ahpC in resistance to peroxynitrite and stage-specific survival in macrophages. Microbiology 148:3139–3144

Masutani H, Ueda S, Yodoi J (2005) The thioredoxin system in retroviral infection and apoptosis. Cell Death Differ 12(Suppl 1):991–998

McDermid JM, Lalonde RG, Gray-Donald K, Baruchel S, Kubow S (2002) Associations between dietary antioxidant intake and oxidative stress in HIV-seropositive and HIV-seronegative men and women. J Acquir Immune Defic Syndr 29:158–164

Miesel L, Weisbrod TR, Marcinkeviciene JA, Bittman R, Jacobs WR Jr (1998) NADH dehydrogenase defects confer isoniazid resistance and conditional lethality in Mycobacterium smegmatis. J Bacteriol 180:2459–2467

Morris D, Guerra C, Donohue C, Oh H, Khurasany M, Venketaraman V (2012) Unveiling the mechanisms for decreased glutathione in individuals with HIV infection. Clin Dev Immunol 2012:734125

Murray CJ, Rosenfeld LC, Lim SS, Andrews KG, Foreman KJ, Haring D, Fullman N, Naghavi M, Lozano R, Lopez AD (2012) Global malaria mortality between 1980 and 2010: a systematic analysis. Lancet 379:413–431

Nakamura H, de Rosa S, Roederer M, Anderson MT, Dubs JG, Yodoi J, Holmgren A, Herzenberg LA (1996) Elevation of plasma thioredoxin levels in HIV-infected individuals. Int Immunol 8:603–611

Narsaria N, Mohanty C, Das BK, Mishra SP, Prasad R (2011) Oxidative stress in children with severe malaria. J Trop Pediatr 58(2):147–150

Newton GL, Fahey RC, Cohen G, Aharonowitz Y (1993) Low-molecular-weight thiols in streptomycetes and their potential role as antioxidants. J Bacteriol 175:2734–2742

Newton GL, Arnold K, Price MS, Sherrill C, Delcardayre SB, Aharonowitz Y, Cohen G, Davies J, Fahey RC, Davis C (1996) Distribution of thiols in microorganisms: mycothiol is a major thiol in most actinomycetes. J Bacteriol 178:1990–1995

Newton GL, Av-Gay Y, Fahey RC (2000) A novel mycothiol-dependent detoxification pathway in mycobacteria involving mycothiol S-conjugate amidase. Biochemistry 39:10739–10746

Newton GL, Buchmeier N, Fahey RC (2008) Biosynthesis and functions of mycothiol, the unique protective thiol of Actinobacteria. Microbiol Mol Biol Rev 72:471–494

Ng VH, Cox JS, Sousa AO, Macmicking JD, McKinney JD (2004) Role of KatG catalase-peroxidase in mycobacterial pathogenesis: countering the phagocyte oxidative burst. Mol Microbiol 52:1291–1302

Nicholas GM, Eckman LL, Newton GL, Fahey RC, Ray S, Bewley CA (2003) Inhibition and kinetics of mycobacterium tuberculosis and mycobacterium smegmatis mycothiol-S-conjugate amidase by natural product inhibitors. Bioorg Med Chem 11:601–608

Nicholson S, Bonecini-Almeida Mda G, Lapa E Silva JR, Nathan C, Xie QW, Mumford R, Weidner JR, Calaycay J, Geng J, Boechat N, Linhares C, Rom W, Ho JL (1996) Inducible nitric oxide synthase in pulmonary alveolar macrophages from patients with tuberculosis. J Exp Med 183:2293–2302

Nickel C, Trujillo M, Rahlfs S, Deponte M, Radi R, Becker K (2005) Plasmodium falciparum 2-Cys peroxiredoxin reacts with plasmoredoxin and peroxynitrite. Biol Chem 386:1129–1136

Nickel C, Rahlfs S, Deponte M, Koncarevic S, Becker K (2006) Thioredoxin networks in the malarial parasite Plasmodium falciparum. Antioxid Redox Signal 8:1227–1239

Nussenblatt V, Semba RD (2002) Micronutrient malnutrition and the pathogenesis of malarial anemia. Acta Trop 82:321–337

Pabon A, Carmona J, Burgos LC, Blair S (2003) Oxidative stress in patients with non-complicated malaria. Clin Biochem 36:71–78

Pantaleo G, Graziosi C, Fauci AS (1993) New concepts in the immunopathogenesis of human immunodeficiency virus infection. N Engl J Med 328:327–335

Park HD, Guinn KM, Harrell MI, Liao R, Voskuil MI, Tompa M, Schoolnik GK, Sherman DR (2003) Rv3133c/dosR is a transcription factor that mediates the hypoxic response of Mycobacterium tuberculosis. Mol Microbiol 48:833–843

Pastrana-Mena R, Dinglasan RR, Franke-Fayard B, Vega-Rodriguez J, Fuentes-Caraballo M, Baerga-Ortiz A, Coppens I, Jacobs-Lorena M, Janse CJ, Serrano AE (2010) Glutathione reductase-null malaria parasites have normal blood stage growth but arrest during development in the mosquito. J Biol Chem 285:27045–27056

Pereira LA, Bentley K, Peeters A, Churchill MJ, Deacon NJ (2000) A compilation of cellular transcription factor interactions with the HIV-1 LTR promoter. Nucleic Acids Res 28:663–668

Polat G, Aslan G, Eskandari GH, Delialioglu N, Bagdatoglu O, Atik U (2002) The role of nitric oxide and lipid peroxidation in patients with Plasmodium vivax malaria. Parasite 9:371–374

Potter SM, Mitchell AJ, Cowden WB, Sanni LA, Dinauer M, de Haan JB, Hunt NH (2005) Phagocyte-derived reactive oxygen species do not influence the progression of murine blood-stage malaria infections. Infect Immun 73:4941–4947
Price TO, Ercal N, Nakaoke R, Banks WA (2005) HIV-1 viral proteins gp120 and Tat induce oxidative stress in brain endothelial cells. Brain Res 1045:57–63
Pym AS, Domenech P, Honore N, Song J, Deretic V, Cole ST (2001) Regulation of catalase-peroxidase (KatG) expression, isoniazid sensitivity and virulence by furA of Mycobacterium tuberculosis. Mol Microbiol 40:879–889
Rahlfs S, Fischer M, Becker K (2001) Plasmodium falciparum possesses a classical glutaredoxin and a second, glutaredoxin-like protein with a PICOT homology domain. J Biol Chem 276:37133–37140
Rawat M, Newton GL, Ko M, Martinez GJ, Fahey RC, Av-Gay Y (2002) Mycothiol-deficient Mycobacterium smegmatis mutants are hypersensitive to alkylating agents, free radicals, and antibiotics. Antimicrob Agents Chemother 46:3348–3355
Rawat M, Uppal M, Newton G, Steffek M, Fahey RC, Av-Gay Y (2004) Targeted mutagenesis of the Mycobacterium smegmatis mca gene, encoding a mycothiol-dependent detoxification protein. J Bacteriol 186:6050–6058
Reiser K, Francois KO, Schols D, Bergman T, Jornvall H, Balzarini J, Karlsson A, Lundberg M (2012) Thioredoxin-1 and protein disulfide isomerase catalyze the reduction of similar disulfides in HIV gp120. Int J Biochem Cell Biol 44(3):556–562
Rho BS, Hung LW, Holton JM, Vigil D, Kim SI, Park MS, Terwilliger TC, Pedelacq JD (2006) Functional and structural characterization of a thiol peroxidase from Mycobacterium tuberculosis. J Mol Biol 361:850–863
Richard MJ, Guiraud P, Didier C, Seve M, Flores SC, Favier A (2001) Human immunodeficiency virus type 1 Tat protein impairs selenoglutathione peroxidase expression and activity by a mechanism independent of cellular selenium uptake: consequences on cellular resistance to UV-A radiation. Arch Biochem Biophys 386:213–220
Richard D, Bartfai R, Volz J, Ralph SA, Muller S, Stunnenberg HG, Cowman AF (2011) A genome-wide chromatin-associated nuclear peroxiredoxin from the malaria parasite Plasmodium falciparum. J Biol Chem 286:11746–11755
Roederer M, Staal FJ, Raju PA, Ela SW, Herzenberg LA (1990) Cytokine-stimulated human immunodeficiency virus replication is inhibited by N-acetyl-L-cysteine. Proc Natl Acad Sci USA 87:4884–4888
Roederer M, Staal FJ, Osada H, Herzenberg LA (1991) CD4 and CD8 T cells with high intracellular glutathione levels are selectively lost as the HIV infection progresses. Int Immunol 3:933–937
Romani B, Engelbrecht S, Glashoff RH (2010) Functions of Tat: the versatile protein of human immunodeficiency virus type 1. J Gen Virol 91:1–12
Romero-Alvira D, Roche E (1998) The keys of oxidative stress in acquired immune deficiency syndrome apoptosis. Med Hypotheses 51:169–173
Rouse DA, Devito JA, Li Z, Byer H, Morris SL (1996) Site-directed mutagenesis of the katG gene of Mycobacterium tuberculosis: effects on catalase-peroxidase activities and isoniazid resistance. Mol Microbiol 22:583–592
Ruwende C, Hill A (1998) Glucose-6-phosphate dehydrogenase deficiency and malaria. J Mol Med (Berl) 76:581–588
Saitoh M, Nishitoh H, Fujii M, Takeda K, Tobiume K, Sawada Y, Kawabata M, Miyazono K, Ichijo H (1998) Mammalian thioredoxin is a direct inhibitor of apoptosis signal-regulating kinase (ASK) 1. EMBO J 17:2596–2606
Sanni LA, Fu S, Dean RT, Bloomfield G, Stocker R, Chaudhri G, Dinauer MC, Hunt NH (1999) Are reactive oxygen species involved in the pathogenesis of murine cerebral malaria? J Infect Dis 179:217–222
Sareen D, Newton GL, Fahey RC, Buchmeier NA (2003) Mycothiol is essential for growth of Mycobacterium tuberculosis Erdman. J Bacteriol 185:6736–6740

Sarma GN, Savvides SN, Becker K, Schirmer M, Schirmer RH, Karplus PA (2003) Glutathione reductase of the malarial parasite Plasmodium falciparum: crystal structure and inhibitor development. J Mol Biol 328:893–907

Sarma GN, Nickel C, Rahlfs S, Fischer M, Becker K, Karplus PA (2005) Crystal structure of a novel Plasmodium falciparum 1-Cys peroxiredoxin. J Mol Biol 346:1021–1034

Sassetti CM, Boyd DH, Rubin EJ (2003) Genes required for mycobacterial growth defined by high density mutagenesis. Mol Microbiol 48:77–84

Scanga CA, Mohan VP, Tanaka K, Alland D, Flynn JL, Chan J (2001) The inducible nitric oxide synthase locus confers protection against aerogenic challenge of both clinical and laboratory strains of Mycobacterium tuberculosis in mice. Infect Immun 69:7711–7717

Schreck R, Rieber P, Baeuerle PA (1991) Reactive oxygen intermediates as apparently widely used messengers in the activation of the NF-kappa B transcription factor and HIV-1. EMBO J 10:2247–2258

Sherman DR, Sabo PJ, Hickey MJ, Arain TM, Mahairas GG, Yuan Y, Barry CE 3rd, Stover CK (1995) Disparate responses to oxidative stress in saprophytic and pathogenic mycobacteria. Proc Natl Acad Sci USA 92:6625–6629

Shiloh MU, Nathan CF (2000) Reactive nitrogen intermediates and the pathogenesis of Salmonella and mycobacteria. Curr Opin Microbiol 3:35–42

Shiloh MU, Manzanillo P, Cox JS (2008) Mycobacterium tuberculosis senses host-derived carbon monoxide during macrophage infection. Cell Host Microbe 3:323–330

Shor-Posner G, Lecusay R, Miguez MJ, Moreno-Black G, Zhang G, Rodriguez N, Burbano X, Baum M, Wilkie F (2003) Psychological burden in the era of HAART: impact of selenium therapy. Int J Psychiatry Med 33:55–69

Sienkiewicz N, Daher W, Dive D, Wrenger C, Viscogliosi E, Wintjens R, Jouin H, Capron M, Muller S, Khalife J (2004) Identification of a mitochondrial superoxide dismutase with an unusual targeting sequence in Plasmodium falciparum. Mol Biochem Parasitol 137:121–132

Singh A, Guidry L, Narasimhulu KV, Mai D, Trombley J, Redding KE, Giles GI, Lancaster JR Jr, Steyn AJ (2007) Mycobacterium tuberculosis WhiB3 responds to O2 and nitric oxide via its [4Fe-4S] cluster and is essential for nutrient starvation survival. Proc Natl Acad Sci USA 104:11562–11567

Singh A, Crossman DK, Mai D, Guidry L, Voskuil MI, Renfrow MB, Steyn AJ (2009) Mycobacterium tuberculosis WhiB3 maintains redox homeostasis by regulating virulence lipid anabolism to modulate macrophage response. PLOS Pathog 5:e1000545

Singhal N, Austin J (2002) A clinical review of micronutrients in HIV infection. J Int Assoc Physicians AIDS Care (Chic) 1:63–75

Staal FJ, Roederer M, Herzenberg LA (1990) Intracellular thiols regulate activation of nuclear factor kappa B and transcription of human immunodeficiency virus. Proc Natl Acad Sci USA 87:9943–9947

Stehbens WE (2004) Oxidative stress in viral hepatitis and AIDS. Exp Mol Pathol 77:121–132

Steyn AJ, Collins DM, Hondalus MK, Jacobs WR Jr, Kawakami RP, Bloom BR (2002) Mycobacterium tuberculosis WhiB3 interacts with RpoV to affect host survival but is dispensable for in vivo growth. Proc Natl Acad Sci USA 99:3147–3152

Stiebler R, Soares JB, Timm BL, Silva JR, Mury FB, Dansa-Petretski M, Oliveira MF (2011) On the mechanisms involved in biological heme crystallization. J Bioenerg Biomembr 43:93–99

Sturm N, Jortzik E, Mailu BM, Koncarevic S, Deponte M, Forchhammer K, Rahlfs S, Becker K (2009) Identification of proteins targeted by the thioredoxin superfamily in Plasmodium falciparum. PLOS Pathog 5:e1000383

Suresh DR, Annam V, Pratibha K, Prasad BV (2009) Total antioxidant capacity–a novel early bio-chemical marker of oxidative stress in HIV infected individuals. J Biomed Sci 16:61

Sztajer H, Gamain B, Aumann KD, Slomianny C, Becker K, Brigelius-Flohe R, Flohe L (2001) The putative glutathione peroxidase gene of Plasmodium falciparum codes for a thioredoxin peroxidase. J Biol Chem 276:7397–7403

Trujillo M, Mauri P, Benazzi L, Comini M, de Palma A, Flohe L, Radi R, Stehr M, Singh M, Ursini F, Jaeger T (2006) The mycobacterial thioredoxin peroxidase can act as a one-cysteine peroxiredoxin. J Biol Chem 281:20555–20566
Tuteja R (2007) Malaria – an overview. FEBS J 274:4670–4679
Vega-Rodriguez J, Franke-Fayard B, Dinglasan RR, Janse CJ, Pastrana-Mena R, Waters AP, Coppens I, Rodriguez-Orengo JF, Srinivasan P, Jacobs-Lorena M, Serrano AE (2009) The glutathione biosynthetic pathway of Plasmodium is essential for mosquito transmission. PLOS Pathog 5:e1000302
Velez DR, Hulme WF, Myers JL, Weinberg JB, Levesque MC, Stryjewski ME, Abbate E, Estevan R, Patillo SG, Gilbert JR, Hamilton CD, Scott WK (2009) NOS2A, TLR4, and IFNGR1 interactions influence pulmonary tuberculosis susceptibility in African-Americans. Hum Genet 126:643–653
Vilcheze C, Jacobs WR Jr (2007) The mechanism of isoniazid killing: clarity through the scope of genetics. Annu Rev Microbiol 61:35–50
Vilcheze C, Weisbrod TR, Chen B, Kremer L, Hazbon MH, Wang F, Alland D, Sacchettini JC, Jacobs WR Jr (2005) Altered NADH/NAD+ ratio mediates coresistance to isoniazid and ethionamide in mycobacteria. Antimicrob Agents Chemother 49:708–720
Vogt RN, Steenkamp DJ, Zheng R, Blanchard JS (2003) The metabolism of nitrosothiols in the Mycobacteria: identification and characterization of S-nitrosomycothiol reductase. Biochem J 374:657–666
Voskuil MI, Bartek IL, Visconti K, Schoolnik GK (2011) The response of mycobacterium tuberculosis to reactive oxygen and nitrogen species. Front Microbiol 2:105
Washington AT, Singh G, Aiyar A (2010) Diametrically opposed effects of hypoxia and oxidative stress on two viral transactivators. Virol J 7:93
Watt G, Jongsakul K, Ruangvirayuth R (2002) A pilot study of N-acetylcysteine as adjunctive therapy for severe malaria. QJM 95:285–290
Weinberg JB, Lopansri BK, Mwaikambo E, Granger DL (2008) Arginine, nitric oxide, carbon monoxide, and endothelial function in severe malaria. Curr Opin Infect Dis 21:468–475
Wengenack NL, Jensen MP, Rusnak F, Stern MK (1999) Mycobacterium tuberculosis KatG is a peroxynitritase. Biochem Biophys Res Commun 256:485–487
WHO (2011a) Global HIV/AIDS response. Progress Report 2011. Available http://whqlibdoc.who.int/publications/2011/9789241502986_eng.pdf
WHO (2011b) Global tuberculosis control 2011 [Online]. Available http://www.who.int/tb/publications/global_report/en/. Accessed 10 Feb 2012
WHO (2011c) World malaria report 2011 [Online]. Available http://www.who.int/malaria/world_malaria_report_2011/en/. Accessed 14 Feb 2012
Williams CH, Arscott LD, Muller S, Lennon BW, Ludwig ML, Wang PF, Veine DM, Becker K, Schirmer RH (2000) Thioredoxin reductase two modes of catalysis have evolved. Eur J Biochem 267:6110–6117
Wood ZA, Schroder E, Robin Harris J, Poole LB (2003) Structure, mechanism and regulation of peroxiredoxins. Trends Biochem Sci 28:32–40
Xu X, Vilcheze C, Av-Gay Y, Gomez-Velasco A, Jacobs WR Jr (2011) Precise null deletion mutations of the mycothiol synthesis genes reveal their role in isoniazid and ethionamide resistance in Mycobacterium smegmatis. Antimicrob Agents Chemother 55:3133–3139
Yeo TW, Lampah DA, Gitawati R, Tjitra E, Kenangalem E, McNeil YR, Darcy CJ, Granger DL, Weinberg JB, Lopansri BK, Price RN, Duffull SB, Celermajer DS, Anstey NM (2008) Recovery of endothelial function in severe falciparum malaria: relationship with improvement in plasma L-arginine and blood lactate concentrations. J Infect Dis 198:602–608
Zahrt TC, Deretic V (2002) Reactive nitrogen and oxygen intermediates and bacterial defenses: unusual adaptations in Mycobacterium tuberculosis. Antioxid Redox Signal 4:141–159
Zahrt TC, Song J, Siple J, Deretic V (2001) Mycobacterial FurA is a negative regulator of catalase-peroxidase gene katG. Mol Microbiol 39:1174–1185

Zeba AN, Sorgho H, Rouamba N, Zongo I, Rouamba J, Guiguemde RT, Hamer DH, Mokhtar N, Ouedraogo JB (2008) Major reduction of malaria morbidity with combined vitamin A and zinc supplementation in young children in Burkina Faso: a randomized double blind trial. Nutr J 7:7

Zhang Y, Lathigra R, Garbe T, Catty D, Young D (1991) Genetic analysis of superoxide dismutase, the 23 kilodalton antigen of Mycobacterium tuberculosis. Mol Microbiol 5:381–391

Zhang Y, Heym B, Allen B, Young D, Cole S (1992) The catalase-peroxidase gene and isoniazid resistance of Mycobacterium tuberculosis. Nature 358:591–593

Zhang Z, Hillas PJ, Ortiz De Montellano PR (1999) Reduction of peroxides and dinitrobenzenes by Mycobacterium tuberculosis thioredoxin and thioredoxin reductase. Arch Biochem Biophys 363:19–26

Chapter 14
Role of Oxidative Stress in Aging

D. Knoefler, H.L. Tienson, and U. Jakob

Abstract The question of why we age has given rise to many different theories over the last decades. One of the most popular and long-lasting hypothesis is the free radical theory of aging. It postulates that endogenously generated reactive oxygen species (ROS) accumulate over time, causing damage to cellular macromolecules and eventually leading to physiological decline, disease, aging, and death. Over the years, a multitude of correlative evidence has been collected in favor of this aging theory, including the discovery that aging and many age-related diseases are accompanied by substantial cellular oxidative damage. However, genetic manipulation of components of cellular antioxidant defense systems in model organisms, like *Caenorhabditis elegans*, *Drosophila melanogaster* or mice have generated conflicting results and suggested a more complex interplay between endogenous oxidants, antioxidants, and lifespan. The fact that ROS play important roles as second messengers in signaling processes, in hormesis, and during the oxidative burst in innate immune cells, likely contributes to the complexity of this issue. In this chapter, we present an overview of the most crucial experiments conducted to address the free radical theory of aging. Our conclusion is that ROS are major

D. Knoefler
Department of Molecular, Cellular and Developmental Biology, University of Michigan, Ann Arbor, MI, USA

H.L. Tienson
Department of Molecular, Cellular and Developmental Biology, University of Michigan, Ann Arbor, MI, USA

Department of Chemistry and Biochemistry, UCLA, Los Angeles, CA, USA

U. Jakob (✉)
Department of Molecular, Cellular and Developmental Biology, University of Michigan, Ann Arbor, MI, USA

Department of Biological Chemistry, University of Michigan, Ann Arbor, MI, USA
e-mail: ujakob@umich.edu

U. Jakob and D. Reichmann (eds.), *Oxidative Stress and Redox Regulation*,
DOI 10.1007/978-94-007-5787-5_14,

players involved in lifespan and aging but likely not (only) in their role as cytotoxic agents but as regulators of essential physiological processes in the cell.

Keywords Redox regulation • Lifespan • Reactive oxygen species • Caloric restriction • Insulin/IGF signaling • Free radical theory of aging

14.1 The Free Radical Theory of Aging

Max Rubner was the first to suggest that aging might be connected to energy metabolism after he observed that organisms with different lifespans expend the same total amount of energy (Rubner 1908). The idea that organisms have a fixed amount of "vital substances", which, when utilized faster, would shorten lifespan formed the basis of the 'rate-of-living' theory proposed by Raymond Pearl in 1921 (Pearl 1921). Although this theory was never proven to be valid, it drew attention to the concept that oxygen metabolism and lifespan might be connected. When Denham Harman realized that ionizing radiation, which induces the formation of oxygen radicals, causes biological effects that are very similar to the physiological changes that occur during aging, he postulated the 'free radical theory of aging' (Harman 1956). This hypothesis suggested that free radicals, which are generated by cells themselves, accumulate over time, leading to increased cell and tissue damage and eventually causing physiological decline and death. The suggestion that harmful reactive oxygen species (ROS) are endogenously produced was initially received with skepticism but gained acceptance with the discovery of superoxide dismutase (SOD), an enzyme whose sole function is the specific removal of superoxide from cells and organisms (McCord and Fridovic 1969). The 'free radical theory of aging' was later modified to the 'mitochondrial free radical theory of aging' (Harman 1972) to take into account the fact that mitochondria are the major source and also the major target of ROS. To acknowledge the involvement of other non-radical oxygen species, like hydrogen peroxide, Harman's theory underwent a final re-definition and is now often referred to as the 'oxidative stress hypothesis of aging' (Yu and Yang 1996).

Since the inception of the free radical theory of aging, numerous studies have been conducted providing convincing evidence that cells constantly produce ROS, not only during mitochondrial respiration but also during host defense, cell signaling and many other physiological and pathological events (Trachootham et al. 2008; Droge 2002). To counteract free oxygen radicals, aerobic organisms have evolved a number of highly efficient antioxidant defense systems, which include ROS detoxifying enzymes, small molecules ROS scavengers and oxidoreductases. These systems appear to work together to maintain a crucial balance of pro-oxidants and antioxidants within cells and sub-cellular compartments, a process commonly referred to as redox homeostasis (Finkel and Holbrook 2000). Shifting the equilibrium towards more oxidizing conditions (i.e., oxidative stress) either by increasing the levels of pro-oxidants or by decreasing the cell's antioxidant capacity, leads to the toxic accumulation of ROS, which damage cellular macromolecules, including nucleic acids, lipids and proteins. Oxidative damage has been associated with aging as well as many age-related conditions, including cancer, diabetes, artherosclerosis, cardiovascular diseases and a variety of neurodegenerative diseases (Barnham et al. 2004; Ceriello and Motz 2004; Victor et al. 2009; Reuter et al. 2010). Yet, despite the wealth of studies that have been conducted to test the free radical theory of aging, the jury is still out on whether radical formation is the primary cause of aging or represents a secondary effect of aging and age-related diseases. This is in part due to the recent realization that ROS are not toxic *per se*. In fact, it is now clear that cells need to maintain certain levels of oxidants to be able to differentiate, develop and to overall function properly (Finkel and Holbrook 2000). Many physiological processes, including cell signaling (Finkel 2011b; D'Autreaux and Toledano 2007; Ghezzi et al. 2005), protein folding (Kakihana et al. 2012; Margittai and Banhegyi 2010), development (Hernandez-Garcia et al. 2010), and immune response require the presence of certain levels of oxidants (Finkel 2011a). These findings imply that while shifting the redox balance towards pro-oxidants is clearly toxic to the cell, shifting the redox balance towards antioxidants might possible not be beneficial either, as it will interfere with the physiological role that low ROS levels play in cells and organisms. In the following chapter, we will summarize the current view on the role of ROS in aging and age-related diseases, attempting to provide a balanced assessment of the most popular aging theory postulated thus far.

14.2 Interplay of Oxidants and Antioxidants

14.2.1 Physiological Occurrence of Reactive Oxygen Species

14.2.1.1 Oxidant Generation in the Mitochondrial Electron Transport Chain (ETC)

The electron transport chain (ETC) in mitochondria is generally considered to be the major source of ROS in the eukaryotic cell (Cadenas and Davies 2000). Mitochondria produce the energy to oxidatively phosphorylate ADP, utilizing an electrochemical proton gradient, which is generated by a series of redox reactions located in the inner membrane. In a stepwise reaction catalyzed by four enzyme complexes (I–IV), electrons are passed from NADH to the more electronegative electron acceptor oxygen. Three of the complexes (I, III, and IV) also function as proton pumps, which utilize the energy released from the electron transport chain to transfer protons from the matrix into the intermembrane space. The proton gradient is subsequently utilized by complex V (ATP synthase) to drive ATP production. Over 95% of inhaled oxygen is used in this process (Cadenas and Davies 2000). Although very efficient and tightly regulated, the electron transport chain can lead to

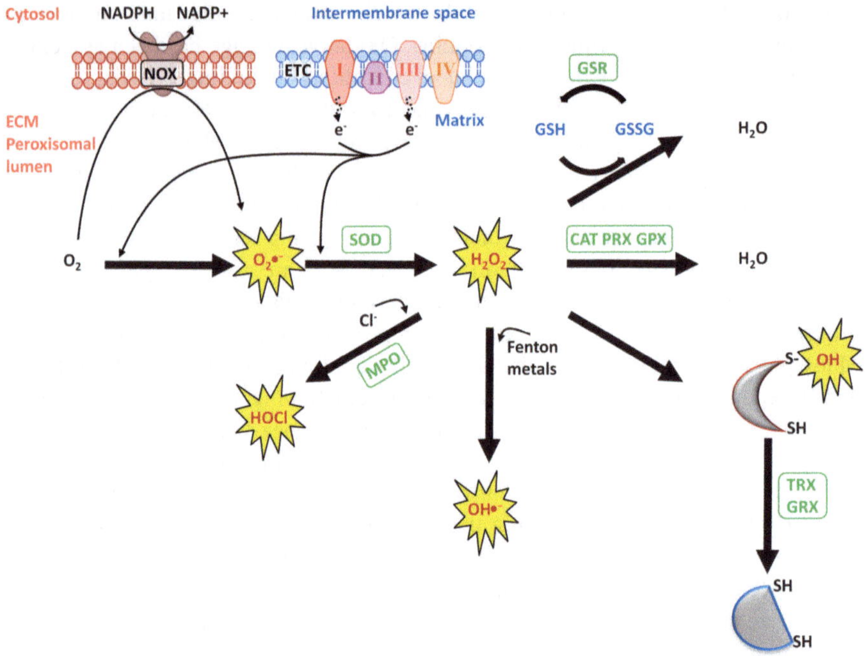

Fig. 14.1 Interplay between oxidants and antioxidants. Reactive oxygen species (ROS) are produced by members of the electron transport chain (ETC), located in the inner membrane of the mitochondria, and by transmembrane NADPH oxidases (NOXs), located in plasma and peroxisomal membranes. Electrons, which constantly leak during the electron transport chain, react with molecular oxygen to form superoxide or hydrogen peroxide (H_2O_2). NADPH oxidases utilize cytosolic NADPH to generate superoxide ($O_2•^-$) either in peroxisomes or the extracellular matrix (ECM). Superoxide is rapidly dismutated to the slow-acting hydrogen peroxide (H_2O_2) in a process that is catalyzed by superoxide dismutase (SOD). H_2O_2 can react with chloride ions to generate the very potent oxidant hypochlorous acid (HOCl), in a process that is catalyzed by myeloperoxidases (MPO) within phagocytes. HOCl is a fast acting oxidant, targeting sulfur-containing amino acids and causing widespread protein aggregation *in vivo*. In the presence of Fenton metals (i.e., iron, copper), peroxide rapidly forms highly reactive hydroxyl radicals ($OH•^-$), which react with and potentially destroy all cellular macromolecules in their vicinity. To detoxify H_2O_2, cells utilize a combination of enzymatic clearance systems, consisting of catalase (CAT), peroxiredoxin (PRX), and glutathione peroxidase (GPX), as well as non-enzymatic small molecule scavengers. One of these scavengers is the small tripeptide glutathione (GSH), which becomes oxidized to GSSG in the process. Regeneration of GSH is achieved by glutathione reductase (GSR). Other peroxide scavengers are surface thiols in proteins, which undergo sulfenic acids (SOH) formation (Hansen et al. 2009; Murphy 2012). Sulfenates are either directly reduced by the thioredoxin (TRX) system or undergo S-glutathionylation, which is reversed by the glutaredoxin (GRX) system

mono- or bivalent reduction of oxygen under physiological conditions, giving rise to superoxide anions and hydrogen peroxide, respectively (Cadenas and Davies 2000; Klotz and Sies 2009) (Fig. 14.1). It is estimated that up to 2% of the molecular oxygen used in mitochondria escapes in form of superoxide anion radicals (Chance and Williams 1956), with complex I and III considered to be the main superoxide producers (Turrens 1997). Not surprisingly, the generation of superoxide and hydrogen peroxide is thus dependent on the mitochondrial metabolic state. Excess of dietary substrates or decreased ATP production due to lack of ADP will stall the flow of electrons through the ETC, which increases electron leakage and hence ROS formation. In contrast, decreasing the metabolic rate by reducing the amount of substrate intake is thought to reduce electron leakage and thus to minimize superoxide and peroxide generation in mitochondria (Heilbronn and Ravussin 2003).

14.2.1.2 Oxidant Production by NADPH Oxidases and Dual Oxidases

NADPH oxidases (NOX) and dual oxidases (DUOX), which are universally distributed in cells and organisms, generate ROS upon exposure to a variety of stimuli, including growth factors, cytokines or bacterial invasion (Lambeth 2004) (Fig. 14.1). During the innate immune response, for instance, invading microorganisms are engulfed by phagocytes. This process triggers the activation of phagocytic NADPH oxidases, which produce large quantities of superoxide anions by transferring electrons from NADPH to oxygen (i.e., respiratory burst). Superoxide radicals are then converted to hydrogen peroxide, which is either directly released into the phagosome of phagocytes or used by myeloperoxidases in neutrophils (a subgroup of phagocytes) to form the potent antimicrobial hypochlorous acid (the active ingredient of household bleach). Dysfunction of phagocytic NADPH oxidase has been implicated in a number of inheritable immunodeficiency's, such as chronic granulomatous disease (Bedard and Krause 2007), which is characterized by the inability of the innate immune system to kill invading pathogens due to a failure to produce sufficient amounts of ROS (Lambeth 2004).

In addition to NADPH oxidases in phagocytic cells, isoforms of NADPH oxidases are involved in a host of other physiological processes. Growth factors, such as angiotensin II, platelet-derived growth factor (PDGF), or vascular endothelial growth factor (VEGF) utilize NOX-mediated ROS signaling to regulate angiogenesis and blood pressure, among other processes (Lambeth 2004; Ushio-Fukai and Nakamura 2008). Thyroidal NADPH dual oxidases, in contrast, provide hydrogen peroxide for thyroid hormone synthesis (Nauseef 2008; Dupuy et al. 1991). Typically membrane-bound, NADPH oxidases utilize cytosolic NADPH to generate superoxide either in the extracellular matrix or the lumen of intracellular organelles (Fig. 14.1). While superoxide itself is not membrane permeable, it is either transported to other cell compartments by ion channels or converted into the highly diffusible hydrogen peroxide. Many NADPH oxidases are ubiquitously expressed and are thought of being capable of generating higher ROS levels in a regulated manner than those continuously produced during respiration (Krause 2007). It is thus not surprising that increased expression and/or activity of several NOX family members has been implicated to play a key role in a number of age-related diseases, including cancer, cardiovascular diseases and neurodegenerative disorders (Bedard and Krause 2007; Krause 2007).

14.2.1.3 Oxidants as By-Products of Biochemical Reactions

In addition to reactions catalyzed by proteins of the ETC and by NADPH oxidases, many other cellular reactions have been shown to produce ROS. In peroxisomes, for example, electrons generated during the β-oxidation of long fatty acids are transferred onto molecular oxygen instead of components of the ETC, thereby producing hydrogen peroxide. Oxidative deamination of aromatic (dietary) amines and monoamine neurotransmitters, such as serotonin and dopamine, is catalyzed by monoamine oxidases (MAO) in a process that leads to the production of potentially neurotoxic by-products, including ammonia and hydrogen peroxide (Bortolato et al. 2008). Other major endogenous ROS producers belong to the heme-containing cytochrome P450 protein superfamily. Members of this family are involved in oxidizing endogenous substrates as well as a broad range of exogenous compounds, including drugs, carcinogens and other xenobiotics. Since the monoxygenation of these substrates is inefficiently coupled to the electron transfer from NADPH to cytochrome P450, it causes a continuous leakage of electrons, resulting in ROS formation even in the absence of substrates (Zangar et al. 2004). Some xenobiotic compounds such as alcohol or drugs can further increase the P450-uncoupling reaction, thereby increasing ROS generation even more. The need to maintain low intracellular ROS level has apparently resulted in the development of feedback mechanisms as the presence of high ROS levels was recently found to decrease cytochrome P450 levels (Zangar et al. 2004).

14.2.2 Antioxidants – Maintaining the Balance

14.2.2.1 Detoxification Systems and ROS Scavengers

Superoxide anions are known to spontaneously dismutate to hydrogen peroxide (H_2O_2) at a slow rate. *In vivo*, this process is massively accelerated by the presence of superoxide dismutases (SODs), which are located in the cytosol, the mitochondrial intermembrane space and matrix, as well as in the extracellular space (Fridovic 1972; Zelko et al. 2002). While H_2O_2 is less reactive and more stable than other ROS (Giorgio et al. 2007), it reacts rapidly with free ferrous (Fe^{2+}) iron in the Fenton reaction, which generates hydroxyl radicals, one of the most reactive oxygen species known. Hence, removal of peroxide is of utmost importance to avoid widespread oxidative damage. Enzymatic clearance of hydrogen peroxide is performed by catalases, glutathione (GSH) peroxidases, and peroxiredoxins (Fig. 14.1). Catalases are mainly found in the cytosol and peroxisomes, as well as in the mitochondrial matrix of some highly metabolically active tissues, such as heart and liver (Radi et al. 1991; Salvi et al. 2007). They catalyze the decomposition of two hydrogen peroxide molecules to oxygen and two water molecules, using one of the fastest turnover rates known for any enzyme (Nicholls et al. 2001). While catalases contribute to hydrogen peroxide decomposition at high hydrogen peroxide concentrations, selenocysteine-containing GSH peroxidases work efficiently at low peroxide levels, suggesting that they serve as the predominant peroxide scavengers under physiological H_2O_2 concentrations (Makino et al. 1994). GSH peroxidases catalyze the reduction of hydrogen peroxide to water by utilizing reduced glutathione (GSH) as electron donor. GSH, a highly abundant tripeptide is oxidized in this process to disulfide-bonded GSSG, and subsequently regenerated by GSH reductase, a NADPH-dependent oxidoreductase. The third group of peroxide-detoxifying enzymes is constituted by peroxiredoxins, which compensate for their slow reaction rates with extremely high cellular concentrations, making them one of the most abundant enzymes in many organisms. The structure, function and role of peroxiredoxins in response to oxidative stress is expertly reviewed in Chap. 4 of this book.

In addition to the various antioxidant enzymes that clear reactive oxygen species, organisms have evolved various small molecules such as glutathione, metallothioneins and vitamins, which are capable of scavenging oxygen radicals. While the non-protein thiol γ-L-glutamyl-L-cysteinyl-glycine (glutathione) can act as reductant for peroxide (Fig. 14.1) and free radicals (Orrenius and Moldeus 1984), metallothioneins, which are low molecular weight metal-containing proteins, are capable of scavenging hydroxyl radicals and superoxide (Thornalley and Vasak 1985). The water-soluble ascorbate (vitamin C) scavenges oxygen radicals in aqueous solution whereas α-tocopherol (vitamin E) protects membranes from radical formation (Niki 1987).

14.2.2.2 Maintaining and Restoring Redox Homeostasis

One of the most ROS-sensitive and reactive target in the cell is the sulfur-containing amino acid cysteine. Many cysteine thiols in proteins have been shown to rapidly react with peroxide, HOCl and/or NO, thereby forming a variety of different oxidative modifications, including sulfenic acid and disulfide bond formation, mixed disulfides with glutathione (S-glutathionylation) or S-nitrosylation (Winterbourn and Hampton 2008). Because of their high sensitivity towards oxidation, cysteine thiols are also the amino acids of choice for those proteins, whose function is regulated by the redox conditions of the environment. Redox sensitive proteins are found to play roles in the majority of cellular functions, ranging from signal transduction (e.g., phosphatases and kinases) and gene expression (e.g., p53) to metabolism (e.g., GapDH) and proteostasis (e.g., Cdc-48) (Brandes et al. 2009; Kumsta et al. 2011). Oxidative modification of redox sensitive cysteines leads to the transient activation (e.g., OxyR, Hsp33, Nrf2) or inactivation (e.g., phosphatases) of a protein's function, making thiol-modifications uniquely able to fine-tune cellular pathways and response systems to the cellular redox environment.

Two highly conserved enzymatic systems, the thioredoxin system and the glutaredoxin system are responsible for maintaining redox homeostasis and reducing most forms of oxidative thiol modifications in proteins (expertly reviewed in Chap. 3). The thioredoxin system consists of the small oxidoreductase thioredoxin, which uses direct thiol-disulfide exchange reactions to reduce sulfenic acids, disulfide bonds or S-nitrosylated cysteines in proteins (Collet and Messens 2010). Thioredoxins are then reduced by thioredoxin reductase, a selenocysteine-containing enzyme in eukaryotes, which utilizes NADPH as the ultimate electron donor. The second redox system consists of the small redox enzymes glutaredoxins, which directly interact with oxidized protein thiols. In contrast to thioredoxins, which are reduced by thioredoxin reductase, glutaredoxins are non-enzymatically reduced by reduced glutathione. Oxidized glutathione is subsequently regenerated by glutathione reductase, which, like thioredoxin reductase, uses NADPH as an electron donor (Holmgren et al. 2005).

As outlined above, both thioredoxin and glutaredoxin systems depend on reduced NADPH as electron source, making both systems and hence the cellular redox status ultimately dependent on the cellular NADPH/$NADP^+$ ratio (Schafer and Buettner 2001). The major source of NADPH within the cell is the pentose phosphate pathway, which generates two molecules of NADPH for every oxidized glucose-6-phosphate molecule. The strict dependence of the cellular redox status on NADPH explains the need for efficient re-routing of glucose from glycolysis to the pentose phosphate pathway under conditions of oxidative stress (Godon et al. 1998; Grant 2008). Oxidative modification and inactivation of key enzymes of glycolysis seem to contribute to these changes in glucose flux, illustrating how redox-sensitive metabolic enzymes play an active part in the oxidative stress defense of organisms (Brandes et al. 2011). Deficiency in glucose-6-phosphate dehydrogenase, which catalyzes the first step of the pentose phosphate pathway, results in lower intracellular NADPH/$NADP^+$ ratios and increased oxidative stress and has been associated with premature cell senescence and a number of different disease conditions (Ho et al. 2007).

14.3 Evidence in Support of the 'Free Radical Theory of Aging'

14.3.1 Oxidative Damage Increases with Age

As outlined above, the free radical theory of aging states that progressive accumulation of ROS during the lifespan of an organism contributes and potentially causes the decline in fitness and might even determine longevity (Finkel and Holbrook 2000). A large amount of correlative evidence has been collected demonstrating a significant increase in oxidative damage to nucleic acids, lipids and proteins as organisms age, using a variety of different model systems ranging from bacteria to human (Golden et al. 2002; Sohal et al. 2002; Muller et al. 2007). By far the most commonly used assay to assess oxidative damage in cells and organisms is the analysis of protein carbonylation. This irreversible oxidative modification occurs primarily on the side chains of prolines, arginines, lysines and threonines in response to oxidative stress (Dalle-Donne et al. 2003). It is commonly assayed by a variety of semi-quantitative techniques, which rely on the detection of carbonylated side chains upon *ex vivo* derivatization (Moller et al. 2011). Unfortunately, none of the available assays are able to determine what proportion of a protein population is affected to what extent, making the effects of protein carbonylation on the physiology of an organism difficult to assess. No distinction can be drawn whether an increased signal results from an increase in the number of carbonylation sites within a few protein molecules, or from carbonylating an increasing proportion of a protein population at a single, physiologically important site. Nevertheless, it is a convenient method to monitor trends in protein oxidation and has been extensively used in aging research. Tissues of mice fed *ad libitum* showed increasing levels of protein carbonylation with age (Sohal et al. 1994) and tissues derived from individuals with premature aging diseases (e.g., progeria) exhibited higher level of protein carbonyls compared to age-matched healthy humans (Stadtman 1992). The extent of protein carbonylation has also been shown to increase during the last third of the lifespan in tissues from different organisms, including house flies, *C. elegans*, rats and humans, summarized in (Levine 2002).

14.3.2 (Some) Oxidants Accumulate with Age

In addition to evaluating the effects of oxidative stress on proteins, lipids and DNA as a read-out for intracellular oxidative damage, assays have been developed that directly measure the concentration of ROS, such as superoxide or peroxide in aging organisms. The obtained results were not always consistent with the free radical theory of aging as it was shown, for instance, that the concentration of hydrogen peroxide in *Drosophila* homogenates increases during the first trimester of their life but remains stable during the remainder of the lifespan (Sohal et al. 1990). In contrast, mitochondrial matrix hydrogen peroxide was observed to increase during aging in *Drosophila* (Cocheme et al. 2011). Peroxide levels also increased in aging *C. elegans* population as was recently demonstrated by using chromosomally encoded peroxide-specific sensor proteins (Back et al. 2012) whereas microsomal superoxide anion production actually declined from reproductive to senescent age, with long-lived mutant animals (i.e., age-1) exhibiting higher superoxide anion levels than the age-matched control animals (Vanfleteren 1993). In mice, both mitochondrial superoxide and hydrogen peroxide release from heart, brain and kidney tissues increased with the age of the animals (Sohal et al. 1994).

14.3.3 (Some) Antioxidant Levels Decrease with Age

Age-induced oxidative damage can either be caused by increased ROS production, decreased detoxification, or a combination thereof. Many studies have been conducted to assess the activity of ROS-detoxifying enzymes in young and old organisms, with the goal to either confirm or rule out the model that older animals have lower levels of ROS-detoxifying activity than young animals, hence the accumulation of oxidative damage. As exemplified below, while such correlation does seem to exist for certain antioxidant systems in some tissues and model systems, it does not generally apply to all ROS or model systems, making a more differentiated discussion necessary.

14.3.3.1 Superoxide Dismutases

Comparative analysis of superoxide dismutase activity in kidney, brain and heart tissue of young and old mice did not reveal any significant alteration in total SOD activity (Sohal et al. 1994). Similarly, activity of Cu/ZnSOD in liver homogenates of mice between 4, 12, or 18 months of age appeared unchanged while analysis of MnSOD activity revealed even an increase of SOD activity with age (Andziak et al. 2005). These results suggested that SOD activity levels in mice do not change with age. Cu/ZnSOD activity in brain tissues of aging rats, however, showed a gradual decline in activity, which appeared to be caused by a decrease in SOD expression levels (Semsei et al. 1991).

Studies in other model systems were consistent with the results obtained in mice and showed that SOD activity levels either remained unaltered during the lifespan (i.e., *C. elegans*) or linearly increased with age (i.e., *Drosophila* lysates) (Vanfleteren 1993; Sohal et al. 1990). Expression levels of Cu/ZnSOD, as determined by mRNA and steady state SOD analysis, remained relatively constant in aged flies (Radyuk et al. 2004). These results ruled out the possibility that a significant decrease in SOD activity and/or level was directly responsible for the oxidative damage observed in aging organisms.

14.3.3.2 Catalases, Glutathione Peroxidases and Peroxiredoxins

Studies assessing the activity of peroxide-detoxifying enzymes during the lifespan revealed a relatively consistent trend, indicating that the peroxide detoxifying capacity of organisms does indeed decrease with age. Analysis of the catalase activity in liver samples of mice, for example, showed a decline with the age of the animals (Perichon and Bourre 1995). A significantly decreased level of catalase and glutathione peroxidase activity was also observed in liver homogenates of 18 months old mice in comparison to 12 months old animals (Andziak et al. 2005), a finding that was independently confirmed for catalase in heart tissue and for glutathione peroxidase in kidney (Sohal et al. 1994). Brain tissue of aged mice, however, exhibited increased catalase and glutathione peroxidase activity (Sohal et al. 1994). This result was in contrast to brain samples of rats, which exhibited a gradual decline in catalase activity with age that coincided with a decrease in catalase mRNA (Semsei et al. 1991).

Analysis of non-rodent model systems were overall also more consistent, revealing kinetics of catalase activity in *Drosophila* that seem to follow a bell-shaped curve with higher levels of catalase in young animals as compared to older animals

(Sohal et al. 1990; Durusoy et al. 1995). In a subsequent more thorough analysis, catalase expression was shown to be both time- and tissue-specific, coinciding with pulses of ecdysteroid synthesis during development followed by a small decline as flies aged (Klichko et al. 2004). Studies in young *C. elegans* adults revealed a similar initial increase in the catalase activity and a decline as the worms aged (Vanfleteren 1993). Dramatic changes in expression level were also observed for peroxiredoxin 5 in *Drosophila*, which showed the highest expression level during embryogenesis, followed by a decline during aging (Radyuk et al. 2009). These results are largely consistent with the idea that the activity levels of peroxide-detoxifying enzymes decrease as animals age, potentially leading to the accumulation of peroxide in aging tissue.

14.3.3.3 Glutaredoxin, Thioredoxin and NADPH

To assess whether changes in the activity of the cellular redox systems contribute to the oxidative damage observed in aging organisms, expression analyses were conducted to monitor the activity of the thioredoxin and glutaredoxin system. An early study focusing on the thioredoxin system in rat kidneys reported decreasing levels of both thioredoxin and thioredoxin reductase with the age of the animals (Cho et al. 2003). The same study also found decreased levels of reduced glutathione and glutathione reductase activity in older rat kidneys as compared to young animals. These results were independently confirmed in aged rat muscles, where expression levels of both mitochondrial thioredoxin reductase and cytosolic thioredoxin were significantly reduced (Rohrbach et al. 2006). In contrast, however, levels of mitochondrial thioredoxin appeared to increase with age (Rohrbach et al. 2006). Moreover, comparative analysis of the glutathione and thioredoxin system in the heart muscle of young and old rats did not reveal any significant changes but did reveal an increase in oxidized GSSG levels, indicative of a pro-oxidant shift in the glutathione reduction potential (Jacob et al. 2010). A pro-oxidant shift in the glutathione pool has also been reported to occur in multiple other tissues of aging mice and rats, generally caused by an increase in oxidized glutathione and sometimes accompanied by a decrease in reduced glutathione. These changes tend to be most significant in liver tissues (Rebrin and Sohal 2008). Given that both glutaredoxin and thioredoxin systems rely heavily on reduced NADPH to maintain redox homeostasis, this shift in redox potential might be partially explained by a decrease in cellular NADPH levels, which has been observed to occur in the neurons of aging rats (Parihar et al. 2008). Studies in invertebrates confirmed some of the results obtained in rodents. It was shown, for instance, that the reduced glutathione pool sharply declines in older flies (Sohal et al. 1990). Moreover, caloric restriction, one of the few near-universal life prolonging measures (see Sect. 14.5.2) has been shown to partially reverse the detected changes in redox potential (Someya et al. 2010; Cho et al. 2003; Rohrbach et al. 2006). In summary, these studies provide convincing evidence that a combined decline in the cellular antioxidant capacity occurs with the age of the animal, which likely contributes to the accumulation of oxidative damage.

14.4 Do Manipulations of the Cell's Antioxidant Capacity Affect Lifespan?

After years of correlative studies, big hopes were spawned with the development of methods that enable genetic manipulations of model organisms, as they should allow direct and unambiguous testing of the validity of the free radical theory of aging. If correct, modulating endogenous ROS levels either by deleting or overexpressing specific antioxidant systems should have clear effects on lifespan. Below is a summary of the current state of affairs, based on genetic manipulation studies conducted in mice, *Drosophila* and *C. elegans*. An overview of the published results can be found in Tables 14.1 and 14.2.

14.4.1 Effects of Manipulating Superoxide Levels on Lifespan of Model Organisms

14.4.1.1 Effects of SOD Deletion

Deletion of either cytosolic Cu/ZnSOD1 or mitochondrial MnSOD2 in mice was found to significantly reduce their lifespan, illustrating the importance of preventing toxic ROS accumulation in mammals (Elchuri et al. 2005; Li et al. 1995). Deletion of cytosolic Cu/ZnSOD1 was also associated with increased oxidative damage and led to age-related loss of skeletal muscle mass and an increased incidence of liver cancer (Muller et al. 2006; Elchuri et al. 2005). Deletion of mitochondrial MnSOD2 had even more severe effects and led to early neonatal death with many animals dying within the first 16 days after birth (Li et al. 1995; Huang et al. 2001). This mortality rate increased further with higher oxygen concentrations (i.e., hyperoxia) (Asikainen et al. 2002). Yet the severity of the phenotype suggested that the decreased lifespan might be not purely due to accelerated aging. Partially depleting the levels of mitochondrial MnSOD2 by making $sod2^{+/-}$ heterozygous mice did not affect their resistance to hyperoxia (Asikainen et al. 2002; Tsan et al. 1998) but did increase their sensitivity towards oxidative stressors (Van Remmen et al. 2003, 2004). These animals showed increased oxidative damage of mitochondrial and nuclear DNA and a significant increase in tumor incidence and age-associated biomarkers, like cataracts and immune response. Yet, their lifespan was unaffected by the mutation (Van Remmen et al. 2003). Homozygous deletion of extracellular SOD (EcSOD) in mice was found to increase their sensitivity to hyperoxia but the lifespan of the animals appeared to not be affected (Carlsson et al. 1995; Sentman et al. 2006). These results provided experimental evidence that deletion of either cytosolic or mitochondrial SOD causes increased oxidative damage and early death in mammals, although it is still inconclusive whether these two phenotypes are directly related.

Table 14.1 Effects of deletion of antioxidant genes on stress resistance and lifespan in model organisms

Antioxidant system	Species	Stress resistance	Lifespan	Refs
Cu/Zn Superoxide dismutase				
Δsod1	Mm		↓	1,2
Δsod1	Dm	↓	↓	3
Δsod1	Ce	↓	↓↔[a]	4–7
Δsod1 + Δsod2; Δsod3; Δsod4; Δsod5	Ce	↓	↓↔	6, 7
ΔECsod	Mm	↓	↔	8, 9
Mn Superoxide dismutase				
Δsod2[Het,b]	Mm	↓	↔	10, 11
Δsod2	Dm	↓ [Het]	↓[b]	12
Δsod2	Ce	↓↔	↑↔	4, 6, 7
Δsod2 + Δsod1; Δsod3; Δsod4; Δsod5	Ce	↓	↑↔	6, 7
Δsod1Δsod2Δsod3Δsod4Δsod5	Ce	↓	↔[c]	13
Catalase				
Δcat	Mm	↔	↔[d]	14
Δcat	Dm	↓	↓	15, 16
Δctl1	Ce		↔	17
Δctl2	Ce		↓	17
Peroxiredoxin				
Δprdx1	Mm		↓	18
Δprdx2[e]	Dm	↓	↓	19
Δprdx2	Ce	↓	↓	20, 21
Δprdx5	Dm	↓	↓[f]	22
Glutathione peroxidase				
Δgpx1	Mm	↓↔	↔	23–25
Δgpx4 [Het,g]	Mm		↑	26
Thioredoxin				
Δtrx1	Ce	↓	↓	27, 28
Δtrx2 [Het,g]	Mm		↔	2
Δtrx2	Dm	↓	↓	29, 30

1 Elchuri et al. 2005, 2 Perez et al. 2009a, 3 Phillips et al. 1989, 4 Yen et al. 2009, 5 Yanase et al. 2009, 6 Doonan et al. 2008, 7 Van Raamsdonk and Hekimi 2009, 8 Carlsson et al. 1995, 9 Sentman et al. 2006, 10 Van Remmen et al. 2003, 11 Van Remmen et al. 2004, 12 Duttaroy et al. 2003, 13 Van Raamsdonk and Hekimi 2012, 14 Ho et al. 2004, 15 Mackay and Bewley 1989, 16 Griswold et al. 1993, 17 Petriv and Rachubinski 2004 18 Neumann et al. 2003, 19 Lee et al. 2009a, 20 Kumsta et al. 2011, 21 Olahova et al. 2008, 22 Radyuk et al. 2009, 23 Ho et al. 1997, 24 Fu et al. 1999, 25 Zhang et al. 2009, 26 Ran et al. 2007, 27 Miranda-Vizuete et al. 2006, 28 Jee et al. 2005, 29 Tsuda et al. 2010, 30 Svensson and Larsson 2007

Mm *M. musculus*, Dm *D. melanogaster*, Ce *C. elegans*, Het heterozygous

[a]Sod1 deletion reduces hypothermia-induced lifespan extension

[b]Homozygous deletion has severe health effects (Li et al. 1995; Huang et al. 2001)

[c]Maximum lifespan extended

[d]Observation up to 1 year of age

[e]Null mutation or RNAi knockdown, stress resistance assessed in RNAi knock-down

[f]Accelerated death during development, maximum lifespan normal

[g]Homozygous deletion embryonic lethal (Yant et al. 2003; Nonn et al. 2003)

Table 14.2 Effects of overexpression of antioxidant genes on stress resistance and lifespan in model organisms

Antioxidant system	Species	Stress resistance	Lifespan	Refs
Cu/Zn SOD overexpression				
sod1 +/− sod2	Mm		↔	1–3
sod1 (motorneurons)	Dm	↑	↑	4
sod1 (ubiquitous)	Dm	↑	↑	5
sod1 +/− sod2	Dm	↔	↔↑[a]	6–8
sod1	Ce	↓	↑	9, 10
Mn SOD overexpression				
sod2 +/− cat	Dm	↑	↑↔↓[b,c]	7, 11–14
sod2	Ce		↑	10
Catalase overexpression				
cat (mitochondria)	Mm		↑	15
cat (peroxisomes)	Mm		↔	2, 15
cat	Dm	↑	↓↔[d]	6, 11, 16
ctl1 + ctl2 + ctl3	Ce		↓[e]	9
Peroxiredoxin overexpression				
prx2 (neurons)	Dm	↑	↑	17
prx5	Dm	↑	↑	18
Thioredoxin overexpression				
trx1	Mm	↑	↑↔[f]	19, 20
trx1(neurons)	Ce		↑	21
trx2	Dm	↑	↑	22, 23

1 Huang et al. 2000, 2 Perez et al. 2009b, 3 Rando et al. 1998, 4 Parkes et al. 1998, 5 Reveillaud et al. 1991, 6 Sun and Tower 1999, 7 Sun et al. 2004, 8 Seto et al. 1990, 9 Doonan et al. 2008, 10 Cabreiro et al. 2011, 11 Mockett et al. 2010, 12 Mockett et al. 1999, 13 Bayne et al. 2005, 14 Sun et al. 2002, 15 Schriner et al. 2005, 16 Griswold et al. 1993, 17 Lee et al. 2009a, 18 Radyuk et al. 2009, 19 Mitsui et al. 2002, 20 Perez et al. 2011, 21 Miranda-Vizuete et al. 2006, 22 Svensson and Larsson 2007, 23 Seong et al. 2001

[a]Simultaneous overexpression of *sod1* and *sod2* increased the lifespan additively

[b]MnSOD has slightly decreased lifespan (4–5%)

[c]Simultaneous overexpression of *sod2* and catalase in mitochondria decreased lifespan by 43%

[d]Two *cat+* strains were tested; while one strain did not show a change in lifespan, the other strain had a significantly decreased lifespan

[e]Deaths by internal hatching

[f]While overexpression of *trx1* in male mice significantly extended earlier part of life (maximum lifespan was unaffected), female mice showed no significant change in lifespan

Very similar results were obtained in *Drosophila*, where deletion of either cytosolic *sod1* or mitochondrial *sod2* significantly decreased lifespan (Phillips et al. 1989; Duttaroy et al. 2003). Again deletion of the mitochondrial isoform caused more severe defects, as demonstrated by the finding that a large fraction of flies died within the first 24 h after eclosion (Duttaroy et al. 2003). Genetic manipulation of SOD levels in *C. elegans*, however, yielded a significantly more ambiguous picture, contributing to a substantial amount of controversy in the field (Doonan et al. 2008). At this point, it is unclear to what extent the presence of additional cytoplasmic and mitochondrial SOD homologues (SOD-5; SOD-3) might have affected the outcome of some of these studies. Initially it was reported that deletion of the major cytosolic Cu/ZnSOD-1 increases the sensitivity of *C. elegans* to superoxide-inducing paraquat and shortens *C. elegans'* lifespan while deletion of the mitochondrial MnSOD-2 also increased sensitivity towards oxidative stressors but did not affect lifespan (Yen et al. 2009; Yanase et al. 2009; Doonan et al. 2008). An independent study in which all five SOD isoforms of *C. elegans* were manipulated either individually or in combination also reported increased sensitivity towards oxidative stressors like paraquat and juglone for *sod-1*, *sod-2*, and *sod-3* mutants but did not confirm reduced lifespan for any of the tested mutants (Van Raamsdonk and Hekimi 2009). In fact, deletion of mitochondrial MnSOD-2 (*sod-2*) alone or in combination with other SOD isoforms was found to even extend *C. elegans*' maximum lifespan by 5–19 days (Van Raamsdonk and Hekimi 2009). This result, however, requires further confirmation as an independent study observed an increase in oxidative stress sensitivity for *sod-2* deletion mutants but did not observe any beneficial effects on lifespan (Doonan et al. 2008). In an attempt to ultimately address the question regarding the significance of SOD in the lifespan of *C. elegans*, Van Ramsdoonk and Hekimi went through the effort to construct a quintuple mutant strain, which lacked all five SOD-homologues. Importantly, this mutant strain showed the same mean lifespan as wild type worms but a significantly increased maximum lifespan (Van Raamsdonk and Hekimi 2012). The authors found that the quintuple mutant strain was highly sensitive to superoxide-inducing paraquat, heat shock and osmotic stress. Yet, the strain was more resistant to exogenous peroxide treatment than wild type strains. This increased peroxide resistance was apparently due to the massive upregulation of catalases, PRDX-2 and thioredoxin reductase 1, which seem to serve as compensatory mechanisms to deal with ROS. In summary, these results suggest that deletion of certain superoxide dismutase isoforms affects the capacity of organisms to deal with oxidative stressors, and causes a decline in fitness. It remains to be determined, however, to what extent superoxide-toxicity truly constitutes an aging factor.

14.4.1.2 Effects of SOD Overexpression and SOD Mimetics

Many gene deletions have been shown to affect the health and/or cause early death in organisms without being directly involved in regulating aging. Thus, a by far more desirable and interpretable phenotype is that of lifespan extension, which, if the free radical theory is indeed correct, should be achieved by increasing the ability of organisms to detoxify otherwise harmful ROS. This should reduce oxidative damage and hence extend lifespan. Much has been done to address this question and stimulation of superoxide detoxification can now either be achieved genetically by overexpressing select SOD isoforms, or chemically by application of SOD mimetics, such as EUK-8 or EUK-134, which detoxify superoxide *in vivo*.

Studies performed in mice showed that the lifespan of mice was unaffected even when both mitochondrial MnSOD and cytosolic Cu/ZnSOD were simultaneously overexpressed (Huang et al. 2000; Perez et al. 2009b). It is of note, however, that Cu/ZnSOD overexpressing mice showed increased lipid peroxidation, suggesting that SOD overexpression might lead to the accumulation of peroxide in cells and tissues, hence potentially off-setting the beneficial effects of enhanced superoxide detoxification (Rando et al. 1998).

Studies in *Drosophila* were highly conflicting with results ranging from (i) 40% lifespan extension upon overexpression of human Cu/ZnSOD in fly motoneurons (Parkes et al. 1998), to (ii) mild lifespan-enhancing effect in flies overexpressing bovine Cu/Zn SOD under the control of the ubiquitous actin 5C gene promoter (Reveillaud et al. 1991), to (iii) no changes in lifespan upon overexpressing of additional copies of either *Drosophila* Cu/ZnSOD or MnSOD (Seto et al. 1990; Mockett et al. 2010), to even (iv) a reduction in lifespan in strains overexpressing mitochondrial MnSOD (Mockett et al. 1999), even when co-overexpressed with catalase (Bayne et al. 2005). It is conceivable that the different effects of SOD overexpression on *Drosophila* lifespan are due to differences in cultivation conditions. Indeed, it has been shown that administration of antioxidants does not cause an extension of mean or maximum lifespan if the control group has an optimal lifespan, while the survival could be extended by antioxidant treatment when the lifespan of the control group was sub-optimal (Kohn 1971). A beneficial effect of Cu/ZnSOD overexpression seems to occur in strains which are rather short-lived, implying that optimal SOD level already exist in wild type strains exhibiting a normal lifespan (Orr and Sohal 2003). It is also possible that expression of antioxidant enzymes affects lifespan in a tissue and/or stage-dependent fashion (Mockett et al. 1999). For instance, overexpression of either cytosolic Cu/ZnSOD or mitochondrial MnSOD initiated during adulthood using a heat shock-inducible expression system extended both mean and maximum lifespan (Sun et al. 2002; Sun and Tower 1999). However, while the inducible expression system enables researchers to work in identical genetic strain backgrounds (Sun and Tower 1999), initiation of transgene expression is based on a heat pulse, which might itself have some negative effects on those strains not overexpressing SOD (Mockett et al. 1999).

Studies in *C. elegans* revealed that despite the fact that overexpression of cytosolic Cu/ZnSOD-1 increases the levels of cellular protein oxidation and enhances the sensitivity of animals towards paraquat-induced oxidative stress, it does increase *C. elegans* lifespan (Doonan et al. 2008; Cabreiro et al. 2011). Overexpression of mitochondrial MnSOD-2 was found to also extend the lifespan of worms by approximately 25% (Cabreiro et al. 2011). A set of independent studies agreed with these results by showing that treatment of worms with the SOD/catalase mimetics EUK-8, EUK-134 or platinum nanoparticles resulted in a significantly increased mean and maximum lifespan of wild type worms, and restored wild type-like lifespan in a short-lived *mev-1* mutant (Kim et al. 2008; Melov et al. 2000). Yet these studies were questioned by reports that showed that treatment with the same SOD mimetic (i.e., EUK-8) causes a dose-dependent shortening of lifespan, particularly at higher doses, suggesting that the effects of antioxidant interventions might be dependent on culture conditions (Keaney and Gems 2003). Moreover, while EUK-8 treatment protected worms against paraquat-toxicity, it did not extend lifespan in non-paraquat treated worms (Keaney et al. 2004). Note that treatment of worms with lower concentrations of the superoxide-generating compounds paraquat or juglone has recently been found to cause lifespan extension in worms (Heidler et al. 2010; Lee et al. 2010), suggesting that superoxide levels need to be carefully monitored in worms before final conclusions can be drawn about the efficiency of SOD mimetics in worms.

In summary, changing superoxide dismutase activity in mice, *Drosophila* or *C. elegans* either by deletion or overexpression of the respective enzymes did not reveal a clear picture about the effects of superoxide on lifespan. The results reflect a much more complex correlation between the amount of superoxide produced at distinct stages in life, the levels of antioxidants present, and the lifespan. It remains to be determined to what extent superoxide-toxicity might have contributed to the observed lifespan reduction and early deaths observed in mitochondrial MnSOD null mutants in fly and mice, and if this truly constitutes superoxide-mediated aging. Another intriguing question is whether overexpression of superoxide dismutases actually adds to cellular oxidative stress since higher SOD activity inevitably increases the hydrogen peroxide pool in the cells.

14.4.2 *Effects of Altering the Hydrogen Peroxide Clearance System on Lifespan of Model Organism*

The hydrogen peroxide clearance system typically consists of catalases, peroxiredoxins and glutathione peroxidases. In the following section we will summarize the effects that genetic manipulation of the individual branches of the hydrogen peroxide clearance system has on the lifespan of mice, *Drosophila* and *C. elegans*.

14.4.2.1 Effects of Catalase Deletion and Overexpression

Catalase-deficient mice develop normally and are healthy at least up to 1 year of age. They do not reveal any increased sensitivity towards hyperoxia or photochemically induced oxidative stress even though tissue slices from catalase knockout mice show slower rates in decomposing extracellular H_2O_2 (Ho et al. 2004). Mice expressing human catalase under the control of the endogenous promoter and regulatory elements, which results in peroxisome-specific overexpression of catalase similar to endogenous gene expression (Chen et al. 2003), did not show any lifespan extension even when Cu/ZnSOD was co-overexpressed (Perez et al. 2009b). In contrast, mice overexpressing the human catalase gene targeted to the mitochondria showed an improved rate of peroxide clearance as well as reduced H_2O_2 production, decreased oxidative damage, and delayed development of cardiac pathology. Most importantly, these mice showed an extension of both median and maximum lifespan by 5 and 5.5 months, respectively (Schriner et al. 2005). These results provide first evidence that increased peroxide clearance from mitochondria might have beneficial effects on mammalian lifespan.

Drosophila mutants with catalase activities of less than 2% showed severe viability effects (Mackay and Bewley 1989). An independent study confirmed these findings, and demonstrated that the lack of catalase activity in *Drosophila* results in decreased viability and lifespan (Griswold et al. 1993). Increasing the levels of catalase in *Drosophila,* however, had either no effect on lifespan (Griswold et al. 1993; Mockett et al. 2010; Sun and Tower 1999) or caused a moderately reduced lifespan, depending in part on the insertion site of the transgene and the magnitude of catalase overexpression (Mockett et al. 2010; Sun and Tower 1999). Only when combined with the simultaneous overexpression of Cu/ZnSOD, was overexpression of catalase found to be beneficial for lifespan (Orr and Sohal 1994). This study was, however, contradicted by another group, who reported that the combined overexpression of catalase and Cu/ZnSOD did not result in a benefit beyond the lifespan expanding effect observed for Cu/ZnSOD overexpressing flies (Sun and Tower 1999).

Similar to the results in *Drosophila* where deletion of catalase significantly reduced lifespan, deletion of peroxisomal catalase 2 (*ctl-2*) in *C. elegans* was found to cause progeric phenotypes and a shortened lifespan. In contrast, deletion of the cytosolic *ctl-1* isoform had no effect on the lifespan of worms (Petriv and Rachubinski 2004). To test whether overexpression of catalase increased *C. elegans* lifespan, the *ctl-1*, *ctl-2*, *ctl-3* gene cluster was overexpressed, generating tenfold higher catalase activity (Doonan et al. 2008). Lifespan measurements did not reveal any significant effect of catalase overexpression in *C. elegans.* In fact, it increased mortality rate by internal hatching, an event that could be suppressed by the simultaneous overexpression of Cu/ZnSOD-1.

These studies suggest that animals lacking catalase accumulate high levels of hydrogen peroxide, which might contribute to a decline in fitness and causes early death at least in some model systems. Reducing peroxide levels by overexpressing catalase in mitochondria appears to have beneficial effects on mammalian lifespan but does not seem to enhance the lifespan of invertebrates.

14.4.2.2 Effects of Peroxiredoxin Deletion and Overexpression

Mice lacking cytosolic peroxiredoxin 1 (*prdx1*) exhibit increased DNA damage and a higher incidence of hemolytic anemia and cancer starting at 9 months of age, which effectively reduces their lifespan (Neumann et al. 2003). Deletion of peroxiredoxin 2 (*prdx2*) also results in hemolytic anemia and increased levels of oxidative stress but the mutant mice seem otherwise healthy and fertile, and their lifespan has not been determined (Lee et al. 2003b). Deletion of the 1-Cys peroxiredoxin 6 (*prdx6*) results in elevated ROS level and increased oxidative damage and heightens the animal's susceptibility towards oxidative stress-mediated damage (Wang et al. 2003). These results confirm the important role that peroxiredoxins play on peroxide detoxification. To our knowledge, only one study presented lifespan data of peroxiredoxin deletion mice so far (Neumann et al. 2003). Adenovirus-mediated overexpression of PRDX6 in the lungs of mice was found to promote resistance to hyperoxia-induced oxidative stress. So far, no lifespan experiments have been reported with these mice.

RNAi knock down of *prx2* in flies resulted in an increased sensitivity to paraquat-mediated oxidative stress. Additionally, flies with a reduction in PRX2 activity, either through RNAi knockdown or a loss of function mutation, exhibited a shortened lifespan (Lee et al. 2009a). The deletion of peroxiredoxin 5 (*Prx5*) in *Drosophila* also leads to a slightly shortened mean lifespan, which appears to be mainly caused by accelerated dying during day 5–10 after eclosion. These results suggest that PRX5 might be important in proliferating tissues and maybe required for embryonic development (Radyuk et al. 2009). After development, the mortality rate of *prx5* null mutants was found to be similar to the mortality rate of wild type flies, resulting in the same maximum lifespan for both wild type and *prx5* mutants (Radyuk et al. 2009). The modest lifespan defect of *prx5* null flies was, however, severely enhanced under oxidative stress conditions (Radyuk et al. 2009). Neuronal expression of either human PRX2 or the *Drosophila* PRX2 homolog as well as ubiquitous expression of PRX5 was found to enhance the resistance of flies to oxidative stress and to extend their lifespan, providing excellent evidence that in flies, detoxification of peroxide through peroxiredoxins prolongs life (Radyuk et al. 2009; Lee et al. 2009a).

Peroxiredoxin 2 (PRDX-2), which is one of the most abundant proteins in *C. elegans*, plays a crucial role for hydrogen peroxide clearance in worms (Kumsta et al. 2011; Olahova et al. 2008). Animals lacking PRDX-2 are much more sensitive to sublethal hydrogen peroxide treatments than wild type *C. elegans,* fail to recover from exogenous peroxide treatment, and exert progeric phenotypes as well as symptoms of chronic oxidative stress (Kumsta et al. 2011; Olahova et al. 2008). Most importantly, *prdx-2* deletion worms show a severe lifespan defect, which seems most pronounced at lower temperatures (Kumsta et al. 2011; Olahova et al. 2008).

These studies show that deletion of peroxiredoxin has some of the most severe effects on the lifespan of model organisms, illustrating its importance in hydrogen peroxide clearance. Importantly, overexpression of peroxiredoxin in *Drosophila* significantly extends lifespan, which is very similar to the effects that overexpression of mitochondrial catalase has in mice. These results suggest that enhanced detoxification of peroxide might indeed be a lifespan-extending factor.

14.4.2.3 Effects of Glutathione Peroxidase Deletion and Overexpression in Mice

In addition to catalases and peroxidases, mice also contain several different glutathione peroxidases, which appear to play a prominent role in detoxifying low levels of peroxide. Mice lacking glutathione peroxidase 1 (*gpx1*) display a normal and healthy development, an unaltered capacity to decompose hydrogen peroxide, a wild type-like protein and lipid oxidation pattern and the same sensitivity to hyperoxia (Ho et al. 1997). While the mutant mice do show an increased sensitivity towards diaquat-induced oxidative stress (Fu et al. 1999) and increased cataract occurrence (Wolf et al. 2005), their lifespan is not affected by the absence of *gpx1* even in combination with a $sod2^{+/-}$ heterozygous mutation (Zhang et al. 2009). In contrast, glutathione peroxidase 4 (*gpx4*) null mutants were found to be embryonic lethal (Yant et al. 2003). Expression of human GPX-4 was able (i) to rescue the embryonic lethality of $gpx4^{-/-}$ mice, (ii) to reduce apoptosis and (iii) to increase cell survival after oxidative insult (Ran et al. 2004). Heterozygous $gpx4^{+/-}$mice showed an increased median lifespan but neither a change in maximum lifespan nor in age-related mortality rate, indicating that aging itself is not slowed in these animals (Ran et al. 2007).

14.4.3 Manipulation of the Thioredoxin/Glutaredoxin System

Genetic manipulations of the thioredoxin system are limited, yet the studies conducted so far all seem to point to the fact that maintainance of cellular redox homeostasis is crucial for development and lifespan. Deletion of either cytosolic thioredoxin 1 or mitochondrial thioredoxin 2 in mice is embryonic lethal (Nonn et al. 2003; Matsui et al. 1996). Heterozygous $trx2^{+/-}$ mice have increased ROS generation and oxidative damage compared to wild type mice but their lifespan is unchanged (Perez et al. 2008, 2009a). Overexpression of TRX-1 was found to extend lifespan of mice (Mitsui et al. 2002). This result was, however, put into question as the lifespan of the control mice was significantly shorter than expected. A more recent study by Perez et al. reported that overexpression of mouse *trx1* causes an increase in the mean lifespan of male mice without affecting their maximal lifespan. These results suggest that the observed decrease in oxidative damage is beneficial to male mice early in life but might not aid in extending lifespan later in life (Perez et al. 2011). Studies in *Drosophila* showed that flies lacking *trx-2* are short-lived while flies overexpressing *trx-2* show enhanced survival upon exposure to hydrogen peroxide or paraquat (Tsuda et al. 2010; Svensson and Larsson 2007). Moreover, a gain-of-function-mutation of *trx* conferred longevity to *Drosophila* (Seong et al. 2001). Very similar results were also obtained in *C. elegans*, where overexpression of *trx-1* was found to extend the mean lifespan of *C. elegans* (Miranda-Vizuete et al. 2006) while deletion of *trx-1* resulted in a reduced resistance to oxidative stress as well as in a shortened mean and maximum lifespan (Miranda-Vizuete et al. 2006; Jee et al. 2005).

14.4.4 *Where Are We Now, What Should We Think?*

The many conflicting results obtained with genetic manipulations of antioxidant enzymes in a variety of different organisms over the last few years clearly illustrate the complexity of redox homeostasis and its role in lifespan. In general, deletion of antioxidant enzymes appears to have one of two outcomes. One being so serious that the organism is severely affected – thus decreased lifespan may not be a direct result of premature aging. The other one having little to no effect at all, suggesting either significant redundancy with other antioxidant enzymes or implying that their function is so highly specialized as to not affect longevity under "normal" conditions. Hence, it is probably unwise to draw significant conclusions from deletion studies. Overexpression studies could be viewed as a more direct approach to analyze the influence of antioxidant systems on aging. However, confusion arose from studies where the same genetic manipulations revealed different effects in different labs. This suggests that specific growth conditions and/or differences in the strain background might be additional sources that influence lifespan and need to be carefully controlled and monitored.

14.5 Lifespan Extension and the Free Radical Theory of Aging – What Works and Why

14.5.1 *Lifespan Extension by Manipulating the Electron Transport Chain*

Manipulation of endogenous ROS production has been used as an alternative approach to assess the role of ROS in aging and lifespan. Screens for mutants with increased sensitivity towards the superoxide generator paraquat in *C. elegans* resulted initially in the identification of the mitochondrial *mev-1* mutant (Ishii et al. 1990). *Mev-1* encodes a subunit of succinate dehydrogenase cytochrome b, a component of mitochondrial complex II (Ishii et al. 1998). Mutants in *mev-1* were found to exert elevated levels of superoxide anions (Senoo-Matsuda et al. 2001), reduced Cu/ZnSOD activity levels and a significantly shortened lifespan (Ishii et al. 1990), which could be reversed by treatment with superoxide dismutase/catalase mimetics (Melov et al. 2000). Similarly, Drosophila mutants of succinate dehydrogenase subunit b (complex II) also exhibited increased levels of mitochondrial hydrogen peroxide production, enhanced sensitivity towards hyperoxia and a shorter lifespan (Walker et al. 2006). Furthermore, *C. elegans* mutants of complex I (i.e., *gas-1*) were found to have elevated levels of superoxide anion and also displayed a shortened lifespan (Kondo et al. 2005; Hartman et al. 2001). These findings seemed to support the free radical theory of aging as they suggested that the decrease in lifespan, which is observed in mitochondrial mutants is caused by increased oxidant levels (Senoo-Matsuda et al. 2001; Sedensky and Morgan 2006). More recent reports indicated, however, that the model that a compromised mitochondrial ETC causes increased oxidant production and hence shorter lifespan is far more complex than previously anticipated. In fact, worms carrying a mutation in *isp-1*, an iron-sulfur protein of complex III, showed increased resistance towards oxidative stress, reduced oxygen consumption, and an extended lifespan, indicating that a reduction in ETC function might also positively affect the lifespan of *C. elegans* (Feng et al. 2001). RNAi-mediated screens for longevity genes further confirmed these results by finding a tenfold overrepresentation of genes involved in mitochondrial function, whose knock-down improved H_2O_2 tolerance (although paraquat tolerance was reduced) and extended lifespan (Lee et al. 2003a). Selective targeting of individual components of the ETC, including proteins of complex I, III, IV and V causes a significant extension in lifespan in *Drosophila* and *C.elegans* (Dillin et al. 2002; Copeland et al. 2009). It is of note that reducing ETC function had to be initiated during *C. elegans* development to achieve lifespan extension whereas reduction of ETC components during adulthood of *C. elegans* resulted in lowered ATP level but no changes in lifespan (Dillin et al. 2002). These results imply that the rate of mitochondrial respiration during development is at least partly responsible for adjusting *C. elegans'* growth rate, development, and adult lifespan. The developmental window during which the intervention seems to be successful ends by the third or early fourth larval stage of *C. elegans*, indicating that an event occurring during larval development might set the clock for lifespan (Rea et al. 2007). One mechanism, which seems to play a role in the lifespan extension mediated by ETC reduction is the mitochondria-specific unfolded protein response (UPR), which can be induced in a cell-non-autonomous way, meaning that signals from one tissue can trigger or control processes in other tissues (Durieux et al. 2011) (Fig. 14.2).

It is of note that many long-lived strains with mutations in the ETC, such as *nuo-6, isp-1* and *clk-1* showed elevated ROS level, which speaks clearly against the original idea that accumulation of ROS is toxic *per se* (Yang and Hekimi 2010; Lee et al. 2010). In fact, treatment of *nuo-6* and *isp-1* mutants with the antioxidant N-acetylcysteine (NAC) abolished the observed lifespan extension, suggesting that (transient) accumulation of oxidants might actually be required for the lifespan-prolonging phenotype of those mutants (Yang and Hekimi 2010). In a similar way, Yang and co-workers showed that treatment with low concentrations of the superoxide generator paraquat caused a transient superoxide boost, which significantly increased the lifespan in *C. elegans* despite an apparent increase in oxidative damage as assessed by protein carbonylation (Yang and Hekimi 2010). These results are in good agreement with the recently postulated mitohormesis theory, which will be discussed in detail later in this chapter.

Deletion of *clk-1* has been shown to increase ROS levels (Lee et al. 2010), UV-resistance (Murakami and Johnson 1996) and lifespan in worms (Wong et al. 1995). A heterozygous deletion mutant of the *clk-1* homologue in mice (*mclk1*) was found to also accumulate higher mitochondrial oxidative protein damage and, most importantly, to extend lifespan in mice (Liu et al. 2005; Lapointe and Hekimi 2008). What appears to contribute to this lifespan extending effect is the mitochondrial ROS-mediated activation of the transcription factor HIF-1α, which in

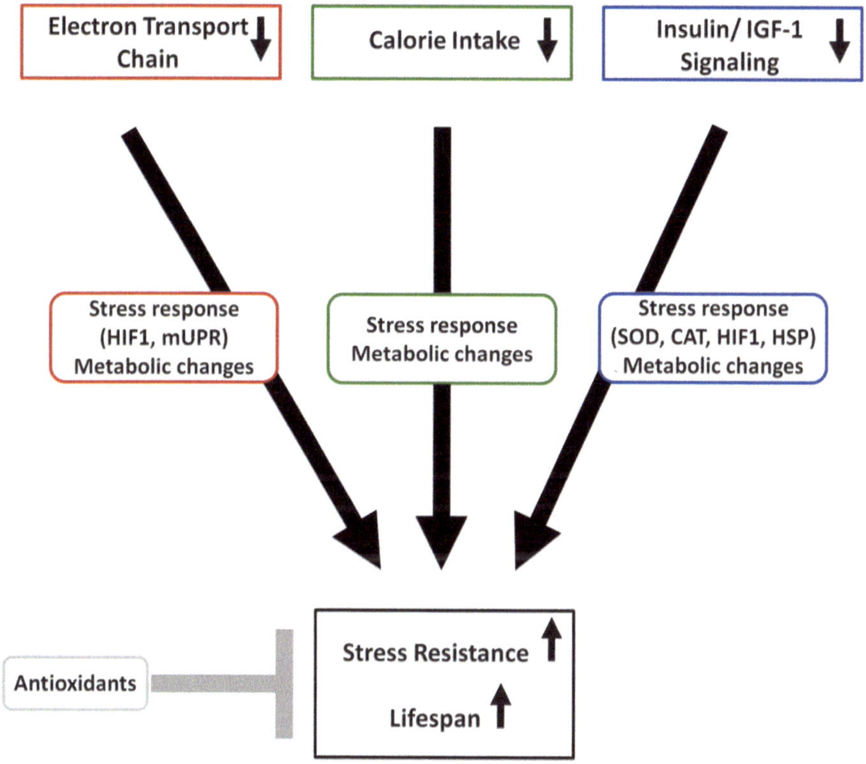

Fig. 14.2 Interventions that extend lifespan. Reduction of components of the electron transport chain (ETC), a decrease in the caloric intake, or a decline in Insulin/IGF-1 signaling extends lifespan in a variety of different organisms. One factor that might contribute to the observed lifespan extension is an increased stress resistance. The role that oxidants play in these lifespan-extending interventions is not fully elucidated yet. However, it has been demonstrated that antioxidants typically interfere with the observed lifespan extensions. Note: Some ETC mutants can also be short-lived

mice is involved in regulating the inflammatory immune response (Wang et al. 2010) (Fig. 14.2). This result led to speculations that the increased lifespan of *mclk1*$^{+/-}$ mice results from an elevated resistance to pathogenic infection especially in aging mice (Wang et al. 2010). Observations in *C. elegans* suggested that compromised respiration could be involved in the ROS-mediated activation of HIF-1, which is indeed responsible for the lifespan extension observed in *C. elegans clk-1* and *isp-1* mutants (Lee et al. 2010). Similarly, the lifespan enhancing effect observed in *C. elegans* upon administering low levels of the superoxide-generator paraquat might also work through the ROS-induced activation of HIF-1, suggesting that the HIF-1 mediated lifespan extension might be evolutionarily conserved (Lee et al. 2010).

Taken together, these studies suggest that lifespan extension achieved by reducing mitochondrial respiration is not simply caused by minimizing the output of harmful reactive oxygen species and decreasing oxidative damage. It rather seems to involve the activation of complex pathways, including the unfolded protein response (UPR) (Durieux et al. 2011), cell-cycle checkpoint control (Rea et al. 2007), changes in HIF-1-mediated gene expression (Lee et al. 2010) and possibly a switch in energy metabolism (Rea and Johnson 2003). That respiratory mutants do not mediate their longevity by one unifying feature was also suggested by the finding that reduction of HIF-1 (either by mutation or RNAi) significantly reduced the extended lifespan of *isp-1* and *clk-1* mutants but only partially affected the long-lived phenotype of other mitochondrial mutants, such as *cyc-1* or *cco-1* (subunits of complex III and IV) (Lee et al. 2010). This finding might also explain the apparently controversial results concerning the role of ROS in lifespan extension of mitochondrial mutants. While antioxidant treatment was found to not affect lifespan extension of some mitochondrial RNAi mutants (Durieux et al. 2011), other studies reported that superoxide is in fact required to mediate lifespan extension (Yang and Hekimi 2010).

14.5.2 Lifespan Extension by Caloric Restriction

Reduction of the daily caloric intake by 30% (dietary or caloric restriction) routinely extends lifespan up to 50% in a variety of different model organisms, including yeast, flies, *C. elegans*, mice and primates (McCay et al. 1935; Fontana et al. 2010). Calorically restricted mice have been found to have lower mitochondrial generation of superoxide and hydrogen peroxide than animals fed *ad libitum* (Sohal et al. 1994; Lambert and Merry 2004; Bevilacqua et al. 2004). Similarly, animals with reduced caloric intake show lower levels of age-accompanying oxidative DNA damage and oxidative protein damage as measured by protein carbonylation, and reveal higher levels of NADPH, specifically in the mitochondria of brain, liver, heart, kidney, ear and eyes of mice (Dubey et al. 1996; Sohal et al. 1994; Someya et al. 2010). Hence, caloric restriction appears to decrease oxidative damage by shifting the glutathione pool to a more reducing redox potential relative to age-matched controls. Whereas earlier findings seemed to indicate that caloric restriction resulted in a lowered metabolic rate, hence the decrease in oxidative damage, more recent studies that were corrected for body mass suggested quite the opposite (Houthoofd et al. 2002). In fact, caloric restriction appears to enhance mitochondria biogenesis (Lopez-Lluch et al. 2006) and to increase the rate of respiration (Lin et al. 2002; Nisoli et al. 2005) (Fig. 14.2). These results, although initially counterintuitive, are fully consistent with recent studies in *C. elegans*, which showed that 2-deoxyglucose (DOG)-mediated glucose restriction during adulthood increased mitochondrial respiration and ROS production, and significantly extended the lifespan (Schulz et al. 2007). Interestingly, pre-treatment of these animals with antioxidants, such as NAC or vitamin E abolished the beneficial effect of glucose restriction on lifespan (Schulz et al. 2007). These findings led to the model of mitohormesis, in which generation of slightly elevated levels of oxidants through increased respiration during a defined time in life enhances expression of antioxidant genes and with that the capacity of organisms to detoxify ROS. Contrary to glucose-restriction, glucose-supplementation of *C. elegans'* diet prevented longevity of *daf-2* worms, a mutant of the Insulin signaling pathway (described in 14.5.3), and shortened the lifespan of wild type animals by inhibiting the transcription factor DAF-16 (Lee et al. 2009b; Schlotterer et al. 2009). Interestingly, increased ROS generation was found at day 15 in worms fed on high-glucose-diet, suggesting that the duration of increased ROS level (or the magnitude) might make a difference in lifespan determination (Schlotterer et al. 2009).

Although it is now generally accepted that caloric restriction increases respiratory rates, the jury is still out whether ROS levels increase, stay unchanged or decrease. Several studies in calorically restricted yeast, flies or rodents observed either no change or even a decrease in ROS production accompanying increased cellular respiration (Barros et al. 2004; Bevilacqua et al. 2004; Miwa et al. 2004; Cocheme et al. 2011; Ash and Merry 2011). So while caloric restriction clearly increases respiratory rates, it might also lead to more efficient respiration possibly through mild uncoupling between electron transport and oxidative phosphorylation (Barros et al. 2004; Bevilacqua et al. 2004; Lopez-Lluch et al. 2006). A mild uncoupling of the two processes could result in an increase of ETC function, which could lead to less opportunity for electron leakage, and hence more efficient mitochondria function and less ROS formation. This conclusion of course argues against a hormetic effect in lifespan extension. However, since CR can be considered as a mild stress condition, it could lead to the observed increase in antioxidant defenses by other means.

14.5.3 Manipulation of the Insulin/IGF-1 Signaling (ILS) Pathway

Genetic manipulation of the ILS pathway has been shown to reproducibly modulate lifespan in *Drosophila*, *C. elegans* and mice (Longo and Finch 2003). As outlined below, interference with the insulin signaling pathway appears to affect levels of both oxidants and antioxidants in organisms, providing additional support for a role of ROS in lifespan determination. Insulin/IGF-1 signaling is a highly conserved pathway, which has been implicated in a multitude of physiological processes, including stress response, diapause, reproduction, metabolism, growth, and aging (Tatar et al. 2003). Signaling through the Insulin/IGF-1 receptor occurs via a phosphorylation cascade, which ultimately causes phosphorylation of the forkhead transcription factor FOXO, and prevents FOXO from its translocation into the nucleus. Conversely, lack of the IGF-1 receptor or disruption of the kinase cascade promotes FOXO's translocation into the nucleus and allows the transcriptional regulator to induce the expression of a variety of stress-related genes (Kenyon 2005). FOXO-controlled genes encode for heat shock proteins, for proteins involved in pathogen resistance, metabolism (e.g. β-oxidation of fatty acids and gluconeogenesis), transcriptional repression and protein degradation as well as for antioxidant enzymes, such as superoxide dismutase, catalase and glutathione S-transferases (reviewed by Murphy 2006) (Fig. 14.2). These findings suggested that the lifespan extension observed in mutants with compromised insulin signaling might be, at least in part, due to increased oxidative stress protection.

Studies in *C. elegans* also revealed that mutants lacking the insulin/IGF-1 receptor *daf-2* or the phosphoinositide kinase PI3K (*age-1*) show increased level of SOD and catalase activity, significantly increased oxidative stress resistance and exhibit very extended lifespans (Honda and Honda 1999; Johnson 1990; Kenyon et al. 1993; Brys et al. 2007). Similar results were observed in *Drosophila*, where deletion of either the insulin receptor or CHICO, the fly insulin receptor substrate, increases both oxidative stress resistance and lifespan (reviewed by (Giannakou and Partridge 2007)). The lifespan-extending features were found to strictly depend on the presence of the FOXO transcription factor *daf-16* as *C. elegans* mutants lacking DAF-16 show increased sensitivity towards paraquat-mediated oxidative stress and are severely short-lived (Yanase et al. 2002; Lin et al. 2001). *Daf-16* mutant worms also show significantly increased protein damage as measured by protein carbonylation, providing further confirmation that these animals experience increased levels of oxidative stress (Yanase et al. 2002).

That the insulin signaling pathway not only influences the lifespan of invertebrates but also of vertebrates became clear when female mice carrying a heterozygous mutation of the insulin-like growth factor type 1 receptor gene *igf-1r*$^{+/-}$ (homozygous null mutation of IGF-1R has been found to be embryonic lethal) were found to be more resistant to oxidative stress and were long-lived (Holzenberger et al. 2003). These results were consistent with earlier studies on the IGF-1R substrate p66^{Shc}, whose mutation was found to increase oxidative stress resistance and extend lifespan (Migliaccio et al. 1999). Subsequent studies with tissue-specific knock-outs of the IGF-1R in the adipose tissue of mice further confirmed these results and showed that these mice have an extended lifespan (Bluher et al. 2003). At this point it is still unclear which of the many FOXO-regulated genes are ultimately responsible for the observed lifespan extension in worms and other organisms, and what exact role(s) ROS play in the ILS-mediated lifespan regulation. A RNAi-mediated knock-down of FOXO-target genes, including glutathione transferase, cytosolic *ctl-1*, peroxisomal *ctl-2*, or mitochondrial superoxide dismutase *sod-3* was found to individually reduce the long lifespan of *daf-2* mutants while deletion of the cytosolic superoxide dismutase *sod-1* had no effect on the lifespan of *daf-2* mutants. Moreover, loss of the extracellular superoxide dismutase *sod-4* further extended *daf-2*-mediated lifespan (Murphy et al. 2003; Ayyadevara et al. 2005; Doonan et al. 2008). That lifespan-extension by compromised Insulin/IGF-1-signaling is not due to a reduction of oxidants and decreased oxidative damage became obvious when long-lived *daf-2* mutants were found to have higher respiratory rates, and exhibit increased levels of mitochondrial ROS level (Houthoofd et al. 2005; Zarse et al. 2012). Moreover, their lifespan was slightly reduced upon treatment with antioxidants (Yang and Hekimi 2010; Brys et al. 2007; Zarse et al. 2012). These results were highly reminiscent of studies conducted with some long-lived

ETC mutants, which also showed increased ROS level that seemed necessary for lifespan extension (see Sect. 14.5.1). When Insulin/IGF-1 signaling was reduced during adulthood of worms, an increase in respiration was observed that caused a transient increase in ROS, which eventually led to increased activity of catalase and SOD (Zarse et al. 2012). The observed lifespan extension was diminished when antioxidants were supplemented (Zarse et al. 2012). The observation that treatment of worms with the superoxide generator juglone caused nuclear translocation (i.e., activation) of DAF-16 while exposure to hydrogen peroxide lead to phosphorylation (i.e., inactivation) of DAF-16 (Weinkove et al. 2006; Nemoto and Finkel 2002) suggested that the type of oxidant and possibly its sub-cellular accumulation might affect signaling processes, oxidative stress resistance and ultimately the lifespan of organisms.

14.6 Is the Free Radical Theory of Aging Still Valid?

Taken all the published studies together, it is now clear that both the expression levels of antioxidants and the production of oxidants are optimized for individual organisms and finely balanced. Any drastic change in oxidant or antioxidant level brings about the possibility to negatively affect tightly regulated processes, such as development, differentiation, signaling, host defense, metabolism, and ultimately lifespan (Fig. 14.3). Future studies aimed to further manipulate the interplay between endogenous oxidants and antioxidants will provide, at most, more correlative evidence. However, to ultimately prove or disprove the free radical theory of aging, we need to temporally and spatially dissect both positive and negative effects of distinct ROS-levels on individual processes during the lifetime of an organism to define which levels, timings and kinetics are required for proper functioning of the organism. Only when we understand the effects of ROS *in vivo* can we begin to manipulate their levels with a predictable outcome. The development of quantitative *in vivo* ROS sensors and read-outs of cellular targets of distinct ROS levels provides us now with the opportunity to make clear distinctions between onset, duration and magnitude of oxidant formation, and to assign the role of ROS in development, proliferation, host defense, aging and disease.

Over the past decades, our view on reactive oxygen species has changed from branding them as undesired, potentially harmful by-products of aerobic life to acknowledging their importance in many physiological processes. The necessity of certain levels of oxidants became even more obvious by the observations that antioxidants were capable of reversing beneficial effects on physiology and lifespan, and by the discovery that many transcription factors and signaling pathways can be influenced by reactive oxygen species (Hamanaka and Chandel 2010; Trachootham et al. 2008). Hence, we need to now move beyond the free radical theory of aging to acknowledge the fact that ROS are major players in lifespan and aging but likely not (only) in their function as cytotoxic agents but as regulators of essential physiological processes in the cell.

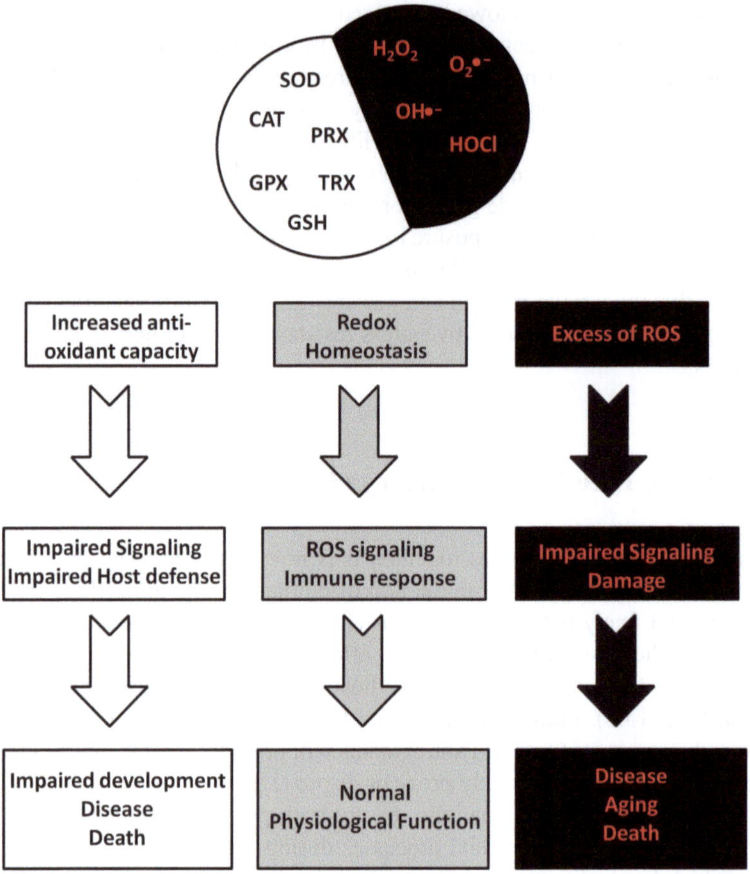

Fig. 14.3 Maintaining redox homeostasis – Essential for the physiology of organisms. Maintaining the proper redox balance between oxidants, such as hydrogen peroxide (H_2O_2), superoxide ($O_2\bullet^-$), hydroxyl radicals ($OH\bullet^-$), hypochlorous acid (HOCl) and antioxidants, like catalase (CAT), peroxiredoxin (PRX), glutathione peroxidase (GPX), superoxide dismutase (SOD), thioredoxin (TRX), glutathione (GSH) is essential for correct physiological processes. Shift towards more oxidizing conditions (i.e., oxidative stress) or increase in the antioxidant capacity of the organism will result in pathophysiological conditions

References

Andziak B, O'Connor TP, Buffenstein R (2005) Antioxidants do not explain the disparate longevity between mice and the longest-living rodent, the naked mole-rat. Mech Ageing Dev 126:1206–1212

Ash CE, Merry BJ (2011) The molecular basis by which dietary restricted feeding reduces mitochondrial reactive oxygen species generation. Mech Ageing Dev 132:43–54

Asikainen TM, Huang TT, Taskinen E, Levonen AL, Carlson E, Lapatto R, Epstein CJ, Raivio KO (2002) Increased sensitivity of homozygous Sod2 mutant mice to oxygen toxicity. Free Radic Biol Med 32:175–186

Ayyadevara S, Dandapat A, Singh SP, Benes H, Zimniak L, Reis RJS, Zimniak P (2005) Lifespan extension in hypomorphic daf-2 mutants of Caenorhabditis elegans is partially mediated by glutathione transferase CeGSTP2-2. Aging Cell 4:299–307

Back P, de Vos WH, Depuydt GG, Matthijssens F, Vanfleteren JR, Braeckman BP (2012) Exploring real-time in vivo redox biology of developing and aging Caenorhabditis elegans. Free Radic Biol Med 52:850–859

Barnham KJ, Masters CL, Bush AI (2004) Neurodegenerative diseases and oxidative stress. Nat Rev Drug Discov 3:205–214

Barros MH, Bandy B, Tahara EB, Kowaltowski AJ (2004) Higher respiratory activity decreases mitochondrial reactive oxygen release and increases life span in Saccharomyces cerevisiae. J Biol Chem 279:49883–49888

Bayne ACV, Mockett RJ, Orr WC, Sohal RS (2005) Enhanced catabolism of mitochondrial superoxide/hydrogen peroxide and aging in transgenic Drosophila. Biochem J 391:277–284

Bedard K, Krause KH (2007) The NOX family of ROS-generating NADPH oxidases: physiology and pathophysiology. Physiol Rev 87:245–313

Bevilacqua L, Ramsey JJ, Hagopian K, Weindruch R, Harper ME (2004) Effects of short- and medium-term calorie restriction on muscle mitochondrial proton leak and reactive oxygen species production. Am J Physiol Endocrinol Metab 286:E852–E861

Bluher M, Kahn BB, Kahn CR (2003) Extended longevity in mice lacking the insulin receptor in adipose tissue. Science 299:572–574

Bortolato M, Chen K, Shih JC (2008) Monoamine oxidase inactivation: from pathophysiology to therapeutics. Adv Drug Deliv Rev 60:1527–1533

Brandes N, Schmitt S, Jakob U (2009) Thiol-based redox switches in eukaryotic proteins. Antioxid Redox Signal 11:997–1014

Brandes N, Reichmann D, Tienson H, Leichere LI, Jakob U (2011) Using quantitative redox proteomics to dissect the yeast redoxome. J Biol Chem 286:41893–41903

Brys K, Vanfleteren JR, Braeckman BP (2007) Testing the rate-of-living/oxidative damage theory of aging in the nematode model Caenorhabditis elegans. Exp Gerontol 42:845–851

Cabreiro F, Ackerman D, Doonan R, Araiz C, Back P, Papp D, Braeckman BP, Gems D (2011) Increased life span from overexpression of superoxide dismutase in Caenorhabditis elegans is not caused by decreased oxidative damage. Free Radic Biol Med 51:1575–1582

Cadenas E, Davies KJA (2000) Mitochondrial free radical generation, oxidative stress, and aging. Free Radic Biol Med 29:222–230

Carlsson LM, Jonsson J, Edlund T, Marklund SL (1995) Mice lacking extracellular superoxide dismutase are more sensitive to hyperoxia. Proc Natl Acad Sci USA 92:6264–6268

Ceriello A, Motz E (2004) Is oxidative stress the pathogenic mechanism underlying insulin resistance, diabetes, and cardiovascular disease? The common soil hypothesis revisited. Arterioscler Thromb Vasc Biol 24:816–823

Chance B, Williams GR (1956) The respiratory chain and oxidative phosphorylation. Adv Enzymol Relat Subj Biochem 17:65–134

Chen XL, Mele J, Giese H, van Remmen H, Dolle MET, Steinhelper M, Richardson A, Vijg J (2003) A strategy for the ubiquitous overexpression of human catalase and CuZn superoxide dismutase genes in transgenic mice. Mech Ageing Dev 124:219–227

Cho CG, Kim HJ, Chung SW, Jung KJ, Shim KH, Yu BP, Yodoi J, Chung HY (2003) Modulation of glutathione and thioredoxin systems by calorie restriction during the aging process. Exp Gerontol 38:539–548

Cocheme HM, Quin C, McQuaker SJ, Cabreiro F, Logan A, Prime TA, Abakumova I, Patel JV, Fearnley IM, James AM, Porteous CM, Smith RAJ, Saeed S, Carre JE, Singer M, Gems D, Hartley RC, Partridge L, Murphy MP (2011) Measurement of H(2)O(2) within living Drosophila during aging using a ratiometric mass spectrometry probe targeted to the mitochondrial matrix. Cell Metab 13:340–350

Collet JF, Messens J (2010) Structure, function, and mechanism of thioredoxin proteins. Antioxid Redox Signal 13:1205–1216

Copeland JM, Cho J, Lo T Jr, Hur JH, Bahadorani S, Arabyan T, Rabie J, Soh J, Walker DW (2009) Extension of Drosophila life span by RNAi of the mitochondrial respiratory chain. Curr Biol 19:1591–1598
D'Autreaux B, Toledano MB (2007) ROS as signalling molecules: mechanisms that generate specificity in ROS homeostasis. Nat Rev Mol Cell Biol 8:813–824
Dalle-Donne I, Rossi R, Giustarini D, Milzani A, Colombo R (2003) Protein carbonyl groups as biomarkers of oxidative stress. Clin Chim Acta 329:23–38
Dillin A, Hsu AL, Arantes-Oliveira NA, Lehrer-Graiwer J, Hsin H, Fraser AG, Kamath RS, Ahringer J, Kenyon C (2002) Rates of behavior and aging specified by mitochondrial function during development. Science 298:2398–2401
Doonan R, McElwee JJ, Matthijssens F, Walker GA, Houthoofd K, Back P, Matscheski A, Vanfleteren JR, Gems D (2008) Against the oxidative damage theory of aging: superoxide dismutases protect against oxidative stress but have little or no effect on life span in Caenorhabditis elegans. Genes Dev 22:3236–3241
Droge W (2002) Free radicals in the physiological control of cell function. Physiol Rev 82:47–95
Dubey A, Forster MJ, Lal H, Sohal RS (1996) Effect of age and caloric intake on protein oxidation in different brain regions and on behavioral functions of the mouse. Arch Biochem Biophys 333:189–197
Dupuy C, Virion A, Ohayon R, Kaniewski J, Deme D, Pommier J (1991) Mechanism of hydrogen-peroxide formation catalyzed by nadph oxidase in thyroid plasma-membrane. J Biol Chem 266:3739–3743
Durieux J, Wolff S, Dillin A (2011) The cell-non-autonomous nature of electron transport chain-mediated longevity. Cell 144:79–91
Durusoy M, Diril N, Bozcuk AN (1995) Age-related activity of catalase in different genotypes of Drosophila-melanogaster. Exp Gerontol 30:77–86
Duttaroy A, Paul A, Kundu M, Belton A (2003) A Sod2 null mutation confers severely reduced adult life span in Drosophila. Genetics 165:2295–2299
Elchuri S, Oberley TD, Qi W, Eisenstein RS, Jackson Roberts L, Van Remmen H, Epstein CJ, Huang TT (2005) CuZnSOD deficiency leads to persistent and widespread oxidative damage and hepatocarcinogenesis later in life. Oncogene 24:367–380
Feng JL, Bussiere F, Hekimi S (2001) Mitochondrial electron transport is a key determinant of life span in Caenorhabditis elegans. Dev Cell 1:633–644
Finkel T (2011a) Signal transduction by mitochondrial oxidants. J Biol Chem 287(7):4434–4440
Finkel T (2011b) Signal transduction by reactive oxygen species. J Cell Biol 194:7–15
Finkel T, Holbrook NJ (2000) Oxidants, oxidative stress and the biology of ageing. Nature 408:239–247
Fontana L, Partridge L, Longo VD (2010) Extending healthy life span-from yeast to humans. Science 328:321–326
Fridovic I (1972) Superoxide radical and superoxide dismutase. Acc Chem Res 5:321–326
Fu YX, Cheng WH, Ross DA, Lei XG (1999) Cellular glutathione peroxidase protects mice against lethal oxidative stress induced by various doses of diquat. Proc Soc Exp Biol Med 222:164–169
Ghezzi P, Bonetto V, Fratelli M (2005) Thiol-disulfide balance: from the concept of oxidative stress to that of redox regulation. Antioxid Redox Signal 7:964–972
Giannakou ME, Partridge L (2007) Role of insulin-like signalling in Drosophila lifespan. Trends Biochem Sci 32:180–188
Giorgio M, Trinei M, Migliaccio E, Pelicci PG (2007) Hydrogen peroxide: a metabolic by-product or a common mediator of ageing signals? Nat Rev Mol Cell Biol 8:722–728
Godon C, Lagniel G, Lee J, Buhler JM, Kieffer S, Perrot M, Boucherie H, Toledano MB, Labarre J (1998) The H2O2 stimulon in Saccharomyces cerevisiae. J Biol Chem 273:22480–22489
Golden TR, Hinerfeld DA, Melov S (2002) Oxidative stress and aging: beyond correlation. Aging Cell 1:117–123
Grant CM (2008) Metabolic reconfiguration is a regulated response to oxidative stress. J Biol 7:1
Griswold CM, Matthews AL, Bewley KE, Mahaffey JW (1993) Molecular characterization and rescue of acatalasemic mutants of Drosophila-melanogaster. Genetics 134:781–788

Hamanaka RB, Chandel NS (2010) Mitochondrial reactive oxygen species regulate cellular signaling and dictate biological outcomes. Trends Biochem Sci 35:505–513
Hansen RE, Roth D, Winther JR (2009) Quantifying the global cellular thiol-disulfide status. Proc Natl Acad Sci USA 106:422–427
Harman D (1956) Aging – a theory based on free-radical and radiation-chemistry. J Gerontol 11:298–300
Harman D (1972) Biologic clock – mitochondria. J Am Geriatr Soc 20:145–147
Hartman PS, Ishii N, Kayser EB, Morgan PG, Sedensky MM (2001) Mitochondrial mutations differentially affect aging, mutability and anesthetic sensitivity in Caenorhabditis elegans. Mech Ageing Dev 122:1187–1201
Heidler T, Hartwig K, Daniel H, Wenzel U (2010) Caenorhabditis elegans lifespan extension caused by treatment with an orally active ROS-generator is dependent on DAF-16 and SIR-2.1. Biogerontology 11:183–195
Heilbronn LK, Ravussin E (2003) Calorie restriction and aging: review of the literature and implications for studies in humans. Am J Clin Nutr 78:361–369
Hernandez-Garcia D, Wood CD, Castro-Obregon S, Covarrubias L (2010) Reactive oxygen species: a radical role in development? Free Radic Biol Med 49:130–143
Ho YS, Magnenat JL, Bronson RT, Cao J, Gargano M, Sugawara M, Funk CD (1997) Mice deficient in cellular glutathione peroxidase develop normally and show no increased sensitivity to hyperoxia. J Biol Chem 272:16644–16651
Ho YS, Xiong Y, Ma WC, Spector A, Ho DS (2004) Mice lacking catalase develop normally but show differential sensitivity to oxidant tissue injury. J Biol Chem 279:32804–32812
Ho HY, Cheng ML, Chiu DT (2007) Glucose-6-phosphate dehydrogenase–from oxidative stress to cellular functions and degenerative diseases. Redox Rep 12:109–118
Holmgren A, Johansson C, Berndt C, Lonn ME, Hudemann C, Lillig CH (2005) Thiol redox control via thioredoxin and glutaredoxin systems. Biochem Soc Trans 33:1375–1377
Holzenberger M, Dupont J, Ducos B, Leneuve P, Geloen A, Even PC, Cervera P, le Bouc Y (2003) IGF-1 receptor regulates lifespan and resistance to oxidative stress in mice. Nature 421:182–187
Honda Y, Honda S (1999) The daf-2 gene network for longevity regulates oxidative stress resistance and Mn-superoxide dismutase gene expression in Caenorhabditis elegans. FASEB J 13:1385–1393
Houthoofd K, Braeckman BP, Lenaerts I, Brys K, de Vreese A, van Eygen S, Vanfleteren JR (2002) Axenic growth up-regulates mass-specific metabolic rate, stress resistance, and extends life span in Caenorhabditis elegans. Exp Gerontol 37:1371–1378
Houthoofd K, Fidalgo MA, Hoogewijs D, Braeckman BP, Lenaerts I, Brys K, Matthijssens F, de Vreese A, van Eygen S, Munoz MJ, Vanfleteren JR (2005) Metabolism, physiology and stress defense in three aging Ins/IGF-1 mutants of the nematode Caenorhabditis elegans. Aging Cell 4:87–95
Huang TT, Carlson EJ, Gillespie AM, Shi Y, Epstein CJ (2000) Ubiquitous overexpression of CuZn superoxide dismutase does not extend life span in mice. J Gerontol A Biol Sci Med Sci 55:B5–B9
Huang TT, Carlson EJ, Kozy HM, Mantha S, Goodman SI, Ursell PC, Epstein CJ (2001) Genetic modification of prenatal lethality and dilated cardiomyopathy in Mn superoxide dismutase mutant mice. Free Radic Biol Med 31:1101–1110
Ishii N, Takahashi K, Tomita S, Keino T, Honda S, Yoshino K, Suzuki K (1990) A methyl viologen-sensitive mutant of the nematode Caenorhabditis-elegans. Mutat Res 237:165–171
Ishii N, Fujii M, Hartman PS, Tsuda M, Yasuda K, Senoo-Matsuda N, Yanase S, Ayusawa D, Suzuki K (1998) A mutation in succinate dehydrogenase cytochrome b causes oxidative stress and ageing in nematodes. Nature 394:694–697
Jacob MH, Janner Dda R, Araujo AS, Jahn MP, Kucharski LC, Moraes TB, Dutra Filho CS, Ribeiro MF, Bello-Klein A (2010) Redox imbalance influence in the myocardial Akt activation in aged rats treated with DHEA. Exp Gerontol 45:957–963

Jee C, Vanoaica L, Lee J, Park BJ, Ahnn J (2005) Thioredoxin is related to life span regulation and oxidative stress response in Caenorhabditis elegans. Genes Cells 10:1203–1210
Johnson TE (1990) Increased life-span of age-1 mutants in Caenorhabditis-elegans and lower gompertz rate of aging. Science 249:908–912
Kakihana T, Nagata K, Sitia R (2012) Peroxides and peroxidases in the endoplasmic reticulum: integrating redox homeostasis and oxidative folding. Antioxid Redox Signal 16:763–771
Keaney M, Gems D (2003) No increase in lifespan in Caenorhabditis elegans upon treatment with the superoxide dismutase mimetic EUK-8. Free Radic Biol Med 34:277–282
Keaney M, Matthijssens F, Sharpe M, Vanfleteren J, Gems D (2004) Superoxide dismutase mimetics elevate superoxide dismutase activity in vivo but do not retard aging in the nematode Caenorhabditis elegans. Free Radic Biol Med 37:239–250
Kenyon C (2005) The plasticity of aging: insights from long-lived mutants. Cell 120:449–460
Kenyon C, Chang J, Gensch E, Rudner A, Tabtiang R (1993) A C-Elegans mutant that lives twice as long as wild-type. Nature 366:461–464
Kim J, Takahashi M, Shimizu T, Shirasawa T, Kajita M, Kanayama A, Miyamoto Y (2008) Effects of a potent antioxidant, platinum nanoparticle, on the lifespan of Caenorhabditis elegans. Mech Ageing Dev 129:322–331
Klichko VI, Radyuk SN, Orr WC (2004) Profiling catalase gene expression in Drosophila melanogaster during development and aging. Arch Insect Biochem Physiol 56:34–50
Klotz LO, Sies H (2009) Cellular generation of oxidants: relation to oxidative stress. In: Redox signaling and regulation in biology and medicine. Wiley-VCH, Weinheim, pp 45–61
Kohn RR (1971) Effect of antioxidants on life-span of C57bl mice. J Gerontol 26:378–380
Kondo M, Senoo-Matsuda N, Yanase S, Ishii T, Hartman PS, Ishii N (2005) Effect of oxidative stress on translocation of DAF-16 in oxygen-sensitive mutants, mev-1 and gas-1 of Caenorhabditis elegans. Mech Ageing Dev 126:637–641
Krause KH (2007) Aging: a revisited theory based on free radicals generated by NOX family NADPH oxidases. Exp Gerontol 42:256–262
Kumsta C, Thamsen M, Jakob U (2011) Effects of oxidative stress on behavior, physiology, and the redox thiol proteome of Caenorhabditis elegans. Antioxid Redox Signal 14:1023–1037
Lambert AJ, Merry BJ (2004) Effect of caloric restriction on mitochondrial reactive oxygen species production and bioenergetics: reversal by insulin. Am J Physiol Regul Integr Comp Physiol 286:R71–R79
Lambeth JD (2004) NOX enzymes and the biology of reactive oxygen. Nat Rev Immunol 4:181–189
Lapointe J, Hekimi S (2008) Early mitochondrial dysfunction in long-lived Mclk1(+/−) mice. J Biol Chem 283:26217–26227
Lee SS, Lee RYN, Fraser AG, Kamath RS, Ahringer J, Ruvkun G (2003a) A systematic RNAi screen identifies a critical role for mitochondria in C-elegans longevity. Nat Genet 33:40–48
Lee TH, Kim SU, Yu SL, Kim SH, Park DS, Moon HB, Dho SH, Kwon KS, Kwon HJ, Han YH, Jeong S, Kang SW, Shin HS, Lee KK, Rhee SG, Yu DY (2003b) Peroxiredoxin II is essential for sustaining life span of erythrocytes in mice. Blood 101:5033–5038
Lee KS, Iijima-Ando K, Iijima K, Lee WJ, Lee JH, Yu K, Lee DS (2009a) JNK/FOXO-mediated neuronal expression of fly homologue of peroxiredoxin II reduces oxidative stress and extends life span. J Biol Chem 284:29454–29461
Lee SJ, Murphy CT, Kenyon C (2009b) Glucose shortens the life span of C. elegans by downregulating DAF-16/FOXO activity and aquaporin gene expression. Cell Metab 10:379–391
Lee SJ, Hwang AB, Kenyon C (2010) Inhibition of respiration extends C. elegans life span via reactive oxygen species that increase HIF-1 activity. Curr Biol 20:2131–2136
Levine RL (2002) Carbonyl modified proteins in cellular regulation, aging, and disease. Free Radic Biol Med 32:790–796
Li Y, Huang TT, Carlson EJ, Melov S, Ursell PC, Olson JL, Noble LJ, Yoshimura MP, Berger C, Chan PH, Wallace DC, Epstein CJ (1995) Dilated cardiomyopathy and neonatal lethality in mutant mice lacking manganese superoxide dismutase. Nat Genet 11:376–381

Lin K, Hsin H, Libina N, Kenyon C (2001) Regulation of the Caenorhabditis elegans longevity protein DAF-16 by insulin/IGF-1 and germline signaling. Nat Genet 28:139–145

Lin SJ, Kaeberlein M, Andalis AA, Sturtz LA, Defossez PA, Culotta VC, Fink GR, Guarente L (2002) Calorie restriction extends Saccharomyces cerevisiae lifespan by increasing respiration. Nature 418:344–348

Liu XX, Jiang N, Hughes B, Bigras E, Shoubridge E, Hekimi S (2005) Evolutionary conservation of the clk-1-dependent mechanism of longevity: loss of mclk1 increases cellular fitness and lifespan in mice. Genes Dev 19:2424–2434

Longo VD, Finch CE (2003) Evolutionary medicine: from dwarf model systems to healthy centenarians? Science 299:1342–1346

Lopez-Lluch G, Hunt N, Jones B, Zhu M, Jamieson H, Hilmer S, Cascajo MV, Allard J, Ingram DK, Navas P, de Cabo R (2006) Calorie restriction induces mitochondrial biogenesis and bioenergetic efficiency. Proc Natl Acad Sci USA 103:1768–1773

Mackay WJ, Bewley GC (1989) The genetics of catalase in Drosophila-melanogaster – isolation and characterization of acatalasemic mutants. Genetics 122:643–652

Makino N, Mochizuki Y, Bannai S, Sugita Y (1994) Kinetic studies on the removal of extracellular hydrogen peroxide by cultured fibroblasts. J Biol Chem 269:1020–1025

Margittai E, Banhegyi G (2010) Oxidative folding in the endoplasmic reticulum: towards a multiple oxidant hypothesis? FEBS Lett 584:2995–2998

Matsui M, Oshima M, Oshima H, Takaku K, Maruyama T, Yodoi J, Taketo MM (1996) Early embryonic lethality caused by targeted disruption of the mouse thioredoxin gene. Dev Biol 178:179–185

McCay CM, Crowell MF, Maynard LA (1935) The effect of retarded growth upon the length of life span and upon the ultimate body size. J Nutr 10:63–79

McCord JM, Fridovic I (1969) Superoxide dismutase an enzymic function for erythrocuprein (Hemocuprein). J Biol Chem 244:6049–6055

Melov S, Ravenscroft J, Malik S, Gill MS, Walker DW, Clayton PE, Wallace DC, Malfroy B, Doctrow SR, Lithgow GJ (2000) Extension of life-span with superoxide dismutase/catalase mimetics. Science 289:1567–1569

Migliaccio E, Giorgio M, Mele S, Pelicci G, Reboidl P, Pandolfi PP, Lanfrancone L, Pelicci PG (1999) The p66(shc) adaptor protein controls oxidative stress response and life span in mammals. Nature 402:309–313

Miranda-Vizuete A, Gonzalez JCF, Gahmon G, Burghoorn J, Navas P, Swoboda P (2006) Lifespan decrease in a Caenorhabditis elegans mutant lacking TRX-1, a thioredoxin expressed in ASJ sensory neurons. FEBS Lett 580:484–490

Mitsui A, Hamuro J, Nakamura H, Kondo N, Hirabayashi Y, Ishizaki-Koizumi S, Hirakawa T, Inoue T, Yodoi J (2002) Overexpression of human thioredoxin in transgenic mice controls oxidative stress and life span. Antioxid Redox Signal 4:693–696

Miwa S, Riyahi K, Partridge L, Brand MD (2004) Lack of correlation between mitochondrial reactive oxygen species production and life span in Drosophila. Strateg Eng Neglig Senescence Why Genuin Control Aging May Be Foreseeable 1019:388–391

Mockett RJ, Orr WC, Rahmandar JJ, Benes JJ, Radyuk SN, Klichko VI, Sohal RS (1999) Overexpression of Mn-containing superoxide dismutase in transgenic Drosophila melanogaster. Arch Biochem Biophys 371:260–269

Mockett RJ, Sohal BH, Sohal RS (2010) Expression of multiple copies of mitochondrially targeted catalase or genomic Mn superoxide dismutase transgenes does not extend the life span of Drosophila melanogaster. Free Radic Biol Med 49:2028–2031

Moller IM, Rogowska-Wrzesinska A, Rao RSP (2011) Protein carbonylation and metal-catalyzed protein oxidation in a cellular perspective. J Proteomics 74:2228–2242

Muller FL, Song W, Liu Y, Chaudhuri A, Pieke-Dahl S, Strong R, Huang TT, Epstein CJ, Roberts LJ 2nd, Csete M, Faulkner JA, Van Remmen H (2006) Absence of CuZn superoxide dismutase leads to elevated oxidative stress and acceleration of age-dependent skeletal muscle atrophy. Free Radic Biol Med 40:1993–2004

Muller FL, Lustgarten MS, Jang Y, Richardson A, van Remmen H (2007) Trends in oxidative aging theories. Free Radic Biol Med 43:477–503
Murakami S, Johnson TE (1996) A genetic pathway conferring life extension and resistance to UV stress in Caenorhabditis elegans. Genetics 143:1207–1218
Murphy CT (2006) The search for DAF-16/FOXO transcriptional targets: approaches and discoveries. Exp Gerontol 41:910–921
Murphy MP (2012) Mitochondrial thiols in antioxidant protection and redox signaling: distinct roles for glutathionylation and other thiol modifications. Antioxid Redox Signal 16:476–495
Murphy CT, McCarroll SA, Bargmann CI, Fraser A, Kamath RS, Ahringer J, Li H, Kenyon C (2003) Genes that act downstream of DAF-16 to influence the lifespan of Caenorhabditis elegans. Nature 424:277–284
Nauseef WM (2008) Biological roles for the NOX family NADPH oxidases. J Biol Chem 283:16961–16965
Nemoto S, Finkel T (2002) Redox regulation of forkhead proteins through a p66shc-dependent signaling pathway. Science 295:2450–2452
Neumann CA, Krause DS, Carman CV, Das S, Dubey DP, Abraham JL, Bronson RT, Fujiwara Y, Orkin SH, van Etten RA (2003) Essential role for the peroxiredoxin Prdx1 in erythrocyte antioxidant defence and tumour suppression. Nature 424:561–565
Nicholls P, Fita I, Loewen PC (2001) Enzymology and structure of catalases. Adv Inorg Chem 51(51):51–106
Niki E (1987) Interaction of ascorbate and alpha-tocopherol. Ann N Y Acad Sci 498:186–199
Nisoli E, Tonetto C, Cardile A, Cozzi V, Bracale R, Tedesco L, Falcone S, Valerio A, Cantoni O, Clementi E, Moncada S, Carruba MO (2005) Calorie restriction promotes mitochondrial biogenesis by inducing the expression of eNOS. Science 310:314–317
Nonn L, Williams RR, Erickson RP, Powis G (2003) The absence of mitochondrial thioredoxin 2 causes massive apoptosis, exencephaly, and early embryonic lethality in homozygous mice. Mol Cell Biol 23:916–922
Olahova M, Taylor SR, Khazaipoul S, Wang JL, Morgan BA, Matsumoto K, Blackwell TK, Veal EA (2008) A redox-sensitive peroxiredoxin that is important for longevity has tissue- and stress-specific roles in stress resistance. Proc Natl Acad Sci USA 105:19839–19844
Orr WC, Sohal RS (1994) Extension of life-span by overexpression of superoxide-dismutase and catalase in Drosophila-melanogaster. Science 263:1128–1130
Orr WC, Sohal RS (2003) Does overexpression of Cu, Zn-SOD extend life span in Drosophila melanogaster? Exp Gerontol 38:227–230
Orrenius S, Moldeus P (1984) The multiple roles of glutathione in drug-metabolism. Trends Pharmacol Sci 5:432–435
Parihar MS, Kunz EA, Brewer GJ (2008) Age-related decreases in NAD(P)H and glutathione cause redox declines before ATP loss during glutamate treatment of hippocampal neurons. J Neurosci Res 86:2339–2352
Parkes TL, Elia AJ, Dickinson D, Hilliker AJ, Phillips JP, Boulianne GL (1998) Extension of Drosophila lifespan by overexpression of human SOD1 in motorneurons. Nat Genet 19:171–174
Pearl R (1921) The biology of death. Utica, N.Y.: Science Press
Perez VI, Lew CM, Cortez LA, Webb CR, Rodriguez M, Liu Y, Qi W, Li Y, Chaudhuri A, van Remmen H, Richardson A, Ikeno Y (2008) Thioredoxin 2 haploinsufficiency in mice results in impaired mitochondrial function and increased oxidative stress. Free Radic Biol Med 44:882–892
Perez VI, Bokov A, van Remmen H, Mele J, Ran QT, Ikeno Y, Richardson A (2009a) Is the oxidative stress theory of aging dead? Biochim Biophys Acta Gen Subj 1790:1005–1014
Perez VI, van Remmen H, Bokov A, Epstein CJ, Vijg J, Richardson A (2009b) The overexpression of major antioxidant enzymes does not extend the lifespan of mice. Aging Cell 8:73–75
Perez VI, Cortez LA, Lew CM, Rodriguez M, Webb CR, van Remmen H, Chaudhuri A, Qi WB, Lee S, Bokov A, Fok W, Jones D, Richardson A, Yodoi J, Zhang Y, Tominaga K, Hubbard GB,

Ikeno Y (2011) Thioredoxin 1 overexpression extends mainly the earlier part of life span in mice. J Gerontol A Biol Sci Med Sci 66:1286–1299
Perichon R, Bourre JM (1995) Peroxisomal beta-oxidation activity and catalase activity during development and aging in mouse-liver. Biochimie 77:288–293
Petriv OI, Rachubinski RA (2004) Lack of peroxisomal catalase causes a progeric phenotype in Caenorhabditis elegans. J Biol Chem 279:19996–20001
Phillips JP, Campbell SD, Michaud D, Charbonneau M, Hilliker AJ (1989) Null mutation of copper-zinc superoxide-dismutase in Drosophila confers hypersensitivity to paraquat and reduced longevity. Proc Natl Acad Sci USA 86:2761–2765
Radi R, Turrens JF, Chang LY, Bush KM, Crapo JD, Freeman BA (1991) Detection of catalase in rat heart mitochondria. J Biol Chem 266:22028–22034
Radyuk SN, Klichko VI, Orr WC (2004) Profiling Cu, Zn-superoxide dismutase expression in Drosophila melanogaster–a critical regulatory role for intron/exon sequence within the coding domain. Gene 328:37–48
Radyuk SN, Michalak K, Klichko VI, Benes J, Rebrin I, Sohal RS, Orr WC (2009) Peroxiredoxin 5 confers protection against oxidative stress and apoptosis and also promotes longevity in Drosophila. Biochem J 419:437–445
Ran QT, Liang HY, Gu MJ, Qi WB, Walter CA, Roberts LJ, Herman B, Richardson A, van Remmen H (2004) Transgenic mice overexpressing glutathione peroxidase 4 are protected against oxidative stress-induced apoptosis. J Biol Chem 279:55137–55146
Ran Q, Liang HY, Ikeno Y, Qi WB, Prolla TA, Roberts LJ, Wolf N, Vanremmen H, Richardson A (2007) Reduction in glutathione peroxidase 4 increases life span through increased sensitivity to apoptosis. J Gerontol A Biol Sci Med Sci 62:932–942
Rando TA, Crowley RS, Carlson EJ, Epstein CJ, Mohapatra PK (1998) Overexpression of copper/zinc superoxide dismutase: a novel cause of murine muscular dystrophy. Ann Neurol 44:381–386
Rea S, Johnson TE (2003) A metabolic model for life span determination in Caenorhabditis elegans. Dev Cell 5:197–203
Rea SL, Ventura N, Johnson TE (2007) Relationship between mitochondrial electron transport chain dysfunction, development, and life extension in Caenorhabditis elegans. PLOS Biol 5:2312–2329
Rebrin I, Sohal RS (2008) Pro-oxidant shift in glutathione redox state during aging. Adv Drug Deliv Rev 60:1545–1552
Reuter S, Gupta SC, Chaturvedi MM, Aggarwal BB (2010) Oxidative stress, inflammation, and cancer how are they linked? Free Radic Biol Med 49:1603–1616
Reveillaud I, Niedzwiecki A, Bensch KG, Fleming JE (1991) Expression of bovine superoxide-dismutase in Drosophila-melanogaster augments resistance to oxidative stress. Mol Cell Biol 11:632–640
Rohrbach S, Gruenler S, Teschner M, Holtz J (2006) The thioredoxin system in aging muscle: key role of mitochondrial thioredoxin reductase in the protective effects of caloric restriction? Am J Physiol Regul Integr Comp Physiol 291:R927–R935
Rubner M (1908) Das problem der lebensdauer und seine beziehungen zu ernährung. Oldenbourg, München
Salvi M, Battaglia V, Brunati AM, la Rocca N, Tibaldi E, Pietrangeli P, Marcocci L, Mondovi B, Rossi CA, Toninello A (2007) Catalase takes part in rat liver mitochondria oxidative stress defense. J Biol Chem 282:24407–24415
Schafer FQ, Buettner GR (2001) Redox environment of the cell as viewed through the redox state of the glutathione disulfide/glutathione couple. Free Radic Biol Med 30:1191–1212
Schlotterer A, Kukudov G, Bozorgmehr F, Hutter H, Du XL, Oikonomou D, Ibrahim Y, Pfisterer F, Rabbani N, Thornalley P, Sayed A, Fleming T, Humpert P, Schwenger V, Zeier M, Hamann A, Stern D, Brownlee M, Bierhaus A, Nawroth P, Morcos M (2009) C-elegans as model for the study of high glucose-mediated life span reduction. Diabetes 58:2450–2456

Schriner SE, Linford NJ, Martin GM, Treuting P, Ogburn CE, Emond M, Coskun PE, Ladiges W, Wolf N, van Remmen H, Wallace DC, Rabinovitch PS (2005) Extension of murine life span by overexpression of catalase targeted to mitochondria. Science 308:1909–1911

Schulz TJ, Zarse K, Voigt A, Urban N, Birringer M, Ristow M (2007) Glucose restriction extends Caenorhabditis elegans life span by inducing mitochondrial respiration and increasing oxidative stress. Cell Metab 6:280–293

Sedensky MM, Morgan PG (2006) Mitochondrial respiration and reactive oxygen species in mitochondrial aging mutants. Exp Gerontol 41:237–245

Semsei I, Rao G, Richardson A (1991) Expression of superoxide-dismutase and catalase in rat-brain as a function of age. Mech Ageing Dev 58:13–19

Senoo-Matsuda N, Yasuda K, Tsuda M, Ohkubo T, Yoshimura S, Nakazawa H, Hartman PS, Ishii N (2001) A defect in the cytochrome b large subunit in complex II causes both superoxide anion overproduction and abnormal energy metabolism in Caenorhabditis elegans. J Biol Chem 276:41553–41558

Sentman ML, Granstrom M, Jakobson H, Reaume A, Basu S, Marklund SL (2006) Phenotypes of mice lacking extracellular superoxide dismutase and copper- and zinc-containing superoxide dismutase. J Biol Chem 281:6904–6909

Seong KH, Ogashiwa T, Matsuo T, Fuyama Y, Aigaki T (2001) Application of the gene search system to screen for longevity genes in Drosophila. Biogerontology 2:209–217

Seto NOL, Hayashi S, Tener GM (1990) Overexpression of Cu-Zn superoxide-dismutase in Drosophila does not affect life-span. Proc Natl Acad Sci USA 87:4270–4274

Sohal RS, Arnold L, Orr WC (1990) Effect of age on superoxide-dismutase, catalase, glutathione-reductase, inorganic peroxides, Tba-reactive material, Gsh Gssg, Nadph Nadp+ and Nadh Nad + in Drosophila-melanogaster. Mech Ageing Dev 56:223–235

Sohal RS, Ku HH, Agarwal S, Forster MJ, Lal H (1994) Oxidative damage, mitochondrial oxidant generation and antioxidant defenses during aging and in response to food restriction in the mouse. Mech Ageing Dev 74:121–133

Sohal RS, Mockett RJ, Orr WC (2002) Mechanisms of aging: an appraisal of the oxidative stress hypothesis. Free Radic Biol Med 33:575–586

Someya S, Yu W, Hallows WC, Xu JZ, Vann JM, Leeuwenburgh C, Tanokura M, Denu JM, Prolla TA (2010) Sirt3 mediates reduction of oxidative damage and prevention of age-related hearing loss under caloric restriction. Cell 143:802–812

Stadtman ER (1992) Protein oxidation and aging. Science 257:1220–1224

Sun JT, Tower J (1999) FLP recombinase-mediated induction of Cu/Zn-superoxide dismutase transgene expression can extend the life span of adult Drosophila melanogaster flies. Mol Cell Biol 19:216–228

Sun JT, Folk D, Bradley TJ, Tower J (2002) Induced overexpression of mitochondrial Mn-superoxide dismutase extends the life span of adult Drosophila melanogaster. Genetics 161:661–672

Sun JT, Molitor J, Tower J (2004) Effects of simultaneous over-expression of Cu/ZnSOD and MnSOD on Drosophila melanogaster life span. Mech Ageing Dev 125:341–349

Svensson MJ, Larsson J (2007) Thioredoxin-2 affects lifespan and oxidative stress in Drosophila. Hereditas 144:25–32

Tatar M, Bartke A, Antebi A (2003) The endocrine regulation of aging by insulin-like signals. Science 299:1346–1351

Thornalley PJ, Vasak M (1985) Possible role for metallothionein in protection against radiation-induced oxidative stress. Kinetics and mechanism of its reaction with superoxide and hydroxyl radicals. Biochim Biophys Acta 827:36–44

Trachootham D, Lu WQ, Ogasawara MA, Valle NRD, Huang P (2008) Redox regulation of cell survival. Antioxid Redox Signal 10:1343–1374

Tsan MF, White JE, Caska B, Epstein CJ, Lee CY (1998) Susceptibility of heterozygous MnSOD gene-knockout mice to oxygen toxicity. Am J Respir Cell Mol Biol 19:114–120

Tsuda M, Ootaka R, Ohkura C, Kishita Y, Seong KH, Matsuo T, Aigaki T (2010) Loss of Trx-2 enhances oxidative stress-dependent phenotypes in Drosophila. FEBS Lett 584:3398–3401

Turrens JF (1997) Superoxide production by the mitochondrial respiratory chain. Biosci Rep 17:3–8
Ushio-Fukai M, Nakamura Y (2008) Reactive oxygen species and angiogenesis: NADPH oxidase as target for cancer therapy. Cancer Lett 266:37–52
van Raamsdonk JM, Hekimi S (2009) Deletion of the mitochondrial superoxide dismutase sod-2 extends lifespan in Caenorhabditis elegans. PLOS Genet 5:e1000361
van Raamsdonk JM, Hekimi S (2012) Superoxide dismutase is dispensable for normal animal lifespan. Proc Natl Acad Sci 109:5785–5790
van Remmen H, Ikeno Y, Hamilton M, Pahlavani M, Wolf N, Thorpe SR, Alderson NL, Baynes JW, Epstein CJ, Huang TT, Nelson J, Strong R, Richardson A (2003) Life-long reduction in MnSOD activity results in increased DNA damage and higher incidence of cancer but does not accelerate aging. Phys Genomics 16:29–37
van Remmen H, Qi WB, Sabia M, Freeman G, Estlack L, Yang H, Guo ZM, Huang TT, Strong R, Lee S, Epstein CJ, Richardson A (2004) Multiple deficiencies in antioxidant enzymes in mice result in a compound increase in sensitivity to oxidative stress. Free Radic Biol Med 36:1625–1634
Vanfleteren JR (1993) Oxidative stress and aging in Caenorhabditis-elegans. Biochem J 292:605–608
Victor VM, Rocha M, Sola E, Banuls C, Garcia-Malpartida K, Hernandez-Mijares A (2009) Oxidative stress, endothelial dysfunction and atherosclerosis. Curr Pharm Des 15:2988–3002
Walker DW, Hajek P, Muffat J, Knoepfle D, Cornelison S, Attardi G, Benzer S (2006) Hypersensitivity to oxygen and shortened lifespan in a Drosophila mitochondrial complex II mutant. Proc Natl Acad Sci USA 103:16382–16387
Wang XS, Phelan SA, Forsman-Semb K, Taylor EF, Petros C, Brown A, Lerner CP, Paigen B (2003) Mice with targeted mutation of peroxiredoxin 6 develop normally but are susceptible to oxidative stress. J Biol Chem 278:25179–25190
Wang DT, Malo D, Hekimi S (2010) Elevated mitochondrial reactive oxygen species generation affects the immune response via hypoxia-inducible factor-1 alpha in long-lived Mclk1(+/-) mouse mutants. J Immunol 184:582–590
Weinkove D, Halstead JR, Gems D, Divecha N (2006) Long-term starvation and ageing induce AGE-1/PI 3-kinase-dependent translocation of DAF-16/FOXO to the cytoplasm. BMC Biol 4:e1741–e7007
Winterbourn CC, Hampton MB (2008) Thiol chemistry and specificity in redox signaling. Free Radic Biol Med 45:549–561
Wolf N, Penn P, Pendergrass W, van Remmen H, Bartke A, Rabinovitch P, Martin GM (2005) Age-related cataract progression in five mouse models for anti-oxidant protection or hormonal influence. Exp Eye Res 81:276–285
Wong A, Boutis P, Hekimi S (1995) Mutations in the Clk-1 gene of Caenorhabditis-elegans affect developmental and behavioral timing. Genetics 139:1247–1259
Yanase S, Yasuda K, Ishii N (2002) Adaptive responses to oxidative, damage in three mutants of Caenorhabditis elegans (age-1, mev-1 and daf-16) that affect life span. Mech Age Dev 123:1579–1587
Yanase S, Onodera A, Tedesco P, Johnson TE, Ishii N (2009) SOD-1 deletions in Caenorhabditis elegans alter the localization of intracellular reactive oxygen species and show molecular compensation. J Gerontol A Biol Sci Med Sci 64:530–539
Yang W, Hekimi S (2010) A mitochondrial superoxide signal triggers increased longevity in Caenorhabditis elegans. PLOS Biol 8:e1000556
Yant LJ, Ran QT, Rao L, van Remmen H, Shibatani T, Belter JG, Motta L, Richardson A, Prolla TA (2003) The selenoprotein GPX4 is essential for mouse development and protects from radiation and oxidative damage insults. Free Radic Biol Med 34:496–502
Yen K, Patel HB, Lublin AL, Mobbs CV (2009) SOD isoforms play no role in lifespan in ad lib or dietary restricted conditions, but mutational inactivation of SOD-1 reduces life extension by cold. Mech Ageing Dev 130:173–178

Yu BP, Yang R (1996) Critical evaluation of the free radical theory of aging – a proposal for the oxidative stress hypothesis. Ann N Y Acad Sci 786:1–11

Zangar RC, Davydov DR, Verma S (2004) Mechanisms that regulate production of reactive oxygen species by cytochrome P450. Toxicol Appl Pharmacol 199:316–331

Zarse K, Schmeisser S, Groth M, Priebe S, Beuster G, Kuhlow D, Guthke R, Platzer M, Kahn CR, Ristow M (2012) Impaired insulin/IGF1 signaling extends life span by promoting mitochondrial L-proline catabolism to induce a transient ROS signal. Cell Metab 15:451–465

Zelko IN, Mariani TJ, Folz RJ (2002) Superoxide dismutase multigene family: a comparison of the CuZn-SOD (SOD1), Mn-SOD (SOD2), and EC-SOD (SOD3) gene structures, evolution, and expression. Free Radic Biol Med 33:337–349

Zhang YQ, Ikeno Y, Qi WB, Chaudhuri A, Li Y, Bokov A, Thorpe SR, Baynes JW, Epstein C, Richardson A, van Remmen H (2009) Mice deficient in both Mn superoxide dismutase and glutathione peroxidase-1 have increased oxidative damage and a greater incidence of pathology but no reduction in longevity. J Gerontol A Biol Sci Med Sci 64:1212–1220

Chapter 15
Oxidative Stress in Cancer

Peter Storz

Abstract Increased oxidative stress is a common feature observed in many different types of cancer. Depending on the radical formed, its concentration, and cellular location where its generation occurs, reactive oxygen species (ROS) have multiple functions within tumor cells. ROS-induced macromolecule damage can contribute to tumor initiation. Low levels of ROS can initiate cellular signaling pathways that mediate tumor cell proliferation, survival and tumor progression to a metastatic phenotype. High levels of ROS initiate signaling pathways that mediate tumor cell death, but also contribute to formation of cancer stem cells that induce tumor recurrence. Understanding the multitude and complexity of ROS-regulated pathways in cancer cells and targeted modulation of intracellular ROS levels using antioxidants or chemotherapy at different stages of tumor progression may be an effective strategy for combination therapy.

Keywords Oxidative stress • Reactive oxygen species • Antioxidants • Tumor • Cancer

Abbreviations

2DG	2-deoxyglucose
4-HNE	4-Hydroxy-noneal
Ask1	apoptosis signal-regulating kinase-1
ATM	ataxia telangiectasia mutated
BSO	L-S, R-buthionine sulphoximine
CSC	cancer stem cells

P. Storz (✉)
Department of Cancer Biology, Mayo Clinic, 4500 San Pablo Road,
Jacksonville, FL 32224, USA
e-mail: storz.peter@mayo.edu

U. Jakob and D. Reichmann (eds.), *Oxidative Stress and Redox Regulation*,
DOI 10.1007/978-94-007-5787-5_15,

DR	death receptor
DRL	death receptor ligand
EGF	epidermal growth factor
EGF-R	epidermal growth factor-receptor
Erk1/2	extracellular-regulated kinases 1/2
FOXO	forkhead homeobox type O
GF	growth factor
GF-R	growth factor receptor
GSH	glutathione
GSSG	glutathione disulphide
GPX	glutathione peroxidase
GST	Glutathione S-transferase
H_2O_2	hydrogen peroxide
HIF-1	hypoxia-inducible factor-1
JNK	c-Jun N-terminal Kinase
MAPK	mitogen-activated protein kinase
MMP	matrix metalloproteinase
MPTP	mitochondrial permeability transition pore
mROS	mitochondrial ROS
NF-κB	nuclear factor κ-B
NOX	NADPH oxidase
PDK1	phosphoinositide-dependent kinase 1
PDGF	platelet-derived growth factor
PDGF-R	platelet-derived growth factor-receptor
PI3-K	phosphatidylinositol 3-kinase
PKC	protein kinase C
PKD	protein kinase D
ROS	reactive oxygen species
Prdx	peroxiredoxin
PTEN	phosphatase and tensin homologue
SOD	superoxide dismutase
TGFβ	transforming growth factor β
TIMP	tissue inhibitor of metalloproteinases
TNFα	tumor necrosis factor α
VEGF	vascular epithelial growth factor.

15.1 Oxidative Stress in Cancer Cells

Oxidative stress occurs when an imbalance between pro-oxidant and anti-oxidant molecules alters the redox potential in a cell with a net effect of increasing ROS levels. Accumulation of intracellular oxidative stress can lead to the conversion of normal cells to cancer cells, and increased levels of oxidative stress are hallmarks of many cancers (Storz 2005; Trachootham et al. 2006). Oxidative stress-induced

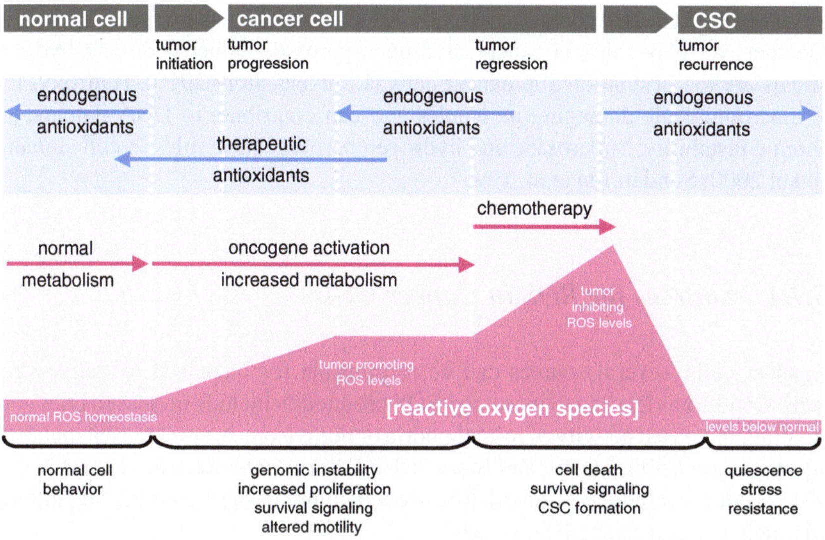

Fig. 15.1 Relations between cellular or therapy-induced oxidative stress and cellular or therapeutic antioxidants at different stages of tumor biology. While in normal cells antioxidants and ROS generation are in balance (normal homeostasis), increased ROS levels are hallmarks of tumor cells. In tumors increased ROS generation can lead to genomic instability, increased cell proliferation, survival signaling and increased cell motility. At this stage, lowering intracellular ROS levels by therapeutic antioxidants may have regulatory effects on tumor progression. Exuberant increase in ROS can be achieved by chemotherapy and can mediate tumor regression by inducing irreparable cell damage and tumor cell death. However, some tumor cells can develop stress resistance by upregulating antioxidant systems. Such cancer stem cells (*CSCs*) are highly resistant to stresses and after clonal expansion can be responsible for tumor recurrence

macromolecule damage and ROS-regulated signaling events have been reported to affect all aspects of tumor formation and progression, including genomic instability and mutagenesis, energy production, cell proliferation, survival and chemoresistance, increased cell motility and metastasis, stemness and recurrence, and angiogenesis (Fig. 15.1).

However, when produced in excess or when cancer cell antioxidant systems fail, they can contribute to tumor cell death. Tumor cells react to increases in oxidative stress by upregulating antioxidant systems, with the outcome that a balance is established that allows beneficiary effects of ROS, but avoids damaging effects that induce cell death. The cellular signaling pathways, by which such fine tuning is accomplished are little known, and some of the ROS-sensing proteins or molecules that contribute to such signaling are discussed in this chapter. Moreover, ROS-sensitive signaling pathways are constitutively switched on in many cancer cells and they participate in all aspects of tumor biology.

The sources for intracellular oxidative stress in cancer cells are manifold and different reactive oxygen species are generated. These include radicals such as superoxide ($O_2 \cdot^-$), nitric oxide (NO·), or hydroxyl radicals (·OH), as well as non-radicals such as hydrogen peroxide (H_2O_2). Among the variety of cellular ROS that can play roles in cancer, hydrogen peroxide, superoxide and hydroxyl radicals are the best studied in cancer cells (Liou and Storz 2010). Hydroxyl ions are macromolecule damaging molecules and can contribute to DNA damage and genomic instability. Superoxide and hydrogen peroxide have roles in cell signaling (Finkel 2000; Sundaresan et al. 1995).

15.1.1 Sources for ROS in Cancer Cells

In cancer cells, several sources can be responsible for increased oxidative stress levels. Conditions leading to increased ROS production include increased oncogenic signaling, increased activity of mitochondria or peroxisomes, or increased metabolic and enzymatic activity (Liou and Storz 2010; Babior 1999; Szatrowski and Nathan 1991). Additional inducers of oxidative stress in tumor tissue are infiltrating immune cells such as macrophages (Storz 2005).

Growth factors and cytokines such as tumor necrosis factor α (TNFα), platelet-derived growth factor (PDGF), epidermal growth factor (EGF), transforming growth factor β (TGFβ), or insulin can induce intracellular ROS production in cancer cells. Most of them induce the formation of superoxide, which then is converted to hydrogen peroxide (Storz 2005; Sundaresan et al. 1995; Bae et al. 2000; Lo and Cruz 1995; Meier et al. 1989; Ohba et al. 1994; Tiku et al. 1990). Additionally, oncogenic mutations of molecules downstream of growth factor receptors such as K-ras lead to increased superoxide production (Minamoto et al. 2000, 2002). Another downstream effector of many growth factor receptors, including epidermal growth factor-receptor (EGF-R) and c-Met, is the small RhoGTPase Rac-1 (Ferraro et al. 2006). Both, oncogenic K-ras and Rac-1 have been shown to induce superoxide production either via NADPH oxidases (NOX) or at the mitochondria (Liou and Storz 2010; Chiarugi and Fiaschi 2007).

Increased superoxide generation at the mitochondria is also induced by mitochondrial dysfunction (Cadenas 2004; Dayal et al. 2009; Pelicano et al. 2009). Mitochondrial respiration produces superoxide as a byproduct of oxidative phosphorylation at complexes I (NADH-ubiquinone oxidoreductase) and III (ubiquinol-cytochrome c oxidoreductase). Superoxide is released into the mitochondrial matrix and the inter-membrane space, but can be released into the cytoplasm via the mitochondrial permeability transition pore (MPTP) (Crompton 1999; Storz 2006). Superoxide usually has a very short half-life and rapidly is dismutated into hydrogen peroxide. This step is mediated by superoxide dismutase (SOD) enzymes. Its breakdown product hydrogen peroxide is a *bona fide* second messenger since it is highly diffusible and can regulate the activity of signaling molecules by direct and reversible oxidation (Liou and Storz 2010; Chiarugi and Fiaschi 2007; Rhee et al. 2000).

Other respiratory organelles that can contribute to increased superoxide and hydrogen peroxide levels in cancer cells are the peroxisomes (Dansen and Wirtz 2001). Peroxisomes generate ROS through acyl-CoA oxidase and xanthine oxidase (del Rio et al. 1992; Singh 1996). Additionally contributing to increased intracellular ROS levels in cancer cells are increased activities of metabolic enzymes such as oxidases, cyclo-oxygenases, lipoxygenases, or thymidine phosphorylase (reviewed in Liou and Storz 2010).

Macrophages or other inflammatory cells (i.e. neutrophils, eosinophils) infiltrating the tumor tissue can be a source of external oxidative stress. Macrophages infiltration into normal tissues may contribute to tumor initiation since infections or inflammation (i.e. pancreatitis) have been identified as a critical risk factor for tumor development (Balkwill 2009). Macrophages can produce and release reactive oxygen species such a superoxide, nitric oxide or hydrogen peroxide (respiratory burst), and released components such as nitric oxide and superoxide can react with each other to generate the highly toxic peroxinitrite radicals (Babior 1999; Cui et al. 1994). Moreover, macrophages can secrete cytokines such as TNFα to induce ROS production in target cells (Babior 1999; Balkwill 2009).

15.1.2 Endogenous Antioxidants in Cancer Cells

Increases in oxidative stress in normal cells can lead to mutations and initiate cancer. In proliferating cancer cells increased ROS levels that are induced by above described mechanisms need to be kept in check to facilitate optimal growth conditions. Cancer stem cells (CSC) often have upregulated antioxidant systems that make them unresponsive to chemotherapy. The detoxification from ROS in cancer cells can be mediated by several antioxidant enzymes.

Depletion of cancer cells from superoxide is occurs through upregulation of superoxide dismutase enzymes. These include MnSOD, Cu/ZnSOD and FeSOD. All are metallo-enzymes that utilize divalent metal ions as cofactors to dismutate superoxide to hydrogen peroxide. Different isoforms of SODs are located at distinct compartments in the cell. For example, MnSOD is located in the mitochondria matrix and Cu/ZnSOD in the cytosol (Liou and Storz 2010). SOD enzymes are encoded by nuclear genes and signaling pathways exist by which increases in mitochondrial oxidative stress are sensed and translated to nuclear gene induction (Storz 2006, 2007; Storz et al. 2005a). The breakdown product of superoxide in SOD-catalyzed dismutase reactions is hydrogen peroxide.

Multiple enzymes detoxify cells from hydrogen peroxide by generating water and oxygen. Most prominent are catalase enzymes located in the cytosol and peroxisomes (Bendayan and Reddy 1982; Hashimoto and Hayashi 1990; Litwin et al. 1987). Peroxiredoxins (Prdx) are thioredoxin peroxidases that catalyze the reduction of hydrogen peroxide, organic hydroperoxides and peroxynitrite (Hofmann et al. 2002; Rhee et al. 1999; Wood et al. 2003). Disregulation of enzymes that detoxify cells from hydrogen peroxide can have dramatic effects on cancer development or progression. For example, the knockout of the Prdx1 gene in mice leads to dramatically increased levels of oxidative stress and animals prematurely die of cancer (Neumann et al. 2003).

Glutathione peroxidases (GPX) mediate the breakdown of hydrogen peroxide and organic hydroperoxides (Brigelius-Flohe 1999). Glutathione (GSH) protects cellular proteins from oxidative stress by reducing disulphide bonds to cysteins. During this process GSH gets oxidized to glutathione disulphide (GSSG). Glutathione reductase enzymes then recycle GSSG to GSH (Beutler 1969). These enzymes are constitutively active in many tumor cells. Glutathione S-transferases (GST) are overexpressed in a variety of tumors and catalyze the conjunction of GSH to electrophilic compounds (Sharma et al. 2004; Townsend and Tew 2003). Glutathione S-transferases have been shown to contribute to chemoresistance, as well as to regulate cellular signaling pathways (Townsend and Tew 2003).

15.2 Altered ROS Levels – Consequences for Cancer Cells

15.2.1 ROS in Macromolecule Damage and Tumor Initiation

Increased oxidative stress in cells can induce the damage of macromolecules including DNA, proteins and lipids. This can have a role in tumor initiation (DNA damage and mutation). Additionally, altered proteins and lipids could serve as biomarkers.

DNA damaging effects of ROS are mostly induced by hydroxyl ions since hydrogen peroxide for example is not very reactive towards DNA. Hydroxyl ions are radicals that are highly diffusible. They can initiate the formation of DNA lesions through mediating single-strand or double-strand breaks or by oxidation of DNA bases and formation of DNA adducts (Maynard et al. 2009; Wiseman and Halliwell 1996). If cells harboring such DNA modifications escape programmed cell death they can continue to proliferate, raising the likelihood for cancerous growth.

Oxidation of proteins can alter their function, including inactivation or constitutive activation, which can contribute to oncogenic growth. Examples are direct oxidation of signaling molecules that lead to their activation (i.e. K-ras, Src) (Lander et al. 1997; Nakashima et al. 2002; Sun and Kemble 2009) or their inactivation (i.e. phosphatases, DNA repair enzymes) (Meng et al. 2002). In contrast to such specific ROS-regulated mechanisms, oxidative stress can induce less-specific modifications including formation of crosslinked and glycated proteins, increased protein carbonylation, nitration of amino-acid residues and protein degradation (Levine 2002; Squier and Bigelow 2000; Wells-Knecht et al. 1997).

Finally, by reacting with polyunsatureated or polydesaturated fatty acids ROS can induce lipid peroxidation (Gardner 1989; North et al. 1994). Lipid peroxidation generates several genotoxic molecules such as reactive aldehydes that can modify proteins and DNA. As an example, 4-Hydroxy-noneal (4-HNE) is the most cytotoxic lipid peroxidation product. 4-HNE can diffuse away from its membrane productions site and covalently modify proteins to alter their function. Clinical studies show that lipid peroxidation events induced by ROS can be used as tumor markers in the serum of patients (Lauschke et al. 2002).

15.2.2 ROS in Tumor Cell Proliferation

Tumor cell proliferation can be induced by increased production of superoxide at the mitochondria (Burdon 1995; Burdon et al. 1990; Parkash et al. 2006). The increased expression of MnSOD depletes superoxide levels and mediates quiescence (Wang et al. 2005). For example, stable expression of MnSOD in pancreatic cancer cells, which are known for high intracellular oxidative stress levels, reduced their proliferation rate (Cullen et al. 2003). This led to the hypothesis that mitochondrial superoxide drives proliferation of cells, whereas SOD-induced conversion into hydrogen peroxide drives quiescence, identifying MnSOD as a mitochondrial ROS switch that determines cell fate (Sarsour et al. 2008).

Tumor cell proliferation is regulated by ROS-mediated alterations in cell cycle control and induction of signaling pathways. Cellular signaling mechanisms that are used by superoxide to induce cell proliferation are not well defined. One *bona fide* signaling pathway that can regulate tumor cell proliferation and may be activated by superoxide, but also its breakdown product hydrogen peroxide, is the mitogen-activated protein kinase (MAPK) cascade (Irani et al. 1997; Kumar et al. 2008; McCubrey et al. 2007; Roberts and Der 2007). Extracellular-regulated kinases 1/2 (Erk1/2) signaling is activated by growth factors and K-ras activating mutations (Roberts and Der 2007; Khavari and Rinn 2007). Mechanisms of how Erk1/2 is activated by oxidative stress are either direct oxidation and activation of Ras (Lander et al. 1997), ROS-induced inactivation of negatively-regulatory phosphatases (Chan et al. 2008), or activation of other upstream kinases such as $p90^{RSK}$ (McCubrey et al. 2007). Other ROS-sensing signaling proteins are Akt, protein kinase C (PKC) or protein kinase D (PKD) enzymes, and the phosphatidylinositol 3-kinase (PI3-K)/Akt as well as the PKC-PKD pathways can induce cell proliferation in various cancer cell lines (Storz 2005; Rozengurt 2011). However, above signaling molecules were shown to be activated by hydrogen peroxide rather than superoxide (Liou and Storz 2010; Song et al. 2009; Storz and Toker 2003; Prasad et al. 2000).

Loss of redox control or increased oxidative stress can also mediate cell cycle progression by upregulating gene transcription of cyclins (Felty et al. 2005; Menon et al. 2005; Ruiz-Ramos et al. 2009). Another cell cycle-related protein activated by ROS is ataxia telangiectasia mutated (ATM) (Ditch and Paull 2012). Animal models lacking ATM have high levels of ROS and increased oxidative stress-induced damage (Browne et al. 2004). Similar is found in patients lacking ATM (Reichenbach et al. 2002).

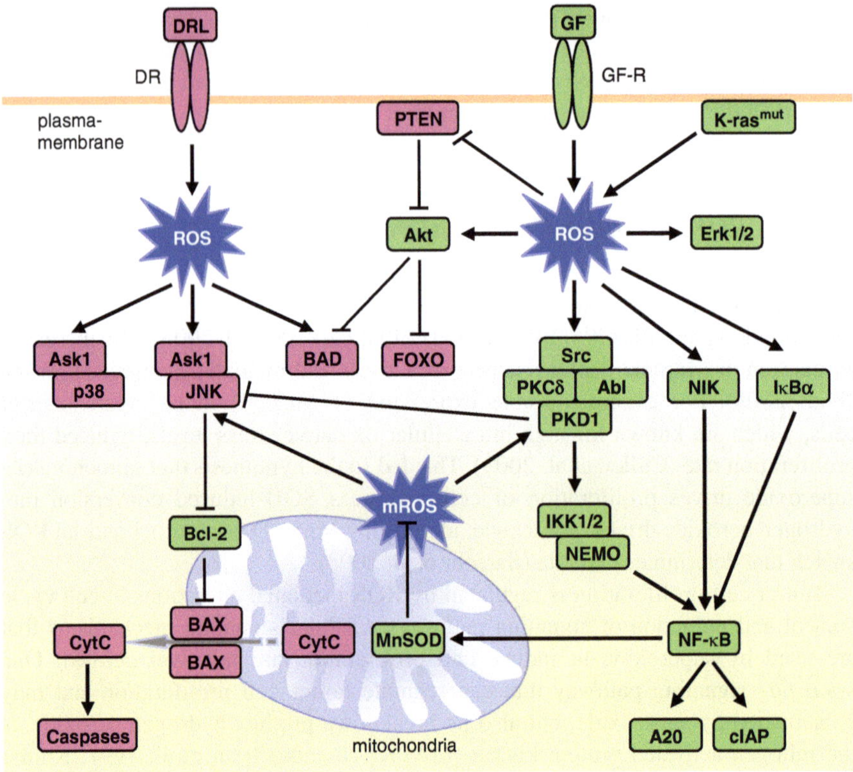

Fig. 15.2 ROS signaling in tumor cell survival and apoptosis. Reactive oxygen species (*ROS*) either generated at the plasma membrane through growth factor receptors (*GF-R*) or death receptors (*DR*) or at the mitochondria (*mROS*) can initiate both protective and cell death signaling pathways. The activation of these pathways is also dependent on radical formed and its concentration. Shown in *green* are protective signaling molecules and their pathways of activation. Shown in *red* are signaling molecules that contribute to cell death in response to ROS

15.2.3 ROS in Tumor Cell Survival and Apoptosis

Dependent on the radical formed, cellular location where this occurs, as well as its concentration, reactive oxygen species can have divergent effects on cell survival and cell death pathways (Fig. 15.2).

Mitochondria-generated oxidative stress and treatment of cells with hydrogen peroxide can lead to the recruitment and activation of a signaling complex consisting of Src, Abl, PKCδ, and PKD1 that initiates activation of the transcription factor nuclear factor-κB (NF-κB) via the canonical IKK complex and phosphorylation and degradation of IκBα, an inhibitory protein for NF-κB (Storz 2006; Storz et al. 2004a, b, 2005a; Song et al. 2009; Storz and Toker 2003; Chiu et al. 2007; Mihailovic et al. 2004). Target genes for NF-κB in this signaling pathway encode anti-apoptotic proteins such as cIAP and A20, as well as antioxidant proteins including MnSOD (Storz 2007; Storz et al. 2005a, b). Inhibition of this pathway in tumor cells renders them susceptible to ROS-mediated cell death (Storz and Toker 2003). This survival pathway is also activated when cells loose their anchorage (Cowell et al. 2009). Loss of anchorage in non-transformed cells induces an apoptotic cell death program called anoikis (Chiarugi and Fiaschi 2007). In tumor cells survival to this mechanism allows tumor metastasis to distant sites.

NF-κB in cancer cells is a redox sensor and many cancers show increased expression and activity of NF-κB (Rayet and Gelinas 1999; Li and Karin 1999; Schreck et al. 1992). Besides the PKC/PKD pathway, its activation by ROS can be caused through several other signaling mechanisms. These include hydrogen peroxide-mediated activation of NIK (Wang et al. 2007) through oxidative inhibition of NIK-regulating phosphatases, resulting in downstream signaling of NIK to the IKK complex (Li and Engelhardt 2006). Finally, NF-κB activation in response to oxidative stress can occur in an IKK-independent fashion. For example, it was shown that ROS can lead to tyrosine phosphorylation of IκBα. Tyrosine phosphorylation of IκBα can lead to its release from NF-κB, without inducing its degradation (Imbert et al. 1996). Besides mediating tumor cell survival, NF-κB was implicated in tumor cell proliferation, increased tumor cell motility, and development of drug resistance during therapy (chemoresistance) (Ahmed et al. 2006).

Mitochondrially-generated ROS are increased by the tumor suppressor protein p53 and $p66^{shc}$ through various pathways (reviewed in Pani and Galeotti 2011). p53-induced ROS production and activation of p53 target genes participate in the induction of cell growth arrest, apoptosis and senescence (Johnson et al. 1996). Dependent on its acetylation status, p53 also promotes the expression of antioxidant genes, which decreases its effects on senescence (Lu and Finkel 2008).

Growth factor receptors such as EGF-R or platelet-derived growth factor-receptor (PDGF-R) as well as activating mutations of K-ras can lead to oxidative stress-mediated activation of Akt. Akt is activated by hydrogen peroxide, either directly or through ROS-induced activation of its upstream kinase phosphoinositide-dependent kinase 1 (PDK1) (Prasad et al. 2000; Higaki et al. 2008). Moreover, the tumor suppressor phosphatase and tensin homologue (PTEN), a negative regulator of Akt signaling, can be inactivated by hydrogen peroxide, further potentiating Akt functions (Lee et al. 2002). Inactivation of PTEN additionally-increases cellular ROS levels due to deregulation of antioxidant enzymes (Huo et al. 2008). Akt can increase tumor cell survival through phosphorylation and inactivation of pro-apoptotic proteins including forkhead homeobox type O (FOXO) transcription factors, Bad, BimEL and Bax (Pastorino et al. 1999; Qi et al. 2006; Xin and Deng 2005). FOXO transcription factors are redox-sensing proteins and oxidative stress can mediate their activation (Storz 2011). This leads to induction of genes that mediate cell cycle arrest, induce apoptosis (i.e. FasL), or encode antioxidant genes (Storz 2011).

Another kinase that was implicated in ROS-induced cell survival and resistance to oxidative stress is Erk1/2, but in some cancers Erk1/2 seems to sensitize to ROS-induced apoptosis (Chan et al. 2008; Lee et al. 2005; Ostrakhovitch and

Cherian 2005). Besides Erk1/2, other members of the MAPK family such as p38 and c-Jun N-terminal Kinase (JNK), when activated by oxidative stress, are pro-apoptotic. Both p38 and JNK are activated by apoptosis signal-regulating kinase-1 (Ask1). Ask1 is regulated in its activity by interacting with thioredoxin, a redox-regulated protein (Saitoh et al. 1998; Takeda et al. 2003). JNK activation, for example, can occur in response to mitochondrial, as well as NADPH oxidase-induced oxidative stress (Storz 2005, 2006, 2007; Simon et al. 2000; Xu et al. 2002). When activated by ROS, JNK mediates phosphorylation and downregulation of Bcl-XL and Bcl-2, two proteins that protect from ROS-mediated apoptosis (Cadenas 2004; Gottlieb et al. 2000). JNK also increases the expression of pro-apoptotic proteins such as Bax to facilitate mitochondrial cytochrome c release (Shim et al. 2007; Zhang et al. 2008).

15.2.4 ROS in Tumor Cell Motility

The intracellular ROS state in cells governs all crucial mechanisms contributing to tumor cell metastasis, including degradation of extracellular matrix, increased potential to migrate and invade, intravasation and protection from anoikis.

Increased oxidative stress can contribute to increased cell motility, resulting in tumor expansion and metastasis. For example, subpopulations of breast cancer cell lines that show higher levels of endogenous ROS production than their origin counterpart, showed increased motility. Orthotopic tumors generated with these cells were metastatic, whereas parental cells were not (Pelicano et al. 2009). Another example is that carcinoma cells, when treated with hydrogen peroxide, prior to intravenous injection into mice show enhanced metastasis (Kundu et al. 1995). On the cellular levels it was shown that highly invasive tumor cell lines express rather high levels of intracellular ROS, probably due to downregulation of MnSOD (Hitchler et al. 2006, 2008).

Above we discussed the importance of the PI3-K/Akt pathway for tumor cell survival and the transcription factor FOXO3a as one of the primary targets of Akt to mediate this. Dependent on the tumor cell type and isoform expressed Akt also can decrease cell motility (Yoeli-Lerner and Toker 2006; Yoeli-Lerner et al. 2005). It was shown that stress conditions such as increased oxidative stress or nutrient depletion can activate the transcription factor FOXO3a (Storz et al. 2009). When activated in cells that grow under suboptimal conditions (where Akt is not active), FOXO3a induces an increase in their motility, potentially allowing their escape from stress or nutrient depletion (Storz et al. 2009). This is facilitated by upregulation of the matrix metalloproteinases (MMP) MMP-9 and MMP-13 (Storz et al. 2009). MMPs reorganize the extracellular matrix, degrade proteins that compose the basal membrane and have been implicated in many aspects of tumor progression (Radisky et al. 2005). Besides FOXO3a other ROS-induced transcription factors such as NF-κB have been implicated as regulators of MMP expression (Brenneisen et al. 2002; Kheradmand et al. 1998). Moreover, it was shown that hydrogen peroxide and not superoxide is responsible for MMP induction (Wenk et al. 1999). Hydrogen peroxide induces MMP expression via activation of the MAP kinases Erk1/2, p38 and JNK (Nelson and Melendez 2004). Increases in ROS can also lead to activation of MMPs by direct oxidation or downregulate tissue inhibitor of metalloproteinases (TIMP), proteins that inactivate MMPs (Rajagopalan et al. 1996). Treatment of tumor cells with MMP-3 can induce mitochondrial ROS production, initiating epithelial-to-mesenchymal transition (EMT) (Radisky et al. 2005). One of the drivers of this process is the RhoGTPase Rac1b (Radisky et al. 2005).

Another member of the Rac family, Rac-1, activates the NADPH oxidase NOX1 and increases superoxide production at cellular membranes (Werner and Werb 2002). NOX1 expression not only can lead to cell transformation, it also can maintain the transformed state (Mochizuki et al. 2006; Tobar et al. 2008). Further it is necessary for formation of invadopodia, which are actin-based structures needed for tumor cells to invade tissue and blood vessels (Diaz et al. 2009). In this context Rac-1/NOX signaling can mediate actin cytoskeleton rearrangements. For example, Rac-1 can activate the actin severing enzyme cofilin to increase cell migration (Kim et al. 2009; Sundaresan et al. 1996). Similar to Rac-1, Src can regulate NOX1-mediated induction of ROS to mediate cell migration (Gianni et al. 2008). Moreover, ROS can regulate the transcriptional repressor Snail to decrease expression of E-cadherin, contributing to EMT and increased cell motility (Wu 2006).

The Rac-1/Nox pathway can act through NF-κB to modulate cell adhesion (Tobar et al. 2008). In endothelial cells this can lead to a loss of cell-cell adhesions, loosen the integrity of the endothelium, and result in increased vascular permeability and intravasation of cancer cells (Cheng et al. 2004; van Wetering et al. 2002). Besides NOX-induced ROS, mitochondrially-induced ROS have been shown to regulate cell adhesion (Chiarugi and Fiaschi 2007). This is probably mediated through Src and FAK, both kinases located to focal contacts, actin-based structures that contribute to cell spreading, cell migration and prevention of anoikis. Interestingly, Rac-1 may be involved in this process too, since growth factors and integrins can induce a Rac-1-caused increase in mitochondrial oxidative stress, leading to Src activation, cell adhesion and cell spreading (Bae et al. 2000; Werner and Werb 2002).

15.2.5 ROS in Cancer Stem Cells

Following radiotherapy or chemotherapy, recurrence of tumors can be initiated by a small subpopulation of surviving cancer cells. These cells are highly stress- and drug-resistant, express markers of stem cells, and therefore were named cancer stem cells (Diehn et al. 2009). CSCs are capable of self-renewal and differentiation and therefore have features to initiate tumors. Their adaption to increased ROS and promotion of cell survival and stress resistance is mediated by an increase in their antioxidant capacity (Tanno and Matsui 2011). For example, mammary epithelial CSCs contain low levels of superoxide when compared to the more mature progeny or even normal epithelia cells (Diehn et al. 2009). This is caused by increased expression of a variety of genes encoding ROS scavenging enzymes or regulating glutathione synthesis (Diehn et al. 2009; Trachootham et al. 2009). Further, keeping ROS at such low levels is critical for maintaining their stem cell phenotype.

Increased levels of ROS are critical mediators of many chemotherapeutics and ionizing radiation therapy leading to DNA damage and tumor cell death (Shackleton et al. 2006; Ward 1985). The expression of antioxidants in CSCs prevents oxidative damage of DNA and cell death, and CSC-enriched tumor cell populations accumulate fewer DNA strand breaks or mutations after irradiation.

Important regulators of stress resistance in CSCs are FOXO transcription factors (Storz 2011). These transcription factors in cancer have been implicated in regulation of migration, proliferation, DNA repair, cell cycle arrest and cell death. They are less expressed in the bulk of tumor cells, but are critical for the maintenance of leukemic stem cells. For example FOXO1 confers stress resistance by upregulating catalase and superoxide dismutase genes (SOD) (Tothova et al. 2007). Consequently, treatment of CSCs with L-S, R-buthionine sulphoximine (BSO), a compound that pharmacologically depletes the ROS scavenger glutathione significantly sensitizes them to radiotherapy (Diehn et al. 2009).

15.2.6 Hypoxia, Angiogenesis and HIF-1

In primary tumors an increasing tumor mass faces cycles of oxygen depletion (hypoxia) and reoxygenation (Dewhirst et al. 2008; Hockel and Vaupel 2001). If prolonged, such limitations in oxygen supply can be damaging to the tumor cells. An escape mechanism for tumor cells to such conditions is their metabolic switch to anaerobic glycolysis (Pani et al. 2009). While normal cells only switch to glycolysis when adequate oxygen supply is not ensured or when mitochondria function is suppressed (Ganapathy et al. 2009), tumor cells can constantly use glycolysis as an energy source under normoxia (Warburg effect) to become independent of such conditions (Dang and Semenza 1999). This switch can be induced by mitochondrial dysfunction, oncogenic transformation or loss of tumor suppressor genes (Hsu and Sabatini 2008). The transcription factor hypoxia-inducible factor-1 (HIF-1) regulates glycolysis-related genes and inhibits mitochondrial respiration (Pani et al. 2009), which in sum results in adaption of tumor cells to hypoxic conditions (Pouyssegur et al. 2006) and the development of an aggressive tumor phenotype (Harris 2002).

With increased tumor mass, nutrient and oxygen support is limited to cells in the tumor center. Therefore, more nascent blood vessels are required to ensure supply to the tumor center (Claffey et al. 1996; Senger et al. 1994). Hypoxic conditions stimulate blood vessel development (angiogenesis) through induction of intracellular oxidative stress. For example, an increase in intracellular oxidative stress levels can be mediated by hypoxia-induced expression of VEGF (Jo et al. 2011). Antioxidants effectively inhibit or decrease angiogenesis since they can modify densitometry of microvessels and proliferation of endothelial cells (Rabbani et al. 2009). Blood flow in newly developed vessels often is instable, leading to periods of high oxygen and hypoxia, and causing additional oxidative stress (Brown and Bicknell 2001). Nutrient deprivation and hypoxia both can increase expression of vascular epithelial growth factor (VEGF) (Spitz et al. 2000). This is mediated through HIF-1 and its co-factor p300 (Liou and Storz 2010).

HIF-1 is composed of the two subunits HIF-1α and HIF-1β (Harris 2002). Under normoxia conditions HIF-1α is proteasomally degraded (Kaelin and Ratcliffe 2008), whereas under conditions of hypoxia, formation of superoxide and hydrogen peroxide lead to accumulation of HIF-1α (Wang et al. 2005). HIF can be induced by increased production of both reactive oxygen species in response to oncogenic signaling through the PTEN/PI3-K/Akt signaling pathway, or to mitochondria-generated oxidative stress (Huo et al. 2008; Liu et al. 2006; Xia et al. 2007). Due to its function as an inducer of VEGF signaling and contribution to angiogenesis, increased HIF-1 expression was shown to correlate with poor prognosis and increased metastasis. HIF-1 also impacts tumor cell metastasis by impacting cell motility and EMT (Erler et al. 2006). This is mediated through HIF-1α regulation of metastasis-related genes including lysyl oxidase and MMPs. For example, MMP-9 and MMP-13 can directly mediate tumor cell invasion and metastasis when expressed by tumor cells (Storz et al. 2009). As a consequence of its effects on preventing cellular acidification, HIF-1 can contribute to cell invasion by increasing the formation of lactate and CO_2, both of which favor the degradation of extracellular matrix (Pouyssegur et al. 2006; Rofstad et al. 2006). MMPs are important mediators of vessel growth within the tumor microenvironment and MMP-induced formation of capillary-like structures occurs through upregulation of ROS.

15.3 Modulation of ROS Levels as Therapeutic Strategy

ROS can be beneficial for tumor cells allowing acquisition and maintenance of tumorigenic characteristics, but can be damaging when in excess. Many tumor tissues have elevated levels of oxidative stress, which in most normal tissues would induce cell death (Storz 2005). The upregulation of intracellular antioxidant systems allows tumor cells not only to keep ROS levels in check, but also to keep them at levels where they contribute to beneficiary signaling. Therefore, tumor cells can thrive under oxidative stress, inducing increased tumor cell proliferation, anti-apoptotic signaling and increasing cell motility (Liou and Storz 2010).

There are several therapeutic strategies aiming to increase intracellular ROS levels in cancer cells, with the goal that oxidative stress reaches a level that induces senescence or even cell death. This may be achieved in cancer cells by compounds that inhibit antioxidant systems or downregulate cellular pathways that mediate expression of antioxidants. Decreasing their antioxidant capacity can dramatically increase intracellular ROS levels in cancer cells, leading to irreparable damage. Normal cells may not be affected by such a strategy since they generally have lower basal ROS levels than cancer cells and are less dependent on antioxidant systems. A caveat with this approach is that a threshold of cytotoxicity may not be reached in cancer cells. An advantage is that cancer stem cells may also be targeted by this approach, since they show increased expression of antioxidant systems to keep their intracellular ROS levels low (Trachootham et al. 2009).

Another strategy is to exuberantly-increase oxidative stress in tumor cells to levels at which they induce cell death. This can be achieved by radiation or chemotherapy (Trachootham et al. 2009). Increased ROS levels in tumor cells can be obtained by blocking glucose metabolism, which is increased in tumor cells and compensates excess metabolic production of ROS. It was suggested that inhibition of glucose metabolism may provide a mechanism by which cancer cells can be specifically targeted. This can be achieved with glucose analogues that can not be metabolized such as 2-deoxyglucose (2DG) and is even more pronounced when combined with mitochondrial electron chain blockers that mediate additional induction of ROS (Aykin-Burns et al. 2009; Coleman et al. 2008). Strategies to increase intracellular ROS generation in tumor cells may be even more effective when antioxidant systems are depleted at the same time (Trachootham et al. 2009). However, high levels of ROS can drive tumor cells into a stem cell-like phenotype, which is stress resistant and responsible for recurrence of tumors. Thus, therapies that increase intracellular ROS in cancer cells may represent a good strategy to target the bulk of proliferating tumor cells, but may drive a subpopulation of tumor cells into a dormant state or a CSC phenotype. These cells then need to be targeted with a separate strategy that considers their unique redox status.

Depleting ROS levels with therapeutic antioxidants may have tumor preventive functions. For example, oncogenic mutations in K-ras have been linked to ROS formation leading to premalignant lesions in pancreatic cancer (Li et al. 2002). Moreover, ROS generated by macrophages have been implicated in the initiation and progression of many tumor types (Keibel et al. 2009; Qian and Pollard 2010). Depleting oxidative stress at early stages therefore may be effective in prevention or delaying tumor development. Examples of antioxidants that are tested for tumor therapy are NOV-002, a mimetic of glutathione disulfide, or EUK-134, a mimetic of superoxide dismutase (SOD) (Townsend and Tew 2003; Bechtel and Bauer 2009). It becomes obvious that depending on the strategy used for therapy, the use of antioxidants in combination therapy could have adverse effects on chemotherapeutic drugs that kill cells by increasing intracellular ROS levels. Therefore, combination therapy with antioxidants and apoptosis-inducing agents may only be effective when these compounds can mediate cell death via pathways independent of ROS. Another explanation of why many tumor therapies in which antioxidants were used were quite ineffective is that at treatment the tumor cells may already have acquired a mutational state in which they are independent of ROS as a tumorigenic factor.

In summary, specific modulation of oxidative stress levels or regulation of specific production of certain reactive oxygen species in tumor cells could be a powerful tool to enhance therapeutic outcomes. However, it may require combination therapy with inhibitors of other pathways, specifically tailored to the signaling pathway or tumor type that needs to be targeted. One example where this type of therapy may be effective is the use of chemotherapy in combination with a specific strategy to target antioxidant systems of CSCs, with the goal to kill the bulk of tumor and additionally prevent recurrence.

Acknowledgements The author would like to thank Heike Döppler and Jenni Bachhofer for critical reading of the manuscript. Work in the Storz laboratory is supported by the NIH grants GM86435 and CA140182.

References

Ahmed KM, Cao N, Li JJ (2006) HER-2 and NF-kappaB as the targets for therapy-resistant breast cancer. Anticancer Res 26(6B):4235–4243

Aykin-Burns N et al (2009) Increased levels of superoxide and H_2O_2 mediate the differential susceptibility of cancer cells versus normal cells to glucose deprivation. Biochem J 418(1):29–37

Babior BM (1999) NADPH oxidase: an update. Blood 93(5):1464–1476

Bae YS et al (2000) Platelet-derived growth factor-induced H(2)O(2) production requires the activation of phosphatidylinositol 3-kinase. J Biol Chem 275(14):10527–10531

Balkwill F (2009) Tumour necrosis factor and cancer. Nat Rev Cancer 9(5):361–371

Bechtel W, Bauer G (2009) Modulation of intercellular ROS signaling of human tumor cells. Anticancer Res 29(11):4559–4570

Bendayan M, Reddy JK (1982) Immunocytochemical localization of catalase and heat-labile enoyl-CoA hydratase in the livers of normal and peroxisome proliferator-treated rats. Lab Invest 47(4):364–369

Beutler E (1969) Effect of flavin compounds on glutathione reductase activity: in vivo and in vitro studies. J Clin Invest 48(10):1957–1966

Brenneisen P, Sies H, Scharffetter-Kochanek K (2002) Ultraviolet-B irradiation and matrix metalloproteinases: from induction via signaling to initial events. Ann N Y Acad Sci 973:31–43

Brigelius-Flohe R (1999) Tissue-specific functions of individual glutathione peroxidases. Free Radic Biol Med 27(9–10):951–965

Brown NS, Bicknell R (2001) Hypoxia and oxidative stress in breast cancer. Oxidative stress: its effects on the growth, metastatic potential and response to therapy of breast cancer. Breast Cancer Res 3(5):323–327

Browne SE et al (2004) Treatment with a catalytic antioxidant corrects the neurobehavioral defect in ataxia-telangiectasia mice. Free Radic Biol Med 36(7):938–942

Burdon RH (1995) Superoxide and hydrogen peroxide in relation to mammalian cell proliferation. Free Radic Biol Med 18(4):775–794

Burdon RH, Gill V, Rice-Evans C (1990) Oxidative stress and tumour cell proliferation. Free Radic Res Commun 11(1–3):65–76

Cadenas E (2004) Mitochondrial free radical production and cell signaling. Mol Aspects Med 25(1–2):17–26

Chan DW et al (2008) Loss of MKP3 mediated by oxidative stress enhances tumorigenicity and chemoresistance of ovarian cancer cells. Carcinogenesis 29(9):1742–1750

Cheng GC et al (2004) Oxidative stress and thioredoxin-interacting protein promote intravasation of melanoma cells. Exp Cell Res 300(2):297–307

Chiarugi P, Fiaschi T (2007) Redox signalling in anchorage-dependent cell growth. Cell Signal 19(4):672–682

Chiu TT et al (2007) Protein kinase D2 mediates lysophosphatidic acid-induced interleukin 8 production in nontransformed human colonic epithelial cells through NF-kappaB. Am J Physiol Cell Physiol 292(2):C767–C777

Claffey KP et al (1996) Expression of vascular permeability factor/vascular endothelial growth factor by melanoma cells increases tumor growth, angiogenesis, and experimental metastasis. Cancer Res 56(1):172–181
Coleman MC et al (2008) 2-deoxy-D-glucose causes cytotoxicity, oxidative stress, and radiosensitization in pancreatic cancer. Free Radic Biol Med 44(3):322–331
Cowell CF et al (2009) Loss of cell-cell contacts induces NF-kappaB via RhoA-mediated activation of protein kinase D1. J Cell Biochem 106(4):714–728
Crompton M (1999) The mitochondrial permeability transition pore and its role in cell death. Biochem J 341(Pt 2):233–249
Cui S et al (1994) Activated murine macrophages induce apoptosis in tumor cells through nitric oxide-dependent or -independent mechanisms. Cancer Res 54(9):2462–2467
Cullen JJ et al (2003) The role of manganese superoxide dismutase in the growth of pancreatic adenocarcinoma. Cancer Res 63(6):1297–1303
Dang CV, Semenza GL (1999) Oncogenic alterations of metabolism. Trends Biochem Sci 24(2):68–72
Dansen TB, Wirtz KW (2001) The peroxisome in oxidative stress. IUBMB Life 51(4):223–230
Dayal D et al (2009) Mitochondrial complex II dysfunction can contribute significantly to genomic instability after exposure to ionizing radiation. Radiat Res 172(6):737–745
del Rio LA et al (1992) Metabolism of oxygen radicals in peroxisomes and cellular implications. Free Radic Biol Med 13(5):557–580
Dewhirst MW, Cao Y, Moeller B (2008) Cycling hypoxia and free radicals regulate angiogenesis and radiotherapy response. Nat Rev Cancer 8(6):425–437
Diaz B et al (2009) Tks5-dependent, nox-mediated generation of reactive oxygen species is necessary for invadopodia formation. Sci Signal 2(88):ra53
Diehn M et al (2009) Association of reactive oxygen species levels and radioresistance in cancer stem cells. Nature 458(7239):780–783
Ditch S, Paull TT (2012) The ATM protein kinase and cellular redox signaling: beyond the DNA damage response. Trends Biochem Sci 37(1):15–22
Erler JT et al (2006) Lysyl oxidase is essential for hypoxia-induced metastasis. Nature 440(7088):1222–1226
Felty Q, Singh KP, Roy D (2005) Estrogen-induced G1/S transition of G0-arrested estrogen-dependent breast cancer cells is regulated by mitochondrial oxidant signaling. Oncogene 24(31):4883–4893
Ferraro D et al (2006) Pro-metastatic signaling by c-Met through RAC-1 and reactive oxygen species (ROS). Oncogene 25(26):3689–3698
Finkel T (2000) Redox-dependent signal transduction. FEBS Lett 476(1–2):52–54
Ganapathy V, Thangaraju M, Prasad PD (2009) Nutrient transporters in cancer: relevance to Warburg hypothesis and beyond. Pharmacol Ther 121(1):29–40
Gardner HW (1989) Oxygen radical chemistry of polyunsaturated fatty acids. Free Radic Biol Med 7(1):65–86
Gianni D et al (2008) The involvement of the tyrosine kinase c-Src in the regulation of reactive oxygen species generation mediated by NADPH oxidase-1. Mol Biol Cell 19(7):2984–2994
Gottlieb E, Vander Heiden MG, Thompson CB (2000) Bcl-x(L) prevents the initial decrease in mitochondrial membrane potential and subsequent reactive oxygen species production during tumor necrosis factor alpha-induced apoptosis. Mol Cell Biol 20(15):5680–5689
Harris AL (2002) Hypoxia – a key regulatory factor in tumour growth. Nat Rev Cancer 2(1):38–47
Hashimoto F, Hayashi H (1990) Significance of catalase in peroxisomal fatty acyl-CoA beta-oxidation: NADH oxidation by acetoacetyl-CoA and H_2O_2. J Biochem 108(3):426–431
Higaki Y et al (2008) Oxidative stress stimulates skeletal muscle glucose uptake through a phosphatidylinositol 3-kinase-dependent pathway. Am J Physiol Endocrinol Metab 294(5):E889–E897
Hitchler MJ et al (2006) Epigenetic regulation of manganese superoxide dismutase expression in human breast cancer cells. Epigenetics 1(4):163–171

Hitchler MJ, Oberley LW, Domann FE (2008) Epigenetic silencing of SOD2 by histone modifications in human breast cancer cells. Free Radic Biol Med 45(11):1573–1580
Hockel M, Vaupel P (2001) Tumor hypoxia: definitions and current clinical, biologic, and molecular aspects. J Natl Cancer Inst 93(4):266–276
Hofmann B, Hecht HJ, Flohe L (2002) Peroxiredoxins. Biol Chem 383(3–4):347–364
Hsu PP, Sabatini DM (2008) Cancer cell metabolism: Warburg and beyond. Cell 134(5):703–707
Huo YY et al (2008) PTEN deletion leads to deregulation of antioxidants and increased oxidative damage in mouse embryonic fibroblasts. Free Radic Biol Med 44(8):1578–1591
Imbert V et al (1996) Tyrosine phosphorylation of I kappa B-alpha activates NF-kappa B without proteolytic degradation of I kappa B-alpha. Cell 86(5):787–798
Irani K et al (1997) Mitogenic signaling mediated by oxidants in Ras-transformed fibroblasts. Science 275(5306):1649–1652
Jo M et al (2011) Oxidative stress is closely associated with tumor angiogenesis of hepatocellular carcinoma. J Gastroenterol 46(6):809–821
Johnson TM et al (1996) Reactive oxygen species are downstream mediators of p53-dependent apoptosis. Proc Natl Acad Sci USA 93(21):11848–11852
Kaelin WG Jr, Ratcliffe PJ (2008) Oxygen sensing by metazoans: the central role of the HIF hydroxylase pathway. Mol Cell 30(4):393–402
Keibel A, Singh V, Sharma MC (2009) Inflammation, microenvironment, and the immune system in cancer progression. Curr Pharm Des 15(17):1949–1955
Khavari TA, Rinn J (2007) Ras/Erk MAPK signaling in epidermal homeostasis and neoplasia. Cell Cycle 6(23):2928–2931
Kheradmand F et al (1998) Role of Rac1 and oxygen radicals in collagenase-1 expression induced by cell shape change. Science 280(5365):898–902
Kim JS, Huang TY, Bokoch GM (2009) Reactive oxygen species regulate a slingshot-cofilin activation pathway. Mol Biol Cell 20(11):2650–2660
Kumar B et al (2008) Oxidative stress is inherent in prostate cancer cells and is required for aggressive phenotype. Cancer Res 68(6):1777–1785
Kundu N, Zhang S, Fulton AM (1995) Sublethal oxidative stress inhibits tumor cell adhesion and enhances experimental metastasis of murine mammary carcinoma. Clin Exp Metastasis 13(1):16–22
Lander HM et al (1997) A molecular redox switch on p21(ras). Structural basis for the nitric oxide-p21(ras) interaction. J Biol Chem 272(7):4323–4326
Lauschke H et al (2002) Lipid peroxidation as additional marker in patients with colorectal cancer. Results of a preliminary study. Eur Surg Res 34(5):346–350
Lee SR et al (2002) Reversible inactivation of the tumor suppressor PTEN by H_2O_2. J Biol Chem 277(23):20336–20342
Lee WC et al (2005) Role of ERK in hydrogen peroxide-induced cell death of human glioma cells. Neurochem Res 30(2):263–270
Levine RL (2002) Carbonyl modified proteins in cellular regulation, aging, and disease. Free Radic Biol Med 32(9):790–796
Li Q, Engelhardt JF (2006) Interleukin-1beta induction of NFkappaB is partially regulated by H_2O_2-mediated activation of NFkappaB-inducing kinase. J Biol Chem 281(3):1495–1505
Li N, Karin M (1999) Is NF-kappaB the sensor of oxidative stress? FASEB J 13(10):1137–1143
Li D et al (2002) DNA adducts, genetic polymorphisms, and K-ras mutation in human pancreatic cancer. Mutat Res 513(1–2):37–48
Liou GY, Storz P (2010) Reactive oxygen species in cancer. Free Radic Res 44(5):479–496
Litwin JA et al (1987) Immunocytochemical localization of peroxisomal enzymes in human liver biopsies. Am J Pathol 128(1):141–150
Liu LZ et al (2006) Reactive oxygen species regulate epidermal growth factor-induced vascular endothelial growth factor and hypoxia-inducible factor-1alpha expression through activation of AKT and P70S6K1 in human ovarian cancer cells. Free Radic Biol Med 41(10):1521–1533
Lo YY, Cruz TF (1995) Involvement of reactive oxygen species in cytokine and growth factor induction of c-fos expression in chondrocytes. J Biol Chem 270(20):11727–11730

Lu T, Finkel T (2008) Free radicals and senescence. Exp Cell Res 314(9):1918–1922
Maynard S et al (2009) Base excision repair of oxidative DNA damage and association with cancer and aging. Carcinogenesis 30(1):2–10
McCubrey JA et al (2007) Roles of the Raf/MEK/ERK pathway in cell growth, malignant transformation and drug resistance. Biochim Biophys Acta 1773(8):1263–1284
Meier B et al (1989) Human fibroblasts release reactive oxygen species in response to interleukin-1 or tumour necrosis factor-alpha. Biochem J 263(2):539–545
Meng TC, Fukada T, Tonks NK (2002) Reversible oxidation and inactivation of protein tyrosine phosphatases in vivo. Mol Cell 9(2):387–399
Menon SG et al (2005) Differential susceptibility of nonmalignant human breast epithelial cells and breast cancer cells to thiol antioxidant-induced G(1)-delay. Antioxid Redox Signal 7(5–6):711–718
Mihailovic T et al (2004) Protein kinase D2 mediates activation of nuclear factor kappaB by Bcr-Abl in Bcr-Abl + human myeloid leukemia cells. Cancer Res 64(24):8939–8944
Minamoto T, Mai M, Ronai Z (2000) K-ras mutation: early detection in molecular diagnosis and risk assessment of colorectal, pancreas, and lung cancers – a review. Cancer Detect Prev 24(1):1–12
Minamoto T, Ougolkov AV, Mai M (2002) Detection of oncogenes in the diagnosis of cancers with active oncogenic signaling. Expert Rev Mol Diagn 2(6):565–575
Mochizuki T et al (2006) Inhibition of NADPH oxidase 4 activates apoptosis via the AKT/apoptosis signal-regulating kinase 1 pathway in pancreatic cancer PANC-1 cells. Oncogene 25(26):3699–3707
Nakashima I et al (2002) Redox-linked signal transduction pathways for protein tyrosine kinase activation. Antioxid Redox Signal 4(3):517–531
Nelson KK, Melendez JA (2004) Mitochondrial redox control of matrix metalloproteinases. Free Radic Biol Med 37(6):768–784
Neumann CA et al (2003) Essential role for the peroxiredoxin Prdx1 in erythrocyte antioxidant defence and tumour suppression. Nature 424(6948):561–565
North JA, Spector AA, Buettner GR (1994) Cell fatty acid composition affects free radical formation during lipid peroxidation. Am J Physiol 267(1 Pt 1):C177–C188
Ohba M et al (1994) Production of hydrogen peroxide by transforming growth factor-beta 1 and its involvement in induction of egr-1 in mouse osteoblastic cells. J Cell Biol 126(4):1079–1088
Ostrakhovitch EA, Cherian MG (2005) Inhibition of extracellular signal regulated kinase (ERK) leads to apoptosis inducing factor (AIF) mediated apoptosis in epithelial breast cancer cells: the lack of effect of ERK in p53 mediated copper induced apoptosis. J Cell Biochem 95(6):1120–1134
Pani G, Galeotti T (2011) Role of MnSOD and p66shc in mitochondrial response to p53. Antioxid Redox Signal 15(6):1715–1727
Pani G et al (2009) Redox-based escape mechanism from death: the cancer lesson. Antioxid Redox Signal 11(11):2791–2806
Parkash J, Felty Q, Roy D (2006) Estrogen exerts a spatial and temporal influence on reactive oxygen species generation that precedes calcium uptake in high-capacity mitochondria: implications for rapid nongenomic signaling of cell growth. Biochemistry 45(9):2872–2881
Pastorino JG, Tafani M, Farber JL (1999) Tumor necrosis factor induces phosphorylation and translocation of BAD through a phosphatidylinositide-3-OH kinase-dependent pathway. J Biol Chem 274(27):19411–19416
Pelicano H et al (2009) Mitochondrial dysfunction and reactive oxygen species imbalance promote breast cancer cell motility through a CXCL14-mediated mechanism. Cancer Res 69(6):2375–2383
Pouyssegur J, Dayan F, Mazure NM (2006) Hypoxia signalling in cancer and approaches to enforce tumour regression. Nature 441(7092):437–443
Prasad N et al (2000) Oxidative stress and vanadate induce tyrosine phosphorylation of phosphoinositide-dependent kinase 1 (PDK1). Biochemistry 39(23):6929–6935

Qi XJ, Wildey GM, Howe PH (2006) Evidence that Ser87 of BimEL is phosphorylated by Akt and regulates BimEL apoptotic function. J Biol Chem 281(2):813–823

Qian BZ, Pollard JW (2010) Macrophage diversity enhances tumor progression and metastasis. Cell 141(1):39–51

Rabbani ZN et al (2009) Antiangiogenic action of redox-modulating Mn(III) ortho-tetrakis-N-ethylpyridylporphyrin, MnTE-2-PyP(5+), via suppression of oxidative stress in a mouse model of breast tumor. Free Radic Biol Med 47:992–1004

Radisky DC et al (2005) Rac1b and reactive oxygen species mediate MMP-3-induced EMT and genomic instability. Nature 436(7047):123–127

Rajagopalan S et al (1996) Reactive oxygen species produced by macrophage-derived foam cells regulate the activity of vascular matrix metalloproteinases in vitro. Implications for atherosclerotic plaque stability. J Clin Invest 98(11):2572–2579

Rayet B, Gelinas C (1999) Aberrant rel/nfkb genes and activity in human cancer. Oncogene 18(49):6938–6947

Reichenbach J et al (2002) Elevated oxidative stress in patients with ataxia telangiectasia. Antioxid Redox Signal 4(3):465–469

Rhee SG et al (1999) A family of novel peroxidases, peroxiredoxins. Biofactors 10(2–3):207–209

Rhee SG et al (2000) Hydrogen peroxide: a key messenger that modulates protein phosphorylation through cysteine oxidation. Sci STKE 2000(53):pe1

Roberts PJ, Der CJ (2007) Targeting the Raf-MEK-ERK mitogen-activated protein kinase cascade for the treatment of cancer. Oncogene 26(22):3291–3310

Rofstad EK et al (2006) Acidic extracellular pH promotes experimental metastasis of human melanoma cells in athymic nude mice. Cancer Res 66(13):6699–6707

Rozengurt E (2011) Protein kinase D signaling: multiple biological functions in health and disease. Physiology (Bethesda) 26(1):23–33

Ruiz-Ramos R et al (2009) Sodium arsenite induces ROS generation, DNA oxidative damage, HO-1 and c-Myc proteins, NF-kappaB activation and cell proliferation in human breast cancer MCF-7 cells. Mutat Res 674(1–2):109–115

Saitoh M et al (1998) Mammalian thioredoxin is a direct inhibitor of apoptosis signal-regulating kinase (ASK) 1. EMBO J 17(9):2596–2606

Sarsour EH et al (2008) Manganese superoxide dismutase activity regulates transitions between quiescent and proliferative growth. Aging Cell 7(3):405–417

Schreck R, Albermann K, Baeuerle PA (1992) Nuclear factor kappa B: an oxidative stress-responsive transcription factor of eukaryotic cells (a review). Free Radic Res Commun 17(4):221–237

Senger DR et al (1994) Vascular permeability factor, tumor angiogenesis and stroma generation. Invasion Metastasis 14(1–6):385–394

Shackleton M et al (2006) Generation of a functional mammary gland from a single stem cell. Nature 439(7072):84–88

Sharma R et al (2004) Antioxidant role of glutathione S-transferases: protection against oxidant toxicity and regulation of stress-mediated apoptosis. Antioxid Redox Signal 6(2):289–300

Shim HY et al (2007) Acacetin-induced apoptosis of human breast cancer MCF-7 cells involves caspase cascade, mitochondria-mediated death signaling and SAPK/JNK1/2-c-Jun activation. Mol Cells 24(1):95–104

Simon HU, Haj-Yehia A, Levi-Schaffer F (2000) Role of reactive oxygen species (ROS) in apoptosis induction. Apoptosis 5(5):415–418

Singh I (1996) Mammalian peroxisomes: metabolism of oxygen and reactive oxygen species. Ann N Y Acad Sci 804:612–627

Song J et al (2009) PKD prevents H_2O_2-induced apoptosis via NF-kappaB and p38 MAPK in RIE-1 cells. Biochem Biophys Res Commun 378(3):610–614

Spitz DR et al (2000) Glucose deprivation-induced oxidative stress in human tumor cells. A fundamental defect in metabolism? Ann N Y Acad Sci 899:349–362

Squier TC, Bigelow DJ (2000) Protein oxidation and age-dependent alterations in calcium homeostasis. Front Biosci 5:D504–D526

Storz P (2005) Reactive oxygen species in tumor progression. Front Biosci 10:1881–1896
Storz P (2006) Reactive oxygen species-mediated mitochondria-to-nucleus signaling: a key to aging and radical-caused diseases. Sci STKE 2006(332):re3
Storz P (2007) Mitochondrial ROS–radical detoxification, mediated by protein kinase D. Trends Cell Biol 17(1):13–18
Storz P (2011) Forkhead homeobox type O transcription factors in the responses to oxidative stress. Antioxid Redox Signal 14(4):593–605
Storz P, Toker A (2003) Protein kinase D mediates a stress-induced NF-kappaB activation and survival pathway. EMBO J 22(1):109–120
Storz P, Doppler H, Toker A (2004a) Protein kinase Cdelta selectively regulates protein kinase D-dependent activation of NF-kappaB in oxidative stress signaling. Mol Cell Biol 24(7):2614–2626
Storz P, Doppler H, Toker A (2004b) Activation loop phosphorylation controls protein kinase D-dependent activation of nuclear factor kappaB. Mol Pharmacol 66(4):870–879
Storz P, Doppler H, Toker A (2005a) Protein kinase D mediates mitochondrion-to-nucleus signaling and detoxification from mitochondrial reactive oxygen species. Mol Cell Biol 25(19):8520–8530
Storz P et al (2005b) Functional dichotomy of A20 in apoptotic and necrotic cell death. Biochem J 387(Pt 1):47–55
Storz P et al (2009) FOXO3a promotes tumor cell invasion through the induction of matrix metalloproteinases. Mol Cell Biol 29(18):4906–4917
Sun G, Kemble DJ (2009) To C or not to C: direct and indirect redox regulation of Src protein tyrosine kinase. Cell Cycle 8(15):2353–2355
Sundaresan M et al (1995) Requirement for generation of H_2O_2 for platelet-derived growth factor signal transduction. Science 270(5234):296–299
Sundaresan M et al (1996) Regulation of reactive-oxygen-species generation in fibroblasts by Rac1. Biochem J 318(Pt 2):379–382
Szatrowski TP, Nathan CF (1991) Production of large amounts of hydrogen peroxide by human tumor cells. Cancer Res 51(3):794–798
Takeda K et al (2003) Roles of MAPKKK ASK1 in stress-induced cell death. Cell Struct Funct 28(1):23–29
Tanno T, Matsui W (2011) Development and maintenance of cancer stem cells under chronic inflammation. J Nippon Med Sch 78(3):138–145
Tiku ML, Liesch JB, Robertson FM (1990) Production of hydrogen peroxide by rabbit articular chondrocytes. Enhancement by cytokines. J Immunol 145(2):690–696
Tobar N et al (2008) RAC1 activity and intracellular ROS modulate the migratory potential of MCF-7 cells through a NADPH oxidase and NFkappaB-dependent mechanism. Cancer Lett 267(1):125–132
Tothova Z et al (2007) FoxOs are critical mediators of hematopoietic stem cell resistance to physiologic oxidative stress. Cell 128(2):325–339
Townsend DM, Tew KD (2003) The role of glutathione-S-transferase in anti-cancer drug resistance. Oncogene 22(47):7369–7375
Trachootham D et al (2006) Selective killing of oncogenically transformed cells through a ROS-mediated mechanism by beta-phenylethyl isothiocyanate. Cancer Cell 10(3):241–252
Trachootham D, Alexandre J, Huang P (2009) Targeting cancer cells by ROS-mediated mechanisms: a radical therapeutic approach? Nat Rev Drug Discov 8(7):579–591
van Wetering S et al (2002) Reactive oxygen species mediate Rac-induced loss of cell-cell adhesion in primary human endothelial cells. J Cell Sci 115(Pt 9):1837–1846
Wang M et al (2005) Manganese superoxide dismutase suppresses hypoxic induction of hypoxia-inducible factor-1alpha and vascular endothelial growth factor. Oncogene 24(55):8154–8166
Wang Y et al (2007) The endogenous reactive oxygen species promote NF-kappaB activation by targeting on activation of NF-kappaB-inducing kinase in oral squamous carcinoma cells. Free Radic Res 41(9):963–971
Ward JF (1985) Biochemistry of DNA lesions. Radiat Res Suppl 8:S103–S111

Wells-Knecht MC et al (1997) Age-dependent increase in ortho-tyrosine and methionine sulfoxide in human skin collagen is not accelerated in diabetes. Evidence against a generalized increase in oxidative stress in diabetes. J Clin Invest 100(4):839–846

Wenk J et al (1999) Stable overexpression of manganese superoxide dismutase in mitochondria identifies hydrogen peroxide as a major oxidant in the AP-1-mediated induction of matrix-degrading metalloprotease-1. J Biol Chem 274(36):25869–25876

Werner E, Werb Z (2002) Integrins engage mitochondrial function for signal transduction by a mechanism dependent on Rho GTPases. J Cell Biol 158(2):357–368

Wiseman H, Halliwell B (1996) Damage to DNA by reactive oxygen and nitrogen species: role in inflammatory disease and progression to cancer. Biochem J 313(Pt 1):17–29

Wood ZA et al (2003) Structure, mechanism and regulation of peroxiredoxins. Trends Biochem Sci 28(1):32–40

Wu WS (2006) The signaling mechanism of ROS in tumor progression. Cancer Metastasis Rev 25(4):695–705

Xia C et al (2007) Reactive oxygen species regulate angiogenesis and tumor growth through vascular endothelial growth factor. Cancer Res 67(22):10823–10830

Xin M, Deng X (2005) Nicotine inactivation of the proapoptotic function of Bax through phosphorylation. J Biol Chem 280(11):10781–10789

Xu YC et al (2002) Involvement of TRAF4 in oxidative activation of c-Jun N-terminal kinase. J Biol Chem 277(31):28051–28057

Yoeli-Lerner M, Toker A (2006) Akt/PKB signaling in cancer: a function in cell motility and invasion. Cell Cycle 5(6):603–605

Yoeli-Lerner M et al (2005) Akt blocks breast cancer cell motility and invasion through the transcription factor NFAT. Mol Cell 20(4):539–550

Zhang R et al (2008) In vitro and in vivo induction of apoptosis by capsaicin in pancreatic cancer cells is mediated through ROS generation and mitochondrial death pathway. Apoptosis 13(12):1465–1478

Wells-Knecht MC et al (1997) Age-dependent increase in ortho-tyrosine and methionine sulfoxide in human skin collagen is not accelerated in diabetes. Evidence against a generalized increase in oxidative stress in diabetes. J Clin Invest 100(4):839–846

[illegible]

Chapter 16
Redox Pathways as a Platform in Drug Development

Danyelle M. Townsend and Kenneth D. Tew

Abstract Redox homeostasis is frequently aberrantly regulated in human diseases such as cancer and neurological disorders. Partly as a consequence, there is optimism in validating and extending redox controlling pathways as a platform for the discovery/development of drugs, particularly in cancer. As the primary redox buffer, cellular thiols have been variously therapeutically targeted. N-acetylcysteine is the simplest pharmaceutical version of a bioavailable redox equivalent. It has uses in a number of disparate human pathologies. Other agents have redox active centers primarily as a function of nucleophilic centers associated with the variable valence states of sulfur. Redox homeostasis is aberrantly regulated in cancer cells and this has provided an opportunity to advance treatment concepts that attempt to produce a beneficial therapeutic index. A component of the approaches to prevent cancer is based upon the possibility that thiols provide a way of detoxifying environmental electrophiles prior to enacting damage to DNA that could progress a cell towards a cancerous phenotype. Further, therapies designed to enhance myeloproliferation, hematopoietic progenitor cell mobilization and immune response also have a foundation in modulation of redox pathways within the bone marrow compartment. As a consequence of these principles, a number of "redox modulating" drugs are under development and progressing towards FDA review.

D.M. Townsend (✉)
Department of Pharmaceutical and Biomedical Sciences, Medical University of South Carolina, 173 Ashley Ave, Charleston, SC 29425-5050, USA
e-mail: townsed@musc.edu

K.D. Tew
Departments of Cell and Molecular Pharmacology and Experimental Therapeutics, Medical University of South Carolina, Charleston, SC 29425, USA

U. Jakob and D. Reichmann (eds.), *Oxidative Stress and Redox Regulation*,
DOI 10.1007/978-94-007-5787-5_16,

16.1 Introduction

Imbalance of redox homeostasis is a characteristic of a number of human disease pathologies. In particular, cancer and neurological disorders are characterized by alterations in a variety of pathways involving reduced glutathione (GSH). GSH (γ-glutamyl-cysteinyl-glycine) is the most abundant non-protein thiol in biological systems. GSH participates as a redox donor in many aspects of metabolism, gene expression, signal transduction, proliferation and apoptosis and as a cofactor for redox modulating enzymes such as glutathione S-transferases (GSTs), peroxidases, peroxiredoxins and thiol reductases (Tew 1994). Aberrant redox homeostasis may be cause: effect related to altered regulation of signal transduction pathways that lead to uncontrolled cell growth (cancer) and/or cell death (neurodegenerative disorders; Tew et al. 2011). At least one classical view of redox status is exemplified by the dynamic equilibrium between the production of reactive oxygen or nitrogen species (ROS; RNS) and the detoxification of these species by a variety of antioxidants. Under ROS/RNS accumulating conditions, the reversible formation of mixed disulfides between redox-sensitive protein thiol groups and Cys (S-thiolation) and/or GSH (S-glutathionylation) can occur. These represent reversible redox regulation events important in many aspects of cell signaling and protein function. Cell redox status may be defined by thiol balance and the equilibrium of free thiols:disulfides. The ratio of GSH and its disulfide GSSG are primary variables used to estimate the reduction equivalents of a cell (sometimes expressed as the Nernst potential) and redox dysregulation is linked with initiation and progression of cancer. Constitutively elevated intrinsic oxidative stress in cancer tissues has been linked to increases in ROS generation and accumulation (Szatrowski and Nathan 1991), disruption of thiol and non-radical circuits and expression of antioxidant enzymes including superoxide dismutase (SOD), glutathione peroxidase, glutathione S-transferases (GST) and peroxiredoxins (Tew 1994; Li and Oberley 1997). Chronic oxidative stress can activate redox sensitive transcription factors such as nuclear factor κB (NF-κB) and Nrf2, oncogenes and signaling pathways. Downstream of these events, expression of survival factors, such as BCL2 and MCL1 can influence expression of pro-apoptotic pathways mediated by caspases. Cancer progression can occur via regulation of redox dependent gene expression in pathways involving cell proliferation, senescence, metastasis, and angiogenesis.

Expression of a number of oncogenes and tumor suppressors is also impacted by redox conditions. For example, constitutive up-regulation of Ras is associated with elevated cellular oxidative stress, as is the chimeric BCR-ABL tyrosine kinase (Trachootham et al. 2008). p53 has a critical role in regulating a number of genes associated with oxidative stress, including SOD and glutathione peroxidases (Drane et al. 2001). Its inactivation increases oxidative stress and enhances tumor growth, both of which could be suppressed via antioxidants (Sablina et al. 2005). During cancer initiation the S-glutathionylation of p53 prevents DNA binding (Velu et al. 2007), exemplifying the importance of redox sensitive cysteines in these pathways. In many instances, these pathways provide a fertile platform for the discovery and development of novel therapeutic agents. A particularly fertile platform for drug discovery in cancer has come from the observations that GST (especially GSTP) is expressed at high levels in many tumors (e.g. ovarian, non-small cell lung, breast, colon, pancreas and lymphomas) and in a wide range of drug resistant cell lines and tumors (Tew 1994). Mechanistically, structural properties of GSTP are critical in facilitating GSH dependent catalysis. Its N-terminal domain adopts a topology similar to several proteins from a thioredoxin fold super-family, which binds GSH with high affinity. This glutathione binding site (often referred to as the G-site; Mannervik 1985) is situated in a cleft formed between intra-subunit domains. Domain 1 is connected to domain 2 by a short linker sequence at the C-terminus of the linker sequence. The C-terminal domain together with a loop from the N-terminal domain forms the substrate-binding site (H-site) that is hydrophobic and is adjacent to the G-site, permitting proper orientation of the bound reactants. Different amino acids in the H site of GST isozymes account for distinct substrate specificities. The homodimeric structure is common for proteins containing thioredoxin folds and structural interactions at the inter-subunit interface are crucial for complex assembly and stability. The domains at the inter-subunit surface of the GSTP homodimer are dominated by hydrophobic interactions between residues from domain 1 of one subunit and domain 2 of the other. Aromatic residues play major roles in these interactions. The interface is approximately 25–35 Å and at a height of 25 Å, diverges to create a V-shaped crevice that is solvent-accessible (Sinning et al. 1993). In addition, the active site cleft in GSTP is shallow, while in other GST isozymes it is larger and more open (Sheehan et al. 2001).

There are drugs that have been developed for cancer that impact specific redox pathways and this review will summarize such examples.

16.1.1 Sulfur as a Nucleophile

Maintenance of sulfur homeostasis is crucial in determining cellular redox homeostasis and response to a variety of chemical and physical stimuli/stresses. Sulfur has unique variable valence states, which in cysteines can empower biological flexibility permitting certain proteins to participate in sensing and/or signaling reactions that can influence cell fate. There are >200,000 cysteines encoded by the human genome (Jones 2008), implying restricted incorporation into proteins. There is perhaps some link with selective evolutionary status, since organism cysteine content appears to correlate with degree of biological complexity (Miseta and Csutora 2000). The major non-protein thiol in cells is glutathione (GSH), where the cysteine can exist as a thiolate with a pKa of 9.65, accounting for its nucleophilic reactivity. GSH homeostasis is maintained in cells by a complex series of balanced pathways including the γ-glutamyl cycle and recycling through the cleavage activity of membrane-associated γ-glutamyl transpeptidase. A characteristic of the transformed phenotype is that cancer cells have millimolar concentrations of GSH combined with aberrant redox homeostasis. Balanced oxidation and reduction reactions determine cellular redox homeostasis and play essential roles in regulating signaling cascades, including those associated with proliferation, inflammatory responses, apoptosis, and senescence.

16.1.2 Selenium vs. Sulfur

Selenium (Se) is a rare element found in the form of selenocysteine (the 21st amino acid) coded for by the UGA codon in 25 proteins (Kryukov et al. 2003). The absence of Se from the mammalian diet leads to death, and indeed, there have been preclinical studies and extensive clinical trials with Se supplements in a chemopreventive setting, particularly in high-risk prostate cancer patients. Since earlier clinically positive results in the mid 1990s, (Clark et al. 1998) there has been much debate as to the value of Se as a preventive supplement (Marshall 2001). There is also debate over which chemical form of selenium is most bioavailable and whether this influences outcomes of selenium based trials. While the therapeutic efficacy of Se will await final scrutiny when many large-scale trials reach completion in the next 5 years, there is value in analyzing how the chemistry/biochemistry of Se complements and/or differs from that of sulfur. Moreover, from an evolutionary standpoint, recoding of the UGA stop codon into selenocysteine sense codon involves a significant expenditure of biological energy (Forchhammer et al. 1989), implying that the amino acid fulfills some critically important biological role that sulfur containing cysteines cannot. Such a conclusion has initiated a variety of theories as to what properties of selenocysteine facilitate its maintenance in the genome/proteome. It has been argued that the enhanced nucleophilicity of Se (with a pKa of 5.2 (Wessjohann et al. 2007)) compared to S and the lower Pka of selenol compared to thiol may convey enhanced catalytic functionality. However, the enhanced nucleophilicity is a factor of <10 (Songstad and Pearson 1967, 1968), quite modest by the rate constant changes that would be imbued. In an eminently readable article, Hondal et al. (Hondal and Ruggles 2011) discuss three chemico-biological properties (with particular reference to thioredoxin reductase (Lothrop et al. 2009)) that may be involved in selenocysteine functionality. By considering three contingencies, (i) selenium is a better nucleophile than sulfur; (ii) selenium acts as a better leaving group; (iii) selenium has better electrophilic characteristics; they reach the conclusion that electrophilicity rather than nucleophilicity is the advantage maintained by natural selection. Se, they argue, is needed for its superior electrophilicity since this property is shared with all small molecule substrates that are reduced by the N-terminal redox center. To account for the selective pressure on the genome that is needed to maintain the insertion machinery Se also has the property of being highly resistant to irreversible oxidation, making Se-containing enzymes resistant to inactivation by oxidation. The other physicochemical properties of Se and S are quite similar, but the chemical rate of recycling of the oxidized selenocysteine residue back to its parent form is very large compared to the same cycling of an oxidized Cys residue (Cys-SO_2– to Cys-SH). This ability to resist irreversible oxidation can explain the biological pressure to maintain the selenocysteine-insertion machinery (Hondal and Ruggles 2011).

16.2 Redox, Thiols and the Bone Marrow Environment: Myeloproliferative Drugs

In humans, hematopoietic stem cells (HSCs) in the bone marrow give rise to all differentiated hematopoietic cell lineages, including erythroid, myeloid, lymphoid, and megakaryocytes. Functionally and spatially characterized niches within the marrow cavity are determinant factors in stem cell fate (Schofield 1978). The osteoblastic niche maintains HSC quiescence while the vascular niche microenvironment promotes HSC proliferation and differentiation (Wilson et al. 2008; Calvi et al. 2003). Adhesion proteins, cytokine and chemokine signaling molecules direct cells to their respective niches. Ca2+ gradients in the endosteum enable calcium-sensitive HSCs to sense and migrate appropriately (Adams et al. 2006). The chemokine CXCL12 and its receptor CXCR4 are regulators of HSC quiescence and niche retention (Nie et al. 2008). Metallo- and/or serine proteases are involved in CXCL12 and CXCR4 inactivation and mobilization (Christopherson et al. 2003). The balance between proteases and their inhibitors is critical in the steps involved in HSC mobilization. Given that a number of proteases and their inhibitors contain Cys residues that are subject to oxidative regulation, redox conditions will regulate protease activity. In addition, serine protease inhibitors are also regulated by S-glutathionylation, a property that conveys a further level of regulatory control on this process (Fig. 16.1).

Hematopoiesis operates optimally when the marrow compartment is between 1 and 3% oxygen (Cipolleschi et al. 1993), but hypoxic niches and gradients are required for HSC maintenance. A hypoxic environment in the osteoblastic niche encourages quiescence in HSCs and movement to the more oxygenated vascular niche promotes HSC differentiation (Iwasaki and Suda 2009). Regions of low oxygen may protect quiescent HSCs from toxic oxygen free radicals and a ROS gradient may regulate HSC-niche interactions (Parmar et al. 2007). N-acetyl cysteine (NAC) reverses ROS suppression of N-cadherin-mediated HSC adhesion to osteoblasts, inducing cell migration (Hosokawa et al. 2007). Gradients of ROS may affect elements of the niche structure and serve to regulate HSC within the marrow microenvironment. Redox modulation is a regulator of cell self-renewal and differentiation in glial cells, where cysteine pro-drugs alter extracellular redox and enhance self-renewal of progenitor cells while buthionine sulfoximine (BSO), by depleting GSH, promotes differentiation (Smith et al. 2000). These types of observations imply that redox gradients and their pharmacological manipulation have a role in influencing HSC migration and differentiation and myeloproliferation.

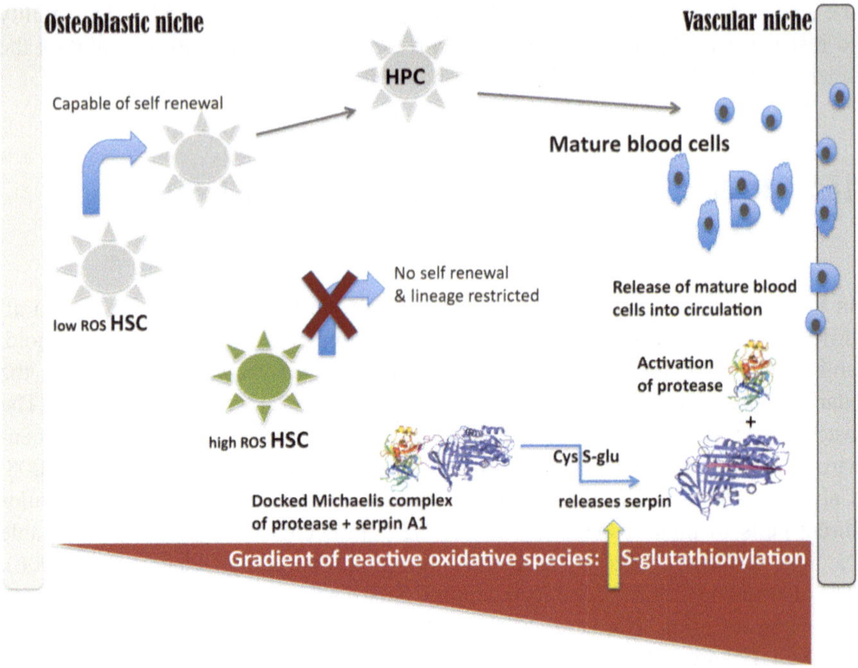

Fig. 16.1 Representation of how S-glutathionylation of a critical cysteine residue in serpin A1 may influence release of mature blood cells from the bone marrow compartment. Various gradients help to define the osteoblastic and vascular niches of the marrow compartment. Hematopoietic stem (*HSC*) and progenitor (*HPC*) cells traverse this space and mature towards the variety of blood cells that present the circulating blood lineages. Two types of HSC exist. Depicted are low and high ROS populations. The former are characterised by cell surface markers that are consistent with their existence in the osteoblastic niche. High ROS HSC lack the capacity to self-renew and are lineage restricted, although treatment with agents such as NAC can reverse this (Jang and Sharkis 2007). Eventual release of mature cells from the marrow involves some form(s) of proteolytic action that result in their entry into the vascular niche and general circulation

Over a half-century ago cystine and cysteine were shown to be important for maintaining the required balance between HSC quiescence, self-renewal, and lineage commitment in human bone marrow (Baldini and Sacchetti 1953). Subsequently it became clear that thiol conditions inversely correlated with the capacity of ROS to attenuate HSC self-renewal differentiation. Bone marrow has two HSC populations defined by low and high intracellular ROS. Low ROS populations have markers (N-cadherin, Notch1, p21, and calcium-sensing receptor) indicating osteoblast association and retained self-renewal capacity in serial transplantation. High ROS cells lacked this repopulating capacity and were lineage restricted (Fig. 16.1). NAC treatment restores functional activity to this high ROS population. (Jang and Sharkis 2007). Such information would suggest that therapeutic strategies that aim to convert high ROS cells into low may be useful in treatments involving HSC transplantation and engraftment.

Excess accumulation of ROS can occur as a consequence of cancer chemotherapy and this can cause phosphorylation of p38 MAPK in HSC contributing to a failure to maintain quiescence and self-renewal capabilities (Ito et al. 2006). The forkhead box-O (FoxO) family of redox-sensitive transcription factors up-regulates ROS detoxifying genes such as MnSOD, catalase, and GADD45 in HSC. Triple FoxO knockout animals develop a fatal bone marrow defect characterized by loss of quiescence, increased apoptosis and elevated intracellular ROS in HSC. Supplementation of thiol pools by NAC rescues these HSC, but increased ROS levels can trigger differentiation of HSCs to myeloid progenitors (Tothova et al. 2007). NAC treatment during granulocyte colony-stimulating factor (G-CSF) induced mobilization suppresses HPC mobilization (Tesio et al. 2011). NAC induced changes in thiols and redox state increases transcription of cytokines such as IL-1, IL-4, IL-6, IL-7, IL-8 and TNF-α (Rovin et al. 1997; Gosset et al. 1999) and directional migration of HSC in response to stromal cell-derived factor-1 (SDF-1) define a requirement for GSH. These studies emphasize how pharmacological manipulation by agents like NAC can control fluctuations of ROS and thiol balance and maintain HSC self-renewal, proliferative capacity and lineage fate.

16.3 GSH and GSTP as Drug Platforms

16.3.1 Direct Manipulation of GSH

In the context of drugs that impact redox pathways there are examples of agents that achieve this through direct targeting of thiol maintenance pathways and there are drugs that have indirect effects, frequently as a consequence of their capacities to produce electrophilic centers. A number of standard anticancer drugs are known to influence redox homeostasis in both cancer and normal cells. Even drugs that do not have an apparently obvious electrophilic character will cause chain reactions that alter intracellular ROS levels. Metabolism in cardiac sarcoplasmic reticulum and mitochondria of quinone based anticancer drugs such as the anthracycline antibiotics is linked to cardiac toxicity (Doroshow 1983). The efficacy/toxicity of a number of alkylating agents and platinum drugs with established electrophilic centers is also linked with ROS induced damage (Wondrak 2009). Others, such as paclitaxel, cytarabine, vinblastine, campotothecin, or histone deacetylase inhibitors result in production of ROS that is unrelated to direct drug properties (Rahmani et al. 2005; Fawcett et al. 2005). Arsenic trioxide has found a niche in the treatment of promyelocytic leukemia and its therapeutic activity stems in part from the enhanced sensitivity of the cancer cells to the thiol active components of arsenic (Pelicano et al. 2003) and other metabolites that regulate components of the NADPH oxidase complex (Chou et al. 2004). Combining arsenic trioxide with the superoxide dismutase inhibitor 2-methoxyestradiol has shown promise in targeting chronic lymphocytic leukemia cells (CLL) (Zhou et al. 2003). Motexafin gadolinium is a thioredoxin (Trx) reductase inhibitor that causes apoptosis in malignant cells through induction of ROS (Evens et al. 2005). Agents that cause a reduction in glutathione levels, such as phenyl-isothiocyanate can also effectively kill resistant CLL and chronic myeloid leukemia (CML) cells (Trachootham et al. 2008). Use of thiol modulating agents has also been successfully adapted as a modality for the prevention of drug toxicities (Wondrak 2009). As an example, TDZD-8 (4-benzyl-2-methyl-1,2,4-thiazoline-3,5-dione) selectively induces apoptosis through the depletion of free thiols of leukemic stem cells (LSC), but not hematopoietic stem and progenitor cells (Guzman et al. 2007). Inhibition of NF-κB with niclosamide produces elevation of ROS in AML cells without targeting normal bone marrow (Jin et al. 2010). Nevertheless, many drugs have toxic effects through their impact on myeloid progenitor stem cells. It is clear that the bone marrow compartment is a tissue with complex associations with thiols and redox balance.

16.3.2 Biologically Available Cysteine: NAC, GSH Ethyl Ester

The simplest form of "bioavailable sulfur reducing equivalents" is the n-acetylated derivative of L-cysteine (NAC) that serves as a precursor for cysteine, enhances the de novo synthesis of GSH and acts as a scavenger of ROS such as OH· and H_2O_2 (Aruoma et al. 1989). As a pharmaceutical, NAC has a broad usage in a number of unrelated pathologies. In each case, the delivery of thiol equivalents is linked with the beneficial impact of the drug. A major therapeutic usage is in the reversal of the potentially lethal liver damage associate with acetaminophen overdose (Kanter 2006). However, the agent has utility in each of the following conditions (for review see Kelly 1998): reversal of carbon monoxide poisoning; unstable angina; amyotrophic lateral sclerosis (ALS); bile duct blockage in infants; Alzheimer's disease; allergic reactions to the anti-seizure drug phenytoin, amelioration of symptoms from ocular infection (keratoconjunctivitis). It is also used for reducing levels of a certain type of cholesterol and homocysteine thereby influencing risk factors for heart disease in patients with serious kidney disease. Additional therapeutic uses include: the agent has mucolytic activity by reducing mucus viscosity by splitting disulfide bonds in various mucoproteins. Inhaled NAC is indicated as an adjuvant in other respiratory pathologies with excessive and/or thick mucus production (e.g. emphysema, bronchitis, chronic obstructive pulmonary disease). Although there is some debate, high doses of oral NAC have been found to modulate inflammation in cystic fibrosis patients (Tirouvanziam et al. 2006) Mechanistically, the data indicate that NAC provides thiol equivalents that compensate redox imbalance that is implicated in inflammatory imbalances in cystic fibrosis. In a similar vein there are studies that support the utility of NAC in the prevention of radiocontrast induced nephropathy, however this indication is not entirely supported by other clinical trial testing studies (Marenzi et al. 2006).

At a cellular level, NAC can impact cell survival and apoptosis through the activation of cell surface signal-regulated kinase pathways. NAC increases mitogenesis and suppresses apoptosis of B-lymphocytes in p53-deficient mice (Martin et al. 2000). NAC prevents activation of c-Jun N-terminal kinase, p38 MAP kinase and redox-sensitive AP-1 and NFκB transcription factor activities, all of which have roles in T cell proliferation (Zafarullah et al. 2003). Administration of NAC in HIV patients restores GSH levels and ameliorates viral replication, diminishing infection and disease progression (Nakamura et al. 2002), enhances the antibody-dependent cellular toxicity of neutrophils (Roberts et al. 1995) and attenuates CD4+ T lymphocyte decline (Akerlund et al. 1996). In cancer patients, NAC has been used both as a chemoprotectant and in the setting of chemoprevention. NAC mediated GSH restoration has been used to ameliorate chemotherapy-induced immunity disorders and to prevent cancer development in ROS-susceptible Atm- and p53-deficent mice (Reliene and Schiestl 2006). In a rat brain tumor model, NAC treatment prevents myelosuppression without compromising treatment efficacy (Neuwelt et al. 2004). With the chloroethylnitrosourea BCNU, pretreatment with NAC prevents drug associated redox dysregulation and myelotoxicity. Through carbamoylation of active cite cysteines, BCNU can inactivate glutathione reductase and other enzymes linked with GSH homeostasis (Tew et al. 1985). In this regard, the drug's myeloprotective activity may be mediated through the protection of some or all of these pathways enabling normal GSH synthesis and maintenance of intracellular GSH balance.

Although considered quite benign from a pharmaceutical perspective, adverse effects have been reported for NAC. In a mouse model extremely high doses can cause cardiac and pulmonary damage. Primarily in the liver, NAC is metabolized to *S*-nitroso-*N*-acetylcysteine that can create a hypoxia like response and lead to pulmonary artery hypertension (Palmer et al. 2007). Even though the mouse studies used doses of NAC that were significantly higher than those used in therapeutic applications in man, these toxicities are relevant.

Other bioavailable GSH precursors include glutathione monoesters in which the carboxyl group of the glycine residue is esterified. Unlike glutathione itself, such esters are effectively transported into various types of cells and can be converted intracellularly into glutathione (Levy et al. 1993). Glutathione diethyl ester is rapidly split to the glutathione monoethyl ester by mouse plasma glutathione diester α-esterase. Both the glutathione mono- and diesters have similar effects on cellular glutathione levels in mice. However, human plasma lacks glutathione diester α-esterase and as a consequence human cells transport glutathione diethyl ester much more effectively than the corresponding monoethyl (glycyl) ester. Human cells rapidly convert glutathione diethyl ester to the monoester and can provide the cells with a means of producing glutathione over a protracted period of time. Thus the glutathione diethyl ester is an effective delivery agent for glutathione monoester and thus for glutathione.

16.3.3 NOV-002

Viral diseases frequently cause depletion of GSH and treatment of HIV patients with precursors of GSH such as N-acetylcysteine have a positive impact upon survival (Herzenberg et al. 1997). GSH levels in antigen presenting cells can influence their T helper1 and Th2 cytokine response patterns in immune response and explains for the differences in immune responses in "Th1" and "Th2" mouse strains (Peterson et al. 1998). These effects may be mediated by extracellular redox changes that can also impact intracellular redox signaling and proliferation in peripheral blood mononuclear cells (Sahaf et al. 2005). Such factors are consistent with some aspects of the pharmacology of NOV-002, an agent that has myeloproliferative properties and causes transient oxidation of cell surface proteins, reflected by time sensitive alterations in intracellular redox homeostasis. NOV-002 (Fig. 16.2, panel A) is a formulation of glutathione disulfide (GSSG), containing a trace amount of cisplatin, (1,000:1molar ratio). The cisplatin component does not contribute to efficacy via cytotoxicity. A 60 mg/day dose of NOV-002 administered over 6 months would result in a cumulative total dose of cisplatin equivalent to <2% of a typical chemotherapeutic single dose of cisplatin. Animal pharmacokinetic studies suggest that the addition of cisplatin serves to increase the C_{max} and $t_{1/2}$ for GSSG bioavailability (Townsend et al. 2008).

Unusually for a drug in preclinical and clinical development, the availability of clinical data preceded full preclinical evaluation of NOV-002. A significant number of clinical trials were generated from earlier Russian investigations indicating an unusual pharmacological profile for oncology indications. The data suggested that for lung cancer patients addition of NOV-002 to existing chemotherapeutic regimens produced higher patient survival statistics and improved quality of life parameters. Unfortunately, further Phase III clinical trials carried out under the auspices of the US Food and Drug Administration failed to confirm these positive efficacy results. As a consequence, further development of the drug is now on hold. Nevertheless, a number of interesting features related to the mechanism of action of the drug did emerge from these studies. For example, NOV-002 has little or no cytotoxicity even at high doses, but has the capacity to modulate redox conditions at the cell surface with subsequent intracellular changes in the redox status and cell signaling pathways. Certain cell surface protein thiols can act as sensors of extracellular redox stress and their oxidative modification has been linked to regulation of cell signaling and other functions (Dominici et al. 1999). NOV-002 treatment of cells serves to oxidatively modify cell surface proteins with commensurate changes in a number of characteristics of membrane structure. Subsequent to these changes hydrogen peroxide levels increase resulting in a process that was temporally consistent with the time and concentration dependent increase in the phosphorylation of three kinases, JNK, p38 and ERK. In bone marrow these kinases can play direct roles in the regulation of cell proliferation. Moreover AKT, a kinase also implicated in marrow proliferation acting in concert with JAK2 and STAT5 was also impacted by NOV-002. The JAK-STAT signaling pathway is intimately involved in governing the

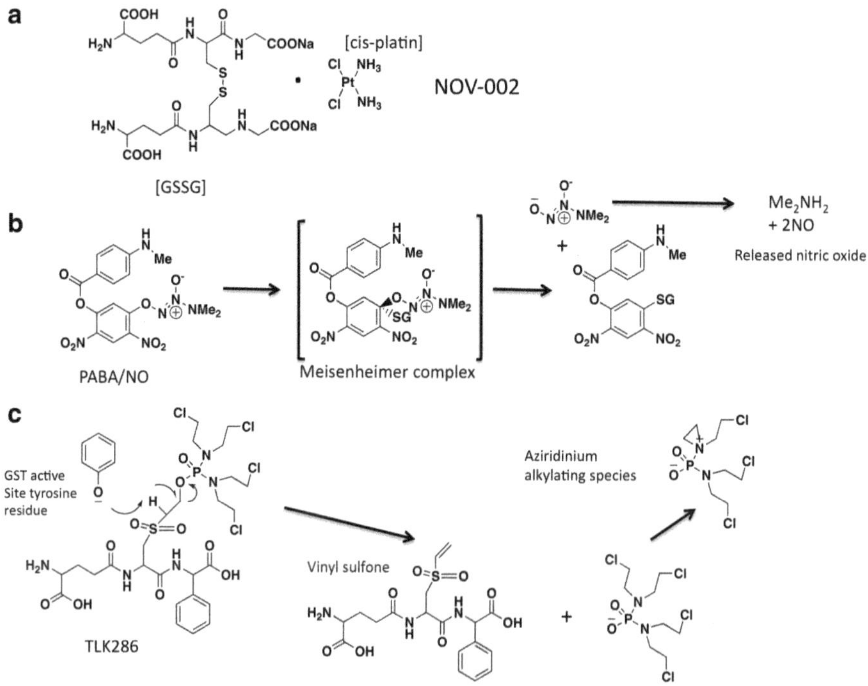

Fig. 16.2 Structures and metabolism of the diglutathione mimetic NOV-002 (*panel A*) and two GSTP activated pro-drugs. For PABA/NO (*panel B*), release of nitric oxide is achieved through an intermediate Meisenheimer complex formed through GSH complex formation under the catalytic activity of GST. *Panel C* shows TLK286 metabolism where proton abstraction can activate the pro-drug to release a vinyl sulfone and tetra-functional alkylating species that become aziridinium species after the chlorines act as leaving groups

response of cells to cytokines and growth factors (O'Shea et al. 2002) and activation of this pathway by NOV-002 indicates that the drug is affecting those pathways that control hematopoiesis and myeloproliferation.

16.3.4 Mesna

Mesna (Mesnex) is the sodium salt of 2-mercaptoethane sulfonate. It is a synthetic sulfhydryl compound used as an adjuvant to reduce the complications arising from hemorrhagic cystitis and hematuria as a consequence of treatment with alkylating agents such as cyclophosphamide or ifosfamide. Each of these cancer drugs is metabolized through the liver cytochrome p450 system to release acrolein that has an electrophilic vinyl group but can be detoxified by interaction with cysteine and GSH. Mesna provides the biologically available thiol that can carry out this reaction in the bladder. In some countries, Mesna is also used as a mucolytic agent, with indications and properties characteristic of NAC (Brock et al. 1984).

16.3.5 Tavocept

Tavocept is 2,2′-dithio-bis-ethane sulfonate a water-soluble disulfide with some characteristics similar to NOV-002. Cellular reducing equivalents provided by glutathione and/or the Grx/Trx systems can reduce Tavocept into the free thiol form, 2-mercaptoethane sulfonate mesna (Verschraagen et al. 2004), the disulfide remains predominant in the oxidative environment of the plasma (Shanmugarajah et al. 2009). Preferential uptake of Tavocept in the kidneys leads to reduction to mesna and results in protection against cisplatinum-induced nephrotoxicity. Tavocept has also been shown to protect against chemotherapy-induced myelosuppression (Hausheer et al. 1998). Mechanisms of action may include pharmacological modulation of physiological thiol and disulfide levels (Pendyala et al. 2003), or the direct conjugation of mesna with platinum species. The γGGT-mediated formation of platinum-SG adducts is prevented by the premature formation of mesna-platinum adducts (Hausheer et al. 1998). Mesna conjugation with GSH also inhibits γGGT and prevents further activation of GS-platinum species. Unreduced Tavocept does provide disulfide equivalents that can influence the Trx and Grx systems and such features do seem to be important in the therapeutic utility of Tavocept in treating lung cancer patients. A drug-induced shift towards oxidized Trx and Grx may impact the signaling pathways that are mediated by these proteins. With most of the thiol:disulfide drug entities, their interface with the Trx system is a critical aspect of their mechanism of action. Trx in particular is involved directly in mediating a series of events critical to cell proliferation or apoptosis. For example, loss of Trx blocks T lymphocyte activation and proliferation and circulating Trx-1 can act like a cytokine as a chemo-attractant for monocytes, neutrophils and lymphocytes in addition to inhibiting neutrophil migration into inflammatory sites (Nakamura et al. 2001; Bertini et al. 1999). Tavocept may impact Trx activity in a manner similar to thioredoxin-interacting protein (TXNIP) that is involved in regulation of HSC quiescence and mobilization (Jeong et al. 2009). Tavocept mediated modulation of the ratio of oxidized and reduced Trx may be important to the myeloproliferative activities associated with the drug. Other redox-cycling antioxidants such as sodium thiosulfate, diethyldithiocarbonate and tempol have some actions that overlap with Tavocept.

16.3.6 Amifostine (WR-2721; Ethyol)

Amifostine (WR-2721) is an agent that emerged from early research by scientists at Walter Reed Army Medical Center in the 1970s. It is an thiophosphate pro-drug subject to activation by blood and tissue alkaline phosphatase to the thiol WR-1065 (Korst et al. 1997). WR-1065 may function mechanistically through the prevention and/or repair of DNA damage induced by ROS, induction of cellular hypoxia, inhibition of apoptosis, as well as the direct covalent binding and inactivation of cytotoxic drugs. More recent studies indicate additional roles in the regulation of transcription factors and involvement in apoptosis, cell cycle regulation and cell differentiation (Khodarev et al. 2004). The rationale for this lies in the fact that normal tissues have relatively high levels of alkaline phosphatase creating the expectation that more of the prodrug will be activated and protect normal tissues compared to tumor. The therapeutic utility of amifostine is in combination with chemo- or radiation therapies and in 1999 US Bioscience received full FDA approval for Ethyol for the indication of reduction of moderate-to-severe, post-operative radiation-induced xerostomia (dry mouth) in patients undergoing radiation treatment for head and neck cancer.

Amifostine also protects HPCs during chemotherapy and reduces engraftment periods after autologous bone marrow transplantation and stimulates the formation of multipotent and erythroid bone marrow progenitor cell lineages (List et al. 1997) and is used in patients receiving bone marrow transplants, as well as in the treatment of a range of pathologies characterized by deficient hematopoiesis resulting from increased cell death of bone marrow precursors. In MDS, amifostine stimulates hematopoietic recovery (Schanz et al. 2009), selectively induces apoptosis in MDS cell lines but protects non-malignant HPC from cytotoxic drugs (Ribizzi et al. 2000). Mechanistically, such protection has been attributed to inhibition of stress-induced apoptosis in normal HPCs via activation of NF-κB/Rel transcription factors (Romano et al. 1999).

Another organic thiol, vitalethine (Knight et al. 1994) also abrogates apoptotic signals and potentiates hematopoiesis. Aminothiols essentially increase levels of intracellular thiols and thereby abrogate apoptotic signaling initiation, alter DNA replication and cell cycle progression and influence the binding of redox-sensitive transcription factors to target DNA sequences. As with other thiols, amifostine binding to NF-κB, AP-1, and p53 enhances the affinity of these proteins to target regulatory sequences with subsequent impact on the transactivation of downstream genes (Shen et al. 2001). Within this mechanism of action, there is commonality with how thioredoxins influence the binding of the redox-transcription factors, *C-fos* and *C-jun* to DNA. There is a common potentiation of responses to proliferative signals and activation of hematopoietic growth factors, such as the stem cell factor kit ligand (KL) (Abate et al. 1990). The myeloid growth promoting property of amifostine may coincide or even compete with that of KL (List et al. 1998). Furthermore, amifostine can regulate cell cycle progression and apoptosis through the induction of p53 through a JNK-dependent signaling pathway and subsequent expression of the cyclin-dependent kinase inhibitor p21 (Pluquet et al. 2003). Amifostine has recently been identified as a potent hypoxia-mimetic that induces HIF-1 expression and accumulation (Koukourakis et al. 2004). Because the bone marrow niches are defined by gradients of oxygen tension activation of hypoxia-sensitive genes may be required to promote events such as HSC mobilization and differentiation. In all, the thiol delivery system serves a number of distinct and overlapping functions in influencing normal tissue proliferative pathways.

16.3.7 TLK199 (Telintra)

In the mid 1990s, because increased expression of GST isozymes was frequently associated with drug resistant cells, their inhibition was considered a rationale approach to improve therapeutic response. For example, ethacrynic acid, although an effective inhibitor of various GST isozymes was not successful in the clinical management of cancer patients. The dose-limiting toxicity of fluid imbalance was triggered by the diuretic properties of the drug (O'Dwyer et al. 1991). One consequence of this platform was the conceptual design of a peptidomimetic inhibitor of GSTP, TLK199 (γ–glutamyl–S–(benzyl)cysteinyl R(−)phenyl glycine diethyl ester) now called Telintra or ezatiostat. While designed to sensitize drug-resistant tumors that overexpress GSTP (Morgan et al. 1996), preclinical mechanism of action studies with this agent revealed an unanticipated myeloproliferative activity in animals (Gate et al. 2004; Ruscoe et al. 2001). Given the link between GSTP and the kinase pathways, a model for how TLK199 can produce proliferative effects in the marrow compartment was proposed (Fig. 16.3). Increased myeloproliferation in GSTP deficient mice compared with wild type animals (Ruscoe et al. 2001) was characterized by higher levels of white blood cells in GSTP knock out mice. Spleen cell counts were higher in knock out animals, and this was associated with a twofold increase in B-lymphocytes, whereas T lymphocytes and NK cell counts were similar in both strains. In contrast, no difference in thymocyte counts and thymus subset composition was observed. Red blood cell and platelet counts were also higher in GSTP null mice. As a whole, these data inferred that the absence of GSTP expression potentiates hematopoiesis by influencing the proliferation and/or the differentiation of hematopoietic progenitor cells. TLK199 stimulated colony formation (i.e. proliferation) of most major progenitor cells in wild type, but not knock out animals. The JNK inhibitor SP600125 decreased marrow colonies produced by cytokine treatment of knock out animals and JNK phosphorylation was endogenously elevated in bone marrow cells from GSTP null animals. Such data are consistent with GSTP acting as a physiological inhibitor of JNK, where TLK199 disassociates GSTP from JNK allowing kinase phosphorylation and subsequent activation of the kinase cascade (Fig. 16.3; Adler et al. 1999). Figure 16.3 also stylizes the formation of GSTP oligomeric structures. Such structures are frequently observed during gel separation of proteins, even in the presence of reducing conditions and imply that disulfide bonds can inactivate GSTP in either a reversible or irreversible manner (Adler et al. 1999; Townsend et al. 2006) Inhibition of JNK abrogates the increased phosphorylation of this kinase observed in the presence of TLK199 and consequently reduces the myeloproliferative effects suggesting that JNK plays a role in both the genetic and pharmaceutical enhanced myeloproliferation. This is consistent with reports indicating a possible role for JNK in control of proliferation. Discrimination between the survival and apoptotic functions of JNK seems to correlate with the level and duration of the enzyme activation. A strong and sustained activation is associated with apoptosis, whereas a weaker and transient phosphorylation is correlated with proliferation (Shaulian and Karin 2001). For example, in mouse hematopoietic BaF3 cells, JNK activity was

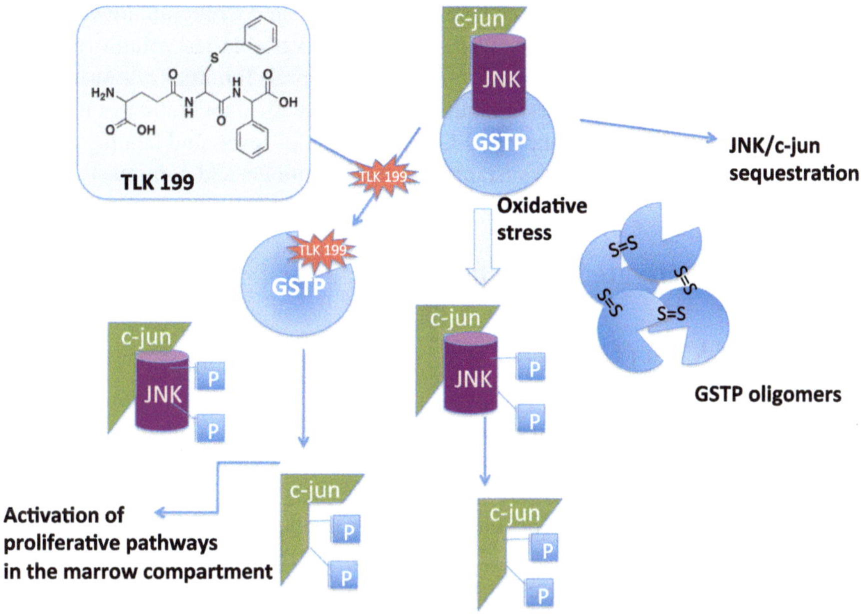

Fig. 16.3 Representative model of signaling cascade regulated by the complex formed between GSTP and proteins of c-jun NH2-terminal kinase. The cartoon illustrates how TLK199 can cause myeloproliferation by activation of JNK mediated proliferative pathways leading to phosphorylation of both JNK and c-jun in the bone marrow compartment. A common theme in regulation of myeloproliferation is the activation potential of ROS and this is mimicked by TLK199. Subsequent drug induced dissolution of the complex can lead to the formation of higher order GSTP structures such as the oligomers shown

three times lower when cells were exposed to mitogenic concentrations of IL–3 than to cytotoxic concentrations of anisomycin (Terada et al. 1997). From such reports, regulation of JNK activity by GSTP should be a viable target for drug intervention. Consequently, Telik Inc. has underwritten a number of clinical trials to test the drug in myelodysplastic syndrome (MDS) patients. A liposomal formulation was developed for intravenous administration and used in Phase 1–2A studies in MDS patients (Raza et al. 2009a). An oral dosage formulation (tablet) was subsequently developed and a Phase 1 trial completed suggesting that the drug is well-tolerated and showed clinical activity (Raza et al. 2009b). Thereafter, the tablet formulation was studied in a Phase II clinical trial setting (Raza et al. 2012). Results from all *Telintra* clinical trials suggested that the agent has a favorable safety profile and provided clinical benefit to some patients and while clinical testing of *Telintra* is still ongoing, a recent case report on a patient with idiopathic chronic neutropenia (ICN; is a heterogeneous group of hematologic diseases characterized by low circulating neutrophil levels often associated with recurrent fevers, chronic mucosal inflammation, and severe systemic infections) was encouraging (Lyons et al. 2011). A 64-year-old female with longstanding rheumatoid arthritis developed ICN with frequent episodes of sepsis over a 4 year period was treated with granulocyte colony stimulating factors (G-CSF) but had a delayed, highly variable and volatile response. She was enrolled in a clinical trial evaluating the oral Telintra (ezatiostat). She responded by the end of the first month of treatment with stabilization of her ANC (despite tapering and then stopping G-CSF), clearing of fever, and healing of areas of infection. This response was sustained for over 8 months and continues.

16.4 GST Activated Prodrugs

16.4.1 TLK286 (Telcyta)

An improvement of the therapeutic index has always been a major goal in the design and development of new anticancer drugs. Trends towards targeted therapies have resulted in new drugs with particular specificities (e.g., geftinib, erlotinib and imatinib) and prior to the development of drug resistance, improved treatment successes. The first GST activated prodrug to emerge from preclinical studies was TLK-286, now called Telcyta. Its design strategy was based on the principle that proton-abstracting sites at the active site of GST could initiate a cleavage reaction converting an inactive prodrug into a cytotoxic species (Lyttle et al. 1994). The presence of a histidine residue in proximity to the G binding site of GSTP was integral to the removal of the sulfhydryl proton from the GSH co-substrate, generating a nucleophilic sulfide anion. This is more reactive with electrophiles in the absence of GSH and crystallographic analysis of GSTP assigns the tyrosine hydroxyl residue as the proton abstraction moiety (Reinemer et al. 1992). For the scheme shown in Fig. 16.2 (panel B), the proton abstraction deprotonates the α-carbon to yield the sulfone, that subsequently undergoes β -elimination to give the active alkylating species. Unlike other standard anticancer nitrogen mustard drugs (e.g. melphalan, chlorambucil) TLK-286 contains a tetrakis (chloroethyl) phosphorodiamidate moiety. Other compounds with this structure are more cytotoxic than a similar structure with a single bis-(chloroethyl) amine group. As with other mustards, the chlorines can act as leaving groups, creating aziridinium ions with electrophilic characteristics. Although the precise sequential formation of the four possible chlorine leaving events is not known, these species presumably possess the capacity to alkylate target nucleophiles such as DNA bases. The four chlorines could provide the reactivity to form crosslinks with bonding distances greater than those for bifunctional agents. However, the absence of some form of myelosuppression indicates that organ toxicities are not similar to those standard bifunctional alkylating agents studied to date. In Phase I and II trials, Telcyta did not show any significant hematological or cumulative toxicities despite drug treatment periods that were >1 year (Rosen et al. 2004). Furthermore, despite being a micromolar inhibitor of DNA-dependent protein kinase, TLK-286 causes minimal direct damage to DNA (Townsend et al. 2002). For TLK-286, activation to the alkylating species also results in the production of a vinyl sulfone derivative of the

GSH backbone (Fig. 16.2, panel C). Although the biological importance of this species to the pharmacology of the drug is not clear, it could be a factor in the chain reactions leading to lipid peroxidation and even production of hydrogen peroxide. Of relevance to this observation, catalase is over-expressed in cells selected for resistance to TLK-286 (Rosario et al. 2000). In terms of human pharmacology, an optimized two-compartment model describes the pharmacokinetic characteristics of TLK-286 in humans. This model only applies when structural parameters are not normalized for size, but instead for body surface area and body weight. In this model, the clearance and volume of distribution at steady state decrease with increasing dose. The decrease in blood clearance with increasing drug dose caused the dose-normalized area under the blood concentration versus time curve to increase with dose. The combined effect of dose-related decreases in both the clearance and volume of distribution is that increasing doses also increase the distribution half-life but decrease the elimination half-life (Rosen et al. 2004).

The drug's favorable therapeutic index is gained from selective activation by GSTP, an enzyme that is widely over-expressed in tumors compared with normal tissues and in various drug-resistant diseases. The preclinical data support the prediction of the mechanism of activation. Published Phase I and Phase II clinical results were encouraging from the standpoint of tolerance and toxicity. Toxicities were mild, non-cumulative and well tolerated, especially in older patient populations. However, Phase III clinical studies in platinum-refractory or -resistant ovarian cancer have not been conclusive. At this time, the company has somewhat diffuse plans to have further Phase III studies, particularly in NSCLC and ovarian cancer. Development of the drug towards FDA registration will by necessity await further indications of positive enhancements in disease response and/or survival.

16.4.2 PABA/NO

NO is involved in a diverse number of physiological processes characterized in many previous publications, but can also have cytotoxic consequences. As a consequence, directed delivery (tumor cells with high GSTP) of a therapeutic concentration of nitric oxide is a relevant approach to drug design. A strategy to derivatize the 0_2 position of a diazeniumdiolate with protective groups has been used to convert them into substrates for GST (Saavedra et al. 2006). The resulting inactive prodrug can become cytotoxic when localized in a cell that has high GSTP concentrations. PABA/NO (O_2-[2,4-dinitro-5-[4- (N-methylamino)benzoyloxy]phenyl]1-(N,N-dimethylamino)diazen-1-ium-1,2-diolate) has N-methyl-p-aminobenzoic acid bound via its carboxyl oxygen as a 5-substituent on the 2,4-dinitrophenyl ring (Saavedra et al. 2001). PABA/NO belongs to the O_2-aryl diazeniumdiolates, electrophiles shown to transfer their aryl groups to attacking nucleophiles with simultaneous production of ions that release NO at physiological pH (Xinhua 2008). In the presence of GSH PABA/NO is activated by GSTP (Saavedra et al. 2006).

This reaction results in the formation of a Meisenheimer-complex intermediate, and subsequently the leaving group of the reaction generates two moles of NO according to the scheme:

$$\text{PABA/NO} + \text{GSH} \rightarrow \underset{\text{Meisenheimer complex}}{[\text{PABA/NO-SG}]} \rightarrow \text{PABA-SG} + 2\text{NO} + \text{HNMe}_2.$$

Elevated NO levels lead to cytotoxic effects by forming RNS/ROS intermediates. PABA/NO-induced nitrosative stress results in limited levels of protein nitrosylation but high levels of S-glutathionylation and activates endoplasmic reticulum stress mediated cell death pathways (Findlay et al. 2004). PABA/NO is presently in preclinical development, however, early stage clinical trials with agents of this class are under consideration.

16.5 Sulfur Based Agents in Cancer Chemoprevention

For rational reasons and perhaps because pharmacological cures have been slow to emerge, general approaches to cancer prevention have gained mainstream status over the last few decades. While there are many principles underlying these efforts, improving the detoxification environment by administration of reasonably non-toxic compounds has been a major focus. Connections with redox can be classified into two domains. First, many inductive chemicals influence Phase II metabolic enzyme expression with the outcome that levels of GSH and associated enzyme systems are increased with commensurate protective effects. Second, many small molecules that induce protection have active sulfur moieties that influence the process. Common to each is the indication that the small molecule will have one or more electrophilic centers capable of reacting by one of two basic mechanisms. Bifunctional inducers are large planar or polycyclic aromatics that elevate both Phase II and Phase I (mostly cytochromes p450) genes or monofunctional inducers (a wide diversity of chemical structures) that elevate Phase II enzymes selectively (Spencer et al. 1990). An in depth analysis of those chemical characteristics that contribute to protective enzyme induction can be found elsewhere (Dinkova-Kostova et al. 2004), however, some common characteristics do apply, their reactivity with thiol groups and their impact upon the Nrf2 transcription complex.

Although a wide array of chemical structures have the capacity to induce Phase II detoxification enzymes, close scrutiny of the chemical classes reveals that these compounds have characteristics that place them into approximately ten classes. Further breakdown identifies commonalities with respect to thiols and to induction of genes through the Nrf2 transcription factor complex. For example, Table 16.1 shows some of the chemical structures of small molecules that either have active sulfur centers or impact through thiol-mediated routes on downstream gene expression pathways. These agents can include, curcumin, ethacrynic acid, cinnamates, crotonates, acrylates and a variety of indole agents that have Michael reaction acceptor characteristics. Reactivity of Michael acceptors with nucleophiles

Table 16.1 Examples of agents with chemopreventive properties

Drug name	Type of agent	Structure
Curcumin	Double Michael addition	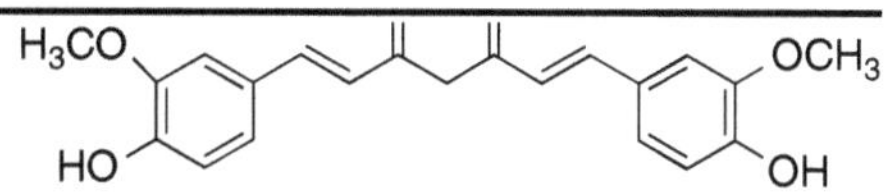
Ethacrynic acid	Michael addition	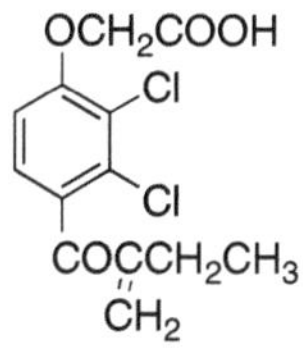
Sulforaphane	Isothiocyanate	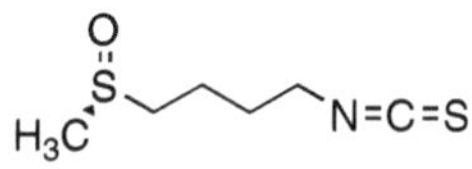
4 phenyl-1, 2-dithiole-3-thione	Dithiol thione	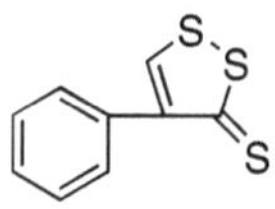
3-cyclohexenyl-4-methyl-1,2 oxathiol-3-ene-2-oxide	Oxathiolene oxide	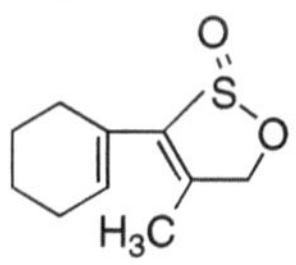
Diallyl sulfide	Alkyl sulfide	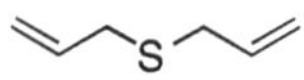
Phenylarsine oxide	Trivalent arsenical	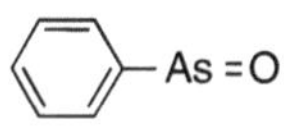
Cumene hydroperoxide	Organic peroxide	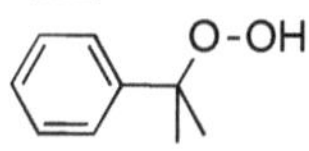
1,2 ethanedithiol	Vicinal mercaptans	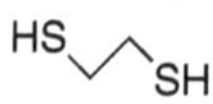
Lipoic acid	Vicinal dithiol	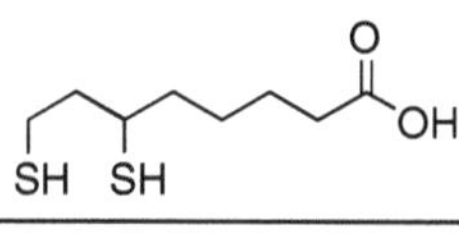

is contingent on the strength of the electron-withdrawing function and the potency of induction of Phase II enzymes in a prevention setting correlates with the strength of the electron withdrawing group and subsequently the nucleophilicity of the electrophilic carbon center.

These react quite readily with thiol nucleophiles particularly GSH and the thioether conjugates can have biological impact on a number of functional processes. For example, for many years ethacrynic acid has been used as a diuretic and as a reversible and irreversible inhibitor of GST isozymes. Michael addition reactions tend to be quite non-specific and this equates to a broad spectrum of "off-target" effects that complicate the use of this type of drug in a prevention setting and as mentioned earlier, its therapeutic activity in cancer treatment was restricted by the diuretic properties.

The N=C=S functional group of all isothiocyanates reacts at its highly electrophilic central carbon atom with biological thiols to give dithiocarbamates. Those isothiocyanates, thio-, dithiocarbamates and related sulfur compounds shown in Table 16.1 all induce expression of detoxification enzymes. Some isothiocyanates are found in cruciferous vegetables (e.g. sulforaphane is in broccoli) as glucoside precursors. The consumption of these vegetables exposes them to myrosinase that catalyzes their conversion to the reactive isothiocyanates. The accumulation of sulforaphane in various tissues contributes to its high potency as an inductive agent and intracellular accumulation can occur following conjugation with cellular glutathione under catalysis by GST's. In mice, 1,2-dithiole-3-thione derivatives can protect against toxicity and carcinogenicity. The anti-schistosomal agent oltipraz has been developed as a chemopreventive agent against liver cancer in China, where aflatoxin contamination of stored food can produce DNA adducts. More recently, a variety of polysulfides from Allium plants have been developed as preventive agents, with diallyl disulfide being the most potent. Hydroperoxides, including hydrogen and cumene hydroperoxide produce species that react rapidly with non-protein thiols such as GSH. Trivalent arsenicals, particularly those with vicinal thiols, for example phenylarsine oxide, are potent inducers of Phase II enzymes. In a similar manner, vicinal dimercaptans and dihydrolipoic acid also have closely spaced thiol groups.

Analysis of the common aspects of the chemistry of these inducers points to the critical nature of the thiol reactive centers. In general, each inducer has the potential to interact with the Nrf2:Keap1 transcription machinery in a manner that results in widespread over-expression of Phase II metabolism genes. Whether the events are likely to have long-term value as a cancer prevention strategy remains to be established, but the mechanism of this process is discussed in more detail in the next section.

16.5.1 Transcriptional Regulation and Drug Response – Nrf2 and ARE's

Reactive oxygen and nitrogen species (ROS; RNS) are byproducts of aerobic metabolism and contributors to cellular redox balance. Upstream of the many components of GSH homeostasis, the transcription factor nuclear factor-erythroid 2-related factor 2 (Nrf2) is one master switch that regulates the expression of stress response genes critical in mounting a cellular defense against electrophiles or ROS/RNS. Nrf2 can impact expression of genes that have one or more antioxidant response elements (ARE) in their promoter regions. One domain of NRF2 negatively regulates its activity through protein–protein interactions with Kelch-like ECH-associated protein 1 (KEAP1), a member of a metalloprotein family with between 25 and 27 conserved cysteine residues, essentially half of which are in basic regions, and maintain the thiolate state under physiological conditions. Such cysteines can fulfill the criteria required of redox sensors. They also provide a platform for cancer chemoprevention by which long-term exposure to low doses of inductive natural product (dietary) electrophiles can induce protective phase II enzyme systems.

There are complications associated with the approach of enhancing Nrf2 as a mechanism of chemoprevention (Hayes and McMahon 2006, 2009). For any of the (primarily) electrophilic agents discussed in the preceding section, their activities are contingent upon long term, low dose regimens that constitute chronic "low key" stress. Over a period, these chemical events could prove to be problematic. For example, somatic mutations that alter Nrf2/Keap1 interactions can cause constitutive activation of the Nrf2 transcription complex with potentially oncogenic consequences. This concept has been addressed by a report (DeNicola et al. 2011) using primary murine cells, where expression of endogenous oncogenic alleles such as *Kras, Braf* and *Myc* has an impact upon Nrf2 and the downstream gene clusters that regulate cellular redox balance. In cancer cells, constitutively elevated levels of Nrf2 can occur through both diminished Nrf2 turnover and increased levels of Nrf2 mRNA. Conditional endogenous oncogenes K-Ras^{G12D}, B-Raf^{V619E} and Myc^{ERT2} can each stably elevate basal Nrf2 through enhanced transcription, activating the cluster of genes involved in ROS response and lowering intracellular levels of ROS (a more reduced intracellular environment). Oncogene-directed, enhanced expression of Nrf2 may be a mechanism for the activation of the Nrf2 antioxidant program and genetic targeting of the Nrf2 pathway may impair K-Ras^{G12D}-induced proliferation and tumorigenesis *in vivo*. In such circumstances, the Nrf2 antioxidant and cellular detoxification program may be a (Hayes and McMahon 2006) mediator of oncogenesis. This interpretation may be contextual since oncogenic signaling mediated Nrf2 activation stimulates not only the antioxidant battery but also transporters, proteasome subunits, heat shock proteins, growth factors and other Phase II metabolizing gene products (Hayes and McMahon 2009). Moreover, glutathione S-transferases are among the genes regulated by NRF2 and the importance of the link with S-glutathionylation may prove important (Townsend et al. 2009). Alterations in Nrf2 activity could shift the balance of GST involvement in the forward S-glutathionylation reaction, perhaps producing changes in post-translational patterns that in turn promote proteins that regulate oncogenesis. In the end, these results do not directly suggest that chemoprevention is a flawed strategy, since dietary supplements would need to be viewed through a requirement for existing predisposing genetic abnormalities, at least one of which would involve Nrf2. Nevertheless, they do support the principles that tumor cells have aberrant methods for equilibrating ROS and these pathways could prove to be viable therapeutic targets. A schematic representation of the impact of oncogene signaling on Nrf2 functions is shown in Fig. 16.4.

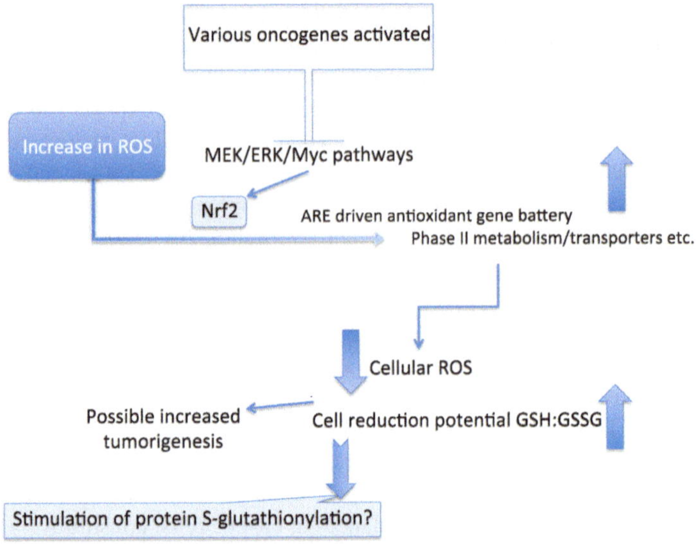

Fig. 16.4 Nrf2 is one of the primary transcriptional controls governing antioxidant response to stress. Its activation by electrophilic natural products is one approach to the therapeutic approaches involved in chemoprevention. As discussed in the narrative, there are possible inherent complications to this as a chronic or long-term strategy to enhancing cellular protective enzymes. The scheme summarizes the influence of Nrf2 and oncogene stimulation on downstream redox pathways

16.6 Summary and Outlook

Evolution has acted over many millennia to fine tune adaptations to an oxidative environment. As a consequence, cellular pathways to handle ROS and RNS are complex, redundant and multi-layered. Perhaps because of their inherent genetic plasticity and instability, many cancer cells have abnormal redox balance. Consequently, those pathways that regulate redox homeostasis have been a focus for therapeutic intervention. Along the way, it has become apparent that the regulation of bone marrow cell proliferation is also sensitive to redox changes and a number of small molecule myeloproliferative drugs are at various stages of preclinical and clinical development. Success to date in gaining FDA approval for such agents has been limited (see summary in Table 16.2), but the pipeline for redox targeted drug discovery and development remains robust. Moreover, the growing interest in the importance of manipulating redox in other human pathologies, particularly neurodegenerative, traumatic brain injury and cardiovascular diseases provides further opportunities to design and test small molecule agents that may gain therapeutic advantage.

Table 16.2 Redox active drugs at various stages of development

Drug name	Target and/or mechanism	Development status
N-acetyl cysteine (NAC)	Cysteine precursor. Enhances GSH synthesis	Many clinical applications (see text) particularly rescue from acetaminophen toxicity
Mesna	Provides available thiol equivalents	Used to prevent side effects of alkylating agent therapies and as mucolytic agent
Tavocept	Can be reduced to provide thiol equivalents	Pharmacology similar to sodium thiosulfate, diethyldithiocarbonate and tempol
Amifostine (WR-2721; Ethyol)	Thiophosphate pro-drug	1999 FDA approval for Ethyol for reduction of radiation-induced xerostomia (dry mouth)
NOV-002	Glutathione disulfide mimetic alters cellular redox	Phase III trial in combination with cancer drugs in lung cancer failed 2010. Some clinical trials continue
Telintra	Peptidomimetic inhibitor of GSTP	Ongoing Phase II trials in myelodysplastic syndrome
Telcyta	GSTP activated pro-drug	Phase III trials in ovarian cancer were negative, but other Phase II/III trials continue
PABA/NO	GST activated pro-drug releasing NO	Preclinical development but some early Phase I studies with similar prodrugs are beginning

Acknowledgements This work was supported by grants from the National Institutes of Health (ES017453, CA08660 and CA117259) and support from the South Carolina Centers of Excellence program. We thank the Drug Metabolism and Pharmacokinetics and Proteomics Core Facilities at the Medical University of South Carolina. This work was conducted in a facility constructed with support from the National Institutes of Health, Grant Number C06 RR015455 from the Extramural Research Facilities Program of the National Center for Research Resources.

References

Abate C, Patel L, Rauscher FJ 3rd, Curran T (1990) Redox regulation of fos and jun DNA-binding activity in vitro. Science 249:1157

Adams GB et al (2006) Stem cell engraftment at the endosteal niche is specified by the calcium-sensing receptor. Nature 439:599

Adler V et al (1999) Regulation of JNK signaling by GSTp. EMBO J 18:1321

Akerlund B et al (1996) Effect of N-acetylcysteine(NAC) treatment on HIV-1 infection: a double-blind placebo-controlled trial. Eur J Clin Pharmacol 50:457

Aruoma OI, Halliwell B, Hoey BM, Butler J (1989) The antioxidant action of N-acetylcysteine: its reaction with hydrogen peroxide, hydroxyl radical, superoxide, and hypochlorous acid. Free Radic Biol Med 6:593

Baldini M, Sacchetti C (1953) Effect of cystine and cysteine on human bone marrow cultured in medium deficient in amino acids. Rev Hematol 8:3

Bertini R et al (1999) Thioredoxin, a redox enzyme released in infection and inflammation, is a unique chemoattractant for neutrophils, monocytes, and T cells. J Exp Med 189:1783

Brock N, Hilgard P, Pohl J, Ormstad K, Orrenius S (1984) Pharmacokinetics and mechanism of action of detoxifying low-molecular-weight thiols. J Cancer Res Clin Oncol 108:87
Calvi LM et al (2003) Osteoblastic cells regulate the haematopoietic stem cell niche. Nature 425:841
Chou WC et al (2004) Role of NADPH oxidase in arsenic-induced reactive oxygen species formation and cytotoxicity in myeloid leukemia cells. Proc Natl Acad Sci USA 101:4578
Christopherson KW 2nd, Cooper S, Broxmeyer HE (2003) Cell surface peptidase CD26/DPPIV mediates G-CSF mobilization of mouse progenitor cells. Blood 101:4680
Cipolleschi MG, Dello Sbarba P, Olivotto M (1993) The role of hypoxia in the maintenance of hematopoietic stem cells. Blood 82:2031
Clark LC et al (1998) Decreased incidence of prostate cancer with selenium supplementation: results of a double-blind cancer prevention trial. Br J Urol 81:730
DeNicola GM et al (2011) Oncogene-induced Nrf2 transcription promotes ROS detoxification and tumorigenesis. Nature 475:106
Dinkova-Kostova AT, Fahey JW, Talalay P (2004) Chemical structures of inducers of nicotinamide quinone oxidoreductase 1 (NQO1). Methods Enzymol 382:423
Dominici S et al (1999) Redox modulation of cell surface protein thiols in U937 lymphoma cells: the role of gamma-glutamyl transpeptidase-dependent H_2O_2 production and S-thiolation. Free Radic Biol Med 27:623
Doroshow JH (1983) Effect of anthracycline antibiotics on oxygen radical formation in rat heart. Cancer Res 43:460
Drane P, Bravard A, Bouvard V, May E (2001) Reciprocal down-regulation of p53 and SOD2 gene expression-implication in p53 mediated apoptosis. Oncogene 20:430
Evens AM, Balasubramanian L, Gordon LI (2005) Motexafin gadolinium induces oxidative stress and apoptosis in hematologic malignancies. Curr Treat Options Oncol 6:289
Fawcett H, Mader JS, Robichaud M, Giacomantonio C, Hoskin DW (2005) Contribution of reactive oxygen species and caspase-3 to apoptosis and attenuated ICAM-1 expression by paclitaxel-treated MDA-MB-435 breast carcinoma cells. Int J Oncol 27:1717
Findlay VJ et al (2004) Tumor cell responses to a novel glutathione S-transferase-activated nitric oxide-releasing prodrug. Mol Pharmacol 65:1070
Forchhammer K, Leinfelder W, Bock A (1989) Identification of a novel translation factor necessary for the incorporation of selenocysteine into protein. Nature 342:453
Gate L, Majumdar RS, Lunk A, Tew KD (2004) Increased myeloproliferation in glutathione S-transferase pi-deficient mice is associated with a deregulation of JNK and Janus kinase/STAT pathways. J Biol Chem 279:8608
Gosset P, Wallaert B, Tonnel AB, Fourneau C (1999) Thiol regulation of the production of TNF-alpha, IL-6 and IL-8 by human alveolar macrophages. Eur Respir J 14:98
Guzman ML et al (2007) Rapid and selective death of leukemia stem and progenitor cells induced by the compound 4-benzyl, 2-methyl, 1,2,4-thiadiazolidine, 3,5 dione (TDZD-8). Blood 110:4436
Hausheer FH et al (1998) Modulation of platinum-induced toxicities and therapeutic index: mechanistic insights and first- and second-generation protecting agents. Semin Oncol 25:584
Hayes JD, McMahon M (2006) The double-edged sword of Nrf2: subversion of redox homeostasis during the evolution of cancer. Mol Cell 21:732
Hayes JD, McMahon M (2009) NRF2 and KEAP1 mutations: permanent activation of an adaptive response in cancer. Trends Biochem Sci 34:176
Herzenberg LA et al (1997) Glutathione deficiency is associated with impaired survival in HIV disease. Proc Natl Acad Sci USA 94:1967
Hondal RJ, Ruggles EL (2011) Differing views of the role of selenium in thioredoxin reductase. Amino Acids 41:73
Hosokawa K et al (2007) Function of oxidative stress in the regulation of hematopoietic stem cell-niche interaction. Biochem Biophys Res Commun 363:578
Ito K et al (2006) Reactive oxygen species act through p38 MAPK to limit the lifespan of hematopoietic stem cells. Nat Med 12:446

Iwasaki H, Suda T (2009) Cancer stem cells and their niche. Cancer Sci 100:1166

Jang YY, Sharkis SJ (2007) A low level of reactive oxygen species selects for primitive hematopoietic stem cells that may reside in the low-oxygenic niche. Blood 110:3056

Jeong M et al (2009) Thioredoxin-interacting protein regulates hematopoietic stem cell quiescence and mobilization under stress conditions. J Immunol 183:2495

Jin Y et al (2010) Antineoplastic mechanisms of niclosamide in acute myelogenous leukemia stem cells: inactivation of the NF-kappaB pathway and generation of reactive oxygen species. Cancer Res 70:2516

Jones DP (2008) Radical-free biology of oxidative stress. Am J Physiol Cell Physiol 295:C849

Kanter MZ (2006) Comparison of oral and i.v. acetylcysteine in the treatment of acetaminophen poisoning. Am J Health Syst Pharm 63:1821

Kelly GS (1998) Clinical applications of N-acetylcysteine. Altern Med Rev 3:114

Khodarev NN, Kataoka Y, Murley JS, Weichselbaum RR, Grdina DJ (2004) Interaction of amifostine and ionizing radiation on transcriptional patterns of apoptotic genes expressed in human microvascular endothelial cells (HMEC). Int J Radiat Oncol Biol Phys 60:553

Knight GD, Laubscher KH, Fore ML, Clark DA, Scallen TJ (1994) Vitalethine modulates erythropoiesis and neoplasia. Cancer Res 54:5623

Korst AE, Eeltink CM, Vermorken JB, van der Vijgh WJ (1997) Pharmacokinetics of amifostine and its metabolites in patients. Eur J Cancer 33:1425

Koukourakis MI et al (2004) Amifostine induces anaerobic metabolism and hypoxia-inducible factor 1 alpha. Cancer Chemother Pharmacol 53:8

Kryukov GV et al (2003) Characterization of mammalian selenoproteomes. Science 300:1439

Levy EJ, Anderson ME, Meister A (1993) Transport of glutathione diethyl ester into human cells. Proc Natl Acad Sci USA 90:9171

Li JJ, Oberley LW (1997) Overexpression of manganese-containing superoxide dismutase confers resistance to the cytotoxicity of tumor necrosis factor alpha and/or hyperthermia. Cancer Res 57:1991

List AF et al (1997) Stimulation of hematopoiesis by amifostine in patients with myelodysplastic syndrome. Blood 90:3364

List AF, Heaton R, Glinsmann-Gibson B, Capizzi RL (1998) Amifostine stimulates formation of multipotent and erythroid bone marrow progenitors. Leukemia 12:1596

Lothrop AP, Ruggles EL, Hondal RJ (2009) No selenium required: reactions catalyzed by mammalian thioredoxin reductase that are independent of a selenocysteine residue. Biochemistry 48:6213

Lyons RM, Wilks ST, Young S, Brown GL (2011) Oral ezatiostat HCl (Telintra(R), TLK199) and idiopathic chronic neutropenia (ICN): a case report of complete response of a patient with G-CSF resistant ICN following teatment with ezatiostat, a glutathione S-transferase P1-1 (GSTP1-1) inhibitor. J Hematol Oncol 4:43

Lyttle MH et al (1994) Glutathione-S-transferase activates novel alkylating agents. J Med Chem 37:1501

Mannervik B (1985) The isoenzymes of glutathione transferase. Adv Enzymol Relat Areas Mol Biol 57:357

Marenzi G et al (2006) N-acetylcysteine and contrast-induced nephropathy in primary angioplasty. N Engl J Med 354:2773

Marshall JR (2001) Larry Clark's legacy: randomized controlled, selenium-based prostate cancer chemoprevention trials. Nutr Cancer 40:74

Martin KR, Kari FW, Barrett JC, French JE (2000) N-acetyl-L-cysteine simultaneously increases mitogenesis and suppresses apoptosis in mitogen-stimulated B-lymphocytes from p53 haploinsufficient Tg.AC (v-Ha-ras) mice. In Vitr Mol Toxicol 13(Winter):237

Miseta A, Csutora P (2000) Relationship between the occurrence of cysteine in proteins and the complexity of organisms. Mol Biol Evol 17:1232

Morgan AS, Ciaccio PJ, Tew KD, Kauvar LM (1996) Isozyme-specific glutathione S-transferase inhibitors potentiate drug sensitivity in cultured human tumor cell lines. Cancer Chemother Pharmacol 37:363

Nakamura H et al (2001) Circulating thioredoxin suppresses lipopolysaccharide-induced neutrophil chemotaxis. Proc Natl Acad Sci USA 98:15143
Nakamura H, Masutani H, Yodoi J (2002) Redox imbalance and its control in HIV infection. Antioxid Redox Signal 4:455
Neuwelt EA, Pagel MA, Kraemer DF, Peterson DR, Muldoon LL (2004) Bone marrow chemoprotection without compromise of chemotherapy efficacy in a rat brain tumor model. J Pharmacol Exp Ther 309:594
Nie Y, Han YC, Zou YR (2008) CXCR4 is required for the quiescence of primitive hematopoietic cells. J Exp Med 205:777
O'Dwyer PJ et al (1991) Phase I study of thiotepa in combination with the glutathione transferase inhibitor ethacrynic acid. Cancer Res 51:6059
O'Shea JJ, Gadina M, Schreiber RD (2002) Cytokine signaling in 2002: new surprises in the Jak/Stat pathway. Cell 109(Suppl):S121
Palmer LA et al (2007) S-nitrosothiols signal hypoxia-mimetic vascular pathology. J Clin Invest 117:2592
Parmar K, Mauch P, Vergilio JA, Sackstein R, Down JD (2007) Distribution of hematopoietic stem cells in the bone marrow according to regional hypoxia. Proc Natl Acad Sci USA 104:5431
Pelicano H et al (2003) Inhibition of mitochondrial respiration: a novel strategy to enhance drug-induced apoptosis in human leukemia cells by a reactive oxygen species-mediated mechanism. J Biol Chem 278:37832
Pendyala L et al (2003) Modulation of plasma thiols and mixed disulfides by BNP7787 in patients receiving paclitaxel/cisplatin therapy. Cancer Chemother Pharmacol 51:376
Peterson JD, Herzenberg LA, Vasquez K, Waltenbaugh C (1998) Glutathione levels in antigen-presenting cells modulate Th1 versus Th2 response patterns. Proc Natl Acad Sci USA 95:3071
Pluquet O et al (2003) The cytoprotective aminothiol WR1065 activates p53 through a non-genotoxic signaling pathway involving c-Jun N-terminal kinase. J Biol Chem 278:11879
Rahmani M et al (2005) Coadministration of histone deacetylase inhibitors and perifosine synergistically induces apoptosis in human leukemia cells through Akt and ERK1/2 inactivation and the generation of ceramide and reactive oxygen species. Cancer Res 65:2422
Raza A et al (2009a) Phase 1-2a multicenter dose-escalation study of ezatiostat hydrochloride liposomes for injection (Telintra, TLK199), a novel glutathione analog prodrug in patients with myelodysplastic syndrome. J Hematol Oncol 2:20
Raza A et al (2009b) Phase 1 multicenter dose-escalation study of ezatiostat hydrochloride (TLK199 tablets), a novel glutathione analog prodrug, in patients with myelodysplastic syndrome. Blood 113:6533
Raza A et al (2012) A phase 2 randomized multicenter study of 2 extended dosing schedules of oral ezatiostat in low to intermediate-1 risk myelodysplastic syndrome. Cancer 118(8):2138–2147
Reinemer P et al (1992) Three-dimensional structure of class pi glutathione S-transferase from human placenta in complex with S-hexylglutathione at 2.8 A resolution. J Mol Biol 227:214
Reliene R, Schiestl RH (2006) Antioxidant N-acetyl cysteine reduces incidence and multiplicity of lymphoma in Atm deficient mice. DNA Repair 5:852
Ribizzi I, Darnowski JW, Goulette FA, Sertoli MR, Calabresi P (2000) Amifostine cytotoxicity and induction of apoptosis in a human myelodysplastic cell line. Leuk Res 24:519
Roberts RL, Aroda VR, Ank BJ (1995) N-acetylcysteine enhances antibody-dependent cellular cytotoxicity in neutrophils and mononuclear cells from healthy adults and human immunodeficiency virus-infected patients. J Infect Dis 172:1492
Romano MF et al (1999) Amifostine inhibits hematopoietic progenitor cell apoptosis by activating NF-kappaB/Rel transcription factors. Blood 94:4060
Rosario LA, O'Brien ML, Henderson CJ, Wolf CR, Tew KD (2000) Cellular response to a glutathione S-transferase P1-1 activated prodrug. Mol Pharmacol 58:167
Rosen LS et al (2004) Phase 1 study of TLK286 (Telcyta) administered weekly in advanced malignancies. Clin Cancer Res 10:3689
Rovin BH, Dickerson JA, Tan LC, Fassler J (1997) Modulation of IL-1-induced chemokine expression in human mesangial cells through alterations in redox status. Cytokine 9:178

Ruscoe JE et al (2001) Pharmacologic or genetic manipulation of glutathione S-transferase P1-1 (GSTpi) influences cell proliferation pathways. J Pharmacol Exp Ther 298:339
Saavedra JE et al (2001) The secondary amine/nitric oxide complex ion R(2)N[N(O)NO](−) as nucleophile and leaving group in S9N)Ar reactions. J Org Chem 66:3090
Saavedra JE et al (2006) PABA/NO as an anticancer lead: analogue synthesis, structure revision, solution chemistry, reactivity toward glutathione, and in vitro activity. J Med Chem 49:1157
Sablina AA et al (2005) The antioxidant function of the p53 tumor suppressor. Nat Med 11:1306
Sahaf B, Heydari K, Herzenberg LA (2005) The extracellular microenvironment plays a key role in regulating the redox status of cell surface proteins in HIV-infected subjects. Arch Biochem Biophys 434:26
Schanz J et al (2009) Amifostine has the potential to induce haematologic responses and decelerate disease progression in individual patients with low- and intermediate-1-risk myelodysplastic syndromes. Leuk Res 33:1183
Schofield R (1978) The relationship between the spleen colony-forming cell and the haemopoietic stem cell. Blood Cells 4:7
Shanmugarajah D et al (2009) Analysis of BNP7787 thiol-disulfide exchange reactions in phosphate buffer and human plasma using microscale electrochemical high performance liquid chromatography. J Chromatogr B Analyt Technol Biomed Life Sci 877:857
Shaulian E, Karin M (2001) AP-1 in cell proliferation and survival. Oncogene 20:2390
Sheehan D, Meade G, Foley VM, Dowd CA (2001) Structure, function and evolution of glutathione transferases: implications for classification of non-mammalian members of an ancient enzyme superfamily. Biochem J 360:1
Shen H et al (2001) Binding of the aminothiol WR-1065 to transcription factors influences cellular response to anticancer drugs. J Pharmacol Exp Ther 297:1067
Sinning I et al (1993) Structure determination and refinement of human alpha class glutathione transferase A1–1, and a comparison with the Mu and Pi class enzymes. J Mol Biol 232:192
Smith J, Ladi E, Mayer-Proschel M, Noble M (2000) Redox state is a central modulator of the balance between self-renewal and differentiation in a dividing glial precursor cell. Proc Natl Acad Sci USA 97:10032
Songstad J, Pearson RG (1967) Application of the principle of hard and soft acids and bases to organic chemistry. J Am Chem Soc 89:1827
Songstad J, Pearson RG (1968) Nucleophilic reactivity constants toward methyl iodide and trans-[Pt(py)2Cl2]. J Am Chem Soc 90:319
Spencer SR, Wilczak CA, Talalay P (1990) Induction of glutathione transferases and NAD(P)H:quinone reductase by fumaric acid derivatives in rodent cells and tissues. Cancer Res 50:7871
Szatrowski TP, Nathan CF (1991) Production of large amounts of hydrogen peroxide by human tumor cells. Cancer Res 51:794
Terada K, Kaziro Y, Satoh T (1997) Ras-dependent activation of c-Jun N-terminal kinase/stress-activated protein kinase in response to interleukin-3 stimulation in hematopoietic BaF3 cells. J Biol Chem 272:4544
Tesio M et al (2011) Enhanced c-Met activity promotes G-CSF-induced mobilization of hematopoietic progenitor cells via ROS signaling. Blood 117:419
Tew KD (1994) Glutathione-associated enzymes in anticancer drug resistance. Cancer Res 54:4313
Tew KD, Kyle G, Johnson A, Wang AL (1985) Carbamoylation of glutathione reductase and changes in cellular and chromosome morphology in a rat cell line resistant to nitrogen mustards but collaterally sensitive to nitrosoureas. Cancer Res 45:2326
Tew KD et al (2011) The role of glutathione S-transferase P in signaling pathways and S-glutathionylation in cancer. Free Radic Biol Med 51:299
Tirouvanziam R, Conrad CK, Bottiglieri T, Herzenberg LA, Moss RB (2006) High-dose oral N-acetylcysteine, a glutathione prodrug, modulates inflammation in cystic fibrosis. Proc Natl Acad Sci USA 103:4628
Tothova Z et al (2007) FoxOs are critical mediators of hematopoietic stem cell resistance to physiologic oxidative stress. Cell 128:325

Townsend DM, Shen H, Staros AL, Gate L, Tew KD (2002) Efficacy of a glutathione S-transferase pi-activated prodrug in platinum-resistant ovarian cancer cells. Mol Cancer Ther 1:1089

Townsend DM et al (2006) A glutathione S-transferase pi-activated prodrug causes kinase activation concurrent with S-glutathionylation of proteins. Mol Pharmacol 69:501

Townsend DM, Pazoles CJ, Tew KD (2008) NOV-002, a mimetic of glutathione disulfide. Expert Opin Investig Drugs 17:1075

Townsend DM et al (2009) Novel role for glutathione S-transferase pi. Regulator of protein S-Glutathionylation following oxidative and nitrosative stress. J Biol Chem 284:436

Trachootham D, Lu W, Ogasawara MA, Nilsa RD, Huang P (2008) Redox regulation of cell survival. Antioxid Redox Signal 10:1343

Velu CS, Niture SK, Doneanu CE, Pattabiraman N, Srivenugopal KS (2007) Human p53 is inhibited by glutathionylation of cysteines present in the proximal DNA-binding domain during oxidative stress. Biochemistry 46:7765

Verschraagen M et al (2004) Possible (enzymatic) routes and biological sites for metabolic reduction of BNP7787, a new protector against cisplatin-induced side-effects. Biochem Pharmacol 68:493

Wessjohann LA, Schneider A, Abbas M, Brandt W (2007) Selenium in chemistry and biochemistry in comparison to sulfur. Biol Chem 388:997

Wilson A et al (2008) Hematopoietic stem cells reversibly switch from dormancy to self-renewal during homeostasis and repair. Cell 135:1118

Wondrak GT (2009) Redox-directed cancer therapeutics: molecular mechanisms and opportunities. Antioxid Redox Signal 11:3013

Xinhua J (2008) Structure-based design of anticancer prodrug PABA/NO. Drug Des Devel Ther 2:123

Zafarullah M, Li WQ, Sylvester J, Ahmad M (2003) Molecular mechanisms of N-acetylcysteine actions. Cell Mol Life Sci 60:6

Zhou Y, Hileman EO, Plunkett W, Keating MJ, Huang P (2003) Free radical stress in chronic lymphocytic leukemia cells and its role in cellular sensitivity to ROS-generating anticancer agents. Blood 101:4098

Index

U. Jakob and D. Reichmann (eds.), *Oxidative Stress and Redox Regulation*,
DOI 10.1007/978-94-007-5787-5,

MIX
Papier aus verantwortungsvollen Quellen
Paper from responsible sources
FSC® C105338

If you have any concerns about our products,
you can contact us on
ProductSafety@springernature.com

In case Publisher is established outside the EU,
the EU authorized representative is:
Springer Nature Customer Service Center GmbH
Europaplatz 3, 69115 Heidelberg, Germany

Printed by Libri Plureos GmbH
in Hamburg, Germany